MATERIAL SCIENCE AND ENGINEERING

PROCEEDINGS OF THE 3RD ANNUAL 2015 INTERNATIONAL CONFERENCE ON MATERIAL SCIENCE AND ENGINEERING (ICMSE2015), GUANGZHOU, GUANGDONG, CHINA, 15–17 MAY 2015

Material Science and Engineering

Editor

Ping Chen
Department of Electrical and Computer Engineering, Boise State University, USA

CRC Press
Taylor & Francis Group
Boca Raton London New York Leiden

CRC Press is an imprint of the
Taylor & Francis Group, an **informa** business

A BALKEMA BOOK

CRC Press/Balkema is an imprint of the Taylor & Francis Group, an informa business

© 2016 Taylor & Francis Group, London, UK

Typeset by V Publishing Solutions Pvt Ltd., Chennai, India

Published by: CRC Press/Balkema
P.O. Box 11320, 2301 EH Leiden, The Netherlands
e-mail: Pub.NL@taylorandfrancis.com
www.crcpress.com – www.taylorandfrancis.com

ISBN: 978-1-138-02936-1 (Hbk)
ISBN: 978-1-315-63888-1 (eBook PDF)

Material Science and Engineering – Chen (Ed.)
© 2016 Taylor & Francis Group, London, ISBN 978-1-138-02936-1

Table of contents

Mechanical engineering and manufacturing technology

Other engineering topics

Material Science and Engineering – Chen (Ed.)
© 2016 Taylor & Francis Group, London, ISBN 978-1-138-02936-1

Preface

The 2015 International Conference on Material Science and Engineering (ICMSE2015) was successfully held in Guangzhou, China, from May 15 to May 17, 2015. The ICMSE2015 is becoming one of the leading international conferences for presenting novel and fundamental advances in the fields of material science and engineering in China. It provides a forum for scientists, scholars, engineers and students from the universities all around the world and the industry to present ongoing research activities, and hence to foster research relations between the universities and the industry. This conference provides opportunities for the delegates to exchange new ideas and application experiences face to face, to establish business or research relations and to find global partners for future collaboration. The first ICMSE (ICMSE2013) was successfully held during October 4–6, 2013 in Guilin, China. The second ICMSE (ICMSE2014) was successfully held during August 8–9, 2014 in Xi'an, China.

This proceedings tends to collect the up-to-date, comprehensive and worldwide state-of-art knowledge on material science and engineering. All of accepted papers were subjected to strict peer-reviewing by 2–4 expert referees. The papers have been selected for this proceedings based on originality, significance, and clarity for the purpose of the conference. The selected papers and additional late-breaking contributions made an exciting technical program on conference. The conference program was extremely rich, featuring invited keynote speeches, panel discussions, high-impact oral and poster presentations. The keynote speaker is internationally recognized leading expert in the research field of material science and engineering, who have demonstrated outstanding proficiency and have achieved distinction in their profession. We hope this conference will not only provide the participants a broad overview of the latest research results on material science and engineering, but also provide the participants a significant platform to build academic connections.

We would like to express our sincere gratitude to all the members of Technical Program Committee and organizers for their enthusiasm, time, and expertise. Our deep thanks also go to many volunteers and staffs for the long hours and hard work they have generously given to ICMSE2015. Finally, we would like to thank all the authors, speakers, and participants of this conference for their contributions to ICMSE2015, and also looks forward to welcoming you to the ICMSE2016.

ICMSE2015 Organizing Committee

Metallic materials and applications

Material Science and Engineering – Chen (Ed.)
© 2016 Taylor & Francis Group, London, ISBN 978-1-138-02936-1

Effect of the cooling rate and final cooling temperature on the microstructure of GCr15 bearing steel

X.D. Huo & Z.Z. Tian
School of Material Science and Engineering, Jiangsu University, Zhenjiang, China

L.J. Li
School of Mechanical and Automotive Engineering, South China University of Technology, Guangzhou, China

N.F. Liu
Shaoguan Iron and Steel Group Co. Ltd., Shaoguan, China

ABSTRACT. The effect of cooling parameters on the transformation and microstructure of GCr15 bearing steel was investigated by using Gleeble-3800 thermo-simulation system. Under the lower cooling rate, the transformed microstructure is composed of pearlite colony and proeutectoid carbide precipitated on the grain boundary. As the cooling rate is higher than $8°C \cdot s^{-1}$, partial austenite transforms to martensite below 187°C. The final cooling temperature is an important parameter and obviously influences the microstructure at room temperature. If appropriate cooling rate and final cooling temperature are adopted, carbide network can be avoided and no martensite appears.

Keywords: bearing steel; cooling rate; final cooling temperature; transformation

1 INTRODUCTION

GCrl5 steel is a bearing steel with high contents of carbon and chromium. It is widely used in the bearing manufacturing industry. About 80% of the bearing steel products in China are GCrl5 steel [1–3].

GCrl5 is a hypereutectoid steel and proeutectoid carbide precipitated first from austenite when it is cooled slowly after high-temperature rolling. Secondary carbide grows up along the austenite grain boundary and forms the network. The carbide network can cause a decrease in the lifetime of bearings, so it should be avoided. An appropriate cooling after hot forming can lead to the elimination of this defect, and, besides, it can prepare a suitable initial structure for subsequent annealing. Unfortunately, relevant work is insufficient [4, 5].

In this paper, the Gleeble-3800 thermo-simulation system was used to study the effect of cooling rates on the transformation of GCr15 bearing steel. The dynamic CCT diagram of GCr15 steel was established, and the continuous cooled microstructure was investigated. On this basis, the effect of the final cooling temperature on the microstructure of GCr15 bearing steel at a cooling rate of $3°C \cdot s^{-1}$ was analyzed.

2 EXPERIMENTAL PROCEDURE

A commercial GCr15 high-chromium bearing steel produced by a steel company was investigated in this paper. The chemical composition of the GCr15 steel used in this study is listed in Table 1. The experiments were conducted on the Gleeble-3800 thermo-simulation system. The specimens were taken from a 50 mm diameter bar and were machined to the shape and dimension as shown in Figure 1.

The corresponding specimens were first heated to 1100°C at the heating rate of $10°C \cdot s^{-1}$ and austenitized for 5 min, and were subsequently cooled to 980°C at the rate of $10°C \cdot s^{-1}$ and deformed by approximately 40%. The samples were continuously cooled to 860°C and deformed by approximately 30%, and then cooled to room temperature at the cooling rates of 0.5, 1, 1.5, 2, 3,

Table 1. Chemical composition of the GCr15 steel investigated [wt%].

C	Si	Mn	S	P	Cr	Ni	Cu
0.98	0.24	0.35	0.001	0.013	1.47	0.009	0.01

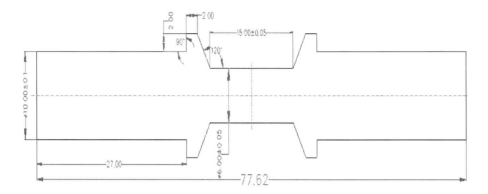

Figure 1. The specimens for the thermal simulation study.

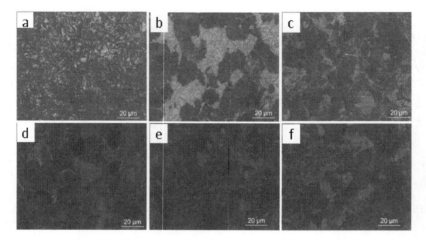

Figure 2.   The optical micrographs for the GCr15 steel investigated at different cooling rates. (a) 0.5°C · s⁻¹, (b) 2°C · s⁻¹, (c) 3°C · s⁻¹, (d) 5°C · s⁻¹, (e) 8°C · s⁻¹, (f) 10°C · s⁻¹.

5, 8, 10, 15, and 20°C · s⁻¹. The microstructure was investigated and the dynamic CCT diagram was constructed on the basis of observation by light optical microscopy and the dilatometric analysis.

In order to study the effect of the final cooling temperature on the carbide network, the above parameters were adopted. After hot compressive deformation, the samples were cooled to different final cooling temperatures (T = 750, 700, 650, 600°C) with the cooling rate of 3°C · s⁻¹. Subsequently, one group was cooled to room temperature with the cooling rate of 1°C · s⁻¹, and another group was quenched in water for keeping the microstructure at a higher temperature.

The specimens for the metallographic examination were mechanically ground, polished and etched with a 3% Nital solution, and then observed with a LEICA DM 2500 M optical microscope.

3   RESULTS AND DISCUSSION

3.1   *The dynamic CCT diagram and microstructure at room temperature*

Figure 2 shows the microstructures of GCr15 bearing steel formed as a result of various cooling rates. When the cooling rate changed from 0.5 to 2°C · s⁻¹, the transformed microstructure composed of a large lamellar pearlite and white carbide network along the grain boundary. When the cooling rate increased to 5°C · s⁻¹, the diameter of the pearlite colony became smaller gradually, the rod or line proeutectoid carbide precipitated on the grain boundary, and the network was not obvious. With the further increase in the cooling rate, the transformed microstructure was dominated by pearlite and martensite, and a small amount of fine spot carbide was distributed in the matrix. The primary microstructure was

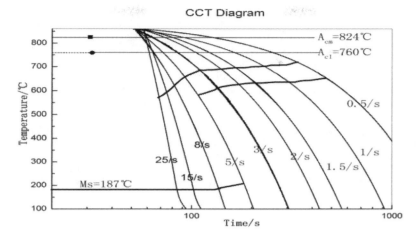

Figure 3.   The dynamic CCT diagram of the GCr15 steel investigated.

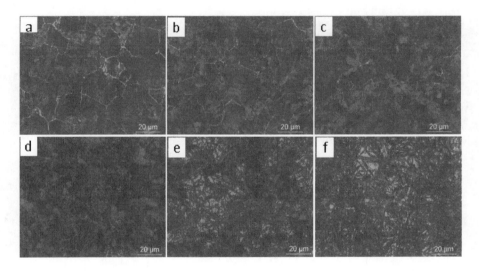

Figure 4.   Microstructure at different final cooling temperatures with the cooling rate of $3°C \cdot s^{-1}$: (a) 700°C, quenching; (b) 650°C, quenching; (c) 600°C, quenching; (d) 700°C, air cooling; (e) 650°C, air cooling; (f) 600°C, air cooling.

martensite and some retained austenite as the cooling rate increased above $10°C \cdot s^{-1}$, and the proeutectoid carbide could not be observed.

The dynamic CCT curves are shown in Figure 3. When the cooling rate was lower than $5°C \cdot s^{-1}$, the formed microstructure primarily included proeutectoid carbide and pearlite. The initial and final transformation temperatures gradually decreased as the cooling rate increased. When the cooling rate varied from 8 to $10°C \cdot s^{-1}$, the initial transformation temperature dramatically decreased and the line of final temperature disappeared. Over-cooled austenite subsequently transformed to martensite when the temperature dropped to 187°C. When the

cooling rate was above $10°C \cdot s^{-1}$, proeutectoid carbide and pearlite transformation were inhibited, and partial austenite transformed to martensite after the temperature declined to the martensite phase transition. The primary microstructure was martensite and partially retained austenite.

3.2   *Effect of the final cooling temperature on the microstructure*

Figure 4 shows the optical micrographs at different final cooling temperatures with the cooling rate of $3°C \cdot s^{-1}$. The initial and final temperatures of the sample cooled by $3°C \cdot s^{-1}$ were 681°C and 605°C, respectively. Air cooling was simulated as $1°C \cdot s^{-1}$, whose initial and final temperatures were 702°C and 640°C, respectively. When the final cooling temperature was 700°C, pearlite transformation had not yet started. The primary microstructure of the sample quenched was martensite, and that of the sample cooled by $1°C \cdot s^{-1}$ composed of large lamellar pearlite and white carbide networks along the grain boundary. When the final cooling temperature was 600°C, pearlite transformation had finished, and there was no appreciable distinction between the microstructure of the samples cooled by the different methods employed. When the final cooling temperature was 650°C, pearlite transformation was not yet complete. A small amount of retained austenite transformed to martensite in the quenching sample; however, retained austenite subsequently completed pearlite transformation in the sample cooled by air cooling. It can be seen that the final cooling temperature obviously influenced the microstructure at room temperature. If an appropriate cooling rate and final cooling temperature are adopted, the carbide network can be avoided and no martensite appears.

4 CONCLUSIONS

When the cooling rate was lower than $5°C \cdot s^{-1}$, the transformed microstructure composed of pearlite colony and proeutectoid carbide precipitated on the grain boundary. The diameter of pearlite colony and the pearlite lamellar spacing became smaller as the cooling rate was increased.

When the cooling rate exceeded $8°C \cdot s^{-1}$, the final temperature line disappeared and over-cooled austenite transformed to martensite as the temperature became lower than 187°C. At a higher cooling rate, the primary microstructure was martensite and partially retained austenite.

The final cooling temperature obviously influenced the microstructure at room temperature. If an appropriate cooling rate and final cooling temperature are adopted, the carbide network can be avoided and no martensite appears.

REFERENCES

[1] Y.K. Sun and D. Wu: J. Iron Steel Res. Int. Vol. 16 (2009), p. 61.
[2] J.G. Zhang, D.S. Sun, H.S. Shi, et al: Mater. Sci. Eng. Vol. A326 (2002), p. 20.
[3] N.Q. Peng, G.B. Tang, J. Yao, et al: J. Iron Steel Res. Int. Vol. 20 (2013), p. 50.
[4] F. Yin, L. Hua, H.J. Mao, et al: Materials and Design, Vol. 43 (2013), p. 393.
[5] C.X. Yue, L.W. Zhanga, S.L. Liao, et al: Mater. Sci. Eng. Vol. A499 (2009), p. 177.

Material Science and Engineering – Chen (Ed.)
© 2016 Taylor & Francis Group, London, ISBN 978-1-138-02936-1

Effect of different flakiness ratios of two-liquid bimetal casting on the technology and microstructure

Y.C. Zhu
School of Materials Science and Engineering, Harbin Institute of Technology, Harbin, China
Institute of Materials Science and Engineering, Jiamusi University, Jiamusi, Heilongjiang, China

Z.J. Wei
School of Materials Science and Engineering, Harbin Institute of Technology, Harbin, China

S.F. Rong, C.L. Ma & Y.H. Wu
Institute of Materials Science and Engineering, Jiamusi University, Jiamusi, Heilongjiang, China

ABSTRACT: The two-liquid bimetal casting technique, which is being widely researched and used, provides a feasible way to the large plate composite materials among dissimilar metals. A prerequisite to realize the complete metallurgical bonding between dissimilar bimetals was proposed in view of the bimetal casting categories. The bimetal casting can be divided into two categories, namely plate casting and rod casting. The crusher hammer and wear plate were chosen to prepare the two-liquid bimetal, and the low alloy steel upside surface was solid or liquid state due to the difference in the flakiness ratio, which determined the timing of the poured high chromium white cast iron.

Keywords: Two-liquid bimetal; Metallurgical bonding; Plate casting and rod casting

1 INTRODUCTION

Traditional single wear-resisting materials were unable to meet the requirement of sustained load conditions [1, 2]. Two-liquid bimetal casting technique has been widely researched and used, and which provides a feasible way to the large plate composite materials among dissimilar metals [3, 4]. Existing two-liquid bimetal casting techniques cannot be definitely classified based on two kinds of the molten metallic solution before the gravity casting [5, 6]. So, many studies and applications were built around a certain problem or product, and enough attention was not paid to their correlation and differentiation. How to define the two-liquid bimetal was an absolute essential, which will establish a foundation platform for advanced research [7].

During the initial research period, the bimetal casting can be divided into two categories, namely plate casting and rod casting [8]. A prerequisite to realize the complete metallurgical bonding between dissimilar bimetals can be proposed in view of the bimetal casting categories. According to the fundamental theory, two-liquid bimetal can be strictly defined.

2 EXPERIMENTAL METHODS

2.1 Material contents and preparation

In accordance with the different performance demand in different parts, two-liquid bimetal was prepared between the high chromium white cast iron (HCWCI) and the low alloy steel (LAS). Two kinds of alloy chemical contents are listed in Table 1.

Table 1. Two kinds of alloy chemical contents (wt%).

Contents	C	Cr	Mn	Mo	Si
LAS	0.3	0.5	0.6	≥0.2	≤0.4
HCWCI	3.0	18.0	1.5	≥0.2	≤0.4

Table 2. Poured temperatures (°C) of the LAS and HCWCI.

Casting type	LAS	HCWCI
Rod casting	1436	1550
Plate casting	1420	1550

For better controlling the poured temperature, two medium-frequency furnaces are necessary to melt the HCWCI and LAS. First, the melted LAS was poured into the CO_2 sodium silicate-bonded sand mold cavity. When the LAS casting surface was in a solid mixed state or a complete liquid state, the HCWCI was immediately poured into the rest of the mold cavity at a certain temperature. The actual temperature of the LAS in the sand mold cavity and the poured temperature of the HCWCI are listed in Table 2. The LAS casting surface phase has a remarkable effect on the microstructure and the metallurgical bonding quality.

2.2 Experimental methods

Two KGPS-800 medium-frequency furnaces with a melting capacity of 150 kg and 250 kg were, respectively, used for melting the LAS and HCWCI. The temperature field was detected by the Sixteen Channel Tour-inspector. The microstructure was observed by using the OLYMPUS-GX71 electronic microscope and the scanning electron microscope. The alloy element contents and distribution were analyzed by energy dispersive X-ray spectroscopy along the line.

3 EXPERIMENTAL RESULTS AND ANALYSIS

3.1 Technology and microstructure of the rod casting bimetal

The flakiness ratio of the crusher hammer is not over 5, in line with the characteristic dimensions of the rod casting. So, it is adopted for preparing and studying the bimetal compound technology and microstructure, as shown in Figure 1a, b.

When the LAS is kept in the solid state at 1436°C by the Tour-inspector, the HCWCI is poured into the sand cavity at 1550°C, as shown in Figure 1a. The crusher hammer is shown in Figure 1b. The

LAS begins to melt at the upside surface under the large quantity heat. With the evolution of heat, the interface between the LAS and the HCWCI gradually solidifies and forms the composition layer. From Figure 1c, it can be seen that the composite layer is clear and distinguishable. Moreover, there is not any oxidative inclusion content in casting. Between the composition layer and the LAS, there is a transition region, and for the same reason, there is a transition region between the composition and the HCWCI. The above-mentioned features are attributed to the elements diffusion that is divided into two phases: the elements diffusion of the alloy process and the elements diffusion of the solidification process. The former causes the composition layer evolution, but the latter directly leads to the transition region on both sides of the composition layer.

As shown in Figure 2a, the composition layer and the transition region can be seen clearly. The LAS microstructure is the ferrite and pearlite, and the HCWCI microstructure is the $(Cr, Fe)_7C_3$ and austeinitic. The particle of the carbide is present in the composition layer matrix, and the interface between the particles and the matrix is clearly seen in Figure 2b. These particles take on circle or oval. The EDAX is used to analyze the position A, and was found that these particles are the disperse carbide.

3.2 Technology and microstructure of the plate casting bimetal

The flakiness ratio of the wear plate is over 5, and conforms to the characteristic dimensions of the plate casting. So, the wear plate is adopted for preparing and studying compared with the rod casting, as shown in Figure 3a, b.

To ensure the LAS in the liquid state on the upside, the LAS is tested at 1420°C by the Tour-inspector, and then the HCWCI is poured into the sand cavity at 1550°C, as shown in Figure 3a. So, the technology method can be called the Liquid Film

a   b   c

Figure 1.   Crusher hammer and composition layer microstructure: a) sand mold; b) crusher hammer; c) composition layer microstructure.

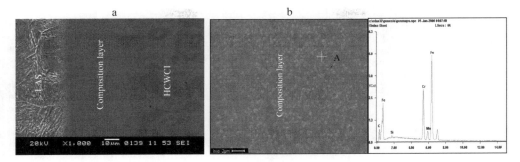

Figure 2. Composite layer and EDAX analysis: a) composition layer; b) energy spectrum of point A.

Figure 3. Wear plate and composition layer microstructure: a) sand mold; b) wear plate; c) composition layer microstructure.

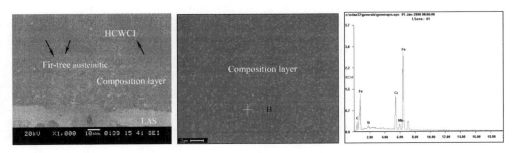

Figure 4. Composite layer and EDAX analysis: a) composition layer; b) energy spectrum of point B.

Bonding (LFB). The wear plate is shown in Figure 3b, and the flakiness ratio of the wear plate is over 10. As shown in Figure 3c, the composition layer is clear and straight, but there are vertical fir-tree crystals growth along the transition region. The LAS liquid film under the scour action of HCWCI is diluted, and the alloying element from the near to the distant shows the continual gradient change from low to high. So, the amount of eutectic blocks and eutectic carbide increases from the adjacent parts of the transition region, and beyond is the $(Cr, Fe)_7C_3$ and $(Cr, Fe)_{23}C_6$. The LAS microstructure is also the ferrite and pearlite.

The constitutional supercooling theories related to the formation of the microstructure are the major cause of the fir-tree austeinitic. There is a large mount of disperse granular carbide in austenitic and the composition layer, as shown in Figure 4a, and the chemical contents are similar to that shown in Figure 2b. Thus, the LAS upside surface is the solid or liquid state, which directly concerns with the microstructure of the transition region of the composition layer.

4 CONCLUSIONS

A prerequisite to realize the complete metallurgical bonding between dissimilar bimetals was proposed in view of the bimetal casting categories.

The bimetal casting can be divided into two categories, namely plate casting and rod casting, and the LAS upside surface was the solid or liquid state due to the difference in the flakiness ratio, which determined the timing of the poured HCWCI.

Two types of composition layers had the similar disperse granular carbide, but the transition region of the wear plate had the fir-tree austeinitic, and the amount of eutectic blocks and eutectic carbide increased from the adjacent parts of the transition region.

ACKNOWLEDGMENTS

This work was financially supported by the National Natural Science Foundation of China Project under Grant No. 51371090, and the Science and Technology Support Program of 12th Five-Year Plan under Grant No. 2011BAD20B03010401, and the Educational Department Surface Project of Heilongjiang Province under Grant No. 12521519, and the Cultivation Plan of the New Century Excellent Talents of Heilongjiang Province under Grant No. 1155-NCET-017, and the College Student Science and Technology Innovation of Heilongjiang Province under Grant No. 201410222037, and the College Student Science and Technology Innovation of Jiamusi University under Grant No. xsld2014–002, and the Graduate Student Science and Technology Innovation of Jiamusi University LZR2014_007.

REFERENCES

[1] Xiong. G.R, Zheng. M.Z, Zhao. L.Z, Research state on technology of metal matrix composites prepared by casting process. Casting Technology, J. 2006, (4): 563–565.

[2] Rong. S.F, Liu C, Guo. J.W, Influence of casting technique on the mechanical properties of bimetal composites. Journal of Iron and Steel Research International, J. 2011, (31): 5–6–2.

[3] An. J.Y, Qi. J.Y, Microstructure and properties of bimetal composite on high chromium cast iron and carbon steel, J. Water Conservancy and Electric Machine, 1992, (8): 49–53.

[4] Asta M, Beckermann C, Karma A, Solidification microstructures and solid-state parallels: Recent developments, future direction, J. Acta Materialia, 2009, (57): 941–971.

[5] Liang H, Xie. Z. Wu. Y, C, Research on continuous core-filling casting forming process of copper-clad aluminum bimetal composite material, J. Acta Metall. Sin. (Engl. Lett.), 2010, (23): 206–214.

[6] Xiao. X.F, Ye. S.P, Yin. W.X, High Cr white cast iron/ carbon steel bimetal liner by lost foam casting with liquid-liquid composite process, J. China Foundry, 2012, (2): 136–142.

[7] Prasad. B.K, Modi. O.P, Patwardhan. A.K, Effects of some material and experimental variables on the slurry wear characteristics of Zinc-Aluminum Alloys, J. Mater Eng. Perform, 2001, (10): 75–80.

[8] Rong. S.F, Zhu. Y.C, Study on lining board with bimetal liquid composite casting, J. Advanced Materials Research, 2011, (317): 158–161.

Material Science and Engineering – Chen (Ed.)
© *2016 Taylor & Francis Group, London, ISBN 978-1-138-02936-1*

Comprehensive research of carbide indexable toll edge cutting simulation

F. Li

Avic Chengdu Engine (Group) Co. Ltd., Chengdu, Sichuan, China

ABSTRACT: The orthogonal cutting force was established in the precision carbide chip in the light of classical model in this paper. The model reveals the relationship between the blade edge and the force, and shows the natural characteristic of plastic deformation in the cutting area of the workpiece. This paper also discusses about the effect of edge preparation of the cutting tool (round edge and chamfer edge) on cutting force and tool wear in orthogonal as determined with Finite Element Method (FEM) simulations. The results obtained in this paper provide a fundamental understanding of the process mechanics for cutting with realistic cutting tool edges and may assist in the tool edge design.

Keywords: orthogonal cutting force; blade edge; cutting characteristic

1 INTRODUCTION

In the process of metal cutting, the cutting edge has an important influence on the cutting state. At the beginning, the cutting edge contacts the workpiece, and the tool cutting part is mainly concentrated on the edge region and its adjacent area, and the mechanical strength of the edge and the adjacent area. Geometric parameters play a crucial role in the process of cutting force-thermal characteristic and tool wear. During the cutting process, the tool–workpiece contact area increases gradually compared with the tool–chip contact area. The extrusion and friction between the tool and the workpiece plays a key role in the cutting process, and it is more obvious that the different cutting edges of the blades have an effect on the cutting state.

2 CUTTING MECHANISM OF CARBIDE CUTTING TOOLS

The radius of the transition circle of the cutting tool edge is one of the main parameters in the characterization of the blade. The size of the edge radius reflects the degree of sharp cutting edges and affects the cutting process. According to the difference in the transfer form at the cutting edges, the blade shape can be subdivided: sharp blade, blunt round blade, and chamfer edge, which are, respectively, establish the corresponding cutting force models. It is assumed that the cutting edges are absolutely sharp cutting models on the basis of which the following hypothesis is proposed:

The deformation of the workpiece material is in plane strain state;
The workpiece material is isotropic, homogeneous continuous and incompressible;
Cutting process is a steady state process, and there is no obvious product cut tumor;
The temperature rise caused by the wear of the back blade is small compared with that caused by the shear cutting, and its impact on the shear zone material should be ignored.

2.1 *Blunt round blade 2d orthogonal cutting force model*

The place around the cutting edge will generate a sliding phenomenon because of the existence of a blunt edge, similar to a negative rake angle acting on the material of the workpiece and part of the material of the workpiece is squeezed and discharged through the flank face of the tool. This is the slip effect that the cutting tool edge has on the material of the workpiece. Under the action of the slip effect, a force on the edge blade, which is defined as $\vec{F_e}$, and the shearing force $\vec{F_r}\,'$ constitute the cutting force, as shown in Figure 1.

The blunt round radius is r_n and the actual rake angle at the diversion point o is γ_0'. The resultant force at the blade edge is the force that is produced by the chip when flowing to the rake face of the tool. The expressions of the blunt edge cutting force are given as follows:

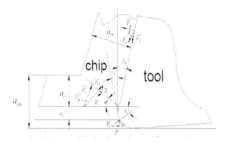

Figure 1. Blunt round blade orthogonal cutting force model.

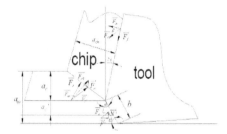

Figure 2. Chamfer edge orthogonal cutting force model.

$$
\begin{cases}
\overrightarrow{F_{\triangle}} = \dfrac{\pi a_w r_n \sigma}{180} + \dfrac{k a_c a_w \cos(\beta - \gamma_0)}{\sin\varphi \cos(\varphi + \beta - \gamma_0)} \\
\overrightarrow{N_{\triangle}} = \left(-\dfrac{\pi a_w r_n \sigma}{180}\right) + \dfrac{k a_c a_w \sin(\beta - \gamma_0)}{\sin\varphi \cos(\varphi + \beta - \gamma_0)}
\end{cases}
\tag{1}
$$

2.2 Chamfer edge 2d orthogonal cutting force model

At the chamfer edge, the cutting force can be decomposed into two forces, namely $\overrightarrow{F}'$ which is parallel to the main cutting force and $\overrightarrow{N}'$ which is perpendicular to the workpiece surface. The cutting model is shown in Figure 2.

The chamfer width is b, its angle is ϕ, and the expressions of the cutting force are given as follows:

$$
\begin{cases}
\overrightarrow{F_{\triangle}} = a_w \sigma \sin^2\theta \cdot \dfrac{b^2}{2} + \dfrac{k a_c a_w \cos(\beta - \gamma_0)}{\sin\varphi \cos(\varphi + \beta - \gamma_0)} \\
\overrightarrow{N_{\triangle}} = a_w \sigma \sin\theta \cdot \cos\theta \cdot \dfrac{b^2}{2} + \dfrac{k a_c a_w \sin(\beta - \gamma_0)}{\sin\varphi \cos(\varphi + \beta - \gamma_0)}
\end{cases}
\tag{2}
$$

3 DIFFERENT BLADE SHAPE CUTTING FINITE ELEMENT MODELS

Three different blade shape models are established in the deform software, and the parameter settings are as follows: sharp blade—the rake angle and

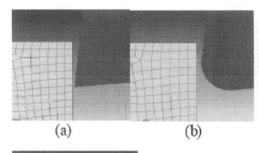

(a)   (b)

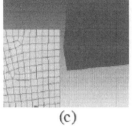

(c)

Figure 3. Three different finite element models.

the clearance angle are 5°, the transition radius is 0.01 mm; blunt—the rake angle and clearance angle are 5°, the transition radius is 0.1 mm; chamfer—the rake angle and clearance angle are 5°, the chamfer angle is 10°, the chamfer width is 0.1 mm. The actual rake angle is positive in the sharp blade cutting process; on the contrary, the blunt blade and the chamfering edge are negative, as shown in Figure 3(a, b, c). Because the contact with the workpiece between the blunt edge and the chamfer edge contact is in a different way and the flow of material edge is also in a different way, two factors together result in the difference of the size of the cutting force and the distribution of the temperature field.

The simulation model material and the cutting condition set are presented in Table 1.

4 RESULT OF THE SIMULATION ANALYSIS

4.1 Cutting force analysis

The generation of the cutting force is one of the most important phenomena in the cutting process, which directly affects the cutting temperature and tool wear and the machined surface quality. The blunt edge takes its different edge radii, i.e. $r_{n1} = 0.05$ mm, $r_{n2} = 0.1$ mm, $r_{n3} = 0.2$ mm. The cloud images of the cutting force stress are shown in Figure 4.

We set the friction angle $\beta = \arctan \mu = 16.7°$ and the shear angle $\varphi = \pi/4 - \beta/2 + \gamma_0/2 = 28.3°$, and estimate $r_n = (0.05$ mm, 0.1 mm, 0.2 mm) in the blunt cutting force model (2–1) as a function of

Table 1. Cutting model parameters.

Tool material	Workpiece material	Friction type/ coefficient	Cutting thickness (mm)	Cutting speed (m/min)	Feed (mm)	Origin temperature (°C)
WC	45 steel	Shear friction/0.5	0.2	100	0.25	20

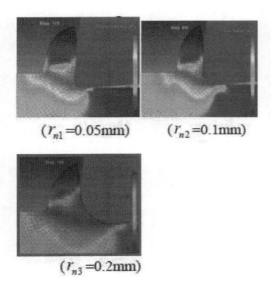

$(r_{n1} = 0.05\text{mm})$ $(r_{n2} = 0.1\text{mm})$

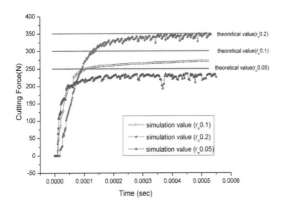

$(r_{n3} = 0.2\text{mm})$

Figure 4. Cutting force stress cloud images of the blunt edge.

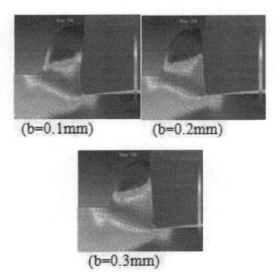

$(b = 0.1\text{mm})$ $(b = 0.2\text{mm})$

$(b = 0.3\text{mm})$

Figure 6. Cutting force stress cloud images of the chamfer edge with different widths.

Figure 5. The comparison of theoretical values and simulation values.

time, and calculate the theoretical value compared with the simulation values. The results are plotted in Figure 5.

From the stress cloud images, we can see that the stress region increases with the increase in the blunt round radius. This is due to the increased

contact area between the rake face of the tool and the chip. Compared with theoretical values and simulation values of cutting force, the conclusion is that with the increase in the blunt round radius value, the theoretical result is slightly larger than the simulation one, with the same trend.

Chamfer edge is determined by two parameters, namely angle (ϕ) and width (b), so the finite element simulation can be divided into two groups. One group is angle unchanged ($\phi = 10°$), in which we change the width of the chamfer edge to 0.1 mm, 0.2 mm, and 0.3 mm. The stress cloud image is shown in Figure 6

We set $\phi = 10°$ and b = (0.1 mm, 0.2 mm, 0.3 mm) into the cutting model (2–2), and calculate the theoretical values compared with the simulation values, as shown in Figure 7.

The other group is width unchanged (b = 0.2 mm), in which we change the angle of the chamfer edge to 5°, 10°, and 20°. The stress cloud images are shown in Figure 8.

We set b = 0.2 mm and $\phi = (5°, 10°, 20°)$ into the cutting model (2–2), and calculate the theoretical values compared with the simulation values, as shown in Figure 9.

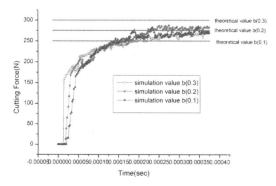

Figure 7. The comparison of theoretical values and simulation values.

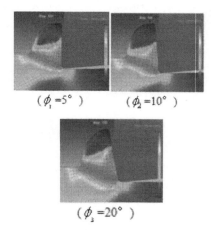

$(\phi_1 = 5°)$ $(\phi_2 = 10°)$

$(\phi_3 = 20°)$

Figure 8. Cutting force stress cloud images of the chamfer edge with different angles.

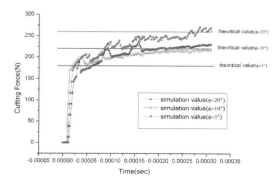

Figure 9. The comparison of theoretical values and simulation values.

From Figure 6–Figure 9, we can see that the stress concentration expands with the increase in the chamfer width and angle. This is mainly because when the chamfer width increases, the con-

tact length between the cutter chamfer and the chip increases, and the friction between the tool and the chip makes the cutting force to increase. Furthermore, when the chamfer angle increases, the negative rake angle increases, the process of squeeze between the cutting tool and the workpiece deepens, and then promotes the improvement of the cutting force. We can also see that with the increase in the width of the chamfer, the cutting force increases to a smaller extent. Compared with the chamfer angle's effect on the cutting force, the influence of the chamfer width on the cutting force is small.

By comparison with the simulation values and the theoretical values, we can find that under the same condition of the material, the cutting tool angle, the cutting speed, the feed and other related cutting parameters, the cutting force of the blunt blade increases with the increasing radius, and the cutting force of the chamfer edge increases with the increase in the width and angle of edge. The simulation and theoretical values are consistent during the simulation process; however, most of the simulation values compared with the theoretical values are somewhat smaller, so the simulation value can be seen as a safe value and can be used in the actual cutting process.

4.2 Tool wear analysis

In the cutting process, carbide cutting tools are prone to collapse or micro collapse for various reasons, so it is necessary to study the tool edge wear of different blade cuttings. From the finite element simulation of tool wear of three different edges, we can see that the chamfer edge tool wear value is 0.00834 mm, the sharp edge tool wear value is 0.0303 mm and the blade abrasion value is 0.0645 mm, which is the largest as shown in Figure 10. On the one hand, the existence of blunt strengths and the intensity of the blade prolongs the tool life; on the other hand, the increase in cutting deformation results in higher temperature, and reduces the tool life. Therefore, it is necessary to choose a suitable blunt round radius.

We conduct the finite element wear analysis on the geometry parameters of the blunt edge and the chamfer according to Section 3.1. The different wear values are shown in Figure 8, from which we can see based on the same cutting condition that different blunt edges of the tool have a big different wear value. When the radius is 0.1 mm, the wear has a relative minor value, and there is a minimum theoretical value of tool wear when the chamfer edge angle sets a big value, while the width sets a small one from Figure 11. So, the choice of different geometric dimensions of the blunt edge and the chamfer edge has a great influence on the strength and life of the tool.

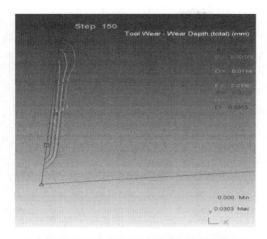

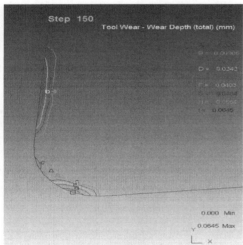

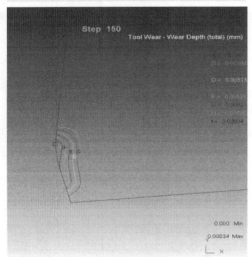

Figure 10. Cutting tool wear cloud images of different finite element models.

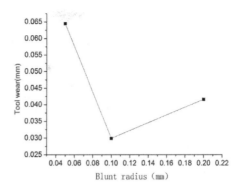

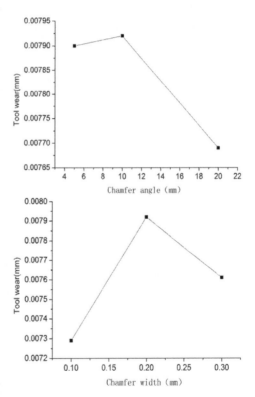

Figure 11. The comparison of tool wear of different finite element models.

5 CONCLUSION

1. This paper draws the 2d orthogonal cutting force model of the blunt round edge and the chamfer edge. Based on the 2d orthogonal cutting force model of the sharp edge, three theoretical formulas of the cutting force are presented.

2. The simulation cutting force values of the blunt edge and the chamfer edge are obtained by

finite element simulation. To a certain extent, the simulation value matches the theoretical value, and then the effect of the blade shape on the force is analyzed.
3. The wear values of the blunt edge and the chamfer edge are obtained by finite element simulation, which show the method to reduce the wear by contrast analyzing, and prepare for the further calculation of tool life.

Studies have shown that under the given cutting conditions, the cutting edge blunt round can strengthen the cutting edge, but generate more heat, with more tool loss. So, it exists a best theory value for the transition radius of the cutting edge, making the cutting performance and tool life in a state of equilibrium. Whether models based on two-dimensional can accurately reflect the state of three-dimensional cutting needs further experiments.

REFERENCES

[1] Yung-Chang Yen, Anurag Jain, Taylan Altan. A finite element analysis of orthogonal machining using different tool edge geometries. Journal of Material Processing Technology 146 (2004) 72–81.
[2] J.M. Zhou, H. Walter, M. Andersson, J.E. Stahl. Effect of chamfer angle on wear of PCBN cutting tool. International Journal of Machine Tools & Manufacture 43 (2003) 301–305.
[3] Zhang Weiji. The theory of metal cutting and cutting tools. Zhejiang University Press, 2005.6.
[4] Shi Hanmin. Metal Cutting Theory and Practice— A New Perspective. Huazhong University of Science and Technology Press 2003.12.
[5] Yang Shucai. Precision cutting of titanium alloy Ti6Al4V cutter blade action mechanism and application research Harbin. Ph.D. Dissertation, university of science and technology 2011.8.
[6] Chen Riyao Metal cutting principle (2nd Ed) China Machine Press 2002.1.

Material Science and Engineering – Chen (Ed.)
© 2016 Taylor & Francis Group, London, ISBN 978-1-138-02936-1

Microstructure of plasma surface iron alloy coating reinforced by pre-coated SiC$_p$

X.D. Du

School of Material Science and Engineering, Hefei University of Technology, Hefei, Anhui, China

Z.L. Song

Technical Center, Anhui Jianghuai Automobile Co. Ltd., Hefei, Anhui, China

Z. Dan, J. Shen & G.F. Liu

School of Material Science and Engineering, Hefei University of Technology, Hefei, Anhui, China

ABSTRACT: In this work, the iron alloy coating reinforced by pre-coated SiC particles was coated on the substrate of medium carbon steels by means of the plasma cladding process. The microstructure of coating and the influence of SiC particles were analyzed. The results show that pre-coating can protect the SiC particles from decomposition during the coating process. The undissolved SiC particles distribute dispersedly and uniformly in the coating. The microstructure of the coating matrix can be influenced by adding the SiC particles and becomes more complicated, which is quite different from the top to the bottom of the coating. It consists of the cellular crystal at the bottom part, the dendrite crystal at the middle part and the hypoeutectic crystal at the top part.

Keywords: coated SiC$_p$; iron alloy coating; microstructure; pre-coated; plasma; composite; particle reinforce; surface; hardness; microstructure; iron alloy

1 INTRODUCTION

Particle-Reinforced Metal Matrix Composites (PRMMCs) have the properties of both metals (good ductility and toughness) and ceramics (high strength and Young's modulus) [1~5]. So, these are attractive candidates for the components that need good abrasion resistance and impact wear resistance. For improving the surface properties of metallic components, PRMMC coatings are often obtained by surface modification techniques. Ceramic particles, such as WC, SiC and Cr$_2$C$_3$, are usually added into the molten pool to form a PRMMC coating on the metal substrate. There are many attempts in this respect. For example, the WC$_p$ (WC particles)/Fe alloy coating [6, 7], the WC$_p$/Ti6-Al-4V coating [8], the WC$_p$/Ti coating [9], the SiC$_p$/Al-Si alloy coating [10], and the SiC$_p$/Ti–6Al-4V coating [11] have already been carried out. These can greatly improve the wear resistance and prolong the service life of the components. It is well known that SiC particles have high hardness, high Young's modulus, good wear resistance and low cost, and are often used for reinforcing the coating of aluminum alloy [10] or titanium alloy [11]. However, these components are unsuitable for reinforcing iron alloys because of the strong reaction with the iron alloy in the liquid state during the coating process because of the high dissolution of SiC$_p$. J. D. Majumdar [12, 13] investigated the laser cladded SiC$_p$+AISI316 L stainless steel, and found that the SiC particles were dissolved mostly. G. Abbas prepared the coating with the mixed powder containing 20 wt% SiC and stainless steel powders by laser, and pronounced that the SiC particles were dissolved completely [14]. M. H. Guo [15] found that the SiC$_p$ particles with small size were dissolved completely after injecting into the molten pool of low carbon steel. So, it is hard to form the PRMMC coat reinforced by SiC particles without pre-coating. In this paper, the Powder Immersion Reaction-Assisted Coating Process (PIRAC) [16] was used to form a protective layer around the SiC particle by reacting with Cr powders for protecting the SiC$_p$ in the molten iron alloy. The pre-coated SiC particles were then injected into the molten pool to form the composite coating. The microstructure was analyzed subsequently.

2 EXPERIMENTAL DETAILS

The medium carbon steel was used as substrate material, which were cut into 100 mm × 50 mm × 10 mm

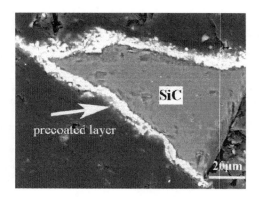

Figure 1. Morphology of coated SiC_p.

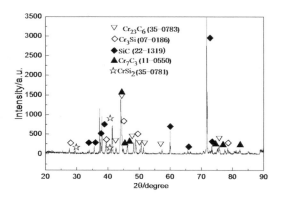

Figure 2. XRD pattern of pre-coated SiC_p shown in Figure 1.

and followed by grounding and degreasing. The self-fluxing FeCrBSi alloy powders (45–109 μm) with the composition of 0.1C-15Cr–1.0B-1.0Si-bal Fe (wt.%) and the angular SiC particles with an average size of 100 μm were prepared for binding or reinforcing. The SiC particles were pre-coated with Cr by means of PIRAC.

A plasma surfacing equipment (PTA–400-D1–ST) was employed to manufacture the coating. The main part is the plasma torch, whose schematic drawing is shown in Figure 1. The parameters are listed as follows: the current was 130 A, the travel velocity was 90 mm/min, the swing width was 22 mm, the torch swing frequency was 22 min^{-1}, the feeding gas (Ar) flow was 5 L/min, the plasma gas (Ar) flow was 5 L/min, and the protective gas (Ar) flow was 13 L/min.

Optical Microscope (OM) and scanning electron microscope (SEM, Card Zeiss EvoMA15) utilizing secondary and backscattered electron imaging (SEI and BEI, respectively) were used for examining the microstructure. The XRD (D/MAX2500VL/PC) with Cu-K$_\alpha$ radiation was used for analyzing the phases in the coating.

3 EXPERIMENTAL RESULTS

3.1 Pre-coated SiC particles

SiC particles were changed from green to gray during the precasting process, whose structure is shown in Figure 1. There is a bright layer around the SiC particle. The XRD analysis (Fig.2) reveals that some new phases such as Cr_3Si, $CrSi_2$, Cr_7C_3, $Cr_{23}C_6$ are formed and constitutes the protective layer. Because carbon atoms diffuse easily compared with silicon atoms, it can be inferred that the microstructure of the pre-coated layer consists of carbides in the sequence of $CrSi_2/Cr_3Si/Cr_3Si+Cr_7C_3/Cr_7C_3/Cr_{23}C_6$ from the inner to the outer layer.

P.T.A. PROCESS plasma transferred arc

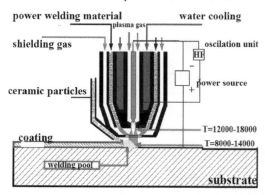

Figure 3. Schematic diagram of the plasma surface process.

3.2 Microstructure of coating

During the coating process, the plasma beam with high temperature can be formed between the workpiece and the tungsten electrode, as shown in Figure 3. The metallic powders are fed into the plasma beam and then melted to form a molten pool. By moving the plasma torch, the pool goes forward and forms a tail vat. Meanwhile, the SiC particles are injected into the pool tail vat. If the SiC particles do not dissolve completely during the process, the coating will trap them after rapid solidification.

The photographs of the upper of the coating and the cross section show that the SiC particles are embedded in the coating (Fig. 4a, b). At the upper of the coating, the SiC particles are not dissolved significantly and are angular in shape, being very similar to the their original shape (Figure 4c). The severe dissociation does not occur even for the interior of the coating, as shown in Figure 4d. The cross-section's photograph (Figure 4b) shows that the SiC particles are distributed dispersedly and uniformly. There are about 2045 particles per square centimeter, as shown in Figure 4b. Conversely, the SiC particles without pre-coating are dissolved entirely, i.e. none of the SiC particles existed, as shown in Figure 4e. It indicates that the pre-coating can prevent the SiC particles from dissolving and give them a better distribution.

For comparison, we investigated the iron alloy coating without the addition of SiC. The XRD

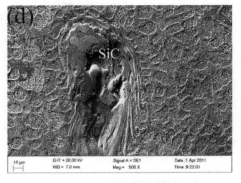

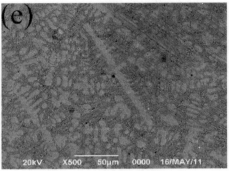

Figure 4. Distribution and morphology of retained SiC particles in coating: (a) distribution of pre-coated SiC shown from the top view of the coating; (b) distribution of pre-coated SiC shown from the cross-section; (c) single pre-coated SiC morphology in the coating surface; (d) single pre-coated SiC morphology in the interior of the coating; (e) microstructure of the cross-section reinforced by uncoated SiC particles.

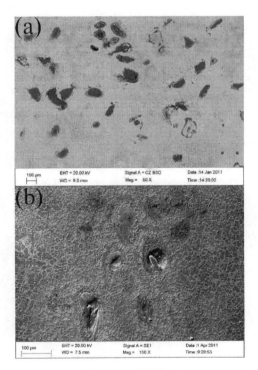

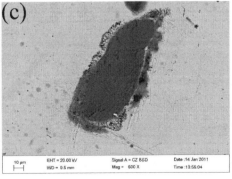

Figure 4. (*Continued*).

result, as shown in Figure 5, reveals that the main phases consist of α-(Fe, Cr), γ-(Fe, Cr) solid solution and M_7C_3 carbide. Figure 6 shows the microstructure in the upper and the bottom parts of the iron alloy coating. It consists of the primary phase and the eutectic mixture, and exhibits the cellular crystal in growth morphology (Figure 6b). However, there are no dendrite in the two parts.

Figure 7 displays the micrographs of different regions in the coating added with SiC particles, and shows that the addition of SiC can influence the matrix microstructure significantly. At the bottom of the coating, as shown in Figure 7a, the microstructure consists of primary crystal and eutectic, and is similar to that without the addition of SiC. It consists of hypoeutectic crystal at the middle of the coating, and some dendrites exist as a typical distinction (Figure 7b). In the upper part, it consists of finer hypereutectic crystals, which differ from the former greatly (Figure 7c).

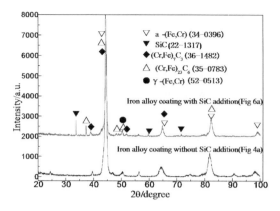

Figure 5. XRD patterns of two kinds of coatings.

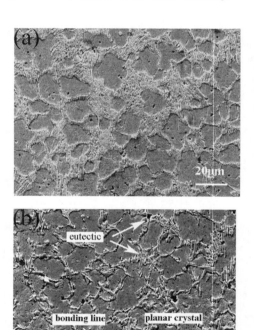

Figure 6. Microstructure of iron alloy coating without the addition of SiC particles: (a) microstructure of the upper part of the cross section; (b) bottom part of the cross-section.

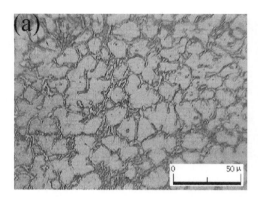

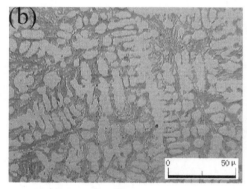

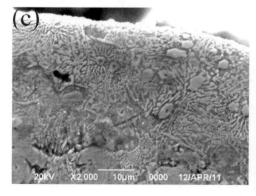

Figure 7. Microstructure of coating with the addition of pre-coated SiC particles: (a) microstructure of the bottom part of the cross section (OM); (b) microstructure of the middle part of the cross section (OM); (c) microstructure of the upper part of the cross section (SEM).

4 DISCUSSION

The majority of ceramic/metal couples are non-equilibrium systems due to the presence of the chemical potential gradient across the interface. It means that the diffusion or the chemical reac-tions can easily take place at the interface. Unde-sirably, a severe interfacial reaction results in a significant dissolution of reinforcement and for-mation of the reaction products, which are harm-ful to the properties. It is well know that SiC is easy to react with the iron alloy binder, in par-ticular when the binder phase is in liquid state. Therefore, how to control the interfacial reaction is always the key factor for preparing the ceramic/metal composites. Generally, diffusion welding processes can be used as the most common methods to obtain chemical bonding for SiC/iron alloy composites. In this process, the reaction will prolong up to several hours. The thickness of the reaction layer depends on time and can be estimated by the formula $X^2 \propto t$. Although the plasma surface process, in which the temperature in molten pool is very high and the reaction time is much shorter, cannot be controlled easily, it turns out to be feasible for decreasing the dissolution of SiC particles because of the shorter heating time. In this investigation, the SiC particles remain in the iron alloy matrix, which is different from that reported in a previous study [15] conducted with uncoated SiC particles. It indicates that pre-coating may be feasible for protecting the SiC particles. As the pre-coated SiC particles are fed into the molten pool, the outer layer consists of Cr_7C_3, $Cr_{23}C_6$ carbides contacting with the liquid metal initially and preventing the SiC particles from dissolving to some degree. However, the outer layers could also be damaged finally, because the Cr_7C_3, $Cr_{23}C_6$ carbides would be decomposed into Cr and C atoms above 1323 K in the iron or steel matrix [16].

When the coated SiC particles are fed into the pool tail, some that have contact with the plasma arc are decomposed completely, some are left on the molten surface, and others penetrate into the pool and dissolve partly. According to J. A. Vreeling [11], only when the initial velocity of particles was higher than v_{min}, could the particles overcome the surface barrier and penetrate into the pool. Subsequently, as moving down into the molten pool, these particles are slowed down by stokes' force and buoyancy. SiC particles must get sufficient kinetic energy to overcome the barrier. However, it is considerably difficult for SiC_p due to its lower density. Therefore, the particles need a much higher velocity to reach the bottom of the pool. This is the reason why there are few SiC particles at the bottom of the coating.

At the top of the molten pool, the interfacial reaction is less intensive due to the lower temperature and the higher velocity of solidification. Hence, some SiC particles at the top of the coating are not destroyed and keep the original shape (Figure 4c). However, as discussed above, some particles will be decomposed partly at the top of the molten pool. These partially dissolved SiC particles move down continually with the outer layer of the pre-coating particles, and dissolve and stay at the top region of the molten pool [17]. It changes the matrix composition of the top region. The microstructures change from the hypoeutectic structure to the hypereutectic one.

In addition, the solidification could be speeded up by the Ar gas flow. These all lead to the formation of the finer hypoeutectic structure at the top part (Figure 7c).

When penetrating further into the interior of the molten pool, the undissolved SiC particles move in a slower rate. So, the amount of the SiC_p particles reaching the interior of the pool decreases sharply, which indicates that the concentration of carbon and silicon is lower in this region. As a result, the composition below the top part is different from that at the top region. The middle part of the molten pool consists mainly of hypoeutectic and could be found in dendrite shape obviously. It might be caused by the dissolution of SiC particles. The increase in the concentrations of carbon and silicon promotes the constitutional supercooling. Thus, the crystal tends to grow up in the form of dendrite. Few SiC particles could reach the bottom, so the matrix composition does not change obviously, and the structure is similar to that in the coating without SiC addition.

5 CONCLUSIONS

The pre-coated SiC particles reacting with Cr powders can form protective layers on their surface. The layers can prevent the SiC particles from dissolving during the coating. It is an effective method to protect the SiC particles during the plasma cladding process of the PRMMC coating.

At the top of the molten pool, the cooling rate is higher. The SiC particles cannot be deteriorated severely and can keep their original morphology. Due to the lower density, the SiC particles are not easy to penetrate into the molten pool. Therefore, the amount of the SiC particles is lower in the interior than that at the top of the coating

In the interior of the coating, the addition of SiC_p influences the microstructure of the iron alloy matrix. At the middle part, severe dissociation cannot occur and the composition is still in the hypereutectic range. The crystal tends to form dendrite. Few SiC particles can arrive at the bottom of the coating and change the composition of the matrix. So, the microstructure at the bottom is similar to that without SiC addition.

ACKNOWLEDGMENT

The author acknowledges the financial support from the State Key Development Program of Basic Research of China (No. 2011CB013402), and the Natural Research Fund of Education Bureau of Anhui Province (No. KJ2012A232).

REFERENCES

[1] I. Sahin, A.A. Eker, Analysis of microstructures and mechanical properties of particle reinforced AlSi7 Mg2 matrix composite materials, J. Mater. Eng. Perf. 20 (2011) 1090–1096.

[2] A.E. Karantzalis, A. Lekatou, E. Georgatis, H.T. Tsiligiannis, Mavros, Solidification observations of dendritic cast Al alloys reinforced with TiC particles, J. Mater. Eng. Perf. 19 (2010) 1268–1275.

[3] K. Venkateswarlu, S. Saurabh, V. Rajinikanth, R.K. Sahu, A.K. Ray, Synthesis of TiN reinforced aluminium metal matrix composites through microwave sintering, J. Mater. Eng. Perf. 19 (2010) 231–236.

[4] S.J. Fu, H. Xu, Microstructure and wear behavior of (Ti,V)C reinforced ferrous composite, J. Mater. Eng. Perf. 19 (2010) 825–828.

[5] C. Sun, M. Song, Z.W. Wang, Y.H. He, Effect of particle size on the microstructures and mechanical properties of SiC-reinforced pure aluminum composites, J. Mater. Eng. Perf. 20 (2011) 1606–1612.

[6] D.J. Liu, L.Q. Li, F.Q. Li, and Y.B. Chen, WC$_p$/Fe metal matrix composites produced by laser melt injection, Surf. Coat. Technol. 202 (2008) 1771–1777.

[7] G. Liu, M.H. Guo, H.L. Hu, Improved wear resistance of low carbon steel with plasma melt injection of WC particles, J. Mater. Eng. Perf. 19 (2010) 848–851.

[8] Y.T. Pei, V. Ocelik, and J.Th.M. De Hosson, Ti–6 Al–4V strengthened by laser melt injection of WCp particles, Acta Mater. 50 (2002) 2035–2051.

[9] S. Harsha, D.K. Dwivedi, A. Agarwal, Performance of flame sprayed Ni-WC coating under abrasive wear conditions, J. Mater. Eng. Perf. 17 (2008) 104–110.

[10] R. Anandkumar, A. Almeida, R. Colaço, R. Vilar, V. Ocelik, and J. Th. M. De Hosson, Microstructure and wear studies of laser clad Al-Si/SiC(p) composite coatings, Surf. Coat. Technol. 201 (2007) 9497–9505.

[11] J.A. Vreeling, V. Ocelik, and J.T.M. De Hosson, SiCp/Ti6 Al4V functionally graded materials produced by laser melt injection, Acta Mater. 50 (2002) 4913–4924.

[12] J.D. Majumdar, AjeetKumar, and L. Li, Direct laser cladding of SiC dispersed AISI 316 L stainless steel, Tribol. Int. 42 (2009) 750–753.

[13] J.D. Majumdar and L. Li, Studies on Direct Laser Cladding of SiC Dispersed AISI 316 L Stainless Steel, Metall. Mater. Trans. A. 40 (2009) 3001–3008.

[14] G. Abbas and U. Ghazanfar, Two-body abrasive wear studies of laser produced stainless steel and stainless steel + SiC composite clads, Wear. 258 (2005) 258–264.

[15] M.H. Guo, A.G. Liu, M.H. Zhao, H.L. Hu, and Z.J. Wang, Microstructure and wear resistance of low carbon steel surface strengthened by plasma melt injection of SiC particles, Surf. Coat. Technol. 202 (2008) 4041–4046.

[16] W.M. Tang, Z.X. Zheng, H.F. Ding, and Z.H. Jin, The interfacial stability of the coated-SiC/Fe couple, Mater. Chem. Phys. 77 (2002) 236–241.

[17] Y.B. Chen, D.J. Liu, F.Q. Li, and L.Q. Li, WCp/Ti-6Al-4V graded metal matrix composites layer produced by laser melt injection, Surf. Coat. Technol. 202 (2008) 4780–4787.

Material Science and Engineering – Chen (Ed.)
© 2016 Taylor & Francis Group, London, ISBN 978-1-138-02936-1

Characterization of Al-Mn particles in Mg-6Al-xMn magnesium alloys

L. Zhang, Z. Wu & J.H. Wang
School of Mechanical Engineering, Qinghai University, Xining, China

ABSTRACT: Mg-6Al-xMn magnesium alloy was prepared by the gravity casting process. The Al-Mn particles were observed in Mg-6Al-xMn alloy. These particles were characterized by Scanning Electron Microscope (SEM), Electron Probe Microanalyzer (EPMA), metallographic analysis software and X-Ray Diffractometer (XRD). It was found that Mn containing particles in Mg-6Al-0.3Mn specimens consist of Al_8Mn_5 and $Al_{11}Mn_4$, some of them contained iron and form $Al_8(Mn,Fe)_5$ inside which iron replaces some manganese atoms. These Mn containing particles exhibit three distinctive morphologies, the majority of the particles have angular block morphology; a few of them display a bar-like morphology and the balance exhibit fine granular morphology. Their quantity and size depended on the Mn content and solidification conditions. As the Mn content increased and solidification rate decreased, the amount and size of Al-Mn particles increase.

Keywords: Mg-6Al-xMn magnesium alloy; Al-Mn particles; microstructure; particle type

1 INTRODUCTION

Mg-Al-based alloys are among the most popular cast Mg alloys with wide commercial applications. It is well known, addition of manganese to magnesium alloys will change their corrosion resistance due to the formation of $Al_x(Mn,Fe)_y$ particles during solidification [1–2]. The composition of Al-Mn particles affects the cathode density, which dictates their electrochemical behavior. For example, particles with high Al concentration like Al_6Mn and Al_4Mn show a relatively low cathode current density, while those containing high Mn concentrations such as Al_8Mn_5 have a continuously high cathode current density [3–4]. Hiroshi M et al. [5] observed that the corrosion rate depended on the impurity concentration in the AM50 alloy and increased with Fe/Mn ratio. On the other hand, Al-Mn particles also play a role in dictating the mechanical behavior of Mg alloys. Small cracks resulted from Al-Mn intermetallic compounds on the fatigue crack surface of rolled AM60 may be responsible for oscillatory crack growth and crack arrest [6].

Many studies mentioned the effect of manganese on grain refinement of Mg-Al based alloys such as grain refinement with increasing Mn concentration. The reason is Al-Mn particles could act as heterogeneous nucleation sites leading to grain refinement [7–9]. It was observed that the metastable hexagonal ε-Al-Mn phase, which was added to the melt, can act as a nucleation in Mg-Al alloys [9]. Easton et al. [10] observed that Mn addition to Mg-Al alloys resulted in an increase in grain size when Al concentration was above 2.0 wt.% and the reason was probably due to the formation of less potent Al-Mn-carbides.

In cast Mg-Al alloys, particles containing aluminum and manganese may have various types, sizes and morphologies. Among them, $Al_{11}Mn_4$, Al_8Mn_5, Al_6Mn, Al_6Mn, Al_4Mn, AlMn were already reported [2, 4, 11–16]. The size of Al-Mn particles usually ranges from 0.1 to 30 μm [2, 11–14]. Various morphologies including needles, crosses, flower and angular blocky structures were reported [2, 13–19]. In addition, these particles can be observed within the grains and at boundaries [11, 15–16, 20]. For Mg-6Al-xMn cast alloys, the amounts of Al-Mn particles change with Mn content and locate preferably in the middle of ingots rather than on the surfaces. Unfortunately, this phenomenon has not been well understood. In this work, detailed characterization of the Al-Mn particles in Mg-6Al-xMn cast magnesium alloy was conducted, with a focus on the distribution, morphology and composition of Al-Mn particles.

2 EXPERIMENTAL

Mg-6Al-xMn alloys were prepared from pure Mg, Al and MnCl2 and their composition is shown Table 1. The alloys were melted under a protective gas cover (N_2+0.5%SF_6) in a stainless steel crucible, held for 30 min, and finally poured into a preheating mould. The casting and mould temperatures were set to 680°C and 180°C, respectively. After solidification, the ingots were removed from the steel mould using dry air.

The distribution of Mn in the ingots was examined. The specimens were polished and etched using a solution mixed with alcohol, acetate picric acid and water. Morphology and chemical composition of Al-Mn particles were examined using Scanning Electron Microscopy (SEM) equipped with an energy dispersive spectrometer. Backscattered electron mode was used to achieve a better contrast. Elemental analysis of the specimens was conducted using an Electron Probe X-ray Microanalyzer (EPMA). X-ray mappings were conducted to examine the distribution of Al, Mn, and Fe elements. Quantity and size of particles was conducted using Leica DMI3000M Metallographic analysis software. The phase constitutions were determined by a D/MAX-2500X-Ray Diffractometer (XRD).

Table 1. Chemical composition of Mg-6Al-xMn magnesium alloy ingots (wt.%).

Mn	Al	Zn	Si	Fe	Ni	Cu	Mg
0.1	5.928	0.18	<0.05	<0.005	<0.001	<0.005	Bal.
0.2	5.983	0.19	<0.05	<0.005	<0.001	<0.005	Bal.
0.3	5.976	0.18	<0.05	<0.005	<0.001	<0.005	Bal.
0.4	5.859	0.17	<0.05	<0.005	<0.001	<0.005	Bal.

3 RESULTS AND DISCUSSION

Figure 1 shows typical SEM images including second electron and and backscattered electron micrographs for the specimen with 0.3 wt.% Mn. The microstructure of the specimens consists of α-Mg solid solution, β-$Mg_{17}Al_{12}$ and Mn containing particles, determined by EDS and XRD as shown in Figure 2 and Figure 3. The island and large block shaped β-$Mg_{17}Al_{12}$ are continuously located at the grain boundary. These Al-Mn particles exhibit

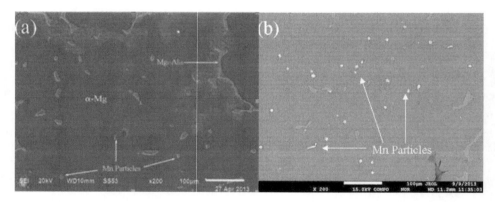

Figure 1. SEM images of a cast specimen with 0.3 wt.% Mn (a) second electron, and (b) backscattered electron images.

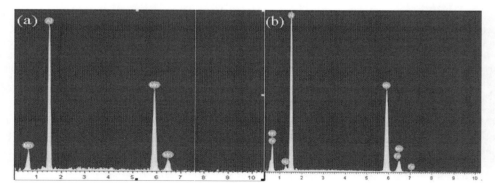

Figure 2. EDS analysis of Mn containing particles.

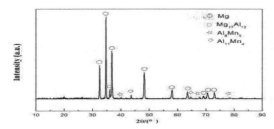

Figure 3. XRD patterns of Mg-6Al-0.3Mn as cast alloy.

Table 2. Chemical composition and morphology of Al-Mn particles.

Particle morphology	Analyzed elemental (atomic fraction %)				
	Mg	Al	Mn	Fe	Al/Mn
Bar-like	31.07	51.35	17.58	–	2.82
Angular blocky	–	58.28	41.72	–	1.40
Fine granular	12.26	27.77	9.97	–	2.71
	1.36	57.10	40.52	1.02	1.41

three distinctive morphologies, most of them have angular block morphology; some display a bar-like morphology and a few exhibit fine granular morphology. These particles locate inside the grains and at grain boundaries (Fig. 1a). Al-Mn particles (white) and α-Mg have a sharp contrast in the backscattered electrons image (Fig. 1b), so that we can easily conduct quantitative metallographic analysis for the Al-Mn particles. Most of these particles contain Mn and Al, some of these particles contain not only Mn and Al, but Fe as well (Figs. 2).

In order to identify the types of Mn-containing particles, XRD and EDS analyses were carried out. Figure 3 provides the XRD patterns of a cast specimen with 0.3 Mn content. It can be seen, the sample with 0.3 wt% Mn contains α-Mg, $Mg_{17}Al_{12}$, $Al_{11}Mn_4$ and Al_8Mn_5. Based on the phase diagram [21–22], the Al-Mn phases is considered to be $Al_{11}Mn_4$ and Al_8Mn_5 in the as-cast Mg-Al-Mn alloy. The typical composition of the bigger particles (6–10 μm) determined by EDS is given in Table 2. The Mg element around Mn-content particle can be evaluated by EDS analysis. To correct the atomic ratio of Al and Mn in Al-Mn particles, Al in α-Mg according to ratio of $Mg_{12}Al_{17}$ should be deducted. The calculated atomic ratio of Mg and Al is 17.63:1. After deducting the excess Al atoms in Al-Mn particles, the atomic ratio of Al and Mn in Al-Mn particles given in Table 2. It is clear that the bar-like and fine granular Al-Mn phase were $Al_{11}Mn_4$ and angular blocky Al-Mn

phase were Al_8Mn_5. Previous studies indicate the absence of iron in Al-Mn particles [10, 14, 20]. However in the present work, iron was detected in some fine Al-Mn particles, in agreement with other works [5, 16]. However, it was difficult to determine the composition of the smaller particles (less than 5 μm). The Al/Mn in these particles is closed to Al_8Mn_5. Therefore it is presumed that they are $Al_8(Mn, Fe)_5$, where Mn was replaced by Fe.

Figures 4 and 5 show the effect of Mn content and position in cast specimens on the quantity, size and morphology of Mn containing particles. The density of particles (number of particle/every square millimeter) increases with Mn content in specimens (Fig. 4), from approximately 20 to 70 conducted by metallographic analysis software in 0.1 and 0.4 Mn content, respectively. This is mainly due to the Mg-Al-Mn alloy melting in an iron crucible, the electronegativity between Mn and Fe is larger than Mn and Al, according to the electronegative principle, Mn react preferentially with Fe, and then excess Mn react with Al to produce Al-Mn compound. Furthermore, the size of particles also increases with Mn content in specimens (Fig. 5a, b and c), from approximately 2 to 10 μm in the 0.1 and 0.4 Mn content, respectively. These results are in the range reported in the literatures for Al-Mn particles (from 0.1 μm to 30 μm). For low Mn content specimens, the particles are smaller and fewer. With increase of Mn content, they grow bigger and more. The Al-Mn particles are more located within the bulk than at the surface of the castings (Fig. 5c and d). These particles precipitate during cooling, since the solubility of Mn in liquid Mg decreases with decreasing temperature and it is further reduced by the presence of Al and Fe in the melt. In Mg-6Al-xMn alloy containing 0.1–0.4 wt.% Mn, supersaturation in Mn occurs below 646 °C. Al-Mn particles can then nucleate and grow in the melt. In the alloy solidification process, as the temperature decreases, the solubility of Mn decreases. Therefore, under equilibrium conditions, the Mn and Fe, initially present, would form compound and be precipitated. Then the excess Mn and Al formed Al-Mn particles and dispersed in the alloy. Consequently, as the Mn content of the specimens increases, the size and quantity of Al-Mn particles increase. In steel moulds, heat is rapidly conducted through the mould/metal interface. Undercooling close to the mould wall is relatively important and a skin composed of fine α-Mg grains usually forms at the surface of castings shortly after mould filling. Nucleation of Al-Mn particles at the surface is likely but when the melt temperature locally reaches 590 °C, nucleation of α-Mg stars and rapid growth occurs [23]. The rapid growth of α-Mg prevents the development of Al-Mn particles. To the

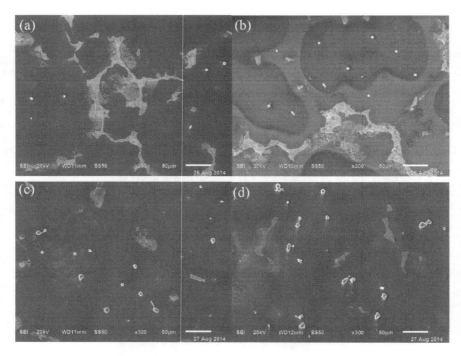

Figure 4.   SEM micrographs of Mg-6Al-xMn as cast alloy with different Mn content (a) 0.1 (b) 0.2 (c) 0.3 (d) 0.4.

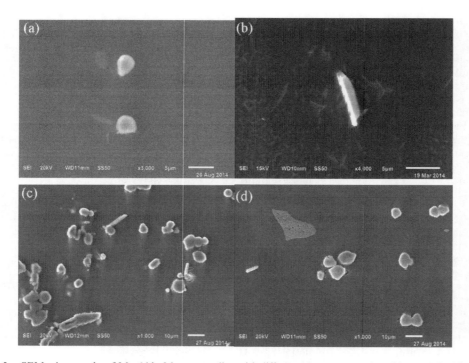

Figure 5.   SEM micrographs of Mg-6Al-xMn as cast alloy with different Mn content and position angular block Mn-particle of 0.1 Mn content in the bulk (b) bar-like Mn-particle of 0.1 Mn content in the bulk (c) 0.4 Mn content in the bulk (d) 0.4 Mn content at the surface.

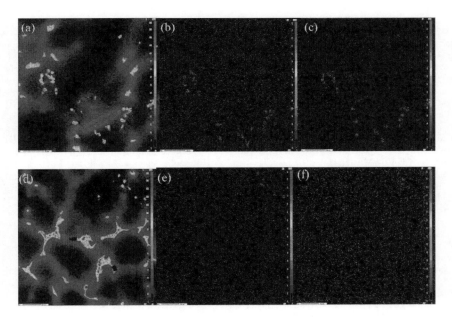

Figure 6. Elemental EPMA mappings of Mg-6Al-xMn showing the distribution of: (a) Al, (b) Mn, and (c) Fe with 0.4 Mn content, and (d) Al, (e) Mn, and (f) Fe with 0.2 Mn content.

contrary, in the bulk of alloy has a lower thermal conductivity. In this case, the early nucleation and growth of α-Mg were prevented. Consequently, there is a relatively longer period of time during which the Al-Mn particles within the bulk can grow.

EPMA was further used to determine the elemental analysis of the specimens. Figure 6 shows typical elemental X-ray mapping of manganese, aluminum and iron at the 0.2 and 0.4 Mn content specimens. These mappings were used to compare quantitatively the distribution of particles in the 0.2 Mn content and 0.4 Mn content castings. Image analysis from these mappings indicates that proportion of Al-Mn particles in 0.4 Mn content specimens is far more than in 0.2 Mn content specimens. From the mappings of the distribution of the element Fe, it can be confirmed that some of Al-Mn particles contain Fe. Small amounts of Fe and Mn, Al reacts to form $Al_8(Mn, Fe)_5$. These particles were not completely precipitated during solidification and remain inside the alloy.

4 CONCLUSIONS

In this work, the size, distribution, morphology and composition of Mn containing particles formed in Mg-6Al-xMn magnesium casting alloy were investigated.

These Mn containing particles were observed more located within the bulk rather than on the surface of the Mg-6Al-xMn castings. As the Mn content of the specimen increases, the amount and size of the particles increase. The types of Mn containing particles in Mg-6Al-0.3Mn specimens were Al_8Mn_5 and $Al_{11}Mn_4$, some of them contained iron and form $Al_8 (Mn, Fe)_5$ inside which iron replaces some manganese atoms. These particles exhibit bar-like, angular blocky and fine granular morphology. The size and quantity of Al-Mn particles depend on the Mn content and solidification conditions. When solidification is slow, more time is allowed for the nucleation and growth of Al-Mn particles within the bulk than the surface of alloys.

ACKNOWLEDGMENTS

This research is supported by Deformation mechanism of low cost and high ductility Mg-Sn-Pb-Zr magnesium alloy at low temperature (Project No: 2014-ZJ-709), Key laboratory of Qinghai Provence light alloy.

REFERENCES

[1] Ghali E. Uhlig's corrosion handbook: magnesium and magnesium alloys [M]. New York: John Wiley & Sons, 2000: 793–800.

[2] Wei Liu-ying, Westengen H, Aune T.K, Albright D. Characterisation of manganese-containing intermetallic particles and corrosion behaviour of die cast Mg-Al-based alloys [C]. Magnesium technology. Nashville, TN. Minerals, Metals and Materials Society. United States: 2000: 153–160.

[3] Mohsen D, Robert M.A, Pellumb J, David W.S, Gianluigi A.B. The cathodic behaviour of Al-Mn precipitates during atmospheric and saline aqueous corrosion of a sand-cast AM50 alloy [J]. Corrosion Science, 2014, 83: 299–309.

[4] Song G, Atrens A. Understanding magnesium corrosion [J]. Adv Eng Master, 2003, 5: 837–858.

[5] Hiroshi M, Yasuhiro I, Kazuaki F, Hiroshi N, Kazunori H. Effect of impurity Fe on corrosion behavior of AM50 and AM60 magnesium [J]. Corrosion Science, 2013, 66: 203–210.

[6] Zeng Rongchang, Han E, Ke Wei, Dietzel W, Kainer K.U, Andrejs A. Influence of microstructure on tensile properties and fatigue crack grow thin extruded magnesium alloy AM60 [J]. International Journal of Fatigue, 2010, 32: 411–419.

[7] Khan S.A, Miyashita Y, Mutoh Y, Sajuri Z.B. Influence of Mn content on mechanical properties and fatigue behavior of extruded Mg alloys [J]. Materials Science and Engineering A, 2006, 5: 837–858.

[8] Laster T, Nurnberg M.R, Janz A, Letzig D, Schmid-Fetzer R, Bormann R. The influence of manganese on the microstructure and mechanical properties of AZ31 gravity die cast alloys [J]. Acta Materialia, 2006, 54: 3033–3041.

[9] Cao P, Qian M, StJohn D.H. Effect of manganese on grain refinement of Mg-Al based alloys [J]. Scr Master, 2006, 54: 1853–1858.

[10] Easton M.A, Schiffl A, Yao J-Y, Kaufmann H. Grain refinement of Mg-Al(-Mn) alloys by SiC additions [J]. Scr Master, 2006, 55: 379–382.

[11] Barbagallo S, Laukli H.I, Lohne O, Cerri E. Divorced eutectic in a HPDC magnesium-aluminum alloy [J]. Alloys Compd, 2004, 378: 226–232.

[12] Wang R.M, Eliezer A, Gutman E.M. An investigation on the microstructure of an AM50 alloy [J]. 2003, 355: 201–207.

[13] Wang R.M, Eliezer A, Gutman E.M. Microstructures and dislocations in the stressed AZ91D magnesium alloys [J]. Mater Sci Eng A, 2003, 344: 279–287.

[14] Gertsman V.Y, Li J, Xu S, Thomson J.P, Sahoo M. Microstructure and second-phase particles in low- and high-pressure die-cast magnesium alloy AM50 [J]. Metall Mater Trans A Phys Metall Mater Sci, 2005, 36: 1989–1997.

[15] Wang Y, Xia M, Fan Z, Zhou X, Thompson G.E. The effect of Al8Mn5 intermetallic particles on grain size of as-cast Mg-Al-Zn AZ91D alloy [J]. Intermetallics, 2010,18: 1683–1689.

[16] Dobrzanski L.A, Tanski T, Cizek L, Brytan Z. Structure and properties of magnesium cast alloys Journal of Materials [J]. Processing Technology, 2007, 192–193: 567–574.

[17] Chang L.L, Cho J.H, Kang S.B. Microstructure and mechanical properties of AM31 magnesium alloys processed by differential speed rolling [J]. Journal of Materials Processing Technology, 2011, 211: 1527–1533.

[18] Kreutzer P, Anton R. TEM investigations on the growth of the icosahedric phase in AlMn films produced by simultaneous deposition of the components [J]. Materials Science and Engineering, 2000, 294–296: 854–858.

[19] Tamura Y, Yagi J, Motegi T, Kono N, Tamehiro H. Manganese bearing particles in liquid AZ91 magnesium alloy [J]. Mater Sci Forum, 2003, 419–422: 703–706.

[20] Kaya AA, Uzan P, Eliezer D, Aghion E. Electron microscopical investigation of as cast AZ91D alloy [J]. Mater Sci Technol, 2000, 16: 1001–1006.

[21] Lindahl B.B, Selleby M. The Al-Fe-Mn system revisited-An updated thermodynamic description using the most recent binaries [J]. Computer Coupling of Phase Diagrams and Thermochemistry, 2013, 43: 86–93.

[22] Ohno M, Mirkovic D, Schmid-Fetzer R. Liquidus and solidus temperatures of Mg-rich Mg-Al-Mn-Zn alloys [J]. Acta Materialia, 2006, 54: 3883–3981.

[23] Sin L.S, Dube D, Tremblay R. Characterization of Al-Mn particles in AZ91D investment castings [J]. Materials Characterization, 2007, 58: 989–996.

Material Science and Engineering – Chen (Ed.)
© 2016 Taylor & Francis Group, London, ISBN 978-1-138-02936-1

Structure, hydrogen storage properties and thermal stability of Mg_{90} $(RE_{0.25}Ni_{0.75})_{10}$ alloys

J.J. Li, Z.M. Wang, P. Lv, P. Liu, J.Q. Deng & Q.R. Yao
School of Materials Science and Engineering, Guilin University of Electronic Technology, Guilin, China

ABSTRACT: Mg_{90} $(RE_{0.25}Ni_{0.75})_{10}$ (RE = La, Pr, Nd) alloys were produced by arc melting + mechanical alloying. Their structure, hydrogen storage properties, and thermal stability were studied by means of XRD, SEM, PCTPro2000 and DSC analysis, respectively. The results indicate that as-prepared Mg_{90} $(RE_{0.25}Ni_{0.75})_{10}$ alloys are all multi-phase alloys, MgH_2, Mg_2NiH_4, $LaH_{2.3}$ (RE = La), $PrH_{2.92}$ (RE = Pr), Nd_2H_5 (RE = Nd) and NdH_3 (RE = Nd) have been formed in related alloys after hydrogenation. Mg_{90} $(La_{0.25}Ni_{0.75})_{10}$ alloy has the maximum hydrogen absorption capacity (3.37 wt.%) at 593 K, the maximum hydrogen desorption capacity (3.00 wt.%) at 623 K, and higher radios of hydrogen-absorption capacity to hydrogen-desorption capacity (98.59% at 573 K, 99.33% at 623 K). Mg_{90} $(La_{0.25}Ni_{0.75})_{10}$ alloy has the lower T_x and higher T_p (DSC curves), Mg_{90} $(NdNi_3)_{10}$ has the lowest TPD temperature (407.1 K).

Keywords: hydrides; intermetallic compounds; crystal structure; thermodynamic; laser annealing

1 INTRODUCTION

Magnesium-based alloys are attractive for hydrogen storage applications due to their high theoretical storage capacity, light weight and low cost. However, the slow H-absorption kinetics and high H-desorption temperature make them difficult for practical application. In general, structure modification and multi-element alloying can be used to improve the hydrogen absorption/desorption kinetics of magnesium-based alloys. Mechanical alloying, melt-spinning have been proved to be extremely appropriate techniques for producing nanocrystalline Mg-based alloys with different compositions at high production rate and low processing cost. Alloying magnesium with Rare-Earth (RE) can improve hydrogen storage properties of Mg-Ni-La, the improvement of its absorption kinetics can be ascribed to the nanosized particles of rare earth metal hydrides and Mg_2Ni which are embedded in Mg matrix after activation, and the increasing of desorption kinetics is attributed to the nano-sized particles of rare earth metal hydrides and Mg_2NiH_4 embedded in MgH_2 matrix after hydrogenation.

In previous work, Glass-Forming Ability (GFA) and hydrogen storage properties of melt-spun $Mg_{70}(Ni_3La)_{30}$ alloys, and thermal stabilities of $Mg_{70}(RE_{0.25}Ni_{0.75})_{30}$ (RE = La, Pr, Nd) amorphous alloys have been investigated. Based on experimental results, a three-dimensional interface reaction process of nucleation and growth was introduced

to explain the initial hydrogenation kinetics, a "geometrical contractionmodel" was employed to explain the dehydrogenation process. Now a series of experiments were done to investigate the effect of RE on microstructure, hydrogen storage properties and thermal stability of Mg_{90} $(RE_{0.25}Ni_{0.75})_{10}$ (RE = La, Pr, Nd) alloys prepared by arc melting + mechanical alloying.

2 EXPERIMENTAL

$RENi_3$ (RE = La, Pr, Nd) intermediate alloys were prepared by arc melting of pure La, Pr, Nd and Ni (purity better than 99.9%), 3 wt.% of RE were excessively added to compensate for the losses of RE during melting. The master alloy Mg_{90} $(RE_{0.25}Ni_{0.75})_{10}$ (RE = La, Pr, Nd) were prepared by mechanical alloying with pure Mg and RENi3 (RE = La, Pr, Nd) powder. The milling process was performed under the protection of pure argon in a QM-1SP planetary ball mill for 100 hr. The weight ratio of steel balls to powder was 10:1 and the speed was 250 rpm.

X-Ray Diffraction (XRD) measurements were employed to determine phase structure (Bruker D8 diffractometer; Cu Kα radiation, 5°/min). The hydrogen absorption/desorption kinetics were measured from 2 MPa to vacuum using a Sieverts-type apparatus at 523 K, 573 K, 593 K and 623 K respectively. Prior to the measurements, the sample was activation-treated by repeated hydriding/dehydriding at 623 K for 3 cycles. In each activation cycle, the sample was first hydrogenated under a hydrogen pressure of 2 MPa for 10 h and then under vacuum evacuated for 1 h. Thermal properties of hydrogenated samples were investigated by differential scanning calorimetry (DSC, NET ZSCH STA 449F3) at different ramping rates (5, 10, 20 K/min) under a continuous argon flow from 303 K to 873 K. Temperature-Programmed Desorption (TPD) is carried out at 5 K/min under vacuum (10^{-3} Pa) from 323 K to 623 K by PCTpro2000.

3 RESULTS AND DISCUSSION

3.1 Structural characterization of Mg_{90} ($RE_{0.25}Ni_{0.75}$)$_{10}$ alloys

XRD patterns of Mg_{90} ($RE_{0.25}Ni_{0.75}$)$_{10}$ (RE = La, Pr, Nd) alloys were shown in Figure 1. It can be seen that as-prepared Mg_{90} ($RE_{0.25}Ni_{0.75}$)$_{10}$ alloys are all multi-phase alloys, $LaNi_5$, La_7Ni_3 (RE = La), $PrMg_2Ni_9$ (RE = Pr) and Nd_2Ni_7 (RE = Nd) phases have been formed besides Mg and Ni phases. XRD

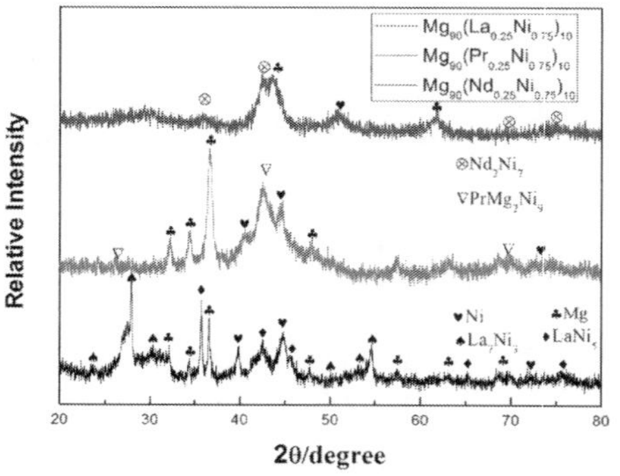

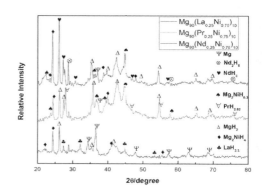

Figure 1. The XRD patterns of Mg_{90} ($RE_{0.25}Ni_{0.75}$)$_{10}$ alloys prepared.

Figure 2. The XRD patterns of hydrogenated Mg_{90} ($RE_{0.25}Ni_{0.75}$)$_{10}$ alloys.

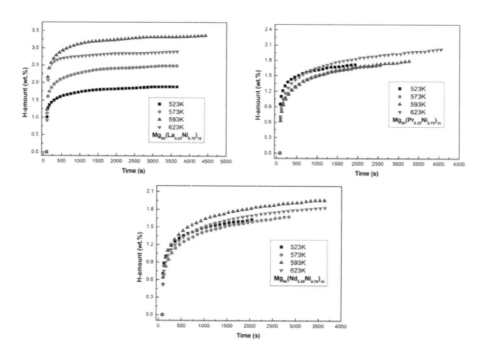

Figure 3. The hydrogen absorption kinetics curve of Mg_{90} ($RE_{0.25}Ni_{0.75}$)$_{10}$ (RE = La, Pr, Nd) alloys.

patterns of hydrogenated $Mg_{90}(RE_{0.25}Ni_{0.75})_{10}$ alloys under a hydrogen pressure of 2 MPa at 573 K are shown in Figure 2. After hydrogenation, MgH_2, Mg_2NiH_4, $LaH_{2.3}$ (RE = La), $PrH_{2.92}$ (RE = Pr), Nd_2H_5 (RE = Nd) and NdH_3 (RE = Nd) hydriding phases have been obtained.

3.2 Hydrogen absorption/desorption kinetics of $Mg_{90}(RE_{0.25}Ni_{0.75})_{10}$ alloys

Hydrogenation of $Mg_{90}(RE_{0.25}Ni_{0.75})_{10}$ alloys were carried out at 623 K under 3 MPa H2 for 20 h and then evacuated to vacuum (10^{-3} Pa) for 1 h. After activation, hydrogen absorption/desorption kinetics properties of $Mg_{90}(RE_{0.25}Ni_{0.75})_{10}$ alloys were measured at 523 K, 573 K, 593 K and 623 K respectively. Hydrogen-absorption kinetic curves of $Mg_{90}(RE_{0.25}Ni_{0.75})_{10}$ alloys are shown in Figure 3. It can be clearly seen that hydrogen-absorption capacity of $Mg_{90}(La_{0.25}Ni_{0.75})_{10}$ and $Mg_{90}(Nd_{0.25}Ni_{0.75})_{10}$ increase in the order of 593 K > 623 K > 573 K > 553 K, and that of $Mg_{90}(Pr_{0.25}Ni_{0.75})_{10}$ alloy increase in this order 623 K > 593 K > 573 K > 553 K. Thinking of the relationship between hydrogenation rate and the addition of RE, it also can be seen that hydrogen-absorption rate increases according to the subsequence of La > Pr > Nd at 523 K, 573 K and 623 K in 250 s, while the subsequence (La > Nd > Pr) is observed at 593 K. Hydrogen-desorption kinetic curves of $Mg_{90}(RE_{0.25}Ni_{0.75})_{10}$ alloys are shown in Figure 4, the maximum hydrogen-desorption capacity of $Mg_{90}(RE_{0.25}Ni_{0.75})_{10}$ alloys increase with the increase of temperature from 523 K to 623 K. The dehydrogenation rate of $Mg_{90}(RE_{0.25}Ni_{0.75})_{10}$ alloy have been improved according to the subsequence of La > Pr > Nd at 523 K and 623 K in 250 s, while the rate increases according to the subsequence of Pr > La > Nd at 573 K and 593 K.

As to the maximum hydrogen absorption/desorption capacity of $Mg_{90}(RE_{0.25}Ni_{0.75})_{10}$ alloys, experimental data are listed in Table 1. $Mg_{90}(La_{0.25}Ni_{0.75})_{10}$ alloy has the maximum hydrogen absorption capacity (3.37 wt.%,) at 593 K, and the maximum hydrogen desorption capacity (3.00 wt.%,) at 623 K. two high radios of hydrogen-absorption capacity to hydrogen-desorption capacity is 98.59% and 99.33% for $Mg_{90}(La_{0.25}Ni_{0.75})_{10}$ alloy at 573 K and 623 K respectively, indicating that La is a good element to improve hydrogen absorption/desorption properties of $Mg_{90}(RE_{0.25}Ni_{0.75})_{10}$ (RE = La, Pr, Nd) alloys.

3.3 Thermal stabilities of $Mg_{90}(RE_{0.25}Ni_{0.75})_{10}$ alloys

DSC behaviors of hydrogenationed $Mg_{90}(RE_{0.25}Ni_{0.75})_{10}$ (RE = La, Pr, Nd) alloys were

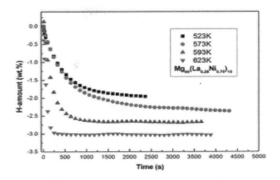

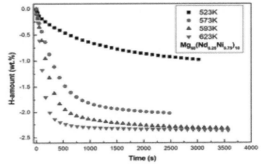

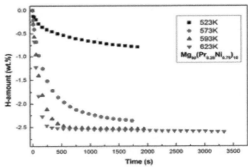

Figure 4. The hydrogen desorption kinetics curve of $Mg_{90}(RE_{0.25}Ni_{0.75})_{10}$ (RE = La, Pr, Nd) alloys.

performed at different temperature rates (5, 10, 20 K/min) under a continuous argon flow from 303 K to 873 K, related DSC curves were shown in Figure 5, here T_x and T_p stand for the start and peak temperature of dehydrogenation respectively.

From Figure 5, some points can be obtained: (1) T_x and T_p rises slightly (10~20 K) as the heating rates increasing in all samples. (2) A proportional relationship between the temperature rate and ΔT ($\Delta T = T_p - T_x$). (3) $Mg_{90}(La_{0.25}Ni_{0.75})_{10}$ alloy has the lower T_x and higher T_p among these alloys. (4) The value of ΔT is about 20 K, 30 K and 60 K for $Mg_{90}(La_{0.25}Ni_{0.75})_{10}$, $Mg_{90}(Nd_{0.25}Ni_{0.75})_{10}$

Table 1. Hydrogen absorption/desorption capacity of Mg_{90} $(RE_{0.25}Ni_{0.75})_{10}$ (RE = La, Pr, Nd) alloys.

Sample/ wt.%	Mg_{90} $(LaNi_3)_{10}$		Mg_{90} $(PrNi_3)_{10}$		Mg_{90} $(NdNi_3)_{10}$	
	Abs	Des	Abs	Des	Abs	Des
523 K	1.92	−1.90	1.72	−0.80	1.63	−0.98
573 K	2.49	−2.37	2.25	−2.21	1.68	−2.01
593 K	3.37	−2.67	1.79	−2.54	1.96	−2.32
623 K	3.02	−3.00	2.02	−2.60	1.83	−2.37

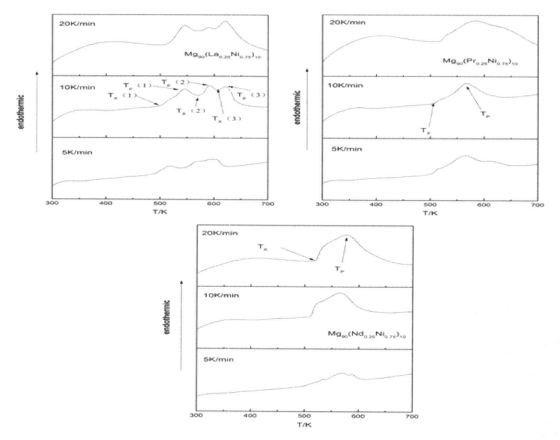

Figure 5.   The DSC kinetics curve of Mg_{90} $(RE_{0.25}Ni_{0.75})_{10}$ (RE = La, Pr, Nd) alloys.

and Mg_{90} $(Pr_{0.25}Ni_{0.75})_{10}$ alloy respectively, indicating that the bigger the temperature difference of the alloy (ΔT) is, the more time needed for the dehydrogenation process from the initial to the end. At the same time, three endothermic peaks have been obviously observed in the DSC curves of hydrogenated Mg_{90} $(La_{0.25}Ni_{0.75})_{10}$ alloy, which were correspondingly related to three hydrides. however, in the cases of Mg_{90} $(Pr_{0.25}Ni_{0.75})_{10}$ and Mg_{90} $(Nd_{0.25}Ni_{0.75})_{10}$ alloys, there are wide endother-

mic peaks due to the overlapping of endothermic peaks of related hydrides, this results has been discussed in previous work.

For commercial production, hydrogen storage alloy must be used at room temperature, such as $LaNi_5$-type hydrogen storage alloys, so the initial hydrogen desorption temperature is a key factor. TPD curves of Mg_{90} $(RE_{0.25}Ni_{0.75})_{10}$ (RE = La, Pr, Nd) alloys have been shown in Figure 6, which were measured at 5 K/min under vacuum (10^{-3} Pa) from

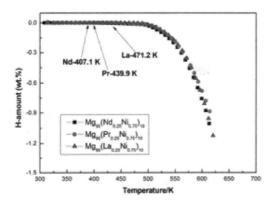

Figure 6. The TPD curve of Mg_{90} $(RE_{0.25}Ni_{0.75})_{10}$ (RE = La, Pr, Nd) alloys.

323 K to 623 K. It can be seen that Mg_{90} $(LaNi_3)_{10}$ alloy has a higher initial desorption temperature (471.3 K) and Mg_{90} $(PrNi_3)_{10}$ alloy has a lower initial desorption temperature (440.0 K). And Mg_{90} $(NdNi_3)_{10}$ has the lowest desorption temperature (407.1 K) among all alloys. According to the existing literature, the desorption temperature of pure magnesium is 523–573 K in a general way. So it can be seen that the addition of rare earth element can effectively reduce the desorption temperature of Mg-based hydrogen storage alloys. And the catalytic effect of decreasing the initial desorption temperature increases according to the subsequence Nd > Pr > La.

4 CONCLUSIONS

1. Mg_{90} $(RE_{0.25}Ni_{0.75})_{10}$ (RE = La, Pr, Nd) alloys were produced by arc melting + mechanical alloying. XRD results indicate that as-prepared Mg_{90} $(RE_{0.25}Ni_{0.75})_{10}$ alloys are all multi-phase alloys, including Mg, Ni, $LaNi_5$, $La_7 Ni_3$ (RE = La), $PrMg_2 Ni_9$ (RE = Pr) and Nd2 Ni7 (RE = Nd) phases. After hydrogenation, MgH2, Mg2 NiH4, LaH2.3 (RE = La), PrH2.92 (RE = Pr), Nd2H5 (RE = Nd) and NdH3 (RE = Nd) have been formed in related alloys.
2. Hydrogen absorption/desorption properties of Mg_{90} $(RE_{0.25}Ni_{0.75})_{10}$ alloys have been measured at 523 K, 573 K, 593 K and 623 K. The results indicate that Mg_{90} $(La_{0.25}Ni_{0.75})_{10}$ alloy has the maximum hydrogen absorption capacity (3.37 wt.%) at 593 K, and the maximum hydrogen desorption capacity (3.00 wt.%) at 623 K. Two high radios of hydrogen-absorption capacity to hydrogen-desorption capacity is 98.59% and 99.33% for Mg_{90} $(La_{0.25}Ni_{0.75})_{10}$ alloy at 573 K

and 623 K respectively, indicating that La is a good element to improve hydrogen absorption/ desorption properties of Mg_{90} $(RE_{0.25}Ni_{0.75})_{10}$ (RE = La, Pr, Nd) alloys.
3. Mg_{90} $(La_{0.25}Ni_{0.75})_{10}$ alloy has the lower T_x and higher T_p (In DSC curves), so good hydrogen-desorption properties is obtained. The catalytic effect of decreasing the initial desorption temperature increases according to the subsequence of Nd > Pr > La, Mg_{90} $(Nd_{0.25}Ni_{0.75})_{10}$ has the lowest desorption temperature (407.1 K).

ACKNOWLEDGEMENTS

This project is financially supported by the National Natural Foundations of China (51261003 and 51471055), the Natural Foundations of Guangxi Province (2012GXNSFGA060002), and Guangxi Experiment Center of Information Science (20130113).

REFERENCES

[1] Bendersky L.A. 2011. Effect of rapid solidification on hydrogen solubility in Mg-rich Mg–Ni alloys. Int. J. Hydrogen Energy 36:5388–99.
[2] Chen H.X. 2013. Hydrogen storage properties and thermal stability of amorphous $Mg_{70}(RE_{25} Ni_{75})_{30}$ alloys. J. Alloys Compd 563:1–5.
[3] Di Chio M. 2008. Effect of microstructure on hydrogen absorption in $LaMg_2$ Ni. Intermetallics 16:102–6.
[4] Friedlmeier G. 1999. Preparation and structural, thermal and hydriding characteristics of melt-spun Mg–Ni alloys. J. Alloys Compd 292:107–17.
[5] Hong T.W. 2000. Dehydrogenation properties of nano-/amorphous Mg_2NiH_x by hydrogen induced mechanical alloying. J. Alloys Compd 312:60–7.
[6] Jain I. 2010. Hydrogen storage in Mg: A most promising material. Int. J. Hydrogen Energy 35:5133–44.
[7] Kalinichenka S. 2009. Structural and hydrogen storage properties of melt-spun Mg–Ni–Y alloys. Int. J. Hydrogen Energy 34:7749–55.
[8] Kalinichenka S. 2011. Hydrogen storage properties and microstructure of melt-spun Mg_{90} Ni_8RE_2 (RE = Y, Nd, Gd). Int. J. Hydrogen Energy 36:10808–15.
[9] Li Q. 2004. Characteristics of hydrogen storage alloy Mg2 Ni produced by hydriding combustion synthesis. J. Mater. Sci. Techno I20:209–12.
[10] Liu J. 2009. Investigation on kinetics mechanism of hydrogen absorption in the $La_2 Mg_{17}$-based composites. Int. J. Hydrogen Energy 34:1951–7.
[11] Løken S. 2007. Nanostructured Mg–Mm–Ni hydrogen storage alloy: Structure–properties relationship. J. Alloys Compd 446:114–20.
[12] Lv P. 2013. Study on glass-forming ability and hydrogen storage properties of amorphous Mg60 Ni30 La10-xCox (x = 0,4) alloys. Mater. Charact 86:200–205.

[13] Orimo S. 1997. Notable hydriding properties of a nanostructured composite material of the Mg_2Ni-H system synthesized by reactive mechanical grinding. Acta Mater 45:331–41.

[14] Ouyang L. 2006. The hydrogen storage behavior of Mg_3 La and Mg_3 $LaNi_{0.1}$. Scripta Mater 55: 1075–8.

[15] Ouyang L. 2007. A new type of Mg-based metal hydride with promising hydrogen storage properties. Int. J. Hydrogen Energy 32:3929–35.

[16] Ouyang L. 2009. Hydrogen storage properties of $LaMg_2$ Ni prepared by induction melting. J. Alloys Compd 485:507–9.

[17] Ouyang L. 2009. Structure and hydrogen storage properties of Mg_3Pr and $Mg_3PrNi_{0.1}$ alloys. Scripta Mater 61:339–42.

[18] Puszkiel J.A. 2010. Synthesis of $Mg_{15}Fe$ materials for hydrogen storage applying ball milling procedures. J. Alloys Compd 495:655–658.

[19] Reilly Jr J.J. 1968. Reaction of hydrogen with alloys of magnesium and nickel and the formation of Mg2 NiH4. Inorg. Chem 7:2254–6.

[20] Renaudin G. 2000. Neodymium trihydride, NdH_3, with tysonite type structure. J. Alloys Compd 313:L10-L4.

[21] Sakintuna B. 2007. Metal hydride materials for solid hydrogen storage: a review. Int. J. Hydrogen Energy 32:1121–40.

[22] Skripnyuk V. 2009. Improving hydrogen storage properties of magnesium based alloys by equal channel angular pressing. Int. J. Hydrogen Energy 34:6320–4.

[23] Song M.Y. 2008. Preparation of Mg–23.5 Ni–10(Cu or La) hydrogen-storage alloys by melt spinning and crystallization heat treatment. Int. J. Hydrogen Energy 33:87–92.

[24] Spassov T. 1999. Hydrogenation of amorphous and nanocrystalline Mg-based alloys. J. Alloys Compd 287:243–50.

[25] Spassov T. 2002. Nanocrystallization and hydrogen storage in rapidly solidified Mg–Ni–RE alloys. J. Alloys Compd 334:219–23.

[26] Spassov T. 2004. Mg–Ni–RE nanocrystalline alloys for hydrogen storage. Mater. Sci. Eng., A 375:794–9.

[27] Tanaka K. 1999. Improvement of hydrogen storage properties of melt-spun Mg–Ni–RE alloys by nanocrystallization. J. Alloys Compd 293:521–5.

[28] Tanaka K. 2009. TEM studies of nanostructure in melt-spun Mg–Ni–La alloy manifesting enhanced hydrogen desorbing kinetics. J. Alloys Compd 478:308–16.

[29] Todorova S. 2009. Mg_6 Ni formation in rapidly quenched amorphous Mg–Ni alloys. J. Alloys Compd 469:193–6.

[30] Yan R. 2011. Kinetic properties of glass transition of amorphous Mg-Ni-La alloys. Zhongnan Daxue Xuebao (Ziran Kexue Ban)/Journal of Central South University (Science and Technology) 42:3686–91.

[31] Yartys V. 1997. Desorption characteristics of rare earth (R) hydrides (R = Y, Ce, Pr, Nd, Sm, Gd and Tb) in relation to the HDDR behaviour of R–Fe-based-compounds. J. Alloys Compd 253:128–33.

[32] Zaluska A. 1999. Nanocrystalline magnesium for hydrogen storage. J. Alloys Compd 288:217–25.

[33] Zhang H.G. 2014. Effect of Mg content on structure, hydrogen storage properties and thermal stability of melt-spun $Mg_x(LaNi_3)_{100-x}$ alloys. Int. J. Hydrogen Energy In Press, Corrected Proof, http://dx.doi.org/10.1016/j.ijhydene.2014.04.044.

[34] Zhang Y. 2009. Hydriding and dehydriding characteristics of nanocrystalline and amorphous $Mg_{20-x}La_xNi_{10}$ (x = 0–6) alloys prepared by melt-spinning. J Rare Earth 27:514–9.

[35] Zhou H.Y. 2012. Effect of rapid solidification on phase structure and hydrogen storage properties of $Mg_{70}(Ni_{0.75} La_{0.25})_{30}$ alloy. Int. J. Hydrogen Energy 37:13178–84.

Material Science and Engineering – Chen (Ed.)
© 2016 Taylor & Francis Group, London, ISBN 978-1-138-02936-1

Precipitation behavior of MnS inclusion for high-speed wheel steel during the solidification process

Y. Tang

CISDI Engineering Co. Ltd., Chongqing, China

ABSTRACT: In order to improve the fatigue life of high-speed wheel steel, the LF-VD secondary refining process is carried out with Al-deoxidation and slag of high basicity, high Al_2O_3 content and low oxidizing property. After the LF-VD process, during the continuous casting process, inclusions in molten steel still change to more complex ones. The transformation process of complex inclusions is investigated in this paper, and it is found that during the solidification process, MnS would precipitate around the existing inclusions in molten steel to form complex inclusions cored by MnS, leading to a lower melting point and better ductility of complex inclusions, and improving the fatigue performance of wheel steel applied to high-speed railway.

Keywords: inclusion; MnS; precipitation; wheel steel

1 INTRODUCTION

High-speed wheel steel suffers from recycling and alternative stress in its service life, leading to fatigue destruction. Especially with the dramatic rise of speed of rail train, fatigue destruction plays a vital role in wheel destruction. In order to meet the demand of the development of a high-speed rail train, high-quality wheel steel is required to have an excellent fatigue performance, which mostly depends on the T [O] content of wheel steel. In other words, ultra-low oxygen content is the major feature for high-speed wheel steel.

However, ultra-low oxygen content is just one of the key points for inclusion control of high-speed wheel steel. Because during the solidification process, with the decrease in temperature, MnS precipitation would take place, forming complex inclusions composed of oxide and sulfide. So, based on this, precipitation behavior of MnS needs to be investigated.

In previous research, many investigations have been made with regard to MnS precipitation rules and some important relations. The formation of oxide–sulfide complex inclusions has been investigated in terms of thermodynamics by S.K. CHOUDHARY [1]. It is found that in order to form Ca, Al oxide other than CaS, low [S] content in molten steel must be guaranteed. Morphology control for MnS inclusion during the solidification process was investigated by Katsunari OIKAWA [2]. It was found that there would be different morphologies of MnS depending on different contents of alloy elements. In addition, interfacial energy would decrease as the MnS content increases. Inclusion precipitation was investigated by Marc WINTZ [3]. It was found that FeS content in sulfides would decrease as the Mn/S ratio increases.

2 EXPERIMENTAL AND RESULTS

2.1 Experimental

Three heats were studied in a plant. Production flow was LD → LF → VD → calcium treatment (No. 1, No. 2 heat) → soft bottom-blown by Ar bubbling → continuous casting. Calcium treatment was used in No. 1 and No. 2 heat in order to make sure of smooth casting, while No. 3 heat was exempted from calcium treatment. Steel and slag samples were fetched throughout the whole process from the beginning of LF to round billets. In this paper, experimental samples were taken from round billets for high-speed wheel. The morphology and composition of inclusions were investigated by SEM-EDS. The melting behavior of MnS inclusion was observed by a confocal laser scanning microscope.

2.2 Results

2.2.1 Typical complex inclusions during the solidification process

Figure 1 shows several kinds of MnS–oxide complex inclusion. As shown in Figure 1, complex inclusion consists of MnS outside and Al, Mg, Ca oxide inside. Actually, in other situations, there

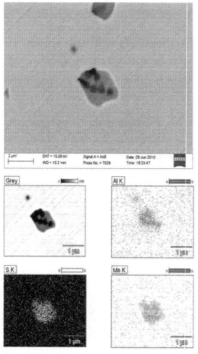

a) MnS–Al$_2$O$_3$ complex inclusion

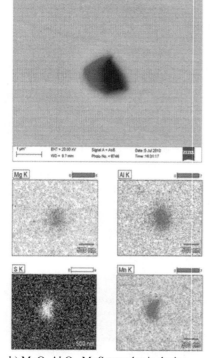

b) MgO–Al$_2$O$_3$–MnS complex inclusion

Figure 1. (*Continued*)

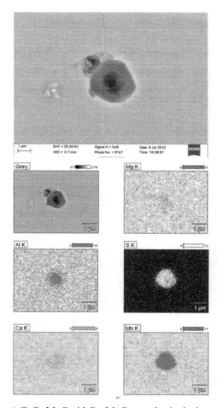

c) CaO–MgO–Al$_2$O$_3$–MnS complex inclusion

Figure 1. MnS–oxide complex inclusion formed during the solidification process.

may be other kinds of complex inclusions whose inside is TiN or other non-oxide inclusion.

2.2.2 *Melting behavior of MnS–oxide complex inclusion*

Figure 2 shows the melting behavior of MnS–oxide complex inclusion during the heating process, observed by the confocal laser scanning microscope from room temperature to 1300°C.

As shown in Figure 2, in the heating process, long strip MnS inclusions change gradually into string inclusions composed of near-spherical inclusions that grow up later. During insulation at 1300°C, MnS inclusion slowly melts into the steel matrix and gets smaller and smaller.

The melting process is actually the reverse process of precipitation. So, if it is imagined in the reverse direction in Figure 2, beginning from 1300°C, then finally to 25°C, this is the MnS precipitation process during the solidification of molten steel. So, it could also be seen from this that MnS precipitates at about 1300°C.

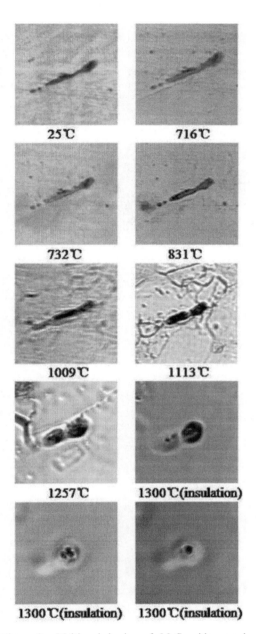

25℃   716℃

732℃   831℃

1009℃   1113℃

1257℃   1300℃(insulation)

1300℃(insulation)   1300℃(insulation)

Figure 2. Melting behavior of MnS–oxide complex inclusion during the heating process.

3 DISCUSSION

After secondary refining, in the process of continuous casting, MnS would precipitate around oxide inclusion in the molten steel during the solidification process.

Under the condition of ultra-low oxygen content and Al deoxidation, the existence of Al_2O_3 and

MgO-Al_2O_3 spinels is inevitable, detrimental to the fatigue performance of high-speed wheel steel. So, this situation must be improved, and, fortunately, MnS could play this important role.

Because MnS has a lower melting point and better formability, it could deform together with the steel matrix under an external force such as rolling, reducing the harmful effect of inclusions with a higher melting point and worse formability such as Al_2O_3 and MgO-Al_2O_3.

In order to optimize the effect caused by MnS precipitation, the content control for Mn and S is very critical, which is know-how and confidential for this paper. This content must be controlled precisely to ensure a proper complex inclusion whose outer layer is MnS and internal layer is oxide inclusions or other non-oxide inclusion.

In general condition, sulfide inclusion in steel consists of MnS and FeS, whose specific composition depends on Mn/S. FeS content in sulfides would decrease as the Mn/S ratio increases. Both MnS and FeS are plastic inclusions with better deformability, so they are positive for high-speed wheel steel.

On the other hand, S content must be kept very low to avoid the formation of CaS inclusion, which is undeformable and detrimental to the fatigue performance of steel. On the other hand, the formation of CaS is greatly related to the content of dissolved Al[4]. In order to avoid CaS formation, maximum allowable content of S decreases as dissolved Al increases.

4 CONCLUSIONS

Under the condition of Al deoxidation and slag of high basicity, high Al_2O_3 content and low oxidizing property, the precipitation behavior of MnS inclusion for wheel steel is investigated. On the basis of the analysis, the following conclusions can be drawn:

1. MnS inclusion forms during the solidification of molten steel. In the confocal experiment, MnS would melt at 1300°C;
2. MnS would precipitate around Al_2O_3 and MgO-Al_2O_3 inclusions, reducing the harm of undeformable inclusions;
3. Proper content of Mn and S should be controlled precisely to ensure the formation of complex inclusions whose outer layer is MnS and internal layer is Al_2O_3, MgO-Al_2O_3, and to avoid the formation of CaS with bad deformability.

REFERENCES

[1] S.K. Choudhary, et al. Thermodynamic Evaluation of Formation of Oxide–Sulfide Duplex Inclusions in Steel. ISIJ International, 2008, 48 (11): 1552–1559.

[2] Katsunari Oikawa, et al. The Control of the Morphology of MnS Inclusions in Steel during Solidification. ISIJ International, 1995, 35 (4): 402–408.

[3] Marc Wintz, et al. Experimental Study and Modeling of the Precipitation of Nonmetallic Inclusions during Solidification of Steel. ISIJ International, 1995, 35 (6): 715–722.

[4] Suito Hideaki, et al. Thermodynamics on Control of Inclusions Composition in Ultra-clean Steels. ISIJ International, 1996, 36 (5): 528–536.

[5] Nishimori H, et al. Bull. Jpn. Inst. Met., 1993, 1: 441.

[6] Hojo Masatake, et al. Oxide Inclusion Control in Ladle and Tundish for Producing Clean Stainless Steel. ISIJ International, 1996, 36(Supplement): S128–S131.

[7] Kim Jong wan, et al. Formation Mechanism of Ca-Si-Al-Mg-Ti-O Inclusions In Type 304 Stainless Steel. ISIJ International, 1996, 36(Supplement): S140–S143.

[8] Jiang Min, et al. Formation of MgO-Al$_2$O$_3$ inclusions in high strength alloyed structural steel refined by CaO-SiO$_2$-Al$_2$O$_3$-MgO slag. ISIJ International, 2008, 48 (7): 885–890.

[9] Okuyama Goro, et al. Effect of Slag Composition on the Kinetics of Formation of Al$_2$O$_3$–MgO Inclusions in Aluminum.

[10] Holappa L.E.K., et al. Inclusion Control in High-Performance Steels. Journal of Materials Processing Technology, 1995, 53: 177–186.

[11] Higuchi Yoshihiko, et al. Inclusion Modification by Calcium Treatment. ISIJ International, 1996, 36 (Supplement): S151–S154.

[12] Kusano Yoshiaki, et al. Calcium Treatment Technologies for Special Steel Bars and Wire Rods. ISIJ International, 1996, 36 (Supplement): S77–S80.

[13] Ye Guozhu, et al. Thermodynamics and Kinetics of the Modification of Al$_2$O$_3$ Inclusions. ISIJ International, 1996, 36 (Supplement): S105–S108.

Material Science and Engineering – Chen (Ed.)
© 2016 Taylor & Francis Group, London, ISBN 978-1-138-02936-1

Mechanism research on the formation and control of MnS inclusion in heavy rail steel

H.G. Li, T.M. Chen & Liang Chen
Panggang Group Research Institute Co. Ltd., China

ABSTRACT: In order to know how MnS precipitate in rail steel and find effective measures to control, this paper has made a thermodynamics calculation for the formation of MnS and calculated by Factsage software to confirm the composition of refining slag, found that the precipite temperature of MnS is about 1365°C, and choose slag system in the area where SiO2: 10%~35%, CaO: 45%~60%, Al2O3: 15%~35% in the ternary slag phase diagram is benefit to control T[O]. Put forward measures to control the precipitate of MnS that [S] at 0.003%~0.004%; low superheat casting and reduce super-cooling degree for molten steel; calcium treatment to promote MnS distribute dispersively and improve the quality evaluation.

Keywords: MnS; precipitation; thermodynamics; calcium treatment

1 INTRODUCTION

Railway is the backbone of the comprehensive transportation system, the development of railway transportation has a great influence for the economic development of our country. In order to improve the transportation efficiency, railway transportation is in the direction of "speeder and heavier" developing. That's put forward a high request for the operation performance of railway steel.

Generally speaking, inclusion have a bad influence for the mechanical property of railway steel. And with railway transportation developing, requirement for the purity of steel is more stringent [2]. It can be found that more stringent requirement for the purity of molten steel in the production of railway steel have been advanced, by comparing TB/T2344—2003 "43 kg–75 kg Hot rolled rail order condition" "Provisional technical conditions of 200 km/h railway line for passenger traffic 60 kg/m" and Provisional technical conditions of 350 km/h railway line for passenger traffic 60 kg/m. The requirements for inclusion A for 250 km/h railway line for passenger traffic is ≤2.5, but ≤2.0 for 350 km/h.

2 EFFECTS OF MNS ON PROPERTY OF RAILWAY STEEL

A type iclusion is mainly about sulfide like MnS, FeS, mainly exists in the form of MnS which is ductile. MnS block the continuity of matrix for railway steel, induce the cracks emergence and spread. Weaken the strength and toughness of railway steel, and shorten the service life eventually.

A comparative for the chemical composition of different countries' railway steel has been shown in Table 1. It can be found that [S] has been controlled at different level: Japan 0.006% US and France and Russia control at 0.012%~0.016%.

Comparing the control of A type inclusion for domestic and international, results have been shown as Table 2. Can be seen that the control for inclusion A, Japan is the best at 1.0 no inclusion D, and comparing with the other countries inclusion morphology and size are more fine and short; for inclusion B Pangang is better, but inclusion A and C occurred 2.0 and 1.5; the another three countries control the level of inclusion A at 1.5~2.0, but no inclusion B, and inclusion C D also occurred 1.5 and 0.5 respectively. Japan control the purity is the best and inclusion control well, Pangang has a certain gap compared with other countries as a whole.

Studies have pointed out that[3]: sulfide can be used to wrap up brittle inclusion like Al_2O_3. But because of the block to the continuity of matrix, come into blowhole segregation, or combine with thermal crack and the segregation of phosphorus and sulfur, lead to internal crack. And MnS is hard to adhesive with matrix after making light pressure to billet, but distribute as belt and strip. Because of the different contraction ratio in the process of cooling and make crack to be possible. Also due to

Table 1. Chemical composition of railway steel %.

	C	Si	Mn	P	S	O	N	H	Alt
US	0.82	0.57	1.18	0.014	0.013	0.0011	0.0040	0.00007	≤0.005
Japan	0.73	0.25	0.85	0.017	0.006	0.0011	0.0021	0.00007	≤0.005
France	0.75	0.55	0.98	0.014	0.016	0.0012	0.0037	0.00007	≤0.005
Russia	0.75	0.34	0.86	0.009	0.012	0.0034	0.0011	0.00010	≤0.005
Pangang	0.75	0.60	0.90	0.010	0.010	0.0014	0.0030	0.00006	

Table 2. Grades of inclusion rating in railway steel.

	Level				
	A	B	C	D	DS
Japan	1.0	0.5	0.5	–	–
US	1.5	–	1.5	0.5	–
France	2.0	–	–	0.5	–
Russia	2.0	–	–	–	–
Pangang	2.0	–	1.5	0.5	–

the reason of uneven cooling, shear stress produced, and makes adjacent cracks connected, internal crack produced while shear stress exceed adhesive power of matrix. Analysis found that inclusion A of 2.5~3.0 levels on the waist of railway steel. And the hydrogen atom from situation gathered at the tip of MnS, make a large pressure, HIC produced and diffused along MnS organization[5].

3 GENERATION AND CONTROL OF MNS IN RAILWAY STEEL

For the control of MnS, the direction is mainly about generation removal and modification treatment.

3.1 Thermodynamic calculation for the generation of MnS

Choose U75V(C: 0.75%, Si: 0.6%, Mn: 0.9%, S: 0.01%, P: 0.01%) of Pangang as railway steel sample to make thermodynamic calculate, using type (1), (2) combine with the chemical composition to get Tl and Ts:

$$T_L = 1538 - 83\,[\%C] - 8\,[\%Si] - 5\,[\%Mn]$$
$$- 30\,[\%P] - 25\,[\%S] \tag{1}$$
$$T_S = 1538 - 175\,[\%C] - 20\,[\%Si] - 30\,[\%Mn]$$
$$- 280\,[\%P] - 575\,[\%S] \tag{2}$$

The calculation results of U75V is that $T_L = 1465°C$ and $T_S = 1359°C$.

The equation of precipitate MnS is shown as the follow:

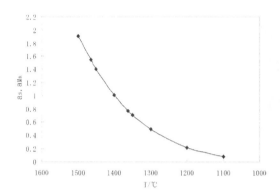

Figure 1. Relationship between T and equilibrium $a_{Mn} \cdot a_S$.

$$[S] + [Mn] = (MnS) \quad \Delta G^0 = -158783 + 95.0T$$
$$\log \frac{a_{MnS}}{a_{Mn} \cdot a_S} = \frac{8293}{T} - 4.96 \tag{3}$$

It have close connection between the value of $a_{MnS} \cdot a_S$ and temperature of molten steel while reaction reach equilibrium and shown as Figure 1.

The relationships between actual values of $a_{MnS} \cdot a_S$ and T of steel molten, and relationship between the solid fraction and T of steel molten are shown as Scheil equation-type (4) and type (5) respectively.

$$a_S = a_{S,o}(1 - f_S)^{k_S - 1} \quad a_{Mn} = a_{Mn,o}(1 - f_S)^{k_{Mn} - 1} \tag{4}$$
$$T = T_m - \frac{T_m - T_L}{1 - f_s \dfrac{T_L - T_S}{T_m - T_S}} \tag{5}$$

And $a_{s,o}$, $a_{Mn,o}$ are initial value of molten steel. a_s, a_{Mn} are the actual value of S and Mn respectively in molten steel; k_S, k_{Mn} are partition coefficients of S and Mn respectively ($k_s = 0.02$, $k_{Mn} = 0.84$); f_S is solid fraction of molten steel and Tm is melting temperature of pure iron and the value is 1538°C

T_L of the steel have been chosen as the point of initial activity, interaction coefficient between each two elements are shown as Table 3 while 1600°C.

40

Table 3. Interaction coefficient between each two elements at 1600°C.

Items	C	Si	Mn	P	S
S	0.11	0.063	−0.026	0.029	−0.028
Mn	−0.07	–	–	−0.0035	−0.048

Table 4. Interaction coefficient between each two elements at 1465°C.

Items	C	Si	Mn	P	S
S	0.1185	0.0679	−0.0280	0.0313	−0.0302
Mn	−0.0754	–	–	−0.0038	−0.0517

And the value of interaction coefficient in the other temperature could be calculated by type (6).

$$e_B^K(T) = \frac{1873}{T} \times e_B^K(1873) \qquad (6)$$

The results of interaction coefficient at T_L were calculated by type (6) and shown as Table 4.

The results of calculation is that $f_S = 1.272$, $f_{Mn} = 0.8767$, and the value of activity is that $a_S = 0.0127$, $a_{Mn} = 0.789$. And the relationship between T and $a_s \cdot a_{Mn}$ were shown as Figure 2. while the content of [S] is 0.010%. And the other values of [S]% are 0.003%, 0.004%, 0.005%, 0.006%, 0.007%, 0.008%, 0.01%, 0.011%, 0.012%, corresponding actual activity product1 to 9, and shown as Figure3.

As Figure 2 shows that equilibrium activity product decrease with temperature decreasing, while value of actual activity product increased, counter trends means that value of activity product would be equal while molten steel solidified, then actual > equilibrium and precipitate MnS, with temperature going down the driving force (D-value between actual and equilibrium activity product) of precipitate MnS enhanced rapidly. As Figure 3 shows that the precipitations are difference with different [S] contents.

3.2 Control MnS

Conclusion could be drawn from the results mentioned above that the start precipitation of MnS is too low to float and remove inclusion, and control MnS is mainly about generation and distribution.

3.2.1 Formation control for MnS

Measures to control the formation of MnS could be found based on thermodynamic results which have been shown as Figure 4, temperature that start precipitate MnS change from 1361°C to 1365°C while [S] change from 0.003% to 0.012%. Found that [S]

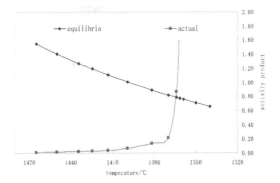

Figure 2. Relationship between $a_S \cdot a_{Mn}$ and T.

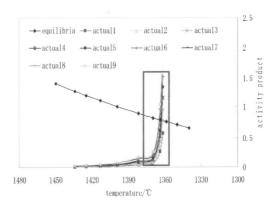

Figure 3. Results of difference [S] contents.

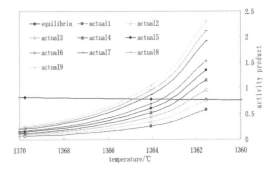

Figure 4. Circumstances of precipitate for different content of [S].

have little bit effect on the temperature that MnS start precipitatE, but a great effect on the driving force of precipitation MnS. And more content of sulfur means MnS precipitate more quick. Compareing with T_S-solidus temperature, conclusion can be reached that MnS precipitate start close to T_S. Three aspects measures can be proposed based on the above.

1. *Control [S]*

   The precipitation thermodynamics calculation results shows that decrease [S] is benefit for decreasing precipitate speed; and the precipitation of MnS can be locked out while [S] have been controlled at 0.003%~0.004%.

2. *Continuous casting control*

   It's very difficult to further decrease [S] while the content it is very low, and the prime cost increased substantially. So control the process of continuous casting is meaningful to control MnS. Control the temperature of molten steel below the temperature of start precipitation and solidified based on the results that calculated by precipitation thermodynamics and MnS could not precipitate. Studies have shown[8] that decrease the temperature gradient of super-cooled liquid steel is helpful to elongate cold zone, and have significant effect on the control of columnar crystals. So casting with low superheat and reduce the degree of super-cooling, not only control MnS but also promote nucleation and growth, and increase solid fraction to prevent the growth of columnar crystals. Combine with the favorable dynamic conditions from electromagnetic stirring, it's helpful to optimize uniformity of liquid steel and weakly the segregation of sulfur to restrain the precipitation of MnS.

3. *Control T[O]*

   MnS inclusion can wrap up oxide inclusion, and weaken the destructive effect for rail. That means control T[o] is the prerequisite for the control of MnS. Calculation results from FACTSAGE shows that control the composition of refining slag in the ternary slag phase diagram that SiO_2: 10%~35% CaO: 45%~60% Al_2O_3: 15%~35% and deoxidize without aluminum, can achieve the control of T[O] effectively.

3.2.2 *MnS distribute control*

Mn is well-distributed in molten steel due to the close radius between Mn atom and Fe atom, but Mn appears co segregation phenomenon caused by S which is easy to segregate in molten steel. Calcium treatment can consume [S] and formed CaS earlier while T[O] is low, and CaS has high-melt-point that can play the role of crystal nucleus. So calcium treatment not only consume [S] but also promote MnS distribute dispersively, and the total mount of MnS can be decreased and MnS formed individual or wrapped in CaS[9], equation between [Ca] and [S], [Ca] and [O] were shown as type (7) and type (8) respectively. Conclusions could be arrived that [Ca] could dioxide deeply and also consume [S][10].

$$[Ca] + [S] = CaS \ \Delta G = -548100 + 103.85T \quad (7)$$
$$[Ca] + [O] = CaO \ \Delta G = -491216 + 146.47T \quad (8)$$

Calculation results shows that with temperature decreasing the requirements for the amount of [Ca] increasing gradually. But the cast-ability turn to be worse while CaS is overmuch. And [Ca] is easy to be oxygenated, the key of calcium treatment is insure appropriate content of [Ca].

4 CONCLUSION

1. Thermodynamic calculation result shows that control [S] at 0.003%~0.004% can inhibit the precipitation of MnS.

2. Low super-heat casting and decreasing the degree of super-cooling can control the precipitation of MnS effectively, elongate cold zone can improve the ratio of equiaxed crystal and optimize the quality of billets.

3. Choose deoxidizing slag system in the ternary slag phase diagram that SiO_2: 10%~35% CaO: 45%~60% Al_2O_3: 15%~35% and can realize the control of T[O] effectively as a prerequisite for the control of MnS.

4. Residual T[O] in molten steel, and ensure an appropriate content of [Ca] can weaken the Co-segregation of S and Mn to realize MnS distribute dispersively.

REFERENCES

[1] Guo Xiaoyang. Research on the development trend of railway transportation for our country [J]. The cooperation of economy and technology, 2012, 433:11.

[2] Li Chunlong, Zhi Jianguo, Chen Jianjun, etc. The production technology and quality level of Bao steel for high speed railway heavy rail steel [J].

[3] Shan Lintian. Inclusion and deoxidation system for heavy rail steel [J]. Iron and Steel, 1997, 32(5): 82.

[4] Wang Quan, Li Zhili, Han Fengying. Research of longitudinal crack for rail web on heavy rail [J]. Baotou steel technology, 2003, 29(4): 67.

[5] Yin Guanghong, Shi Qing, Sun Yuanning. Analysis of factors that influence HIC for line pipes steel [J]. Steel Pipe, 2004, 33(6): 20–23.

[6] Huang Xihu. Theory of iron and steel metallurgy [M]. Beijing: Metallurgical Industry Press, 1990: 55.

[7] Qi Jianghua, Wu Jie, Suo Jinping, etc. Research on MnS inclusions for high speed heavy rail that cut length as 100 meters [C]. 14th Academic Conference of quality and inclusions control, GuiLin Guan Xi, 2010: 110–111.

[8] Min Yi, Liu Chengjun, Wang Deyong, etc. The prediction of center equiaxial crystal rate for cc round billets of 37Mn5 [N]. Journal of iron and steel research, 2011: 23(10): 41–42.

[9] Wu Xiaodong etc. Analysis of the effects of calcium treatment to modificate inclusions for gear steel 20CrMo [J]. Technology of hot work, 2012: 41(23): 22.

[10] Wu Wei etc. Research on the technology of non aluminum deoxidation for the steel of heavy rail [J]. Iron and Steel, 2007: 42(3): 35.

Material Science and Engineering – Chen (Ed.)
© 2016 Taylor & Francis Group, London, ISBN 978-1-138-02936-1

Methodology of the fatigue test for the partial component of the railway vehicle's carbody

Z.Z. Wang
R&D Center, CSR Qingdao Sifang Railway Vehicles Co. Ltd., Qingdao, China

X.W. Wu
State Key Laboratory of Traction Power, Southwest Jiaotong University, Chengdu, China

ABSTRACT: In this paper, a methodology of the fatigue test is proposed to assess the fatigue strength of the underframe of the carbody. In this test, three directions' dynamic loads are taken into consideration. Before the test, the payload is applied on the underframe, which corresponds to the 1/3 weight of the whole carbody. In addition, the inertial loading test method is adopted to generate the vertical dynamic loads based on the 6 DOF vibration test rig of carbody fatigue test bench. Moreover, the lateral and longitudinal force of the center pivot, and the longitudinal force of the coupler, are taken into consideration. Based on the fatigue result, the fatigue test methodology of the underframe can sufficiently identify the dangerous points of the underframe, and also has a good agreement with the FEM model and the operation experience of the railway vehicles.

Keywords: fatigue tests; carbody fatigue test bench; underframe of carbody; inertial loading test method

1 INTRODUCTION

With the development of the railway industry, railway vehicles have become extremely popular in the world, especially in China. According to the government's report, about 18,000 km of passenger lines will be constructed by 2020, and the operation speed would exceed 200 km/h, which means that the operation of vehicles will be more frequent. However, it is well known that the wheel/rail interaction will be worse because of the frequent operation of railway vehicles. Therefore, the fatigue problems of railway vehicles could be caused by the dynamic loads induced by the deteriorative wheel/rail interaction. The occurrence of the fatigue problems for the components would threaten the operation safety of railway vehicles. Consequently, the fatigue characteristic is tremendously important for the component of railway vehicles.

According to the author's knowledge, the fatigue features of the wheelset, the axlebox and the bogie frame have been well validated by the fatigue test in the laboratory. The carbody is a main structure of railway vehicles, which would endure the dynamic loads caused by the vibration of the bogie frame and the equipment under the underframe of the carbody and the longitudinal oscillation between the carbodies. However, the fatigue properties of

the carbody of railway vehicles have not been well documented and investigated by the fatigue test in the existing literature. Most investigations on the fatigue properties of the railway vehicle's carbody have been conducted by using the FEM and the static load tests [1, 23, 4]. However, the carbody of railway vehicles is a large structure. Thus, it is impossible to consider all the details in the FEM model of the carbody, such as the weld joints and the defects caused by the manufacture. Although the static load test can be used to detect some dangerous locations and the local distribution of the stress under the concentrated loads, it cannot take the influence of the flexible vibration of the structure on the fatigue features of the carbody into consideration.

Thus, it is extremely necessary to verify the fatigue characteristics of the carbody by the fatigue test or dynamic load test. However, the huge cost of the full-scale fatigue tests for the carbody and the huge cost for the carbody hinder the designers to conduct the full-scale fatigue tests, which also limits the development of the fatigue tests for the carbody. Currently, only the Traction Power State Key Laboratory of Southwest Jiaotong University can conduct the full-scale fatigue tests for the carbody in China.

Compared with the full-scale fatigue test, the fatigue tests for the partial component cutting from the carbody are economical and accepted by most designers. In this paper, we proposed a method to conduct the fatigue tests for the partial component of the carbody and to evaluate the equivalent loads for the partial component fatigue tests.

This paper is organized as follows. Section 1 introduces the method of fatigue tests for the carbody used in the railway vehicles. Section 2 indicates the partial component fatigue tests for the underframe of carbody. Section 3 depicts the methodology of equivalent loads adopted in the partial component fatigue tests.

2 SURVEY OF FATIGUE TEST METHODS FOR THE CARBODY TEXT AND INDENTING

In order to evaluate the fatigue strength of a rolling stock carbody, it is important to set the testing conditions very similar to the real dynamic load conditions. According to the existing literature, there are two major methods to evaluate the fatigue strength: dynamic concentrated loading test and inertial loading test.

Oomura's [6] dynamic load testing method is the concentrated loading test method, as shown in Figure 1. An actuator generates the dynamic concentrated loads, which are transferred to the underfloor of the carbody through the special jigs. The carbody structure is subjected to four points bending moment. They prove the validity of the testing method by comparing with analysis results and other test measurements. However, bending moment at center shows some difference between the concentrated loading case and the distributed loading case. Also, discontinuity of shear force at loading points causes different stress distributions from the distributed loading case. Jun-Seok Kim [5] also conducted a whole body test for a Korean tilting train by the concentrated loading test method to evaluate the fatigue strength. This method cannot consider the effects of the flexible vibration of the carbody on the fatigue strength.

Sung II SEO proposed another dynamic loading test method for the carbody based on the inertial loading test, as shown in Figure 2. Four supporting beams at the position of air springs of the actual bogie system are fixed on the test bed. The carbody sits on the four supporting beams through the coil spring. Basic loads corresponding to the weight of passengers and the equipment are distributed on the underfloor. Taking into account the fact that dynamic load is transferred from the bogie under the body bolster. Spring between the carbody and the supporting beams plays the same role as the secondary suspensions of the actual bogie system. The actuators move the carbody to vibrate at a constant acceleration. Before beginning the dynamic fatigue load test, a pre-test to determine the magnitude of exciting forces of the actuators is conducted. The actuators below the bolsters excite the carbody on the spring at the frequency of 5 Hz, and the exciting force is determined so that the amplitude of acceleration at the quarter of the carbody may be 0.2 g, which is required by the standard specifications for urban transit units.

The inertial loading test method can effectively simulate the dynamic loads induced by the motions of the carbody. However, it can be seen that the dynamic load tests, mentioned above, just considered the vertical loads. According to the author's knowledge, the railway carbody would endure three directions' dynamic loads. For the vertical direction, the vertical loads are transferred to the carbody by the secondary suspension and the mount points of the equipment under the underfloor of the carbody. The lateral dynamic forces, which are induced by the lateral vibration when the vehicle passes through the curve line, are applied on the center pivot. The longitudinal dynamic forces

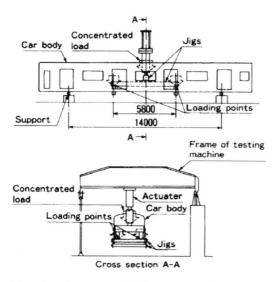

Figure 1.   Concentrated loading test method.

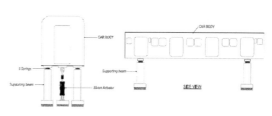

Figure 2.   Inertial loading test method.

caused by the longitudinal oscillation between two carbodies are applied to the position of the coupler. The amplitudes of the acceleration in the three directions are ±0.15 g, ±0.15 g and ±0.15 g, respectively, for the high-speed train, which are given by the EN 12663-1-2010[7]. Therefore, it is necessary to assess the fatigue strength of the carbody, considering three directions' dynamic loads. In this paper, we propose a method to evaluate the fatigue strength of the carbody with the consideration of three directions' dynamic loads.

3 FATIGUE TEST METHOD FOR AN UNDERFRAME OF THE CARBODY

Figure 3 indicates the underframe of the carbody used in the fatigue test, which is cut from an urban transit unit. The specimen is 1/3 of the whole underframe of the carbody, which would endure the lateral, vertical and longitudinal forces. To consider the three directions' dynamic forces in the fatigue tests, Traction Power State Key Laboratory of Southwest Jiaotong University proposed a completely new method for the fatigue tests of the carbody based on the fatigue test bench of the carbody, as shown in Figure 4.

It can be seen from Figure 4 that the main facility of the test bench is the 6 DOF vibration test rig, which can move in 6 directions with the help of 4 hydraulic actuators. Therefore, the 6 DOF vibration test rig is adopted to apply the vertical forces of air spring, the lateral and longitudinal forces of the center pivot. The longitudinal force of the coupler is applied by one longitudinal hydraulic actuator. A long longitudinal beam, which is connected to the underframe by ball joints, is used to release the residual accelerations. The basic load on the underframe of the carbody corresponds to the 1/3 weight of the carbody. The vertical supporting beams are used to support the underframe in the non-working condition. When the fatigue tests

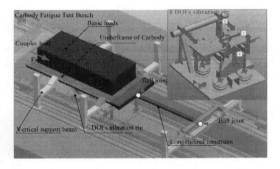

Figure 4. Fatigue test for the underframe of the carbody.

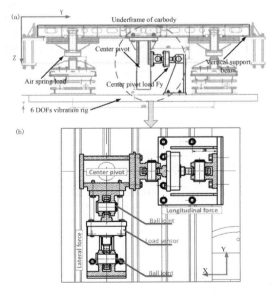

Figure 5. Connection set-ups between the underframe and the 6 DOF vibration rig.

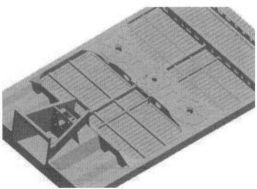

Figure 3. Underframe of the carbody.

begin, the underframe would separate from the vertical supporting beams, and all the vertical loads would be transferred to the underframe by the air spring loading set-up, as shown in Figure 5(a). The vertical air spring loading set-up is fixed on the 6 DOF vibration test rig, which can move in the vertical direction without the influences of the lateral and longitudinal motions.

For the lateral and longitudinal force of the center pivot, two set-ups were fixed on the 6 DOF vibration test rig to transfer the lateral and longitudinal motions, as shown in Figure 5(b).

4 FATIGUE LOADS OF THE FATIGUE TESTS

In order to verify the fatigue strength of the underframe, all the dynamic loads that exist in the operation of vehicles should be included. According to the EN 12663-1-2010, the amplitude of the acceleration for three directions should be ±0.15 g. The vertical and lateral motions are adopted to simulate the dynamic loads induced by the track. The lateral motions are mainly caused by the curing track. The longitudinal motions are induced by the traction and braking system. Moreover, 107 load cycles are suggested to evaluate the fatigue strength of the carbody by the EN 12663-1-2010 standard.

In this fatigue test, the inertia loading test method is used for the vertical motion. Before the test, the payload that corresponds to the 1/3 weight of the whole carbody is applied on the underframe. During the fatigue tests, the underframe moves in the vertical direction with a constant amplitude of the acceleration through the 6 DOF vibration rig. To assess the fatigue strength of the mounting position of coupler, the longitudinal acceleration is converted to the longitudinal force. Meanwhile, the lateral acceleration is also converted to the load force to simulate the lateral force of bumstop when the railway vehicle negotiates the curving track. The profile of the fatigue loads is designed to account for the process of the traction and braking and the process of negotiating curving, as shown in Figure 6.

It can be seen that the load/unload for the vertical loads is taken into consideration to simulate the fluctuation of the carbody due to the track irregularities. In addition, the direction of the lateral forces is changed every 20 cycles to represent the right and left curve line. Meanwhile, the longitudinal load of the coupler and the center pivot switch the direction every 40 cycles and 20 cycles, respectively, to simulate the process of the traction and braking. Figure 7 shows the underframe of the carboy on the test bench.

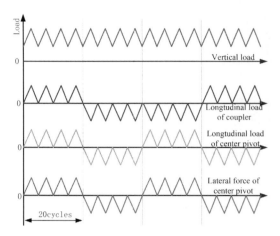

Figure 6.   Profile of the fatigue loads.

Figure 7.   Underframe of the carbody on the test bench.

5   DISCUSSION

The method of the fatigue test for the carbody can sufficiently identify the critical location of the carbody structure. However, dynamic loads generated by the payloads on the underframe are much bigger than normal conditions, which is the main cause of the crack. It can be seen from Figure 9 that the amplitude of the stress under the fatigue loads (±0.15 g) for the full-scale carbody is about 5.1 Mpa at the location of Node 2. However, the dynamic stress measured at the location of Node 2 during the fatigue test is obviously larger than the full-scale carbody condition, as shown in Figure 10. It shows that the maximum of the dynamic stress reaches to about 25 Mpa, which increases the growth rate of crack initiation and propagation.

The underframe is a partial component of the full-scale carbody, which does not contain the

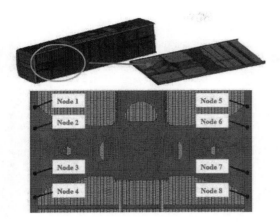

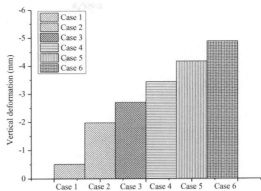

Figure 10.   Comparison for the vertical stiffness of the full-scale carbody and the underframe with different load cases.

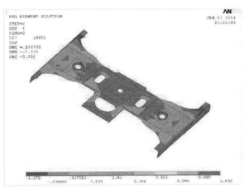

tical deformation for the full-scale carbody. Case 2 to Case 6 represent the underframe with different payloads, which are, respectively, 0 t, 1 t, 2 t, 3 t and 4 t. It can be seen that the stiffness of the underframe is smaller than the full-scale carbody. However, the dynamic loads suggested by the EN standard are based on the full-scale carbody. Therefore, the loads for the full-scale carbody should be equivalent to the partial component conditions before the fatigue tests.

Figure 8.   Fatigue assessment for the underframe of the full-scale carbody.

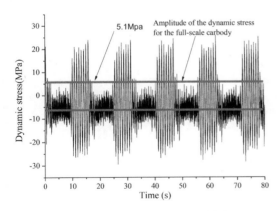

Figure 9.   Dynamic stress measured in the fatigue tests.

side wall of the carbody. Therefore, the stiffness decreases to a large extent compared with the full-scale carbody. Figure 11 illustrates the comparison for the vertical stiffness of the full-scale carbody and the underframe of the carbody with different load cases. In Figure 11, Case 1 indicates the ver-

6   CONCLUSIONS

In this paper, a methodology of the fatigue test for the underframe of the carbody is proposed. Before the test, the payload is loaded on the underframe of the carbody, and the inertial loading method is adopted to apply the vertical load. Moreover, the lateral and longitudinal force of the center pivot, and the longitudinal force of the coupler are taken into consideration. Based on the fatigue result, the fatigue test methodology of the underframe can sufficiently identify the dangerous points of the underframe. However, the stiffness of the underframe is smaller than the full-scale carbody condition; thus, fatigue loads of the full-scale carbody should be equivalent to the partial component condition. However, there has been no effectively equivalent method. Therefore, further investigation on the equivalent method should be conducted in the future.

ACKNOWLEDGMENTS

This work was supported by the National "863" Program (No.2012AA112001), the Research Foundation of National Railway Co. (No. 2014J012-C),

and the Science and Technology Development Program of Sichuan Province (No. 2012GZX0082).

REFERENCES

[1] Ridnou J.A. 2003. Methodology for evaluating vehicle fatigue life and durability. *PhD thesis, The University of Tennessee, Knoxville.*

[2] Hyun-Kyu Jun & Hyun-Seung Jung, et al. 2010. Fatigue crack evaluation on the underframe of EMU carbody. *Procedia Engineering*: 893–900.

[3] Clormann U.H. 1986. Local stresses of welded joints for fatigue assessment. *TH Darmstadt, Diss.*

[4] B.R. Miao. 2006. Simulation research of locomotive carbody structure fatigue based on multibody dynamics and finite element method. *PhD thesis, southwest jiaotong university*, Chengdu.

[5] Sung II Seo, ChoonSoo Park & Ki Hwan Kim. 2005. Fatigue Strength Evaluation of the Aluminum Carbody of Urban Transit Unit by Large Scale Dynamic Load Test [J]. *JSME International Journal* 48 (1).

[6] Oomura K, Okumo S, Kawai S, Masai K & Kasai Y. 1992. Fatigue Test of an Actual Car Body Structure. *Trans, Japan, Soc. Mech* 58 (545, A): 20–25.

[7] EN 12663-1: 2010 Railway applications—Structural requirements of railway vehicle bodies—Part 1: Locomotive and passenger rolling stock [s].

Material Science and Engineering – Chen (Ed.)
© 2016 Taylor & Francis Group, London, ISBN 978-1-138-02936-1

Iron slag as a flame retardant for flexible Poly(Vinyl Chloride)

M. Gao, H. Wang, Y.H. Wang & C.B. Guo
School of Environmental Engineering, North China University of Science and Technology, Yanjiao, Beijing, China

ABSTRACT: Iron slag as a flame retardant was used to PVC, and the mechanical and flame retardant properties of the samples were studied. The resultant data show that iron slag had a better effect on the mechanical properties of the sample, especially tensile strength and impact strength, and 1% of iron slag obtained good flame retardance. PVC samples containing iron slag showed a high limiting oxygen index, low maximum smoke densities, and high thermal stability, which indicated that the flame retardance of the treated PVC was improved.

Keywords: iron slag; Sb_2O_3; flame retardant; PVC

1 INTRODUCTION

Poly(Vinyl Chloride) (PVC), a well-known and economical polymer, is used for many electrical applications. The high chlorine content (56.8%) of PVC makes it relatively resistant to ignition and burning. For many applications of rigid or semi-rigid PVC, additions of flame-retardant additives are not required. However, to make it easy to process, semi-rigid and flexible PVC compounds always contain a large volume of plasticizers such as DOP [di(2-ethylhexyl) phthalate], which can deteriorate flame retardation and smoke suppression properties. When the PVC products contain 45 parts DOP, the Limiting Oxygen Index (LOI) would decrease to about 24 and the PVC would thus become a high flammable material. However, plasticized PVC products can still have good fire performance, particularly if additionally fire-retarded (Chen & Wang, 2010).

Many additives, such as alloys, organic substances, and inorganic compounds including tin, zinc, copper, iron, and molybdenum, have been used in the flame retardation and smoke suppression of PVC (Xu et al. 2005). Because of the high loading, it is essential that a good degree of flame retardancy is obtained, but mechanical properties decrease obviously. Using coupling agents and synergists are good ways to solve this problem (Qu et al. 2005, Pi et al. 2003). Iron slag after flotation, an industrial solid waste, contains many kinds of metal oxides (Levchik & Weil, 2005), which may have an effect of flame retardation and smoke suppression on PVC.

The purpose of the present study is to study the mechanical and flame retardant properties of samples treated with combinations of iron slag, and find another good way to use iron slag after flotation.

2 EXPERIMENTAL

2.1 Materials

PVC, SG2; Dioctyl Phthalate (DOP); Tribasic lead Sulfate (TS), Dibasic lead Phosphate (DP), commercially available; Stearic Acid (SA), Sb_2O_3, Iron Slag (IS), industrial grade, were supplied by Tianjin Fuchen Chemical Reagent Factory (Tianjin, China).

2.2 Instrumentation

LOI values were determined in accordance with ASTM D2863–70 by means of a General Model HC-1 LOI apparatus. The vertical burning test was conducted by a CZF-II horizontal and vertical burning tester (Jiang Ning Analysis Instrument Company, China). The mechanical properties were tested according to the GB/T 1040.2–2006 standard with an LJ-5000 tensile testing machine (Chengde Experimental Factory). Thermogravimetry (TG) was carried out on a DTA-2950 thermal analyzer (Dupont Co. USA) under a dynamic nitrogen (dried) atmosphere at a heating rate of 10°C min^{-1}. The Maximum Smoke Density (MSD) evaluations of the samples, in the form of plates measuring $25.3 \times 25.3 \times 3$ mm^3, were made in accordance with ASTM D 2843–1993 by means of a General Model JCY-1 instrument (Nanjing Jiangning Analysis Instrument Factory).

Table 1. The composition of the samples.

No.	PVC (g)	DOP (g)	TS (g)	DP (g)	ZS (g)	Sb$_2$O$_3$ (g)	IS (g)
A	100	30.0	2.0	2.0	0.5	7.0	0.0
B	100	30.0	2.0	2.0	0.5	7.0	0.5
C	100	30.0	2.0	2.0	0.5	7.0	1.0
D	100	30.0	2.0	2.0	0.5	7.0	2.0
E	100	30.0	2.0	2.0	0.5	7.0	3.0
F	100	30.0	2.0	2.0	0.5	7.0	4.0
G	100	30.0	2.0	2.0	0.5	7.0	5.0
H	100	30.0	2.0	2.0	0.5	7.0	6.0

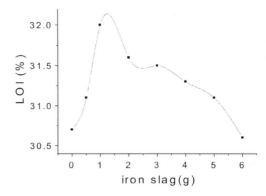

Figure 1. LOI curves of the PVC samples containing different contents of iron slag.

2.3 *Preparation of flame-retardant PVC samples*

Formulations according to predetermined material (including PVC, three salts, salts of stearic acid, DOP, hydroxide, chloride, and coal) were mixed in a mixer for 3 to 5 minutes at 35–45°C. Then, the mixture was plastified on a two-roll mill at 165°C for 6–8 min, and compressed at 180°C to form sheets of 100 mm × 10 mm × 3 mm. The compositions are listed in Table 1.

3 RESULTS AND DISCUSSION

3.1 *Flame retardancy of Sb$_2$O$_3$ and iron slag*

Sb$_2$O$_3$ is good for the flame-retardant effect of PVC material, it is in the gas phase and condensed phase, so its better flame-retardant effect, used in the plastics and rubber industry, has attracted widespread attention.

From the oxygen index values shown in Figure 1, the flexible PVC including only Sb$_2$O$_3$ has good flame retardancy. To get better flame retardancy for PVC, iron slag was added. The data are provided in Table 1. The LOI of samples (A-C) increased with the addition of iron slag, with the added of iron slag, whose LOI increased from 30.7% to 32.0%. When more iron slag was added, the LOI of the samples decreased, so 1.0 g iron slag was optimal. Iron slag and Sb$_2$O$_3$ can be a good flame retardant for flexible PVC.

From Figure 2, we can see that the MSD values of the samples containing iron slag were lower than those of the PVC. Moreover, the maximum smoke densities of the PVC samples decreased with the content of iron slag. However, it increased when the content was more than 4.0 g.

3.2 *Effect of iron slag on the mechanical properties*

Mechanical properties of the samples such as tensile strength and impact strength were measured, and shown in Figures 3 and 4.

Figure 3 shows the effect of the iron slag on the tensile strain of the samples, which shows that the

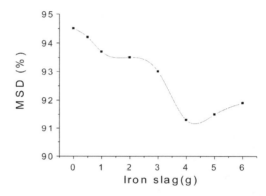

Figure 2. MSD curves of the PVC samples containing different contents of iron slag.

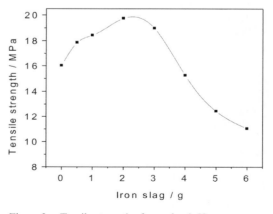

Figure 3. Tensile strength of samples A-H.

tensile strain increases when the added iron slag is less than 2.0 g, but decreases when the content is more than 2.0 g.

Figure 4 shows the effect of the iron slag on the impact strength of the samples. When the added

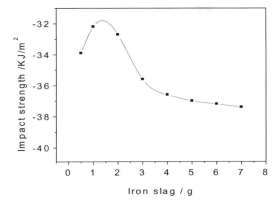

Figure 4.　Impact strength of the samples.

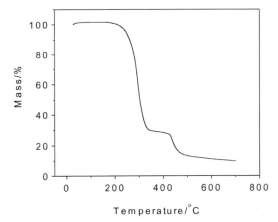

Figure 5.　Thermal analysis of sample A.

iron slag is more than 1.0 g, with the iron slag added, the impact strength of the samples decreases.

So, in a certain range, with the added impact strength of the samples, the tensile strength and impact strength of the samples improved, and the application scope of the samples was not affected.

3.3　Degradation of flexible PVC

The simultaneous DTG and TG curves of sample A and sample F were generated in dynamic nitrogen from ambient temperature to 750°C, and are shown in Figures 3 and 4. The Initial Decomposition Temperature (IDT) determined by 5% of weight loss, the Integral Procedure Decomposition Temperature (IPDT) determined by 50% of weight loss, and char yield at 700°C were measured, and are listed in Table 2.

From Figures 5 and 6, it can be seen that PVC decomposes in three weight losses in total weight loss (Tian et al. 2003). The first stage is mainly due

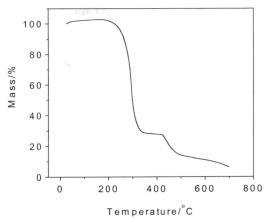

Figure 6.　Thermal analysis of sample F.

Table 2.　Thermal data of the samples from thermogravimetric analysis.

Sample	IDT (°C)	IPDT (°C)	Char yield (%)
A	246.7	283.8	9.5
F	250.0	300.0	6.1

to the emission of hydrogen chloride and the degradation of DOP. The second stage is where the carbonaceous backbone suffers chain scission, and thus a large number of low-molecular-weight compounds (and smoke) are produced. The third stage tends to be, just as for many other polymers, a very slow reaction.

The first stage is mainly due to the emission of hydrogen chloride and the degradation of DOP. At this stage, all the samples lose mass by about 60–70%. In fact, for the samples of plasticized polymer without fire-retardant additives, just over 70% is lost, with somewhat smaller losses for the samples that were treated with flame retardants. The second stage is where the carbonaceous backbone suffers chain scission, and thus a large number of low-molecular-weight compounds (and smoke) are produced. The third stage tends to be, just as for many other polymers, a very slow reaction. The rate of which is relatively uniform for a wide temperature range. As regards to the second and third weight stages, the only possible systematic statement is that both the rates and the percentages of weight loss are much smaller than those at the first stage.

From Table 2, it can be seen that for sample F, compared with sample A, the mass loss rate is decreased at a temperature above 400°C. The decrease significantly of mass loss rates lowers the amount and rate of the release of combustible products from the flexible PVC's decomposition, consequently suppressing the resins' flammability. The increase in char yields agrees with the mechanism of flame retardant. The introduction of flame retardants leads to more char formed at the expense of flammable volatile products of thermal degradation, thus suppressing the combustion and increasing the LOI.

3.4 *Thermal stability of flexible PVC*

The thermal stability of the PVC samples is assessed with two parameters: IDT and IPDT. IDT indicates the apparent thermal stability of PVC, the failure temperature of the resins in processing and molding. On the other hand, IPDT exhibits the resins' inherent thermal stability, i.e. the decomposition characteristics of the resins' volatile composition. From Table 2, it can be seen that PVC/iron slag shows a relatively higher IDT than do the PVC resin, showing that the apparent thermal stability of PVC increases. On the other hand, the existence of flame retardants (iron slag) exhibits a higher IPDT than the PVC, retarding the mass loss rate of the PVC at high temperatures. The high IPDT implies the PVC resins' potential application in highly anti-thermal coatings and thermal insulating materials.

4 CONCLUSIONS

Iron slag can act as a synergistic agent added to the flexible PVC to achieve a good flame-retardant performance. 1.0 g Iron slag and 7.0 g Sb_2O_3 were added to obtain an LOI of 32.0%, and the maximum smoke densities of PVC samples were decreased with the content of iron slag. In a certain range, iron slag has a little effect on the mechanical properties of the sample. Proper iron slag also can improve the decomposition temperature of PVC, and increase the apparent thermal stability inherent thermal stability of PVC.

ACKNOWLEDGMENT

The work was supported by the fundamental research funds for the Central Universities (3142013102).

REFERENCES

[1] Chen, L. & Wang, Y. 2010. A review on flame retardant technology in China. Part I: development of flame retardants. Polymers for Advanced Technologies, 21(1), 1–26.
[2] Xu, J.Z. Zhang, C.Y. & Qu, H.Q. 2005. Zinc hydroxystannate and zinc stannate as flame-retardant agents for flexible poly(vinyl chloride). J. Appl. Polym. Sci. 98, 1469–1475.
[3] Qu, H.Q. Wu, W.H. & Jiao, Y.H. 2005. Thermal behavior and flame retardancy of flexible poly (vinyl chloride) treated with Al (OH) 3 and ZnO. J. Polym. Int. 54, 1469–1473.
[4] Pi, H. Guo, S. & Ning, Y. 2003. Mechanochemical improvement of the flame-retardant and mechanical properties of zinc borate and zinc borate–aluminum trihydrate-filled poly(vinyl chloride). J. Appl. Polym. Sci. 89(3), 753–762.
[5] Levchik, S.V. & Weil, E.D. 2005. Overview of the recent literature on flame retardancy and smoke suppression in PVC. Polymers for advanced technologies. 16(10), 707–716.
[6] Tian, C.M. Wang, H. & Liu, X.L. 2003. Flame retardant flexible poly(vinyl chloride) compound for cable application. J. Appl. Polym. Sci., 89, 3137–3142.

Material Science and Engineering – Chen (Ed.)
© 2016 Taylor & Francis Group, London, ISBN 978-1-138-02936-1

Research on dissimilar metal fusion welding-brazing of titanium alloy and aluminum alloy

H. Li

State Nuclear Power Engineering Co. Ltd., Shanghai, P.R. China

J.J. Dai

Qingdao Binhai University, Qingdao, P.R. China

ABSTRACT: Ti/Al double metal structure can reduce component quality, reduce costs, and meet specific integrated performance, and therefore has broad application prospects. Fusion welding-brazing has become a hot-spot research method of dissimilar metal welding of titanium alloy to aluminum alloy. This paper describes the basic principles and difficulties of fusion welding-brazing of titanium alloy to aluminum alloy, and discusses the interface behavior of fusion welding-brazing of titanium alloy to aluminum alloy.

Keywords: titanium alloy; aluminum alloy; fusion welding-brazing

1 INTRODUCTION

The rapid development of the aerospace and automotive industry has continuously proposed higher requirements on the overall performance of material structure, a single alloy structure has been unable to fully meet the needs of modern industrial development; lightweight, composite structures and low cost are the inevitable trend of development. Titanium alloy is characterized by high specific strength, corrosion resistance, and stable intermediate temperature performance and is widely used in aerospace, chemical, automotive fields. Aluminum alloy is characterized by low density, high specific modulus, corrosion resistance and good fracture toughness, and widely used in a variety of welded structures and products. Ti/Al double metal structure can reduce component mass, decrease costs, and meet specific integrated use performance, therefore it has broad application prospects [1–2]. However, due to the large differences in physical and chemical properties such as melting point, thermal conductivity, thermal expansion coefficient, crystal type between titanium alloy and aluminum alloy, their weldability is poor when using conventional welding methods, and it is easy to form continuous intermetallic compound at the weld, seriously affecting the quality of welded joints. Fusion welding-brazing, with features of both fusion welding and brazing, has become a hot research method for dissimilar alloy welding of titanium alloy and aluminum alloy [2].

2 DIFFICULTIES IN TITANIUM ALLOY AND ALUMINUM ALLOY WELDING

Difficulties in titanium alloy and aluminum alloy welding are mainly the following aspects [3–4]:

1. Strong chemical activity Titanium alloy and aluminum alloy are highly chemically active metals, it is easy to generate high-melting dense TiO_2 and Al_2O_3 on their surface, the oxide material is not conducive to the interfacial bonding between the two, and the oxide will make it easy to produce slag in the weld, reducing the strength of welded joints. At high temperature, hydrogen, nitrogen, oxygen have large solubility in titanium alloy and aluminum alloy, and are easy to form a brittle phase with titanium; their solubility decreases at low temperatures, and is easy to generate porosity, reducing the toughness and ductility of the weld.
2. Large differences in the melting point Pure aluminum's melting point is about 660°C, while pure titanium's melting point about 1670°C. If using fusion welding, in the welding process, when the molten pool temperature reaches the melting point of titanium, it will cause burning and evaporation of the low-melting aluminum and its alloying elements and the non-uniform chemical composition of the weld, and thus the joint is difficult to weld.
3. Large difference in thermal conductivity and thermal expansion coefficients. Thermal

conductivity and thermal expansion coefficient of pure aluminum are respectively 16 times and three times of those of pure titanium, causing large welding deformation and tendency for hot cracking under the welding stresses.

4. Small mutual solubility According to Al-Ti binary alloy phase diagram, aluminum and titanium have a very small mutual solubility, and therefore, titanium and aluminum liquids react with each other to form various brittle Ti-Al intermetallic compounds during the welding.

3 STRUCTURE AND MECHANICAL PROPERTIES OF TITANIUM ALLOY AND ALUMINUM ALLOY FUSION WELDING-BRAZING

3.1 Joint structure of titanium alloy and aluminum alloy fusion welding-brazing

In the aluminum alloy and titanium alloy fusion welding-brazing test, with the proper process parameters, welds with a good profile and without obvious non-wetting and cracks can be achieved. Fusion welded-brazed joints of titanium alloy and aluminum alloy has a dual nature of fusion welding and brazing, its structure is divided into three areas and seven parts, as shown in Figure 1.

Area 1: brazed joint area, which includes titanium alloy base metal zone (Part A), heat affected zone (Part B) and brazing interface zone (Part C).

Area 2: the weld area, Part D as shown in Figure 1.

Area 3: welded joint area, which includes aluminum alloy base metal zone (Part G), heat affected zone (Part F) and the fusion zone (Part E).

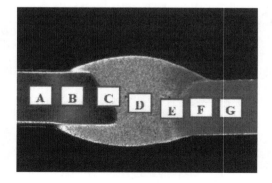

Figure 1. Titanium alloy and aluminum alloy fusion welding-brazing joint morphology [5].

3.2 Microstructure of fusion welded joints

The study found that aluminum alloy's fusion welded joint is composed of overheating zone and fusion zone, and its microstructure is mainly columnar grain and equiaxed grain. When the grain growth direction coincides with the direction of the fastest cooling, coarse columnar grains will come into being. When the columnar grain structure grows to a certain extent, the whole molten pool may have been in the supercooled state, and then the fine equiaxed grain zone will be formed inside the molten pool.

3.3 Microstructure of brazing interface

As Part C of Figure 1 shown, the melting amount of the base metal titanium is small, base metal's original interface retains completely. Studies have indicated that the brazing interface between the base metal titanium alloy and weld metal is mainly composed of the diffusion zone and the interface transition layer zone, wherein the diffusion zone is mainly produced when the low-melting aluminum alloy elements diffuse into the high-melting titanium alloy, and the interface transition layer is mainly produced when a small amount of molten titanium alloy elements interact with elements in the molten pool and cool down. Further research indicates that the diffusion zone is mainly composed of solid solution, and the interface transition layer is mainly composed of Ti-Al intermetallic compounds, and the thickness and the microstructure of the interface transition layer composed of intermetallic compounds are affected by the interface temperature and reaction time and other factors.

3.4 Mechanical properties of the welded joint

Tensile tests of titanium alloy-aluminum alloy dissimilar metal fusion welded-brazed joints show that the fracture position of the joint is closely related to the tensile strength of the joint. If the joint strength is high, mostly the joint fractures at the weld while if the joint strength is low, mostly the joint fractures at the interface, and the tensile strength of joints are mainly affected by the process parameters. Reference[7] points out that when the welding heat input is small or excessive heat focuses on the aluminum base material area, there will be non-metallurgical bonding area at the titanium alloy base metal interface, resulting that the joint strength at the titanium alloy brazing interface is low and the joint partially fractures at the interface; when the heat input is huge or heat distribution is appropriate, alloy elements at the titanium alloy interface react fully with each other, and form a good metallurgical bonding, the joint is firm, and its tensile strength is high, fracture often occurs at the weld, this is an ideal welded joint.

4 BRAZING INTERFACIAL REACTION AND GROWTH MECHANISM

Brazing interface achieves connection by forming an interface reaction layer. In the welding process of aluminum alloy and titanium alloy, the liquid molten weld metal wets and spreads on the solid titanium alloy base material surface and interacts with the base material. The reaction process of brazing interface is divided into the following phases [6–9]:

1. Dissolving of the titanium alloy base material. The liquid phase low-melting aluminum contacts the solid phase titanium alloy base material interface, the titanium alloy base metal contact surface layer is dissolved into the molten weld metal, then the main alloying element atoms, driven by the concentration gradient, diffuse mutually and form diffusion boundary layer at the forefront of the interface.
2. The interaction of the elements.
   ① At the micro area with high Ti concentration, Ti first reacts with weld component Al to generate intermetallic compounds $TiAl_3$. $TiAl_3$ further reacts with other elements o liquid phase weld metal to generate other intermetallic compounds.
   ② At the area near the interface, supersaturated solid solutions are formed.
   ③ Weld component Al diffuses to the solid base material to form a concentration gradient, the reaction generates different types of intermetallic compounds of titanium and aluminum.
3. Solid-state phase transformation. With decreasing molten pool temperature, the mutual solubility between the alloying elements is decreased, and the solute in the supersaturated solution diffuses outward, and precipitates intermetallic compounds with the solid interface as the substrate. Finally, the molten pool solidifies as a whole, the solid phase transformation does not continue, an intermetallic compound layer is formed at the interface.

The intermetallic compound layer generated at the brazing interface is mainly composed of a non-continuous layer and a continuous layer. The non-continuous intermetallic compound layer is formed when the titanium alloy base material components dissolves to the weld and reacts, and the continuous intermetallic compound layer is formed when the weld components diffuse to the titanium base metal and react.

Intermetallic compound interface layers at different brazing interfaces vary in thickness, the intermetallic compound layer at the top of the joint is thick, the intermetallic compound layer at the upper part of the interface becomes thinner gradually, while there are only few non-continuous metallic compound layers at the lower part of the interface. Analysis indicates that morphological differences of the intermetallic compound layer at different positions of the same interface is generated due to different interface reaction degrees, the key factors that determine the degree of reaction of the interface are the interfacial reaction temperature and time. The fundamental cause is differences in heating of different positions of the joint interface in the welding process. The titanium alloy base metal surface at the interface top has a high temperature, the rates at which the base metal elements dissolve to the molten weld metal and the rates at which the weld metal elements diffuse to the solid base metal are both great, the intermetallic compounds controlled by diffusion grow rapidly; the higher the temperature, the longer the cooling time, the longer the reaction effect lasts, the thicker the generated intermetallic compound layer; the surface temperature of titanium alloy base metal at the bottom of the interface is dependent upon the heat transfer of the molten weld metal, and therefore its interface temperature is low, the diffusion process is slow, and the generated intermetallic compound layer is small [6–10].

5 CONCLUSIONS

Dissimilar welding of titanium alloy to aluminum alloy is one of the research hot spots of dissimilar metal welding. Fusion welding-brazing has certain advantages against brazing or fusion welding alone, but there are issues yet to be resolved, such as the brittle phase and the weak mechanical properties of the joint. Therefore, it is necessary to continue in-depth study of the formation mechanism and distribution of brittle phases, and their effect on the mechanical properties in the fusion welding-brazing joint of titanium alloy to aluminum alloy in order to lay the theoretical foundation for improving the quality of welded joints.

REFERENCES

[1] Yajiang Li, Juan Wang & Peng Liu. 2004. *Welding and Application of Dissimilar Metals Difficult to Weld*. Beijing: Chemical Industry Press. (in Chinese).

[2] Shixiong Lv, Xiaojun Jing, Yongxian Huang, Yuanrong Wang, Shufei Li & Yongqiang Xu. 2012. Dissimilar Fusion Welding-Brazing Joint Temperature Field and Interface Microstructure. *Welding Journal* 33(7): 13–18 (in Chinese).

[3] Ryabov B P. 1990. *Welding of Aluminum and Aluminum Alloys to Other Metals*. Beijing: Astronautics Press. (in Chinese).

[4] Shuhai Chen, Liqun Li &Yanbin Chen. 2008. Al/Ti Dissimilar Alloys Laser Fusion Welding- Brazing Joint Interface Properties. *The Chinese Journal of Nonferrous Metals* 18(6): 13–18. (in Chinese).

[5] Shixiong Lv, Tao Yang & Yongxian Huang. 2013. Ti/Al TIG Micro-Fusion Welding-Brazing Interface Behavior and Joint Fracture Behavior. *Rare Metal Materials and Engineering* 42(3): 478–182. (in Chinese).

[6] Shuhai Chen. 2009. *Ti/Al Dissimilar Alloys Laser Fusion Welding-Brazing Process and Connection Mechanism*. Harbin: Harbin Institute of Technology PhD Thesis. (in Chinese).

[7] Peng Dong. 2011. *Dissimilar Alloys Laser Deep Penetration Brazing and Technology Research*. Beijing: Beijing University of Technology PhD Thesis. (in Chinese).

[8] Baohua Zhu. 2006. *Al/Ti Dissimilar Alloys Laser Fusion Welding-Brazing Process and Joint Organizational Performance*. Harbin: Harbin Institute of Technology Master's Degree Thesis. (in Chinese).

[9] Shuhai Chen, Liqun Li & Yanbin Chen. 2013. Joining mechanism of Ti/Al dissimilar alloys during laser welding-brazing process. *Journal of Alloys and Compounds* 509: 891–898.

[10] Jiaming Ni, Liqun Li & Yanbin Chen. 2007. Dissimilar Alloys Laser Fusion Welding-Brazing Joint Properties. *The Chinese Journal of Nonferrous Metals* 17(4): 617–622. (in Chinese).

Material Science and Engineering – Chen (Ed.)
© 2016 Taylor & Francis Group, London, ISBN 978-1-138-02936-1

Analyzing the structure strength of tracked vehicles' wheel box based on the virtual prototyping technology

F. Yang & Q.G. Shang
The Mechanical Engineering Department of Engineering Technology Institute, Changchun, China

X. Jin
Yanbian University, Jilin, China

ABSTRACT: The rigid–flexible model of a tracked vehicle is established on the basis of the FEA model of the tracklayer wheel box. The structure strength of the wheel box under the typical condition is tested based on this model. The stress and the weak units are achieved after the testing. The application of the method testifies that the virtual prototyping technology can provide technical support during the design, and improve the process.

Keywords: wheel box; structure strength; virtual prototyping

1 INTRODUCTION

In the new equipment design and the process of improving the old equipment, the design of the feasibility and reliability of components are extremely crucial. In the past, the physical prototype was usually examined to verify these capabilities. It took very long time and the cost was very high. Using the virtual prototyping technology, the technical indices and the project of the design can be examined and forecasted prior to the physical prototype.

In recent years, with the development of virtual prototyping technology, using co-simulation modeling technology and rigid–flexible coupling of the components, the structural strength analysis methods have been applied to an increased number of mechanical product designs and development processes. Using the virtual prototyping technology not only can get rid of the dependence on the physical prototype, but also will save a lot of manpower, material resources and financial resources. It has an important practical application significance. Virtual prototyping technology is widely used in some developed countries. For examples, Boeing Corporation used the virtual prototype technology on the 777 aircraft to test the design and the performance of the plane. It made the aircraft connect successfully in the absence of drawings. It not only saved dozens billion dollars, but also shortened the development cycle by half. Caterpillar, the largest tractor Corporation of the world, fundamentally changed the design and test steps by using the virtual prototyping technology. They shortened the

development cycle for a few months and saved nearly 40% of the cost.

Tracked vehicles wheel box is an important component of the transmission system. This paper is based on the theories of modal analysis and multibody system dynamics, and establishes the wheel box rigid–flexible coupling model of a certain type tracked vehicles by the virtual prototyping technology. Besides, it then examines the structural strength of the wheel box through a virtual test. These techniques provide technical support to the structure design of the box-type in the transmission system.

2 VIRTUAL MODEL OF WHEEL BOX SYSTEM BUILDING

2.1 FEA model of the wheel box and the bearing model building

This paper takes a wheel box design of a certain type tracklayer. The design is adopt during the improvement of the vehicle. First, the 3D model of the box is built based on the size of the wheel box. Using this model of the wheel box, the finite element meshes are plotted out. The material properties of the box are listed in Table 1. The entire total is divided into 76,151 tetrahedral units. There are 22,262 nodes between them. During this process, the joint between the finite element model and the driving shaft must be taking into account. It is a joint between a rigid body and a flexible body. The shafts of the wheel box are joined with the box through the

Table 1. Material properties of the box.

Material name	Aluminum alloy
Poisson's Ratio	0.31
Density kg/m³	2.66e3
Youngs Modulus N/m²	7.7e7
Ultimate Tensile Strength Mpa	176.4

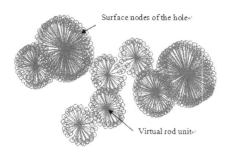

Figure 1. Bearing model.

Figure 2. Finite element model of the wheel box.

bearings. This paper applies a multi-point constraint in the finite element model to simulate the bearing role. The multi-point constraint (MPC, Multi-point Constraint) is a constraint to the nodes. The nodes will be dependent on a certain degree of freedom. The freedom is defined as a number of other independent nodes freedom function. Multi-point constraints can be used to transfer the load between incompatible units. It can show the specific characterization of the physical phenomena, such as the rigid joint, hinged, and sliding. A virtual rod unit is established at the axle hole's center of the finite element model, as shown in Figure 1. The freedoms of the surface nodes of the hole depend on the corresponding virtual rod unit. The joints between the shafts and the wheel box are also established at the corresponding virtual rod unit. The forces of the shaft act on the wheel box through the corresponding virtual unit and multi-point constraint. The wheel box finite element model with the bearing model is shown in Figure 2.

2.2 Gear transmission system model building

This paper will assume the gear transmission system as the stiffness fixed multi-body system. The gear meshing relationship between the tooth surfaces is defined as the contact. In the multi-body dynamics theory, the contact of expression based on Hertz's collision theory can be described as follows:

$$F_n = \begin{cases} k(x_1 - x)^e - C_{\max}\dot{x}f(x,d) & q < q_1 \\ 0 & q \geq q_1 \end{cases}, \quad (1)$$

where:
x—generalized distance used for contact calculation;
$\dot{x}$—generalized velocity;
x_1—distance above which the contact will occur;
k—contact stiffness coefficient;
e—contact force non-linear coefficient;
$C_{\max}$—maximum of damping coefficient; and
d—penetration deepness.

$f(x)$ is defined as a step function, which makes the damping force equal to zero when the contact occurs, and the damping force is the maximum when the transient penetration exceeds the critical depth of penetration.

The above formula denotes when $x \geq x_1$, the two objects do not come into contact, and the contact force is zero. When $x < x_1$, the two objects come into contact, and the size of the contact force is related to the stiffness coefficients, damping, contact force non-linear coefficient and penetration deepness.

The friction between the gears is defined as Coulomb friction. It is simplified as follow: ① the friction is unrelated to the contact area; ② the direction of the friction between the two objects is opposite to the relative slip velocity direction; and ③ the size of the friction between the two objects is inversely proportional to the size of the pressure.

By formula (1) and the above simplification, we can express the contact force during the collision exactly. Then, the input load will pass on step by step.

2.3 Rigid–flexible coupling model building

To couple the finite element model of the wheel box and the gear transmission model, which have been built before together, we must take the finite element model into a model calculation to produce a neutral modal, which can characterize its inherent characteristics of the frequency. Modal analysis is, in essence, a kind of coordinate transformation. Its aim is to put the response vector that is described in the physical coordinate system into the modal coordinate system. Each basal vector of this coordinate system is just a feature vector of the vibration system. In this paper, we use the method of Craig–Bampton to take the modal analysis on the wheel box. The box freedom can be denoted as follows:

$$\mathbf{u} = \left\{ \begin{matrix} \mathbf{u}_B \\ \mathbf{u}_I \end{matrix} \right\} = \left[\begin{matrix} \mathbf{I} & \mathbf{0} \\ \mathbf{\Phi}_{IC} & \mathbf{\Phi}_{IN} \end{matrix} \right] \left\{ \begin{matrix} q_C \\ q_N \end{matrix} \right\}, \qquad (2)$$

where:
$\mathbf{u}_B$—boundary DOF;
$\mathbf{u}_I$—interior DOF;
$\mathbf{I}, \mathbf{0}$—identity and zero matrices, respectively;
$\mathbf{\Phi}_{IC}$—physical displacements of the interior DOF in the constraint modes;
$\mathbf{\Phi}_{IN}$—physical displacements of the interior DOF in the normal modes;
q_C—modal coordinates of the constraint modes; and
q_N—modal coordinates of the fixed-boundary normal modes.

At this point, the corresponding modal stiffness matrix and quality matrix of the wheel box can be denoted as follows:

$$\hat{\mathbf{K}} = \mathbf{\Phi}^T \mathbf{K} \mathbf{\Phi}$$
$$= \left[\begin{matrix} \mathbf{I} & \mathbf{0} \\ \mathbf{\Phi}_{IC} & \mathbf{\Phi}_{IN} \end{matrix} \right]^T \left[\begin{matrix} \mathbf{K}_{BB} & \mathbf{K}_{BI} \\ \mathbf{K}_{IB} & \mathbf{K}_{II} \end{matrix} \right] \left[\begin{matrix} \mathbf{I} & \mathbf{0} \\ \mathbf{\Phi}_{IC} & \mathbf{\Phi}_{IN} \end{matrix} \right]$$
$$= \left[\begin{matrix} \hat{\mathbf{K}}_{CC} & \mathbf{0} \\ \mathbf{0} & \hat{\mathbf{K}}_{NN} \end{matrix} \right], \qquad (3)$$

$$\hat{\mathbf{M}} = \mathbf{\Phi}^T \mathbf{M} \mathbf{\Phi}$$
$$= \left[\begin{matrix} \mathbf{I} & \mathbf{0} \\ \mathbf{\Phi}_{IC} & \mathbf{\Phi}_{IN} \end{matrix} \right]^T \left[\begin{matrix} \mathbf{M}_{BB} & \mathbf{M}_{BI} \\ \mathbf{M}_{IB} & \mathbf{M}_{II} \end{matrix} \right] \left[\begin{matrix} \mathbf{I} & \mathbf{0} \\ \mathbf{\Phi}_{IC} & \mathbf{\Phi}_{IN} \end{matrix} \right]$$
$$= \left[\begin{matrix} \hat{\mathbf{M}}_{CC} & \hat{\mathbf{M}}_{NC} \\ \hat{\mathbf{M}}_{CN} & \hat{\mathbf{M}}_{NN} \end{matrix} \right], \qquad (4)$$

where: the subscripts I, B, N, C denote the internal DOF, boundary DOF, normal mode and constraint mode, respectively. The caret on $\hat{\mathbf{M}}$ and $\hat{\mathbf{K}}$ denotes that this is generalized mass and stiffness.

The load on the wheel box put in by the drive shaft through the bearings can be simplified as the distributed load. The kinematics equations

$$M \ddot{x} + Kx = F \qquad (5)$$

Figure 3. Rigid-flexible coupling model of the wheel box.

can be transferred to the modal coordinates using the modal switch matrix. This can be represented in the simplified form as follows:

$$\hat{M} \ddot{q} + \hat{K} q = \Phi^T F = f \qquad (6)$$

where f—is the load vector projected to the modal coordinates, i.e. the modal force.

In the rigid–flexible coupling model, roller bearings are simplified as revolute joints, and ball bearings are simplified as spherical joints. Two middle axes are fixed on the wheel box. In the simplified model, we add six-direction generalized forces between the wheel box and the middle axes to simulate the actions. The translational stiffness of the generalized forces is 1e10 N/mm and the rotation stiffness is 1e6 $N \cdot mm$/deg. It will reflect the deformation and the transmission of force between the middle axes and the wheel box during the course of working factually. Then, the gear transmission system and the wheel box model are coupled together by these three types of constraints. The trunk of the wheel box is fixed on the bottom deck of the power bay through four bolt holes. We also use the six-direction generalized forces whose translational stiffness and rotation stiffness are both too large to describe the constraint relations. The rigid–flexible coupling model is shown in Figure 3.

3 VIRTUAL TEST RESULTS AND ANALYSIS

3.1 Wheel box modal analysis

First, the modal analysis of the wheel box is taken under the virtual environment. The results are shown in Figure 4. On this basis, the modal analysis of the rigid–flexible coupling model is also taken. The results are shown in Figure 5.

First order model   Second order model

Third order model   Forth order model

Figure 4.   Model of the wheel box.

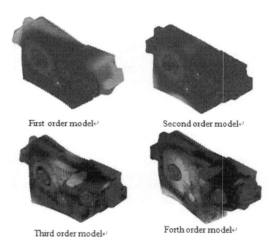

First order model   Second order model

Third order model   Forth order model

Figure 5.   Model of rigid-flexible coupling mode.

From the table, it can be obtained that the modal parameters of the system without the constraints between the gear transmission system and the wheel box are quite different from the ones of the system with the constraints. The frequencies fall wholly. This is because the frequencies calculated from the model without the gear transmission system are the parameters of the trunk of the wheel box, but not the ones of the whole wheel box system. So, the gear transmission system and the constraints between the gear transmission system and the wheel box are necessary when analyzing the frequencies of the model.

Because the trunk of the wheel box is fixed on the bottom deck of the power bay and the gear transmission system is connected with the engine's output shaft and the gear box input shaft, it must be influenced by the external vibration excitation. At this time, if the frequency of the external excitation is similar to the inherent frequency, the resonance will occur. This probability must be avoided during the design process. Otherwise, life and reliability of the box will be affected; the whole system may not work.

Reference [1] shows that the inherent frequency ranges between 16.7 Hz and 33.3 Hz when the engine rotating speed is between 1000 and 2,000 rpm. The mesh frequencies of each gear is shown as follows: the frequency of the first gear and the reverse gear is 296 Hz; that of the second gear is 494 Hz; that of the third gear is 592 Hz; that of the fourth gear is 691 Hz; and that of the fifth gear is 802 Hz. The external excitation frequency bringing from the road surface when running is usually no more than 100 Hz. It can be seen that the mesh frequency of the fifth gear is the same as the inherent frequency of the first-order model. In this way, when the vehicle is running at the fifth gear, the wheel box may resonate. It will debase the reliability of the vehicle. The results show that there are certain defects in the wheel box design, which is still needed to be further improved to make the

Table 2.   The vibration frequency of each order model.

Frequency of the wheel box model with the gear transmission system

Order	1	2	3	4	5
Frequency/Hz	801.93	1317.47	1630.17	1825.54	2078.63
Order	6	7	8	9	10
Frequency/Hz	2204.90	2715.56	2907.80	3091.84	3146.86

Frequency of the wheel box model without the gear transmission system

Order	1	2	3	4	5
Frequency/Hz	781.37	1184.46	1394.34	1549.95	2053.27
Order	6	7	8	9	10
Frequency/Hz	2192.13	2295.53	2389.97	2561.92	2720.97

inherent frequency to keep away from the external excitation frequency.

3.2 Strength analysis of the wheel box

The wheel box is at the most atrocious working condition when the tracked vehicle takes the largest vertical slope at the first gear and the maximum rotating speed of the engine. The structural strength analysis is taken under this condition.

This paper sets the speed of the vehicle at this time as v, the track efficiency as η_X, the traction as F, and the resistant torque acting on the passive-axis as M_{qb}. Based on the parameters of the vehicle, formulas (7), (8), and (9) can be obtained as follows:

$$Fv = W_f \cdot \eta_{ch} \cdot \eta_X, \tag{7}$$

$$\eta_X = 0.975 - 0.000075v^2, \tag{8}$$

$$M_{qb} \cdot \omega_{qb} \cdot \eta_{ch} \cdot \eta_X = F \cdot v, \tag{9}$$

We can obtain $M_{qb} = 1\ 383.3\ \text{N} \cdot \text{m}$. The rotating speed of the drive axis is 2 000 r/min.

Therefore, a rotating speed of 2 000 r/min is added on the drive axis of the wheel box rigid–flexible coupling model, and a resistant torque of 1 383.3 $N \cdot m$ is added on the passive axis. Then, the load under this condition can be denoted exactly.

The measured working stress and ultimate stress distribution are shown in Figures 6 and 7. As shown in the figures, the stress distribution areas are mainly concentrated around the axle hole of the passive axis and the middle axis. The maximal stress distributions are near the screw hole on the diagonal directions of the passive axis hole. The stress distribution around the passive axis is the

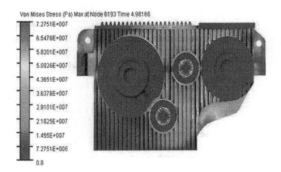

Figure 7. Ultimate stress distribution of the wheel box.

most concentrated, with the peak value reaching 60~70 MPa. Comparing with the ultimate strength of 176.4 MPa for aluminous alloy, the safety factor is more than 2.5. Thus, the design requirement is met.

4 CONCLUSIONS

The tracked vehicle wheel boxes' finite element model and the rigid–flexible coupling model are established in this paper using the virtual prototyping technology. The stress distribution of the wheel box under the utmost condition is analyzed by using the models. The result is similar to that of the actual process. The feasibility of the method is verified.

REFERENCES

[1] Cao Feng-li, Analysis on dynamic and static characteristics of Tracked vehicle gearbox system, Beijing, 2005.
[2] Yu Yu-chun, Analysis on Gear Box Modes for Tracked Vehicles, ISSN 1672-1497, 2006(6).
[3] Du Han-ping, The Virtual Prototype Technology of Crawler Crane Based on the Rigid-Flexible Multi-body Dynamic, Dalian University of Technology, 2006.
[4] Yang Guang-wu, Simulation of Prediction of Fatigue Life for Component of Railway Vehicle, Southwest Jiaotong University, 2005.
[5] Chen Wei-ping, Zhang Yun-qing, Dynamic Analysis of Mechanical Systems and Application Guide ADAMS, Tsinghua University Press, 2005.

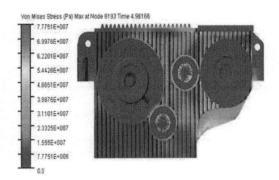

Figure 6. Working stress distribution of the wheel box.

Material Science and Engineering – Chen (Ed.)
© 2016 Taylor & Francis Group, London, ISBN 978-1-138-02936-1

Characterization of precipitate phases in a Mg-Gd-Y-Nd-Zr alloy

J.C. Xiong, Z.W. Du, T. Li & X.L. Han
National Analysis and Testing Center for Nonferrous Metals and Electronic Materials,
General Research Institute for Nonferrous Metals, Beijing, China

Y.J. Li, X.G. Li & K. Zhang
State Key Laboratory for Fabrication and Processing of Nonferrous Metals, General Research Institute
for Nonferrous Metals, Beijing, China

ABSTRACT: The crystallography of the precipitates in a Mg-7Gd-5Y-1Nd-0.5Zr alloy was investigated using Transmission Electron Microscopy (TEM) and Scanning Transmission Electron Microscopy (STEM). The β′ precipitate, which has a b.c.o. structure, was present in a significant fraction in 483 K/225 h- and 483 K/300 h-aged samples. Atomic-resolution High-Angle Annular Dark-Field (HAADF) images were taken to characterize the structure of the β′ phase. The zigzag arranged arrays in the high-angle annular dark-field images demonstrate that the β′ phase has a Mg_7RE-type structure. Some small precipitates with different structures were found at the neck of β′ precipitates in the 300 h-aged alloys. *In situ* observations on the 300 h-aged sample were conducted during the aging treatment at 538 K. The nucleation of β precipitates was found to take place at the area of those small precipitates that has a higher concentration of rare-earth elements.

Keywords: Mg-7Gd-5Y-1Nd-0.5Zr alloy; precipitates; microstructure; HAADF-STEM; *in situ* observation

1 INTRODUCTION

Magnesium Rare-Earth (RE) alloys with a high yield strength and creep resistance at both elevated temperature and room temperature have received considerable attention due to their potential applications in the aerospace, aircraft and automotive industries. Enormous efforts have been made on improving the mechanical properties of Mg-RE alloys. The addition of rare-earth alloys such as Gd, Y, Dy, and Nd can significantly enhance the mechanical properties of magnesium alloys. The key reason to this strengthening behavior is the precipitation of the Mg-RE phase during isothermal aging according to previous research on magnesium alloys. The Mg-Gd system is one of the most promising candidates due to the remarkable age-hardening response and very good thermal stability of the main strengthening phase (Smola, B. et al. 2002, Rolhlin, L.L. 2003, Lorimer, G.W. et al.1987, Lorimer, G.W. et al. 2003).

The isothermal aging precipitation sequence varies from the different kinds of solute elements. In Mg-Gd alloys, the precipitation sequence has been proposed previously as follows: S.S.S.S-β″-β′-β (Peng, Q.M. et al. 2012). The first precipitated β″ phase has a DO19 structure with a lattice parameter of a = 0.64 nm and c = 0.52 nm.

The precipitated β′ precipitate phase has a b.c.o. structure with a lattice parameter of a = 0.64 nm, b = 2.22 nm, and c = 0.52 nm, and it has been proved to be the main strengthening phase in peak-aging samples (Gao, X. et al. 2006, Zhu, Y. et al. 2006, Li, T. et al. 2013). The equilibrium β phase has a f.c.c. structure with a = 2.22 nm. Another f.c.c. structural intermediate precipitate with a lattice parameter of a = 0.74 nm, designated as β1, has been found to form between β′ and β (Nie, J.F. et al. 2000; Gao, X. et al. 2006).

Previous research (Li, Y.J. et al. 2010, Zhang, K. et al. 2008) has shown that the Mg-7Gd-5Y-1Nd-0.5Zr (EW75) alloy exhibits high strength at both room temperature and elevated temperature. The ultimate tensile strength and yield strength of the alloy at T5 temper can be more than 415, 340 MPa. Yet, some details of the precipitates are still unclear. In the present paper, HAADF-STEM was used to characterize the crystal structure of precipitates in the EW75 alloy.

2 EXPERIMENTAL PROCEDURE

EW75 alloy with a nominal composition of Mg-7Gd-5Y-1Nd-0.5Zr (wt%) was used for the present study. The ingot was melted in a medium-frequency induction furnace under RJ-2 flux refining. A two-step solution treatment (793 K for 16 h at 808 K for 20 h) was carried out to eliminate the segregation in the as-cast alloy. The solution-treated ingot was then quenched in water and aged at 483 K.

To make transmission electron microscope samples, the ingot was cut into 3 mm in diameter and 1 mm-thick slices and ground to about 90 u, and then the samples were twin-jet electro-polished in a solution of 10 ml perchloric acid and 190 ml ethanol at 243 K and 45 V.

The observation and analysis were performed on a JEM-2010 transmission electron microscope and a Tecnai G^2 F20 scanning transmission electron microscope operating at 200 keV. The *in situ* observations were made on a Tecnai G^2 F20 microscope with a Gatan double tilt heating holder. To reduce the time needed for the process of phase transition, the sample temperature was set at 543 K.

3 RESULTS AND DISCUSSION

3.1 *Microstructure of β′ precipitates*

A common microstructural feature of the 483 K peak-aged samples is the distribution of the β′ precipitate. Figure 1 shows the morphology and Selected Area Electron Diffraction (SAED) pattern of the 483 K/225 h-aged sample. The SAED pattern consists of three different oriented β′ variant patterns and a Mg $[0001]_\alpha$ pattern. The HRTEM image (Figure 1b) reveals that the β′ precipitates are coherent with the Mg matrix.

The microstructure of the aged alloy is composed mainly of a dispersion of an oval-shaped particle formed on the {11–20} plane of the Mg matrix. These precipitates were proved by the electron diffraction to be β′ precipitate, which has a b.c.o. structure, and its isomorphic with α-Mg, which has a h.c.p. structure. The SAED pattern of 225 h-aged samples, shown in Figure 1, demonstrates that the β′ precipitates and the α-Mg matrix have a fixed orientation relationship of $[001]_\beta$// $[0001]_\alpha$, $[100]_\beta$//$[11–20]_\alpha$.

3.2 *STEM observations of precipitates*

Due to the limitation of TEM technology, the RE atoms and Mg atoms in the β′ precipitate were difficult to distinguish. In order to obtain a further understanding of the atom occupation in these β′ precipitates, HAADF-STEM images with the incident beam parallel to the $[0001]_\alpha$ direction were taken using a FEI Tecnai G2 F20 transmission electron microscope, as shown in Figure 2.

The HAADF-STEM technique forms the images from high-angle scattered electrons and produces a contrast proportional to the atomic number square (Zhu, Y. et al. 2006). So, in HAADF-STEM images, columns with heavier elements have a brighter contrast. As shown in Figure 2, the β′ precipitates should have a high concentration of rare-earth elements than the matrix as they have a brighter contrast. The contrast in some of the joint part of the β′ precipitates was even brighter (indicated by the black circles in Figure 2). These areas should have different structures from the β′ precipitates.

High-magnification HAADF-STEM images are shown in Figure 3, in which the brighter spots represent the columns with rare-earth atoms, and the darker spots represent the columns of Mg atoms. In the β′ precipitates, the rare-earth atoms were distributed zigzagged, and every two zigzag

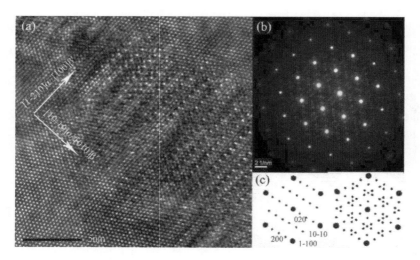

Figure 1. (a) HRTEM image of 483 K/225 h-aged samples; (b) corresponding SAED pattern; (c) schematic pattern for the superimposed β′ phase and matrix. Incident beam is parallel to the $[0001]_\alpha$ direction.

arrays of RE atoms were equally distanced apart by 1.1 nm, indicated by the broken lines displayed in Figure 3(b). This kind of structure is consistent with the ideal β′ structure model proposed by Ninshijima, M. et al. (2007), as displayed in Figure 3c, which was also proved by Li, T. et al. (2013). This kind of β′ phase has a chemical composition of Mg_7RE, close to the measured chemical composition in Mg-Gd alloys (Honma, T. et al. 2007).

As revealed by the recorded HAADF-STEM image, the b.c.o. structural β′ phase

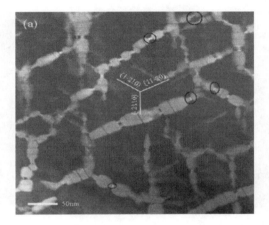

Figure 2. HAADF-STEM image of 210°C/300 h-aged samples showing the three variants of the β′ precipitates and the small platelets at the neck of the β′ precipitates. Incident beam is parallel to the $[0001]_\alpha$ direction.

has a lattice parameter of $a_{\beta'} = 2\ a_0 = 0.64$ nm, $b_{\beta'} = 4\sqrt{3}a_0 = 2.22$ nm, and $c_{\beta'} = c_0 = 0.52$ nm. While the lattice parameter of α-Mg is $a_0 = 0.321$ nm, the actual lattice parameter of the β′ phase is measured to be $a_{\beta'} = 0.664$ nm, slightly two times larger than a_0, as an effect of the Mg atoms partly being replaced by RE atoms. Chen, P. et al. (2009) and Liu. H et al. (2013) conducted a series of first-principles calculations on the structure and chemical composition of the β′ phase, and their results also agree with the Mg_7RE structure model. Honma, T. et al. (2005) also suggests a chemical composition of Mg_7RE.

The ellipsoid β′ phase in over-aged alloys is often connected by some smaller platelets. The structure of these smaller platelets is different from β′ bulk particles, as shown in Figure 4. As with the proposed structure model demonstrated, the recorded HAADF image shows that the RE atoms in β′ bulk particles are arranged in a zigzag fashion, and the length of a period in the $[010]_{\beta'}$ direction is 2.22 nm, eight times of the plane spacing of the $\{10\text{–}10\}_\alpha$ plane. However, the smaller platelets at the horns have a different periodicity along the $[010]_{\beta'}$ direction. In β′ bulk particles, every two adjacent arrays of the zigzag-arranged RE atoms are staggered and equally distanced apart by 1.1 nm (Fig. 3b). However, as the broken lines and arrowheads in Figure 4b present, one kind of the zigzag arrays in the platelets (marked as type A) at the joint part of β′ bulk particles has the same arrangement, and the other kind of zigzag arrays (marked as type B) sticks to each other, not separated by three Mg atom layers. These two kinds of structure are similar to that of

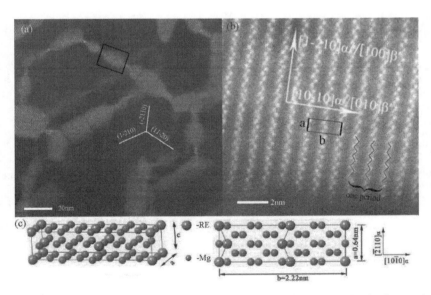

Figure 3. HAADF-STEM images of the sample aged at 483 K for 300 h; (a) low-magnification image taken along the $[0001]\alpha$ zone axis; (b) high-resolution image taken along the $[0001]_\alpha$ zone axis; (c) structure model for the β′ precipitates and atomic arrangement projected along the $[001]_{\beta'}$ axis.

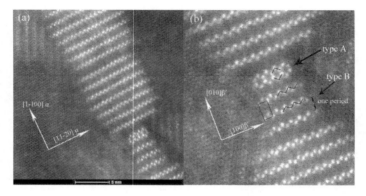

Figure 4.   (a) Atomic-scaled HAADF-STEM images of the platelets formed at the joint part of the β′ precipitates; (b) local amplification corresponding to the selected area in (a). Incident beam is parallel to [0001]$_\alpha$ direction.

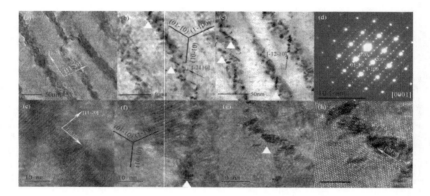

Figure 5.   *In-situ* observed results of the EW75 alloy during aging treatment; (a) TEM image of the 483 K/300 h-aged sample; (b) TEM image of the 483/300 h + 538 K/10 min-aged sample; (c) TEM image of the 483/300 h + 538 K/1 h-aged sample. (d) Corresponding SAED pattern; (e) HRTEM image of the 483 K/300 h-aged sample; (f) HRTEM image of the 483/300 h + 538 K/15 min-aged sample; (g) HRTEM image of the 483/300 h + 538 K/1 h-aged sample; (h) HRTEM image of the precipitates formed during aging treatment at 538 K. Incident beam is parallel to the [0001]$_\alpha$ direction.

an intermediate phase reported for the Mg-Nd alloy (Saito, K. & Hiraga, K. 2011).

No other diffraction spots were visible in the SAED pattern, and these two kinds of nanoscale precipitates only exist in the joint part of the β′ precipitates in the over-aged EW75 alloy. According to the study by Gao, X. et al. (2006), the β′ phase would turn into the β$_1$ phase or the β phase during the aging treatment. In addition, the precipitate phase transition process often starts at the joint part of the β′ precipitates (Nie, J.F. & Muddle, B.C. 2000). So, the production of these two kinds of structure might be the beginning of the β′ phase transition.

3.3   *In situ observations of the phase transition process*

In order to determine the role of the small platelets (types A and B in Figure 4b) in the process of the phase transition during the aging treatment, *in situ* observations were made. To reduce the time needed for the process of the phase transition, the 483 K/300 h-aged sample was then aged at 538 K. The results are shown in Figure 5.

When the sample was aged at 538 K, the β′ phase decomposed and a kind of plate-shaped precipitated with a habit plane parallel to the {10–10}$_\alpha$ nucleated at the neck of the β′ precipitates. At the first step of the aging process, the plate-shaped precipitates grew very fast. After 10 min aging, the length along the [1–210]$_\alpha$ direction and the width along the [10–10]$_\alpha$ direction of the precipitates grew up to 20 nm and 2.5 nm, respectively (Figure 5f). In the next 1h of the aging process, the size of the precipitates barely changed (Fig. 5 g). The size and the volume fraction under the condition was not enough to obtain the diffraction patterns, but the features of this kind of precipitates were the same as the observed β precipitates in many Mg-RE alloys (Li, T. et al. 2013, Gao, X. et al. 2006).

The concentration of the RE element in the β precipitate with a nominal composition of Mg$_5$RE was higher than that in the α-Mg matrix and β' phase with a nominal composition of Mg$_7$RE. While the neck of the β' precipitates contained a large proportion of rare-earth atoms, the nucleation of the β precipitate would take place at the neck to obtain lower nucleation energy.

4 CONCLUSIONS

The microstructure of the EW75 alloy in the over-aging stage was investigated using transmission electron microscopy and scanning transmission electron microscopy. The 483 K/300 h- and 483 K/225 h-aged samples were mainly composed of the α-Mg matrix and β' phase. The β' phase is coherent with the α-Mg matrix and has a b.c.o. structure with a lattice parameter of a$_{\beta'}$ = 0.64 nm, b$_{\beta'}$ = 2.22 nm, and c$_{\beta'}$ = 0.52 nm. Atomic-scaled HAADF images with the incident beam parallel to the [0001]$_\alpha$ direction agree with a Mg$_7$RE-type β' phase.

The β' precipitates in the 483 K/300 h-aged sample were connected by some platelet-shaped precipitates. These smaller precipitates have a higher concentration of rare-earth elements than the β' precipitate. The rare-earth atoms in these precipitates have a different periodicity along the [010]$_{\beta'}$ direction from the β' bulk particles. During the aging treatment, the Mg$_5$RE-type β phase with a habit plane parallel to the {10–10}$_\alpha$ would nucleate in these smaller precipitates.

REFERENCES

[1] Chen, P. Li, D.L. Yi, J.X. Wen, L. Tang, B.Y. Peng, L.M. and Ding, W.J. 2009. Structural, elastic and electronic properties of β' phase precipitate in Mg-Gd alloy system investigated via first-principles calculation. *Solid State Science* 11 (2009): 2156–2161.

[2] Gao, X. He, S. Zeng, X.Q. Peng, L.M. Ding, W.J. & Nie, J.F. 2006. Microstructure evolution in a Mg-15Gd-0.5Zr (wt.%) alloy during isothermal aging at 250°C. *Mater Sci Eng A*, 2006, 431: 322.

[3] Honma, T. Ohkubob, T. & Hono, K. 2005. Chemistry of nanoscale precipitates in Mg-2.1Gd-0.6Y-0.2Zr (at.%) alloy investigated by the atom probe technique. *Mater Sci Eng A*, 2005, 395: 301.

[4] Honma, T. Ohkubo, T. Kamado, S. & Hono, K. 2007. Effect of Zn additions on the age hardening of Mg-2.0Gd-1.2Y-0.2Zr alloys. *Acta Mater*, 2007, 55: 4137.

[5] He, S.M. Zeng, X.Q. Peng, L.M. Gao, X. Nie, J.F. & Ding, W.J. 2007. Microstructure and strengthening mechanism of high strength Mg-10Gd-2Y-0.5Zr alloy. *J. Alloys Compd.* 2007 (427): 316.

[6] Li, Y.J. Zhang, K. Li, X.G. Ma, M.L. Wang, H.Z. & He, L.Q. 2010. Influence of extrusion on microstructures and mechanical properties of Mg-5.0Y-7.0Gd-1.3 Nd-0.5Zr magnesium alloy. *Chin. J. Nonferrous Met.* 20: 1692.

[7] Li, T. Du, Z.W. Zhang, K. Li, X.G. Yuan, J.W. Li Y.J. Ma, M.L. Shi, G.L. Fu, X. and Han, X.L. 2013. Characterisation of precipitates in a Mg-7Gd-5Y-1Nd-0.5Zr alloy aged to peak-ageing plateau. *Journal of Alloys and Compounds* 2013, 574: 174–180.

[8] Li, T. Zhang, K. Du, Z.W. Li, X.G. Li, Y.J. Ma, M.L. & Yuan, J.W. 2013. Characterization of β precipitate phase in Mg-7Gd-5Y-1Nd-0.5Zr alloy. *Journal of rare earths*, 2013, 4(31): 410–414.

[9] Liu, H. Gao, Y. Liu, Y. Zhu, J.Z. Wang, Y.M. Wang, Y. & Nie, J.F. 2013. A simulation study of the shape of precipitates in Mg-Y and Mg-Gd alloys. *Acta Materialia* 61 (2013): 453–466.

[10] Lorimer, G.W. Baker, C. Lorimer, G.W. & Unsworth, W. 1987. Proceedings of the London conference on magnesium technology. *London: The Institute of Metals*: 47–53.

[11] Lorimer, G.W. Apps, P.J., Karimzadeh, H. & King, J.F. 2003. Improving the performance of Mg-rare alloys by the use of Gd or Dy additions. *Mater. Sci. Forum*, 419: 279.

[12] Nie, J.F. & Muddle, B.C. 2000, Characterisation of strengthening precipitate phases in a Mg-Y-Nd alloy. *Acta Mater*, 48: 1691–1703.

[13] Nishijima, M. Yubuta, K. & Hiraga, K. 2007. Characterization of β' Precipitate Phase in Mg2-at%Y Alloy Aged to Peak Hardness Condition by High-Angle Annular Detector Dark-Field Scanning Transmission Electron Microscopy (HAADF-STEM). *Materials Transactions* 48: 84–87.

[14] Peng, Q.M. Ma, N. & Li, H. 2012. Gadolinium solubility and precipitate identification in Mg-Gd binary alloy. *J. Rare Earths*, 2012, 30(10): 1064.

[15] Rokhlin, L.L. 2003. *Magnesium Alloys Containing Rare Earth Metals*. London: Taylor and Francis.

[16] Saito, K. & Hiraga, K. 2011. The Structures of Precipitates in an Mg-0.5at%Nd Age-Hardened Alloy Studied by HAADF-STEM Technique. *Mater Trans* 2011, 52: 1860.

[17] Smola, B. Stulikova, I. Vonbuch, F. & Mordlike, B.L. 2002. Structure aspects of high performance Mg alloys design. *Mater. Sci. Eng. A*, 324: 113.

[18] Xu, Z. Weyland, M. & Nie, J.F. 2014. Shear transformation of coupled β$_1$/β' precipitates in Mg-RE alloys: A quantitative study by aberration corrected STEM. *Acta Materialia* 81: 58–70.

[19] Zhang, K. Li, X.G. Li, Y.J. & Ma, M.L. 2008. Effect of Gd content on microstructure and mechanical properties of Mg-Y-RE-Zr alloys. *Trans. Nonferrous Met. Soc. Chin.* 18: s12.

[20] Zhu, Y. Niewczas, M. Couillard, M. & Botton, G.A. 2006. Single atomic layer detection of Ca and defect characterization of Bi-2212 with EELS in HAADF STEM. *Ultramicroscopy* 106: 1076–1081.

Material Science and Engineering – Chen (Ed.)
© 2016 Taylor & Francis Group, London, ISBN 978-1-138-02936-1

Molecular dynamics simulation of the crack propagation of aluminum with void under various conditions

P. Chen, X.B. Liu & Z. Fang
School of Aeronautic Manufacturing Engineering, Nanchang Hangkong University, Nanchang, China

ABSTRACT: The molecular dynamics simulation model of aluminum with void was set up. The crack propagation behavior of aluminum with void was studied by using the embedded atom method. The effects of the void size and loading rate on the crack propagation behavior of aluminum with void were investigated. The results show that the bigger the void, the shorter the time of the system being split. In addition, the details of atomic motion around the void are much more difficult to display. Within the same amount of time, the higher the loading rate, the fewer the details of atomic motion in the system are shown, and the time of the system being split is significantly shortened. Meanwhile, the edge shape of the system after the split becomes more complicated.

Keywords: crack propagation; molecular dynamics simulation; energy evolution; void; loading rate

1 INTRODUCTION

Aluminum alloy will inevitably have tissue defects such as void, which are inclusions when being manufactured and used. The internal crack propagation of materials occurs normally accompanied with the evolution of a series of void. Nucleation, growth and accumulation of void are widely considered to be the main reasons for the ductile damage of plastic materials as well as to be the decisive factors of material strength [1]. Computer science, which has been continuously developed in recent years, provides a powerful tool for researching the micro characteristics of materials. The molecular dynamics simulation could research the system dynamic evolution process in the microscale and accurately grasp the law of microscopic movement such as void evolution and crack propagation. Domestic and foreign scholars have conducted a lot of research in molecular dynamics simulation of crack propagation of aluminum alloy without void and with void [2–7].

Most researchers have merely explored the crack propagation behavior by studying crack or void and tended to focus on the regular pattern of crack propagation. However, there are no particular discussions about the effects of void size and loading rate on the crack propagation. Hence, this paper utilized the LAMMPS molecular dynamics software to investigate the effects of void size and loading rate on the crack propagation behavior of aluminum with void.

2 THE ESTABLISHMENT OF THE MODEL

Aluminum is a kind of face-centered cubic metal. Its structure characteristics and physical characteristics are given in Table 1.

Figure 1 shows the initial molecular dynamics model with prefabricated void. The size of this model is $40a_0 \times 40a_0 \times 0.25a_0$, where a_0 is a lattice constant under a certain temperature. The center of void is set at the point (30, 20) on the XY plane. The void shows a space size that is formed after removing 13 atoms. So, the total number of atoms in this system is 3268. On the XY plane, the size of the region that corresponds to the upper layer and the lower layer is $40a_0 \times 2a_0$; the size of the region that corresponds to the left upper layer and the left lower layer is $10a_0 \times 18a_0$; and the

Table 1.   The structure characteristics and physical characteristics of single crystal Al.

Material	Single crystal Al
Type of structure	FCC
Lattice constant a_0 (nm)	0.405
Minimum distance between atoms (nm)	0.286
Minimum quality of atoms M (kg)	4.254×10^{-26}
Binding energy coefficient A (eV)	0.2703
Potential energy curves of the gradient coefficient α (nm)	11.650

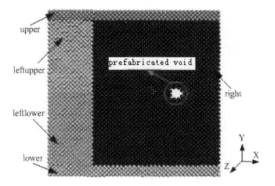

Figure 1. The initial molecular dynamics model with prefabricated void.

size of the region that corresponds to the right layer is $30a_0 \times 36a_0$. The crack length between the left upper layer and the left lower layer is set to $10a_0$. The acting force of the atom both on the upper layer and the lower layer is set to zero. The initial temperature is set to 1 K in order to avoid the thermal excitation effect. This model uses the embedded atom method to carry out the molecular dynamics simulation. The boundary conditions of the X and Y direction are set as aperiodic, while those of the Z direction are set as periodic. After being fully relaxed, the system achieves the balance. The initial rates of the upper layer, the left upper layer, the left lower layer and the right layer on the Y direction are set to 30 m/s, and those on the X and Z direction are set to zero. The atoms on the lower layers are assumed to be fixed. The time step of the simulation is set to 0.001 ps. The microcanonical ensemble (NVE) is employed in the whole system.

3 THE POTENTION FUNCTION AND ALGORITHM OF SIMULATION

The object of the molecular dynamics is a particle system, and the interaction between atoms in the system is described by a potential function. Thus, the EMA [8] potential function was adopted to simulate and calculate the interaction between aluminum atoms. The total potential energy of the system can be expressed as follows:

$$U = \sum_i F_i(\rho_i) + \frac{1}{2}\sum_{j \neq i} \phi_{ij}(r_{ij}) \qquad (1)$$

where F_i is the embedded energy function; ρ_i is the summation of electron cloud density generated by all atoms (except the i-th atom) at the position of the i-th atom; ϕ_{ij} is a pair of potential functions between the i-th atom and the j-th atom; r_{ij} is the distance between the i-th atom and the j-th atom.

The velocity Verlet algorithm was used for the calculation and simulation in this work.

4 SIMULATION RESULTS AND DISCUSSION

4.1 *The effect of void size on crack propagation behavior*

To explore the effect of void size on crack propagation behavior, two different sizes of prefabricated void were engaged in the original model. The first engaged size was formed by removing 5 atoms, and another was formed by removing 23 atoms. The results of the simulation were drawn into time steps--total energy curve. Figure 2 shows the system energy evolution of different void sizes.

Curve b in Figure 2 shows the energy evolution of the original model. Curve a shows the energy evolution of the model, in which 5 atoms were removed. A short period of energy fluctuation occurred in curve a. With the crack propagation, the system energy declined steadily. The atomic motion shows that the system has been split at the 43000th step. Curve c shows the energy evolution of the model, in which 23 atoms were removed. The shape of curve c is similar to that of curve a and curve b. The curve shows that the system energy increased at first and then declined steadily, which means a slight energy fluctuation occurs. However, the time when the fluctuation appeared is later than that of curve a and curve b. The system has the maximum energy at about the 20000th step. From the atomic motion diagram, the biggest difference between the original model and the curve c model is the crack tip and the surrounding morphology. Figure 3 shows the instantaneous atomic motion diagram in curve c when the program ran up to 21000 steps. The contact area between the upper layer and the right layer has no obvious tearing trace, and the system is split at about the 38000th step.

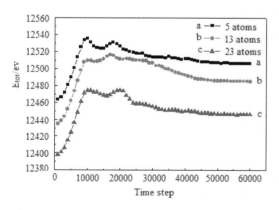

Figure 2. System energy evolution of different void sizes.

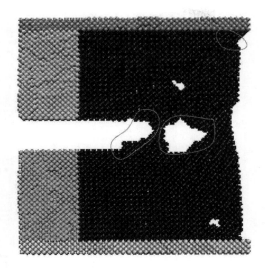

Figure 3. Instantaneous atomic motion diagram in curve c when the program ran up to 21000 steps.

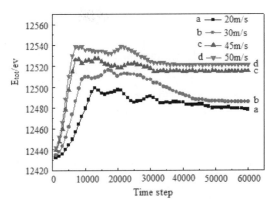

Figure 4. The energy evolution diagram under different loading rates.

4.2 The effects of loading rate on the crack propagation behavior of the system with prefabricated void

Different loading rates along the Y direction were set, which include v = 20 m/s, 45 m/s and 50 m/s, respectively, to investigate the effect of different loading rates on the crack propagation behavior of the system with prefabricated void. These, together with the original model rate of 30 m/s, are drawn into the time step-energy evolution diagram when other conditions remain unchanged. Figure 4 shows the energy evolution diagram under different loading rates.

Curve b in Figure 4 corresponds to the energy evolution of the original model. Curve a corresponds to the energy evolution with a loading rate of 20 m/s, and the system energy in this model from the beginning to the balance stage is smaller than the energy in the original model. This is because all the conditions of the system were unchanged, except the loading rate, so the loading rate is the only factor to affect the system energy. That is, the kinetic energy reflects the change of total energy. The degree of fluctuation in curve a is much greater than that in curves b, c and d. When the loading rate is relatively low, the motion details of the atoms are difficult to clearly present at the same time step. The atomic motion trajectory shows that the system is split at about the 58000th step. Curves c and d correspond to the energy evolution with a loading rate of 45 m/s and 50 m/s, respectively. Their trends are roughly the same. In the balance stage, the system energy in curves c and d is obviously

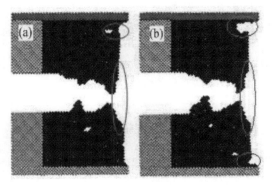

Figure 5. The marginal morphology after the system split.

higher than that in curve b. With the increase in the loading rate, the motion of the system fasted and kinetic energy increased as well as the energy of the system in the balance stage. The atomic motion trajectory shows that the atomic motion in curves c and d is much stronger than that in curves a and b. From curve c, it can be observed that the system was split at about the 30000th step. However, from curve d, the system was split at about the 29000th step. In addition, curve d shows that the degree of tearing in the contact area between the upper layer and the right layer is the largest among those of all the curves. The main reason is the increase in the rate. The violent motion of the atoms leads to a complicated marginal morphology when the crack traverses through the system. The marginal morphology after the system split is shown in Figure 5. Figure 5 (a) and Figure 5 (b) are the marginal morphology corresponding to curves c and d, respectively. The areas marked by the red circle are the main notable morphology.

5 CONCLUSION

The molecular dynamics simulation model of single crystal aluminum with void is set up. Discussions about the main factors in influencing crack propagation behavior are proposed, and some conclusions could be drawn as follows:

The bigger the void, the shorter the time system split, and the details of the atomic motion around the void are much more difficult to display. The increase in the loading rate leads to the aggravation of the atoms movement in the system, resulting in fewer details of the atomic motion being displayed at the same time. The time of the system split is significantly shortened, and the marginal morphology after the split is more complicated.

ACKNOWLEDGMENTS

This study was principally supported by the Natural Science Foundation of Jiangxi Province (2010GQC0803) and the Science and Technology Project of Jiangxi Province Education Department (KJLD12073).

REFERENCES

[1] Zhang Jun-shan. 2004, Strength of materials [M]. Harbin: Harbin Institute of Technology press: 107–109.

[2] W.P. Wu, Z.Z. Yao. 2013, Molecular dynamics simulation of stress distribution and microstructure evolution ahead of a growing crack in single crystal nickel [J]. Theoretical and Applied Fracture Mechanics 62:67–75.

[3] Zhu Zhi-xiong, Zhang Hong, Liu Chao-feng, Yi Dan-qing, Li Zhi-cheng. 2009, Molecular dynamics simulation for solidification process of Ni-Al alloys [J]. Chinese Journal of nonferrous metals 19 (8):1409–1416.

[4] D. Terentyev, E.E. Zhurkin, G. Bonny. 2012, Emission of full and partial dislocations from a crack in BCC and FCC metals: An atomistic study [J]. Computational Materials Science (55):313–321.

[5] Zeng Xiang-guo, Xu Shu-sheng, Chen Hua-yan. 2010, Molecular dynamics simulation on plasticity deformation mechanism and failure near void for magnesium alloy [J]. Trans. Nonferrous Met. Soc. China (20):519–522.

[6] Aude Simar, Hyon-Jee Lee Voigt, Brian D. Wirth. 2011, Molecular dynamics simulations of dislocation interaction with voids in nickel [J]. Computational Materials Science (50):1811–1817.

[7] Tapan G. Desai, Paul Millett, Michael Tonks, et al. 2010, Atomistic simulations of void migration under thermal gradient in UO_2 [J]. Acta Materialia (58): 330–339.

[8] Ackland G.J, Vitek V. 1990, Many-body potentials and atom-scale relaxations in noble-metal alloys [J]. Physical Review B41(15):10324–10333.

Material Science and Engineering – Chen (Ed.)
© *2016 Taylor & Francis Group, London, ISBN 978-1-138-02936-1*

Effects of the ytterbium modification on the microstructure and mechanical properties of the Mg-3Al-1Si magnesium alloy

X.L. Wei, K. Jia, L. Li, W.B. Yu & H.Z. Zhao
School of Materials Science and Engineering, Southwest University, Chongqing, China

ABSTRACT: The microstructures and mechanical properties of the Mg-3 Al-1Si (AS31) alloy with Ytterbium (Yb) addition at ambient and high temperatures were investigated by means of metallurgical phase analysis, X-Ray Diffraction (XRD) analysis and Scanning Electron Microscopy (SEM) analysis. The experimental results indicate that an appropriate amount of Yb (X = 0.2%, 0.5%, wt%) in the alloy can notably refine the Chinese script-shaped eutectic Mg_2Si phases. With the introduction of 0.5% Yb, the amount of needle-shaped phases $Mn_2Si_2Yb_1$ is formed, resulting in the decrease of stress concentration and mechanical properties. After the solid solution treatment, the Chinese script-shaped Mg_2Si phases change from the Chinese script shape to a short pole or a round shape, which is more evident with Yb addition, thus the mechanical properties are increased.

Keywords: Mg_2Si phase; casting; solid solution treatments; mechanical properties

1 INTRODUCTION

Magnesium alloys, one of the lightest metal structural materials, with high specific intensity, high specific rigidity, low density, nice machinability, damping capacity, excellent thermal conductivity and electromagnetic shielding, and recyclability, have been extensively paid attention to be applied in the automobile industry during the past decades, in order to reduce fuel consumption and gas emission, which are considered as eco-structural metal material in the 21st century and have attracted many researchers [1–4].

AS (Mg-Al-Si) series of magnesium alloys have an excellent creep resistant due to the grain boundaries that are suppressed by Mg_2Si particles, exhibiting a low density of $1.99 \times 103\,kg\cdot m^{-3}$, high elastic modulus of 120 Gpa, high melting temperature of 1085°C, high hardness of $4.5 \times 109\,Nm^{-2}$ and low thermal expansion coefficient of $7.5 \times 10\text{–}6\,K^{-1}$ [5]. Generally, the Mg_2Si phase has three morphologies, namely the brittle Chinese script eutectic Mg_2Si, polygon—or dendritic-shaped hypereutectic Mg_2Si [6]. The mechanical properties of the alloys are influenced by their shape, size and distribution. A fine and dispersive distribution of Mg_2Si particles will remarkably improve the mechanical properties of the alloys, while large Mg_2Si phases will worsen the alloy's strength and ductility [7]. To date, the shape, size and distribution of Mg_2Si particles in AS series alloys have not been controlled satisfactorily, and the mechanical properties of the alloys are still low considerably. Therefore,

the modification of the Mg_2Si phase is a main approach to improve the mechanical properties of the alloys due to its low cost. Previous investigations have indicated that the addition of some elements has positive effects on the refinement of the microstructure and on the improvement of Mg_2Si particles, such as Sb [8–12], Ca [13, 14], KBF4 [15], Y [9], P [16, 17], and RE [18–22]. However, less work has been carried on the modification of Mg_2Si by ytterbium in the AS series of magnesium alloys. The special electronic configuration of Yb atoms gives rise to its atomic radius bigger than the others in the Y series rare-earth metals and out of the regularity. The maximum solid solubility of Yb in the magnesium alloy is still much smaller than other heavy rare-earth metals [1].

Therefore, the refinement and modification effects of Yb on the AS31 alloys were investigated in the present study, and the modification mechanism is also discussed. The purpose of this preliminary experiment is to find a simplified and effective casting process route to produce a relatively fine Mg_2Si phase-reinforced magnesium alloy.

2 EXPERIMENTAL

The starting materials were commercial AS31 and Mg−15.05wt% Yb master alloys in this investigation (Mass fraction, the same below unless otherwise specified). The AS31 alloy preheated at 300°C was melted at 720°C in a graphite crucible by an electrical resistance furnace under flux protection.

Table 1. Chemical compositions of studied alloys (mass fraction, wt.%).

Alloy no.	Al	Si	Mn	Zn	Yb	Mg
1 AS31+0 Yb	3.21	1.05	0.24	0.11	0	Base
2 AS31+0.2% Yb	3.21	1.07	0.26	0.11	0.16	Base
3 AS31+0.5% Yb	3.15	1.04	0.36	0.11	0.46	Base

Then, the Mg-Yb master alloy as a modifier agent was added to the melt at 750°C, stirring for about 2 min. After holding at 720°C for 20 min, the melt was poured into a steel mold preheated at 200°C to produce $40 \times 120 \times 20$ mm ingots. For comparison, the alloys with different contents of Yb were prepared at the same conditions. The actual chemical compositions of the investigated alloys are listed in Table 1. The solution treatments were carried out at 420°C for different holding times, respectively, and quenched in the cool water.

All the metallographic specimens were etched by 4% Nital and observed by optical microscopy. The phase characteristics and qualitative analysis of the microstructure were conducted using Scanning Electron Microscopy (SEM, JSM-6610, Japan), equipped with an Energy-Dispersive Spectrometer (EDS, Inca X-Max), and X-Ray Diffraction (XRD, Shimadzu XRD-7000). The ambient and high-temperature (150°C) tensile tests were conducted by using an electronic universal testing machine (WDW3050) at a constant cross-head speed of 1 mm/min. The fracture surfaces of the tensile test specimens were also examined with SEM.

3 RESULTS AND DISCUSSION

3.1 Microstructure and phase analysis of the as-cast alloys

The XRD patterns of the alloys with different Yb contents are shown in Figure 1. According to the XRD patterns, the phase components of both the unmodified AS31 alloy and the alloys modified with different contents of Yb are α-Mg and Mg_2Si phases. However, the added Yb and Al are not detected by XRD due to the low content of Yb, and Al atoms have a large solid solubility in magnesium [23].

According to the Mg-Al-Si ternary phase diagram [23], the as-cast microstructure of the AS31 alloy without Yb addition should be composed of the α-Mg matrix, the coarse Chinese script-type eutectic Mg_2Si and the β-$Mg_{17}Al_{12}$ phase. Figure 2a shows the representative optical micrograph of the unmodified AS31 alloy. It can be seen from Figure 2a that the microstructure of the unmodified as-cast AS31 alloy contains three constituents:

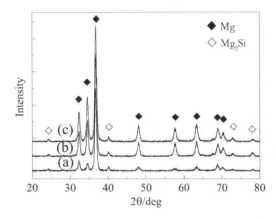

Figure 1. XRD patterns of the alloys with different Yb contents: (a) 0 Yb; (b) 0.2% Yb; (c) 0.5% Yb.

α-Mg matrix, coarse Chinese script-type eutectic Mg_2Si embedded in the α-Mg matrix and gray β-$Mg_{17}Al_{12}$ phase. Due to the large solid solubility of Al in Mg and low content, the $Mg_{17}Al_{12}$ phases mainly precipitated discontinuously in the matrix. This experimental result is in good agreement with previous investigations [11, 24].

The typical micrographs of AS31 alloys with 0.2% and 0.5% Yb addition are shown in Figure 2b, c, respectively. After 0.2% Yb addition, the $Mg_{17}Al_{12}$ phase did not change obviously. While comparing Figure 2a with Figure 2b, it can be found that the size of the Chinese script shaped eutectic Mg_2Si phase was significantly refined by the addition of 0.2% Yb. When the Yb content increased to 0.5%, as shown in Figure 2a, the size of the eutectic Mg_2Si phase did not change obviously. Meanwhile, some needle-shaped or lamellar shaped (in three-dimensional space) phases were also observed. It is hard to identify this phase by the XRD pattern due to its low intensity. Figure 3a, b shows the SEM BEC images of the EDS point analysis of the needle-shaped or lamellar-shaped phases in the 0.5% Yb alloy. The results of the composition examinations by the EDS detector are summarized in Table 2. It reveals that the new phase is composed of Mn, Si and Yb elements, implying the Mn-Si-Yb intermetallic compound. By the analysis of the EDS results, it can be found that the stoichiometry ratio of Mn, Si and Yb was close to 2:2:1. Therefore, the phase was the probable $Mn_2Si_2Yb_1$ compound.

As mentioned above, the microstructure of the as-cast AS31 alloy can be refined by the addition of Yb elements. It causes a decrease in the particle size and an increase in the number of particles, implying an increase in the number of nuclei during solidification. Therefore, it can be suggested

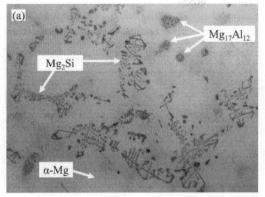

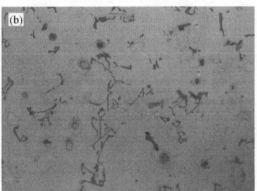

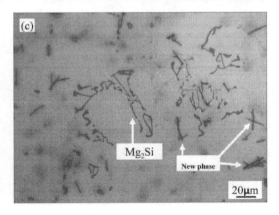

Figure 2. Optical microstructures of the as-cast alloys with different Yb additions: (a) 0 Yb, (b) 0.2% Yb, (c) 0.5% Yb.

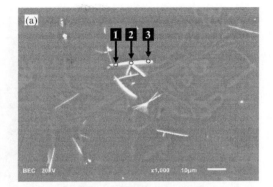

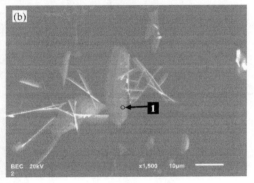

Figure 3. EDS point analysis of the 0.5% Yb alloy.

Table 2. EDS results of the needle-shaped or lamellar-shaped phases in the 0.5% Yb alloy.

| Positions | Elements (at.%) | | | | Total |
	Mg	Mn	Si	Yb	%
Figure 3(a)-1	71.42	11.68	11.61	5.29	100
Figure 3(a)-2	65.57	13.83	13.74	6.85	100
Figure 3(a)-3	82.7	6.4	7.6	3.3	100
Figure 3(b)-1	93.86	2.96	2.28	0.89	100

that Yb can enhance the nucleation of Mg_2Si in the Mg melt. A possible explication is that, under the non-equilibrium solidification condition, the Yb atoms are easily absorbed onto the growth front of the crystal or solid-liquid interface, so they change both the solid-liquid interfacial energy and the surface energy of the crystal, reduce the critical nucleation radius and increase the nucleation number [9, 10, 22]. Finally, the crystals of the alloys are refined. Meanwhile, the rare-earth metal as a surface active element, easily adsorbed on the crystal plane, will change the surface energy of the Mg_2Si crystal by lattice distortion, since the atomic radius of the Yb atom is larger than that of Mg and Si atoms [1, 9, 15]. This effectively suppresses the growth steps of the Mg_2Si crystal.

3.2 Microstructure after the heat treatment

Figure 4 shows the microstructures of the investigated alloys after the solution treatment at 420°C

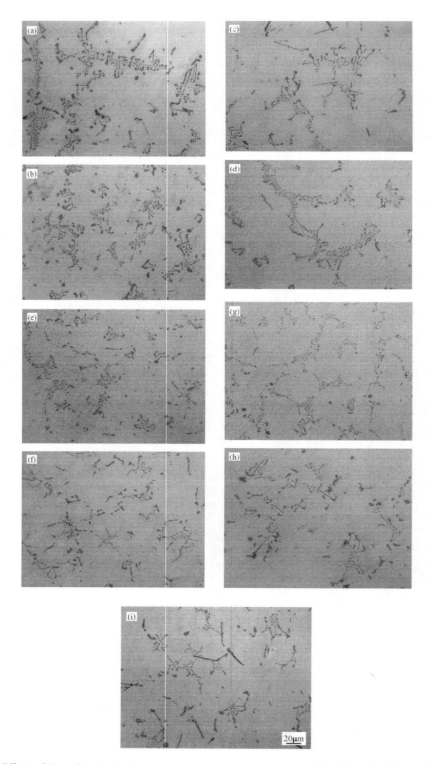

Figure 4.   Effects of the solid solution time on the microstructure of alloys: (a) 0 Yb-24 h, (b) 0.2% Yb-24 h, (c) 0.5% Yb-24 h, (d) 0 Yb-34 h, (e) 0.2% Yb-34 h, (f) 0.5% Yb-34 h, (g) 0 Yb-48 h, (h) 0.2% Yb-48 h, (i) 0.5% Yb-48 h.

for 24, 34 and 48 hours followed by water quenching, respectively. The gray $Mg_{17}Al_{12}$ phases can easily dissolve into the matrix, so it is hardly found in Figure 4 [7]. It can be seen from Figure 4a, b, c that after the 24-hour heat treatment, the Chinese script-shaped Mg_2Si phases are partly dissolved and broken. As shown in Figure 4d, e, f, after the 34-hour solution treatment, most of the Chinese script-shaped Mg_2Si phases are dissolved and parts of Mg_2Si phases granulated, which is more obvious with Yb modification. When increasing the holding time to 48 hours, the Mg_2Si phases in the primary AS31 alloy are mostly fully granulated, while the sizes of Mg_2Si phases with Yb modification tend to increase, as shown in Figure 4 g, h, i. That is to say, adding the trace Yb to AS31 alloys can enhance the dissolution process of the Mg_2Si phase during the solid solution treatment. This is maybe because the Mg_2Si is smaller with the introduction of Yb.

As revealed by the thermodynamic analysis, with a certain volume of Mg_2Si, only the spherical has the smallest surface energy. Therefore, the sheet-like Mg_2Si transforming to spherical is a spontaneous process to reduce the surface energy, depending on Si and Mg diffusivity. The alloy element diffusivity in the matrix is affected by its solubility in the matrix alloy. Furthermore, the solubility of Si atoms in the Mg matrix is only about 0.003% (mole fraction) [6, 25]. This means that the diffusion of Si atoms into the Mg matrix is difficult. However, the vibration energy of the atomic diffusion coefficient is increased by the solution treatment. In addition, the defects such as vacancy, dislocation and sub-boundary will also increase the diffusivity of Si atoms into the Mg matrix. These factors accelerate the diffusion of Si atoms into the Mg matrix, as well as an easier diffusion along the grain boundaries and interfaces. Therefore, the Mg_2Si/Mg interface is the possible and the only diffusion way for Si atoms [25].

From the dynamic conditions analysis, the Gibbs–Thomson equation can be expressed as follows:

$$C_\alpha(r) = C_\alpha(\infty)\exp(2\sigma v_B/K_B Tr) \qquad (1)$$

Here, $C_\alpha(r)$ is the Si element concentration of the interface when the radius of curvature is r; $C_\alpha(\infty)$ is the element concentration of the Si atoms at the parallel interface; σ is the surface energy; v_B is the volume of the Si atom; K_B is the shape factor; T is the temperature; and r is the radius of curvature of the interface. According to Equation (3.2–1), the atomic concentration of Si atoms decreases with the increased radius of curvature. In addition, there always exist concaves and convexities on the particles along the interface. Therefore, during the solution treatment, the high concentration of Si atoms of the interface will diffuse to a low concentration with a lower curvature. Then, the equilibrium of Si concentration at this point is destroyed. In order to maintain the balance of Si concentration, the low radius of curvature of Mg_2Si will be dissolved to compensate for the occurrence of deficiencies. The α-Mg matrix in a plane position with the over-saturation of Si atoms will precipitate the Mg_2Si phase and change the morphology of the Mg_2Si phase. Then, the concaves will become more sunken while the convexities become smoother. Finally, the Chinese script-shaped Mg_2Si dendrite will be broken and become granulated [7, 25]. Notably, the morphology evolution of the Chinese script-shaped Mg_2Si particles is not simultaneous. Therefore, parts of the Mg_2Si phases granulate, while others are not.

3.3 Mechanical properties

Table 3 outlines the mechanical properties of the investigated AS31 alloys with and without Yb modification under the as-cast and solution treatment conditions. From the results, it can be seen that with 0.2% Yb modification, the ambient tensile strength and elongation of the alloys increased from 195.69 MPa and 8.59% to 207.49 MPa and 10.66%, respectively; however, the performance at the high temperature is not altered greatly.

Table 3. Tensile properties of the as-cast experimental alloys.

Alloy	State	Room temperature		150°C	
		σ_b/MPa	δ/%	σ_b/MPa	δ/%
Alloy 1	F-as-cast	195.69	8.59	184.48	12.45
	T4+420°C+48 h	212.27	11.25	183.94	13.12
Alloy 2	F-as-cast	207.49	10.66	184.02	15.45
	T4+420°C+34 h	230.46	12.69	183.77	16.16
Alloy 3	F-as-cast	203.65	9.77	179.83	13.44
	T4+420°C+34 h	210.93	12.14	176.08	13.38

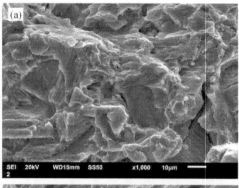

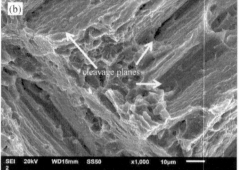

Figure 5. SEM images of the tensile fracture surface of the alloys: (a) 0 Yb, (b) 0.5% Yb.

With the introduction of 0.5% Yb, the tensile strength and elongation are not much improved. After the solid solution treatment, the ambient tensile strength and elongation of the alloys are evidently improved, while the high temperature tensile strengths of the alloys are decreased obviously. Although the Mg$_2$Si particles in the 0.5% Yb alloy are refined, the mechanical properties of the alloy are dropped by the stress concentration caused by the needle-shaped or lamellar-shaped phases. Comparing the fractograph of 0 Yb with 0.5% Yb, as shown in Figure 5, we can find that there exist some huge cleavage planes in the 0.5% Yb alloy caused by the needle-shaped or lamellar-shaped phases.

The mechanical properties are mainly related to the morphology of the eutectic Mg$_2$Si particles, because the micro-cracks are easily diffused along the boundary between the Chinese script-shaped Mg$_2$Si phases and the α-Mg matrix under a lower stress [24]. In this case, the coarse Mg$_2$Si phases and the α-Mg grains are refined by Yb additions, and the solution treatment will greatly reduce the expansion of the micro-cracks. Thus, the mechanical properties and particularly elongation are improved [7, 25].

4 CONCLUSIONS

1. The addition of trace Yb to AS31 alloys has a good modification effect on the eutectic Mg$_2$Si phase. Meanwhile, some needle-shaped or lamellar-shaped intermetallics phases are also formed, when over 0.5% Yb is introduced.
2. The solution treatment can notably improve the morphologies and distribution of the coarse eutectic Mg$_2$Si particles. In addition, the effect of the solution treatment on the Mg$_2$Si phase is enhanced by Yb addition.
3. With an appropriate solution treatment holding time, both the tensile strength and elongation of the AS31 alloys are increased. In particular, Yb modification can evidently shorten the granulating time of the eutectic Mg$_2$Si phases..

REFERENCES

[1] Yu, W.B. 2007. High-strength wrought magnesium alloy with dense nano-scale spherical precipitate. *Chinese Science Bulletin* 52:1867–1871.
[2] Chen, X. 2005. Influence of Ca addition on microstructure and mechanical properties of in-situ Mg$_2$Si/ZM5 magnesium matrix composite. *The Chinese Journal of Nonferrous Metals* 15:410–414.
[3] Wang, H.Y. 2005. Modification of Mg$_2$Si in Mg-Si alloys with K2TiF6, KBF4 and KBF4+K2TiF6. *Journal of Alloys and Compounds* 12:105–108.
[4] Farshid, M. & Mahmoud, M. 2012. High temperature tensile properties of modified Mg/Mg$_2$Si in situ composite. *Materials and Design* 33:557–562.
[5] Emamy, M. 2013. The influence of Ni addition and hot-extrusion on the microstructure and tensile properties of Al-15% Mg$_2$Si composite. *Materials and Design* 46:381–390.
[6] Song, P.W. 2007. Spheroidization of Mg$_2$Si particles in Mg-4Al-2Si alloys during solution treatment process. *Journal of Materials Engineering* 3:34–37.
[7] Liu, J. 2011. Morphology of Mg$_2$Si particles and its effects on tensile properties of Mg-Al-Zn-Si alloy. *Founder Technology* 32:1291–1295.
[8] Chen, K. 2014. Effect of co-modification by Ba and Sb on the microstructure of Mg$_2$Si/Mg-Zn-Si composite and mechanism. *Journal of Alloys and Compounds* 592:196–201.
[9] Jiang, Q.C. 2005. Modification of Mg$_2$Si in Mg-Si alloys with yttrium. *Materials Science and Engineering A* 392:130–135.
[10] Yang, M.B. 2001. Comparison of Sb and Sr on modification and refinement of Mg$_2$Si phase in AZ61–0.7Si magnesium alloy. *Transactions of Nonferrous Metals Society of China* 19:287–292.
[11] Guo, X.H. 2010. Effect of Sb on modification of primary Mg$_2$Si crystal in hypereutectic Mg-Si alloy. *The Chinese Journal of Nonferrous Metals* 20:24–29.
[12] Ren, B. 2010. Effect of Sb on microstructure and mechanical properties of Mg$_2$Si/Al-Si composites. *Transactions of Nonferrous Metals Society of China* 20:1367–1373.

[13] Cong, M.Q. 2005. Effect of Ca on the microstructure and tensile properties of Mg-Zn-Si alloys at ambient and elevated temperature. *Journal of Alloys and Compounds* 539:168–173.

[14] Moussa, M.E & Wal, M.A & El-Sheikh, A.M. 2014. Combined effect of high-intensity ultrasonic treatment and Ca addition on modification of primary Mg_2Si and wear resistance in hypereutectic Mg-Si alloys. *Journal of Alloys and Compounds* 615: 576–581.

[15] Wang, H.Y. 2008. Influence of the amount of KBF_4 on the morphology of Mg_2Si in Mg-5Si alloys. *Materials Chemistry and Physics* 108:353–358.

[16] Li, C. & Liu, X.F. & Wu, Y.Y. 2008. Refinement and modification performance of Al–P master alloy on primary Mg_2Si in Al-Mg-Si alloys. *Journal of Alloys and Compounds* 465:145–150.

[17] Qin, Q.D. 2007. Effect of phosphorus on microstructure and growth manner of primary Mg_2Si crystal in Mg_2Si/Al composite. *Materials Science and Engineering A* 447:186–191.

[18] Zheng, N. 2007. Modification of primary Mg_2Si in Mg-5Si alloys with Y2O3. *Transactions of Nonferrous Metals Society of China* 17:440–443.

[19] Zhao, Y.G. 2005. Microstructure of the Ce-modified in situ Mg_2Si/Al-Si-Cu composite. *Journal of Alloys and Compounds* 389:L1–L4.

[20] Huang, Z.H. 2004. Effects of cerium on microstructure and mechanical properties of as-cast magnesium alloy AZ91D. *Chinese Journal of Rare metals* 28:683–686.

[21] Wu, X.F & Wang, Z. 2012. Influence of Nd addition on microstructure, tensile properties and fracture behavior of cast Al-18 Mg_2Si alloy. *Journal of Liaoning University of Technology* 32:241–246.

[22] Wang, L.P. & Guo, E.J. & Ma, B.X. 2008. Modification effect of lanthanum on primary phase Mg_2Si in Mg-Si alloys. *Rare Earths* 26:105–109.

[23] Zhou, J.X. 2011. Effects of Sr addition on as cast microstructure of AS31 alloy. *Advanced Structural Materials* 686:125–128.

[24] Nasiri, N. & Emamy, M. & Malekan, A. 2012. Microstructural evolution and tensile properties of the in situ Al–15% Mg_2Si composite with extra Si contents. *Materials and Design* 37:215–222.

[25] Peng, L. 2011. Influence of solution treatment on microstructure and properties of in-situ $Mg_2Si/$ AZ91D composites. *Transactions of Nonferrous Metals Society of China* 21:2365–2371.

[26] Ye, L.Y. 2013. Modification of Mg_2Si in Mg-Si alloys with gadolinium. *Material Characterization* 79:1–6.

Material Science and Engineering – Chen (Ed.)
© 2016 Taylor & Francis Group, London, ISBN 978-1-138-02936-1

Surface characteristics of TiN, TiAlN and AlCrN coated tungsten carbide

P. Jaritngam, C. Dumkum & V. Tangwarodomnukun

Department of Production Engineering, King Mongkut's University of Technology Thonburi, Bangkok, Thailand

ABSTRACT: The surface characteristics of coating materials were examined in this study, where the single layer coatings of TiN and TiAlN and the bilayer coatings of AlCrN/TiN and AlCrN/TiAlN on tungsten carbide substrate were performed by using the arc evaporation process. The surface morphology of coating layer(s) were observed and analyzed in terms of thickness, surface roughness, hardness and chemical composition. The scratch test was also performed to determine the adhesion properties of coating layer on the substrate. The investigations revealed that although the bilayer coatings caused a slightly rough surface compared to the single layer coatings, they can provide a good resistance to crack due to the high hardness of AlCrN. The critical loads for each coating material were also reported and analyzed in this paper.

Keywords: coating; adhesion; arc evaporation; surface roundness; tungsten carbide

1 INTRODUCTION

Surface coating processes have been increasingly developed and employed to modify and improve the surface properties of workpiece for protection in aggressive environments. The coating of cutting tool surface is a renowned process to enhance the tool resistance to wear during the machining operation. TiN, TiCN, TiAlN and Al2O3 are typically used as the coating materials for gaining a better oxidation and corrosion resistance of cutting tool surface (Dobzañski & Golombek 2005, Mo et al. 2007, Liew et al. 2013). In general, the cutting tool material can be oxidized under an elevated temperature in the machining process; particularly in the hard and dry cutting operations. When the tool material is quickly oxidized and worn out during the process, the cutting performance and tool life become low accordingly. In order to prolong the tool life, the coating material and its characteristics should be well analyzed and selected. Kumar et al. (2014a) found that the AlCrN-coated cutting tool can provide a good oxidation resistance in the machining process. This is due to the chromium and aluminum contained in the AlCrN that can form a strong protective oxide film to strengthen the tool surface at high temperature. By depositing the AlCrN film on the typical coating materials such as TiN and TiAlN, the coated tool could be further extended its functionality in a broadened spectrum. Hence, this research is to bring out the characteristics of single layer coatings of TiN and TiAlN and the bilayer coatings of AlCrN/TiN and AlCrN/TiAlN, where their micro-structure, surface roughness, hardness and adhesion to the substrate surface were investigated and analyzed. The findings of this research would be useful for improving the surface properties of cutting tool and the machining performance accordingly.

2 EXPERIMENT

Tungsten carbide was selected as a substrate material in this study. The sample surface was ground and then polished by a mechanical polisher with the diamond compound whose grain size of 60000. The surface was polished until the average surface roughness (R_a) of approximately 0.12 μm was achieved. Arc evaporation process was employed to form the single layer coatings of TiAlN and TiN on the substrate. The bilayer coatings of AlCrN/TiN and AlCrN/TiAlN on the tungsten carbide were also made where the TiN and TiAlN layers were used as the interlayer. The roughness of substrate and coating surfaces was evaluated by Carl Zeiss, Surfcom 480 A. The micro-hardness of coating layer(s) was examined under the applying load of 50 gf. The thickness of coatings was measured by the aid of Scanning Electron Microscope (SEM). Furthermore, the scratch test was performed by using the revet test machine (CSM Instrument) equipped with a diamond stylus whose tip radius of 200 μm. A point load for scratch test was progressively applied from 0 to 100 N with the rate of 100 N/min and scan speed of 10 mm/min. The scratching distance was kept constant at 10 mm for all tests. After the scratch test, the top surface morphology of coating was observed to realize the scratching marks as well as the delamination and failure of coating(s).

3 RESULTS AND DISCUSSION

3.1 Surface characteristics

The micrographs of single layer and bilayer coatings are presented in Figure 1, highlighting that the thickness of single layer coatings is approximately 2 μm while that of bilayer coatings is 5 μm. Figure 2 shows the top surface morphology of coatings. The arithmetic surface roughness (R_a) of TiN and TiAlN coatings was found to be varied in between 0.13–0.15 μm while the WC substrate surface was 0.12 μm. This can be indicated that the surface roughness was slightly increased after a coating layer was formed. In addition, the roughness of AlCrN or the top coat layer was further increased to be 0.19–0.2 μm. This would be due to the large size and great number of micro-droplets of coating material occurring at the high processing temperature that can introduce a rough film surface on the sample (Kumar et al. 2014a, Ali et al. 2010, Huang et al. 2011).

The hardness of TiN and TiAlN coatings was 1600 and 2700 HV, respectively, while that of bilayer coatings was approximately 3200–3300 HV.

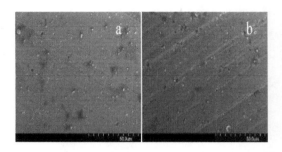
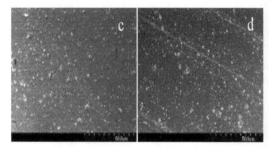

Figure 2. Surface morphology of (a) TiN, (b) TiAlN, (c) AlCrN/TiN and (d) AlCrN/TiAlN coatings.

Thus, it is evident that a harder surface can be achieved when the AlCrN was applied as the top coat layer. The chemical compositions detected by using the Energy Dispersive Spectroscopy (EDS) are shown in Figure 3, where the weight% of relevant elements was measured. The compositions of AlCrN growing on the different interlayers (TiN and TiAlN in this study) were found to be similar for both cases. This could be interpreted that the chemical composition of AlCrN was not significantly altered by the substrate composition.

3.2 Scratch test analyses

The scratch test results are plotted in Figure 4, in which the normal force, friction coefficient, frictional force, acoustic emission and penetration depth were read during the test. Since the normal force was linearly increased with the scratching distance, the status of coating layer was changed with respect to the force applied. Under a certain load, a crack was initiated on the coating film. Such amount of force is called the 1st critical load or L_{C1}. The 2nd stage or L_{C2} can be subsequently appeared when the film chipping or partial delamination were presented. Until the normal force reached the 3rd critical load (L_{C3}), the coating became failure or total delamination. Based on these results, the film adhesion of coatings can be evaluated accordingly. Table 1 concludes the critical loads of the four coatings obtained from the scratch test. From the table, it can be noted that the bilayer

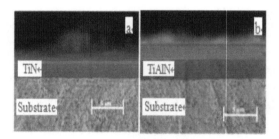
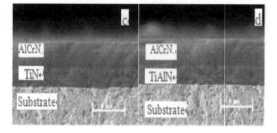

Figure 1. Cross-sectional view of (a–b) single layer coatings and (c–d) bilayer coatings on WC substrate.

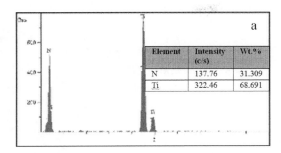

Element	Intensity (c/s)	Wt.%
N	137.76	31.309
Ti	322.46	68.691

a

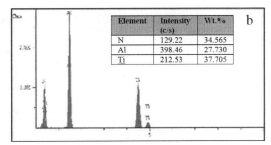

Element	Intensity (c/s)	Wt.%
N	129.22	34.565
Al	398.46	27.730
Ti	212.53	37.705

b

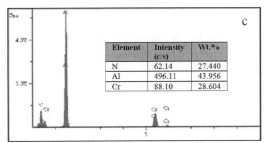

Element	Intensity (c/s)	Wt.%
N	62.14	27.440
Al	496.11	43.956
Cr	88.10	28.604

c

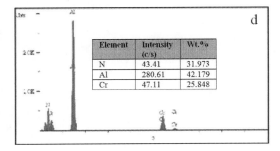

Element	Intensity (c/s)	Wt.%
N	43.41	31.973
Al	280.61	42.179
Cr	47.11	25.848

d

Figure 3. EDS spectrum and chemical composition associated with the top coat layer of (a) TiN, (b) TiAlN, (c) AlCrN/TiN and (d) AlCrN/TiAlN coatings.

coating of AlCrN/TiN and AlCrN/TiAlN can resist the formations of crack and chipping better than the other two single layer coatings. However, the single layer seems to effectively withstand the delamination better than the bilayer. The fiction coefficient (μ) was progressively increased with the normal force and film damage. This finding was also similar to that of Kumar et al. (2014a, b).

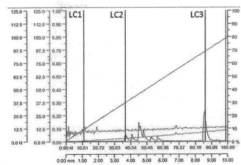

(a)

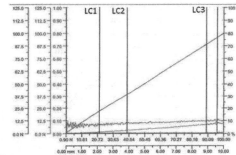

(b)

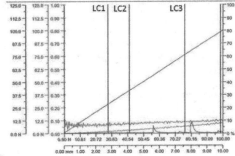

(c)

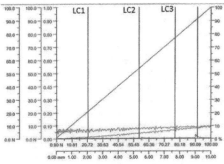

(d)

Figure 4. Scratch test results and critical loads for: (a) TiN, (b) TiAlN, (c) AlCrN/TiN and (d) AlCrN/TiAlN coatings.

83

Table 1. Results for scratch test under the progressive load of 1 to 100 N.

Coating	Critical load (N)		
	L_{C1}	L_{C2}	L_{C3}
TiN	12	37	86
TiAlN	22	39	89
AlCrN/TiN	28	41	76
AlCrN/TiAlN	21	53	77

4 CONCLUSIONS

TiN, TiAlN, AlCrN/TiN and AlCrN/TiAlN coatings were deposited on the tungsten carbide substrate by using the arc evaporation technique. The surface characteristics and adhesion of coatings were tested and presented in this study. The average surface roughness increased after the AlCrN film was formed on top of the TiN or TiAlN layer. This is plausibly subject to the size and amount of AlCrN micro-droplets depositing on the sample surface. The micro-hardness of bilayer coatings was approximately 3200–3300 HV, which basically corresponded with the hardness of AlCrN. Since the AlCrN hardness is greater than the TiN and TiAlN, the bilayer coatings have proven themselves in the scratch test in resisting the crack and chipping formations. In addition, TiAlN was found to have a higher adhesive strength than the TiN film. It can be further remarked that the selection of coating material and potential use of AlCrN for the bilayer coating are able to be justified, specifically in the improvement of cutting tool performance.

REFERENCES

[1] Ali, M., Hamzah, E., Qazi, I.A. & Toff, M.R.M. 2010. Effect of cathodic arc PVD parameters on roughness of TiN coating on steel substrate. *Current Applied Physics* 10: 471–474.

[2] Dobzañski, L.A. & Golombek, K. 2005. Structure and properties of selected cemented carbides and cermet covered with TiN/(Ti, Al, Si)N/TiN coating obtained by the cathodic arc evaporation process. *Materials Research* 8: 113–116.

[3] Huang, R.X., Qi, Z.B., Sun, P., Wang, Z.C. & Wu, C.H. 2011. Influence of substrate roughness on structure and mechanical property of TiAlN coating fabricated by cathodic arc evaporation. *Physics Procedia* 18: 160–167.

[4] Kumar, T.S., Prabu, S.B., Manivasagam, G. & Padmanabham, K.A. 2014. Comparison of TiAlN, AlCrN and AlCrN/TiAlN coatings for cutting-tool applications. *International Journal of Minerals, Metallurgy and Materials* 21: 796.

[5] Kumar, T.S., Prabu, S.B. & Manivasagam, G. 2014. Metallurgical characteristics of TiAlN/AlCrN coating synthesized by the PVD process on a cutting insert. *Journal of Materials Engineering and Performance* 23: 2877.

[6] Liew, W.Y.H., Jie, J.L.L., Yan, L.Y., Dayou, J., Sipaut, C.S. & Madlan, M.F.B. 2013. Frictional and wear behavior of AlCrN, TiN, TiAlN single-layer coating, and TiAlN/AlCrN, AlN/TiN nano-multi-layer coatings in dry sliding. *Procedia Engineering* 68: 512–517.

[7] Mo, J.L., Zhu, M.H., Lei, B., Leng, Y.X. & Huang, N. 2007. Comparison of tribological behaviours of AlCrN and TiAlN coatings deposited by physical vapor deposition. *Wear* 263: 1423–1429.

Material Science and Engineering – Chen (Ed.)
© 2016 Taylor & Francis Group, London, ISBN 978-1-138-02936-1

Influence of V and N addition on the microstructure and properties of low-alloy steel containing 1% chromium

X.M. Dong
School of Materials Science and Engineering, Shanghai University, Shanghai, P.R. China
Baosteel Research Institute, Baoshan Iron and Steel Co., Shanghai, P.R. China

Z.H. Zhang
Baosteel Research Institute, Baoshan Iron and Steel Co., Shanghai, P.R. China

Y.X. Chen
School of Materials Science and Engineering, Shanghai University, Shanghai, P.R. China

ABSTRACT: The effect of V and N addition on the microstructure and mechanical properties of low-alloy steel containing 1% Cr (1 Cr) is investigated. The coarsening of austenite grain and transformation of bainite are inhibited due to the occurrence of the undissolved V(C, N) particles in the microstructure. The transformation of ferrite was improved by the undissolved V(C, N) particles as the nucleation of proeutectoid ferrite, promoting the transformation of ferrite and inhibiting the transformation of bainite. Subsequently, the ferrite and pearlite microstructure with a good corrosion-resistant property is formed, and its strength reaches up to 80ksi(552MPa). With the content of V and N increasing, the undissolved V(C, N) favors the increasing ferrite phase content and reducing pearlite and bainite phase, making the strength of the material lower and greatly improving the toughness.

Keywords: low-alloy steel; V(C, N) precipitates; ferrite and pearlite

1 INTRODUCTION

CO_2 corrosion could cause the failure of pipes and equipment, resulting in a huge economic loss and catastrophic accidents. Moreover, the leakage of oil due to pipe failure would cause fire accidents, environmental pollution. Therefore, CO_2 corrosion has been one of the most common corrosion problems in the oil and gas industry [1–4]. The addition of the Cr element can effectively improve the CO_2 corrosion-resistant property of the material, so carbon steel with low grade and the material containing 1% Cr is used for producing tubing products in the oil industry [1–3]. Besides, many studies have found that the microstructure of the material could influence the corrosion-resistant property. For example, the corrosion-resistant property of the material with ferrite+pearlite is better than that of the material with martensite and bainite. It has been suggested that Fe_3C in pearlite has a positive potential with respect to ferrite, which forms a galvanic couple between ferrite and Fe_3C. Thus, ferrite dissolves preferentially as an anodic phase and lamellar Fe_3C in pearlite accumulates on the surface, making the protective corrosion scales to form easily on the steel with the ferritic-pearlitic microstructure in a CO_2-containing solution, which can improve the corrosion resistance of the steel [5, 6].

However, the strength of the steel with the ferrite and pearlite microstructure is low, which is no more than 55 ksi, so the pipe produced with a low-strength material is not suitable for application in the oil well deeper than 3000 m, leading to the strict restriction of usable range. In order to increase the strength, microalloy elements (Ti, Nb, V) are added to improve the strength of the steel due to the fine grain strengthening and precipitation strengthening [7–9]. However, Cr could raise the hardenability of the steel, preventing the formation of proeutectoid ferrite and pearlite during hardening [10], promoting the formation of bainite and martensite. Thus, it is difficult to design the material containing 1% Cr with the ferrite–pearlite microstructure and high strength reaching up to 80 ksi, and related research work is rare.

In this paper, the influence of V and N on the microstructure and mechanical properties of low-alloy steel is studied to obtain a high-strength ferrite–pearlite microstructure a with corrosion-resistant property in the material containing 1% Cr.

Table 1. Chemical composition of the tested steels (wt%).

Sample	C	Si	Mn	S	P	Cr	V	N
V1	0.35	0.20	0.6	≤0.005	≤0.012	1.05	0.08	0.004
V2	0.35	0.20	0.6	≤0.005	≤0.012	1.05	0.13	0.01
V3	0.35	0.20	0.6	≤0.005	≤0.012	1.05	0.18	0.016

2 EXPERIMENTAL PROCEDURE

Three medium-carbon steels, containing various contents of V and N, were vacuum-melted, casted into ingots, press-forged and hot-rolled to a 15 mm-thick plate. The chemical compositions of the tested steels are given in Table 1. The test pieces were reheated to 900°C holding for 30 min, and cooled in still air. The still air-cooling rate was estimated to be between 1.15 and 1.35°C/s. The test pieces were then mechanically polished and etched in 4% Nital. Microstructural evaluations of the tested steels were carried out by Optical Microscopy (OM) and Transmission Electron Microscopy (TEM) equipped with an Energy-Dispersive X-ray spectroscope (EDX). The ferrite grain volume fractions of ferrite, pearlite and bainite were determined by using the mean linear intercept and point counting methods on etched metallographic specimens at appropriate magnifications. Tensile test pieces with a round shape (L0 = 40 mm, d0 = 8 mm) and Charpy V-notch impact test pieces were taken from the center of the plate.

3 RESULTS

3.1 Microstructure

Figure 1 shows the microstructure of all the samples with different V and N additions. The volume fractions of ferrite, pearlite, bainite and pearlite grain size are summarized in Table 2. As can be seen, the microstructure of V1 heated to 900°C is characterized as ferrite, pearlite and bainite, and proeutectoid ferrite appears as a thin, continuous network at prior austenite grains (Figure 1a). The volume fraction of proeutectoid ferrite is about 10% (Table 2).

The microstructure of V2 at 900°C consists of polygonal ferrite idiomorphs, nucleated at austenite grain boundaries, surrounded by pearlite (Figure 1b), and proeutectoid ferrite appears more evenly distributed in the microstructure. The volume fraction of proeutectoid ferrite increased to about 30%. The microstructure of V3 consists of polygonal ferrite and pearlite without bainite, and ferrite also nucleates at austenite grain boundaries or within the former austenite grains, surrounded by pearlite (Figure 1c). Furthermore, the volume fraction of proeutectoid ferrite increased to about 40%. It can be found that an increased proeutectoid ferrite volume fraction is accompanied by simultaneous increases in V and N

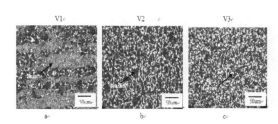

Figure 1. Microstructures of (a) V1, (b) V2 and (c) V3 at an austenizing temperature of 900°C.

Table 2. Volume fractions of ferrite, pearlite and bainite, and mean linear intercept grain sizes of V1,V2 and V3.

Sample	Ferrite content (%)	Pearlite content (%)	Bainite content (%)	Grain size (µm)
V1	10	40	50	40
V2	30	70	0	25
V3	40	60	0	15

concentrations, and the fraction of bainite decreases and the ferrite–pearlite microstructure becomes the dominant morphology. Moreover, the grain sizes of V1, V2 and V3 are finer with more addition of V and N, and the grain intercepts are 40 µm, 25 µm and 15 µm, respectively (Table 2).

3.2 Mechanical properties

The mechanical properties of different V and N additions are shown in Figure 2. It is clear from Figure 3 that the yield strength and tensile strength of V1, V2 and V3 steels reach up to the requirements of 80ksi grade pipe. The yield strength and tensile strength of V1, V2 and V3 at the same austenizing temperature decrease slightly, and elongation and toughness are improved dramatically, showing that more addition of V and N leads to a decrease in the yield strength and tensile strength, which is contradictory to some research results showing that the strength of the material increases with increasing contents of V and N [10].

As for the toughness, the Charpy impact value of V1 is about 20 J at the austenizing temperature of 900°C. However, the impact values of V2 and V3 at the same austenizing temperature reach up to 40 J and 75 J, respectively, which are twice and four times than that of V1.

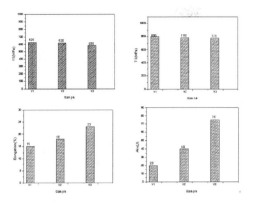

Figure 2. A comparison of the mechanical properties of different normalizing temperatures for V1, V2 and V3.

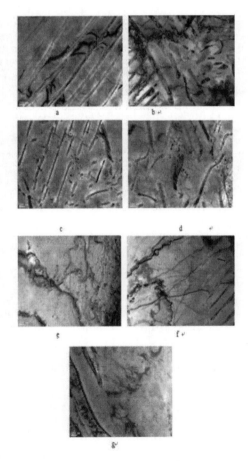

Figure 3. TEM micrographs of V(C,N) precipitation within pearlite and ferrite at 900°C; (a-c) TEM micrographs of pearlite of V1, V2 and V3, respectively, (d) another micrograph of pearlite of V3; (e,f) TEM micrographs of ferrite of V1, V2 and V3, respectively; (g) corresponding EDX patterns and diffraction patterns.

3.3 TEM analysis

TEM micrographs of V1-V3 at the austenizing temperature of 900°C are shown in Figure 3. Figure 3(a-c) displays the TEM micrograph of pearlite of V1, V2 and V3, respectively. The TEM observations of V1 steel do not reveal any dispersive precipitates due to the low concentrations of carbide—and carbonitride-forming element (Figure 3a). The investigated V2 and V3 steel containing more V and N has structures consisting mostly of pearlite, proeutectoid ferrite and V(C, N) carbide or carbonitride particles (Figure 3b).

Figure 3(e, f) shows the TEM micrograph of ferrite of V1, V2 and V3, respectively, showing that the content of V(C, N) precipitates increases with more addition of V and N. The energy-dispersive X-ray spectroscope (EDX) taken from this kind of particles (Figure 3g) reveals that these particles are V-rich precipitation, and the diffraction pattern (Figure 3g) shows that these particles are mainly V(C, N).

4 DISCUSSION

4.1 Microstructure

The chemical composition of the material is an important factor in developing the final microstructure of the steel. The addition of Cr strongly influences the formation of proeutectoid ferrite. If the proeutectoid ferrite is formed during austenite transformation, it decorates the prior austenite grains, disabling them from their role as nucleation sites for bainite formation. Subsequently, ferrite intragranular nucleation is the primary mechanism of austenite transformation. Therefore, Cr could induce the hardenability of the steel, preventing the formation of proeutectoid ferrite and pearlite during hardening [10], and promoting the formation of bainite and martensite. Therefore, bainite is formed in sample V1 due to the Cr addition and low V addition, explaining why it is difficult to obtain the ferrite and pearlite microstructure in the material containing Cr.

It is well known that the precipitates are complex due to the solubility in nitride—and carbide-forming vanadium carbonitrides V(C, N). From the solubility product data based on Narita [13], it can be seen that the solubility of VN is much lower than VC. It is clear that VN is much more stable than VC at 900°C. Thus, VN precipitates are the key factor to the microstructure and mechanical properties.

The equilibrium temperatures for the complete dissolution of VN in austenite are calculated according to Equation (1)[12], and the corresponding results are given in Table 3:

$$\text{Log } [V] \cdot [N] = -7840/T_{VN} + 3.02 \qquad (1)$$

Table 3. Calculated temperature for the complete dissolution of VN.

Sample	T_{VN}
V1	930
V2	1054
V3	1446

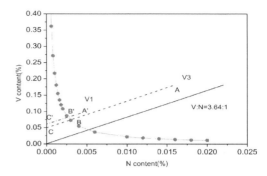

Figure 4. The solubility curve of steel with different V and N contents.

It is indicated from Table 3 that the content of V and N greatly influences the complete dissolution temperature of VN.

According to Equation (1), the solubility curve of VN is calculated, as shown in Figure 4. It is indicated that with the austenizing temperature increasing, the solubility of VN is improved. The full line represents the stoichiometry line of V and N. The V and N proportion line of sample V3 (point A) is parallel to the stoichiometry line, intersecting with the solubility curve of VN of 900 °C at B, and AB represents the content of undissolved VN at 900°C, and BC represents the content of VN precipitated during the phase transformation.

Similarly, the V and N proportion line of sample V1 (point A') is also parallel to the stoichiometry line, intersecting with the solubility curve of VN of 900 °C at B', and A'B' represents the content of undissolved VN at 900°C, and B'C' represents the content of VN precipitated during the phase transformation.

Based on the above discussion, it can be seen that with the content of V and N increasing, the concentration of undissolved VN increases, which is consistent with the results observing by TEM.

At 900°C, a large amount of vanadium in V1 steel is dissolved, and V segregate at grain boundaries favors the formation of grain boundary bainite. With the addition of V and N in V2 and V3 steel increasing, V and N present as VN, because their complete dissolution occurring at temperatures is higher than that of V1 steel (Table 3). Therefore, a less amount of V is available to segregate at grain boundaries, resulting in the formation of grain boundary ferrite in comparison with the grain boundary bainite, while the intragranular nucleation of ferrite is promoted (Figure 1(a, b, c)).

It is well known that V(C, N) precipitates restricted the prior austenite grain size and enhanced the proeutectoid ferrite volume because the grain boundaries represent sites for proeutectoid ferrite nucleation [13]. Comparing the microstructure of V1-V3 steel, it is obvious that ferrite and pearlite grain size becomes smaller with the content of V and N increasing. As shown in Figure 4, the increasing V and N content leads to more precipitates of undissolved V(C, N), which restricted the growth of austenite grain boundaries and inhibited the grain growing during the austenization process. Therefore, these precipitates (vanadium carbonitrides and nitrides) reduce the ferrite–pearlite grain size obtained by the decomposition of the austenite in samples V1-V3 [14].

Additionally, V(C, N) is also attributed to the intragranular nucleation of ferrite, promoting the formation of proeutectoid ferrite [10,15]. It is assumed that the formation of an allotriomorphic ferrite layer on the prior austenite grain boundaries inhibits the formation of bainite and contributes indirectly to the intragranular nucleation of ferrite [16–18].

As shown in Figure 4, the undissolved V(C, N) favors the nucleation of proeutectoid ferrite, forming a thin, continuous network at prior austenite grains of V1 and V2 steel during the phase transformation. With the contents of V and N increasing in samples V2-V3, the concentration of undissolved V(C, N) decreases, contributing to the increasing fraction of proeutectoid ferrite and intragranular ferrite (Table 2).

4.2 Mechanical properties

The main alloy element in the microalloyed medium carbon forging steels is V. It is assumed that V promotes strengthening by the precipitation of V(C, N) particles, which precipitate during and/or after austenite-ferrite transformation[10]. However, as shown in Figure 2, the addition of V and N in V1-V3 leads to the decrease in the yield strength and tensile strength at the same austenizing temperature, which is contradictory to the results mentioned above [11]. It is known that strength is also quite sensitive to the pearlite content, which is explained by the fact that there is a linear relationship between work hardening and the pearlite content, and pearlite work hardens much more rapidly than ferrite. From Figure 1 and Table 2, it can be seen that the concentration of ferrite increases gradually from V1 to V3 and that of pearlite reduces consequently, attributing to the reduced strength. Meanwhile, the strain hardening rate of bainitic sheaves is very high in comparison with the ferritic–pearlitic microstructure[18], resulting in a higher ultimate tensile strength and enhancing the strength of V1 and V2 at the higher austenizing temperature due to the high content of bainite.

As shown from Figure 2, the toughness of V1-V3 varies with different V and N contents, resulting from the different fractions of bainite and ferrite. Bainitic sheaves have a low capacity of stopping the propagation of brittle crack, so the formation of bainite dramatically reduces the toughness of the material.

The proeutectoid ferrite and intragranular ferrite, which attributes to the ferrite nucleation of V(C, N), are better suited to deflect the propagation of cleavage cracks and therefore more desirable from the toughness point of view in comparison with bainitic sheaves [18]. Therefore, the undissolved V(C, N) contributes to the higher concentration of ferrite, which helps in the improvement of toughness (Figure 2d).

5 CONCLUSION

1. Cr could prevent the formation of proeutectoid ferrite and pearlite, promoting the formation of bainite. The moderate addition of V and N contributes to the formation of proeutectoid ferrite and intragranular ferrite, inhibiting the transformation of bainite and obtaining the ferrite-pearlite microstructure with a good corrosion-resistant property.
2. V(C, N) precipitates restrict the prior austenite grain size and enhance the proeutectoid ferrite volume, and it is also attributed to the intragranular nucleation of ferrite, promoting the formation of proeutectoid ferrite.
3. In contrast to the former research, the addition of V and N lowers the strength of the steel. V(C, N) serves as the nucleation of proeutectoid ferrite and promotes the transformation of ferrite, which increases the concentration of ferrite and reduces that of pearlite, leading to strength reduction and toughness improvement.

ACKNOWLEDGMENT

This research was supported by the Key Projects in the National Science & Technology Pillar Program (No. 2011BAE25B00).

REFERENCES

[1] Paolinelli L.D. & Perez T. 2008. Electrochemical and sulfide stress corrosion cracking behaviors of tubing steels in a H_2S/CO_2 Annular Environment, *Corros. Sci.* 50:2456–2464.

[2] Kermani M.B. & Morshed A. 2003. Carbon dioxide corrosion in oil and gas production—a compendium, *Corrosion,* 59(8):659–683.

[3] Linter B.R. & Burstein G.T. 1999. Reactions of pipeline steels in carbon dioxide solution. *Corros. Sci.* 41:117–139.

[4] Xia Z. & Chou K.C. 1989. Effect of chromium on the pitting resistance of oil tube steel in a carbon dioxide corrosion system. *Corrosion* 45:636–642.

[5] Palacios C.A. & Shadley J.R. 1991. Characteristics of corrosion scales on steels in a CO_2 saturated NaCl brine. *Corrosion* 47:122–127.

[6] Crolet J.L. & Olsen S. & Wilhelmsen W. 1994. Influence of a layer of undissolved cementite on the rate of CO_2 corrosion of carbon steel. *Corrosion NACE International,* Houston, TX:4.

[7] Show B.K. & Veerababu R. 2010. Effect of vanadium and titanium modification on the microstructure and mechanical properties of a microalloyed HSLA steel. *Mater. Sci. Eng. A* 527:1595–1604.

[8] Zhao J. & Lee J.H. & Kim Y.W. & Jiang Z. 2013. Effects of Tungsten Addition on the Microstructure and Mechanical Properties of Microalloyed Forging Steels. *Mater Sci Eng A,* 559:427–35.

[9] Matlock D.K. & Krauss G. 2001. Microstructures and properties of direct-cooled microalloy forging steels. *J Mater Process Technol.* 117:324–328.

[10] Siwecki T. & Eliasson J. & Lagneborg R. 2010. Vanadium Microalloyed Bainitic Hot Strip Steels. *ISIJ Int.,* 2010, 50:760–767.

[11] Narita K. 1975. Physical chemistry of the groups IVa (Ti, Zt), Va (V, Nb, Ta). *Trans ISIJ.* 15:145–52.

[12] Sandberg O. & Westerhult P. & Roberts W. 1985. Vanadium-Modified alloy steels for quench-and-temper applications requiring high hardenability. *Journal of Heat Treating.* 4(2):184–193.

[13] Ollilainen V. & Kasprzak W. & Holappa L. 2003. Fatigue performance evaluation of forged versus competing process technologies. *J Mater Process Technol.* 134:405–412.

[14] Babakhani A. & Ziaei S. 2010. Investigation on the effects of hot forging parameters on the austenite grain size of vanadium microalloyed forging steel (30MSV6). *J. Alloys Compd.* 490:572–575.

[15] Garcia-Mateo C. & Capdevila C. & Cabalero F. 2008. Influence of V Precipitates on Acicular Ferrite Transformation Part 1: The Role of Nitrogen. *ISIJ Int.* 2008, 50:1270–1275.

[16] Capdevila C. & Ferrer J. & Garcia-Mateo G. 2006. Influence of Deformation and Molybdenum Content on Acicular Ferrite Formation in Medium Carbon Steels. *ISIJ Int.* 46: 1093.

[17] Madariaga I. & Guttierez I. 1999. Role of the particle–matrix interface on the nucleation of acicular ferrite in a medium carbon microalloyed steel. *Acta Mater.* 47:951.

[18] de Andres C.G. & Capdevila C. & Madariaga I. & Gutierrez I. 2001. Role of molybdenum in acicular ferrite formation under continuous cooling in a medium carbon microalloyed forging steel. *Scripta Mater.* 45:709.

Material Science and Engineering – Chen (Ed.)
© 2016 Taylor & Francis Group, London, ISBN 978-1-138-02936-1

Effect of heat treatment on the banded structure of a hot-rolled 30CrMo steel

M. Wang
The State Key Laboratory of Refractories and Metallurgy, Hubei Collaborative Innovation Center for Advanced Steels, Wuhan University of Science and Technology, Wuhan, China
State Key Laboratory of Development and Application Technology of Automotive Steels (Baosteel Group), Shanghai, China

G. Xu & Y. Zhang
The State Key Laboratory of Refractories and Metallurgy, Hubei Collaborative Innovation Center for Advanced Steels, Wuhan University of Science and Technology, Wuhan, China

L. Wang
State Key Laboratory of Development and Application Technology of Automotive Steels (Baosteel Group), Shanghai, China

ABSTRACT: The effect of annealing, normalizing and quenching+tempering on the banded structure of a hot-rolled 30CrMo steel was studied by optical microscopy. The results show that the microstructure of the hot-rolled 30CrMo steel is ferrite and pearlite. Element segregation is the main reason for the banded structure. After annealing and normalizing, the banded structure became thinner, while it was eliminated by quenching + tempering. In the production process, it was suggested to increase the tapping temperature, preservation temperature and holding time for the uniformity of alloying elements. The hot-rolled 30CrMo steel should be quenched and tempered to eliminate the banded structure.

Keywords: hot-rolled 30CrMo steel; microstructure; banded structure; heat treatment

1 INTRODUCTION

30CrMo steel belongs to the CrMo series steel (Tan et al. 2014). It is widely used in machinery manufacturing industry, chemical industry, steam turbine and boiler manufacturing industry for its good mechanical properties, hardenability and mechanical processability (Zheng & LI 2013, Yuan et al. 2011). However, hot-rolled 30CrMo steel in a steel company has some problems such as composition segregation and banded structure. Banded structure may reduce the mechanical properties, cutting properties and plastic working properties of the steel. Workpieces machined using steel with a banded structure usually crack in the phase boundaries of ferrite and pearlite (Wang et al. 2013). In this paper, the microstructure of a hot-rolled 30CrMo steel was studied. To reduce or eliminate the banded structure, different heat treatment processes were carried out. This paper attempts to study the effect of heat treatment on the banded structure, and provides a theoretical basis for determining the heat treatment technology of the hot-rolled 30CrMo steel.

2 EXPERIMENTAL MATERIAL AND SCHEME

The experimental material was cut from the hot-rolled 30CrMo steel. The chemical composition of the steel is given in Table 1. The tapping temperature of the molten steel in the converter was 1660~1690°C, and the superheat temperature was 15~25°C. Casting billets were rolled to 4 mm in thickness by seven-pass rolling on a 4-high reversal mill with a final rolling temperature of 872°C. After rolling, the steel was cooled to 618°C by laminar cooling and coiled.

To study the effect of heat treatment on the banded structure, different heat treatment routes were carried out. Schematic diagrams are shown in Figure 1. For the annealing treatment, the sample was heated

Table 1. Chemical composition of the 30CrMo steel (wt%).

C	Mn	Si	P	S	Cr	Ti	Mo	N
0.32	0.60	0.25	0.014	0.003	1.10	0.0335	0.189	0.008

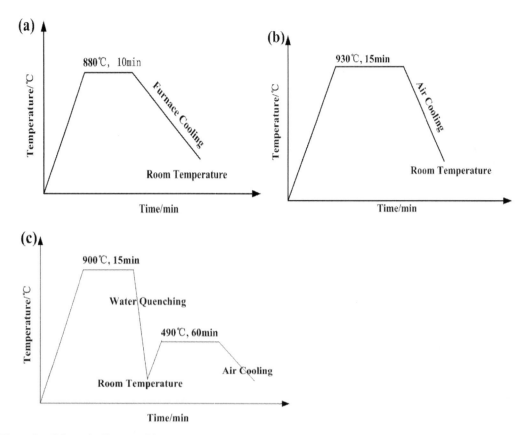

Figure 1. Schematic diagram of heat treatment: (a) annealing; (b) normalizing; (c) quenching + tempering.

to 880°C for 10 min, followed by furnace cooling to room temperature. For normalizing the treatment, the sample was heated to 930°C and held for 15 min, followed by air cooling to room temperature. In the quenching and tempering process, the sample was heated to 900°C for 15 min and quenched to room temperature in water. After quenching, the sample was tempered at 490°C for 60 min and subsequently cooled to room temperature in air. The samples taken from the hot-rolled steel and steels after heat treatment were mechanically grinded, polished, followed by etching in a solution containing 4% nitric acid and 96% ethanol. Microstructures were observed by a Zeiss optical microscope.

3   EXPERIMENTAL RESULTS

Microstructures of the hot-rolled steel after heat treatment are shown in Figure 2. It is observed that the microstructure of the hot-rolled 30CrMo steel consists of ferrite and pearlite. A coarse pearlite band can be clearly seen in the hot-rolled steel

in Figure 2a. Figure 2b illustrates that the banded structure becomes thinner after annealing and normalizing treatment compared with the hot-rolled steel. However, the banded structure still exists in the microstructure. The banded structure is alleviated after normalizing treatment (Fig. 2c). However, it should be noted in Figure 4d that the banded structure is eliminated after quenching and tempering.

4   DISCUSSION

During the freezing process of liquid steel, alloying elements are distributed unevenly and segregate in dendrites. Coarse dendrites are elongated along the direction of deformation in the rolling process. After rolling, elongated dendrites form solute-rich zones, while solute-depleted zones exist between solute-rich zones. In the subsequent slow cooling process, proeutectoid ferrite forms in solute-depleted zones, followed by partitioning carbon into solute-rich zones and promoting the formation of the pearlite band (Zhang 2014, Liu 2000).

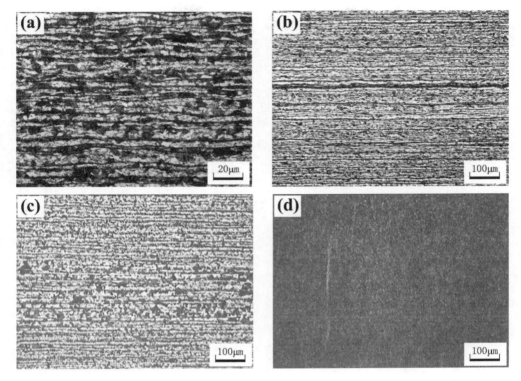

Figure 2. Microstructure of the 30CrMo steel under different treatments: (a) hot rolled; (b) annealing; (c) normalizing; (d) quenching + tempering.

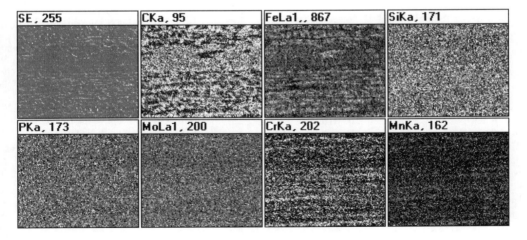

Figure 3. Electron probe analysis of the coarse pearlite band.

Electron probe analysis of the coarse pearlite band is shown in Figure 3. Carbon presents positive segregation, while Cr, Mn and Fe present negative segregation in the pearlite band. This means that element segregation is the main reason for the banded structure. In the production process, it is necessary to increase the tapping temperature, preservation temperature and holding time for the uniformity of alloying elements. However, the banded structure cannot be eliminated by using these methods. Fast cooling can suppress the uneven distribution of C and thus reduce the banded structure (Thompson & Howell 1992, Grossterlinden et al. 1992). Cai et al. studied the effects of cooling rate on the banded structure, and found that the difference on A_{r3} temperatures for solute-depleted and solute-rich zones in the matrix is a prerequisite for the banded structure (Cai et al. 2012). With the increase in the cooling rate, the diffusion distance of C decreases and the banded structure is relieved.

To relieve or eliminate the banded structure of the hot-rolled steel, different heat treatment processes were carried out according to the routes shown in Figure 1. It can be seen that annealing and normalizing can relieve the banded structure of the hot-rolled 30CrMo steel, while quenching + tempering can eliminate the banded structure. Liu studied the reason of formation, harmful effect and removal of the banded structure in low-carbon alloy steel, and found that conventional heat treatment cannot eliminate alloying element segregation in the banded structure (Liu 2000).

In normalizing and annealing processes, although composition was more uniform after heating to high temperature and kept for a certain time, the heating temperature was not high enough for the alloying element to be fully redistributed. So, normalizing and annealing relieved the banded structure. The diffusion of C was hindered during quenching. Residual C in solute-depleted zones after ferrite formation cannot be partitioned into solute-rich zones for pearlite transformation. The banded structure was eliminated after quenching + tempering. The microstructure after quenching + tempering consists of tempered sorbite and ferrite, which meets the user's demands.

5 CONCLUSION

The effect of annealing, normalizing and quenching + tempering on the banded structure of a hot-rolled 30CrMo steel was studied by optical microscopy. The results show that element segregation is the main reason for the banded structure. After annealing and normalizing, the banded structure became thinner, while it was eliminated by quenching + tempering. In the production process, it was suggested to increase the tapping temperature, preservation temperature and holding time for the uniformity of alloying elements. Thus, the hot-rolled 30CrMo steel should be quenched and tempered to eliminate the banded structure.

ACKNOWLEDGMENTS

The authors are grateful to the financial support from the National Natural Science Foundation of China (NSFC) (No. 51274154), the National High Technology Research and Development Program of China (No. 2012AA03A504), the State Key Laboratory of Development and Application Technology of Automotive Steels (Baosteel Group), and the Key Project of Hubei Education Committee (No. D20121101).

REFERENCES

[1] Tan, W. et al. 2014. An analysis of microstructure & mechanical properties of 30CrMo steel produced by CSP line. *Journal of Wuhan Engineering Institute* 26(3): 59–61.
[2] Zheng, Y. & Li, W. 2013. Microstructure and mechanical properties of 30CrMo strip steel produced by continuous casting and rolling. *Hot Working Technology* 42(9): 143–146.
[3] Yuan, Y.Q. et al. 2011. Development of thick alloy structural steel plate 30CrMo. *Research on Iron and Steel* 39(6): 46–49.
[4] Wang, F. et al. 2013. Current research situation of alleviating or eliminating band structure in steel. *Hot Working Technology* 42(5): 52–57.
[5] Zhang, Y.H. et al. 2014. Research status of banded structure in steel. *Steel Rolling* 31(3): 45–47.
[6] Liu, Y.X. 2000. Reason of formation, harmful effect and removal of band structure in low carbon alloy steel. *Heat Treatment of Metals* (12): 1–3.
[7] Thompson, W.S. & Howell, R.P. 1992. Factors influencing ferrite/pearlite banding and origin of large pearlite nodules in a hypoeutectoid plate steel. *Materials Science and Technology* (8): 777.
[8] Grossterlinden, R. et al. 1992. Formation of pearlite banded structure in ferrite-pearlite steel. *Steel Research* 63(8): 331.
[9] Cai, Z. et al. 2012. Mechanism of Effect of Cooling Rate on Ferrite/Pearlite Banded Structure. *Journal of Iron and Steel Research* 24(6): 25–30.

Material Science and Engineering – Chen (Ed.)
© *2016 Taylor & Francis Group, London, ISBN 978-1-138-02936-1*

Effect of Ti content on microstructures and Charpy impact properties in the Coarse-Grained Heat-Affected Zones of HSLA steels

X. Luo, Z.D. Wang, H. Tang, J. Cai & K.X. Chen
University of Science and Technology of Beijing, Beijing, P.R. China

X.H. Chen
State Key Laboratory for Advanced Metals and Materials, University of Science and Technology Beijing, Beijing, P.R. China

ABSTRACT: This study aimed to determine the effects of Ti content on microstructures and Charpy impact properties in HAZs of two HSLA steels. An original approach for acquiring *in situ* reactive Ti-containing inclusions with micron size was presented. Microstructure evolution and Charpy impact properties were investigated at coarse-grained simulated HAZs for steels with different Ti contents. SEM was used to characterize microstructures and inclusions at HAZs. It was found that inclusions in the steel with less Ti content were titanium oxides with a round shape and smaller sizes, which pinned prior austenite grain boundaries and induced ferrite during the thermal cycle, thus promoting Charpy impact properties. Those in the steel with more Ti content was confirmed to be titanium nitrides with a cubic shape and larger sizes, which acted as crack initiation sites, leading to the decrease in Charpy impact properties. Fractographs of both steels were examined, which is consistent with the microstructure and toughness.

Keywords: Ti content; HAZ; microstructure; inclusion; Charpy impact property

1 INTRODUCTION

HSLA (High-Strength Low-Alloy) steels are required to maintain a good balance between strength, ductility and toughness to fulfill the ever-increasing demand for advanced materials in manufacturing vessels, automotives and other industries. Since they should be welded to make available structures, e.g., shells and frames, welding process is indispensable. During the course of welding, HSLA steels often meet with a problem of low fracture toughness because a Heat-Affected Zone (HAZ) is formed in welded regions. In general, the weld Coarse-Grained HAZ (CGHAZ) adjacent to the fusion line is deemed to be the weakest linkage among the various regions within a HAZ.

During the welding process, the CGHAZ experiences thermal cycles up to a high peak temperature around the melting point and fast cooling. This high peak temperature provides a high driving force for the growth of austenite grain, leading to coarse grains in HAZs. Fast cooling induces relatively brittle microstructures in the weld CGHAZ, such as side plates ferrite, bainite and martensite.

Oxide metallurgy, in which oxides with excellent high temperature stability are used, has been developed recently to prevent the deteriorated toughness of the HAZ. Oxides promote the formation of ferrite during welding because they act as ferrite nucleation sites, while the formation of bainite or martensite is prevented. The pinning effect of fine inclusions also contributes to the toughness. Among the oxides, titanium oxides are given considerable attention due to its thermostability and inducing effect on the nucleation of some favorable microstructures, such as acicular ferrite and intragranular ferrite.

Four suggestions of the mechanism by which inclusions nucleate acicular ferrite were proposed, and it has been reported that Ti_2O_3 absorbs Mn in the vicinity of itself and then promotes the transformation of ferrite. M. Fattahi et al concluded that Ti–Mn complex inclusions are effective nucleation sites for ferrite, and the Mn-depleted zone promotes the temperature and chemical driving force of austenite–ferrite transformation. It has also been reported that fine titanium oxide inclusions are favorable for the nucleation of intragranular ferrite, and the best effect comes with the sizes of the inclusions in the range of 0.2–3 μm.

In the present study, we proposed a novel approach to obtain *in situ* fine titanium oxide inclusions in the melt, and investigated the effect of different Ti contents on the microstructure and toughness of HAZs of two trial-produced steels.

Table 1. Chemical compositions of steels A and B (wt%).

Steel	C	Si	Mn	Ni	Cr	Mo	Cu	V	Nb	Ti
A	0.05	0.03	0.85	4.5	0.45	0.5	0.2	0.03	0.06	0.05
B	0.05	0.03	0.85	4.5	0.45	0.5	0.2	0.03	0.06	0.1

2 EXPERIMENTAL PROCEDURE

In order to obtain Ti-containing inclusions in the steels, Ti was added in a special way. First, Fe, as well as other alloys except Ti, was fully melted at 1600°C in a vacuum melting furnace with vacuum up to about 0.08 Pa, and then a pure Ti wire of diameter 2 mm was fed to the melt. It should be noted that a strong electromagnetic stirring (~4 kHz frequency) was applied in the whole process. After holding for 1 min, the melt was poured into a cast iron mold. The ingot was finally rolled into slabs. The chemical compositions of as-received novel HSLA steels are given in Table 1. For the sake of simplicity, the steels are referred to as "A" and "B" according to the content of Ti, which is the only difference between the two steels.

Metallographic specimens were cut perpendicular to the roll direction from the hot plate and were ground, polished, etched with 4% nitric acid–alcohol or saturated water solution of picric acid, and then observed with 9XB-PC optical microscopy. The microstructure of the specimen was further investigated by using a ZEISS ULTRA 55 thermal field emission scanning electron microscope. Charpy impact tests were performed three times for A and B steels on standard Charpy V-notch specimens (size; 10 × 10 × 55 mm, orientation; perpendicular to the rolling direction, temperature −50°C). Weld thermal cycle simulation tests were conducted on a Gleeble 1500 simulator. Simulated heat input and the applied peak temperature were 30 kJ/cm and 1350°C, respectively. The value of $T_{8/5}$ was calculated by the empirical equation:

$$T_{8/5} = (0.043 - 4.3 \times 10^{-5} T_0) \frac{\eta^2 E^2}{\delta^2}$$

$$\times \left[\left(\frac{1}{500 - T_0} \right)^2 - \left(\frac{1}{800 - T_0} \right)^2 \right] F_2$$

here, T_0 is the preheating temperature (°C); η is the relative thermal efficiency; E is the weld heat input (J/cm); δ is the thickness of the slabs (cm); and F_2 is the two-dimensional heat transfer coefficient. To match the practical MIG welding with a heat input of 30 KJ/cm, the values of T_0, η, E, δ and F_2 were assigned to be 20, 0.7, 30 and 1, respectively. Hence, the $T_{8/5}$ was calculated to be 35 s.

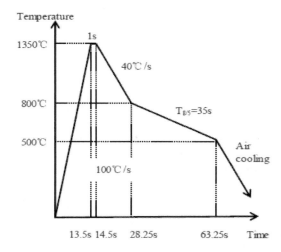

Figure 1. Schematic diagram of weld thermal cycles of the HSLA steels.

The thermal cycle of the weld simulation, which is primarily characterized by the peak temperature and the cooling time from 800°C to 500°C ($T_{8/5}$), is shown in Figure 1. After reaching the peak temperature of 1350°C and holding for 1 s, the specimens were cooled down with $T_{8/5}$ of 35 s.

3 RESULTS AND DISCUSSION

3.1 Microstructure of the HSLA steels

Figure 2(a) and (b) shows the microstructure of as-rolled steels A and B etched by the saturated water solution of picric acid. Both steel A and steel B show elongated grains with inclusions inside. There is a distinguishable difference in the shape and quantity of the inclusions in both steels. Inclusions in steel A are sub-round in shape and small in quantity; however, those in steel B are cubic in shape and large in quantity.

Figure 3 reveals an overview of the microstructure at simulated HAZs for steels A and B. As shown in Figure 3(a) and (d), the microstructure consists of predominantly lath bainite, ferrite side plate, intragranular ferrite and granular bainite, as indicated by the arrow. The EDXS spectra indicate that round or subround inclusions in steel A are

96

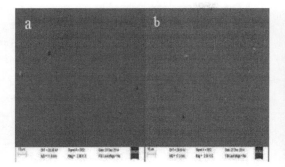

Figure 2. SEM micrographs of the as-rolled (a) A and (b) B steels etched by the saturated water solution of picric acid.

titanium oxides. Meanwhile, cubic inclusions in steel B are indicated to be titanium nitride according to the EDXS spectra.

3.2 Crack propagation and impact fracture

The dominating microstructure in both steels is lath bainite, which has a relatively low crack resistance. Figure 5(a) shows that the microcrack propagates almost vertically in a single lath bainite grain. Figure 5(b) shows that the microcrack propagates through two adjacent lath bainite grains. It is obvious that the propagation direction of the crack can be changed by the lath bainite boundaries. Thus, it can be inferred that lath bainite boundaries impede crack propagation to some degree.

Figure 6(a) shows that fine titanic oxide inclusions induced the nucleation and growth of ferrite in the prior austenite grain of steel A. As shown in Figure 6(b), the microcrack propagated vertically in bainite and the direction changed when the microcrack encountered intragranular ferrite, as indicated by the red arrow. Figure 6(c) and 6(d) shows the pinning effect of the titanic oxide inclusions. It is much evident that fine titanic oxide inclusions do have the advantage in refining the HAZ grains and inhibiting microcrack propagation.

It was found that inclusions, especially TiN inclusions, act as cleavage crack nuclei and have a deleterious effect on the impact toughness. TiN particles further deteriorate the impact toughness in the presence of lath bainite or large effective ferrite grain size. Figure 7 shows that cubic Ti-containing inclusion initiates microcracks in lath bainite, which agrees with the previous research and accounts for the lower Charpy impact toughness in steel B.

Figure 8 shows the typical examples of SEM fractographs for steels A and steel B that have different toughness properties, respectively. As shown in Fig. 8(a), the evidence for a quasi-cleavage

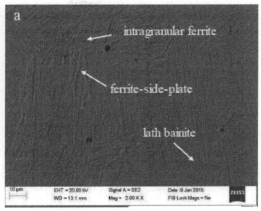

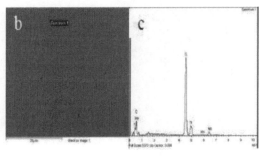

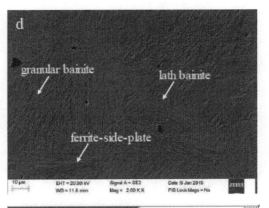

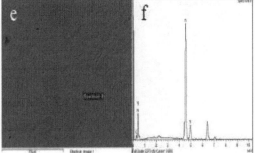

Figure 3. SEM and EDXS spectra of the simulated (a)–(c) A and (d)–(f) B steels.

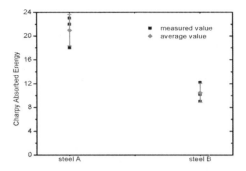

Figure 4. Charpy impact test results of the simulated steel A and B.

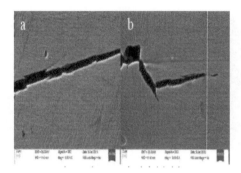

Figure 5. Crack propagation in lath bainite.

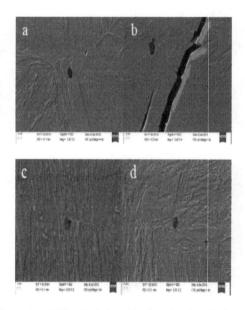

Figure 6. (a) Ferrite induced by fine titanic oxide inclusions (b) ferrite consumes energy for microcrack propagation and (c) and (d) pinning effect of fine titanic oxide inclusions in steel A.

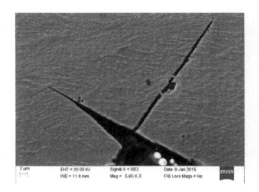

Figure 7. The cubic Ti-containing inclusion in steel B initiates microcracks.

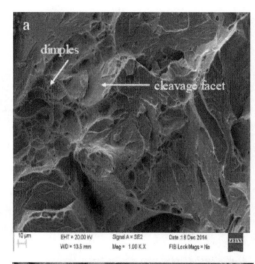

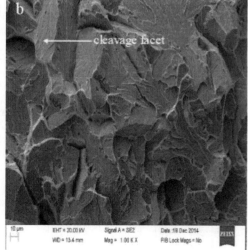

Figure 8. Fracture surface of steels (a) A and (b) B.

fracture is clear for steel A due to the observation of dimples and cleavage facets, which are in accordance with the existence of ductile ferrite and brittle lath bainite. The fracture surface of steel B exhibited a total brittle fracture behavior with a typical transgranular fracture characteristic. The only cleavage facets corresponded to the brittle lath bainite and large cubic inclusions.

4 CONCLUSIONS

1. An innovative approach for acquiring *in situ* reactive Ti-containing inclusions with micron size was explored preliminarily. Steel A with 0.05 wt% Ti exhibited a higher impact toughness at a coarse grain HAZ by virtue of the pinning effect of titanium oxide inclusions and induced nucleation of intragranular ferrite.
2. The cubic inclusions, namely titanium nitrides, are detrimental to the impact toughness at the coarse grain HAZ due to the relatively larger size and stress concentration caused by its angular shape.
3. The relatively low toughness of steel A is attributed to the brittle microstructures, e.g. lath bainite and ferrite side plate, which dominate the matrix. There is still much room for the promotion of the impact toughness when the microstructures become favorable through a preferable component design and deformation mechanism. The innovative approach is promising when nitrogen is first reduced to a certain extent during the melting process..

REFERENCES

[1] Bose-Filho, W.W. & Carvalho, A.L.M. & Strangwood, M. 2007. Effects of alloying elements on the microstructure and inclusion formation in HSLA multipass welds. Materials Characterization 58:29–39.
[2] Byun, J.S. & Shim, J.H. & Cho, Y.W. 2003. Non-metallic inclusion and intragranular nucleation of ferrite in Ti-killed C–Mn steel. Acta Materialia 51:1593–1606.
[3] Fattahi, M. & Nabhani, N. & Hosseini, M. 2013. Effect of Ti-containing inclusions on the nucleation of acicular ferrite and mechanical properties of multipass weld metals. *Micron* 45:107–114.
[4] Fattahi, M. & Nabhani, N. & Vaezi, M.R. 2011. Improvement of impact toughness of AWS E6010 weld metal by adding TiO_2 nanoparticles to the electrode coating Materials Science and Engineering: A 528:8031–8039.
[6] Moon, J. & Kim, S. & Lee, C. 2011. Effect of thermo-mechanical cycling on the microstructure and strength of lath martensite in the weld CGHAZ of HSLA steel. Materials Science and Engineering: A 528:7658–7662.
[7] Qian, L. 2012. Microstructure and mechanical properties of a low carbon carbide-free bainitic steel co-alloyed with Al and Si. Materials and Design 39: 264–268.
[8] Shanmugam, S. & Misra, R.D.K. & Hartmann, J. 2006. Microstructure of high strength niobium-containing pipeline steel. Materials Science and Engineering: A 441:215–229.
[9] Shim, J.H. & Cho, Y.W. & Chung, S.H. & Shin, J.D. & Lee, D.N. 1999. Nucleation of intragralunar ferrite at Ti_2O_3 particles in low carbon steel. Acta Materialia 47:2751–2760.
[10] Sung, H.K. & Shin, S.Y. & Cha, W. 2011. Effects of acicular ferrite on charpy impact properties in heat affected zones of oxide-containing API X80 linepipe steels. Materials Science and Engineering: A 528:3350–3357.
[11] Vargas-Arista, B. & Angeles-Chavez, C. & Albiter, A. 2009. Metallurgical investigation of the aging process on tensile fracture welded joints in pipeline steel. Materials Characterization 60:1561–1568.
[12] Zhang, D. & Terasaki, H. & Komizo, Y. 2010. In situ observation of the formation of intragranular acicular ferrite at non-metallic inclusions in C–Mn steel. Acta Materialia 58:1369–1378.
[13] Zhang, L. & Kannengiesser, T. 2014. Austenite grain growth and microstructure control in simulated heat affected zones of microalloyed HSLA steel. Materials Science and Engineering: A 613:326–335.

Material Science and Engineering – Chen (Ed.)
© 2016 Taylor & Francis Group, London, ISBN 978-1-138-02936-1

Adsorption kinetics of soil-exchangeable Zn and Mn by Fe–Al binary oxide

Y.L. Dong

College of Resources and Environment, Linyi University, Linyi, China

ABSTRACT: The objective of this study was to test the adsorption kinetics between Iron-Aluminium (Fe-Al) binary oxide and soil-exchangeable Zn and Mn. After 14 days, the concentrations of soil-exchangeable Zn and Mn were reduced by 83.58% and 71.61%, respectively; thus Fe–Al binary oxide is an effective adsorbent for soil-exchangeable heavy metals. The Elovich equation was more suitable than the other equations for describing the kinetics of soil-exchangeable Zn. The adsorption kinetics of Fe-Al binary oxide with soil Mn were favorably described by the Elovich equation and the power function equation.

Keywords: Fe–Al binary oxide; soil; Zn; Pb; adsorption

1 INTRODUCTION

Contamination of the environment with toxic heavy metals is an urgent problem worldwide, and removal of these toxic heavy metal ions is often essential. Numerous methods have been used to remove heavy metals, including chemical precipitation, ion-exchange, reverse osmosis, coagulation and flocculation, membrane separation, biosorption, and adsorption. Among these methods, adsorption was the most preferred method for removal of heavy metals due to its simplicity and high effectiveness (Neda et al. 2013; Sdiri et al. 2012). Numerous studies have been conducted to synthesize and develop adsorbents such as kaolinite (Schaller et al. 2009), illite (Ozdes et al. 2011), and other adsorbents (Engates and Shipley 2011) for removing heavy metals from aqueous solutions.

Hydrate Fe and Al oxides are important adsorbents in soil (Trivedi et al. 2001). They have a very large surface area and a high adsorption capacity, enabling them to control the sorption capacity of soils (Geelhoed et al. 1997; Matis et al. 1999). The properties, applications and adsorption qualities of hydrated Fe and Al oxides have been extensively reported (Liu et al. 2009; Ren et al. 2009; Liao et al. 2011), However, these oxides usually exist as Fe–Al binary oxides in the soil.

Fe–Al binary oxide is an adsorbent with a large surface area, and it has the potential to provide a cost-effective solution to some of the most challenging problems in environmental remediation (Granados-Correa and Bulbulian, 2012). Currently, binary oxides are used as effective adsorbents for the removal of metal ions from wastewater (El-Kamash et al. 2007; Zhang et al. 2009; Hong et al. 2010; Hye-Jin et al. 2011). However, due to the complex nature of soil composition, little is known about the ability of Fe–Al binary oxide to adsorb heavy metals in soil. Therefore, it is important for understanding the behavior of Fe-Al binary oxide in the soil.

In this study, Fe–Al binary oxide was added to soil with the objectives: (1) to demonstrate the reduction in soil-exchangeable Zn and Mn by Fe–Al binary oxides, and (2) to describe the adsorption kinetics of soil-exchangeable Zn and Mn using chemical and empirical kinetic equations.

2 MATERIALS AND METHODS

2.1 *Preparation of Fe–Al binary oxide*

Fe–Al binary oxide was synthesized according to the method of Basu et al. (2012). A solution of $FeCl_3 \cdot 6H_2O$ and $AlCl_3 \cdot 6H_2O$ was titrated with 3.0 mol L^{-1} NaOH at a rate of 5 mL min^{-1} to pH 7.4. The suspension was then aged with intermittent stirring for 2 days at 60°C in a polyethylene beaker. The precipitate was separated from the supernatant by vacuum filtration and was washed several times with deionized water, until the conductivity $< 1 \times 10^2 \text{ µs cm}^{-1}$.

2.2 *Soil collection and processing*

Brown soil from Yihe River coast located in the Shandong Province of China (E 118°34′, N 35°06′)

was collected from the 0–20 cm topsoil horizon. The soil was air-dried, stones and plant root debris were removed, and then it was passed through a 2.0-mm sieve prior to the experiment. The soil pH was 7.81, and the concentrations of available Zn and Mn were 7.52 mg kg^{-1} and 22.64 mg kg^{-1}, respectively.

2.3 Pot experiments design

A study was conducted in 20 pots containing a mixture of 5 g Fe–Al binary oxide and 1 kg of soil. The incubation experiment was conducted in a growth chamber at room temperature and the soil moisture was maintained with distilled water. Four replicate pots were harvested each sampling day (0, 3, 5, 7, and 14), and the soils were air-dried.

2.4 Soil analysis

Soil-available heavy metals were measured according to the method of Tessier et al. (1979): After harvest, each 15 g subsample of soil was extracted for 2 h at 25 ± 2°C with 150 mL of 1 mol L^{-1} MgCl$_2$ extraction solution. Extracts were filtered and analyzed by Inductively Coupled Plasma Optical Emission Spectroscopy to determine Zn and Mn.

2.5 Data analysis

Mathematical equations were used to describe the adsorption kinetics of soil-exchangeable heavy metals with the Fe–Al binary oxide according to the method of Zhu et al. (2000). The fitting of experimental data to these equations provided regression coefficients (R^2) and Standard Errors (SE) to determine the best equation to describe the kinetics of heavy-metal adsorption to the Fe–Al binary oxide.

Zero-order kinetic equation: $(1 - X_t/X_0) = -kt + a$;
First-order kinetic equation: $\ln(1 - X_t/X_0) = -kt + a$;
Parabolic diffusion equation: $(1 - X_t/X_0) = -kt^{1/2} + a$;
Elovich kinetic equation: $X_t = a + k\ln t$;
Power function equation: $\ln X_t = \ln a + k\ln t$;

where X_t = amount of available heavy metal adsorbed to the binary oxide at time t; it was obtained by the amount of available heavy metal at time t minus the amount of available heavy metal at t = 0. X_0 = amount of heavy metal at equilibrium. It was evaluated by a graphical extrapolation of X from 1/t to 1/(t = 0). k is an empirical rate constant. a and lna are intercepts.

3 RESULTS

3.1 Influence of soil-exchangeable Zn content by Fe–Al binary oxide

The concentration of soil-exchangeable Zn decreased rapidly between days 4 and 6, but then decreased gradually to day 14 (Fig.1). The concentration of total soil-exchangeable Zn decreased by 79.31% on day 5 and by 83.58% on day 14. Adsorption of soil-exchangeable Zn by Fe–Al binary oxide was best approximated by the Elovich kinetic equation ($R^2 = 0.658$). The R^2 values for the other equations were not significant (Table 1).

3.2 Influence of soil-exchangeable Mn concentration by Fe–Al binary oxide

The quantity of exchangeable heavy metals in the soil reflects the short-term effectiveness of the Fe–Al binary oxides. The concentration of soil-exchangeable Mn rapidly decreased from 22.64 mg kg^{-1} at t = 0 to 11.97 mg kg^{-1} on day 3, but then decreased slowly to day 14 (Fig. 2). The Fe–Al binary oxide decreased

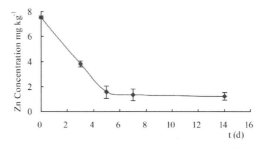

Figure 1. Changes of soil-exchangeable Zn content after Fe-Al oxides added in the soil.

Table 1. Kinetic parameter of the adsorption effect on soil Zn by Fe-Al binary oxide.

Equation	Parameter estimates				
	a	k	R^2	Sig.	SE
Zero order	4.304	0.170	0.445	0.333	0.113
First order	1.441	0.034	0.424	0.349	0.234
Parabolic diffusion	2.807	1.054	0.546	0.261	1.007
Elovich	2.743	1.532	0.658	0.189	0.873
Power	1.122	0.311	0.638	0.201	0.186

Table 2. Kinetic parameter of the adsorption effect on soil Mn by Fe-Al binary oxide.

Equation	Parameter estimates				
	a	k	R^2	Sig.	SE
Zero order	9.921	0.473	0.926	0.037	0.781
First order	2.326	0.035	0.890	0.057	0.072
Parabolic diffusion	6.327	2.712	0.972	0.014	0.485
Elovich	6.724	3.633	0.994	0.002	0.183
Power	2.082	0.273	0.987	0.007	0.025

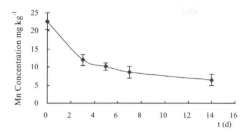

Figure 2. Changes of soil-exchangeable Mn content after Fe-Al oxides added in the soil.

the total soil-exchangeable Mn by 71.6%. The R^2 values for zero-order kinetic, first-order kinetic, parabolic diffusion, Elovich, and power function equations were 0.926, 0.890, 0.972, 0.994, and 0.987 respectively (Table 1). The highest R^2 value was for the Elovich equation, followed by the power function equation. Significant values were also obtained for the zero-order and parabolic diffusion equations.

4 DISCUSSION

Heavy metals sorption on pure mineral and soil surfaces has been extensively studied to evaluate its mobility and fate in soils and surface (Wang et al. 2008; Zeng et al. 2009; Zhao et al. 2011). In most cases, adsorption reactions control the concentration of heavy metals in soil solutions and therefore their bioavailability. Although the extent of adsorption and desorption of various heavy metals by mineral surfaces has been widely studied, but little study about the effect of mineral in the soil was reported. Reaction processes between oxides and heavy metals in the soil are more closed to the fact. Knowledge of sorption reaction rates is useful for providing an insight into reaction mechanisms and the processes occurring on the adsorbing surface. The soil-exchangeable heavy metal concentrations were reduced after incubation with Fe–Al binary oxide for 14 days in the order Zn > Mn. An effective method for decreasing heavy metals leaching from soil is stabilization. Our results confirmed that the Fe–Al binary oxide increased the stability of heavy metals and decreased their extractable fraction. Therefore, Fe–Al binary oxide has the potential to be utilized as an effective adsorbent for reducing the mobility of Zn and Mn in soil.

Adsorption–desorption kinetics can describe the mechanism of binding and release of an adsorbate from the sorbent. It depends on the physical and chemical characteristics of the sorbent as well as on the process of mass transfer (Naiya et al. 2009). To investigate the kinetics of adsorption and desorption, two kinds of mathematical models are commonly used: a chemical kinetics model, such as

the zero-order kinetic equation and the first-order reaction kinetic equation; and an empirical model, such as the Elovich equation and the power function equation. Different equations have a different physical meaning. The parabolic diffusion equation reflects the diffusion transfer mechanism in the process of adsorption–desorption. The Elovich equation was originally developed to describe the kinetics for the chemisorption of gases to the solid surfaces, but later on, it was modified by Chien and Clayton (1980), with the assumption that the activation energy of sorption increases linearly with surface coverage (Cheung et al. 2001). The Elovich equation is often used to describe soil and mineral adsorption reactions that show a decrease in adsorption rate and an increase in adsorption capacity (Mehdi, et al. 2011). The experimental data for soil-exchangeable Mn demonstrated a high degree of correlation with the Elovich kinetic equation and the power function equation, respectively. The Elovich equation also modeled the adsorption of soil-exchangeable Zn better than the other equations. These results indicate that the process for adsorption of soil-exchangeable Mn similar to that for Zn, which may be due to the surface coverage of the adsorbent. The Elovich equation showed the changes of adsorption rate with surface coverage of adsorbate. The results demonstrate that the fixation of Mn and Zn by oxide on the soil was mainly associated with the concentration of them in the soil.

During the experiment period, the soil pH decreased from 7.8 to 7.3, indicating that the concentration of hydrogen ions increased. According to the following surface complexation reaction, heavy metals adsorb to reactive surface-hydroxyl groups on Fe–Al binary oxides and release hydrogen ions in the process.

$$2Fe_xAl_y-OH + M^{2+} = Fe_xAl_y-O-M-O-Fe_xAl_y + 2H^+$$

M = exchangeable heavy metal ions in the soil. This equation may exemplify the adsorption mechanism for exchangeable heavy metals on Fe–Al binary oxide surface in the soil.

ACKNOWLEDGEMENT

This work was supported by the National Natural Science Foundation of China (41201228).

REFERENCES

[1] Basu T., Gupta K., Ghosh U.C. (2012) Performances of As(V) adsorption of calcined synthetic iron (III)–aluminum(III) mixed oxide in the presence of some groundwater occurring ions. Chem. Eng. J. 183:303–314.

[2] Cheung C.W., Porter J.F., McKay G. (2001) Sorption kinetic analysis for the removal of cadmium ions from effluents using bone char. Water Res. 35: 605–612.

[3] Chien S.H., Clayton W.R. (1980) Application of Elovich equation to the kinetics of phosphate release and sorption in soils. Soil Sci. Soc. Am. J. 44: 265–268.

[4] El-Kamash A.M., El-Gammal B., El-Sayed A.A. (2007) Preparation and evaluation of cerium(IV) tungstate powder as inorganic exchanger in sorption of cobalt and europium ions from aqueous solutions. J. Hazard. Mater. 141: 719–728.

[5] Engates K.E., Shipley H.J. (2011) Adsorption of Pb, Cd, Cu, Zn, and Ni to titanium dioxide nanoparticles: effect of particle size, solid concentration, and exhaustion. Environ. Sci. Pollut.Res. 18: 386–395.

[6] Geelhoed J.S., Findenegg G.R., Vanriemsdijk W.H. (1997) Availability to plants of phosphate adsorbed on goethite: experiment and simulation. Eur. J. soil sci. 48:473–481.

[7] Granados-Correa F., Bulbulian S. (2012) Co(II) Adsorption in Aqueous Media by a Synthetic Fe–Mn Binary Oxide Adsorbent. Water Air Soil Pollut. 223: 4089–4100.

[8] Hong H., Farooq W., Yang J. (2010) Preparation and evaluation of Fe-Al binary oxide for arsenic removal: comparative study with single metal oxides. Sep. Sci. Technol. 45: 1095–1981.

[9] Hye-Jin H., Jung-Seok Y., Bo-Kyong K., Ji-Won Y. (2011) Arsenic Removal Behavior by Fe-Al Binary Oxide: Thermodynamic and Kinetic Study. Sep. Sci. Technol. 46: 2531–2538.

[10] Liao S.J., Liu G.L., Zhu D.W., Li Y., Ren L.Y., Cui J.Z. (2011) Characterization and properties of boron-doped aluminum hydroxide for Mn^{2+} adsorption and soil acidification. Environ. Earth Sci. 62: 1047–1054.

[11] Liu G.L., Zhu D.W., Liao S.J., Ren L.Y., Cui J.Z., Zhou W.B. (2009) Solid-phase photocatalytic degradation of polyethylene–goethite composite film under UV-light irradiation. J. Hazard. Mater. 172: 1424–1429.

[12] Matis K.A., Zouboulis A.I., Zamboulis D., Valtadorou A.V. (1999) Sorption of As by goethite particles and study of their flocculation. Water, Air Soil Poll. 111: 297–316.

[13] Mehdi J.S., Ali M., Ataallah S. (2011) Mercury (II) removal from aqueous solutions by adsorption on multi-walled carbon nanotubes. Korean J. Chem. Eng. 28: 1029–1034.

[14] Naiya T.K., Bhattacharya A.K., Mandal S.N., Das S.K. (2009) The sorption of lead(II) ions on rice husk ash. J. Hazard. Mater. 163: 1254–1264.

[15] Neda A., Tahereh K. (2013) A comparison on efficiency of virgin and sulfurized agro-based adsorbents for mercury removal from aqueous systems. Adsorption, 19: 189–200.

[16] Ozdes D., Duran C., Senturk H.B. (2011) Adsorptive removal of Cd(II) and Pb(II) ions from aqueous solutions by using Turkish illitic clay. J. Environ. Manag. 92: 3082–3090.

[17] Ren L.Y., Zhu D.W., Cu, J.Z.i Liao S.J., Geng M.J., Zhou WB, Hamilton D (2009) Plant availability of boron doped on iron and manganese oxides and its effect on soil acidosis. Geoderma, 151: 401–406.

[18] Schaller M.S., Koretsky C.M., Lund T.J., Landry C.J. (2009) Surface complexation modeling of Cd(II) adsorption on mixtures of hydrous ferric oxide, quartz and kaolinite. J. Colloid Interface Sci. 339: 302–309.

[19] Sdiri A., Higashi T., Jamoussi F., Bouaziz S. (2012) Effects of impurities on the removal of heavy metals by natural limestones in aqueous systems. J. Environ. Manage. 93: 245–253.

[20] Tessier A., Campbell P.G.C., Bisson M. (1979) Sequential extraction procedure for the speciation of particulate trace metals. Anal. Chem. 51: 844–851.

[21] Trivedi P., Axe L., Dyer J. (2001) Adsorption of metal ions onto goethite: single-adsorbate and competitive systems. Colloids Sur. A. 191: 107–121.

[22] Wang Y.J., Zhou D.M., Sun R.J., Jia D.A., Zhu H.W., Wang S.Q. (2008) Zinc adsorption on goethite as affected by glyphosate. J. Hazard. Mater. 151: 179–184.

[23] Zeng H., Singh A., Basak S., Ulrich K.U., Sahu M., Biswas P., Catalano J.G., Giammar D.E. (2009) Nanoscale size effects on uranium(VI) adsorption to hematite. Environ. Sci. Technol. 43: 1373–1378.

[24] Zhang G., Liu H., Liu R., Qu J. (2009) Adsorption behavior and mechanism of arsenate at Fe–Mn binary oxide/water interface. J.Hazard.Mater. 168: 820–825.

[25] Zhao X.Y., Guo X.J., Yang Z.F., Liu H., Qian Q.Q. (2011) Phase-controlled preparation of iron (oxyhydr)oxide nanocrystallines for heavy metal removal. J. Nanopart. Res. 13: 2853–2864.

[26] Zhu D.W., Chen X.H., Cheng D.S. (2000) Electrical characteristics and desorption kinetics of soil boron. Pedosphere, 10: 61–68.

Material Science and Engineering – Chen (Ed.)
© 2016 Taylor & Francis Group, London, ISBN 978-1-138-02936-1

Effect of sand-blasted treatment on the oxidation resistance of TP347 stainless steel in high-temperature water vapor

J.W. Chen, Z. Jiang & H. Mu
Guizhou Electric Power Test Research Institute, Guiyang, China

C. Luo, H.D. Liu & B. Yang
School of Power and Mechanical Engineering, Wuhan University, Wuhan, China

ABSTRACT: Oxidation behaviors of TP347 stainless steel after sand-blasted treatment were investigated in high-temperature water vapor. Surface morphologies and surface roughness of the treated samples before the oxidation experiment were studied by Atomic Force Microscopy (AFM). The samples were weighed together with the quartz crucible using a 0.1 mg electric scale. The chemical compositions and microstructures of the oxide scales were characterized by EBSD, SEM and EDX. The results showed that the surface roughness of the samples had increased after sand-blasted treatment compared with the untreated samples. Furthermore, the lower oxidation rate was detected from the treated samples, and a higher Cr content and thinner thickness was found in the oxide scales. As a result, oxidation resistance of the treated TP347 stainless steel was significantly improved by the sand-blasted treatment.

Keywords: sand-blasted treatment; TP347 stainless steel; oxide scale; oxidation resistance

1   INTRODUCTION

Austenite heat-resistant steel is widely used as the material of high-temperature super-heater and reheater tube in a super-critical unit. However, the oxidation on the inner surface of super-heater and reheater tubes may lead to steam flow restriction and blocking of tube bends, which will result in boiler failure from overheating [1–2]. TP347 stainless steel is one of the heat-resistant steels used as materials for boiler super-heater tubes and high-temperature reheater tubes. It tends to form a Cr-rich oxide layer that can protect the steel against further oxidation in dry oxidizing atmospheres; however, the oxidation resistance of the steel is decreased in wet atmospheres at elevated temperatures [3–6]. Therefore, improving the diffusion of Cr from the steel to the surface is a very important factor to improve the oxidation resistant of the steel. Studies [7–8] have shown that it is very easy to form cellular cluster oxides on the surface under the conditions of high surface roughness and high surface residual strain. As the short-circuit diffusion path increases, it is easier for Cr in the process strained layer to form Cr_2O_3 on the surface, which can improve the oxidation resistance of the steel. Sand-blasted treatment can effectively improve the surface roughness, and introduce a lot of dislocations and defects on the surface of the steel, which

can improve the efficiency of the diffusion of Cr at a high temperature and form a dense protective oxide scale on the surface.

Therefore, in order to explore the oxidation mechanism of the stainless steel with the different surface roughness and surface morphologies, this paper is mainly focused on the effect of sand-blasted treatment on the oxidation behavior of TP347 stainless steel at 550–750°C for 25 h in water vapor. The comparison between the untreated and sand-blasted treated samples from the oxidation dynamics and mechanism, the constituents and microstructures of the oxide scale is systematically investigated.

2   EXPERIMENTAL PROCEDURE

The material used in this paper was TP347 stainless steel ($10 \times 10 \times 10$ mm), with the chemical composition (wt%) as follows: C: $\leqq0.10$, Si: $\leqq1.00$, Mn: $\leqq2.00$, Cr: $17.0–20.0$, Ni: $9.0–13.0$, Nb: $8 \times C\%–1.5$, S: $\leqq0.030$, P: $\leqq0.035$, and balance Fe. The stainless steel was ground, and then polished with 400–3000 grit paper prior to the sand-blasted treatment. The samples were treated at 0.3, 0.5, and 0.7 MPa, respectively. The duration time of the sand-blasted process was 120 s. Before the oxidation experiment, Atomic Force Microscopy (AFM) in SPM-9500J3

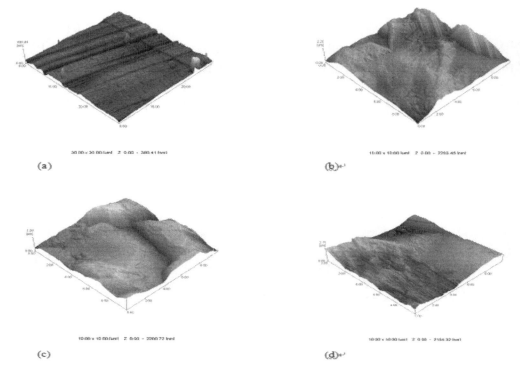

(a)

(b)

(c)

(d)

Figure 1. Surface morphologies of TP347 before the oxidation experiment (AFM): (a) without sand-blasted treatment (b) 0.3 MPa (c) 0.5 MPa (d) 0.7 MPa.

Table 1. Surface roughness of the samples with and without sand blasting (Ra/μm).

Blasting pressure (MPa)	0	0.3	0.5	0.7
Surface roughness (Ra/μm)	0.022	0.242	0.304	0.28

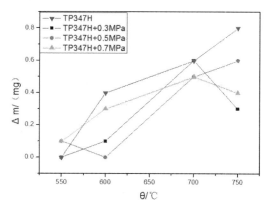

Figure 2. Effect of the sand-blasted treatment on the weight gain of TP347 steel in water vapor at 550–750°C for 25 h.

was performed to detect the surface morphologies and surface roughness. The samples were weighed together with the quartz crucible using a 0.1 mg electric scale.

The samples were oxidized at 550, 600, 650, 700 and 750°C for 25 h in a self-designed tubular horizontal oxidation oven under atmospheric pressure, respectively. The water vapor was produced by a steam generator from deionized water. After the oxidation experiment, the samples were weighed for comparison with the untreated samples. The chemical compositions and microstructures of the oxide scales were observed by EDX, SEM and EBSD.

3 RESULTS AND DISCUSSION

3.1 *Surface morphologies and surface roughness*

The AFM of the surface morphologies of the samples with and without the sand-blasted treatment before the oxidation experiment is shown in Figure 1. As shown in Figure 1, the surface of the sample without the sand-blasted treatment was relatively smooth, which had only a little scratch by sand paper. However, the undulating surface of the sam-

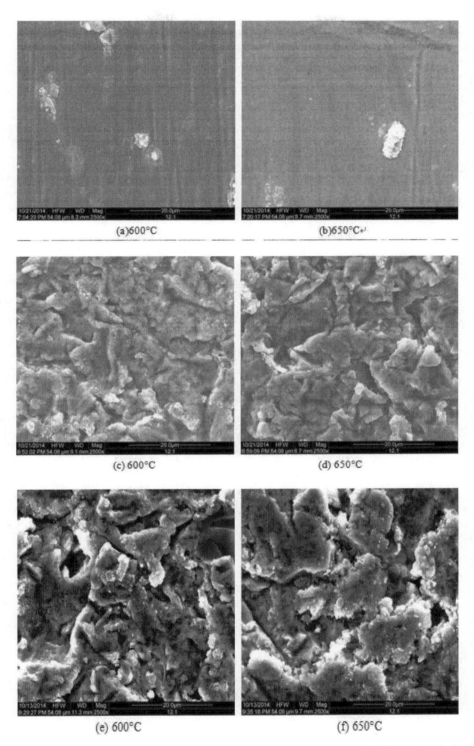

Figure 3. Surface morphologies of TP347 steel after oxidation in water vapor at 600–650°C for 25 h: (a) and (b) untreated samples; (c) and (d) 0.3 MPa; (e) and (f) 0.5 MPa.

ples with the sand-blasted treatment was observed, which was rougher than that of the untreated one. Moreover, the higher surface roughness was observed from the higher-pressure treated samples. However, when the blasting pressure increased to 0.7 MPa, the surface roughness decreased due to the high blasting pressure that smoothens the uneven surface. The surface roughness of the samples with and without the sand-blasted treatment obtained by AFM is presented in Table 1.

3.2 Oxidation dynamics

Figure 2 shows the weight gains of virgin and blasted treated samples oxidized at 550, 600, 650, 700, and 750°C for 25 h in water vapor. As the temperature increases, the weight gains of all the samples showed an upward trend, indicating that the temperature was an important factor in the weight gains of oxidized samples. The treated samples presented higher weight gains at a temperature lower than 600°C. Above 600°C, the samples with sand blasting gradually exhibited an enhanced effect on oxidation resistance, which showed lower weight gains than that of the untreated sample. The sample with the maximum surface roughness at the blasting pressure of 0.5 MPa had the minimum weight gain, which indicated that the surface roughness of the sample could affect the oxidation weight gain, and the larger the surface roughness of the sample, the better the oxidation resistance.

3.3 Surface morphologies and composition analysis of the oxide scales

Figure 3 shows the surface morphologies of the virgin and blasted samples oxidized at different temperatures for 25 h. With the increasing oxidation temperature, both the treated and untreated samples showed aggravated surface oxidation, which indicated that the temperature was the significant factor on the oxide scale. Compared with the untreated samples, morphologies of oxide scales of the treated samples exhibited an intensive regular shape. With the uneven size, spherical oxide distributed on the relatively flat surface of the untreated sample could be drawn as high Fe oxides in the spectrum analysis. Moreover, a comparatively dense oxide scale was formed on the surface of the treated samples at 600°C, and the oxide scale became more compact with increasing temperatures. It was assumed that the diffusion of Cr became easier to form a dense oxide scale on the surface of the treated samples. However, oxides of the untreated samples dispersed on the surface with cavity, and there were more relatively loose oxides in the defects. As a result, the untreated sample could not form a dense oxide scale to suppress the diffusion of O element and prevent further oxidation of the base material.

The oxide scale formed on the surface of the steel consisted of Fe_3O_4 and Cr_2O_3, of which Cr_2O_3 was more effective to prevent further oxidation due to its denser structure than Fe_3O_4. Better oxidation resistance would be obtained in the steel with a higher Cr_2O_3 content oxide scale on the surface. Therefore, the relative content of Cr in the oxide scale was regarded as a standard to judge the oxidation resistance of the steel.

The ratio of Cr to Fe+Cr on the surface of oxide scales formed in water vapor at 600–700°C for 25 h by surface scanning EDX is shown in Figure 4. As can be seen from Figure 4, with the increasing temperature, the relative content of the Cr element of the surface oxide scale was increased. Furthermore, the treated samples had a higher content of Cr, which indicated that the sand-blasted treatment helped to promote the diffusion of Cr into the sample's surface to form a relatively dense oxide scale, thereby enhancing the oxidation resistance of the material.

3.4 Cross-sectional morphologies of the oxide scale

The untreated and treated samples at 0.5 MPa were ground to the metallurgical surface for observing the oxide scale by Electron Back-Scattered Diffraction (EBSD) associated with the Field Emission Scanning Electron Microscope (FESEM). Figure 5 shows the cross-sectional images of the oxide scales formed at 650°C for 25 h. As it can be seen from Figure 5 (a) and (b), the treated sample had a much thinner oxide scale of 6 μm, which was only about 50% of the untreated sample. Clearly, the growth rates of the oxide scales became slower considerably and the thickness of the oxide scales was controlled effectively after the sand-blasted treatment, which indicated that the oxidation

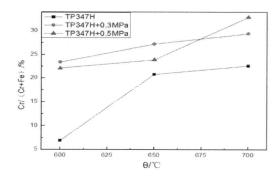

Figure 4.   Fraction of Cr/(Cr+Fe) on the surfaces of the oxide scales in water vapor at 600–700°C for 25 h.

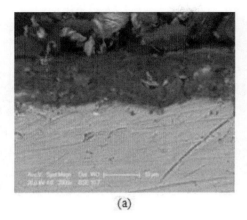 

(a) (b)

Figure 5.   EBSD of cross section of TP347 at 650°C for 25 h: (a) untreated sample; (b) 0.5 MPa.

resistance of the sand-blasted treatment samples in high-temperature water vapor was significantly improved.

4   CONCLUSION

The surface morphologies of TP347 steel were obviously changed by the sand-blasted treatment, which resulted in the increased surface roughness, a state of stress on the surface and the increased density of dislocation and sub-grain boundary defects.

The relative content of Cr had effectively improved and the growth rate of the oxide scale reduced greatly due to high energy state surface, dislocations and other defects generated by the sand-blasted treatment. The thickness of the oxide scale had greatly decreased, which was only about 50% of that of the untreated sample.

The oxidation resistance of TP347 stainless steel in high-temperature water vapor was markedly enhanced after the sand-blasted treatment.

REFERENCES

[1] Dooley, R.B., Wright, I.G., & Tortorelli, P. (2007). Program on technology innovation: oxide growth and exfoliation on alloys exposed to steam. *EPRI Report*, Palo Alto, CA.
[2] Zengwu, Y., Min, F., Xuegang, W., & Xingeng, L. (2012). Effect of shot peening on the oxidation resistance of TP304H and HR3C steels in water vapor. *Oxidation of metals*, 77(1–2), 17–26.
[3] Halvarsson, M., Tang, J.E., Asteman, H., Svensson, J.E., & Johansson, L.G. (2006). Microstructural investigation of the breakdown of the protective oxide scale on a 304 steel in the presence of oxygen and water vapour at 600 C. *Corrosion Science*, 48(8), 2014–2035.
[4] Pint, B.A., & Rakowski, J.M. (2000). Effect of water vapor on the oxidation resistance of stainless steels. *Corrosion* 2000.
[5] Wright, I.G., & Dooley, R.B. (2010). A review of the oxidation behaviour of structural alloys in steam. *International Materials Reviews*, 55(3), 129–167.
[6] Saunders, S.R.J., Monteiro, M., & Rizzo, F. (2008). The oxidation behaviour of metals and alloys at high temperatures in atmospheres containing water vapour: A review. *Progress in Materials Science*, 53(5), 775–837.
[7] Otsuka, N., & Fujikawa, H. (1991). Scaling of austenitic stainless steels and nickel-base alloys in high-temperature steam at 973 K. *Corrosion*, 47(4), 240–248.
[8] Langevoort, J.C., Fransen, T., & Geilings, P.J. (1984). On the influence of cold work on the oxidation behavior of some austenitic stainless steels: High temperature oxidation. *Oxidation of metals*, 21(5–6), 271–284.
[9] Rehn, I.M. (1981). Corrosion problems in coal-fired boiler superheater and reheater tubes: steam-side oxidation and exfoliation. Development of a chromate-conversion treatment. *Final report (No. EPRI-CS-1812). Foster Wheeler Development Corp.*, Livingston, NJ (USA).

Material Science and Engineering – Chen (Ed.)
© 2016 Taylor & Francis Group, London, ISBN 978-1-138-02936-1

Investigation on the magnetic properties of heat-resistant steels and oxides

J.W. Chen, Z. Jiang & H. Mu
Guizhou Electric Power Test Research Institute, Guiyang, China

Q. Wan, H.D. Liu & B. Yang
School of Power and Mechanical Engineering, Wuhan University, Wuhan, China

ABSTRACT: Martensite and austenite heat-resistant steels are currently favored structural materials in the application of steam generators due to their excellent high-temperature mechanical properties. One of the limited factors in the real application is the formation of a large amount of oxides in the steam side. In this paper, the magnetic properties of T23, T91, Super304, TP304, TP347H and the oxides of TP347H were studied by the vibrating sample magnetometer of the Physical Property Measurement System. The results suggested that martensite heat-resistant steels and the oxides of TP347H, which revealed stronger magnetism, were ferromagnetic, while the austenite heat-resistant steels were paramagnetic, which could not be influenced by the applied magnetic field.

Keywords: heat-resistant steels; oxide scale; magnetic properties

1 INTRODUCTION

Recently, the appearance of new generation power plants has increased the operating steam temperature and pressure to achieve higher efficiency and better environmental protection [1].

New heat-resistant steels such as T23, T91, Super304, TP304 and TP347H have been widely used in supercritical and ultra-supercritical power plants [2]. During the service, oxide layers would be formed in the steam side of these steels. The large difference in the thermal expansion coefficient between oxide layers and the heat-resistant steels made the oxide layers to peel off easily when the temperature of the tubes varied greatly during the start–stop process in the boilers [3]. Part of the peeled oxides would accumulate at the bending steels, which would lead to the overheating of tubes. Other peeled oxides would flow with the steam and cause abrasive wear of the tubes and turbine blades [4–6].

Many studies have focused on the oxidation mechanism, microstructures of the oxide layers, and the exfoliation reasons, while little literature related to the magnetic detection technology of the oxide layers has been published [7–9]. The aim of this work is to investigate the magnetic properties of T23, T91, Super304, TP304, TP347H and the oxides of TP347H, which can provide a scientific basis for the development of the application of the magnetic detection technology.

2 EXPERIMENTAL DETAILS

2.1 *Sample preparation*

The samples used in the present work were obtained from the commercial tubes. As shown in Figure 1, the tubes were cut into a dimension of 5*5*2 mm, and the samples of Super304, TP347H, TP304, T23, T91 and oxides of TP347H were labeled as 1#, 2#, 3#, 4#, 5#, 6#, respectively.

2.2 *Characterization*

PPMS (Physical Property Measurement System) is an instrument designed for the detection of the physical properties of materials. VSM (Vibrating Sample Magnetometer), as part of PPMS, can provide the magnetic properties of materials. In this paper, the magnetizing curves and hysteresis loops of the samples prepared in the previous section were obtained to analyze the magnetic properties.

3 RESULTS AND DISCUSSION

3.1 *Magnetizing curves*

Figure 2 shows the magnetizing curves of the heat-resistant steels and oxides. The magnetizing curves of Super304, TP347H, and TP304 with a smaller slope were almost parallel to the X-axis, which

(a)　Samples of heat-resistant steels

(b) Samples of oxides

Figure 1.　Prepared samples.

suggested that Super304, TP347H, and TP304 were paramagnetic. While the intensity of magnetization of T23, T91 and oxides of TP347H increased with the increasing applied magnetic field, and then kept stable after reaching the maximum value.

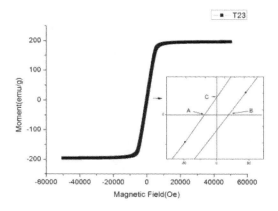

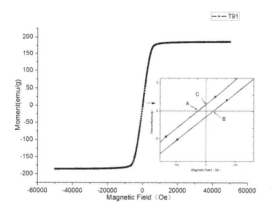

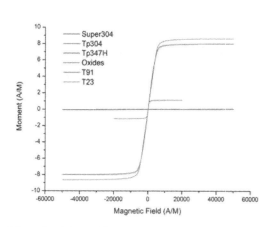

Figure 2.　Magnetizing curves.

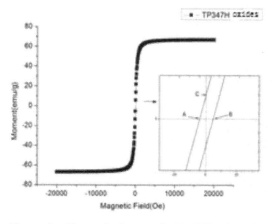

Figure 3. Hysteresis loops of (a) T23, (b) T91, (c) TP347H.

Table 1. Magnetic parameters of austenite steels and oxides.

Samples	Maximum magnetization Ms (emu/g)	Coercivity Hc (A/M)	Residual magnetization Br (Gs)
4#	195.3 (5.2T)	2352.9	122.7
5#	185.8 (6.6T)	2044.5	101.4
6#	66.6 (1.75T)	126.6	42.7

The results suggested that Super304, TP347H, and TP304 were paramagnetic, while T23, T91, and oxides of TP347H were ferromagnetic [10]. Based on the results, austenite steels would not affect the applied magnetic field while the oxides of the austenite could affect the applied magnetic greatly, which suggested that the oxides of the austenite steels could be detected by the magnetic detection technology.

3.2 Hysteresis loops of martensite steel and oxides

The hysteresis loop test was carried out to study the magnetic intensity, coercivity, and residual magnetism of T23, T91 and oxides, as shown in Figure 3.

Points A and B are intersections of the hysteresis loops, and the X-axis correspond to the coercivity −Hc, +Hc. Point C is the intersection of the hysteresis loops, and the Y-axis is an indicator of the residual magnetization. Table 1 lists the magnetic parameters of these three samples. The coercivity and residual magnetization of the oxides were 126.6 and 42.7, respectively, which were the minimum value among the three samples. However, the other two martensite steels revealed a similar value. It could be concluded that the oxides of austenite steels were soft magnetic materials, while martensite steels were hard magnetic materials [11].

4 CONCLUSIONS

The magnetic measurement was conducted on the martensite, austenite steels and oxides. The coercivity and residual magnetization were calculated from the hysteresis loops, and the following conclusions could be obtained:

1. Super304, TP347H and TP304 revealed significant paramagnetism.

2. T23, T91 and oxides of TP347H showed hysteresis. The coercivity and residual magnetization of the oxides were small and should be soft magnetic materials, while T23 and T91 with higher coercivity and residual magnetization were hard magnetic materials.

REFERENCES

[1] Jingbiao Y. 2010. Review on the formation and exfoliation mechanism of steam-side oxidation scale on the boiler tube with high temperature [J]. *Boiler Technology*. 41, (6): 44–49.

[2] Jiancheng Z., Yunfei L., Yuxin W. 2008. Analysis on the falling-off problem of oxide on the inside wall of austenitic stainless steel pipe and its examining method study [J]. *Electrical Equipment*. 9, (2): 58–60.

[3] Bin L., Xuejin C. 2011. Exfoliation Cause Analysis of Oxidation Velamen on the High Temperature Heating Surface Tube of Supercritical Boiler [J]. *Journal of Anhui Electrical Engineering Professional Technique College*. 16 (3): 86–89.

[4] Yupeng W., Wei W., Zhiwu W. 2013. Analysis on accident reason for fall-off of oxide skin of heating [J]. *Guangdong Electric Power*. 26 (3): 91–94.

[5] Wanli J., Lianfeng G. 2009. Technical measure to prevent ESC boiler heating surface pipe blocking from inside pipe oxide peeling-off [J]. *Electric Power Construction*. 30 (10): 71–73.

[6] Jianmin J. Jigang C. 2008. Counter measures against massive exfoliation of oxidation scale on the internal surface of coarse grained 18-8 type stainless steel boiler tubes [J]. *Electric Power*. 1, (5): 37–41.

[7] Jianmin J., Jigang C., Liying T., Hongzhe W., Feng L. 2008. Investigation on Microstructure and Morphology Features of Steam-side Oxidation Scale and Exfoliated Oxide From the Internal Surface of 12X18H12T Tube [J]. *Proceedings of the CSEE*. 28 (17): 43–48.

[8] Huiguo L., Guosheng X., Yi L., Shenzhou M. 2013. Steam-side oxidation mechanism and characterization of oxide scales grown on TP347H steel tube [J]. *Transactions of Materials and Heat Treatment*. 34 (9): 183–188.

[9] Wanli Z., Zheng gang L., Wei W. 2012. Microstructure and formation mechanism of oxide film on the steam side of T91 steel after high temperature operation [J]. *Thermal Power Generation*. 41 (6): 28–31.

[10] Shi T. 2004. Physical Properties of Materials [M]. *Beijing: Press of Beihang University*. 294–295.

[11] Deke S. 2003. Foundation of Materials Science [M]. *Beijing: China Machine Press*. 401–402.

Chemical materials

Material Science and Engineering – Chen (Ed.)
© 2016 Taylor & Francis Group, London, ISBN 978-1-138-02936-1

High-performance UV-cure polyurethane acrylate for UV transparent insulation inks

J.J. Huang, X.C. Zhang, J.T. Liang, Y.C. Feng, W.J. Zhao & Z.H. Yang

College of Materials and Energy, South China Agricultural University, Guangzhou, Guangdong, P.R. China

ABSTRACT: In this work, a series of bio-based polyurethane acrylate oligomers, based on the *in situ* polymerization of different molar ratios of Castor Oil (CO)/Polyethylene Glycol (PEG), were first prepared and characterized by Fourier Transform Infrared (FT-IR) spectroscopy. The UV-cured polyurethane acrylate transparent insulating inks, prepared with the above optimized oligomers, were printed on ITO glass and ITO film substrates for further polymerization under UV irradiation. When the molar ratio of the CO/PEG was 1/6, the UV-PUA films had super combination properties such as excellent electrical insulating, adhesion, and mechanical properties.

Keywords: bio-based; polyurethane acrylate; transparent; insulative inks

1 INTRODUCTION

As governments become increasingly strict about environmental laws and regulations with respect to Volatile Organic Compounds (VOC) emission, chemical industries have to develop and utilize more eco-friendly solvent-free polymerization systems and UV coatings, which are a promising choice (Pappas, S.P. 1978; Sack, M. 1982). The ultraviolet curing technology is based on the polymerization of transforming a liquid multicomponent system, which is induced by an incident UV radiation into a solid three-dimensional network polymer material with some properties of rubber and glass. In addition, this technology has the following features: (1) low energy requirement; (2) very fast and efficient polymerization; (3) cure selectively limiting to the irradiated area; and (4) no environmental pollution by VOC (Allen, N.S. et al. 1989; Chattopadhyay, D.K. & Raju, K. 2007). This makes UV-radiation curable products become a viable alternative to conventional thermal curing of solvent-containing polymer products. Meanwhile, the benefits of UV curing products have been widely utilized for drying inks, coatings, adhesives and other UV-sensitive materials through polymerization. It has been applied to some specific industrial as well as the graphics industry, the automotive industry, the electronic industry with its various varnishing and printing procedures, including screen printing.

Among the oligomers used for UV-curable formulations, polyether acrylates, polyester acrylates, silicone acrylates, polyether urethane acrylates, polyester urethane acrylates and epoxy acrylates are commonly used types of acrylate oligomers

(Meier, W.U. 2007; Eren, T. et al. 2006). Polyurethane acrylate oligomers have gained increasing attention and speedy development due to a wide range of excellent application properties, such as high impact and tensile strength, abrasion resistance and toughness combined with excellent resistance to chemicals and solvents (Schwalm, R. 2007; Nebioglu, A. & Soucek, M.D. 2006). In particular, bio-based polyurethane acrylate oligomers, which are derived from renewable resources such as plant oil and cellulose (Huang, Y.G. & Pang, L.X. 2013; Rengasamy, R. et al. 2013; Wang, C.S. et al. 2011; Patel, M.M. et al. 2009), with their non-toxic, biodegradable and eco-friendly advantages, are gaining extensive research attention. Moreover, there has been a growing trend in modifying bio-derived polymers with Castor Oil (CO) to meet specific application ends or to get balanced properties. Polyurethane acrylate with CO, as a kind of raw material, has shown significant enhancement in thermal stability and mechanical properties (Dieterich, D. 1981; Ha, C. et al. 1996).

In the present work, UV curing polyurethane acrylate oligomers with seven different compositions were synthesized by reacting with Isophorone Diisocyanate (IPDI), Polyester diol (PBA), Polyethylene Glycol (PEG), Castor Oil (CO), Dimethylol Propionic Acid (DMPA), Bisphenol A (BPA) and 2-Hydroxyethyl Acrylate (HEA). Meanwhile, UV curing polyurethane acrylate transparent insulating inks were prepared by the above-synthesized oligomers, Isobornyl Acrylate (IBOA) as the reactive diluent, photoinitiator 184 and PM-2 as the adhesion promoter according to different weight ratios. In addition, relatively high levels of biodegradable

Castor Oil (CO) and Bisphenol A (BPA) were introduced into the polyurethane acrylate oligomers. High curing speed and high insulation were obtained together with little reduction of adhesive force.

2 EXPERIMENTAL PROCEDURE

2.1 *Materials*

Isophorone Diisocyanate (IPDI) was obtained from Bayer AG. Castor Oil (CO) (hydroxyl value = 163 mg KOH/g) was obtained from Guanghua Chemical Factory Co., Ltd, Guangdong. Polyester diol (PBA) (Mn = 1000) was supplied by Qingdao New Yutian Chemical Co., Ltd. Polyethylene Glycol (PEG) (Mn = 1000) was obtained from Shanghai Qiangshun Chemical Factory Co., Ltd. Dimethylol Propionic Acid (DMPA) was obtained from Perstorp Co., Sweden. Dibutyltin Dilaurate (DBTDL) was obtained from Shanghai Lingfeng Chemical Reagent Co., Ltd. Bisphenol A (BPA) was obtained from Sinophorm Chemical regent Co., Ltd. Isobornyl Acrylate (IBOA) was obtained from Guangzhou Swan Chemical Co., Ltd.

2.2 *Synthesis of UV-PUA oligomers*

The synthesis of UV polyurethane acrylate oligomer was by a process of the polyaddition reaction (Dieterich, D. 1981). NCO/OH value of the prepolymer was equal to 1.5, while the weight percentage of DMPA was about 1.5% with respect to the total solid weight. The reactive process is depicted as follows. A constant molar ratio of CO/PEG (A = 1/6) and a certain amount of PBA (5.86 g) were added into a 500 ml four-necked flask equipped with a mechanical stirrer, a drying tube, a nitrogen inlet and a thermometer. The four-necked flask was placed into an oil bath to obtain an even heat distribution. Under continuous stirring, the reactants were heated to 120°C to dehydrate under the condition of vacuum for 3h. After the temperature was cooled down to 70°C, IPDI (22.23 g) was poured into the flask, and then approximately 250 ppm of DBTDL were injected into the reaction vessel. The urethane forming reaction lasted 1.0 h. Next, when the reaction temperature was reduced to 60°C, the DMPA (0.98 g) and BPA (6.95 g) were charged into the reaction vessel to react for over 2 h. The NCO percentage of prepolymer was determined according to the standard of ASTM D2572–97. When the content of NCO reached a theoretical value, the reaction mixture was lowered to 45°C. HEA was then added dropwise into the NCO-terminated prepolymer. Subsequently, the reaction was carried out for 2.5~3 h at a temperature below 45°C. Other polyurethane acrylate oligomers were synthesized through changing the molar ratio of CO/PEG using the similar procedure.

2.3 *Preparation of UV transparent insulation inks*

The newly synthesized UV-PUA oligomers were, respectively, diluted with a reactive monomer (IBOA) in a constant monomer-to-oligomer ratio of 35/65 wt%, and the 3 wt% photo initiator (Irgacure 184) and the 5 wt% adhesion promoter (HEMA) were also added into the mixture systems. Then, these mixtures were stirred slightly above room temperature to achieve homogeneous systems. A series of UV-PUA films were prepared by printing the homogeneous mixing onto a ITO glass and poly-(tetrafluoroethylene) by a 200 mesh screen. Then, UV curing was carried out by exposing the samples to a mercury UV lamp (1000 W/cm) with a wavelength of 250–420 nm for different curing times.

2.4 *Structure characterization and performance test*

Fourier Transform Infrared (FT-IR) spectra of the UV-PUA oligomer and the UV-PUA films with different compositions were obtained using an Excalibur Series FT-IR (Digilab Co., Varian 4100, MA, USA) instrument. Fourier Transform Infrared (FT-IR) spectra of the UV-PUA film were obtained between 4000 and 400 cm^{-1} with a FTIR spectrometer (AVATAR 360, Madison, Nicolet). The mechanical properties including tensile shear strength and elongation at break for all of the specimens were measured on a universal testing machine (SUNS CMT5504, Shenzhen) with triplicate at room temperature at a speed of 5 mm/min according to the standard of ISO 4587:2003. The electrical insulating properties were determined on a megger (ZC36, Shanghai Qiangyue Electric Ltd, Shanghai) according to the standard of JB/T 5466–91 at room temperature. The Shaw hardness was measured with a sclerometer (TH200, Beijing Time High Technology Ltd, Beijing, China) on the basis of GB/T 2411–80. The contact angle of the distilled water on the ITO glass and the coated ITO glass substrates was measured using a manual contact angle goniometer (JCY-2, Shanghai Fangrui Instrument Ltd, Shanghai, China). All measurements had an average value of three replicates of each sample group. The water absorption (Q) measurements of UV-PUA films were determined in distilled water at 25°C on the basis of ASTM D 570.

2.5 *Adhesion property test*

For the adhesion measurements of the UV-PUA films printed on ITO glass substrates or ITO film, the tape adhesion (crosscut adhesion) was measured by following the ASTM D3359 methods, respectively.

3 RESULTS

3.1 *FT-IR spectral analysis*

The typical FT-IR spectra of the UV-cured PUA synthesis process are shown in Figure 1. During the reaction (from a to c in Figure 1), the isocyanate (NCO) bands at 2265–2270 cm^{-1} reduced gradually, and the progressive change of absorption bands at 3350 cm^{-1} and 1720–1730 cm^{-1} stated clearly that NCO reacted with OH and converted to NHCOO. In addition, when the acrylate-terminated reaction was continued for 150 min, no absorption bands were observed at 2265–2270 cm^{-1}, meaning that the NCO groups had been thoroughly reacted.

In addition, the absorption peaks at 1410 cm^{-1} and 810 cm^{-1} appeared and became stronger gradually, implying that the sealing side reaction with HEA occurred and acrylate double bonds were added to the urethane acrylate chain.

3.2 *The mechanical properties of UV transparent insulating inks*

UV transparent insulating inks with different mole ratios of CO/PEG were tested at room temperature, and the results of their mechanical properties including hardness, tensile shear strength and elongation at break are summarized in Table 1. From Table 1, it follows that:

(1) With the increasing mole ratio of CO/PEG, the hardness of the UV-PUA insulating inks increased gradually. This may be attributed to the increasing crosslinking density, the increasing den-

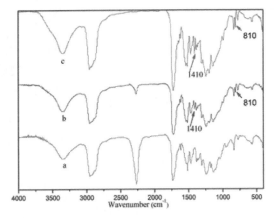

Figure 1. FT-IR spectrum of the UV-PUA3 oligomer synthesis process.

sity of the hard segment and the increasing hydrogen bonding formation in the systems with the increasing CO content. (2) Tensile shear strength increased from 30 to 34.2 MPa with an increasing mole ratio of CO/PEG from 0 to 1/6, but decreased subsequently. Besides, the elongation at the break of UV transparent insulating inks decreased from 8.5% to 2.4% with the increasing mole ratio of CO/PEG from 0 to 1/3. In general, these mechanical properties can be explained by the differences in the chain flexibility of the molecules, cross-linked degree, and crystallinity (Lee, C. et al. 2004; Shen, L. et al. 2006).

3.3 *The surface drying time of UV-PUA transparent insulating inks*

The surface drying time of UV-PUA insulating inks are summarized in Table 1. As shown in Table 1, with the molar ratio of CO/PEG increasing, the surface drying time of UV-PUA insulating inks straightly decreased. Because the molar ratio of CO/PEG increased, the degree of functionality increased in this system, which results in fast curing.

3.4 *Contact angle measurement*

The surface hydrophilicity of the different UV-PUA transparent insulating inks is presented in Table 1. The value of the contact angle of a droplet on a substrate is a sensitive reflection of the surface wetting properties. The distilled water contact angle on the ITO glass was 45°C. However, the distilled water contact angle on ITO glass coated with different urethane acrylate inks was in the range of 90–85°C, reflecting that the UV-PUA inks with different mole ratios of CO/PEG have good wettability on the ITO glass. In addition, when the mole ratio of CO/PEG increased, the contact angle value slightly decreased. This might be explained by the fact that Castor Oil (CO) makes the coating surface more polarized, giving the coating larger surface tension. Therefore, the contact angle of distilled water decreased slightly with the increase in the mole ratio of CO/PEG.

3.5 *The adhesion properties of UV-PUA transparent insulating inks*

UV curing was carried out by exposing the samples under a mercury UV lamp (1000 W/cm) with a wavelength of 250–420 nm for 30 s, 45 s, 60 s and 75 s, respectively. The adhesion property to the ITO glass substrate was investigated by crosscut adhesion (tape adhesion). Figure 2 shows the adhesion strength test results of the UV-PUA inks with different molar ratios of CO/PEG in varied

Table 1.   Physical properties of UV-PUA transparent insulating inks.

Sample item	CO/PEG	Hardness	Contact angle (°)	Surface drying time (s)	Gel content (%)	Tensile strength (MPa)	Elongation at break (%)
UV-PUA1	0	89	90	65	94.4	30.0	8.5
UV-PUA2	1/9	92	89	54	95.0	31.4	7.6
UV-PUA3	1/6	96	87	45	96.9	34.2	4.2
UV-PUA4	1/3	98	85	36	97.8	31.3	2.4

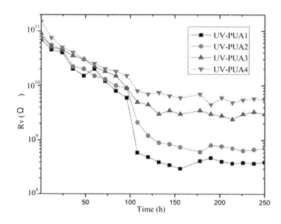

Figure 2.   The results of the adhesive test.

curing time. When the molar ratio of CO/PEG was lower than 1/9, it showed non-ideal adhesion to the ITO glass substrate, due to the fact that the cross-linking density of the polymeric network obtained by UV irradiation was low and the unreacted components remained because of low light reaction rate, even though the light curing time was 75 s. However, the cross-linking density increased with the increasing molar ratio of CO/PEG. This can be attributed to the fact that the light reaction rate increased with the increase in the functionality in a certain amount of light curing time. When the molar ratio of CO/PEG was equal to 1/6, the adhesion strength achieved the most effect (5B). However, the adhesion strength decreased with the continuous increase in the molar ratio of CO/PEG. This can be explained by the fact that shrinkage in this system increases seriously with the further increase of functionality. In addition, when the molar ratio of CO/PEG was low, the adhesion increased with the increasing curing time. However, the increase rate of the adhesion decreased with the increase in curing time, because the amount of unreacted components decreased. When the molar ratio of CO/PEG increased to 1/3, the adhesion showed a trend of decrease with the increasing curing time. This can be attributed to the fact that shrinkage in this system increases seriously with the increase in curing time.

3.6   The electrical insulating properties of UV transparent insulating inks

Figure 3 shows the evolution with time of this parameter for the UV-PUA transparent insulating inks under investigation. It is quite obvious that the general evolution of the electrical resistance properties reduces with time, owing to the interaction between the UV-PUA film and the electrode.

The UV-PUA3 and the UV-PUA4 transparent insulating inks of relatively high mole ratio of CO/PEG maintain very high volume resistance (Rv) values during all the measuring times. Especially, their plateau values (around 10^9–$10^{10}\Omega$) are about one order of magnitude higher than the UV-PUA1 and the UV-PUA2 inks of lower mole ratio of CO/PEG.

In terms of Figure 3, a continuing decrease in the volume resistance (Rv) begins at the first hour and approximately 100 hours later, and the volume resistance (Rv) values decrease sharply. In particular, the UV-PUA1 and the UV-PUA2 show a more apparent reduction. It is possible that the UV-PUA film of low mole ratio of CO/PEG induces swelling or absorbs water when exposed to air for a long time. Then, at approximately 120 h of continuous measurement, the UV-PUA1 and the UV-PUA2 reach a stable value around 10^8–$10^9\Omega$. At the same time, the UV-PUA3 and the UV-PUA4 maintain a plateau value around 10^9–$10^{10}\Omega$. From this analysis, it is possible to state that the UV-PUA film of relatively high mole ratio of CO/PEG improves the volume resistance (Rv) value when compared with the lower one. It is likely that the high values of volume resistance (Rv) are due to the high crosslinking covalent bonds in the UV-PUA film system, and the presence of a low content of hydrophilic chain segments is responsible for the high values of volume resistance.

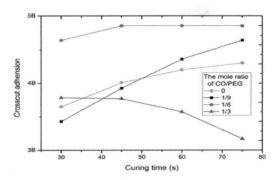

Figure 3. The coating volume resistance (Rv) with time.

4 CONCLUSION

In this study, a series of UV-PUA oligomers with the different molar ratios of CO/PEG were successfully synthesized, and UV-PUA transparent insulating ink formulations were prepared. It was found that the molar ratio of CO/PEG plays an important role in the properties such as hardness and pencil hardness, tensile strength, cross-cut, contact angle, and volume resistance. The experimental results indicated that the optimum molar ratio of the CO/PEG was 1/6. Almost all the UV-PUA films have good hardness, solvent resistance and mechanical properties. It is hopeful that the UV-PUA3 transparent insulating ink can be applied to commercial use in corresponding fields.

ACKNOWLEDGMENTS

This work was supported by the National College Students' Innovative and Entrepreneurial Training Program (201310564122) and the Guangdong Province Science & Technology Program (2012B091100433).

REFERENCES

[1] Huang, Y.G., Pang, L.X. 2013. Synthesis and properties of UV-curable tung oil based resins via modification of Diels-Alder reaction, nonisocyanate polyurethane and acrylates. Prog Org Coat. 76: 654–661.

[2] Rengasamy, R., Senthil, K., Mannari, V. 2013. Development of soy-based UV-curable acrylate oligomers and study of their film properties. Prog Org Coat 76: 78–85.

[3] Wang, C.S., Chen, X.Y., Chen, J.Q. 2011. Synthesis and Characterization of Novel Polyurethane Acrylates Based on Soy Polyols. J Appl Polym Sci 122: 2449–2455.

[4] Patel, M.M., Patel, A.I., Patel, H.B. 2009. Parmar. Iran Polym J 18: 903–915.

[5] Chattopadhyay, D.K., Raju, K. 2007. Prog Polym. Sci 32: 352–418.

[6] Schwalm, R. 2007. UV Coatings: Basics. Recent Developments and New Applications 2: 205–206.

[7] Meier, W.U. 2007. Polyurethanes: Coatings. Adhesives and Sealants.

[8] Nebioglu, A., Soucek, M.D. 2006. J. Coat Technol Res 3: 61–62.

[9] Eren, T., Colak, S., Kusefoglu, S.H. 2006. Preparation of phenol-urea-formaldehyde copolymer adhesives under heterogeneous catalysis. J Appl Polym. Sci 100: 2947–2955.

[10] Shen, L., Wang, L., Liu, T., He, C. 2006. Nanoindentation and morphological studies of epoxy nanocomposites. Macromol Mater Eng 291: 1358–1366.

[11] Ren, Y.B., Pan, H.M., Li, L.S. 2005. Synthesis of polyurethane acrylates by hydrogenated castor oil and dimer-based polyester diol and study on pressure-sensitive adhesive. J Appl Polym Sci 98: 1814–1821.

[12] Lee, C., Iyer, N.P., Han, H. 2004. Polym Phys. J Polym Sci 42: 2202–2214.

[13] Allen, N.S., Edward, L., Edward, M.H. 1989. Photopolymerisation and Photoimaging. Science and Technology 2: 361–363.

[14] Sacks, M. 1982. UV Curing of Coatings. Printing Inks and Adhesives 2: 152–154.

[15] Dieterich, D. 1981. Prog Org Coat 9: 281–340.

[16] Ha, C., Jung, S., Kim, E., Kim, W., Lee, S., Cho, J. 1996 J Appl Polym Sci 25, 1011–1021.

[17] Pappas, S.P. 1978. UV Curing. Science and Technology 2: 307–387.

Material Science and Engineering – Chen (Ed.)
© 2016 Taylor & Francis Group, London, ISBN 978-1-138-02936-1

Fluorescence reabsorption analysis on the laser cooling of Er³⁺-doped fluoroindate glass

Y.H. Jia
Science College, Shanghai Second Polytechnic University, Shanghai, P.R. China

B. Zhong & J.P. Yin
State Key Laboratory of Precision Spectroscopy, East China Normal University, Shanghai, P.R. China

ABSTRACT: Er³⁺-doped CNBZn (CdF_2-$CdCl_2$-NaF-BaF_2-$BaCl_2$-ZnF_2) glass has been investigated as one of the hot materials with the potential for laser cooling. Compared with other rare-earth doping ions such as Yb^{3+} and Tm^{3+}, Er^{3+} has a better cooling prospect. To date, one of the main reasons that restrict the cooling effect is fluorescent reabsorption. In this paper, first, the mechanism of the cooling cycle of Er^{3+} is discussed. Second, using several spectral parameters of Er^{3+}, the reabsorption effects are calculated by using the stochastic model, which is a semi-analytical approach to this problem. Finally, the results show that when the average number of the absorption events increase, the range of the cooling wavelength is strongly restricted because of the redshift of the mean fluorescence wavelength, and the cooling efficiency and cooling power decrease. We find that choosing appropriate shape materials will benefit the fluorescence emission and cooling process.

Keywords: laser cooling; rare-earth ions; fluorescent reabsorption

1 INTRODUCTION

In 1929, Pringsheim suggested the possibility of cooling an object by using anti-stokes fluorescence [1]. In this process, the temperature of the material is decreased by the absorption of low-energy photons and a subsequent emission of photons of greater energy. Later on, many researchers first opposed this idea: Landau proposed that it does not obey the second laws of thermodynamics. It was not until 1995 that the first experimental evidence of laser cooling in a solid was presented: a Yb^{3+}-doped fluorozirconate glass ZBLANP was cooled to 0.3 K below room temperature [2]. Since then, several investigations have been made and great progresses have been achieved. Now, net fluorescent cooling has been observed in many kinds of materials, such as glasses [3], fibers [4], crystals [5], and semiconductors [6]. In 2006, Fernandez reported the first observation of anti-stokes fluorescence cooling in Er^{3+}-doped CNBZn glass and Er^{3+}-doped KPb_2Cl_5 crystal [7], and proved the feasibility of laser cooling in Er^{3+}-doped material, and that they are promising laser cooling material.

2 MECHANISM OF THE COOLING CYCLE

Er^{3+} can be doped with many glasses and crystals, and has several energy levels [8]. Each level in several inhomogeneous broadened levels, as shown in Figure 1, corresponds to a Stark-split manifold. For example, the $^4I_{9/2}$ level has several Stark-split levels, and the $^4I_{15/2}$ level also has several levels. Obviously, it is suitable to choose the $^4I_{15/2}$ and the $^4I_{9/2}$

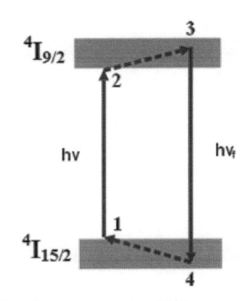

Figure 1. Energy level structure of Er^{3+} ion.

manifolds as the levels of laser cooling, in which the $^4I_{9/2}$ level is the upper level of the laser pumping, and the $^4I_{15/2}$ level is the lower level of the laser pumping. In the cooling cycle, Er^{3+} ions absorb one laser pump photon, and the corresponding frequency of the pumping laser is v; therefore, they are excited from the top of the ground-state manifold to the bottom of the excited-state manifold and the mean fluorescent frequency is v_f. Then, the excitations thermalize within the upper and lower manifolds by absorbing vibrational energy from the host. Because it satisfies the condition of anti-stokes fluorescent emission, $hv_f > hv$, cooling occurs. Compared with Yb^{3+} and Tm^{3+}, the energy level structure of Er^{3+} is very complex: the cooling cycle not only occurs in $^4I_{15/2} \rightarrow ^4I_{9/2}$, but also with transitions associated with the excited states $^4I_{11/2} \rightarrow ^4F_{5/2}$ and $^4I_{13/2} \rightarrow ^2H_{11/2}$.

3 FLUORESCENCE REABSORPTION ANALYSIS

From the above expression, since the heat is taken away by emitted fluorescence, the efficiency of fluorescent emission is very important. Actually, the fluorescent reabsorption is inevitable, and it will weaken the cooling process. In this section, the effect of fluorescent reabsorption on anti-stokes cooling in solids is calculated by using the stochastic model [9]. It is a semi-analytical approach to this problem. This model has recently been used in studies on the spectral characteristics of several fluorescent materials [10]. Using the absorption and fluorescent spectra of Er³:CNBZn as the initial data, first, the absorption coefficient α_R is obtained. For this cooling sample, the absorption coefficient is 0.317 cm⁻¹ [11]. Then, we analyze the random process of fluorescence reabsorption of Er^{3+}:CNBZn.

The procedure for calculating the reabsorption effect is based on the Monte Carlo method. Let us consider a rectangular medium with dimensions $L \times D \times D$ with a laser beam propagating through the center. The random walk process can be described as follows: first, a random number ξ is generated, and then we consider the fluorescence quanta travel a distance l corresponding to a probability for reabsorption of ξ. According to Beer's law,

$$l = \frac{-\ln \xi}{\alpha_R}. \tag{1}$$

Two more random numbers, $0 < \gamma < 1$ and $0 < \phi < 1$, are generated to ensure the position of the photon when it is reabsorbed. The new position can be calculated by

$$r(x_{new}, y_{new}, z_{new}) = r(x_{old}, y_{old}, z_{old}) + \Delta x + \Delta y + \Delta z \tag{2}$$

$$\Delta x = l \cos(2\pi\phi)\cos(2\pi\gamma)$$
$$\Delta y = l \cos(2\pi\phi)\sin(2\pi\gamma) \tag{3}$$
$$\Delta z = l \sin(2\pi\phi)$$

The decision of accepting or rejecting a random step is simply made according to whether the new position falls within the boundaries of the sample cell. If it does, we consider that the reabsorption occur. Then, we can take another step until the new position is outside the boundaries of the sample. Finally, we get the average number of the absorption events N. We can use it to analyze other phenomena in the laser cooling of the solid.

For a certain material, there is an internal quantum efficiency η_0, which is called the exciton-to-photon conversion efficiency. The quantum efficiency that we discussed here is just η_0. For a material that can be laser cooled, it should have a high quantum efficiency. When considering the fluorescent reabsorption, the quantum efficiency will decrease. It can be described as follows:

$$\eta = (\eta_0)^N \tag{4}$$

The cooling efficiency can be defined as the ratio of the cooling power to absorbed power, and by energy considerations [12]

$$\eta_{cooling} = \frac{\lambda - \lambda_{F^*}}{\lambda_{F^*}} \tag{5}$$

where λ and λ_{F^*} are the pump and the effective mean fluorescent wavelengths, respectively. λ_{F^*} can be calculated by mean fluorescent wavelengths

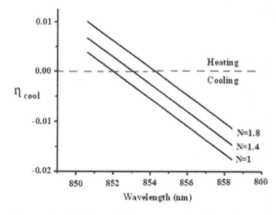

Figure 2. Cooling efficiency as a function of wavelength in different number of absorption events.

124

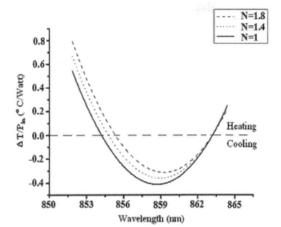

Figure 3. Temperature change as a function of wavelength indifferent number of absorption events.

divided by the quantum efficiency. We take Er^{3+}:CNBZn for example, and the results are shown in Figure 2. It shows the relationship between the cooling efficiency and the pumping wavelength at different numbers of absorption events.

Using the model of the C.W. Hoyt, the normalized change in the temperature of the sample can be expressed as [13]:

$$\frac{\Delta T}{P_{in}} = \kappa \left[\alpha_b + \alpha_r(\lambda)(1-\eta) - \alpha_r(\lambda)\eta \frac{\lambda - \lambda_f}{\lambda_f} \right] \quad (6)$$

We discuss the sensitivity of the cooling process to the average number of the absorption. As shown in Figure 3, with the increase of N, the quantum efficiency decreases, the cooling wavelength range becomes small, and the cooling effect becomes weaker.

4 CONCLUSIONS

In this article, we propose a theoretical model to analyze the cooling cycle. The complicated energy levels with several stark-split manifolds are simplified to two levels, and the mechanism of the cooling cycle is introduced. Afterwards, the effect of fluorescent reabsorption in the solid material is discussed by using the stochastic model. The average

number of the absorption events is obtained, subsequently resulting in the change of cooling efficiency and cooling power. We found that in order to weaken the fluorescence reabsorption effects of adverse effects on the laser cooling of solid materials, the volume of the refrigeration material should not be too large. On the other hand, in order to guarantee the absorption of pump light, choosing an appropriate geometry will benefit the fluorescence emission and cooling process. The material should be well fabricated, and the elongated shape of the cooling element should be more suitable for cooling.

ACKNOWLEDGMENT

This research was financially supported by the National Natural Science Foundation of China and Shanghai Construction of College Experiment Technique Team Plan.

REFERENCES

[1] P. Pringsheim, Z. Phys. 57 (1929), 739–746.
[2] R.I. Epstein, M.I. Buchwald, B.C. Edwards, T.R. Gosnell, and C.E. Mungan, Nature 377 (1995), 500–503.
[3] C.W. Hoyt, M.P. Hasselbeck, M. Sheik-Bahae, R.I. Epstein, S. Greenfield, J. Thiede, J. Distel and J. Valencia, J. Opt. Soc. A m. B 20 (2003), 1066–1074.
[4] X. Luo, M.D. Eisaman, T.R. Gosnell, Opt. Lett. 23 (1998), 639–641.
[5] A. Mendioroz, J. Fernandez, M. Voda, M. Al-Saleh, R. Balda, Opt. Lett. 27 (2002), 1525–1527.
[6] J.B. Khurgin, Phys. Rev. Lett. 98 (2007), 177401.
[7] J. Fernandez, A.J. Garcia-Adeva, R. Balda, Phys. Rev. Lett. 97 (2006), 033001.
[8] Angel J. Garcia-Adeva, Rolindes Baldaa, Joaquin Fernándeza, B, Proc. of SPIE Vol. 6461 (2007), 646102.
[9] B. Heeg, P.A. DeBarber, G. Rumbles, Appl. Opt, Vol. 44, No. 15 (2005), 3117–3124.
[10] B. Heeg and G. Rumbles, J. Appl. Phys. 93 (2003), 1966–1973.
[11] W. LozanoB, C.B. deAraujo, L.H. Acioli, Y. Messaddeq, J. Appl. Phys. 84 (1998), 2263–2267.
[12] R.I. Epstein, J.J. Brown, B.C. Edwards, A. Gibbs, J. Appl. Phys. 90 (2001), 4815–4819.
[13] C.W. Hoyt, M. Sheik-Bahae, R.I. Epstein, B.C. Edwards, and J.E. Anderson, Phys. Rev. Lett. 85 (2000), 3600–3603.

Material Science and Engineering – Chen (Ed.)
© *2016 Taylor & Francis Group, London, ISBN 978-1-138-02936-1*

The first-principle study of structures and properties of Cu$_{n+1}$, PbCu$_n$ (n = 1~8) clusters

L.T. Sun, C.P. Fu, M. Huang & C.Z. Guo

Research Institute for New Materials Technology, Chongqing University of Arts and Sciences, Yongchuan, China

ABSTRACT: In this work, we importantly studied the geometries and energy of Cu$_{n+1}$, PbCu$_n$ (n = 1~8) clusters via the first principle. The stability of the ground state of PbCu$_n$ (n = 1~8) clusters was further discussed by means of binding energy, second-order difference in energy and energy gaps, and density of states. The results show that the stability of PbCu$_n$ (n = 1~8) clusters is induced and controlled by the doping of Pb atoms, resulting in the higher structural stability of PbCu$_3$ and PbCu$_6$ clusters compared with adjacent clusters, and the maximum average binding energy of the PbCu$_3$ cluster with the highest stability. Besides, the energy gaps of Cu$_{n+1}$ and PbCu$_n$ (n = 1~8) clusters present odd–even oscillations with the increasing dimension. The chemical activity of PbCu$_n$ clusters is lower than that of Cu$_{n+1}$ clusters (n = 3, 4, 6~8), whereas the chemical activity of PbCu$_n$ clusters is higher than that of Cu$_{n+1}$ clusters (n = 1, 2, 5). The PbCu$_n$ (n = 1~8) clusters can produce a peak at the distal apart from the Fermi level on the basis of the atomic orbital hybrid of Pb and Cu, in which the peak position differences can influence the physical and chemical properties of the clusters.

Keywords: first principle; clusters; fragmentation energy; fermi energy

1   INTRODUCTION

Clusters are a new hierarchical material structure between the micro and macro levels. Because of the important potential value in research originating from its singular structure and nature, clusters have been drawing increasing attention in the fields of research for the past 20 years [1]. Due to the larger surface area volume ratio of clusters, nano-size clusters have excellent thermal, mechanical and magnetic properties [2–3]. For small size clusters, the current theoretical calculation methods are mainly based on the first principle of quantum mechanics (*ab initio*) method; despite the large calculation workloads, its calculation results are very precise. This computer simulation method can be used to study the physical and chemical properties of the clusters, such as geometric structure, energy and stability. Meanwhile, combining with the experimental observation, it is helpful to better understand the physical and chemical characteristics of clusters on the nanoscale [4].

Cu and Pb play a crucial role in the design of nanoscale electronic devices and the field of catalysis, and have a significant application value and comprehensive prospect. Many theories and experimental research on Cu and Pb clusters have been performed [5–7]. In recent years, mixed clusters with non-noble metal Cu atoms have drawn

researchers' interests [8]. Qian et al. obtained the stable structure of Cu$_{n-1}$ Au (n = 2–10) mixing non-noble metal clusters by the genetic algorithm, and calculated the static susceptibility and absorption spectra of clusters according to the first-principle method based on the static and time-dependent density functional theory, respectively [9]. Feng et al. studied the structure and stability of Cu$_n$ and Cu$_{n-1}$ Ni (n = 3–14) clusters with the density functional theory [10]. Their results have shown that the ground state of Cu$_n$ (n = 3–14) clusters was not the dense structure but the configuration was similar to the dual plane, and the doping of the Ni atom also increased the stability of Cu clusters. Some researchers further investigated the structure and energy properties of Cu$_{12}$ A (A = Fe, Co, Ni) mixed clusters with the density functional theory by using the non-compact low symmetric ground-state geometry of Cu$_{13}$ and four types of 13-atom high-symmetric (Ih, Oh, D5h, D3h) close-packed structures as the initial configuration, and replacing a doping atom in the unequivalent position. It was found that Cu$_{12}$ A (A = Fe, Co) clusters tended to the Ih-replacing geometry as the ground-state structure and replaced the central atoms to form a high-symmetric structure [11]. At present, a large number of studies have focused on transition metal-doped Cu clusters, but Pb-doped Cu clusters lacked systematic research. Therefore, in this work, we carried out the theoretical study on PbCu$_n$ (n = 1–9) clusters and gained the ground-state structure. Our results will help to better understand the formation mechanism and stability rule of small-size clusters and to search for the formation mechanism of large-size clusters.

2 CALCULATION METHODS

Structural optimization was performed by using the spin-polarized density functional theory of the DMol3 program package. During the process of calculations, the Generalized Gradient Approximation (GGA), Becke–Lee–Yang–Parr correlation correction, Double Numeric with Polarization (DNP), and the spin unrestricted approximation (SCF) were chosen to solve the Self-Consistent Field. In the process of geometric structure optimization, the displacement convergence criterion was 0.050 nm, the inter-atomic interaction convergence criterion was 0.020 Hartree/nm, and the energy convergence criterion was 2×10^{-5} Hartree. The self-consistent field convergence criterion was 10^{-6} Hartree. The validity of studying the clusters on the conditions of simulation has been verified previously [12].

For PbCu$_n$ (n = 1~8) clusters, the possible initial configuration and isomers with different components were increased with the increasing Cu atom number. In order to effectively identify the stable structure in the system, we designed all kinds of possible configurations for the given size clusters, and the structure optimization, energy and frequency property calculation were carried out in different spin multiplicities, respectively. For designing the cluster initial configurations, we referred to the copper clusters geometry of previous reports. The atomic space of the initial configuration referred to the atomic space of the pure copper block. After giving the initial configuration, Pb atoms in different positions were optimized with the structure searching method by substitution and hood, and then one original initial configuration became several geometric configurations. Although the workload of research increased, the reliability of the results was ensured. Finally, the lowest energy and no imaginary frequency structure was confirmed as the ground state structure of the clusters. In order to understand the effects of doping the Pb atom into Cu$_n$ clusters on the structural properties, we calculated the ground-state structures of optimized Cu$_n$ (n = 2~9) clusters from the literature [7].

For the purpose of demarcating the calculation method used, we calculated the bond length, vibration frequency and average binding energy and other physical quantities of the Cu$_2$ dimer under the same conditions. The results showed that the bond length of low-energy Cu$_2$ clusters was 0.2278 nm, and binding energy was 1.96 eV, which are in good agreement with the theoretical calculation values (bond length 0.2257 nm, binding energy 2.02 eV) [13] and experimental results (bond length 0.2219 nm, binding energy 2.01 ± 0.08 ev) [14–15].

3 RESULTS AND DISCUSSION

3.1 The stable structure of Cu$_{n+1}$, PbCu$_n$ (n = 1~8) clusters

With total energy as the criterion of judgment, the energy is lower and the structure is more stable. Using the binding energy as the auxiliary reference, the most stable configuration of PbCu$_n$ (n = 1~8) clusters is determined, and its ground state structure is shown in Figure 1. PbCu$_n$ (n = 1~8) ground-state clusters, with the increasing number of the Pb atom, progressively transform from one to two dimensions and then to three dimensions, and become more and more complicated. PbCu is a linear structure, in which the Pb atom and the Cu atom lie in both ends of the linear structure, resulting from the substitution of the Cu atom on the basis of the Cu$_2$ cluster. When increasing one Cu atom, the PbCu$_2$ cluster becomes an isosceles triangle structure of symmetrical C$_{2v}$, which can be regarded as the result of one Cu atom being adsorbed onto the PbCu cluster configuration or one Pb atom replacing the top site of Cu$_3$ clusters. The waistline and hemline length of the PbCu$_2$ cluster structure are 2.636 nm and 2.668 nm, and the vertex angle and the base angle are 60.795° and 59.603°, respectively. The PbCu$_3$ cluster with the symmetry Cs is a planarized structure similar to an abnormal triangle with one hemline out of the vertex. Unlike the PbCu$_3$ structure, PbCu$_4$ is a cap-shape spatial three-dimensional conical structure, where the Pb atom is adsorbed onto the planarized structure consisting of four Cu

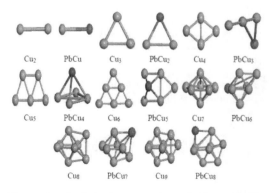

Figure 1. Optimized geometries of Cu$_{n+1}$, PbCu$_n$ (n = 1~8) clusters.

atoms. The $PbCu_5$ cluster with the symmetry Cs is also a spatial three-dimensional structure, where the Pb atom lies in the middle of half round of 5 Cu atoms. Compared with the Cu_6 cluster, its structure is the result of stretching outward caused by substituting the middle atom on the bottom of the Cu_6 cluster structure. The $PbCu_6$ cluster is a three-pyramid structure, where the Pb atom lies in the middle plane of the pyramidal structure. Compared with the double-pyramid Cu_7 structure, $PbCu_6$ results from the outward oblique and upward tensile. The symmetry of more pyramids $PbCu_7$ and $PbCu_8$ clusters is C1 and Cs, $PbCu_7$ and $PbCu_8$ clusters are obtained by absorbing a Cu atom in the corner of the bottom based on $PbCu_6$ and $PbCu_7$ or stretch caused by the substitution of the Pb atom for the Cu atom from the upper end of the pyramid point in Cu_8 and Cu_9.

3.2 The stability of Cu_{n+1}, $PbCu_n$ (n = 1~8) clusters

In order to study the stability of Cu_{n+1}, $PbCu_n$ (n = 1~8) clusters, we calculated the average binding energy E_b, the splitting energy D (n, n-1) and the energy of second-order difference (Δ_{2E}) of Cu_{n+1}, $PbCu_n$ (n = 1~8) clusters. The calculation formulas are as follows:

$$E_b = [E(Pb) + nE(Cu) - E(PbCu_n)] / (n+1) \quad (1)$$

$$D(n, n-1) = E(PbCu_{n-1}) + E(Cu) - E(PbCu_n) \quad (2)$$

$$\Delta_2 E(PbCu_n) = E(PbCu_{n+1}) + E(PbCu_{n-1}) - 2E(PbCu_n) \quad (3)$$

where n in formulas (1), (2) and (3) is the Cu variety number of Cu_{n+1} and $PbCu_n$ (n = 1~8) clusters. In formula (1), E(Cu), E(Pb), E($PbCu_n$), respectively, represent the total energy of the most stable Pb, Cu, $PbCu_n$ cluster structures. The larger the average binding energy is, the more difficult the molecules are decomposed into individual particles. The average binding energy is a good physical quantity that reflects the stability of the clusters. Formulas (2) and (3) represent the change in the value of splitting energy and the energy of second-order difference of Cu_{n+1} and $PbCu_n$ (n = 1~8) clusters as Cu atom numbers change with n. The splitting energy D (n, n-1) and the energy of second-order difference Δ_{2E} can depict the stability of the clusters from another side.

Figure 2 shows the average binding energy curves of Cu_{n+1} and $PbCu_n$ (n = 1~8) clusters as a function of the total number of atoms. It is observed that, as n increases, the average binding of Cu_{n+1} and $PbCu_n$ (n = 1~8) clusters has an overall oscillation change, in which the average binding energy of $PbCu_3$ is

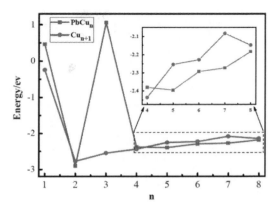

Figure 2. The binding energy per atom as a function of the total number of atoms for Cu_{n+1}, $PbCu_n$ (n = 1~8) clusters.

clearly higher than other clusters. This indicates that $PbCu_3$ has the best stability. When n = 1,3,4, the average binding energy of $PbCu_n$ clusters is greater than $PbCu_{n+1}$ clusters, which indicates that $PbCu_n$ clusters are more stable than $PbCu_{n+1}$ clusters; when n = 2,5,8, the average binding energy of $PbCu_n$ clusters are lower than $PbCu_{n+1}$ clusters, which indicates that $PbCu_n$ clusters are less stable than $PbCu_{n+1}$ clusters. It can thus be seen that doping the Pb atom into Cu_{n+1} clusters has a certain inducing and regulatory effects on the stability of the clusters.

Figure 3 shows the relation graph of the energy of second-order difference Δ_{2E} and the splitting energy D (n, n-1) of $PbCu_n$ (n = 1~8) clusters. For Δ_{2E} and D (n, n-1), the peaks show that the corresponding size clusters are stable than adjacent size clusters. As shown in Figure 3, with the size increasing, Δ_{2E} and D (n, n-1) of $PbCu_n$ (n = 1~8) clusters show consistent volatility, and when n = 3,6, they reach the maximum value, indicating that the clusters corresponding to the cluster size of peaks are more stable than clusters of adjacent size, and $PbCu_3$ is obviously higher than adjacent clusters. It shows a good agreement with the previous conclusion of average binding energy. The average binding energy of $PbCu_3$ clusters is obviously higher than the other clusters, and its stability is the best.

3.3 The energy gap of Cu_{n+1}, $PbCu_n$ (n=1~8) clusters

Figure 4 shows the transformation relation of energy gap of Cu_{n+1}, $PbCu_n$ (n = 1~8) ground-state clusters varying with total atom numbers in clusters. The gap in the solid-state physics refers to the energy gap of the semiconductor or the

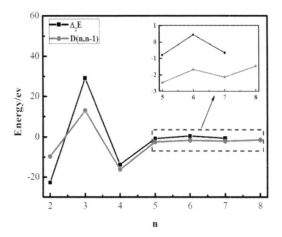

Figure 3. The second-order differences as a function of the total number of atoms for $PbCu_n$ (n = 1~8).

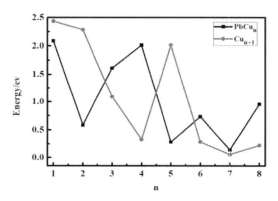

Figure 4. The HOMO-LUMO energy as a function of the total number of atoms for Cu_{n+1}, $PbCu_n$ (n = 1~8) clusters.

insulator from the top of the valence band to the bottom of the conduction band. It is found that the HOMO energy level reflects the strength of losing electrons, whereas the LUMO energy level reflects the strength of gaining electrons. While the gap is a reflection of the ability of an electron transiting from the Highest Occupied Molecular Orbit (HOMO) to the Lowest Unoccupied Molecular Orbit (LUMO), it partly reflects the ability of the molecules involved in chemical reactions.

As shown in Figure 4, the gap of Cu_{n+1}, $PbCu_n$ (n = 1~8) clusters shows an even–odd oscillation with the increasing size. When n = 3, 4, 6, 8, the gap of $PbCu_n$ is broader than Cu_{n+1}. For example, the corresponding chemical activity of $PbCu_n$ is weaker than the Cu_{n+1} cluster. By contrast, when n = 1,2,5, the gap of $PbCu_n$ is narrower than Cu_{n+1}, and its chemical activity is stronger than the Cu_{n+1} cluster.

For $PbCu_n$ (n = 1~8) clusters, the CuPb cluster with the largest gap value, HOMO-LUMO = 2.09 eV, has the worst chemical activity; the $CuPb_7$ cluster with the least gap value, HOMO-LUMO = 0.142 eV, has the best chemical activity. The whole gap of $PbCu_n$ (n = 1~8) clusters shows many properties of the semiconductor and the metal.

3.4 Analysis of state density

To find the reason why the gap even–odd oscillation of $PbCu_n$ (n = 1~8) clusters changes in Figure 4, we choose the state density distribution of $PbCu_4$ and $PbCu_5$ from $PbCu_n$ (n = 1~8) ground-state clusters for comparative purposes.

State density graphs are generated by the Gaussian expansion of discrete energy levels. We take HOMO as the Fermi level, and set this point as the zero of energy. Figure 5 shows the s, p, and d orbits and local density of state of Cu_4 of the Pb atom in the $PbCu_4$ ground-state cluster. From the graph, the leftmost peak of the d orbit of the Pb atom in the $PbCu_4$ cluster is at −0.68 Ha (1 hartree = 110.5×10^{-24} J), where Cu_4 has no peak. However, after the confluence of Cu_4 orbits with each track of Pb, the valence band bandwidth

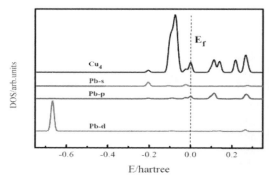

Figure 5. PDOS of the $PbCu_4$ cluster.

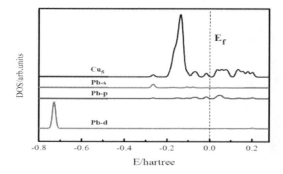

Figure 6. PDOS of the $PbCu_5$ cluster.

grows and the peak appears at −0.68 Ha. As shown in Figure 6, after the orbits of Cu_5 and Pb in $PbCu_5$ clusters mixing, $PbCu_5$ clusters will exhibit a peak value at the far left of the Fermi level, but it peaks at −0.73 Ha, which is different from the left most peak of $PbCu_4$. Therefore, the heterozygosis of the Pb, Cu atomic orbital can make the energy band of $PbCu_n$ (n = 1~8) clusters produce peak far away from the Fermi level. The difference in the peak position will induce the physical and chemical differences in the clusters to illustrate the specific reasons for the gap oscillations.

4 CONCLUSIONS

This paper investigated the geometric structures and properties of Cu_{n+1}, $PbCu_n$ (n = 1~8) clusters via the first principle. We have optimized large amounts of initial structures to obtain several conclusions as follows:

1. Doping the Pb atom into Cu_{n+1} clusters exhibits inducing and regulatory effects on the stability of the clusters, in which $PbCu_3$ and $PbCu_6$ are the most stable to adjacent clusters, and $PbCu_3$ has the maximum average binding energy and best stability.
2. $PbCu_n$ (n = 1~8) clusters show an even–odd oscillation with the increasing size. When n = 3,4,6,8, the chemical activity of $PbCu_n$ is weaker than that of the Cu_{n+1} cluster; when n = 1,2,5, it is stronger than that of the Cu_{n+1} cluster.
3. The heterozygosis of Pb and Cu atomic orbitals can make the energy band of $PbCu_n$ (n = 1~8) clusters produce peaks far away from the Fermi level. The difference in their peak position will induce the physical and chemical differences in the clusters.

ACKNOWLEDGMENTS

This research was financially supported by the Scientific Research Project of Chongqing University of Arts and Sciences (Y2013CJ26), the Frontier and Basic Research Program of Chongqing Municipality (CSTC2014 jcyjA50038), the Talent Introduction Project of Chongqing University of Arts and Sciences and Yongchuan Science & Technology Commission (Ycstc2014nc4002).

REFERENCES

[1] G.H. Wang. Cluster physics, Shanghai Scientific and Technology, Press, Shanghai, 2003.
[2] W. Chen, Preparation of Precious Metals/Monox Mesoporous Assembly with Ultrasound and Environmental Sensitivity of Au, Ag/SiO$_2$, Institute of Solid State Physics, Chinese Academy of Science, Beijing, 2001.
[3] L.D. Zhang, J.M. Mao, Nanomaterials and Nanostructures, Science Press, Beijing, 2001.
[4] J. Li, X. Li, H.J. Zhai, Au$_{20}$: a tetrahedral cluster, Science 299 (2003) 864.
[5] G. Guzman-Ramirez, F. Aguilera-Granja, J. Robles, DFT study of the fragmentation channels and electronic properties of $Cu_n v$ (v = ±1,0,2; n = 3–13) clusters, Eur. Phys. J. D 57 (2010) 335–342.
[6] D. Dai, K. Balasubramanian, Electronic states of Pb$_5$: geometries and energy separations, Chem Phys Lett. 271 (1997) 118–124.
[7] S. Wang, Z.P. Liu, J. Lu, et al, Study on size effects of Cu_n (n ≤ 20) clusters with combined density functional and genetic algorithm methods, Acta chim.sinica. 63 (2007) 1831–1835.
[8] L.H. Dong, B. Yin, L. Zhang, et al, Theoretical study of the effect of nickel and tin doping in copper clusters, Synth. Met. 162 (2012) 119–125.
[9] S. Qian, X.L. Guo, J.J. Wang, First principles study of structures, static polarizabilities and optical absorption spectra of $Cu_{n-1} Au$ (n = 2–10) clusters. Acta Phys. Sin. 62 (2013) 057803.
[10] C.J. Feng, B.Z. Mi, Theoretical Study of Nickel Doping in Copper Clusters, Chin. J. Comput. Phys. 30 (2013) 922–930.
[11] R. Tuersun, Y.Y. Jiang, A. Abulizi, et al, Density functional study of the structural and energetic properties of $Cu_{12} A$ (A = Fe, Co, Ni) clusters, Acta Phys. Sin. 30 (2013) 565–570.
[12] L.T. Sun, D.P. Shi, W.C. Dai, et al, Density functional theory study on the structures and properties of $Eu_m Si_n$ (m = 1–2, n = 1–8) clusters, Acta Phys. Sin. 30 (2013) 98–99.
[13] Y.Q. Jiang, H.M. Duan, Density functional investigations on the neutral and charged Cu_n (n = 2–12) clusters, Acta Phys. Sin. 28 (2011) 656–665.
[14] R.S. Ram, C.N. Jarman, P.E. Bernath. Fourier transform emission spectroscopy of the copper dimmer, J. Mol. Spectrosc. 156(1992) 468.
[15] M.D. Morse, Clusters of transition-metal atoms [J]. Chem. Rev. 86 (1986) 1094.

Material Science and Engineering – Chen (Ed.)
© 2016 Taylor & Francis Group, London, ISBN 978-1-138-02936-1

Research on the influence of species and diameters of the filler on the thermal conductivities of silicone potting glue

J.Y. He & Q. Wang
Beijing University of Chemical Technology, Beijing, China

X.C. Ye
Amoy-BUCT Industrial Bio-Technovation Institute, Xiamen, China

R.Y. Wang, X. Tang, L.C. He & H. He
Beijing University of Chemical Technology, Beijing, China

ABSTRACT: The influence of species and diameters of the filler on the thermal conductivities of silicone potting glue is investigated in this paper. The relationship between the species and diameters of the powder particle and the thermal conductivities of the composite material is explained on a microscopic scale. The experiment shows that the thermal conductivities of the silicone potting material are increased with the increase in the volume fraction of the filler powder. The thermal conductivities of the silicone potting material are influenced significantly by the species, diameters and different particle size complex of the thermal conductive filler when the quantity of powder filling is the same. The thermal conductivity of the potting material can be affected when the filler size is too large or too small. The complexity of the different particle size filler is conducive to improving the thermal conductivity of the potting material.

Keywords: filler powder, diameter, thermal conductivity, silicone, electronic potting glue

1   INTRODUCTION

With the rapid development of microelectronics technology, electronic components have been highly integrated and modularized. In order to ensure the normal work of electronic components against the interference of external dust, moisture, shock, vibration, chemicals and other factors, the electronic components are usually packaged or insulation potting protected [1–3]. The heat dissipation and work reliability of electronic components have been influenced largely by effective potting. The parameters of electronic components can be stabilized, and the service life of electronic components can be greatly improved [4]. Silicone polymer material has a unique climate resistance, aging resistance, excellent mechanical properties and good hydrophobicity because of its special silicon oxygen bond backbone structure [5]. Silicone potting glue has been widely used in the field of sealing and packaging of electronic appliances because of its good temperature resistance, weather resistance, corrosion resistance and electrical properties [6]. Using thermal conductive powder to fill the polymer is the main way to improve thermal conductivity; therefore, the thermal conductivity of potting

glue has been widely studied by many experts and scholars [7–14]. A variety of explanations about the influence of thermally conductive powder on the thermal conductivity of composites have been given, but there are still some shortcomings. The requirements of the integration and modular of electronic components to the stability and life of electrical appliance cannot be satisfied because of the bad thermal conductivity of the pure organic potting glue [15]. Good thermal conductivity of organic potting glue has a great significance to the development of the modern electronic industry [16]. In this paper, the influence of the material microstructure on the thermal conductivity is studied, and the influence of species and diameters of the filler on the thermal conductivity of the composite material is explained.

2   EXPERIMENTAL PROCEDURE

2.1   *Materials and equipment*

501 silicone rubber and M, N components (viscosity: 3.5 Pa·s) were obtained from Zhonghao Chenguang Research Institute.

Alumina powder (5 μm, 40 μm) was purchased from Shandong Lu Xin High-Tech Industry Co., Ltd.

Aluminum nitride powder (5 μm, 40 μm) was obtained from Qingzhou Maite Ke Chong Materials Co., Ltd.

Boron nitride powder was obtained from Qingzhou Maite Ke Chong Materials Co., Ltd.

Kneading machine (ZH-1 type) was purchased from Nanjing Enso Group Ltd.

Thermal conductivity meter (HC-110) was obtained from EKO Company.

Viscometer (DV-III) was obtained from Brookfiled Company.

2.2 Preparation of a two-component mixture

An appropriate amount of thermal conductive powder was added into a 100 portion of the M, N silicone component, and then they were placed in the kneader and kneaded for 0.5 h, respectively, to make the base materials A and B.

2.3 Sample preparation

The base materials A and B were blended, bubbles removed and cured under the condition of 40°C temperature and 50% ± 5% relative humidity. The diameter of the sample is 50 mm, and the height of the sample is 10 mm.

2.4 Performance testing

The viscosity of the base materials A and B was measured according to GB/T 2794–1995 after blending. After the samples were cured, its thermal conductivity was tested using EKO's HC-110 type thermal conductivity meter.

3 RESULTS AND DISCUSSION

3.1 The influence of the filler content on the thermal properties of potting glue

The influence of the filler content on the thermal conductivity of silicone potting glue is shown in Figure 1. It can be seen from Figure 1 that the thermal conductivity of the potting material is improved significantly with the increase in the volume fraction of the thermal conductive filler, while the thermal conductivity of the composite material is not improved significantly at the low packing volume, because the thermal conductivity of the substrate is much lower than the thermal conductivity of the filler. Most of the thermal conductive filler particles could not be contacted directly and their high thermal conductivity cannot work fully

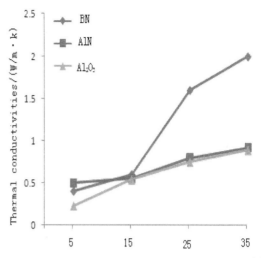

Figure 1. The influence of species of the filler on the thermal conductivities of silicone potting glue.

because of the low amount of filling. The stable thermal network is formed in the resin with the increase in the filler content, so the thermal conductivity of the composite material is increased rapidly. The heat flow can be transmitted along the network, and the high thermal conductivity of the filler is fully played. The thermal conductivity of composite materials is influenced largely by filler particles.

3.2 The influence of species and geometry of the filler on the thermal properties of potting glue

The comparison of the thermal conductivity of three types of potting glue filled with alumina, aluminum nitride and boron nitride, respectively, at the 35% volume fraction of the filler is shown in Figure 2. It can be seen from Figure 2 that the thermal conductivity of silicone potting glue filled with lamellar boron nitride is much higher than that filled with the other two fillers. Because the lamellar boron nitride particle is soft and can deform under pressure, the contact probability between the filler particles can be increased and higher bulk density can be achieved at the same filler volume. The heat flow of the polymer is promoted to dissipate faster through the filler particles because of dense packing and contact of boron nitride particles. Moreover, the intrinsic thermal conductivity of boron nitride is much higher than the other two kinds of fillers. The general trend of the thermal conductivity of the potting material filled with alumina powder, aluminum nitride powder and boron

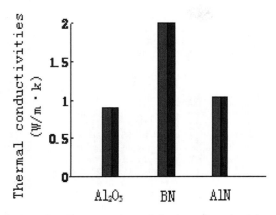

Figure 2. The comparison of the thermal conductivity of silicon potting glue filled with three kinds of inorganic filler particles.

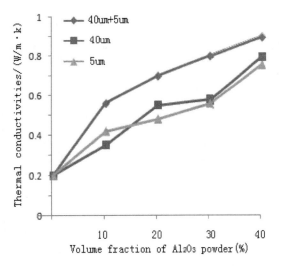

Figure 3. The relationship between the thermal conductivity of composite materials and diameters and volume fraction of the alumina powder.

nitride powder, respectively, is basically consistent with the intrinsic thermal conductivity of the filler particles.

3.3 The influence of diameters of alumina powder on the thermal conductivity of potting glue

The relationship between the alumina powder of different diameters and the thermal conductivity of the composite material is shown in Figure 3. It can be seen from Figure 3 that (1) the thermal conductivity of the composite material is increased with the increase in the filling amount of the alumina powder. (2) Under the same condition of the filler

amount, the thermal conductivity of the composite material filled with a bigger diameter particle is higher than that filled with a smaller diameter particle when the diameter of the alumina particle is bigger than 5 μm. (3) The thermal conductivity of the composite material filled with a complex diameter particle is significantly higher than that filled with a single diameter particle.

4 CONCLUSION

1. The thermal conductivity of the composite material is not influenced significantly by filler species when the content of the filler is low. The thermal conductivity of the silicone potting material is increased with the increasing volume fraction of the filler powder. The high thermal conductivity of the filler is fully played, and the thermal conductivity of composite materials is influenced largely by the filler particles.
2. The general trend of the thermal conductivity of the potting material filled with alumina powder, aluminum nitride powder and boron nitride powder, respectively, is basically consistent with the intrinsic thermal conductivity of the filler particles when the powder filling quantity is the same. The thermal conductivity of the silicone potting material is influenced significantly by the species of the filler particles, diameters of the filler particles and the filler particles' different complex sizes. The thermal conductivity of the potting material can be affected when the filler size is too large or too small. The thermal conductivity of the potting material filled with the mixture of different diameter fillers is bigger than that filled with a single diameter.
3. The thermal conductivity of the potting material is also influenced significantly by the geometry of the thermal conductive filler particles. The larger bulk density of lamellar thermal conductive filler particles can be obtained, and the contact probability between the filler particles can be increased. The heat flow of the polymer is promoted to dissipate faster through the filler particles because of dense packing and contact of thermal conductive filler particles, and the ability of thermal conductivity is increased.

REFERENCES

[1] Zhao Hongzhen, Qi Shuhua, Zhou Wenying, et al. Influence of alumina particles on the thermal properties of silicone rubber [J]. Special Rubber Products. 2007, 28(5):19–22.

[2] Chung Sungil, Im Yonggw an, Kim Hoyoun, et al. Evaluation for micro scale structures fabricated using epoxy-aluminum particle composite and it application [J]. Journal of Materials Processing Technology, 2005, 16:168–173.

[3] Liu S.L., Chen G., Yong M.S. EMC characterization and process study for electronics packaging [J]. Thin Solid Films, 2004, 462–463:454–458.

[4] Wang Ying, Zhou Chunyan. Potting material of high and low temperature impact resistance and insulation [J]. Electronic Technology. 2013, 3:110–113.

[5] Xiong Ting, Yuan Sulan, Wang Youzhi, et al. The research of two-component silicone potting glue of LED optoelectronic display device [J]. Silicone Material, 2011, 25 (2):94–97.

[6] Xing Songmin, Wang Yilu. The synthesis process and product application of silicone [M]. Beijing: Chemical Industry Press, 2000:14–618.

[7] Agari Y., Uno T. Estimation on thermal conductivities of filled polymers [J]. Journal of Applied Polymer Science, 1986, 32:5705–5712.

[8] Liu S.L., Chen G., Yong M.S. EMC characterization and process study for electronics packaging [J]. Thin Solid Films, 2004, 462–463:454–458.

[9] Pascariu G., Cronin P., Crowley D. Next generation electronics packaging utilizing flip chip technology [J]. Electronics Manufacturing Technology Symposium, 2003(16–18):423–426.

[10] Lee Geon-Woong, Park M., Kim Jun-Kyung, et al. Enhanced thermal conductivity of polymer composites filled with hybrid filler [J]. Composites (Part A): Applied Science and Manufacturing, 2006, 37(5):727–734.

[11] Chung Sun-Gil, Im Yong-Gwan, Kim Ho-Youn, et al. Evaluation for micro scale structures fabricated using epoxy-aluminum particle composite and its application [J]. Journal of Materials Processing Technology, 2005, 160(2):168–173.

[12] He Hong, Fu Ren-Li, Han Yan-Chun, et al. High thermal conductivity Si3 N4 particle filled epoxy composites with a novel structure [J]. Journal of Electronic Packaging, 2007, 129 (4):469–472.

[13] Yang Zhibin, Zhang Jianxin, Jiang Ming, et al. Condensation type: CN, 1978579 A [P]. 2004–09–15.

[14] Zhang Yinhua. Two-component condensed type silicone potting glue composition of low hardness and high toughness: CN, 101565600 A [P]. 2009–10–28.

[15] Zhao Nian, Jiang Hongwei. Preparation of high thermal conductivity and flame retardant silicone potting glue [J]. Silicone Material, 2014, (4):243–248.

[16] Qiao Hongyun, Kou Kaichang, Yan Luke, et al. The research progress of silicone potting materials [J]. Journal of Materials Science and Engineering, 2006, 24 (2):321–324.

Material Science and Engineering – Chen (Ed.)
© *2016 Taylor & Francis Group, London, ISBN 978-1-138-02936-1*

Double functional polystyrene beads

X.D. Duan, W.H. Sun, Y.H. Zhang & L.G. Sun
Key Laboratory of Chemical Engineering Process and Technology for High-Efficiency Conversion,
College of Heilongjiang Province, Heilongjiang University, Harbin, China

D.M. Zhao
Department of Food and Environment Engineering, Heilongjiang East University, Harbin, China

ABSTRACT: In this paper, double functional polystyrene beads embedding with CdSe quantum dots and magnetic particles (polystyrene composite beads) were prepared. Magnetism of the Fe_3O_4 particles was characterized by adding the magnetic fluid to an applied magnetic field. The fluorescence, diameter, sphericity of the particles are characterized by fluorescence microscope and optical microscope.

Keywords: microfluidic devices; double functional beads; preparation

1   INTRODUCTION

Microfluidic technology is a new platform in recent years to control liquid drops and prepare polymer beads [1–7]. Sun et al prepared silica colloidal crystals with microfluidic device and it was successfully applied to multiple immune analysis [8], Zhao et al prepared quantum dots which was wrapped by hydrogel and bisexual polymer particle magnetic particles with microfluidic device, these particles showed uniform spectral characteristics and good coding performance, they also can conduct magnetic separation [9], Shepherd et al fabricated polymer microspheres with uniform particle size by adding photoinitiator in the liquid from a synthetic polymers [10]. The microfluidic technology overcomed the disadvantages of traditional methods for preparation of polymer beads. In this paper, we successfully prepared double functional polystyrene beads embedding with CdSe quantum dots and magnetic particles with home-made microfluidic device.

2   EXPERIMENTAL SECTION

Chemicals. Polystyrene, ferrous chloride, ferric chloride, hydrochloric acid, oleic acid, ammonia, sodium hydroxide, polyvinyl alcohol, sodium sulfite, selenium, cadmium acetate, nitrogen, ethanol and toluene were purchased from Sinopharm Chemical Reagent Co., Ltd.

Preparation of the aqueous Fe_3O_4 magnetic particles. Firstly, 25 ml distilled water were added into a 500 ml three-necked flask with nitrogen flow in order to drain the air. After that, 2.0 g $FeCl_2.4H_2O$

and 3.1 g $FeCl_3$ were added into the flask. 0.85 ml HCl was added to make the solids dissolve quickly. After the solid ferric salt and ferrous salt were dissolved, 15.0 g NaOH was dissolved in 250 ml distilled water to gain 1.5 mol/l NaOH aqueous solution. Then, transfer NaOH aqueous solution to the 500 ml flak. After finished dripping, the reaction was complete. Finally, Fe_3O_4 was separated with the external magnetic field and washing three times.

Modification of the aqueous Fe_3O_4 magnetic particles. Firstly, 220 ml distilled water was added into the aqueous Fe_3O_4 magnetic particle with nitrogen flow. Second, raised the temperature to 80°C and added 3 ml oleic acid and 1 ml ammonia into the flak and kept flirring for 30 min. Last, centrifuged the solution and washed 3 times with distill water and ethanol.

Preparation of Cadmium selenide (CdSe) quantum dots. 250 ml deionized water were added into a 500 ml three-necked flask under nitrogen flow. 3.79 g sodium sulfite and 1.97 g selenium were added into the flask to dissolve and react 3 h. 3.5 g solid sodium hydroxide and 1.33 g cadmium acetate solid were dissolved into 50 ml deionized water respectively. 150 ml of anhydrous ethanol and 40 ml oleic acid were poured into 1000 ml conical flask under magnetic stirring. Then the NaOH solution and cadmium acetate were added into the conical flask sequentially, it can be observed the bight yellow color became pale. Finally, 50 ml selenosulfate solution were poured in, the solution changed rapidly deep with orange precipitate. The resulting supernatant was added into autoclave with with PTFE lining in an oven at 180°C for 3 h.

Preparation of polystyrene coated magnetic particles and CdSe quantum dots fluorescent beads. 10.0 g polystyrene granular particles were put into a sealed bottle with 32 ml toluene in an oven at 60°C until polystyrene particles completely dissolved. The 2% magnetic particle and 2%CdSe quantum dots were put into the polystyrene solution according to the mass fraction of polystyrene with the ultrasonic for 15 min. 3 ml polystyrene solution in an 10 ml syringe fixed on a dual channel injection pump and 40 ml 8% aqueous solution of polyvinyl alcohol in an 50 ml syringe on the same pump. Teflon pipe were connected the two syringe with a T-type pass. Set the two channels with an injection amount and the advancing speed respectively to adjust the solution bead size. A 1000 ml pear shaped bottle with 200 ml 8% polyvinyl alcohol solution in 40°C water bath was used to take over the beads flowed out from the PTFE pipe. After 24 h, the solvent was completely evaporated and the polystyrene solution beads became solid polystyrene beads.

3 RESULTS AND DISCUSSION

Fe_3O_4 particles. As shown in Figure 1, the left image is the Fe_3O_4 particles prepared by us. The concentration

is 10 g/l. After ultrasonic, it doesn't precipitate for two days. The right one showed the Fe_3O_4 particles solution under a external magnetic field.

4 CDSE QUANTUM DOTS

Under 365 nm UV light, CdSe quantum dots from the left to the right showed green, yellow, orange and red colors, and the fluorescence is very bright. As Figure 2 shows, from left to right is the reaction time at 1.5 h, 2.5 h, 4 h and 6 h respectively at 150°C, this showed that with the increase of reaction time, the size of CdSe quantum dots increased gradually.

5 THE COMPOSITE BEADS

Figure 3 is the optical microscope image of polystyrene beads which were prepared by wrapping

Figure 2. Photograph of CdSe quantum dots under 365 nm UV light.

Figure 1. The images of Fe_3O_4 particles with (right) and without (left) magnet.

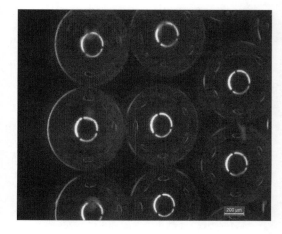

Figure 3. The microscope image of the polystyrene composite beads.

Fe_3O_4 magnetic particles and CdSe quantum dots in polystyrene. It can be inferred that the diameter of the bead is about 680 μm. In this image, polystyrene beads were with uniform particle size and irregular darker spots are distributed in the beads.

Figure 4 is the microscope images of polystyrene composite beads under dark filed and UV filed. The left one is the polystyrene composite beads with orange CdSe quantum dots under dark filed. It can be inferred that the polymer bead's diameter is about 120 μm. The bead was with good sphericity and smooth surface. The right one is the same bead under the UV filed. Quantum dot orange is excited under the irradiation of UV light. According to the bead with the same brightness everywhere, we can draw a conclusion that quantum dots dispersed in polymer uniformly.

Figure 5 is the optical microscope pictures of the fluorescent polystyrene bead which is made of wrapping in Fe_3O_4 magnetic particles and orange CdSe quantum dots. The length of the scale

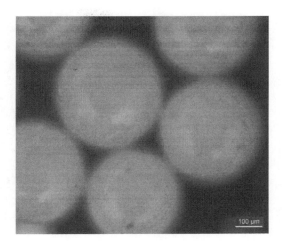

Figure 5. The fluorescence microscope photos of the polymer beads made by polystyrene wrapped orange CdSe quantum dots and magnetic particles.

is 100 μm. The polymer bead's diameter is about 350 μm. We can see Fe_3O_4 magnetic particles black reunion at the outer edge of the polymer beads. Contrast Figure 4, with the increase of the size of the polymer beads, distribution of magnetic particles to the fluorescence distribution of polymer beads' homogeneity effects decreased.

6 CONCLUSION

Double functional polystyrene beads embedding with CdSe quantum dots and magnetic particles have been fabricated by a home-made microfluidic device, the diameter of the beads could be adjusted via the parameter setting of microfluidic device.

ACKNOWLEDGMENTS

The study has been supported by China post doctor foundation (2013M531008) and Heilongjiang Provincial Department of Education (12531521).

REFERENCES

[1] C. Sun, X.W. Zhao, Y.J. Zhao, Fabrication of colloidal crystal beads by a drop-breaking technique and their application as bioassays, Small 4(2008) 592–596.
[2] L. Tao, Y.X. Yan, L. Wang, CdTe quantum dots preparation and application, Journal of Jilin University (Science Edition) 47(2009)1299–1302.

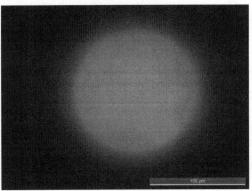

Figure 4. The microscope images of the polystyrene composite beads under dark filed (left) and UV filed.

[3] Y.F. Liu, W.F. He, Y. Sun, The thioglycolic acid stabilized CdTe quantum dots: synthesis and characterization, Journal of Zhengzhou University of Light Industry (Natural Science Edition) 25(2010)28–30.

[4] R.T. Weitz, L. Harnau, S. Rauschenbach, Polymer nanofibers via nozzle-free centrifugal spinning, Nano Letters 8(2008)1187–1191.

[5] X. Brokmann, L. Coolen, J.P. Hermier. Emission properties of single CdSe/ZnS quantum dots close to a dielectric interface, Chemical Physics 318(2005)91–98.

[6] B.G. Wang, H.C. Shum, D.A. Weitz. Fabrication of monodisperse toroidal particles by polymer solidification in microfluidics, ChemPhysChem. 10(2009)641–645.

[7] S. Jeong, J.H. Han, J.T. Jang. Transformative two-dimensional layered nanocrystals, J. Am. Chem. Soc. 133(2011)14500–14503.

[8] Y.J. Zhao, X.W. Zhao, C. Sun. Encoded silica colloidal crystal beads as supports for potential multiplex immunoassay, Analytical Chemistry 80 (2008)1598–1605.

[9] Y.J. Zhao, H.C. Shum, H.S. Chen, Microfluidic generation of multifunctional quantum dot barcode particles, J. Am. Chem. Soc. 133(2011)8790–8793.

[10] R.F. Shepherd, J.C. Conrad, S.K. Rhodes, Microfluidic assembly of homogeneous and janus colloid-filled hydrogel granules, Langmuir 22 (2006) 8618–8622.

Material Science and Engineering – Chen (Ed.)
© 2016 Taylor & Francis Group, London, ISBN 978-1-138-02936-1

Preparation of quaternized cationic waterborne polyurethane and its fixative performance on reactive dyes

M.Y. Zhang
School of Chemistry and Chemical Engineering, Wuhan Textile University, Wuhan, China

J.Q. Liu
Workstation of Wuhan Textile university (LIYUAN), Hubei, Songzi, China

H. Quan
New Textile Chemicals School-Enterprise R&D Center, Hubei Province, Wuhan, China

ABSTRACT: A set of silicon-modified cationic waterborne polyurethane reactive dyes wet rubbing fastness agent having high activity is prepared by isophorone diisocyanate, 1,4-butanediol, N-butyl diethanolamine as hard monomer, mixed polyether as soft monomer, hydroxyl alkyl polysiloxane as modified monomers. The effects that quaternary amination degree of the reactive polyurethane polymer have on the wet rubbing fastness ascend agent are studied. The results prove that quaternary ammonium salt structure make polyurethane macromolecules show excellent resistance to acid and alkali stability, however the polyurethane macromolecules of strong cationic strength are not conducive to improve the fixation textile wet rubbing fastness.

Keywords: cationic polyurethane; quaternary amination; organ silicon; reactive dyes; wet rubbing fastness

1 INTRODUCTION

Reactive dyes has always been one of the important fiber dyeing dye because of its excellent dyeing performance [1]. But reactive dyes have poor wet rubbing fastness in deep color products, which limits its application in textile and fabrics [2]. In recent years, the international market have put forward higher requirements for dyeing textile product of environmental protection and quality. While traditional polymer type of fixing agent not only has formaldehyde to release but also deteriorate the soft feel of textiles seriously. It is imminent to study a new type of green environmental protection textile auxiliaries [3]. The experiment is about cationic waterborne polyurethane modified by organic silicon monomer. Adopt the method of bulk polymerization to synthesis a strong reactivity and moderate quaternary ammonium salt structure organic silicone modified waterborne polyurethane oligomer, and evaluate the main application performance.

2 EXPERIMENT

2.1 Materials and preparation of polyurethane reagents

Isophorone diisocyanate (IPDI), glycol polytetrahydrofuran (PTMG/Mn = 1000), polyether diol (N220/Mn = 2000), polyether triol (ZC330/Mn = 3000), end hydroxyl alkyl polysiloxane (PPC/Mn = 2200), silane coupling agent (KH-550) are industrial products, N—butyl diethanolamine (NBEA), 1, 4—butanediol (BDO), bromobutane are chemical reagent; Acetic acid dibutyl tin laurate (DBTDL) is the analysis of pure reagent.

2.2 Fabric

Pure cotton fabric 13.1 Tex × 13.1 Tex, 120 root/10 cm × 100/10 cm; active red B—3 bf dyed fabric (dye dosage of 5%).

2.3 Instruments

YW-200-A automatic interface tensiometer, TGL—16G centrifuge table high speed, Nanotrac laser particle size analyzer, LFY—208 oscillating softness friction tester, Y571 N scrub fastness tester.

2.4 Preparation of polyurethane fixing agent

Under the condition of 110 °C, –0.01 MPa, put mixed polyether polyol into vacuum dehydration for 1.5–2 hours, after the system temperature cooled to 60 °C, putted in isophorone diisocyanate, catalyst, acetic acid two butyl two tin dilaurate and gradually increased temperature to 85 °C, reaction for 2.5 h, when the system temperature was 40 °C, Multiple homogeneous adding cationic chain extender N-butyl two ethanolamine and non-ion chain extender 1,4-butanediol in 1 h, and heated it up to 60 °C for 1 h, homogeneous adding silane coupling agent in 10 min, when temperature was 40 °C, reaction for 0.5 h in 50 °C, at last, add quaternary amination reagent bromide into it on 55 °C for 0.5 h, neutralized, emulsified and spared.

2.5 Fixing process

Two dip two rolling in working fluid—dehydration—100 °C dried—140 °C baking for 3 minutes.

3 PERFORMANCE TEST

3.1 Acid and alkali stability

In two tubes were taken to 5 mL emulsion, which were dropped into the 10% NaOH and HCl solution, then severe concussion. Observe the change of emulsion, and record the stable pH range.

3.2 Electrolyte stability

Take 16 ml 40 g/l promoting agent emulsion into graduated test tube, and slowly add 4 ml 0.5% MgCl2 solution into it, shake it well, and put it in 100 degree for 1 hour. After that we observe it carefully and record it.

3.3 Emulsion centrifugal stability

Take centrifugal acceleration 3000 r/min for 15 min, and observe the changes of emulsion carefully.

3.4 Determination of wet rubbing fastness

Refer to standard GB/T 8424.3-2001 "test color textile color fastness to calculate"[4].

3.5 The test of interfacial tension

Prepare 40 g/l adjuvant emulsion, and adjust the PH to 5–8, and test the interfacial tension.

3.6 Emulsion particle size and size distribution

Analyze waterborne polyurethane emulsion particle size by laser particle size analyzer.

3.7 Determination of sample color

Refer to standard GB/T 8424.3-2001 "test color textile color fastness to calculate"[5].

3.8 Flexibility test

Take specimens of 20 cm * 5 cm, use the oscillating softness tester to measure the softness, the one end of the specimen rotating rod is fixed on the instrument and natural prolapse, start the power supply to make the turntable began to turn, record the pointer scale when the other end of the fabric break away from the fixed rod, namely softness.

3.9 Fabric hydrophilic test

Dyed fabric of fixation treatment moisture regain for 24 h at room temperature, tile it on the rim of beaker; drop a drop of deionized water at the height of 1 cm and record the time that from water contact the fabric until reflective surface disappear.

4 RESULTS AND DISCUSSION

4.1 Structural design of fixing agent

Introduce silicone chain into polyurethane macromolecular which can give the cationic waterborne polyurethane excellent flexibility and moderate water repellency, at the same time, cap polyurethane macromolecules oligomer with coupling agent, reactive polyurethane macromolecules can be prepared with smaller molecular weight and narrow distribution, which is conducive to the auxiliary latex particles diffusion and adsorption to the surface of fiber.

Table 1. Structural characteristics of quaternary ammonium cationic waterborne polyurethane.

Samples	Q-1	Q-2	Q-3
Soft monomers %	65	65	65
Hard monomers %	35	35	35
Silicone chain %	23	23	23
Polyether chain %	42	42	42
Cationic chain extender %	8.0	8.0	8.0
Quaternization degree %	100	50	0

4.2 The stability of sample emulsion

We can see that electrolyte resistant and centrifugal stability of polyurethane micro emulsion are good and the weakening of quaternization degree has no influence on electrolyte resistant performance and stability of auxiliary.

4.3 Interfacial tension of samples and particle size distribution

We can see that interfacial tension of sample emulsion change a little as the quaternization degree of auxiliary decreases. This is because surface property of emulsion mainly depends on the length and content of organic silicon chain, and has little relation with the cationic strength. At the same time, as the analysis of 2.2: "hydrophilic of auxiliary comes from a higher proportion of polyether chain

Table 2. The stability of sample emulsion.

Auxiliary number	Electrolyte resistance	Centrifugal stability	Acid and alkali stability				
			pH2	pH4	pH6	pH8	pH10
Q-1	No change	No precipitation	Good	Excellent	Good	General	Poor
Q-2			Good	Excellent	Good	General	Poor
Q-3			Good	Excellent	General	General	Poor

Table 3. Interfacial tension of samples and particle size distribution.

	Test the pH value	pH5	pH6	pH7	pH8
Q-1	Interfacial tension (Dyn/cm)	48.3	47.6	46.4	47.3
	Particle size distribution (nm)	34–520	33–402	35–417	35–394
Q-2	Interfacial tension (Dyn/cm)	45.9	44.9	46.8	43.7
	Particle size distribution (nm)	38–376	38–435	35–309	37–418
Q-3	Interfacial tension (Dyn/cm)	43.6	42.3	45.1	43.7
	Particle size distribution (nm)	44–531	32–457	36–417	34–381

Table 4. Main application properties.

Number	Project	Dosage (g/L)			
		0	10	20	40
Q-1	Wet rubbing fastness/level	1–2	1–2	2	2–3
	Aberration ΔE	0	0.62	1.14	1.33
	Hand feeling	Soft	Soft	General soft	General hard
Q-2	Wet rubbing fastness/level	1–2	1–2	2	2
	Aberration ΔE	0	0.18	0.65	1.16
	Hand feeling	Soft	Soft	General soft	General hard
Q-3	Wet rubbing fastness/level	1–2	2	2–3	2–3
	Aberration ΔE	0	0.13	0.50	1.00
	Hand feeling	Soft	Soft	General soft	General hard

and tertiary amines or quaternary ammonium salt structure unit", and this auxiliary is polyurethane high molecular oligomer, its molecular weight is small. So no matter what is the degree of polyurethane macromolecular quaternary ammonium, additive molecules were hydrophilic enough, therefore there is no obvious difference between emulsion particle size and its distribution.

4.4 *Main application properties*

From Table 4, we can see as the auxiliary macromolecule quaternary amination degree gradually reduce, the wet rubbing fastness of fixation dyeing textiles has a tendency to improve; At the same time, the quaternary amination degree of fixing agent for fabric color is smaller and effect is smaller. This is because when the dosage of the tertiary amine chain extender is higher (8.0%), auxiliary macromolecules has enough electropositive, a lot can be adsorbed on the fiber surface. If continue to increase the degree of quaternary ammonium, auxiliary macromolecular may can't adsorb effectively on the fiber surface by the fertilizer larger repulsion between auxiliary molecules, resulting in it's difficult for auxiliary to form a continuous film on the fiber surface.

5 CONCLUSION

Cationic waterborne polyurethane oligomer micro emulsion has good resistance to electrolyte and centrifugal stability, with the decrease of the level agent molecular quaternary ammonium, the emulsion alkali resistance decreases. When tertiary amine type cationic chain extender is higher, as the auxiliary macromolecule quaternary amination degree gradually reduce, the wet rubbing fastness of fixation dyeing textiles has a tendency to improve. Less degree of quaternary ammonium of cationic polyurethane fixing agent has less influence on fabric color.

REFERENCES

[1] Song xinyuan. Reactive dye and its dyeing progress in recent years [J]. Journal of printing and dyeing, 2002, 02:45–49.
[2] Chen rong qi. Improve the wet rubbing fastness of reactive dyes for deep dyed fabric [J]. Journal of printing and dyeing, 2004, 07:20–22.
[3] Ding changbo, Wang xueyan, Zhao zhenhe. Reactive dye wet rubbing fastness of the dyed fabric influencing factors and measures to improve [J]. Journal of textile review, 2014, 02:60–62.
[4] The People's Republic of China national standard. GB/T3921-1997, textile soaping fastness test method [S].
[5] The People's Republic of China national standard. GB/T8424.3-2001, textile fastness test color difference calculation [S].

Material Science and Engineering – Chen (Ed.)
© 2016 Taylor & Francis Group, London, ISBN 978-1-138-02936-1

Effect of unvulcanized polar graft-modified rubbers on mechanical and thermal properties of Polypropylene

X. Li & H.T. Liu

Department of Polymer Material and Engineering, School of Material Science and Engineering, Wuhan Textile University, Wuhan, Hubei Province, China

ABSTRACT: Unvulcanized polar graft-modified rubbers, including Maleic Anhydride Grafted Ethylene Propylene Diene Monomer (MAH-g-EPDM) and 49% Methyl Methacrylate grafted Natural Rubber (MMA-g-NR, i.e. MG49 rubber), were employed to toughen Polypropylene (PP). The results showed that Charpy impact strength of the unnotched PP specimens at room temperature was improved by addition of the graft-modified rubbers. Moreover, Charpy impact strength of the composite increased with increasing content of polar graft-modified rubbers in this study, and the incorporation of MAH-g-EPDM enhanced impact strength of PP more markedly than that of MG49 did. The result of Thermo-gravimetric Analysis (TGA) indicated that the addition of MAH-g-EPDM improved the thermal stability of PP, while the incorporation of MG49 reduced it. The temperatures at the maximum degradation rates of the corresponding PP-based blends increased with the increasing content of MAH-g-EPDM, while they decreased with increasing content of MG49. This work may also provide some information for oil-resistant modification of PP due to the polar groups grafted onto the backbone of rubbers.

Keywords: graft-modified rubber; natural rubber; EPDM; polypropylene; thermal property; mechanical property

1 INTRODUCTION

Low density, low cost and easy processability of Polypropylene (PP) makes it suitable for applications, while PP has to overcome some drawbacks such as the relative low maximum service temperature and thermal oxidation. Graft-modified rubbers have been investigated as effective compatibilizers to improve the adhesion between fillers and rubber-based matrices, incorporating polar groups [1–3]. Figure 1 and Figure 2 show the schematic structures of two typical polar graft-modified rubbers, i.e. maleic anhydride grafted ethylene propylene diene monomer (MAH-g-EPDM) and methyl methacrylate grafted natural rubber (MMA-g-NR, also called MG rubber), respectively. However, to the best of our knowledge, the effect of unvulcanized graft-modified rubbers on mechanical and thermal properties of the graft-modified rubber/plastic blends was seldom reported. In this work, the blending of polystyrene (PP) with MAH-g-EPDM and 49% methyl methacrylate grafted natural rubber (MG49) was performed using a twin screw extruder. The effect of the graft-modified rubbers on mechanical and thermal properties of graft-modified rubber/PP blends was investigated.

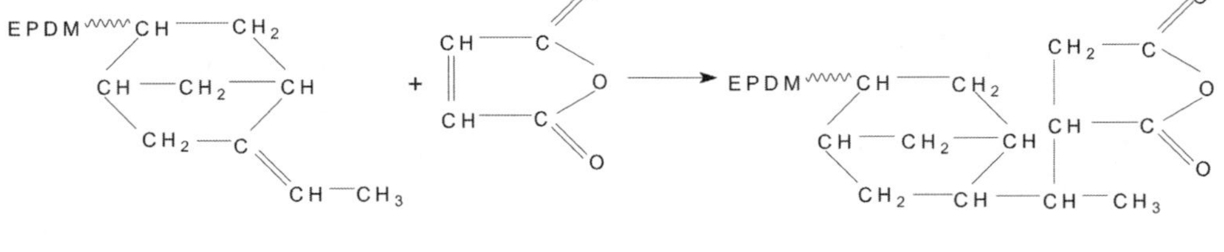

Ethylene Propylene Diene Monomer (EPDM)   Maleic Anhydride (MAH)   MAH-g-EPDM

Figure 1.   Schematic structure of MAH-g-EPDM.

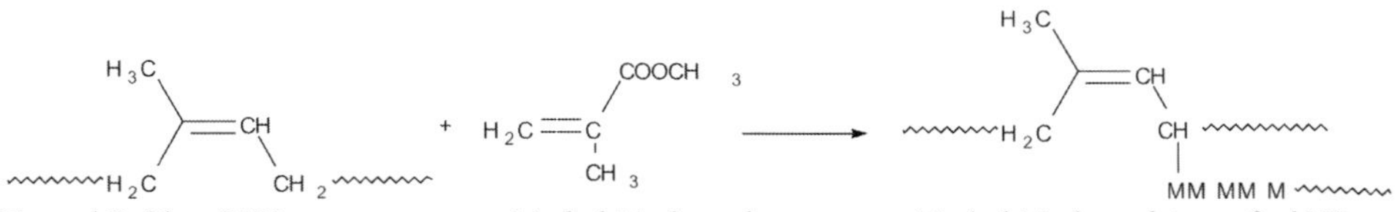

Figure 2. Schematic structure of MG rubber (M represents MMA).

2 EXPERIMENTAL

2.1 *Starting materials*

Commercially available Polypropylene (PP) named PPH-T03 was derived from China Petroleum Chemical Co., Wuhan Branch. Its melt flow rate is 3.5 g/10 min (200°C, 5.0 kg), and the standard is Q/SHPRD253-2009. MAH-g-EPDM (maleic anhydride grafted ethylene propylene diene monomer) with 1% MAH was provided by Shanghai Sunny New Technology Development Co., Ltd. MG 49 (49% methyl methacrylate grafted NR) was obtained from Zhanjiang State Farms, Guangdong, China.

2.2 *Preparation of PP/graft-modified rubber blend specimens*

Graft-modified rubbers (MAH-g-EPDM or MG49)/PP blends with weight ratio 5/95 (5%), 10/90 (10%) and 15/85 (15%) were mixed by means of a dual bolt extruder (Model SHJ-20, Nanjing Jieya Extruder Equipment Co., Ltd.) with an aspect ratio of 32. The blends were extruded as twin laces of 4 mm diameter, which were hauled into a quenching water trough prior to being pelletized. Dried pelletized blends were melt at 170 °C and moulded into the predetermined geometry 120 mm (length) × 120 mm (width) × 4 mm (thickness) by using a plate vulcanizing press machine (Model XLB-D350 × 350 × 2–0.25 MN, Huzhou Dongfang Electrical Equipment Co., Ltd.). To form impact specimens, the cooled PP/graft-modified rubber blend plates were cut into geometry 120 mm (length) × 10 mm (width) × 4 mm (thickness) by using a universal sample preparation machine (Model HY-W, Chengde Dahua Testing Equipment Co., Ltd.).

2.3 *Measurement of Charpy impact strength*

For notched PP-based specimens, Charpy impact strengths of the composites were determined according to the Chinese standard GB/T 1043-93 (idt. DIN EN ISO 179 standard) in the edgewise notched (notch type A) modes, using an XJJ-50 pendulum-type impact tester (Chengde Jinjian

Testing Instrument Co., Ltd., China). The test samples were conditioned at 23 °C and 50% relative humidity for 3 days before testing and all the tests were performed under the same conditions. For each specimen, 12 measurements were carried out and the average values were calculated. Standard deviation was within 1% as well.

2.4 *Measurement of thermal decomposition*

Thermal decomposition of the raw materials and graft-modified rubbers (MAH-g-EPDM and MG49)/PP blends was performed by thermogravimetric analyzer (TG 209 F1, NETZSCH) in nitrogen atmosphere at a scan rate of 20 °C/min from 30 °C to 600 °C.

3 RESULTS AND DISCUSSION

3.1 *Charpy impact strength*

A pendulum-type single-blow impact test in which the specimen is supported at both ends as a simple beam and broken by a falling pendulum was used. The absorbed energy per unit area (kJ/m^2), as determined by the subsequent rise of the pendulum, is a measure of Charpy impact strength. As shown in Table 1, the addition of either MAH-g-EPDM or MG49 improved the impact strength of PP. Moreover, impact strength of PP-based blend increased with increasing graft-modified rubbers content. Unexpectedly, the incorporation of MAH-g-EPDM enhanced impact strength of PP more markedly than that of MG49 did, as natural rubber was considered as a more efficient elasticizer [4–6]. High content (49%) of MMA homopolymer grafted on the main chain of NR and good compatibility between PP and MAH-g-EPDM could be responsible for these mechanical features.

3.2 *Thermal decomposition*

Polymer degradation is generally an undesirable process involving a deterioration of properties. The maximum rate of degradation for PP is normally observed at 430–470 °C in nitrogen [8–9]. As shown in Figure 3, the degradation processes of all the raw materials and blends were one-step processes. Thermograms of the raw materials, including MG49, PP and MAH-g-EPDM respectively, indicate that their thermal stability increased stepwisely following this order. The maximum degradation rates for the graft-modified rubbers/PP blends (15% MAH-g-EPDM/PP and 15% MG49/PP) occurred at the temperatures between those of graft-modified rubbers and PP. This indicated that the addition of

MAH-g-EPDM improved the thermal stability of PP, while the incorporation of MG49 reduced it. Generally, thermal stability of saturated rubbers (including EPDM) is superior to that of unsaturated rubbers (including NR and MG49) [4]. This well-known tendency could be responsible for the aforementioned thermal features.

As shown in Figure 4 and Figure 5, the TGA curves indicated that the temperatures at the

Table 1. Charpy impact strength for the PP-based blend specimens.

PP-based specimens	Charpy impact strength for notched specimens (kJ/m²)*
PP (data from measurement)	4.62
5% MAH-g-EPDM/PP	6.72
10% MAH-g-EPDM/PP	6.85
15% MAH-g-EPDM/PP	7.56
5% MG49/PP	5.78
10% MG49/PP	5.94
15% MG49/PP	6.09
PP (data from MSDS)	4–6
Neat PP [7]	5.38

*For each specimen, 12 measurements were carried out and the average values were calculated. Standard deviation was within 1% as well.

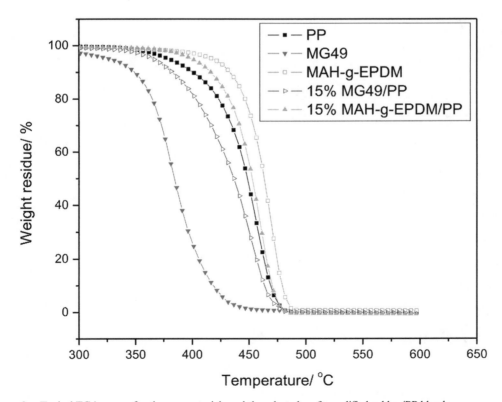

Figure 3.   Typical TGA curves for the raw materials and the selected graft-modified rubber/PP blends.

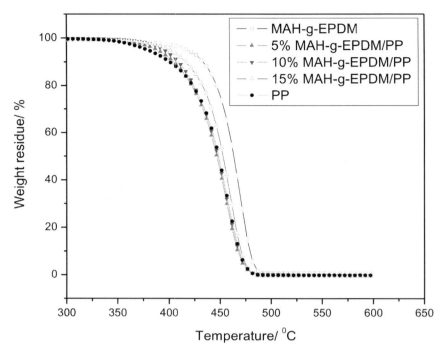

Figure 4.   Typical TGA curves for the selected polar graft-modified rubber/PP blends with different content of MAH-g-EPDM.

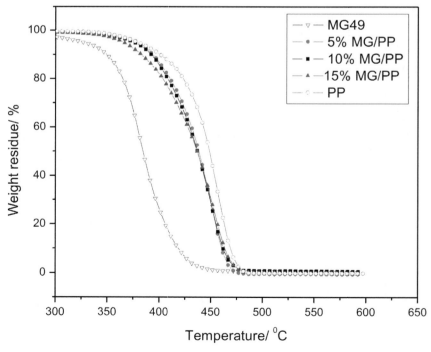

Figure 5.   Typical TGA curves for the selected polar graft-modified rubber/PP blends with different content of MG49.

maximum degradation rates of the corresponding PP-based blends increased with the increasing content of MAH-g-EPDM, while they decreased with increasing content of MG49.

4 CONCLUSIONS

Polar graft-modified rubbers including MAH-g-EPDM and MG49 toughened PP blends were successfully prepared by a melt-blending method. Enhanced impact strength and deviated thermal decomposition of PP were obtained. Charpy impact strength of the notched PP specimens at room temperature was improved by addition of the graft-modified rubbers, which increased with increasing graft-modified rubbers content. Incorporation of MAH-g-EPDM enhanced impact strength of the plastic more markedly than that of MG49 did. TGA measurement indicated that the addition of MAH-g-EPDM improved the thermal stability of PP, while the incorporation of MG49 reduced it. The temperatures at the maximum degradation rates of the corresponding PP-based blends increased with the increasing content of MAH-g-EPDM, while they decreased with increasing content of MG49. This work may also provide some information for oil-resistant modification of PP due to the polar groups grafted onto the backbone of rubbers.

ACKNOWLEDGEMENTS

This work was financially supported by the projects of Hubei Provincial Department of Education (XD2012220 and 201210495002) and the project of Wuhan Textile University (2013CXXL005).

REFERENCES

[1] X. Jiang, Y. Zhang, Y. Zhang, Study of dynamically cured PP/MAH-g-EPDM/epoxy blends. Polym. Test. 23 (2004) 259–266.
[2] G. Hu, B. Wang, X. Zhou, Effect of EPDM-MAH compatibilizer on the mechanical properties and morphology of nylon 11/PE blends. Mater. Lett. 58 (2004) 3457–3460.
[3] H. Zheng, Y. Zhang, Z. Peng, Y. Zhang, Influence of the clay modification and compatibilizer on the structure and mechanical properties of ethylene-propylene-diene rubber/montmorillonite composites. J. Appl. Polym. Sci. 92 (2004) 638–646.
[4] S.H. El-Sabbagh, Compatibility study of natural rubber and ethylene–propylene diene rubber blends. Polym. Test. 22 (2003) 93–100.
[5] H. Huang, J. Yang, X. Liu, Y. Zhang, Dynamically vulcanized ethylene propylene diene terpolymer/nylon thermoplastic elastomers. Eur. Polym. J. 38 (2002) 857–861.
[6] A.P. Mathew, S. Thomas, Izod impact behavior of natural rubber/polystyrene interpenetrating polymer networks. Mater. Lett. 50 (2001) 154–163.
[7] J. Ganster, H.P. Fink, M. Pinnow, High-tenacity man-made cellulose fibre reinforced thermoplastics-Injection moulding compounds with polypropylene and alternative matrices. Compos. A. 37 (2006) 1796–1804.
[8] A.F. Vargas, V.H. Orozco, F. Rault, S. Giraud, E. Devaux, B.L. López, Influence of fiber-like nano-fillers on the rheological, mechanical, thermal and fire properties of polypropylene: An application to multifilament yarn. Compos. A. 41 (2010) 1797–1806.
[9] E. Pārpāritā, M.T. Nistor, M.-C. Popescu, C. Vasile, TG/FT-IR/MS study on thermal decomposition of polypropylene/biomass composites. Polym. Degrad. Stabil. 109 (2014) 13–20.

Material Science and Engineering – Chen (Ed.)
© 2016 Taylor & Francis Group, London, ISBN 978-1-138-02936-1

Properties of medical grade Polypropylene and its as-spun filament

F.Q. Li, X.H. Liu & H.T. Liu
Department of Polymer Material and Engineering, School of Material Science and Engineering,
Wuhan Textile University, Wuhan, Hubei Province, China

ABSTRACT: Charpy impact strength and melt flow rate of a commercially available medical grade polypropylene for injectors and infusion bottles were measured, then the melt-spun filament of the Medical Polypropylene (MPP) was produced for potential hernia repair mesh by a laboratory mixing extruder with a take-up system. The results show that the average Charpy impact strength for the notched MPP plate specimen is 4.4 kJ/m². Melt flow rates for MPP and control PP increased with the increasing temperature. Melt flow rate of MPP is lower than that of control general PP samples. The Young's modulus and sonic velocity orientation angle of the as-spun MPP filament are calculated as 1.94 GPa and 54.2 °, respectively. From the result of this study, the relatively proper condition for the spinning of MPP is at the extrusion temperature of 210 °C, the winding speed of 2.15 m/min and the extrusion rate of 16 r/min.

Keywords: medical polypropylene; thermal property; mechanical property

1 INTRODUCTION

Polypropylene (PP) has been extensively used in the manufacture of medical devices such as tubing, catheters, dialysis units and syringes [1, 2]. With the increasing popularity of mesh use, a wide variety of synthetic meshes (including PP) are now commercially available, making the selection of the most appropriate mesh difficult [3]. However, it is necessary and interesting to make more new product and find the corresponding detailed process. In this preliminary study, we aimed to gain a better understanding of the mechanical properties of a commercially available medical grade polypropylene and its as-spun filament.

2 EXPERIMENTAL

2.1 *Starting materials*

Commercially available medical grade polypropylene (MPP, random copolymer) named GM1600E was derived from Sinopec Shanghai Petrochemical Company Limited. Control general PP samples (model Jin'ao and Luoyang) were also obtained from Sinopec.

2.2 *Preparation of MPP plate specimens*

To form the notched impact specimens, dried MPP granules were melt at 230 °C and injected directly into the predetermined geometry 80 mm (length) × 10 mm (width) × 4 mm (thickness) by using an injection molding machine (Model HBL-1300, Ningbo Haibo Machinery Manufacture Co. Ltd., China).

2.3 *Melt spinning process and as-spun filament of MPP*

To obtain the as-spun filament of Medical Polypropylene (MPP), a laboratory mixing extruder with a take-up system (Dynisco, Franklin MA, USA) was employed. Drawing the MPP melt from the laboratory mixing extruder into fibers, the take-up system pulls the fibers into smaller diameters, wrapping them around a spindle. The winding speed was controlled at 2.15 m/min to match the extrusion rate (15–20 r/min) and produce the desired fiber diameter (0.1–0.2 mm) at the varying extrusion temperatures of 180 °C, 190 °C, 200 °C, and 210 °C, respectively. Figure 1 shows the starting medical polypropylene (left) and its as-spun filament (right).

2.4 *Property measurements*

2.4.1 *Charpy impact strength of MPP plate*
For notched MPP plate specimens, Charpy impact strengths of MPP were determined according to the Chinese standard GB/T 1043-93 (idt. DIN EN ISO 179 standard) in the edgewise notched (notch type A) modes, using an XJJ-50 pendulum-type impact tester (Chengde Jinjian Testing Instrument Co. Ltd., China) [4]. The test samples were conditioned at 23 °C and 50% relative humidity

Figure 1. Starting medical polypropylene (left) and its as-spun filament (right).

for 3 days before testing and all the tests were performed under the same conditions. For each specimen, 12 measurements were carried out and the average values were calculated. Standard deviation was within 1% as well.

2.4.2 *Melt flow rate of MPP*

With a steady load of 2160 g, the Melt Flow Rate (MFR) of MPP and control PP samples (model jinao and luoyang, also from Sinopec) was measured using a melt indexer (model XNR-400 AM, Chengde Dahua Testing Instrument Co. Ltd., China), according to Chinese standard GB/T 3682–2000 at 210 °C, 220 °C, 230 °C, and 240 °C, respectively. The inner diameter of hole of the melt flow is 2.095 ± 0.005 mm. For each specimen, 5 measurements were carried out and the average values were calculated. Standard deviation was within 1% as well.

2.4.3 *Young's modulus and sonic velocity*
orientation angle of as-spun MPP filament
A Young's modulus and sonic velocity orientation angle tester for fibers (model SCY-III, Shanghai Donghua Kaili Chemical Fiber High-Tech Co. Ltd., China) was employed.

3 RESULTS AND DISCUSSION

3.1 *Charpy impact strength of MPP plate*

A pendulum-type single-blow impact test in which the specimen is supported at both ends as a simple beam and broken by a falling pendulum was used. The absorbed energy per unit area (kJ/m²), as determined by the subsequent rise of the pendulum, is a measure of Charpy impact strength. The results show that the average Charpy impact strength for the notched MPP plate specimen is 4.4 kJ/m².

Table 1. Melt flow rates for MPP and control PP samples (g/10 min).

PP samples	210 °C	220 °C	230 °C	240 °C
MPP	10.560	12.774	15.030	18.318
PP (Jin'ao)	12.834	14.631	19.843	22.584
PP (Luoyang)	18.498	24.516	30.120	34.728

3.2 *Melt flow rate of MPP*

Melt flow rates for MPP and control PP samples are shown in Table 1 and Figure 2, indicating that the melt flow rates of all PP samples increased with the increasing temperature. Melt flow rate of MPP is lower than that of control general PP samples.

3.3 *Young's modulus and sonic velocity*
orientation angle of as-spun MPP filament

The Young's modulus and sonic velocity orientation angle tester for MPP fiber were calculated by the following formula:

$$\overline{COS^2\theta} = 1 - \frac{2}{3}\left(\frac{C_u}{C}\right)^2,$$

where θ is the sonic velocity orientation angle (°) for MPP fiber, the sonic velocity in completely amorphous PP, i.e. Cu (PP) = 1.45 km/s, and C is the sonic velocity (km/s) tested in MPP fiber.

$$E = \rho \cdot C^2,$$

where E is the Young's modulus (MPa) for MPP fiber, ρ is the density (910 kg/m³) for MPP fiber, and C is the sonic velocity (km/s) tested in MPP fiber.

The sonic velocity tested in MPP fiber, i.e. C = 1.46 km/s, and the Young's modulus (E) and

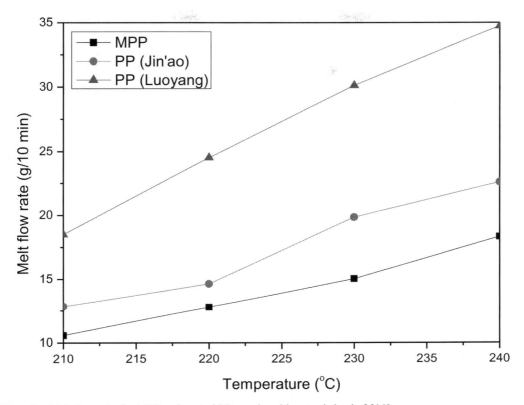

Figure 2.   Melt flow rates for MPP and control PP samples with a steady load of 2160 g.

sonic velocity orientation angle (θ) of the as-spun MPP filament are calculated as 1.94 GPa and 54.2 °, respectively. The value of Young's modulus (E) is consistent with the data of polypropylene reported [5–6], and the θ value is very close to that of completely amorphous PP (54.7 °). This may be due to the as-spun MPP filament was produced without enough high-speed extension. From the result of this study, the relatively proper condition for the spinning of MPP is at the extrusion temperature of 210 °C, the winding speed of 2.15 m/min and the extrusion rate of 16 r/min.

4   CONCLUSIONS

In this paper, Charpy impact strength and melt flow rate of a commercially available medical grade polypropylene for injectors and infusion bottles were measured. The melt-spun filament of the Medical Polypropylene (MPP) was produced for potential hernia repair material by a laboratory mixing extruder with a take-up system. The results show that the average Charpy impact strength for the notched MPP plate specimen is 4.4 kJ/m². Melt flow rates for MPP and control PP increased with

the increasing temperature. Melt flow rate of MPP was lower than that of control general PP samples. The Young's modulus and sonic velocity orientation angle of the as-spun MPP filament are calculated as 1.94 GPa and 54.2 °, respectively. From the result of this study, the relatively proper condition for the spinning of MPP is at the extrusion temperature of 210 °C, the winding speed of 2.15 m/min and the extrusion rate of 16 r/min.

ACKNOWLEDGEMENTS

This work was financially supported by the projects of Hubei Provincial Department of Education (XD2012220 and 201210495002) and the project of Wuhan Textile University (2013CXXL005).

REFERENCES

[1] M. Fuzail, D.J.T. Hill, J. Anwar, M.S. Jahan, L. Rintoul, Effectiveness of DOP mobilizer on the radiolysis of a semi-crystalline ethylene-propylene copolymer, Nulear. Instru. Math. Phys. Res. B 265 (2007) 285–289.

[2] J.L. Aguayo-Albasini, A. Moreno-Egea, J.A. Torralba-Martínez, The labyrinth of composite prostheses in ventral hernias, Cirugía Española (Eng. Ed.) 86 (2009) 139–146.

[3] X. Li, J.A. Kruger, J.W.Y. Jor, V. Wong, H.P. Dietz, M.P. Nash, P.M.F. Nielsen, Characterizing the ex vivo mechanical properties of synthetic polypropylene surgical mesh, J. Mech. Behav. Biomed. Mater. 37 (2014) 48–55.

[4] H. Liu, D. Zuo, H. Liu, L. Li, J. Li, W. Xu, Enhanced impact strength and deviated thermal decomposition of PP and PS toughened with graft-modified rubbers., e-Polym. 134 (2010) 1–5.

[5] P. Mareri, S. Bastide, N. Binda, A. Crespy, Mechanical behaviour of polypropylene composites containing fine mineral filler: Effect of filler surface treatment, Compos. Sci. Technol. 58 (1998) 747–752.

[6] J.O. Iroh, J.P. Berry, Mechanical properties of nucleated polypropylene and short glass fiber-polypropylene composites, Eur. Polym. J. 32 (1996) 1425–1429.

Material Science and Engineering – Chen (Ed.)
© 2016 Taylor & Francis Group, London, ISBN 978-1-138-02936-1

Phenolic Foam modified with butadiene-acrylonitrile rubber as a toughening agent

M. Gao, Z.H. Chai, Y.H. Wang & C.B. Guo
School of Environmental Engineering, North China University of Science and Technology, Yanjiao, Beijing, China

ABSTRACT: Butadiene-Acrylonitrile Rubber (NBR) was used as a toughening agent and mixed with the phenolic resin to prepare the foam. The effects of butadiene-acrylonitrile rubber on the mechanical and flame retardant properties of the phenolic foam were studied by compressive strength, tensile strength, heat stability, Scanning Electron Microscopy (SEM), UL-94 and Limited Oxygen Index (LOI). The apparent density and SEM results showed that the addition of NBR can decrease the apparent density of the phenolic foam. The compressive impact test results showed that the incorporation of NBR into PF can dramatically improve the compressive strength and impact strength, indicating the excellent toughening effect of NBR. The LOI of NBR-modified phenolic foams showed a high value, and the UL-94 results showed that all samples can pass the V-0 rating, indicating that the modified foams still had a good flame retardancy. The thermal properties of the foams were investigated by thermogravimetric analysis.

Keywords: butadiene-acrylonitrile rubber; phenolic foams; flame retardance; flammability

1 INTRODUCTION

Phenolic Foam (PF) has aroused a great deal of interest in recent years due to its low thermal conductivity, excellent fire-proof performance, and low generation of toxic gas during combustion (Lei et al. 2010a,b). Replacing polyurethane foams (Sarier & Onder 2008) and polystyrene foams (Ozkan & Onan 2011), which have high flammability, with phenolic foam for building thermal insulation materials is the development trend. However, the largest weakness of phenolic foams is its brittleness and powdering, which restricts its wide applications (Rangari et al. 2007). Therefore, studying the toughening technology of phenolic foams is particularly necessary. Over the past few decades, different approaches have been developed to toughen phenolic foams (Wei et al. 2012), and these fall into three categories: chemical modification, inert fillers, and fiber reinforcement. Because inert fillers produce much denser and heavier foam, this approach has gradually lost its significance.

In this work, butadiene-acrylonitrile rubber (NBR) was used as a toughening agent and mixed with the phenolic resin to prepare the foam to modify its brittleness.

2 EXPERIMENTAL PROCEDURE

2.1 Materials

Butadiene-acrylonitrile rubber (NBR; molecular weight 1000 g mol^{-1}), *n*-pentane, and *p*-toluic acid was supplied by Tianjin Yongda Chemical Reagent Co., Ltd (Tianjin, China). Phenol, paraformaldehyde, phosphoric acid, and polysorbate 80 were supplied by Tianjin Fuchen Chemical Reagent Factory (Tianjin, China).

2.2 Measurements and characterization

The morphology of foam specimens were observed with a KYKY2800B Scanning Electron Microscopy (SEM) using 15 kV and 60 mA of electric current. Thermogravimetric (TG) analysis was carried out on a HCT-2 thermal analyzer (Beijing Hengjiu Scientific Instrument Factory) under a dynamic nitrogen (dried) atmosphere at a heating rate of 10°C min^{-1}. The vertical burning test was conducted by using a CZF-II horizontal and vertical burning tester (Jiang Ning Analysis Instrument Company, China). The specimens used were 130 × 12.7 × 3 mm^3 according to the UL94 test ASTM D3801 standard. LOI tests were performed at room temperature as per ASTM D2863–97 using a Stanton Redcraft FTA unit. The specimen size for the LOI measurement was 120 × 12 × 12 mm^3 on a JF-3 LOI apparatus. The compressive properties were tested with a WSM-20 KB universal testing machine (Changchun, China) according to GB/T 8813–2008. The impact strength was evaluated with a JJ-20 mnemonic impact tester (XJ-50D). At least three samples were tested to obtain the average values.

Table 1. The composition and apparent density of phenolic foam samples.

No.	PR (g)	FA (g)	CA (g)	FS (g)	NBR (g)	AD (kg m^{-3})
1	20	1	2	1	0	53.7
2	20	1	2	1	1	50.3
3	20	1	2	1	2	47.6
4	20	1	2	1	2.5	46.7
5	20	1	2	1	3	46.4

2.3 Preparation of modified Phenolic Foam samples

A certain percentage of Phenolic Resin (PR), and NBR as a toughening agent was added into a 500-mL plastic beaker at room temperature, and then stirred with a high-speed mechanical mixer for about 30 s. The mixture was then mixed with n-pentane as the Foaming Agent (FA), polysorbate 80 as the Foam Stabilizer (FS) and Curing Agent (CA) prepared with pentaerythritol, phosphoric acid, melamine, dicyandiamide, and formaldehyde as the reference (Gao & Yang, 2009), and stirred at a high speed for 30 s. The resulting viscous mixture was poured into a foaming mold quickly, and cured at 80°C for 2 h. The sample was cut precisely and used for the fire and mechanical testing. The composition and Apparent Density (AD) of phenolic foams are listed in Table 1. Other samples were prepared by following the same procedure.

3 RESULTS AND DISCUSSION

3.1 Apparent density and microstructure of phenolic foams

As is well known, lightweight and good mechanical strength phenolic foam is the ideal foam material in the field of exterior wall insulation material. So, it is important to enhance the physical properties of phenolic foam by modification. As listed in Table 1, the apparent density of phenolic foam samples decreases after the addition of NBR, which is favorable. The microstructure of the phenolic foams was observed by SEM (Figures 1–2). Figure 1 shows the microstructure of the pure phenolic foam, in which the foam cells are not very uniform, and there are many fragments attached to the cell walls. After addition of NBR to the phenolic foam, the shape of foam cells is ellipsoid-like, and the size of the cells is gradually reduced, as shown in Figure 2.

3.2 Compressive strength

The results of compressive strength measurements are shown in Figure 2. The curve shows that the

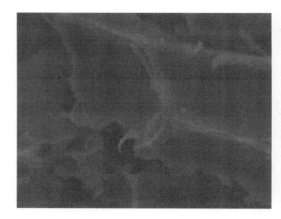

Figure 1. SEM images (×1000) of the pure phenolic foam.

Figure 2. SEM (×1000) images of phenolic foams with 2 g NBR.

compressive strength of the phenolic foam modified with NBR significantly increases by 183% (from 0.12 to 0.37 MPa) compared with that of the neat phenolic foam. This significant improvement may be attributed to the flexible chain of NBR.

3.3 Impact strength

To further explain and support the toughening efficiency of the toughening agent, the impact properties of the phenolic foam samples are investigated. The results are shown in Figure 4. A gradually increasing impact strength trend of the modified phenolic foam samples can be seen obviously with the increasing NBR content, indicating that the introduction of NBR can assuredly improve the shock resistance of the phenolic foam, and further showing the high toughening efficiency of NBR.

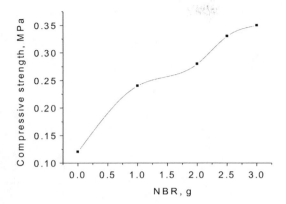

Figure 3.   Compressive strength of PF systems.

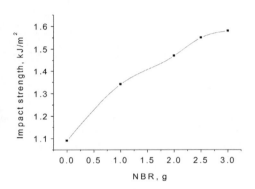

Figure 4.   Impact strength of PF systems.

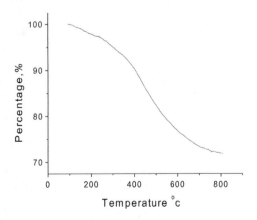

Figure 5.   TG curves of PF foams.

3.4   *Thermal stability*

Figures 5–8 show the TG curves of the pure PF and phenolic foams with various contents of NBR.

From the DG curve, it can be seen that the degradation below 300°C occurs mainly due to

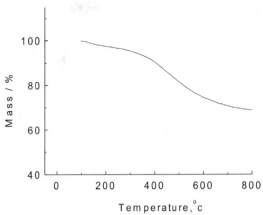

Figure 6.   TG curve of 1.0 g NBR/PF foams.

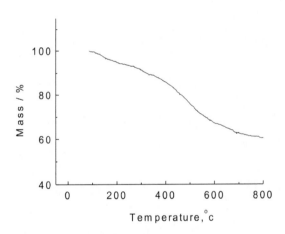

Figure 7.   TG curve of 2.0 g NBR/PF foams.

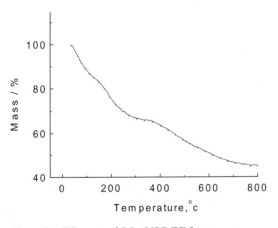

Figure 8.   TG curve of 3.0 g NBR/PF foams.

the evaporation of water, the dehydration of further curing and the degradation of phenolic resin (Streckova, 2012). All the modified phenolic foams present a similar degradation behavior to that of the neat PF, but the main difference is the degradation temperature. Compared with the neat PF, the modified phenolic foams with 1 g NBR have slightly higher degradation temperatures and char yield, while for the modified phenolic foams with 2.0 and 3.0 g NBR, the degradation temperatures and char yield are shifted to a lower temperature. These results indicate that the slight incorporation of NBR can increase the thermal stability of the phenolic foam.

3.5 Flame-retardant behavior

UL-94 vertical burning test and Limiting Oxygen Index (LOI) were used to evaluate the fire-resistant behavior of the PF and PF/NBR systems, as shown in Table 2 and Figure 9.

It is clear that all the samples can pass the V-0 rating, indicating that the modified foams still have a good flame retardancy. The LOI value of the pure PF is 46.8%. The phenolic foam sample modified by 1 g NBR content exhibits the high

Table 2. Flame-retardant properties of EP samples.

Sample no.	PR (g)	NBR (g)	LOI (%)	UL 94 rating
1	20	0	46.8	V-0
2	20	1	43.7	V-0
3	20	2	41.8	V-0
4	20	2.5	38.9	V-0
5	20	3	38.0	V-0

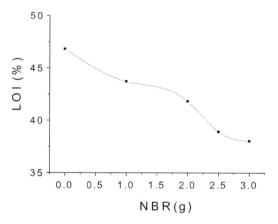

Figure 9. LOI curves of foams with different NBR content.

LOI value of 43.7%. When the content of NBR is 3 g, the LOI values decrease slightly to 38.0%, respectively. Although the NBR is flammable, the PF with a low NBR content still retains its excellent flame-retardant property.

4 CONCLUSIONS

Butadiene-acrylonitrile rubber (NBR) was mixed with the phenolic resin as a toughening agent to prepare a modified foam, which has a lower apparent density. The incorporation of NBR into the PF dramatically improved the compressive strength and the impact strength. The phenolic foam modified by 1–3 g NBR content can pass the V-0 rating, with the high LOI value of 38.0–43.7%, which still retains its excellent flame-retardant property. The thermal stability of the phenolic foam modified with 1 g NBR is increased, while for the modified phenolic foams with 2.5 g NBR, the degradation temperatures and char yield are decreased.

ACKNOWLEDGMENT

This work was supported by the fundamental research funds for the Central Universities (3142013102).

REFERENCES

[1] Lei, S. Guo, Q. Shi, J. & Liu, L. 2010a. Preparation of phenolic-based carbon foam with controllable pore structure and high compressive strength. Carbon, 48, 2644–2646.
[2] Lei, S. Guo, Q. & Zhang, D.Q. 2010b. Preparation and properties of the phenolic foams with controllable nanometer pore structure. J. Appl. Polym. Sci. 117(6), 3545–3550.
[3] Sarier, N. & Onder, E. 2008. Thermal insulation capability of PEG-containing polyurethane foams. Thermochim. Acta 475, 15–21.
[4] Ozkan, D.B. & Onan, C. 2011. Optimization of insulation thickness for different glazing areas in buildings for various climatic regions in Turkey. Appl. Energ. 88(4), 1331–1342.
[5] Rangari, V.K. Hassan, T.A. Zhou, Y.X. & Mahfuz, H. 2007. Cloisite clay-infused phenolic foam nanocomposites. J. Appl. Polym. Sci. 103(1), 308–314.
[6] Wei, D. Li, D. Zhang, L. Zhao, Z. & Ao, Y. 2012. Study on phenolic resin foam modified by montmorillonite and carbon fibers. Procedia Engineering, 27, 374–383.
[7] Gao, M. & Yang, S.S. A novel intumescent flame-retardant epoxy resins system. 2010. J. Appl. Polymer Sci. 115(4), 2346–2351.
[8] Streckova, M. Sopcak, T. Medvecky, L. & Bures, R.J. 2012. Preparation, chemical and mechanical properties of microcomposite materials based on Fe powder and phenol-formaldehyde resin. Chem. Eng. J. 2012, 180, 343–353.

Material Science and Engineering – Chen (Ed.)
© 2016 Taylor & Francis Group, London, ISBN 978-1-138-02936-1

Study on the main affecting factors of the fire-retardant retention on the bamboo bundles of a fire-retardant bamboo scrimber

C.G. Du, J.G. Wei & C.D. Jin
School of Engineering, Zhejiang Agriculture and Forestry University, Linan, Zhejiang, China

ABSTRACT: This paper presents a study on the main affecting factors of the fire-retardant retention on bamboo bundles as the basic unit of a fire-retardant bamboo scrimber. The results showed that when the bamboo bundles impregnated fire-retardant treatments, the fire-retardant retention on the thick bamboo bundles was more than that on the thin bundles, and the composite fire retardant was more than that by the single fire retardant; in addition, the retention of bamboo bundles increased quickly with increasing impregnated time and fire-retardant concentration. The better impregnated time was 120 minutes, and the fire-retardant concentration should not exceed 30%.

Keywords: bamboo scrimber; fire retarding; bamboo bundles; fire-retardant retention; affecting factors

1 INTRODUCTION

In recent years, the frequency, scale and economic losses caused by fire accidents in public places of China present an increasing trend. The large amount of combustible materials for decoration in the high-rise residential buildings is one of the major reasons that the fire spreads quickly [1–3]. At present, the bamboo flooring, bamboo sliced veneer, and other bamboo materials have been increasingly applied in building and decoration fields, and it tends to expand continually. However, the rapid growth of bamboo consumption will inevitably increase the fire occurrence rates because bamboo is a combustible material. Therefore, it is very significant to explore the technology of fire-retardant treatment for bamboo-based materials.

At present, there are few related research reports about bamboo material fire-retardant treatment. Bamboo scrimber, as a new bamboo material, has rapidly developed and has become the mainstream varieties of bamboo industry in nearly 10 years [4]. In addition, it has been widely used in the field of construction and decoration. So, it is very necessary to study the fire-retardant treatment of bamboo scrimber. The common methods of bamboo material fire-retardant treatment include the addition of fire retardant [5]. The treatment results are often represented by the fire-retardant retention as the index, which is more higher, and the inflaming retarding properties are better, and vice versa. There are many complex factors affecting the retention; in fact, it is impossible to tightly control each factor except for the most important factors. As the basic unit of bamboo scrimber, the bundle

has a decisive impact on the manufacturing technology and comprehensive properties of bamboo scrimber. Therefore, the main affecting factors of the fire-retardant retention on the bamboo bundles of the bamboo scrimber are studied, which plays a very vital role in the manufacturing of the bamboo scrimber with excellent properties.

2 MATERIALS AND METHODS

2.1 Materials

Bamboo bunches were selected carbonized bamboo bunches, and their moisture contents were 8–10%. They were purchased from the production enterprises of bamboo scrimber, and named the thick bamboo bundles. The thin bamboo bundles were the thin subdivision from the thick ones via cutting. Diammonium Hydrogen Phosphate (DAP) (≧99.0%) and boric acid (BA) (≧99.5%) were purchased from Xilong Chemical Co., Ltd. Distilled water was used for all the experiments.

2.2 Methods

The experiment was carried out by the single-factor test method. The known quantities of DPA and BA were, respectively, dissolved in distilled water, and then both of them at a 3:2(w/w) ratio were mixed to prepare the Composite Fire Retardant (CFR). According to the requirements for different experiments, the DAP solution prepared was diluted in aqueous solutions at different concentrations as well as in the CFR solution, and bamboo bundles were impregnated in the above solutions at atmospheric pressure, and then they were placed in a drying box for drying until the moisture contents were 8–10%. The fire-retardant retention was expressed as a percentage of the bamboo bundle impregnated fire retardant before and after the quality difference accounting for the quality of the impregnated fire retardant before.

3 RESULTS AND DISCUSSION

3.1 *Effect of bamboo bundles' morphology on fire-retardant retention*

The effects of the morphology with the difference in the treated time or agent concentration on the fire-retardant retention were investigated. Bamboo bundles of different thicknesses were, respectively, impregnated in the aqueous solution of 10% DAP for 0.5 h, 1 h, and 2 h, and the results are shown in Figure 1. Bamboo bundles were impregnated by the DAP solution of 10%, 20%, and 30% for 2 h, and the results are shown in Figure 2.

Figure 1 and Figure 2 show that the morphology of bamboo bundles has a certain impact on the fire-retardant retention. When bamboo bundles were impregnated by fire retardant at the same concentration for different time periods of 30 min, 60 min and 120 min, the fire-retardant retention of the thick bamboo bundles was higher than that of the thin ones, and the rates of the two kinds of bundles increased with the extension of impregnated time. As the treated time was same but the concentration varied, regardless of the agent concentration of 10%, 20% and 30%, the fire-retardant retention of the thick bamboo bundles was higher than that of the thin bamboo bundles, and they increased with the increasing agent concentrations. In general, the fire-retardant retention of the

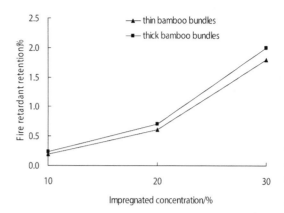

Figure 2. Effect of the bamboo bundles' morphology on retention at different concentrations.

bamboo bundles was higher, and the fire-retardant properties were better. The thick bamboo bundles showed relatively better bundles' morphology than the thin bamboo bundles for manufacturing the fire-retardant bamboo scrimber.

3.2 *Effect of fire-retardant type on fire-retardant retention*

In order to discuss the effects of fire-retardant type and different treated times on the fire-retardant retention, a part of the thick bamboo bundles was placed in the 10% DAP solution and another in the 10% CFR solution, and both were impregnated for 30 min, 60 min and 120 min. The experiment results are shown in Figure 3. Moreover, to study the effects of agent type at different agent concentrations on the fire-retardant retention, the thick bamboo bundles were impregnated with the DAP solution of 10%, 20% and 30% for 2 h as well as with the CFR solution in the same way. The results are shown in Figure 4.

Figure 3 and Figure 4 show that the fire-retardant type has a profound effect on the fire-retardant retention of the thick bamboo bundles. At the same agent concentration but different treated times from 30 min to 120 min, the fire-retardant retention with CFR were higher than that with DPA, and increased with the extension of impregnated time. In addition, at the same treated time but different agent concentrations from 10% to 30%, the changing laws of the fire-retardant retention were the same as above with the increasing agent concentration. Because the fire-retardant retention of the bamboo bundles was higher, the fire-retardant properties of bamboo scrimber were better. Therefore, the composite fire retardant was more appropriate than the single one fire retardant

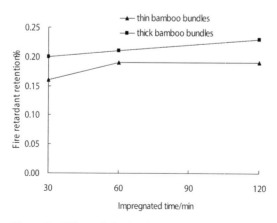

Figure 1. Effect of the bamboo bundles' morphology on retention at different times.

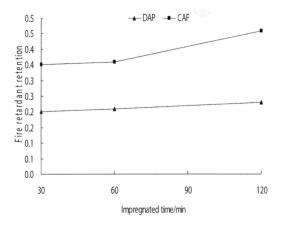

Figure 3. Effect of the fire-retardant type on retention at different times.

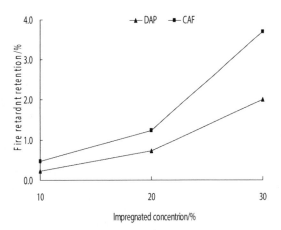

Figure 4. Effect of the fire-retardant type on retention at different concentrations.

for manufacturing the bamboo scrimber with better fire-retardant properties.

3.3 *Effect of impregnated time on fire-retardant retention*

The effects of the different impregnated times on fire-retardant retention were studied. Figure 5 shows the results of the experiment in which the thick bamboo bundles were placed in the 10% CFR solution for 10 min, 30 min, 60 min, 120 min, 240 min and 360 min, respectively.

From the figure, it can be observed that the fire-retardant retention of the bamboo bundles changed with the different treated times, and that the time had a great impact on fire-retardant retention. The fire-retardant retention of the bamboo

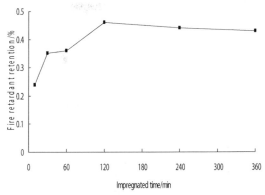

Figure 5. Effect of the impregnated time on fire-retardant retention.

bundles increased quickly with increasing time from 0 to 120 min and the highest was at 120 min. After 120 min, the fire-retardant retention began to decrease slowly. The fire-retardant retention was 45.8% at 30 min, 50.0% at 60 min and 91.7% at 120 min as high as that at 10 min, while it was just 4.2% at 240 min and 6.5% at 360 min as low as that at 120 min. When the concentration of fire retardant was 10%, the fire-retardant retention was not more than 1%, independent of the impregnated time of the bamboo bundles. The reason may be that the bamboo bundles impregnated fire-retardant treatment by the atmospheric pressure method, and the concentration of fire retardant was only 10%. So, it indicates that the agent can easily enter into the bamboo cell cavities, but it is difficult to enter the bamboo cell wall. Another reason is that it is difficult for more agents to enter into the cell cavities filled with water, which prevents the fire-retardant retention from increasing.

3.4 *Effect of impregnated concentration on fire-retardant retention*

Four batches of the thick bamboo bundles were impregnated for 120 min in the solutions of CFR of 10%, 30%, 40% and 50%, respectively, and the experiment results are shown in Figure 6.

Figure 6 shows that the effects of the fire-retardant concentration on the fire-retardant retention of the bamboo bundles are very significant. With the increasing concentration of CFR, the fire-retardant retention of the bamboo bundles increased rapidly. The fire-retardant retention of bundles treated by the 10% agent was, respectively, only 15.4% of that by 20%, 5.0% of that by 30%, 2.2% of that by 40%, and 1.6% of that by 50%. Therefore, the key to obtain a high fire-retardant retention was to increase the agent concentration.

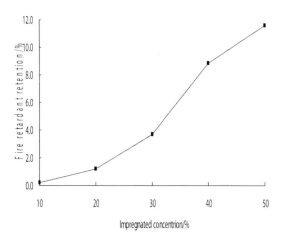

Figure 6. Effect of the impregnated concentration on fire-retardant retention.

However, it was also found in the experiment that the fire retardant appeared and gathered on the surfaces of the bamboo bundles when the agent concentration was higher than 30%, and the amount of precipitation was in proportion to the concentration. This phenomenon may lead to the decrease in fire-retardant properties of the bamboo bundles. Furthermore, fire retardant with a high concentration will make the treatment cost more. Thus, the agent concentration should not exceed 30% when the bamboo bundles are treated by fire retarding.

4 CONCLUSIONS

1. Under different impregnated times and impregnated concentrations, the fire-retardant retention of the thick bamboo bundles was higher than that of the thin ones, and the fire retardant retention of the composite agent was higher than that of the single one.
2. The fire-retardant retention of the bamboo bundles increased quickly with the increasing impregnated time and impregnated concentration. The best impregnated time was 120 min. When the agent concentration exceeded 30%, the fire retardant will precipitate from the surface of the bamboo bundles. So, the fire-retardant concentration should not be higher than 30% when the bamboo bundles are treated by fire retarding.

ACKNOWLEDGMENTS

This work was financially supported by the Natural Science Foundation Project of Zhejiang Province (LY12C16005).

REFERENCES

[1] Yu, M.L. & Jiang, X.T. & Zhang, X. 2014. Forecasting Model based on Regression Neural Network Combination for Significant Fire in Public Places. *Fire Protection Technology and Product Information* (1): 4–7.
[2] Li, C.C. & Yang, X. & Chen, Y. 2009. Investigations and Studies on Major Fires in Public Places. *Industrial Safety and Environmental Protection* 35(5): 48–50.
[3] Chen, J.L. 2014. Analysis of the Preventive Measures and Causes of High-rise Building Fire. *Low temperature construction technology* (2): 149–150.
[4] Yu, W.J. 2012. Current Status and Future Development of Bamboo Scrimber Industry in China. *China Wood Industry* 26(1): 11–14.
[5] Fang, G.Z. 2008. *The Functionality Improvement of Wood.* Beijing: Chemical Publishing House.

Material Science and Engineering – Chen (Ed.)
© *2016 Taylor & Francis Group, London, ISBN 978-1-138-02936-1*

Electrochemical properties of rod-like liquid crystalline molecules based on aromatic aldehyde

M.Y. Zheng, Y.S. Wei, W. Geng & Y.Z. Gu
School of Chemistry and Chemical Engineering, Xianyang Normal University, Xianyang, P.R. China

ABSTRACT: Four rod-like aromatic aldehydes of two series composed of a long rigid core of three six-member rings (cyclohexyl benzene rings or double benzene rings), ester group and terminal aldehyde were prepared in the yield of over 80%. All the target compounds were characterized based on their basic spectral data of Infrared Spectra (IR), Mass Spectra (MS) and ^{1}H NMR. The liquid crystalline properties of the compound were determined by a Differential Scanning Calorimeter (DSC) and a Hot-Stage Polarizing Optical Microscope (HS-POM). All the compounds displayed the mesophase belonging to the nematic phase. The electrochemical property of the compound was found by Cyclic Voltammetry (CV). The Highest Occupied Orbital (HOMO) and the Lowest Unoccupied Orbital (LUMO) and the difference between the frontier molecular orbitals (E_g) of the compounds were calculated. An effect of the terminal straight alkyl chain, the terminal ring system and the polarities of solvents on the frontier molecular orbital and E_g is observed.

Keywords: liquid crystal; frontier molecular orbital; aromatic aldehyde

1 INTRODUCTION

Evaluating the relationship between molecular structures and their properties is an essential task of material chemistry (An Z. W., Zheng M. Y., Wei Y. S., & Li J. L., 2012). It is well known that the molecular architecture is an important factor that governs the liquid crystalline properties of any compound. A rod-like molecular liquid crystalline molecule is one of the thermotropic liquid crystals, generally constructed by a rigid core (e.g. ring system), a rigid bridge bond (e.g. ester) and terminal groups (e.g. alkyls, alkoxy or other polar group) (Hoshino N, Matsuoka Y, Okamoto K, & Yamagishi A 2003; Joo S H, Yun Y K, Jin J I, Kim D C, & Zin W C. 2000; Lehmann M, Gearba R I, Koch M H J, & Ivanov D A. 2004; Cammidge A N, & Crepy K V L. 2003). In this paper, multi-ring aromatic aldehyde liquid crystals were built by a mono-ester bridge, a three six-member ring system, an alkyl and the aldehyde as a polar end.

Our former research designed and synthesized several series of liquid crystals with a terminal alkyl chain and a polar aldehyde end (Zheng M. Y. & An Z. W., 2007), indicating that the rod-like aromatic aldehydes easily exhibited a mesophase with enough thermostability and a wider temperature range of mesophase. Moreover, aldehyde is able to convert into other groups by a suitable reaction, for example aromatic acid (Chouhary V. R., Dumbre D. K., Narkhede V. S., 2012), Schiff base (Su R., Lu L., Zheng S. Z., Jin Y. H. & An S. J., 2015) or

other functional molecules (Zou Z. Q., Deng Z. J., Yu X. H., Zhang M. M., Zhao S. H., Luo T., Yin X., Xu H., & Wang W., 2012). Based on the above ideas, we synthesized four rod-like liquid crystalline molecules of two series bonded with a shorter terminal alkyl chain (n = 3 or n = 5), an ester, a there six-member ring and the aldehyde group. The relationship between the frontier molecular orbitals and their molecular structures was observed.

2 EXPERIMENTAL PROCEDURE

2.1 Materials and characterization

All initial intermediates used in the process were synthesized in our laboratory. The purities of the intermediates were detected by a LC-10 A (Shimadzu) instrument with methanol as the eluent and the flowing rate was 1 mL/min and higher than 99%. Their structures were characterized by IR, GC-MS and ^{1}H NMR methods. Other reagents were obtained from commercial sources and used without further purification.

Elemental analyses were conducted by a PE-2400 analyzer (Perkin Elmer). UV spectra were determined by an Agilent 8453 Spectrometer (Agilent Technologies). IR (KBr) spectra were recorded on a Vertex 70 spectrophotometer (Bruker). Mass data were recorded on a GCMS-QP2010 (Shimadzu) and IE was 70 eV. ^{1}H NMR spectra were obtained on an Avance 500 (Bruker, 500 MHz, solvent

CDCl$_3$). Cyclic voltammetric measurements were performed on an Epsilon (BAS), which was run by a General Purpose Electrochemical System software. The conventional three-electrode system was measured against SCE, Platinum wire and glassy carbon served as the reference, counter and working electrodes, respectively.

2.2 Synthesis

2.2.1 Synthesis of compound 2

Compound **2a** was prepared by heating and stirring a mixture of **1a** (1.23 g, 5 mmol) and SOCl$_2$ (0.59 g, 5 mmol) according to the method described in the literature (Zheng, M. Y. & An, Z. W. 2006), and used the next step without further purifying.

2.2.2 Synthesis of compounds 3

Compound **2a** was synthesized by dissolving **1a** in CH$_2$Cl$_2$ (20 mL) and dropping the mixture into a solution of **2a** according to the method from the literature (Zheng, M. Y. & An, Z. W. 2006). Compound **2a** was a white crystal obtained in 88% yield (two steps). Compounds **2b**, **2c**, and **2d** were also prepared by the same pathway. The synthesis of all the target compounds is outlined in Scheme 1, and their structural data are provided in Table 1. The spectral data are in accordance with the assigned structure.

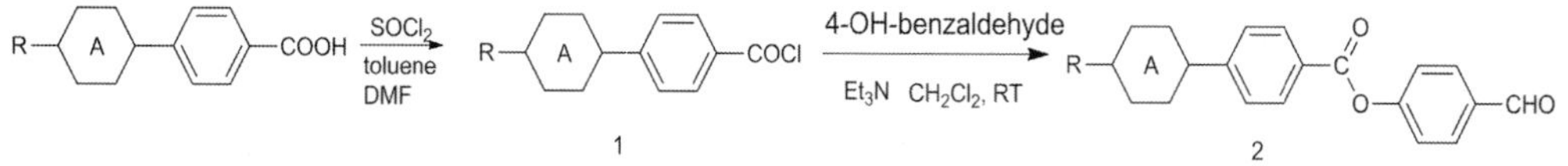

Scheme 1. The synthesis process of resulting compounds.
Series 1, a: R = n-propyl, A = trans-cyclohexyl; b: R = n-pentyl, A = trans-cyclohexyl;
Series 2, c: R = n-propyl, A = phenyl; e: R = n-pentyl, A = phenyl.

Table 1. ^{1}H NMR, MS, appearance, yields, melting points, IR and elemental analysis of the target compounds.

Compd.	Appearance	Yield/ %	m.p./°C	MS ([M+], calcd), m/z	IR (KBr) v/cm^{-1}	1H NMR (500 MHz, CDCl3), δ
2a	White crystal	88	100–101	350 (350.19)	3072, 3043 (m, C-H), 2956, 2920 (s, C-H), 2831, 2746 (vw, H of CHO), 1734 (vs, C = O), 1691 (vs, C = O), 1597, 1502, 1450 (s, HAr), 1265, 1064 (s, C-O-C), 850 (w, 1, 4-Ar)	0.91 (t, J = 6.0 Hz, 3H, CH$_3$), 1.08 (m, 2H, CH$_2$), 1.23 (m, 2H, CH$_2$), 1.35–2.58 (m, 10H, H of cyclohexane), 7.36, 7.38, 7.97, 8.11 (d, J = 8.5, J = 9, J = 8.5, J = 8.5 Hz, 2H of each, 8 PhH), 10.03 (s, 1H, CHO)
2b	White crystal	84	110–111	344 (344.14)	3066 (m, C-H), 2956, 2929 (s, C-H), 2872, 2743 (vw, H of CHO), 1728 (s, C = O), 1701 (s, C = O), 1599, 1504, 1458 (s, HAr), 1273, 1209, 1155, 1070, 1012 (s, C-O-C), 858 (w, 1, 4-Ar)	1.01 (t, J = 7.0 Hz, 3H, CH$_3$), 1.71 (m, 2H, CH$_2$), 2.68 (m, 2H, CH$_2$), 7.32, 7.45, 7.60, 7.76, 8.01, 8.27 (d, J = 8.0, J = 8.5, J = 8.5, J = 8.5, J = 8.0, J = 8.0 Hz, 2H of each, 12 PhH), 10.03 (s, 1H, CHO)
2c	White crystal	81	91–92	378 (378.22)	3010, 3068, 3047 (w, C-H), 2951, 2920 (s, C-H), 2848, 2754 (vw, H of CHO), 1746 (vs, C = O), 1697 (vs, C = O), 1597, 1506, 1446 (s, HAr), 1265, 1205, 1174, 1157, 1062 (s, C-O-C), 852 (w, 1, 4-Ar)	0.92 (t, J = 6.0 Hz, 3H, CH$_3$), 1.09 (m, 2H, CH$_2$), 1.24–1.48 (m, 6H, CH$_2$CH$_2$CH$_2$), 1.51–2.50 (m, 10H, H-of cyclohexane), 7.36, 7.41. 7.98, 8.11 (d, J = 8.0, J = 8.5, J = 8.5, J = 8.0 Hz, 2H of each, 8 PhH), 10.03 (s, 1H, CHO)
2d	White crystal	83	92–93	372 (372.17)	3047 (w, C-H), 2954, 2924 (s, C-H), 2852, 2729 (vw, H of CHO), 1728 (s, C = O), 1705 (s, C = O), 1599, 1556, 1502, 1490 (s, ArH), 1 273, 1070 (s, C-O-C), 842 (w, 1, 4-Ar)	0.91 (t, J = 6.5 Hz, 3H, CH$_3$), 1.36 (m, 4H, CH$_2$CH$_2$), 1.23 (m, 2H, CH$_2$), 1.67 (m, 2H, CH$_2$), 7.30, 7.43, 7.58, 7.74, 7.98, 8.24 (d, J = 8, J = 8.5, J = 8, J = 8, J = 8.5, J = 8.5 Hz, 2H of each 12 PhH), 10.03 (s, 1H, CHO)

3 RESULTS AND DISCUSSION

3.1 Electrochemical analysis

The E_{HOMO} (molecular energy level of HOMO) and E_{LUMO} (molecular energy level of LUMO) can be calculated exactly by the CV method. The E_{HOMO}, E_{LUMO} and the corresponding band gap (Eg) were calculated by using following equations:

$$E_{HOMO} = -e\,(E_{ox}+4.4)(eV);$$

$$E_{LUMO} = -e\,(E_{red}+4.4)(eV);$$

$$E_g = -e\Delta\Phi,\ \Delta\Phi = \Phi_p - \Phi_n$$

(Gao L., Zhang J., He C., Zhang Y., Sun Q. J. & Li Y. F. 2014) where e is the elementary charge; Φ_p is the onset oxidation potential; and Φ_n is the onset reduction potential.

The energy levels and energy gaps of the four compounds were determined by the three-electrode system (SCE, platinum wire and glassy carbon serves as the reference, counter and working electrodes, respectively) with different solvents (THF, CH_2Cl_2 or CH_3CH_2OH with the concentration of 8×10^{-3}) and KCl (0.3 M, water solution) as the supporting electrolyte. The typical voltammogram of 2a is shown in Figure 1. The E_{HOMO}, E_{LUMO} and E_g of the target compound in these solvents are summarized in Table 2, Table 3 and Table 4.

Data from Table 2, Table 3, Table 4 indicate that the E_{HOMO}, E_{LUMO} and E_g of the same compound were different, indicating that these energies were affected by the different solvent.

Taking 2a as example, we find that the values of E_{HOMO} of **2a** were the lowest and E_g was the highest in ethanol, while the values of E_{HOMO} of **2a** were the highest and E_g was the lowest in THF. The E_{HOMO} values of 2b, 2c and 2d had the same changing rule as those of 2a.

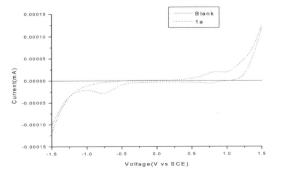

Figure 1. The cyclic voltammogram of 2a in the dilute THF solution with KCl of 0.3 M as the supporting electrolyte. The dotted line shows the CV of blank run under the identical condition.

Table 2. Redox potentials, E_{HOMO}, E_{LUMO} and E_g of the target compounds in THF.

Compd.	$E_o(\Phi_p)$/ eV	$E_{red}(\Phi_n)$/ eV	E_{HOMO} /eV	E_{LUMO} /eV	E_g/ eV
2a	0.48	−0.42	−4.88	−3.98	0.90
2b	0.37	−0.51	−4.77	−3.89	0.88
2c	0.38	−0.49	−4.78	−3.91	0.87
2d	0.36	−0.47	−4.76	−3.93	0.83

Table 3. Redox potentials, E_{HOMO}, E_{LUMO} and E_g of the target compounds in CH_2Cl_2.

Compd.	$E_{ox}(\Phi_p)$/ eV	$E_{red}(\Phi_n)$/ eV	E_{HOMO}/ eV	E_{LUMO}/ eV	E_g/ eV
2a	0.65	−0.47	−5.05	−3.93	1.12
2b	0.62	−0.46	−5.02	−3.94	1.08
2c	0.58	−0.48	−4.98	−3.92	1.06
2d	0.61	−0.41	−5.01	−3.99	1.02

Table 4. Redox potentials, E_{HOMO}, E_{LUMO} and E_g of the target compounds in ethanol.

Compd.	$E_{ox}(\Phi_p)$/ eV	$E_{red}(\Phi_n)$/ eV	E_{HOMO}/ eV	E_{LUMO}/ eV	E_g/ eV
2a	0.69	−0.51	−5.09	−3.89	1.20
2b	0.68	−0.51	−5.08	−3.89	1.19
2c	0.70	−0.48	−5.10	−3.92	1.18
2d	0.69	−0.44	−5.09	−3.96	1.13

However, the E_{LUMO} values of 2a, 2b, 2c and 2d had no clear trend. These solvents differed from each other in their polarities. Their polarities were 3.1 of CH_2Cl_2, 4.0 of THF and 4.3 of C_2H_5OH, indicating that the bigger the polarity of the solvent is, the lower the E_{HOMO} and the higher the E_g of the same compound will be.

There is an effect of the terminal alkyl chain on these energies. Whatever the solvent is, the E_g values of **2a** or **2b** with the terminal n-propyl chain were lower than those of its corresponding compound **2c** or **2d** with the terminal n-pentyl chain. When the terminal alkyl chains were the same, the compound with the terminal cyclohexyl benzene ring structure had lower E_{HOMO} and higher E_g values.

E_g is the energy difference between the frontier molecule orbitals. The transition of the electrons between the frontier molecular orbitals are easier in THF than in the other two solvents, indicating that the polarity of THF can narrow the energy level of the frontier molecular orbitals. Too high or too low polarities of the solvent are not available to decrease the E_g. In other words, the redox reaction of the target molecules can easily occur in THF than in the other two solvents.

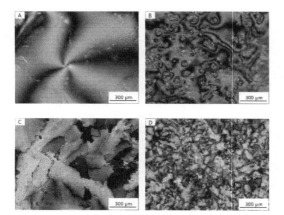

Figure 2. The textures of the target compounds observed under POM during the heating process: A-D are schlieren textures of 2a, 2b, 2c, 2d, taken at 130, 132, 120, and 125°C, respectively; magnification × 300.

Table 5. The DSC data of EPA compounds.

Compd	Tm/°C	ΔH_m(J/g)	Tc/°C	ΔHc(J/g)	Tc-Tm/°C
2a	100	59.53	206	2.57	106
2b	111	64.24	221	1.57	110
2c	93	50.84	202	1.57	109
2d	91	46.84	209	1.84	118

3.2 Liquid property of the target molecules

All the four compounds have the mesophases of schlieren textures (Figure 2) with the temperature ranging from 106 to 118°C (Table 5). The melting points of these rod-like aromatic aldehydes compound were between 91 and 111°C, while their clear points ranged from 202 to 221°C. When the terminal ring system was the same, the melting point of the compound with terminal n-propyl was higher than that with terminal n-propyl, while the clear point of the compound with terminal n-propyl was higher than that with n-pentyl. When the terminal alkyl chain was the same, the compound with terminal cyclohexyl benzene rings had a higher clear point than that with terminal double benzene rings.

4 CONCLUSIONS

The four rod-like liquid crystalline compounds based on the structure of aromatic aldehyde were synthesized in the yield of over 80%. Their structures were characterized by the structural data. Their electrochemical properties were studied in the paper. The result showed that the energy levels of HOMO, LUMO and Eg of the target compounds were influenced by the polarities of the solvent, their

terminal alkyl chain and terminal ring system. Their liquid crystal properties were also affected by the terminal ring system and the terminal alkyl chain. The result will provide a pathway to change the electrochemical property and the temperature range of the mesophase of similar liquid crystal molecules.

ACKNOWLEDGMENTS

We gratefully acknowledge the financial support from the National Natural Science Foundation of China (No. 21102121) and the Natural Science Foundation of Shaanxi Province (No. 2014 JM2–2014).

REFERENCES

[1] An, Z. W. Zheng, M.Y. Wei, Y.S. & Li, J.L. 2012. TEMPO containing esters and their magnetic and liquid crystal properties. Mol. Cryst. Liq. Cryst., 557, 28–34.
[2] Hoshino, N. Matsuoka, Y. Okamoto, K. & Yamagishi, A. 2003. Δ-[Ru(acac)2 L] (L) a mesogenic derivative of bpy) as a novel chiral dopant for nematic liquid crystals with large helical twisting power. J Am Chem Soc, 125(7): 1718–1719.
[3] Joo, S.H. Yun, Y.K. Jin, J.I. Kim, D.C. & Zin, W. C. 2000. Synthesis of liquid crystalline polyesters of various types by acyclic diene metathesis polymerization. Macromolecules, 33(18): 6704–6712.
[4] Lehmann, M. Gearba, R.I. Koch, M.H.J. & Ivanov, D.A. 2004. Semiflexible star-shaped mesogens as nonconventional columnar liquid crystals. Chem Mater, 16(3): 374–376.
[5] Cammidge, A.N. & Crepy, K.V.L. 2003. Application of the Suzuki reaction as the key step in the synthesis of a novel atropisomeric biphenyl derivative for use as a liquid crystal dopant. J Org Chem, 68(17): 6832–6835.
[6] Zheng, M.Y. & An, Z.W. 2006. Rod-like Schiff base magnetic liquid crystals bearing organic radical, Chinese Journal of Chemistry, 24(12): 1754–1757.
[7] Chouhary, V.R. Dumbre, D.K. & Narkhede, V.S. 2012. Solvent-free oxidation of aldehydes to acids by TBHP using environmental-friendly MnO₄–14 -exchanged Mg-Al hydrotalcite catalyst. J. Chem. Sci., 124(4): 835–839.
[8] Su, R. Lu, L. Zheng, S.Z. Jin, Y.H. & An, S.J., 2015. Chem. Synthesis and characterization of novel azocontaining or azoxy-containing Schiff bases and their antiproliferative and cytotoxic activities. Res. Chinese Universities, 31(1): 60–64.
[9] Zou, Z.Q. Deng, Z.J. Yu, X.H. Zhang, M.M. Zhao, S.H. Luo, T. Yin, X. Xu, H. & Wang, W. 2012. A new facile approach to N-alkylpyrroles from direct redox reaction of 4-hydroxy-L-proline with aldehydes. Sci. Chin. Ser. B: Chem., 55(1): 43–49.
[10] Gao, L. Zhang, J. He, C. Zhang, Y. Sun, Q.J. Li, Y.F. Effect of additives on the photovoltaic properties of organic solar cells based on triphenylaminecontaining amorphous molecules. Science China, 2014, 57(7): 966–972.

Material Science and Engineering – Chen (Ed.)
© *2016 Taylor & Francis Group, London, ISBN 978-1-138-02936-1*

Synthesis and crystal structure of ethyl 2-(salicylylideneamino)-4,5,6,7-tetrahydrobenzo[b]thiophene-3-carboxylate

H. Gao & J. Fu
Laboratory of Biologic Resources Protection and Utilization of Hubei Province, Hubei University for Nationalities, Enshi, China

X.J. Song
Laboratory of Biologic Resources Protection and Utilization of Hubei Province, Hubei University for Nationalities, Enshi, China
School of Chemical and Environmental Engineering, Hubei University for Nationalities, Enshi, China

ABSTRACT: The Schiff base ethyl 2-(2-hydroxybenzylideneamino)-4,5,6,7-tetrahydrobenzo[b]thiophene-3-carboxylate ($C_{18}H_{19}NO_3S$, $M_r = 329.40$) was synthesized by the condensation reaction of salicylaldehyde with ethyl 2-amino-4,5,6,7-tetrahydrobenzo[b]thiophene-3-carboxylate, which was efficiently prepared *via* a one-pot Gewald's three-component reaction using cyclohexanone, ethyl 2-cyanoacetate and elemental sulfur as starting materials. The structure of the title compound was determined by single-crystal X-ray diffraction. The crystal belongs to triclinic, space group $P\bar{1}$ with $a = 8.3714(17)$ Å, $b = 10.023(2)$ Å, $c = 10.744(2)$ Å, $\alpha = 82.729(2)°$, $\beta = 67.119(2)°$, $\gamma = 79.631(3)°$, $V = 815.4(3)$ Å^3, $Z = 2$, $D_c = 1.342$ g/cm^3, $\mu = 0.213$ mm^{-1}, $F(000) = 348$, the final $R = 0.0487$ and $wR = 0.1164$ for 2688 observed reflections with $I > 2\sigma(I)$. X-ray diffraction analysis reveals that the title molecule is nearly planar except for the cyclohexene and ester moieties.

Keywords: Gewald reaction; Schiff base; 2-aminothiophene; crystal structure

1 INTRODUCTION

2-Aminothiophene derivatives are an important class of organic intermediates in organic synthesis and commonly used as the scaffold motif of a variety of agrochemicals, dyes, and biologically active products. Thiophenes and their fused hetero-aromatic rings have attracted continuing interest from both synthetic and medicinal chemists over these years due to their various significant pharmacological properties such as antitumor (Liu et al. 2014; Song et al. 2014), antibacterial (Kanawade et al. 2013) and antioxidant (Kotaiah et al. 2012) activities. Schiff bases containing C = N double bond show excellent biological activities, ligand and photoelectric properties. Some of them have been widely used in the fields of drugs (Zhang et al. 2014), catalysis, and functional materials (Jeevadason et al. 2014). In addition, the introduction of an ester group in the compound molecule may improve biological activity because it can effectively regulate lipid-water partition coefficient and increase the fat-solubility of the compound. Therefore, it is worthwhile to investigate the Schiff bases incorporating both

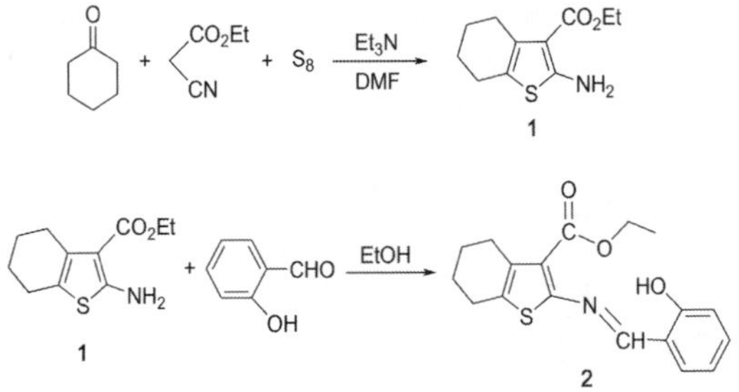

Scheme 1. The synthetic procedure of the title compound 2.

thiophene nucleus and ester group, which have seldom been reported. Herein, the crystal structure of ethyl 2-(salicylylideneamino)-4,5,6,7-tetrahydrobenzo [b] thiophene-3-carboxylate is investigated by X-ray diffraction as part of our ongoing structural studies as well as to provide a basis for consideration for the structure-activity relationships.

The synthetic route of the title compound 2 is outlined in Scheme 1.

2 EXPERIMENTAL

2.1 Instruments and reagents

Melting points were measured with an X-4 digital melting-point apparatus and uncorrected. ^{1}H NMR spectra were obtained on an UNITY INOVA-600 Spectrometer with TMS as internal standard and DMSO-d_6 as the solvent. MS spectra were run on a Agilent 5975 inert Mass Selective Detector using the EI method. Elemental analysis was performed by a Vario EL III analyzer. X-ray diffraction data were collected using a Bruker Smart APEX-CCD diffractometer equipped with a graphite-monochromatized Mo K_α ($\lambda = 0.71073$ Å) radiation. All chemicals used for the preparation were of analytical grade. Solvents were dried by standard methods and distilled prior to use.

2.2 Synthesis of 2-amino-4,5,6,7-tetrahydrobenzo [b]thiophene-3-carboxylate (1)

To a mixture of 4.90 g (0.005 mol) cyclohexanone, 5.66 g (0.05 mol) ethyl cyanoacetate, 1.92 g (0.06 mol) elemental sulfur and 10 mL of anhydrous DMF were dropwise added 3.5 mL of triethylamine with cooling on a ice-bath and stirring (completed within about 15 min). The reaction mixture was stirred at room temperature for a further 4 h, then poured into 15 mL cold water and a large amount of solid precipitate formed. The solid precipitate was filtered off, washed with water, and then recrystallized from 80% ethanol to deliver 2-amino-4,5,6,7-tetrahydrobenzo[b]thiophene-3-carboxylate (1) in 78% yield as yellowish solid. M.p.: 116 °C.

2.3 Synthesis of ethyl 2-(salicylylideneamino)-4,5,6,7-tetrahydrobenzo[b]thiophene-3-carboxylate (2)

A reaction mixture of compound 1 (0.45 g, 2.0 mmol), salicylaldehyde (0.24 g, 2.0 mmol) in 5 mL of anhydrous ethanol was stirred at 50 °C for 6 h. The solvent was removed by evaporation under reduced pressure, and the residue was recrystallized from anhydrous ethanol to give yellow crystal 2 in 76% yield. M.p.:132 ~ 133 °C; ^{1}H NMR (600 MHz, DMSO-d_6) δ: 1.29 (3H, t, $J = 7.2$ Hz, COOCH$_2$CH$_3$), 1.73 (4H, dd, $J = 4.8$ and 4.4 Hz, CH$_2$ at 5 and 6), 2.67 (2H, s, CH$_2$ at 4), 2.70 (2H, s, CH$_2$ at 7), 4.27 (2H, q, $J = 7.2$ Hz, COOCH$_2$CH$_3$), 6.93 ~ 7.65 (4H, m, ArH), 8.75 (1H, s, CH = N), 12.56 (1H, s, OH). IR (KBr, cm^{-1}) v: 3435 (O-H), 1697 (C = O), 1601 (C = N), 1216, 1137 (C-O-C); EI-MS (%): m/z 329.4 (74.6, M$^+$), 150 (100); Anal. Calcd. for C$_{18}$H$_{19}$NO$_3$S: C 65.63, H 5.81, N 4.25; found: C 65.49, H 5.98, N 4.06.

2.4 Crystallographic measurement

A colorless single crystal of compound 2 with dimensions of 0.29 mm × 0.28 mm × 0.23 mm was mounted on a Brucker APEX II diffractometer equipped with a graphite-monochromated Mo$K\alpha$ ($\lambda = 0.71073$ Å) radiation. The intensity data were collected by using a ψ-ω scan mode in the range of $2.06 \leq \theta \leq 26.50°$ at 293(2) K. A total of 4762 reflections were collected and 3311 were independent with $R_{int} = 0.0160$, of which 2688 were observed with $I > 2\sigma(I)$ and used in the succeeding refinements. Absorption correction was not applied. The structure was solved by direct methods with SHELXS-97 (Sheldrick 2008) and expanded using Fourier difference techniques. The non-hydrogen atoms were refined anisotropically, and the hydrogen atoms were added according to theoretical models. Structural refinement was carried out by full-matrix least-squares techniques on F^2 with SHELXL-97 (Sheldrick 2008). The final refinement gave $R = 0.0487$ and $wR = 0.1164$ ($w = 1/[\sigma^2(F_o^2) + (0.0488P)^2 + 0.5764P]$, where $P = (F_o^2 + 2F_c^2)/3$. $S = 0.982$, $(\Delta/\sigma)_{max} = 0.000$, $(\Delta\rho)_{max} = 0.357$ and $(\Delta\rho)_{min} = -0.377$ e/Å^3.

3 RESULTS AND DISCUSSION

3.1 Synthesis

The most efficient and convenient approach for the preparation of 2-aminothiophene derivatives is Gewald reaction. The one-pot cyclocondensation of aldehyde or ketone having α-hydrogen, a cyanomethylene containing an electron-withdrawing group (e.g. cyanoacetate, cyanoacetamide or α-cyano ketone) and elemental sulfur in the presence of organic base (such as morpholine, piperidine, diethylamine or triethylamine, etc.), known as the Gewald reaction, has been one of the most well-studied three-component reactions in recent years. In this paper, ethyl 2-amino-4,5,6,7-tetrahydrobenzo[b] thiophene-3-carboxylate (1) was efficiently prepared through a one-pot Gewald's three-component reaction using cyclohexanone, ethyl 2-cyanoacetate and elemental sulfur as starting materials in the presence of triethylamine, then reacted with salicylaldehyde via the dehydration condensation to afford the product.

3.2 Crystal structure analysis

The selected bond lengths, bond angles and the selected torsion angles are listed in Tables 1, 2 and 3, respectively. The molecular structure and packing diagram of compound 2 are depicted in Figures 1 and 2, respectively. The C–N distances (Table 1) fall between the normal C = N double bond (1.27 Å) and C–N single bond (1.47 Å), show-

Table 1. Selected bond lengths (Å) for compound 2.

Bond	Dist.	Bond	Dist.
O(1)—C(1)	1.349(3)	C(9)—S(1)	1.726(2)
C(7)—N(1)	1.282(3)	C(16)—O(2)	1.199(3)
C(8)—N(1)	1.384(3)	C(16)—O(3)	1.330(3)
C(8)—S(1)	1.741(2)	C(17)—O(3)	1.451(3)

Table 2. Selected bond angles (°) for compound 2.

Angle	(°)
O(1)—C(1)—C(2)	118.4(2)
O(1)—C(1)—C(6)	121.9(2)
N(1)—C(7)—C(6)	122.0(2)
C(11)—C(8)—N(1)	125.67(19)
N(1)—C(8)—S(1)	123.62(16)
C(5)—C(6)—C(7)	119.1(2)
C(10)—C(9)—S(1)	112.17(16)
C(15)—C(9)—S(1)	121.94(18)
O(2)—C(16)—C(11)	123.5(2)
O(2)—C(16)—O(3)	123.5(2)
O(3)—C(16)—C(11)	113.00(19)
O(3)—C(17)—C(18)	107.6(2)
C(11)—C(8)—S(1)	110.71(15)
C(7)—N(1)—C(8)	121.41(18)
C(16)—O(3)—C(17)	116.43(19)
C(9)—S(1)—C(8)	91.62(10)

Table 3. Selected torsion angles (°) for compound 2.

Torsion angle	(°)
O(1)—C(1)—C(2)—C(3)	−179.4(2)
O(1)—C(1)—C(6)—C(5)	179.3(2)
O(1)—C(1)—C(6)—C(7)	3.2(3)
C(1)—C(6)—C(7)—N(1)	−0.6(3)
C(5)—C(6)—C(7)—N(1)	−176.7(2)
S(1)—C(9)—C(10)—C(12)	178.31(18)
S(1)—C(8)—C(11)—C(16)	174.97(17)
N(1)—C(8)—C(11)—C(16)	−4.6(3)
C(8)—C(11)—C(16)—O(2)	−137.1(2)
C(8)—C(11)—C(16)—O(3)	43.5(3)
S(1)—C(8)—N(1)—C(7)	3.0(3)
C(10)—C(11)—C(16)—O(3)	−141.1(2)
O(2)—C(16)—O(3)—C(17)	0.1(3)
C(15)—C(9)—S(1)—C(8)	177.7(2)
C(18)—C(17)—O(3)—C(16)	−167.7(2)
C(11)—C(8)—S(1)—C(9)	1.13(17)
N(1)—C(8)—S(1)—C(9)	−179.25(1)

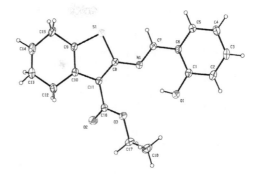

Figure 1. Molecular structure of compound 2 with the atomic labelling.

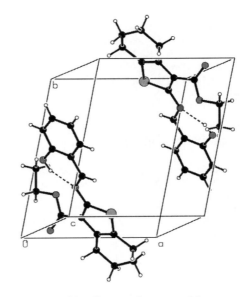

Figure 2. Packing diagram of compound 2.

linkage. All the other bond lengths fall within the expected range. There exists an intramolecular O–H···N hydrogen bond (O–H 0.82 Å, H ... N 1.90 Å, O ... N 2.623(2) Å, O–H ... N 146.5°) to give a planar six-membered ring, as shown in Figure 2. The cyclohexene ring adopts a half-chair conformation and the thiophene ring is essentially planar with the benzene ring with a dihedral angle of only 8.97(11)°.

ing that N(1) atom is partially characterized by sp^2 hybridization. Also, the bond length of C(16)—O(2) is 1.199(3) Å, longer than that of the normal C = O bond (1.119 Å). It is deduced that there exists some degree of electron delocalization around the ester

4 CONCLUSION

The Schiff base ethyl 2-(salicylylideneamino)-4,5,6,7-tetrahydrobenzo[b]thiophene-3-carboxylate was synthesized by condensation reaction of salicylaldehyde with ethyl 2-amino-4,5,6,7-tetrahydrobenzo[*b*]thio-phene-3-carboxylate, which was efficiently prepared *via* a one-pot Gewald's three-component reaction using cyclo-hexanone, ethyl 2-cyanoacetate and elemental sulfur as starting materials. This protocol offered such advantages as mild reaction conditions, simple purification and good yields. The title structure was confirmed by ^{1}H NMR, mass spectroscopy, elemental analysis and X-ray crystallographic studies. X-Ray diffraction analysis reveals that the title molecule is nearly planar except for the cyclohexene and ester moieties.

ACKNOWLEDGEMENTS

This work was financially supported by the Open Fund of Key Laboratory of Biologic Resources Protection and Utilization of Hubei Province, China (PKLHB 1506), the Project of New Strategic Industries for Fostering Talents in Applied Chemistry of Higher Education of Hubei Province and the First-class Discipline of Forestry in Hubei University for Nationalities.

SUPPLEMENTARY MATERIAL

Crystallographic data for compound 2 have been deposited in the Cambridge Crystallographic Data Center (Deposition No. CCDC-855228). These data can be obtained free of charge *via* www.ccdc.cam.ac.uk/conts/ retrieving.html (or from the Cambridge Crystallographic Data Centre, 12, Union Road, Cambridge CB2 1EZ, UK; fax: +44 1223 336033; or deposit@ccdc.cam.ac. uk).

REFERENCES

[1] Jeevadason, A.W.; Murugavel, K.K. & Neelakantan M.A. 2014. Review on Schiff bases and their metal complexes as organic photovoltaic materials. *Renewable and Sustainable Energy Reviews* 36: 220–227.

[2] Kanawade, S.B.; Toche, R.B. & Rajani, J. 2013. Synthetic tactics of new class of 4-aminothieno[2,3-*d*] pyrimidine-6-carbonitrile derivatives acting as antimicrobial agents. *European Journal of Medicinal Chemistry* 64: 314–320.

[3] Kotaiah, Y.; Harikrishna, N. & Nagaraju, K. 2012. Synthesis and antioxidant activity of 1,3,4-oxadiazole tagged thieno[2,3-*d*]pyrimidine derivatives. *European Journal of Medicinal Chemistry* 58: 340–345.

[4] Liu, Z.J.; Wang, Y. & Lin, H.F. 2014. Design, Synthesis and biological evaluation of novel thieno[3,2-*d*] pyrimidine derivatives containing diaryl urea moiety as potent antitumor agents. *European Journal of Medicinal Chemistry* 85: 215–227.

[5] Sheldrick, G.M. 2008. A short history of SHELX. *Acta Crystallogr., Sect. A: Found. Crystallogr.* 2008 64: 112–122.

[6] Song, X.J.; Yang, P.; Gao, H.; Wang, Y.; Dong, X.G. & Tan, X.H. 2014. Facile synthesis and antitumor activity of novel 2-trifluoromethylthieno[2,3-*d*] pyrimidine derivatives. *Chinese Chemical Letters* 25(7): 1006–1010.

[7] Zhang, K.; Wang, P.; Xuan, L.N.; Fu, X.Y.; Jing, F.; Li, S.; Liu, Y.M. & Chen, B.Q. 2014. Synthesis and antitumor activities of novel hybrid molecules containing 1,3,4-oxadiazole and 1,3,4-thiadiazole bearing Schiff base moiety. *Bioorganic & Medicinal Chemistry Letters* 24: 5154–5156.

Material Science and Engineering – Chen (Ed.)
© 2016 Taylor & Francis Group, London, ISBN 978-1-138-02936-1

Magnesium olivine sand as a flame retardant for flexible Poly(Vinyl Chloride)

X.J. Ren, Z.H. Chai, C.G. Song & M. Gao
School of Environmental Engineering, North China University of Science and Technology, Yanjiao, Beijing, China

ABSTRACT: Magnesium olivine sand was used as a flame retardant for PVC, to study the mechanical properties and flame retardancy of samples. The resultant data show that magnesium olivine sand had a better effect on the mechanical properties of the sample, especially the tensile strength and impact strength, and 4% of magnesium olivine sand obtained a good flame retardancy. PVC treated with the flame retardant showed a high limiting oxygen index and a high decomposition temperature, which indicated that the flame retardancy of the treated PVC was improved.

Keywords: magnesium olivine sand; Sb_2O_3; flame retardant; PVC

1 INTRODUCTION

Poly(Vinyl Chloride) (PVC), a well-known and economical polymer, is used for many electrical applications. The high chlorine content (56.8%) of PVC makes it relatively resistant to ignition and burning. For many applications of rigid or semi-rigid PVC, addition of flame-retardant additives is not required. However, to make it easy to process, semi-rigid and flexible PVC compounds always contain a large volume of plasticizers, such as DOP [di(2-ethylhexyl)phthalate], which can deteriorate the flame-retardant and smoke suppression properties. When the PVC products contain 45 parts of DOP, the Limiting Oxygen Index (LOI) would decrease to about 24 and the PVC would thus become a high-flammable material. However, plasticized PVC products can still exhibit a good fire performance, particularly if additionally fire-retarded (Chen & Wang, 2010).

Many additives, such as alloys, organic substances, and inorganic compounds including tin, zinc, copper, iron, and molybdenum, have been used in the flame retardation and smoke suppression of PVC (Xu et al. 2005). Because of the high loading, it becomes essential that a good degree of flame retardancy be obtained, but the mechanical properties decrease obviously. Using coupling agents and synergists are good ways to solve this problem (Qu et al. 2005, Pi et al. 2003). Magnesium Olivine Sand (MOS) contains many kinds of metal oxides, which may have a similar effect.

The purpose of the present study is to study the mechanical properties and flame retardancy of samples treated with a combination of magnesium olivine sand.

2 EXPERIMENTAL PROCEDURE

2.1 *Materials*

PVC, SG2; Dioctyl Phthalate (DOP); Tribasic Lead Sulfate (TLS), Dibasic Lead Phosphate (DLP), commercially available; Stearic Acid (SA), Sb_2O_3, industrial grade, were supplied by Tianjin Fuchen Chemical Reagent Factory (Tianjin, China).

2.2 *Instrumentation*

LOI values were determined in accordance with ASTM D2863-70 by means of a General Model HC-1 LOI apparatus. The vertical burning test was conducted by using a CZF-II horizontal and vertical burning tester (Jiang Ning Analysis Instrument Company, China). The mechanical properties were tested according to the GB/T 1040.2–2006 standard with a LJ-5000 tensile testing machine (Chengde Experimental Factory).

2.3 *Preparation of flame-retardant PVC samples*

Formulation according to the predetermined material (including PVC, three salts, salts of stearic acid, DOP, hydroxide, chloride, and coal) was mixed in a mixer for 3 to 5 minutes at 35–45°C. Then, the mixture was plasticized on a two-roll mill at 165°C for 6–8 min, compressed at 180°C to form sheets of 100 mm × 10 mm × 3 mm. The test specimens were cut from the molded sheets.

3 RESULTS AND DISCUSSION

3.1 Flame retardancy of Sb_2O_3 and magnesium olivine sand

Sb_2O_3 has a good flame-retardant effect on the PVC material. It is in the gas phase, and the condensed phase also plays a role, and thus has a better flame-retardant effect and has been used in the plastics and rubber industry, attracting widespread attention.

From the oxygen index values in Table 1, the Sb_2O_3-treated flexible PVC has good flame retardancy. To achieve better flame retardancy for PVC, magnesium olivine sand was added. The data are presented in Table 1. The LOI of the samples (B-F) increased with the addition of magnesium olivine sand, ranging from 30.7% to 34.0%. When more magnesium olivine sand was added, the LOI of the samples decreased, thus 4.0 g magnesium olivine sand was optimal. Magnesium olivine sand and Sb_2O_3 can be good flame retardants for flexible PVC.

3.2 Effect of magnesium olivine sand on the mechanical properties

Mechanical properties such as tensile strength and impact strength of the samples were measured, which are shown in Figures 1 and 2.

Figure 1 shows the effect of magnesium olivine sand on the tensile strain of the samples, which indicates that the tensile strain decreases when the proportion of magnesium olivine sand is increased.

Figure 2 shows the effect of magnesium olivine sand on the impact strength of the samples. When the added amount of magnesium olivine sand was more than 5.0 g, the sample absorbed less energy and its hardness was much decreased.

So, with the addition of magnesium olivine sand, the tensile strength and impact strength of

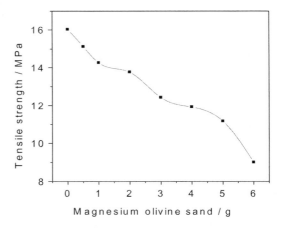

Figure1.　Tensile strength of samples A-H.

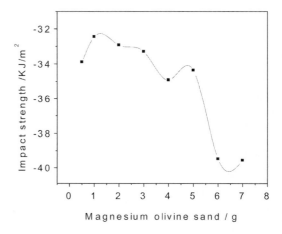

Figure 2.　Impact strength of the samples.

the samples decreased, and the application scope of the samples was affected to a small extent.

3.3 Degradation of flexible PVC

The simultaneous DTG and TG curves of samples A and F were generated in dynamic nitrogen from ambient temperature to 750°C, which are shown in Figures 3 and 4. The Initial Decomposition Temperature (IDT) determined by 5% weight loss, the Integral Procedure Decomposition Temperature (IPDT) determined by 50% weight loss, and the char yield at 700°C were measured, and are listed in Table 2.

From Figures 3 and 4, it can be seen that PVC decomposes in three weight losses in total weight loss (Tian et al. 2003). The first stage is mainly due

Table 1.　The composition of the samples.

Sample	A	B	C	D	E	F	G	H
PVC (g)	100	100	100	100	100	100	100	100
DOP (g)	30.0	30.0	30.0	30.0	30.0	30.0	30.0	30.0
TLS (g)	2.0	2.0	2.0	2.0	2.0	2.0	2.0	2.0
DLP (g)	2.0	2.0	2.0	2.0	2.0	2.0	2.0	2.0
ZS (g)	0.5	0.5	0.5	0.5	0.5	0.5	0.5	0.5
Sb_2O_3 (g)	0.7	0.7	0.7	0.7	0.7	0.7	0.7	0.7
MOS (g)	0.0	0.5	1.0	2.0	3.0	4.0	5.0	6.0
LOI (%)	30.7	32.2	32.7	33.1	33.4	34.0	32.8	31.4

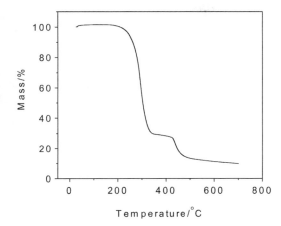

Figure 3.   Thermal analysis of sample A.

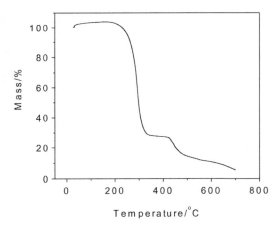

Figure 4.   Thermal analysis of sample F.

Table 2.   Thermal data of the sample from TG analysis.

Sample	IDT (°C)	IPDT (°C)	Char yield (%)
A	246.7	283.8	9.5
F	250.9	298.5	6.2

to the emission of hydrogen chloride and the degradation of DOP. The second stage is where the carbonaceous backbone suffers chain scission, and thus much lower-molecular-weight compounds (and smoke) are produced. The third stage tends to be, just as for many other polymers, a very slow reaction.

The first stage is mainly due to the emission of hydrogen chloride and the degradation of DOP.

At this stage, all the samples lose mass by about 60–70%. In fact, for the samples of plasticized polymer without fire-retardant additives, just over 70% is lost, with somewhat smaller losses for the samples that were treated with the flame retardants. The second stage is where the carbonaceous backbone suffers chain scission, and thus much lower-molecular-weight compounds (and smoke) are produced. The third stage tends to be, just as for many other polymers, a very slow reaction. The rate of which is relatively uniform for a wide temperature range. As regards to the second and third weight stages, the only possible systematic statement is that both the rates and the percentages of weight loss are much smaller than those at the first stage.

From Table 2, it can be seen that for sample F, compared with sample A, the mass loss rate above 400°C is decreased. The significant decrease in the mass loss rates lowers the amount and the rate of the release of combustible products from the flexible PVC's decomposition, consequently suppressing the flammability of resins. The increase in char yield agrees with the mechanism of flame retardancy. The introduction of the flame retardants leads to more char formed at the expense of flammable volatile products of thermal degradation, thus suppressing the combustion and increasing the LOI.

3.4   Thermal stability of flexible PVC

The thermal stability of the PVC material is assessed with two parameters: IDT and IPDT. IDT indicates the apparent thermal stability of PVC, the failure temperature of the resins during processing and molding. On the other hand, IPDT exhibits the resins' inherent thermal stability, i.e. the decomposition characteristics of the resins' volatile composition. From Table 2, it can be seen that PVC/magnesium olivine sand show a relatively higher IDT than does the PVC resin, indicating that the apparent thermal stability of PVC is increased. On the other hand, the existence of the flame retardants (magnesium olivine sand) exhibits a higher IPDT than the PVC, retarding the mass loss rate of the PVC at high temperatures. The high IPDT implies the PVC resins' potential application in highly anti-thermal coatings and thermal insulating materials.

4   CONCLUSIONS

The addition of 4.0 g magnesium olivine sand and 7.0 g Sb_2O_3 obtained a good flame retardancy for flexible PVC, whose LOI reached 34.0%. In a certain range, magnesium olivine sand has a little effect on the mechanical properties of the sample. It also improved the decomposition temperature of the samples, showing a good flame-retardant effect. Magnesium olivine sand can as a synergistic agent for the flame-retardant polymeric materials containing halogen flame retardancy, and the flexible PVC containing magnesium olivine sand has a good processing performance, is cheap and non-toxic, has a very broad market prospect, and a higher application value.

ACKNOWLEDGMENT

The work was supported by the fundamental research funds for the Central Universities (3142013102).

REFERENCES

[1] Chen, L. & Wang, Y. 2010. A review on flame retardant technology in China. Part I: development of flame retardants. Polymers for Advanced Technologies, 21(1), 1–26.

[2] Xu, J.Z. Zhang, C.Y. & Qu, H.Q. 2005. Zinc hydroxystannate and zinc stannate as flame-retardant agents for flexible poly (vinyl chloride). J. Appl. Polym. Sci. 98, 1469–1475.

[3] Qu, H.Q. Wu, W.H. & Jiao, Y.H. 2005. Thermal behavior and flame retardancy of flexible poly (vinyl chloride) treated with Al (OH)$_3$ and ZnO. J. Polym. Int. 54, 1469–1473.

[4] Pi, H. Guo, S. & Ning, Y. 2003. Mechanochemical improvement of the flame-retardant and mechanical properties of zinc borate and zinc borate–aluminum trihydrate-filled poly (vinyl chloride). J. Appl. Polym. Sci. 89(3), 753–762.

[5] Tian, C.M. Wang, H. & Liu, X.L. 2003. Flame retardant flexible poly (vinyl chloride) compound for cable application. J. Appl. Polym. Sci., 89, 3137–3142.

Material Science and Engineering – Chen (Ed.)
© *2016 Taylor & Francis Group, London, ISBN 978-1-138-02936-1*

Preparation of a non-ionic polyacrylamide-grafted starch flocculant and its application in the treatment of coal slurry waste water

Y.J. Sun, Y.L. Yang, F. Zhen & T.J. Zhang
School of Environmental Engineering, North China University of Science and Technology, Yanjiao, Beijing, China

ABSTRACT: A non-ionic polyacrylamide-grafted starch (St-g-NPAM) flocculant for the coal slurry waste water was prepared by using corn starch and Acrylamide (AM) as monomers through solution polymerization. The structure of the synthesized St-g-NPAM was characterized by FT-IR. The effects of the initiator concentration, reaction temperature, and monomer concentration on the percentage of grafting and grafting efficiency were investigated. The results show that the optimal conditions of the polymerization are as follows: monomer-to-substrate ratio 2.3, potassium persulfate 0.6 g/L, reaction time 3 h, and reaction temperature 65°C. The flocculation performance of St-g-NPAM is better than the commercial flocculant PAM. The additional dosage of St-g-NPAM varies between 12 and 18 mg/L to obtain good flocculation capability.

Keywords: polyacrylamide; flocculant; coal slurry waste water; starch; graft

1 INTRODUCTION

Waste water discharge has become one of the most urgent problems for environmental protection. Flocculants are often used in fast solid–liquid separations by an aggregation process of colloidal particles (Lee et al. 2014). The process is termed as flocculation, which has been widely used as an effective water treatment technology (Sen et al. 2011). Flocculants can be divided into two categories: inorganic flocculants and organic flocculants. Among the inorganic flocculants, the salts of multivalent metals such as alum and iron are mostly used, the use of alum usually leads to the problem of residual aluminum (Nasim & Bandyopadhyay 2012). Ferrite flocculants can be expensive, and the resultant excess iron may cause an unpleasant metallic taste, odor, color, corrosion, foaming, or staining. So, synthetic organic flocculants, such as polyacrylamide derivatives, are most frequently used because of their effectiveness.

However, organic flocculants cause environmental problems because they are not readily biodegradable and some of their degraded monomers such as Acrylamides (AMs) are neurotoxic and even show a strong human carcinogenic potential (Das et al. 2013). Starch is one of the most abundant natural polymers in the world. In this paper, starch is introduced into the flocculants because of its biodegradable property and relatively low price. St-g-NPAM was prepared by the reaction with AM as a monomer, which can decrease the toxicity of organic flocculants. Its application in the treatment of coal slurry waste water is discussed.

2 EXPERIMENTAL PROCEDURE

2.1 Materials

Acrylamide (AM), potassium persulfate, hydrochloric acid, sodium hydroxide, acetone, acetic acid glycol, and sodium hydroxide were supplied by Beijing Chemical Plant (Tianjin, China). Corn starch was obtained from Smetana Food, China.

2.2 Preparation of St-g-NPAM flocculant

The reaction was carried out in a 250 ml four-necked round-bottom flask equipped with a stirrer, a thermometer, a condenser, and a nitrogen gas inlet. The flask was heated in a thermostatic water bath. Corn starch and deionized water were poured into the flask and preheated for 1 h at 85°C. After the starch was gelatinized (starch slurry turning to a transparent solution), the flask contents were cooled to 50°C. Then, AM and potassium persulfate were added, and the mixture was allowed to react for 5 h at 60°C under N_2 atmosphere. The product was washed with 200 ml acetone, and then dried and crushed, to obtain the crude product. Then, the crude product was washed for 24 h with the solution of glacial acetic acid and ethylene glycol at a ratio of 6:4, to obtain the refined final product. The synthetic reaction is shown in Scheme 1.

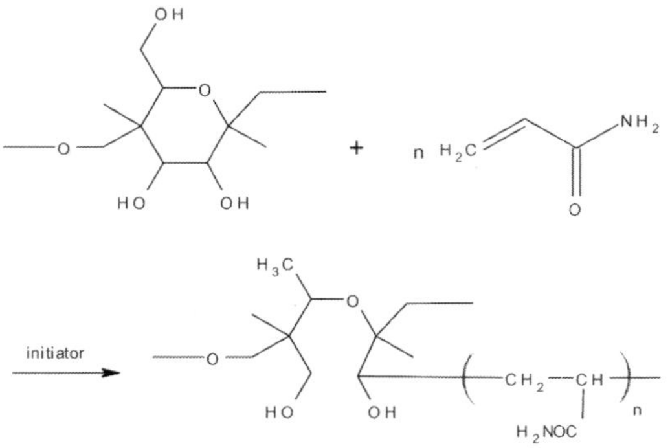

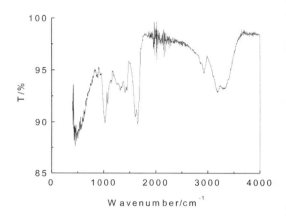

Scheme 1. The synthetic reaction of St-g-NPAM.

Figure 1. FT-IR spectrum of the St-g-NPAM.

2.3 PG and GE properties

The Percentage of Grafting (PG) and Grafting Efficiency (GE) can be calculated by the following formula:

$$PG = \frac{M_2 - M_0}{M_0} \times 100\%, \ GE = \frac{M_2}{M_1} \times 100\%.$$

where M_0 is the weight of the starch; M_1 is the weight of the crude product; and M_2 is the weight of the refined product.

2.4 FT-IR analysis

The chemical structure of the St-g-NPAM flocculant was characterized by the KBr disk method with a NEXUS-470 FTIR (Nicolet) spectrophotometer.

2.5 Flocculation testing

A 30 g coal slurry sample and 1.0 L water were added to a 1.0 L beaker. First, the solution was stirred for 15 min by a mechanical stirrer at 300 rpm/min, and then the speed was adjusted to 120 rpm/min. The flocculants were added and stirred for 1 min, to generate a small flocculation. Finally, the mixture was stirred for 5 min at a low speed of 40 rpm/min, to promote the growth of flocculation. The supernatant was extracted after standing for 15 min, and then its percentage of transmittance (T%) was measured at 550 nm by using a spectrophotometer (model number 722; Shanghai Precise Instrument Plant, China).

3 RESULTS AND DISCUSSION

3.1 IR spectra of PAM

The FT-IR spectra of St-g-NPAM are shown in Figure 1. The broad band at 891 cm^{-1} is due to the stretching mode of the glucose ring. The intense peak at 1030 cm^{-1} is assigned to the glucose ring of C-O stretching. Two peaks at 1430 and 2925 cm^{-1} correspond to the characteristic absorption of the C-C and –CH$_2$– vibrations, respectively. The peak at 3310 cm^{-1} is due to the –NH$_2$– stretching vibration, and the peak at 1675 cm^{-1} is due to the C = O stretching vibration, which is a strong evidence of the incorporation of PAM onto the backbone of the starch.

3.2 Effect of different factors on grafting parameters

3.2.1 Monomer

Figure 2 shows the effect of the monomer-to-substrate ratio on Grafting Parameters (GP). It shows that PG and GE increase when the monomer-to-substrate ratio is less than 2.3. However, they all show decreasing trends when the monomer-to-substrate ratio is more than 2.3. The average number of free radicals increases with the monomer concentration. However, when the total monomer mass increases to a certain value, the polymerization rate is accelerated and the residue monomer increases, which can result in the decrease of PG and GE. The highest value of PG and GE appears at the monomer-to-substrate ratio of 2.3.

3.2.2 Initiator concentration

Figure 3 shows the effect of potassium persulfate on the grafting reaction. The PG and GE values are low when the concentration of potassium persulfate is less. It is suggested the radical initiator is locked inside a cage when it is very less (Das et al. 2012), which may cause a side reaction and consume the initiator. This phenomenon is called the "cage effect." (Wang et al. 2011) When the initiator increases, the radical initiator is sufficient enough to come out from the cage to initiate the grafting process, and therefore PG and GE increase with

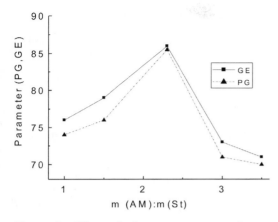

Figure 2. Effect of the monomer to substrate ratio on GP.

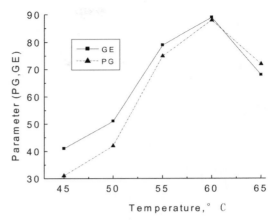

Figure 4. Effect of reaction temperature on GP.

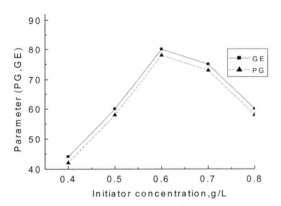

Figure 3. Effect of potassium persulfate on GP.

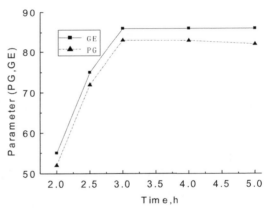

Figure 5. Effect of reaction time on GP.

the initiator. When the concentration of potassium persulfate is 0.6 g/L, the grafting reaction reaches its maximum point with PG and GE values of 80% and 78%, respectively.

3.2.3 *Temperature*

Figure 4 shows that the PG and GE values first increase and then decrease with the reaction temperature. It is suggested the thermal motion of the molecules increases with the increase in the temperature, and more free radicals will be formed, and hence the chain growth will be accelerated, resulting in a higher PG and GE value. However, when the temperature exceeds 60°C, the molecular chains of St-g-NPAM may be broken, and then the chain transfer and chain termination of the reaction and polymerization rate also increase, which causes an obvious decline in the GE value at 60°C. So, 60°C is regarded as the optimal reaction temperature and used in the following discussion.

3.2.4 *Time*

The effect of the reaction time on grafting is shown in Figure 5. As it can be seen in Figure 6, with the reaction time, the PG and GE values increased much before 3 h, and then the PG value increased slowly and the GE value decreased. It is suggested that the concentration of the monomer becomes less gradually after 3 h with the polymerization. So, the optimum reaction time is 3 h.

3.3 *Flocculation characteristics of ST-G-NPAM*

Flocculation performance of St-g-NPAM is determined by the transmittance of the solution and the sedimentation velocity, which are shown in Figure 6 and Figure 7. As it can be seen in Figure 6 and Figure 7, the transmittance of the solution and the sedimentation velocity of St-g-NPAM are much higher than those of PAM, which is a good commercial flocculant. With the increase in

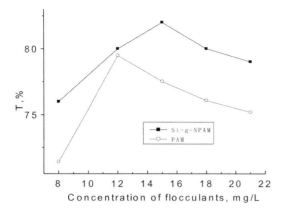

Figure 6.   Flocculation performance of St-g-NPAM.

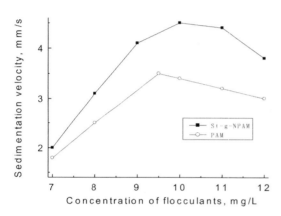

Figure 7.   Effect of St-g-NPAM on flocculation performance.

St-g-NPAM, the T% and sedimentation velocity increase. When the dosages of St-g-NPAM are 15 mg/L and 10 mg/L, the T% and sedimentation velocity reach up to the highest point. However, T% would decrease if the St-g-NPAM dosage continues to increase. Because of less flocculant dosage, its ability of electricity and adsorption bridging need to be enhanced. With the dosage increasing, the capacity of adsorption starts to increase. Nevertheless, the adsorbed particles would be wrapped by the excessive polymer flocculant, so that they could not coagulate but disperse when the dosage

used is excessive. The results of the experiments show that when the additional dosage of St-g-NPAM varies between 12 and 18 mg/L, the flocs are coarse, and the subsidence velocity is fast and water quality is good.

4 CONCLUSIONS

St-g-NPAM can be synthesized by AM and starch with only one-step reaction. The PG and GE of St-g-NPAM are affected by the temperature, the ratio of monomer and starch, the initiator concentration, the reaction time, and the reaction temperature. The PG and GE of St-g-NPAM can reach up to 78% and 80%, respectively, under optimal conditions. The optimal conditions of the polymerization are as follows: monomer-to-substrate ratio 2.3, potassium persulfate 0.6 g/L, reaction time 3 h, and reaction temperature 60°C. The flocculation performance of St-g-NPAM is better than the commercial flocculant PAM. The additional dosage of St-g-NPAM varies between 12 and 18 mg/L to obtain good flocculation capability.

REFERENCES

Das, R., Ghorai, S. & Pal, S. 2013, Flocculation characteristics of polyacrylamide grafted hydroxypropyl methyl cellulose: An efficient biodegradable flocculant. *Chemical Engineering Journal.* 229, 144–152.

Lee, C.S., Robinson J. & Chong, M. F. 2014. A review on application of flocculants in wastewater treatment. *Process Safety and Environmental Protection.* 92, 489–508.

Nasim, T. & Bandyopadhyay, A. 2012. Introducing different poly (vinyl alcohol)s as new flocculant for kaolinated waste water. *Separation and Purification Technology.* 88, 87–94.

Pal, S., Sen, G. & Ghosh, S. 2012. High performance polymeric flocculants based on modified polysaccharides—Microwave assisted synthesis. *Carbohydrate Polymers.* 87, 336–342.

Sen, G., Ghosh, S., Jha, U. & Pal, S. 2011. Hydrolyzed polyacrylamide grafted carboxymethylstarch (Hyd. CMS-g-PAM): An efficient flocculant for the treatment of textile industry wastewater. *Chemical Engineering Journal.* 171, 495–501.

Wang, L.F., Duan, J.C. & Miao, W.H. 2011. Adsorption–desorption properties and characterization of crosslinked Konjac glucomannan-graft-polyacrylamide-co-sodium xanthate. *Journal of Hazardous Materials.* 186, 1681–1686.

Material Science and Engineering – Chen (Ed.)
© 2016 Taylor & Francis Group, London, ISBN 978-1-138-02936-1

Preparation of a cationic polyacrylamide-grafted starch flocculant and its application

Y.J. Sun, Y.L. Yang & Q.L. Xu
School of Environmental Engineering, North China University of Science and Technology, Yanjiao, Beijing, China

ABSTRACT: A cationic polyacrylamide-grafted starch (St-g-AM-DMC) flocculant for the coal slurry waste water was prepared by using corn starch and Acrylamide (AM) as monomers and ethylene methyl propenoyl trimethylammonium chloride (DMC) as a cationic monomer through the solution polymerization. The effects of the initiator concentration, reaction temperature, and monomer concentration on the percentage of grafting and grafting efficiency were investigated. The results show that the optimal conditions of the polymerization are as follows: monomer-to-substrate ratio 2.5, DMC:AM ratio 1:3, potassium persulfate 3×10^{-3} mol/L, and reaction temperature 50°C. The additional dosage of St-g-AM-DMC varies between 15 and 25 mg/L to obtain good flocculation capability.

Keywords: cationic polyacrylamide, flocculant, coal slurry waste water, starch, graft

1 INTRODUCTION

Flocculation is a means of removing organic and inorganic contaminants from waste water and involves the aggregation of dispersed particles into larger flocs that can be separated from the water (Yang et al. 2014). Currently, a wide range of flocculants are extensively used in various industries (Pal et al. 2011). According to the composition, they can be divided into two categories: organic and inorganic flocculants. Both organic and inorganic flocculants are used in various kinds of flocculation phenomena (Liu et al. 2014). Inorganic flocculants are inexpensive, but less effective. Organic flocculants are essentially polymeric in nature. Synthetic polymers are very efficient flocculants. However, they cause environmental problems because they are not readily biodegradable and some of their degraded monomers such as Acrylamides (AMs) are neurotoxic and even show a strong human carcinogenic potential (Yang et al. 2014). To solve these problems, substitutes of natural polymers, such as starch, chitosan, and cellulose, have been investigated as an attractive alternative because natural polymers and their derivatives are biodegradable as well as their degradation intermediates are harmless to humans and the environment.

Starch is one of the most abundant natural polymers in the world, and as an important derivative of starch flocculant, cationic polyacrylamide-grafted starch (St-g-AM-DMC) can be prepared conventionally by a two-step method: first grafting AM to starch, and then adding formaldehyde and dimethylamine through the Mannich reaction, which is complicated and not environmental friendly (Lee & Robinson 2014).

In this article, starch is introduced into the flocculant because of its biodegradable property and relatively low price. St-g-AM-DMC is prepared by a one-step reaction with DMC as a cationic monomer, which can prevent the toxicity of formaldehyde. The application of the treatment of coal slurry waste water is discussed.

2 EXPERIMENTAL PROCEDURE

2.1 Materials

Acrylamide (AM), methyl acrylic acid ethyl trimethyl chloride ammonium (DMC), potassium persulfate, hydrochloric acid, sodium hydroxide, acetone, acetic acid glycol, and sodium hydroxide were supplied by Beijing Chemical Plant (China). Corn starch was obtained from Simeite Food, China.

2.2 Preparation of St-g-AM-DMC flocculant

The reaction was carried out in a 250-mL four-necked round-bottom flask equipped with a stirrer, a thermometer, a condenser, and a nitrogen gas inlet. The flask was heated in a thermostatic water bath. Corn starch and deionized water were poured into the flask and preheated for 1 h at 85°C. After the starch was gelatinized (starch

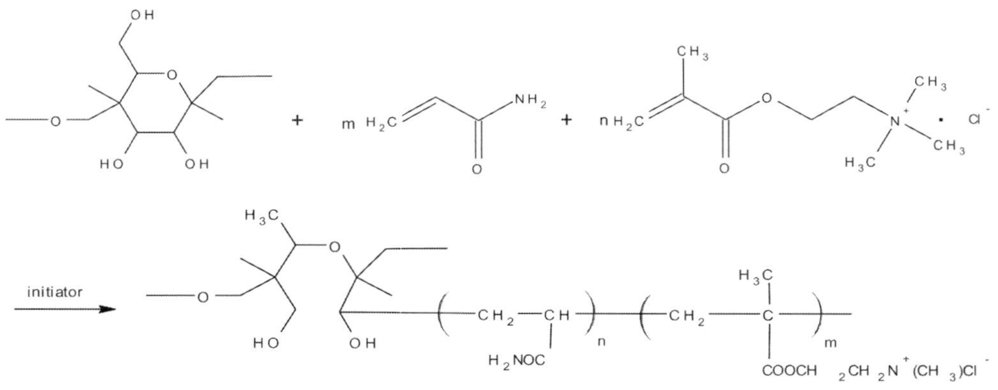

Scheme 1. The synthetic reaction of St-g-AM-DMC.

slurry turning to a transparent solution), the flask contents were cooled to 40°C. Then, AM, DMC, and potassium persulfate were added, and the mixture was allowed to react for 5 h at 55°C under N₂. The product was washed with 200 mL acetone, and then dried and crushed, to obtain the crude product. Then, the crude product was washed for 12 h with the mixture of methanol and ethanol, to obtain the refined final product. The synthetic reaction is shown in Scheme 1.

2.3 PG and GE properties

The Percentage of Grafting (PG) and Grafting Efficiency (GE) can be calculated by the following formula:

$$PG = \frac{M_2 - M_0}{M_0} \times 100\%, GE = \frac{M_2}{M_1} \times 100\%.$$

where M0 is the weight of the starch; M1 is the weight of the crude product; and M2 is the weight of the refined product.

2.4 FT-IR analysis

The chemical structure of the St-g-AM-DMC flocculant was characterized by the KBr disk method with a NEXUS-470 FTIR (Nicolet) spectrophotometer.

2.5 Flocculation testing

A 30 g coal slurry sample and 0.5 L water were added to a 0.5-L beaker. First, the solution was stirred for 15 min by a mechanical stirrer at 300 rpm/min, and then the speed was adjusted to 120 rpm/min. The flocculants were added and stirred for 1 min, to generate a small flocculation. Finally, the mixture was stirred for 5 min at a low speed of 40 rpm/min, to promote the growth of flocculation. The supernatant was extracted after standing for 15 min, then its percentage transmittance (T%) was measured at 550 nm by using a spectrophotometer (model number 722; Shanghai Precise Instrument Plant, China).

3 RESULTS AND DISCUSSION

3.1 IR spectra of CPAM

The FTIR spectra of St and St-g-AM-DMC are shown in Figure 1. It can be seen that the FT-IR spectra of St and St-g-AM-DMC are much different. The broad band at 1000 cm⁻¹ is due to the stretching mode of the glucose ring, and assigned to the glucose ring of C-O stretching. Two peaks at 1400 and 2920 cm⁻¹ correspond to the characteristic absorption of the C-C and -CH2- stretching, respectively. The peak at 3400 cm⁻¹ is due to the -NH2- stretching vibration. The peak of 1120–1450 cm⁻¹ is attributed to trimethyl, which is connected to the N+ of DMC structural units, which is a strong evidence of the incorporation of a cationic moiety onto the backbone of the starch.

3.2 Effect of different factors on grafting parameters

3.2.1 Monomer
Figure 2 shows the effect of the monomer-to-substrate ratio on grafting parameters. It shows that the PG and GE values increase when the monomer-to-substrate ratio is less than 2.5. However, they all show decreasing trends when the monomer-to-substrate ratio is more than 2.5. The average number of free radicals increases with the monomer concentration. However, when the total monomer mass increases to a certain value, the polymerization rate is accelerated and the

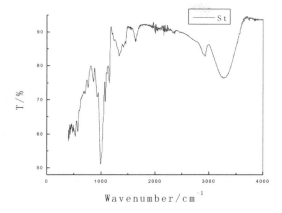

Figure 1.   FT-IR spectrum of St and the St-g-AM-DMC.

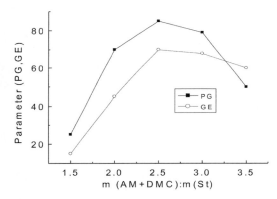

Figure 2.   Effect of the monomer-to-substrate ratio on grafting parameters.

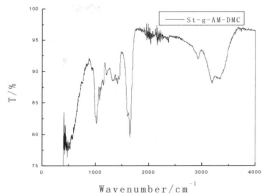

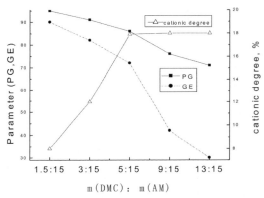

Figure 3.   Effect of the monomer ratio on grafting parameters.

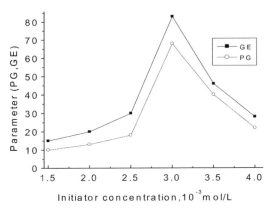

Figure 4.   Effect of potassium persulfate on grafting parameters.

residue monomer increases, which can result in the decrease of the PG and GE values. The highest value of PG and GE appears at the monomer-to-substrate ratio of 2.5.

3.2.2   *Monomer ratio*

Figure 3 shows that the PG and GE values increase with the amount of AM. The reactivity of AM is much higher than that of DMC; therefore, the higher the content of AM is, the higher the PG and GE values are. However, the cationic degree will decrease with the amount of AM. The PG and GE reach their highest values at the DMC:AM ratio of 1:3 (5:15).

3.2.3   *Initiator concentration*

Figure 4 shows the effect of potassium persulfate on the grafting reaction. The reaction is not observed when the concentration of potassium persulfate is 1.5×10^{-3} mol/L. It is suggested the radical initiator is locked inside a cage when it is very less, which may cause a side reaction and consume the initiator. This phenomenon is called the "cage effect". When the initiator increases, the radical initiator is sufficient enough to come out from the cage to initiate the grafting process, and therefore the PG and GE values increase with the initiator. When the concentration of potassium persulfate is 3×10^{-3} mol/L, the grafting reaction reaches its maximum point at the PG and GE values of 68 and 83%, respectively.

3.2.4 *Temperature*

Figure 5 shows that the PG and GE values first increase and then decrease with the reaction temperature. It is suggested the thermal motion of the molecules increases with the increase in the temperature, and more free radicals will be formed, and hence the chain growth will be accelerated, resulting in a higher PG and GE value. However, when the temperature exceeds 50°C, the molecular chains of St-g-AM-DMC may be broken, and the chain transfer and chain termination of the reaction and polymerization rate also increase, which causes a obvious decline in the GE value at 55°C. So, 50°C is regarded as the optimal reaction temperature and used in the following discussion.

3.3 *Flocculation characteristics of St-g-AM-DMC*

The flocculation performance of St-g-NPAM is determined by the transmittance of the solution and the sedimentation velocity, which is shown in Figure 6 and Figure 7. As shown in Figure 6 and Figure 7, the transmittance of the solution and the sedimentation velocity of St-g-AM-DMC are much higher than those of PAM, which is a good commercial flocculant. With the increase in St-g-AM-DMC, the T% and sedimentation velocity increase. When the dosages of St-g-AM-DMC are 20 mg/L and 18 mg/L, the T% and sedimentation velocity reach the highest point. However, T% would descend if the St-g-AM-DMC dosage continues to increase. Because of the less flocculant dosage, its ability of electricity and adsorption bridging need

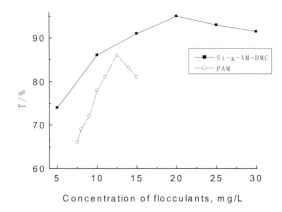

Figure 6. Effect of the reaction time on grafting parameters.

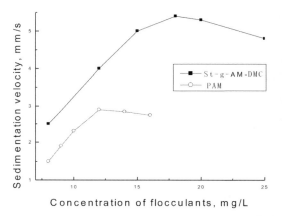

Figure 7. Flocculation performance of St-g-AM-DMC.

to be enhanced. With the dosage increasing, the capacity of adsorption starts to increase. Nevertheless, the adsorbed particles would be wrapped by the excessive polymer flocculant, so that they could not coagulate but disperse when the dosage used is excessive. The results of the experiments show that when the additional dosage of St-g-AM-DMC varies between 15 and 25 mg/L, the flocs are coarse, and the subsidence velocity is fast, and water quality is good.

4 CONCLUSIONS

St-g-AM-DMC can be synthesized by AM, DMC, and starch with only one-step reaction. The PG and GE of St-g-AM-DMC are affected by the temperature, the ratio of monomer and starch, the ratio of AM and DMC, the initiator concentration, the reaction time of 4 h, and the reaction temperature. The PG and GE values of St-g-AM-DMC can

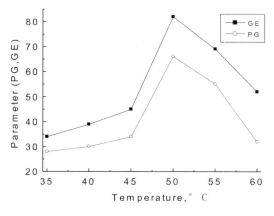

Figure 5. Effect of the reaction temperature on grafting parameters.

reach up to 68 and 83%, respectively, under optimal conditions. The optimal conditions of the polymerization are as follows: monomer-to-substrate ratio 2.5, DMC:AM ratio 1:3, potassium persulfate 3×10^{-3} mol/L, and reaction temperature 50°C. The additional dosage of St-g-AM-DMC varies between 15 and 25 mg/L to obtain good flocculation capability.

REFERENCES

Lee, Ch. Siah. & Robinson, J. 2014. A review on application of flocculants in wastewater treatment. *Process Safety and Environmental Protection.* 92, 489–508.

Liu, H.Y. Yang, X.G. Zhang, Y. & Zhu, H.Ch. 2014. Flocculation characteristics of polyacrylamide grafted cellulose from Phyllostachys heterocycla: An efficient and eco-friendly flocculant. *Water Research.* 59, 165–171.

Pal, S. Ghorai, S. & Dash, M.K. 2011. Flocculation properties of polyacrylamide grafted carboxymethyl guargum (CMG-g-PAM) synthesised by conventional and microwave assisted method. *Journal of Hazardous Materials.* 192, 1580–1588.

Yang, Zh. Li, H.J. Yan, H. Wu, H. & Yang, H. 2014. Evaluation of a novel chitosan-based flocculant with high flocculation performance, low toxicity and good floc properties. *Journal of Hazardous Materials.* 276, 480.

Yang, Zh. Wu, H. & Yuan, B. 2014. Synthesis of amphoteric starch-based grafting flocculants for flocculation of both positively and negatively charged colloidal contaminants from water. *Chemical Engineering Journal.* 244, 209–217.

Material Science and Engineering – Chen (Ed.)
© *2016 Taylor & Francis Group, London, ISBN 978-1-138-02936-1*

Epoxy resins treated with manganese-containing compounds

Y.L. Yang, C.B. Guo & M. Gao
*School of Environmental Engineering, North China University of Science and Technology,
Yanjiao, Beijing, China*

ABSTRACT: A novel cheap Macromolecular Intumescent Flame Retardant (Mn-MIFR) was synthesized and characterized by IR, which indicated that its structure was a caged bicyclic macromolecule containing phosphorus and manganese. Epoxy Resins (EP) were modified with Mn-MIFR to obtain the flame-retardant EP, whose flammability and burning behavior were characterized by UL 94, Limiting Oxygen Index (LOI), dilatation, char yield, Smoke Density Rating (SDR) and Maximum Smoke Density (MSD). The epoxy resins were obtained for the UL 94 V-0 rating at low Mn contents of 4.0%, a LOI of 27.0% and char yield of 18.2%. It was shown that the dilatation, SDR and MSD of EP/Mn-MIFR decreased. The degradation behavior of the EP/Mn-MIFR was studied by the TG and EDX analysis. The experimental results revealed that the Initial Decomposition Temperature (IDT) was decreased, and that the Integral Procedure Decomposition Temperature (IPDT) and the amounts of Mn and P at the residue were increased.

Keywords: Epoxy resins; degradation; flame retardant; synthesis; manganese

1 INTRODUCTION

Recently, halogen-free Intumescent Flame Retardants (IFR) have attracted increased attention from both academic and industrial communities for their multifold advantages including low toxic, low smoke, low corrosion, and non-corrosive gas properties (Wang & Yang 2010, Song, et al. 2008). Three ingredients are necessary for IFR: acid source, carbon source and gas source. Phosphorus-containing compounds are often used as an acid source, while nitrogen-containing compounds are used as a blowing agent. The greatest benefit to be obtained in this way is a dramatic decrease in the heat generated due to the exothermic combustion of polymers. Other advantages include the conservation of the structural integrity of the polymer as a result of the residue of solid carbon and a decrease in the formation of flammable gaseous products (Cullis et al.1984). In the previous work of our group, a novel macro-Molecular IFR (MIFR) containing an acid source, a gas source and a char source was synthesized simultaneously. The thermal degradation and flame retardancy of EP were improved after blending with the MIFR (Sun et al. 2013, Dallas et al. 2011).

However, the conventional IFR additives also have some disadvantages (Poornima 2013, Wang et al. 2007). To achieve a certain flame-retarding level, a higher loading of the IFR additive is needed than that of some halogen-containing flame retardants, at the expense of the mechanical properties of flame-retardant materials. The flame-retardant efficiency of IFR needs to be further improved. Some reports (Dai & Gao 2013) have demonstrated that the presence of some metal-containing compounds with IFR seemed to enhance the flame-retardant action, and exhibit a synergistic effect on the flame retardancy of polymers.

So, in this work, the manganese element was incorporated into MIFR to achieve higher flame retardancy, and the degradation behavior of EP modified with the Mn-MIFR was investigated.

2 EXPERIMENTAL PROCEDURE

2.1 Materials

Pentaerythritol, 85% of phosphoric acid, melamine, 37% of formalin as formaldehyde, urea and manganese oxide were obtained from Beijing Chemical Reagents Co.

2.2 Instrumentation

IR spectra were measured on a NEXUS-470 FT-IR (Nicolet) spectrophotometer using KBr. The elemental analysis was carried out using a Carlo Eroa 1102 Elemental Analyzer. LOI values were determined in accordance with ASTM D2863-70 by means of a General Model HC-1 LOI apparatus. The UL 94 vertical burning classification was obtained using an ATLAS HVUL 2 burning chamber according to FMVSS 302/ZSO3975. The dimension of the sample was 127 mm × 12.7 mm × 3 mm.

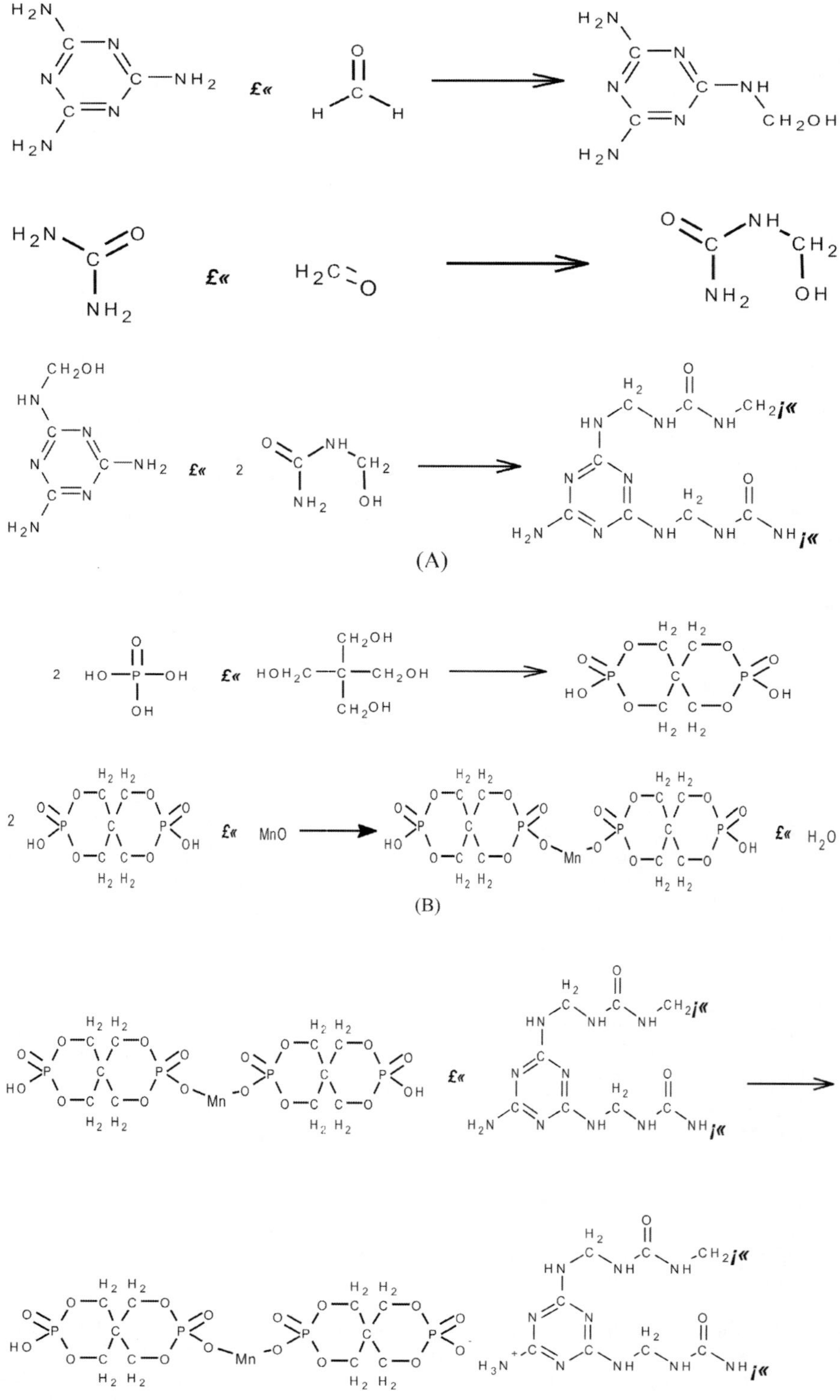

Scheme 1. Synthesis of Mn-MIFR.

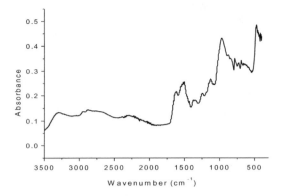

Figure 1. The Fourier transform IR spectra of Mn-MIFR.

Thermogravimetry (TG) was carried out on a DTA-2950 thermal analyzer (Dupont Co. USA) under a dynamic nitrogen (dried) atmosphere at a heating rate of 10°C min⁻¹. Energy-dispersive X-ray spectroscopy (EDX) data were obtained using an EDS2100X system at 20.00 keV.

2.3 Synthesis of MIFR

Formalin (37%)_ as formaldehyde [F] at 1 mol was brought to a pH of 8–8.5 with NaOH and heated. Then, 0.2 mol melamine [M] and 0.5 mol urea [U] were added to the above solution, stirred until dissolved and heated under reflux for 50 min. Heating was stopped and the solution was allowed to cool to obtain MUF prepolymer A.

Phosphoric acid (85%, 1 mol) and pentaerythritol (0.5 mol) were mixed, heated to 120°C, stirred for 4 h until without water distilled to get caged bicyclic pentaerythritol diphosphonate. Then, manganese oxide was doped into under stirring, heated to 80°C, stirred for 1 h to obtain sample B. Sample B was added slowly into sample A under stirring to obtain the Mn-MIFR.

2.4 Characterization of Mn-MIFR

IR (KBr) (see Figure 1), (cm⁻¹): 3335 (-OH, NH), 2934, 2890 (w, CH_2), 1245~1280 (P = O), 1521, 813 (C = N), 1021, 882, 785, 653 (dicyclic P-O-C), 2253 (P-O-Mn).

3 RESULTS AND DISCUSSION

3.1 Flame retardancy of epoxy resins

The flame-retardant properties of EP containing 20% Mn-MIFR are listed in Table 1, and the char and smoke suppression properties of the EP samples are listed in Table 2. From Table 1 and Table 2, we can

Table 1. Flame-retardant properties of EP containing different contents of Mn in the Mn-MIFR.

EP/Mn-MIFR	Mn content in Mn-MIFR (%)					
	0	2.0	3.0	4.0	5.0	6.0
LOI (%)	26.0	26.3	26.2	27.0	26.5	26.5
UL 94	V-1	V-1	V-1	V-0	V-1	V-1

Table 2. Char and smoke suppression properties of the EP samples.

FR	Addition (%)	Char yield (%)	Dilatation (cm³/g)	SDR (%)	MSD (%)
–	0	9.1	6.4	74.35	99.63
MIFR	20	15.1	67.2	61.87	92.67
Mn-MIFR	20	18.2	51.4	64.72	94.25

observe the good flame retardancy of the IFR. With the increase in the Mn content in the Mn-MIFR at the same dosage, the LOI increases. A 20% dosage achieves UL 94 V-0 when the Mn content in the Mn-MIFR exceeds 4.0%. The epoxy resins obtained for the UL 94 V-0 rating at low Mn contents of 4.0% achieve a LOI of 27.0% and a char yield of 18.2%. Such a large increase in the LOI value showed that Mn provided a clear synergistic effect on the LOI values of the EP/MIFR composites. Because metal ions can catalyze the dehydration and oxidation reactions, it enhances the char formation (Wu & Yang 2011) and higher char yield (18.2%) as characterized by the EP/Mn-MIFR. Meanwhile, the dilatation of EP/MIFR increases, more for the Mn-MIFR, which shows that the char is dense and compact, increasing the efficiency of flame retardancy, heat insulation, and protect inner matrix materials. The Smoke Density Rating (SDR) and Maximum Smoke Density (MSD) of the EP samples decrease with the increase in the Mn content in the Mn-MIFR, which show its smoke suppression on EP.

3.2 Degradation of epoxy resins

The simultaneous DTG and TG curves of EP and EP/Mn-MIFR were generated in dynamic nitrogen from ambient temperature to 800°C, which are shown in Figure 2. The Initial Decomposition Temperature (IDT) is defined as the temperature when the weight loss is 5%, and the Integral Procedure Decomposition Temperature (IPDT), the char yield at 800°C, temperatures at the maximum weight loss rate (Tm) and the value of the maximum weight loss rate (Rmax) were measured, and are listed in Table 3.

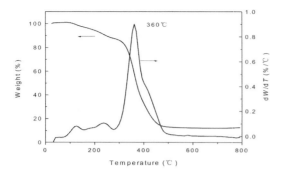

Figure 2. TG and DTG curves of EP.

Table 3. Thermal data of the epoxy resins from the thermogravimetric analysis.

No.	Mn-MIFR (%)	IDT (°C)	IPDT (°C)	Char yield (%)	Tm (°C)	Rmax (%/°C)
EP-1	–	160	422	15.2	354	0.82
EP-2	20	145	465	18.2	320/596	0.28

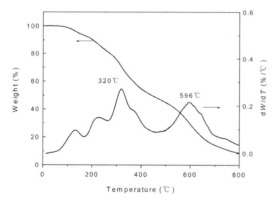

Figure 3. TG and DTG curves of EP/Mn-MIFR.

From Figures 2 and 3, and Table 3, for EP/Mn-MIFR, compared with EP, IDT decreases. It was suggested that the EP/Mn-MIFR was heated and the phosphoric groups were first decomposed, which can catalyze the decomposition of EP to form a carbonaceous char, which changed into a heat-resistant intumescent char by gaseous products such as NH_3 to retard the weight loss rate of EP at high temperatures (Zhao et al. 2011). Moreover, Rmax (0.28%/°C) were decreased, and char yields (18.2%) were increased. This result is supported by the higher LOI values indicated in Table 2.

Table 4. EDX data of the residues of the EP/Mn-MIFR.

Elements	k	ZAF correction	Weight (%)	Atom (%)
C–(Ka)	0.75919	16.9543	4.7670	6.7858
N–(Ka)	0.03338	0.2198	15.3448	18.7312
O–(Ka)	0.07402	0.0914	61.1707	65.3702
P–(Ka)	0.09248	0.6017	13.6533	7.5367
Mn–(Ka)	0.04093	0.7149	5.0642	1.5761

The thermal stability of the epoxy resins is assessed with two parameters: IDT and IPDT. IDT indicates the apparent thermal stability of the epoxy resins, i.e. the failure temperatures of the resins in processing and molding. On the other hand, IPDT exhibits the resins' inherent thermal stability, i.e. the decomposition characteristics of the resins' volatile composition. From Table 3, the EP/Mn-MIFR shows a relatively lower IDT than does the phosphorus-free resin (EP), since phosphorus-groups decompose at low temperatures. On the other hand, the existence of flame retardants (E EP/Mn-MIFR) exhibits a higher IPDT than the EP, retarding the weight loss rate of the polymers at high temperatures. The high IPDT implies the epoxy resins' potential application in highly anti-thermal coatings and thermal insulating materials. The temperatures at the maximum weight loss rate (Tm) are also increased.

3.3 Characterization of the char structure

The EDX analysis of the residues of EP/Mn-MIFR is presented in Table 4. For the EP/Mn-MIFR, the percentage of the carbon atom at the residue is low. It is noted that the amounts of Mn and P at the residue are high. These results imply that Mn and P can accumulate at the surface of the char. Therefore, the presence of the Mn-MIFR can improve the thermal-oxidative stability of the char layers and protect the matrix.

4 CONCLUSIONS

We succeed in synthesizing a novel cheap macromolecular Mn-MIFR with a structure of a caged bicyclic pentaerythritol diphosphonate. The epoxy resins obtained for the UL 94 V-0 rating at low Mn contents of 4.0% achieves a LOI of 27.0% and a char yield of 18.2%. The dilatation, Smoke Density Rating (SDR) and Maximum Smoke Density (MSD) of the EP/Mn-MIFR decrease. In the thermal degradation of the EP/Mn-MIFR, phosphorus groups decompose at a relatively low temperature, and then catalyze the dehydration, decomposition and carbonization of EP to form a heat-resistant char, retarding the weight loss rate of the EP at high temperatures. Mn and MIFR have a synergistic effect on the flame retardancy of EP.

REFERENCES

[1] Wang, G., & Yang, J. (2010). Influences of binder on fire protection and anticorrosion properties of intumescent fire resistive coating for steel structure. Surface and Coatings Technology, 204(8), 1186–1192.

[2] Song, P., Fang, Z., Tong, L., Jin, Y., & Lu, F. (2008). Effects of metal chelates on a novel oligomeric intumescent flame retardant system for polypropylene. Journal of Analytical and Applied Pyrolysis, 82(2), 286–291.

[3] Cullis, C.F., Gad, A.M.M., & Hirschler, M.M. (1984). Metal chelates as flame retardants and smoke suppressants for thermoplastic polymers. European polymer journal, 20(7), 707–711.

[4] Sun, C.F., & Gao, M. (2013, March). Epoxy resins treated with magnesium-containing compounds in material engineering. In Advanced Materials Research (Vol. 648, pp. 73–77).

[5] Dallas, P., Sharma, V.K., & Zboril, R. (2011). Silver polymeric nanocomposites as advanced antimicrobial agents: classification, synthetic paths, applications, and perspectives. Advances in colloid and interface science, 166(1), 119–135.

[6] Poornima, N. (2013). Non-destructive evaluation of photovoltaic materials and solar cells using Photoluminescence.

[7] Wang, D.Y., Liu, Y., Wang, Y.Z., Artiles, C.P., Hull, T.R., & Price, D. (2007). Fire retardancy of a reactively extruded intumescent flame retardant polyethylene system enhanced by metal chelates. Polymer Degradation and Stability, 92(8), 1592–1598.

[8] Dai, Q.J., & Gao, M. (2013). Thermal Degradation of Epoxy Resins Containing Copper Compounds. Advanced Materials Research, 705, 101–105.

[9] Wu, N., & Yang, R. (2011). Effects of metal oxides on in-tumescent flame-retardant polypropylene. Polymers for Advanced Technologies, 22(5), 495–501.

[10] Zhao, X., Sánchez, B.M., Dobson, P.J., & Grant, P.S. (2011). The role of nanomaterials in redox-based supercapacitors for next generation energy storage devices. Nanoscale, 3(3), 839–855.

Material Science and Engineering – Chen (Ed.)
© 2016 Taylor & Francis Group, London, ISBN 978-1-138-02936-1

Chitosan/polyvinylpyrrolidone hydrogels for cimetidine release

X.J. Li, Z.D. Zhou, Y. Huang & G.Y. Li
School of Life and Environmental Sciences, Guilin University of Electronic Technology, Guilin, Guangxi, China

ABSTRACT: Chitosan/polyvinylpyrrolidone-based hydrogels were synthesized using glutaraldehyde as a crosslinking agent. The hydrogels, denoted as CS/PVP, were then characterized using FT-IR spectra and SEM. Swelling studies were investigated at two different pHs. The swelling behaviors of CS/PVP hydrogels were dependent on the pH of the medium and increased faster in an acidic medium than in an alkaline medium. Furthermore, the CS/PVP hydrogels were used for the release of the cimetidine drug. The *in vitro* drug release displayed that cimetidine release from hydrogels was higher in the pH 1.0 solution than in the pH 7.4 solution, which was similar to the swelling results. All these results demonstrated that the CS/PVP hydrogels can be used for its potential capacity as a promising candidate for carrying pharmaceutical substances in the drug delivery system.

Keywords: hydrogels; chitosan; polyvinylpyrrolidone; cimetidine

1 INTRODUCTION

Hydrogels are three-dimensional, hydrophilic, polymeric networks that can retain large amounts of water or biological fluids while remaining insoluble in aqueous solutions, which are similar to a variety of natural living tissues (Dragan, 2014; Xu et al., 2015). Their porosity and response to microenvironments can be useful in biomedical applications, particularly those related to drug delivery. Molecules, even large molecules such as proteins and DNA, can be loaded into the hydrogel structure and later released when the hydrogel responds to a physiological trigger. Therefore, hydrogels are widely applied in the development of novel drug carriers with controlled release characteristics (Cheng et al., 2014; Gao et al., 2014).

Chitosan (CS), poly(1→4)-2-amino-2-deoxy-D-glucan, which is a partially deacetylated biopolymer of acetyl glucosamine obtained after the alkaline deacetylation of chitin, is particularly attractive for pharmaceutical applications because of its high solubilizing capacity, biodegradability, and desired safety profile (Altinisik and Yurdakoc, 2014; Cui et al., 2014). Chitosan-based hydrogels have a good pH sensitivity in an aqueous solution because of abundant of amino and hydroxyl functional groups (Solomko et al., 2014). Based on the excellent gel-forming property, numerous chitosan-based hydrogels have been developed (Alver et al., 2014; Li et al., 2014; Mukhopadhyay et al., 2014; Yang et al., 2013). Yang et al. reported the preparation of novel hydrogels composed of PEG grafted on carboxymethyl chitosan and alginate, and found an improvement of the protein release at pH 7.4 (Yang et al., 2013).

Li et al developed a chitosan/β-glycerophosphate thermosensitive hydrogel loaded with docetaxel for i.t. delivery to enhance therapeutic efficacy and alleviate system toxicity (Li et al., 2014).

Polyvinylpyrrolidone (PVP), a water-soluble biocompatible polymer, was used as a synthetic blood plasma substitute, an additive in drug compositions and a vitreous humor substitute (El Achaby et al., 2014; Zhang et al., 2013). The incorporation of PVP into hydrogels is expected to influence their morphological, swelling, and drug release characteristics (Wei et al., 2014). PVP can form hydrogen bonding with chitosan and increases chitosan hydrogels' mechanical properties and thermal stability (Altinisik and Yurdakoc, 2014; Alver et al., 2014).

In this study, chitosan/polyvinylpyrrolidone-based hydrogel (CS/PVP) was prepared in the presence of glutaraldehyde as a crosslinking agent, and then its applicability as a drug delivery carrier was tested using cimetidine as the model drug. The composition, morphology, swelling kinetics of the hydrogels, and the release profiles of cimetidine from the carrier in a simulated gastric (pH 1.0) and intestinal medium (pH 7.4) were investigated.

2 EXPERIMENTAL PROCEDURE

2.1 Materials

Cimetidine (purity > 99%) was obtained from the Hubei Yuancheng Pharmaceutical Co. Ltd (Wuhan, China). Chitosan (CS, Mw = 4.9×10^5, degree of deacetylation 95%) was obtained from Zhejiang Aoxing Biotechnology Co., Ltd

(Zhejiang, China). PVP K30 (Mw = 50000) and Glutaraldehyde (GA) were obtained from Sinopharm Chemical Reagent Co., Ltd (Shanghai, China). All other reagents and solvents were of analytical grade and used without further purification.

2.2 *Preparation of cimetidine-loaded chitosan/ polyvinylpyrrolidone hydrogels*

A sample of 0.72 g CS was dissolved in 36 mL of 2% acetic acid solution, containing 0.18 g cimetidine and sonicated for 30 min. Then, 2.88 g PVP K30 was gradually added to the CS solution under magnetic stirring (500 rpm, 25°C). After 30 min, 15 mL glycerol was added to the above solution and stirred for 15 min. Finally, 0.12 mL of 25% glutaraldehyde solution were added slowly to the suspension and stirred to assure efficient crosslinking. Cimetidine-loaded Chitosan/Polyvinylpyrrolidone hydrogels (CI-loaded CS/PVP) were collected, and washed with distilled water three times. Hydrogels were dried completely in a vacuum oven at 40°C for 2 days. A schematic representation of the structure of the cimetidine-loaded CS/PVP hydrogel is shown in Figure 1. Unloaded hydrogels were prepared in a similar way without CI to determine the equilibrium swelling values of the hydrogels.

2.3 *Swelling test*

The swelling behavior was detected by immersing the CS/PVP xerogels in a phosphate-buffered saline solution (PBS, 0.1M, pH 7.4) or a HCl solution (0.1M, pH 1.0) for 6 h at room temperature until the equilibrium of swelling had reached. The weights of the swollen samples were measured after the excess surface solution was removed by filter paper. The Swelling Ratio (SR) was calculated using the following equation:

$$SR(wt, \%) = \frac{(W - W_0)}{W_0} \times 100\%$$

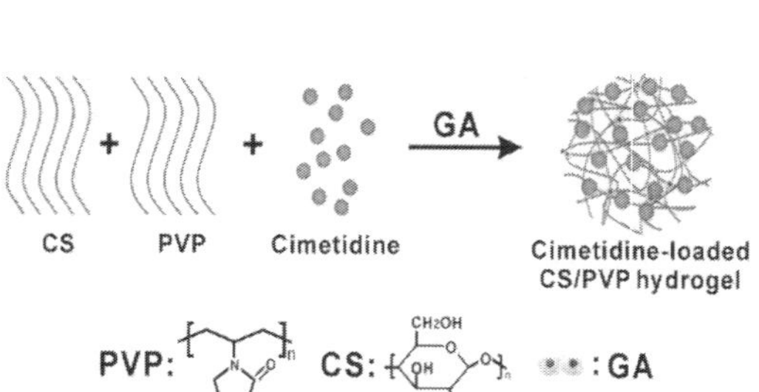

Figure 1. A schematic representation of the structure of CI-loaded CS/PVP hydrogels.

where W and W_0 are the weights of the hydrogels at the swollen state and at the dry state, respectively. Each experiment was repeated three times.

2.4 *In vitro drug release studies*

The CI-loaded CS/PVP xerogel (one disk 1 cm × 1 cm) was distributed into 5 mL of simulated gastric fluid (0.1M HCl, pH 1.0) or simulated intestinal fluid (PBS, pH 7.4, 0.1 M). Then, the solution was introduced into a dialysis membrane (molecular weight cut-off 8000–14000 Da), Afterwards, the dialysis membrane was put into a glass bottle with 95 mL of PBS solution (or HCl solution) and shaken at 100 rpm and 37°C. At a given time interval, a certain amount of the release medium was withdrawn and assayed for the absorbance of cimetidine with a UV/Vis spectrophotometer (Perkin-Elmer Lambda 35, USA) at 218 ± 2 nm using a standard curve with a correlation coefficient $R^2 = 0.999$. The concentrations of the released drug were converted to percent-release by dividing the released quantities over the total loaded drug. Each experiment was repeated three times.

2.5 *Characterization of the hydrogels*

The morphology of the CS/PVP hydrogels and the CI-loaded CS/PVP hydrogels was investigated using a field emission Scanning Electron Microscope (SEM) (Quanta 200 FESEM, FEI, Oregon, USA) in a low-vacuum mode and optic microscope imaging (Microscope LSM5, Carl Zeiss, Germany). Fourier transform infrared (FT-IR) spectra of CS, PVP, CS/PVP hydrogels and the CI-loaded CS/PVP hydrogels were recorded with KBr discs in the range of 4000–400 cm^{-1} on Nicolet AVATAR360 Fourier-transfer infrared.

3 RESULTS AND DISCUSSION

3.1 *The chemical structure of the CI-loaded CS/PV hydrogels*

Figure 2 shows the FT-IR spectra of the CS, PVP, CS/PVP hydrogels and the CI-loaded CS/PVP hydrogels. The FT-IR spectrum of PVP (Figure 2a) indicated the characteristic chemical bonds. The strong absorption band at 1574 cm^{-1} represented the carbonyl group (C = O). Another peak at 1068 cm^{-1} was due to the presence of C–N stretching. The peak at 1374 cm^{-1} was assigned to the vibration of the pyrrolidone ring, and the band at 2862 cm^{-1} to the stretching vibration of the aliphatic—CH$_2$ (Alver et al., 2014). In the spectrum of chitosan (Figure 2b), the basic characteristic peaks of chitosan were found at 3461 cm^{-1} (O-H stretching and N-H stretching, overlap), 2904 and 2854 cm^{-1} (C-H stretching), 1675 cm^{-1}

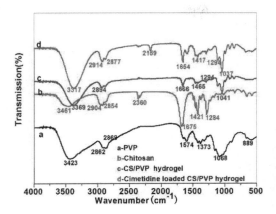

Figure 2. The FT-IR spectra of the PVP (a), CS (b), CS/PVP hydrogels (c) and the CI-loaded CS/PVP hydrogels (d).

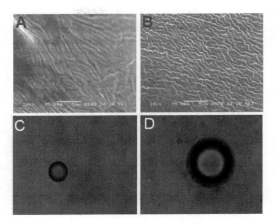

Figure 3. SEM and (400X) optic microscope imaging for CS/PVP (A, C) and CI-loaded CS/PVP hydrogels (B, D).

(NH_2 deformation), 1421 cm^{-1} (CH$_3$ symmetrical deformation) and 1284 cm^{-1} (C-O vibration stretch) (Altinisik and Yurdakoc, 2014; Khan and Ranjha, 2014).

Compared with the spectrum of chitosan and PVP, CS/PVP (Figure 2c) showed variations in the intensity and shifting in the bands. The band at 3461 cm^{-1} shifted to 3369 cm^{-1}, indicating—OH and—NH stretching vibrations. The absorption bands at 1666 cm^{-1} and 1465 cm^{-1} changed due to the hydrogen bond between the carbonyl group of PVP and the NH_2 group of chitosan (Alver et al., 2014) (Solomko et al., 2014). In the spectra of the CI-loaded CS/PVP hydrogel (Figure 2d), when the cimetidine drug was incorporated into the CS/PVP hydrogel, along with all the characteristic bands of the CS/PVP hydrogel, there was no appearance of a new band, which indicated that the CI drug is only loaded into the CS/PVP hydrogel and does not result in the chemical change of the hydrogel.

3.2 Morphology and microstructure of the CI-loaded CS/PVP hydrogels

The surface morphology and microstructure of the CS/PVP and CI-loaded CS/PVP hydrogels were characterized by SEM and optic microscope imaging, which are shown in Figure 3. The SEM observations (Figure 3A and Figure 3B) clearly illustrated that both the CS/PVP and CI-loaded CS/PVP hydrogels have smooth surfaces and a uniform inner structure, which confirms that CS and PVP blended uniformly with a good compatibility without phase separation. Also, all the hydrogels possessed three-dimensional network morphologies with a large number of molecular micropores. These microporous morphologies are very conducive to

the diffusion of water or drug molecules into or out of the hydrogel networks. Optic microscope images of dried CS/PVP and CI-loaded CS/PVP hydrogels are also shown in Figure 3C and Figure 3D. Most of the microspheres were almost spherical in shape. The diameter of the CI-loaded CS/PVP hydrogels is bigger than that of the CS/PVP hydrogels.

3.3 Swelling properties of the CS/PVP hydrogels

Swelling is one of the most important properties of a material affected by various physical and chemical parameters and molecular interactions during the process. The swelling behavior of the CS/PVP hydrogels was measured in a phosphate buffer solution (0.1 M, pH 7.4) or a HCl solution (0.1 M, pH 1.0) for 6 h at 25°C, as shown in Figure 4A. The swelling ratios of the CS/PVP hydrogels were 49.75% and 21.98% at pH 1.0 and 7.4, respectively. The swelling behaviors of the CS/PVP hydrogels were dependent on the pH of the medium and increased faster in an acidic medium than in an alkaline medium. This may be attributed to the—NH_2 groups in chitosan existing as NH_3^+ at strong acidic media (pH 1.0) and NH_2 at pH values above 7.0. This indicated that the intermolecular interaction occurred between the functional groups of chitosan and PVP, which directly interfere with swelling.

3.4 In vitro drug release studies

Hydrogels have been extensively investigated as potential drug delivery carriers due to their excellent physico-chemical properties. Drug release from hydrogel matrices is affected by many factors, such as network porosity, swelling properties, and interactions between the drug and the polymer

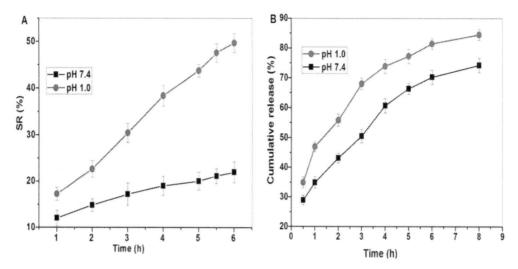

Figure 4. A) Equilibrium swelling ratios of the CS/PVP hydrogels in the pH 7.4 and pH 1.0 solutions at 25°C. B) *In vitro* release profiles of cimetidine from the CI-loaded CS/PVP hydrogels in the pH 7.4 PBS and pH 1.0 HCl media at 37°C. Data are presented as the average standard deviation (n = 3).

chains in the hydrogel network (Wei et al., 2014). In the present work, cimetidine was chosen as a model drug, and the simulated gastric fluid (0.1 M HCl, pH 1.0) and simulated intestinal fluid (PBS, pH 7.4, 0.1 M) solutions were used as modulated media. Figure 4B shows the drug release profiles from the CS/PVP hydrogels *in vitro*. It was clear that the release of cimetidine was dependent on the pH of the medium, and the cumulative percentage of cimetidine released from the hydrogel was higher in the pH 1.0 solution than in the pH 7.4 solution, which was similar to the swelling results. After 8 h, the sample released 84.48% and 74.26% cimetidine at pH 1.0 and 7.4. This parallel behavior is reasonable, since the drug release from the CS/PVP hydrogels into the solution is swelling controlled.

4 CONCLUSIONS

The present study is aimed at developing a polymeric co-network of the CS/PVP hydrogel using glutaraldehyde as a crosslinking agent. The prepared hydrogels were used as a carrier for the cimetidine drug. The presence of CS and PVP in the hydrogel was confirmed by FT-IR spectroscopy. SEM illustrated that both the CS/PVP and CI-loaded CS/PVP hydrogels have smooth surfaces and a uniform inner structure. The swelling behaviors of the CS/PVP hydrogels were dependent on the pH of the medium and increased faster in an acidic medium than in an alkaline medium. The *in vitro* drug release behavior from the CI-loaded CS/PVP hydrogels was dependent on the pH of the medium and was released quickly in the pH 1.0 HCl solution than in

the pH 7.4 PBS solution. In conclusion, the CS/PVP hydrogels can be used for its potential capacity as a promising candidate for carrying pharmaceutical substances in the drug delivery system.

ACKNOWLEDGMENTS

This work was supported by the National Nature Science Foundation of China (No. 81372362, No. 21265003 and No. 81460451) and the Fund of National College Students' Innovative Projects (No. 201310595025).

REFERENCES

[1] Altinisik, A., and Yurdakoc, K. (2014). Chitosan/poly(vinyl alcohol) hydrogels for amoxicillin release. *Polym. Bull.* 71: 759–774.
[2] Alver, E., Metin, A. U., and Ciftci, H. (2014). Synthesis and Characterization of Chitosan/Polyvinylpyrrolidone/Zeolite Composite by Solution Blending Method. *J. Inorg. Organomet. P.* 24: 1048–1054.
[3] Cheng, Y. H., Hung, K. H., Tsai, T. H., Lee, C. J., Ku, R. Y., Chiu, A. W., Chiou, S. H., and Liu, C. J. (2014). Sustained delivery of latanoprost by thermosensitive chitosan-gelatin-based hydrogel for controlling ocular hypertension. *Acta Biomater.* 10: 4360–4366.
[4] Cui, L., Jia, J., Guo, Y., Liu, Y., and Zhu, P. (2014). Preparation and characterization of IPN hydrogels composed of chitosan and gelatin cross-linked by genipin. *Carbohydr. Polym.* 99: 31–38.
[5] Dragan, E. S. (2014). Design and applications of interpenetrating polymer network hydrogels. A review. *Chem. Eng. J.* 243: 572–590.

[6] El Achaby, M., Essamlali, Y., El Miri, N., Snik, A., Abdelouahdi, K., Fihri, A., Zahouily, M., and Solhy, A. (2014). Graphene Oxide Reinforced Chitosan/ Polyvinylpyrrolidone Polymer Bio-Nanocomposites. *J. Appl. Polym. Sci.* 131: 1642–1652.

[7] Gao, X., Cao, Y., Song, X., Zhang, Z., Zhuang, X., He, C., and Chen, X. (2014). Biodegradable, pH-responsive carboxymethyl cellulose/poly(acrylic acid) hydrogels for oral insulin delivery. *Macromol. Biosci.* 14: 565–575.

[8] Khan, S., and Ranjha, N. M. (2014). Effect of degree of cross-linking on swelling and on drug release of low viscous chitosan/poly (vinyl alcohol) hydrogels. *Polym. Bull.* 71.

[9] Li, C., Ren, S., Dai, Y., Tian, F., Wang, X., Zhou, S., Deng, S., Liu, Q., Zhao, J., and Chen, X. (2014). Efficacy, pharmacokinetics, and biodistribution of thermosensitive chitosan/beta-glycerophosphate hydrogel loaded with docetaxel. *AAPS Pharm Sci Tech* 15: 417–424.

[10] Mukhopadhyay, P., Sarkar, K., Bhattacharya, S., Bhattacharyya, A., Mishra, R., and Kundu, P. P. (2014). pH sensitive N-succinyl chitosan grafted polyacrylamide hydrogel for oral insulin delivery. *Carbohydr. Polym.* 112: 627–637.

[11] Solomko, N., Budishevska, O., Voronov, S., Landfester, K., and Musyanovych, A. (2014). pH-sensitive chitosan-based hydrogel nanoparticles through miniemulsion polymerization mediated by peroxide containing macromonomer. *Macromol. Biosci.* 14: 1076–1083.

[12] Wei, Q. B., Fu, F., Zhang, Y. Q., Wang, Q., and Ren, Y. X. (2014). pH-responsive CMC/PAM/PVP semi-IPN hydrogels for theophylline drug release. *J Polym Res* 21: 453–462.

[13] Xu, J. K., Strandman, S., Zhu, J. X. X., Barralet, J., and Cerruti, M. (2015). Genipin-crosslinked catechol-chitosan mucoadhesive hydrogels for buccal drug delivery. *Biomaterials* 37: 395–404.

[14] Yang, J., Chen, J., Pan, D., Wan, Y., and Wang, Z. (2013). pH-sensitive interpenetrating network hydrogels based on chitosan derivatives and alginate for oral drug delivery. *Carbohydr. Polym.* 92: 719–725.

[15] Zhang, Q. G., Hu, W. W., Zhu, A. M., and Liu, Q. L. (2013). UV-crosslinked chitosan/polyvinylpyrrolidone blended membranes for pervaporation. *Rsc Advances* 3: 1855–1861.

Material Science and Engineering – Chen (Ed.)
© 2016 Taylor & Francis Group, London, ISBN 978-1-138-02936-1

Preparation and characterization of a novel desulfurizer, Co$_3$O$_4$/γ-Al$_2$O$_3$-attapulgite, and its optimization for the removal of SO$_2$

T. Zhang, J.Q. Li, Y.N. Wei & M. Ye
School of Petrochemical Engineering, Lanzhou University of Technology, Lanzhou, China

ABSTRACT: A new sorbent catalyst prepared by using γ-Al$_2$O$_3$ and attapulgite as raw materials to support Co$_3$O$_4$, was synthesized for the removal of SO$_2$ at a normal temperature. The optimization of the preparation of the sorbent catalyst and desulfurization experiment was studied under dynamic conditions. The physico-chemical characteristics of the samples were evaluated by various techniques such as XRD, SEM, and FT-IR. Meanwhile, the major factors affecting the preparation of the sample, such as the ratio of ATP and γ-Al$_2$O$_3$, the impregnation concentration, the calcination temperature, and the main factors affecting desulfurization, such as the moisture content, particle size, and column temperature, were investigated. The results showed the optimal synthesis conditions as follows: the γ-Al$_2$O$_3$ content was 35 wt%, the impregnation concentration was 20 wt% and the calcination temperature was 500°C. The moisture content of 30 wt%, the particle size of 2 mm, and the column temperature of 20°C can be achieved at a sulfur tolerance of 5.2 wt%.

Keywords: attapulgite; Co$_3$O$_4$; desulfurization; SO$_2$; adsorption

1 INTRODUCTION

Currently, with the gradual consciousness of the importance of environmental protection, renewable energy and nuclear power have become the world's fastest-growing energy sources. However, fossil fuels continue to supply almost 80 percent of world energy use through 2040 (EIA, 2014). In spite of the environmental issues caused by coal utilization and the perception of being a dirty energy, its position as the most important source of energy presently and in the future remains inevitable with regard to electric power generation due to the fluctuations in crude oil prices and stability of coal prices coupled with its abundance and availability in almost all parts of the world. Consequently, emissions control measures of gas pollutants, especially the removal of SO$_2$, are of utmost necessity and still a hotpot.

There are a variety of technologies to decrease the gas pollution, although the dry Flue Gas Desulfurization (FGD) technology has been argued to be a more advantageous process in terms of capital cost and waste handling (Ogenga et al., 2012; Xu et al., 2006). The related research of the FGD technology had always focused on how to improve the ability of the dry and regenerable sorbents for the past few years. It is well known that granular CuO-CeO$_2$-MnO$_x$/γ-Al$_2$O$_3$ catalysts are synthesized for the selective reduction by the sol-gel method. The experimental results showed that the catalysts maintain a nearly 100% NO conversion at 350°C (Qing et al., 2009). SO$_2$ could be reduced to elemental sulfur by methane and synthesis gas at a low temperature on the Cu/Cr-Al$_2$O$_3$ catalyst (Ismagilov et al., 2010). Two kinds of ZnO porous materials can be synthesized with the rod-shaped morphology to investigate the effect of the texture structure on the desulfurization performance in the Ni/ZnO reactive adsorption desulfurization system (Liu et al., 2013). A high reactive absorbent was prepared by mixing fly ash, lime and a small quantity of KMnO$_4$ for simultaneous desulfurization and denitrification (Zhao et al., 2014). In the field of dry and regenerable adsorbents, active carbon has been devoted considerable attention due to the highly adsorption capacity (Chu et al., 2010; Sumathi et al., 2009). In brief, finding a novel cheap and efficient sorbent is necessary and indispensable for the removal of SO$_2$.

Attapulgite (or commonly called palygoskite) is a specific material that has an excellent adsorption performance due to its specific structure (Bradley, 1940; Cao et al., 2008; Ijagbemi et al., 2009). Many researchers have an intense interest in the use of clays such as attapulgite as a potential sorbent due to its low price and abundant reserves (Zhou, 2011), especially in Jiangsu and Gansu Provinces in China. Apalygorskite/carbon sorbent was synthesized by the hydrothermal carbonization of glucose onto palygorskite under mild conditions; The experimental results showed that the modified palygorskite demonstrates a substantially high adsorption capacity for phenol (Wu et al., 2011). The palygorskite/γ-Fe_2O_3/C nanometer composite material was prepared using palygorskite and spent bleaching earth as the fundamental raw material (Qing et al., 2009). It was noted that when modified by different chemical reagents, the performances and properties of attapulgite would show a difference (Huang et al., 2008; Lilya et al., 2011; Xue et al., 2011; Li et al., 2011; Tang et al., 2011).

In the current study, a novel sorbent catalyst, Co_3O_4/γ-Al_2O_3-attapulgite, was synthesized using attapulgite, γ-Al_2O_3 and cobalt nitrate as raw materials. Major factors such as the ratio of ATP and γ-Al_2O_3, the impregnation concentration and the calcination temperature were investigated and optimized. Furthermore, the efficiency of the samples for the removal of SO_2 was investigated.

2 EXPERIMENTAL PROCEDURE

2.1 Sample preparation and optimization

Attapulgite (purchased from Xuyi, Jiangsu)-ATP was mixed homogeneously with γ-Al_2O_3 at a certain proportion, added water into mud, aged for 12 h, granulated manually, dried at 105°C for 2 h, and roasted for 2 h, to obtain the γ-Al_2O_3-ATP desulfurizer samples. Then, after impregnating the γ-Al_2O_3-ATP desulfurizer samples in the solution of cobalt nitrate for 12 h, dried at 105°C for 2 h, and roasted for 2 h, the Co_3O_4/γ-Al_2O_3-ATP desulfurizer samples were obtained.

Factors influencing the desulphurization efficiency were investigated, For instance, the ratio of ATP and γ-Al_2O_3, the impregnation concentration, and the calcination temperature; furthermore, an orthogonal experiment was conducted to obtain more precise results. The orthogonal experimental method is indicated in Table 1.

2.2 Experimental set-up and method

Figure 1 shows a schematic diagram of the self-designed experimental apparatus. It contains four parts: gas injection system, packed column, as analyzing system, and exhaust gas absorption equipment.

The efficiency of the desulfurization can be calculated by the following equation:

$$E_{ds} = \frac{m_A - m_B}{m_B} \qquad (1)$$

m_A, m_B can be measured in the process of experiment.

m_A—weight of the desulphurizer after desulphurization and

m_B—weight of the desulphurizer before desulphurization.

2.3 Dynamic desulfurization and optimization experiment

Simulated SO_2 gas was introduced into the desulfurization column (20 mm in diameter and 200 mm in length) packed with ATP samples. Exhaust gas flowed to a SO_2 absorber in the alkaline solution. The desulfurization column was placed in a water bath, which adjusted the temperature from 20°C to 80°C. The change in the SO_2 concentration at the outlet of the column was monitored by a flue gas analyzer (MOT500-II).

Factors such as the moisture content, particle size, and column temperature were investigated by the orthogonal experimental method to make the efficiency of the desulfurization optimized.

2.4 Sample characterization

The surface properties and chemical characters of the sample were studied using Scanning Electron Microscopy (SEM) (JSM-701F), X-Ray Diffraction (XRD) (D/Max-2400) and Fourier Transform Infrared Spectrophotometer (FT-IR) (FT-Raman Module made in Nicolet, American).

Table 1. The orthogonal experiment for the preparation of the samples.

	Factor		
Level	A Ratio of ATP and γ-Al_2O_3	B Impregnation concentration/%	C Calcination temperature/°C
1	A_1 = 5.5:4.5	B_1 = 10	C_1 = 400
2	A_2 = 6:4	B_2 = 15	C_2 = 500
3	A_3 = 6.5:3.5	B_3 = 20	C_3 = 600

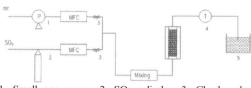

1. Small gas pump; 2. SO_2 cylinder; 3. Check valve; 4. Flue gas analyzer; 5. Exhaust gas absorption equipment.

Figure 1. Schematic diagram of the self-designed experimental apparatus.

3 RESULTS AND DISCUSSION

3.1 *The characterization of the samples*

Figure 2 shows the diagrams of attpulgite modified by three different methods. As can be seen in Figure 2(a), the structure of ATP is layered and rod-shaped crystal, which can form quantities of pore canals, and the surface is uneven. Consequently, attapulgite has a large surface area and a high capacity of absorption. When mixed with activated alumina, another material with the ability of high absorption, the structure of γ-Al_2O_3-ATP, as shown in Figure 2(b), is also rod-shaped crystal, which means that the large surface area and high absorption are still maintained. After loading Co_3O_4, the scattered cluster or agglomerate substance are observed surrounding the rod-shaped crystal structure, as shown in Figure 2(c). Moreover, no significant changes in morphology occurred after the treatment, which shows that the treatment has no apparently effect on the structure.

FT-IR spectral peaks obtained for the samples are shown in Figure 3. The FT-IR spectrum of the attapulgite samples is shown in Figure 3. According to the FTIR spectrum of ATP, the peaks at $3650\sim3200$ cm^{-1} are contributed by the stretching vibration of –OH in combination with the peaks at 3430 cm^{-1} that correspond to the stretching vibration of the Al–OH unit, respectively. The peak at 1630 cm^{-1} corresponds to the bend vibration of zeolite water; the peaks at 1030 and 579 cm^{-1} are attributed to the Si–O–Si bonds; the peak at 827 cm^{-1} may correspond to the stretching vibration of Al–O–Si (Giustetto et al., 2005; Wu et al., 2007). There are few changes in the FT-IR spectrum of γ-Al_2O_3-ATP with the influence of the peaks at $1050\sim400$ cm^{-1} related to the vibration of γ-Al_2O_3. For inorganic metal oxides, the FT-IR spectra of γ-Al_2O_3 and Co_3O_4 are too simple to have a spare bend or stretching vibration peaks of functional groups, even with the effect of the absorbed SO_2, the FT-IR spectra of Co_3O_4/γ-Al_2O_3-ATP have a little change, which means that the functional structures of the three different samples remain after the treatment.

The phase structures of the samples were studied by XRD, and the obtained results are shown in Figure 4. The diffraction peaks at 5.3°, 8.4°, 19.7°, 27.5°, 34.6° and 42.6° relate to the crystal structure of attapulgite (Niu et al., 2009). The diffraction peaks at 27.1°, 35.2° and 67.3° correspond to the crystal structure of γ-Al_2O_3. When mixed with attapulgite, no obvious changes are found in the XRD patterns of the γ-Al_2O_3-ATP sample. Meanwhile, the sample of Co_3O_4/γ-Al_2O_3-ATP shows the characteristic peaks (31.12°, 36.7°, 44.66°, 59.86°, and 65.18°) of the cubic spinel structure known from the bulk CO_3O_4 phase, i.e. Co_3O_4 is loaded on the γ-Al_2O_3-ATP sample.

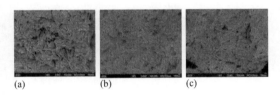

(a)   (b)   (c)

Figure 2.   SEM images of sorbent catalyst samples (a) ATP, (b) γ-Al_2O_3-ATP, (c) Co_3O_4/γ-Al_2O_3-ATP.

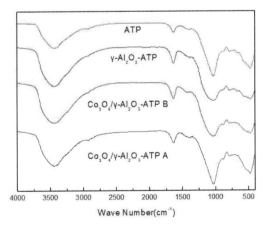

Figure 3.   FT-IR spectral peaks obtained for three different samples (Co_3O_4/γ-Al_2O_3-ATP B means the sample before desulfurization; Co_3O_4/γ-Al_2O_3-ATP A means the sample after desulfurization).

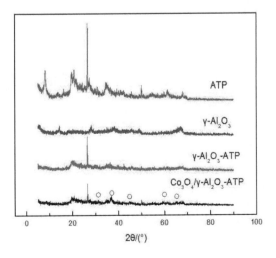

Figure 4.   XRD patterns of the samples.

3.2 *The optimization of the sample preparation and desulfurization*

As shown in Table 2, $R_A > R_B > R_C$, which means that the ratio of ATP and $\gamma\text{-Al}_2O_3$ influences the desulfurization results to a large extent, while the calcination temperature influences the desulfurization results to a small extent. According to the results of the orthogonal experiment, when the ratio of ATP and $\gamma\text{-Al}_2O_3$ is 6.5:3.5, the impregnation concentration is 20% and the calcination temperature is 600°C, as given in Table 2. The sulfur tolerance can be achieved at 5.15%.

As the sulfate efficiency of desulfurization was almost the same when the calcination temperature was 500°C and 600°C, and the calcination temperature of 500°C was chosen to save energy.

Other factors indicated an impact on the dynamic process of desulfurization, such as the moisture content of the samples, as shown in Figure 5. The sulfur tolerance showed a difference under various experimental conditions. Sulfate efficiency increased while the moisture content increased until 30%, and then decreased probably because of the moisture content lost with the simulate flue gas during the desulfurization experiment. So, the moisture content should be about 30 wt%, and the effect on the dynamic process of desulfurization was better.

The column temperature could also influence the desulfurization process; the results are shown in Figure 6. With the increase in the column temperature, the sulfur tolerance decreased, which

Table 2. Results of the orthogonal experiment for the preparation of the samples.

Factors	A	B	C	
Serial number	Ratio of ATP and $\gamma\text{-Al}_2O_3$	Impregnation concentration/%	Calcination temperature/°C	Sulfur tolerance/%
1	5.5:4.5	10	400	4.0354
2	5.5:4.5	15	500	2.5900
3	5.5:4.5	20	600	3.2309
4	6:4	10	500	3.0349
5	6:4	15	600	3.4764
6	6:4	20	400	3.1424
7	6.5:3.5	10	600	4.3958
8	6.5:3.5	15	400	2.9107
9	6.5:3.5	20	500	5.1456
t1	3.2854	3.8220	3.3628	
t2	3.2179	2.9923	3.5902	
t3	4.1507	3.8396	3.7010	
R	0.9328	0.8473	0.3382	
Optimal levels	6.5:3.5	20	600	

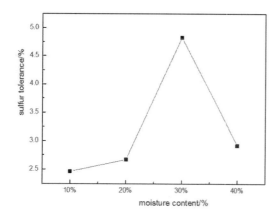

Figure 5. Influences of moisture content.

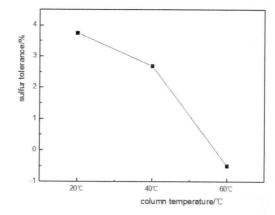

Figure 6. Influences of column temperature.

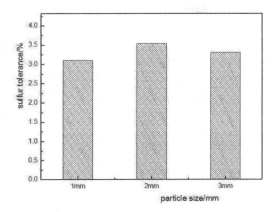

Figure 7.   Influences of particle size.

indicated that the desulfurization process is an exothermic reaction As a result, the column temperature should be about 20°C to obtain a higher sulfur tolerance.

The efficiency of the removal of SO_2 could be affected by the particle size, as shown in Figure 7. When the particle size is about 2 mm, the sulfate efficiency is better. This may be because of the larger specific surface area of the samples in about 2 mm particle size formed in the desulfurization column.

4   CONCLUSIONS

A novel γ-Al_2O_3/attapulgite-based sorbent catalyst was synthesized by using Co_3O_4 upon attapulgite after mixing with γ-Al_2O_3. The optimum synthesizing condition was the ratio of ATP and γ-Al_2O_3 of 6.5:3.5, the impregnation concentration of 20%, and the calcination temperature of 500°C. The moisture content of 30 wt%, the particle size of 2 mm, the column temperature of 20°C, and the sulfur tolerance of the sorbent can be achieved at 5.2%.

ACKNOWLEDGMENTS

This work was financially supported by the National Natural Science Foundation of China (Grant No. 51302123).

REFERENCES

[1] EIA, International Energy Outlook 2013 report. US Department of Energy Washington, D.C. 20585 DOE/EIA-0484 (2013) July 2013. <http:// www.eia. doe.gov/oiaf/ieo/pdf/0484 (2008). pdf> [accessed September, 2008].

[2] D.O. Ogenga, M.M. Mbarawa, K.T. Lee, A.R. Mohamed, I. Dahlan. 2010. Sulphur dioxide removal using South African limestone. J. siliceous materials, Fuel. 89: 2549–2555.

[3] Lusi Xu, Jia Guo, Feng Jin, Hancai Zeng. 2006. Removal of SO_2 from O_2-containing flue gas by activated carbon fiber (ACF) impregnated with NH_3. J. Chemosphere. 62: 823–826.

[4] Zhao Qing-sen, Xiang Jun, Sun Lu-shi, Shi Jin-ming, Su Sheng, Hu Song. 2009. Selective catalytic reduction of NO with NH_3 over sol-gel-derived CuO-CeO_2-$MnOx$/γ-Al_2O_3 catalysts. J. Cent. South Univ. Technol. 16: 0513–0519.

[5] Z.R. Ismagilova, S.R. Khairulina, S.A. Yashnikb, I.V. Ilyukhina, and V.N. Parmona. 2010. Developing New Catalysts and Improving Catalytic Methods for Purifying the Flue Gases of Vanyukov and Flash Smelting Furnaces. J. Catalysis in Industry. 2(4): 353–359.

[6] Liu Yun Qi, She Nannan, Zhao Jinchong, Peng Tingting and Liu Chenguang. 2013. Fabrication of hierarchical porous ZnO and its performance in Ni/ZnO reactive adsorption desulfurization. J. Pet. Sci. 10: 589–595.

[7] Yi Zhao, Tianxiang Guo, Zili Zang. Activity and characteristics of "Oxygen-enriched" highly reactive absorbent for simultaneous flue gas desulfurization and denitrification. J. Front. Environ. Sci. Eng. DOI 10.1007/s11783-014-0636-2.

[8] Chu Ying Hao, Guo Jia Xiu, Liang Juan, Zhang Qiang Bo, Yin Hua Qiang. 2010. Ni supported on activated carbon as catalyst for flue gas desulfurization. J. Science China Chemistry. 53(4): 846–850.

[9] S. Sumathi, S. Bhatia, K.T. Lee, A.R. Mohamed. 2009. Performance of an activated carbon made from waste palm shell in simultaneous adsorption of SOx and NOx of flue gas at low temperature. J. Sci. China Ser E-Tech Sci. 52(1): 198–203.

[10] Bradley, W.F., 1940. The structural scheme of attapulgite. J. Am. Mineral. 25: 405–410.

[11] Cao, J.L., Shao, G.S., Wang, Y., 2008.CuO catalysts supported on attapulgite clay for low-temperature CO oxidation. J. Catal. Commun. 9: 2555–2559.

[12] C.O. Ijagbemi, M.H. Baek, D.S. Kim. 2009. Montmorillonite surface properties and sorption characteristics for heavy metal removal from aqueous solutions. J. Hazard. Mater. 166: 538–546.

[13] Chun Hui Zhou. 2011. An overview on strategies towards clay-based designer catalysts for green and sustainable catalysis. J. Applied Clay Science. 53: 87–96.

[14] Xueping Wu, Wangyong Zhu, Xianlong Zhang, Tianhu Chen, Ray L. Frost. 2011. Catalytic deposition of nanocarbon onto palygorskite and its adsorption of phenol. J. Applied Clay Science. 52:400–406.

[15] Qing Cheng Song, Song Hao, Chen Tian Hu, Wu Xue Ping, Xie Jing Jing. 2009. Preparation and characterization of palygorskite/γ-Fe_2O_3/C nanocomposite materials. J. Chin. Ceram. Soc. 37: 548–553.

[16] Lilya Boudriche, Rachel Calvet, Boualem Hamdi, Henri Balard. 2011. Effect of acid treatment on surface properties evolution of attapulgite clay: An application of inverse gas chromatography. J. Colloids and Surfaces A: Physicochem. Eng. Aspects. 392:45–54.

[17] Ailian Xue, Shouyong Zhou, Yijiang Zhao, Xiaoping Lu, Pingfang Hana. 2011. Effective NH_2- grafting on attapulgite surfaces for adsorption of reactive dyes. J. Journal of Hazardous Materials 194: 7–14.

[18] Min Li, Zhishen Wu, Hongtao Kao. 2011. Study on preparation, structure and thermal energy storage property of capric–palmitic acid/attapulgite composite phase change materials. J. Applied Energy 88:3125–3132.

[19] Jianhua Huang, Yuanfa Liu, Xingguo Wang. 2008. Selective adsorption of tannin from flavonoids by organically modified attapulgiteclay. J. Journal of Hazardous Materials. 160: 382–387.

[20] Yi Tang, Hong Zhang, Xianan Liu, Dongqing Cai, Huiyun Feng, Chunguang Miao, Xiangqin Wang, Zhengyan Wu, Zengliang Yu. 2011. Flocculation of harmful algal blooms by modified attapulgite and its safety evaluation. J. Water Research. 45: 855–2862.

[21] Giustetto, R., Xamena, F.X.L., Ricchiardi, G., Bordiga, S., Damin, A., Gobetto, R., Chierotti, M.R., 2005.Mayablue:a computational and spectroscope study. J. Phys. Chem. B. 109:19360–19368.

[22] Wu, W.S., Fan, Q.H., Xu, J.Z., Niu, Z.W., Lu, S.S. 2007. Sorption–desorption of Th(IV) on attapulgite: effects of pH, ionic strength and temperature. J. Appl. Radiat. Isot. 65:1108–1114.

Electronic materials

Material Science and Engineering – Chen (Ed.)
© 2016 Taylor & Francis Group, London, ISBN 978-1-138-02936-1

Electronic and optical properties of quasi-2D BCN_2

L.N. Jiao, C.M. Li, F. Li, J.P. Wang & Z.Q. Chen
Faculty of Materials and Energy, Southwest University, Chongqing, China

ABSTRACT: We investigate the electronic structures and optical properties of quasi-2D BCN_2 by first-principles calculations, using the plane-wave pseudopotential density functional theory within the Local Density Approximation (LDA). The results on band structures, density of states and optical properties such as dielectric function, refractive index, absorption, reflectivity and energy-loss function are presented and analyzed. It is shown that the structure with more number of B-N bonds is more stable. The calculated results indicate that BCN_2 exhibits metallic properties with zero or small band gaps, which is found to be in good agreement with previous calculations. These findings may provide a theoretical basis for the experimental research.

Keywords: BCN_2; density functional theory; electronic structure; optical property

1 INTRODUCTION

New ternary Boron Carbon Nitrogen (B-C-N) compounds as typical superhard materials have aroused increasing interest for their superior physical and chemical properties such as high hardness, good chemical stability, incompressibility and high bulk modulus. A large amount of literature has reported the experimental synthesis and characterization [1–3] and theoretical studies of B-C-N on crystal structures [4], lattice dynamics [5], electronic [6] and mechanical properties [7].

Mazzoni M. S. C. *et al.* [8] investigated the relative stability and electronic structure of several $B_xC_yN_z$-layered structures using the first-principles calculations. Their results showed that one kind of BCN_2 containing an isolated zig-zag chain of carbons displayed a metallic behavior and high formation energies. Sun Guang [9] synthesized different B-C-N compounds hexagonal BCN_2 and BC_xN (x < 1) by a solvothermal method from $Ca_3B_2N_4$ and CCl_4 at 400 °C. They further studied the electronic structures of hexagonal BCN_2 by the first-principles calculations, drawing a conclusion that the layered BCN_2 showed a metallic behavior and could be a good conducting material. However, most research on B-C-N has focused on the one- and three-dimensional BCN and BC_2N [10–12], and few studies have focused on the two-dimensional BCN_2. In this paper, we present two novel structures of quasi-2D BCN_2 and investigate their electronic and optical properties by the first-principles calculations.

2 CALCULATION METHODS

2.1 *Calculation parameters*

The first-principles calculation is based on the Density Functional Theory (DFT) [13] as implemented in the CASTEP program. We assume that the three-dimensional crystal calculation methods can be adopted to study the quasi-2D BCN_2. The Ceperley–Alder–Perdew–Zunger (CA-PZ) method under a local density approximation (LDA) [14] is used for the electronic exchange-correlation functional. The interaction between the ion core and valence electrons is described by using the UltraSoft PseudoPotential (USPP) [15]. The outer electronic configurations considered in the atomic pseudopotential are as follows: $2s^22p^1$ for B, $2s^22p^2$ for C and $2s^22p^3$ for N. A plane-wave basis set is used with a cut-off energy of 280 eV. The Brillouin zone is sampled by using the Monkors-Park scheme [16] with $3 \times 3 \times 3$ k-points. All geometries are relaxed using the BFGS algorithm [17–20] until the total force and energy on each atom are less than 0.03eV/Å and 10^{-5} eV/atom, and the stress and displacement are less than 0.05 GPa and 0.001 Å, respectively. A vacuum of 10 Å is considered to eliminate the interaction between layers in the direction of the z-axis. Based on these methods, the electronic structures and optical properties are calculated.

2.2 *Structure and stability*

The optimized two unit cells of the BCN_2 monolayer consisting of 32 atoms are shown in Figure 1.

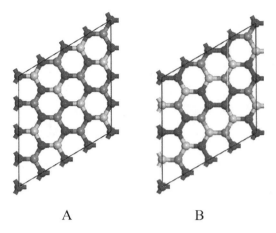

A   B

Figure 1.   Crystal structures of quasi-2D BCN$_2$.

Table 1.   Lattice parameters, volumes, densities, total energy and total bond energy of the two BCN$_2$.

Type	A	B
a/ Å	9.70	9.74
b/ Å	9.72	9.88
c/ Å	2.0	2.0
$\alpha = \beta/°$	90.00	90.00
$\gamma/°$	60.00	60.00
V/Å^3	163.44	166.61
ρ/g $\cdot$ cm^{-3}	4.16	4.08
E$_t$/eV	−6215.74	−6197.12
Total bond energy/eV	−6214.32	−6179.40
ΔE/eV	−1.42	−15.72

Table 2.   Bond energy.

Chemical bond	Bond energy (eV)
C-C	−103.64
B-N	−117.19
C-N	−141.74
B-C	−77.05
C-C	−50.05
N-N	−178.97

Type A has only B-N and C-N bonds, while type B also contains B-C and N-N bonds, except for the above two bonds. The optimized lattice constants, volumes, densities, total energy, and total bond energy are listed in Table 1. The value of the lattice parameter c is taken to be slightly greater than the diameter of the boron atom (~1.9 Å) to ensure the monoatomic layer of the model. The bond energy of each chemical bond is listed in Table 2. As shown in Table 1, the total energy of each structure is lower than the total bond energy of the structure, indicating that the two structures are thermodynamically stable. Type A is obviously more stable, which may be due to the more number of B-N bonds compared with the other.

3   RESULTS AND DISCUSSION

3.1   *Electronic structures*

Figure 2 shows the band structure of the two materials, in which the dashed line represents the Fermi level. The properties of a material, especially the electronic and optical properties, are mainly determined by the properties of the electrons at the Fermi surface. As shown in Figure 2, there is no band gap at E$_f$ for type A, but there exists a large overlapping of the valence and conduction band. This result suggests the existence of free electrons near the Fermi level, which is attributed to the B 2p, C 2p, and N 2p state electrons by analyzing the density of states (Figure 3). Thus, type A is expected to exhibit a metallic behavior, such as good thermal and electrical conductivity, which is consistent with the results reported by Mazzoni M. S. C. and Sun [8,9]. For type B, it also exhibits metallic properties with a small direct band gap of 0.184 eV.

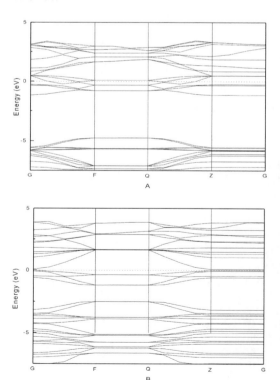

Figure 2.   Band structure of the two BCN$_2$.

206

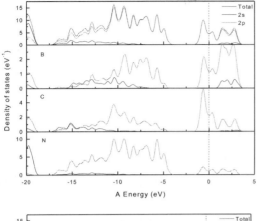

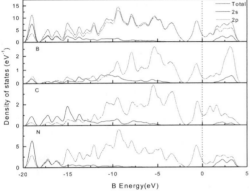

Figure 3. Density of states of the two BCN$_2$.

The Total Density of States (TDOS) and Partial Density of States (PDOS) of the two materials are shown in Figure 3, which reflect the specific compositions of the electronic states in the band structure. For type A, the low-energy band ranging from −20 to −18.7 eV mainly consists of 2s orbit of N atoms, with a small contribution from the 2p orbitals of B and C atoms. The electrons from the B, C, N 2p and 2s atomic orbitals are in a wide energy range between −17.3 and −7.5 eV. The energy bands near the Fermi level are primarily dominated by the 2p orbitals of all atoms, showing a strong hybridization between these electronic states.

In the case of type B, from −20 to −18.4 eV, the densities of states are primarily formed by the N 2s and 2p states, with weak contributions from the 2p and 2s states of B and N atoms. In the energy interval of −18.4 to −4.2 eV, the TDOS are contributed by the 2s and 2p states of B, C and N atoms. Besides, the VBM and CBM are primarily contributed by the 2p states of each atom, which is similar to type A. Simultaneously, the broadening of both the hybrid orbitals is wide and the electronic distribution locality is weak, reflecting the

characteristics of covalent bonds. Both kinds of BCN$_2$ may have metallic conductivity due to the free electrons formed by the 2p state electrons passing through the Fermi level.

3.2 Optical properties

The interaction of photons with electrons will produce the electron transitions between the occupied and unoccupied states, which determines the macroscopic optical properties of solids. The linear optical response function of the system to an external electromagnetic field with a small wave vector is usually described by the dielectric function, which is given by [21]:

$$\varepsilon(\omega) = \varepsilon_1(\omega) + i\varepsilon_2(\omega) \qquad (1)$$

The imaginary part $\varepsilon_2(\omega)$ of the dielectric constant is derived by calculating the electron transition between the occupied and unoccupied orbitals, while the real part $\varepsilon_1(\omega)$ follows from the Kramers-Kronig relationship, namely [21]:

$$\varepsilon_2(\omega) = \frac{4\pi^2}{m^2\omega^2} \sum_{V,C} \int_{BZ} d^3k \frac{2}{2\pi} |e \cdot M_{CV}(K)|^2 \\ \times \delta[E_C(k) - E_V(k) - \hbar\omega] \qquad (2)$$

$$\varepsilon_1(\omega) = 1 + \frac{8\pi e^2}{m^2} \sum_{V,C} \int_{BZ} d^3k \frac{2}{2\pi} \frac{|eM_{CV}(K)|^2}{[E_C(K) - E_V(K)]} \\ \times \frac{\hbar^3}{[E_C(K) - E_V(K)]^2 - \hbar^2\omega^2} \qquad (3)$$

where m and e represent the electron mass and electron charge, respectively; ω is the angular frequency of incident photons; C and V represent the conduction band and valence band, respectively; $E_c(K)$ and $E_v(K)$ imply the intrinsic energy level of the conduction band and the valence band, respectively; BZ represents the first Brillouin zone; K is the reciprocal lattice; and $|eM_{cv}(K)|$ is the momentum transition matrix element. With the knowledge of the complex dielectric function, all other optical constants can be obtained, such as absorption $I(\omega)$, reflectivity $R(\omega)$, and loss function $L(\omega)$, which are expressed as follows:

$$I(\omega) = \sqrt{2}\omega \left[\sqrt{\varepsilon_1^2(\omega) + \varepsilon_2^2(\omega)} - \varepsilon_1(\omega) \right]^{1/2} \qquad (4)$$

$$R(\omega) = \left| \frac{\sqrt{\varepsilon_1^2(\omega) + i\varepsilon_2^2(\omega)} - 1}{\sqrt{\varepsilon_1^2(\omega) + i\varepsilon_2^2(\omega)} + 1} \right|^2 \qquad (5)$$

$$L(\omega) = \text{Im}\left[\frac{-1}{\varepsilon(\omega)} \right] = \frac{\varepsilon_2(\omega)}{\varepsilon_1^2(\omega) + \varepsilon_2^2(\omega)} \qquad (6)$$

The complex refractive index including refractive index $n(\omega)$ and extinction coefficient $k(\omega)$ are given by [22].

$$n' = n(\omega) + ik(\omega)$$
$$n(\omega) = \frac{1}{\sqrt{2}}\left[(\varepsilon_1^2 + \varepsilon_2^2)^{1/2} + \varepsilon_1\right]^{1/2} \qquad (7)$$
$$k(\omega) = \frac{1}{\sqrt{2}}\left[(\varepsilon_1^2 + \varepsilon_2^2)^{1/2} - \varepsilon_1\right]^{1/2}$$

The dielectric function is a very important parameter to connect the microscopic physical transitions between bands to the electronic structure of a solid, which reflects the band structure and other spectral information [23]. From Figure 4, the calculated static dielectric constant $\varepsilon_1(0)$ found to be 1.94 and 14.01 for type A and type B, respectively. For type A, the first peak of the imaginary part $\varepsilon_2(\omega)$ is located at 2.13 eV, which corresponds mainly to the transition from B 2s VB to C 2p CB. For type B, the first peak at 0.88 eV corresponds mainly to the transition from B 2p to C 2p states, and the second peak (3.22 eV) may originate from the intra-band transition from 2s to 2p states of each atom combined with the band structure and density of states. The electrons with energy higher than 15.0 eV are hardly excited.

The calculated results on the refractive index $n(\omega)$ and extinction coefficient $k(\omega)$ are shown in Figure 5. The static refractive indices are 1.39 for type A and 3.75 for type B. When the refractive index increases as the energy increases, the crystals show the properties of normal dispersion, while they show abnormal dispersion properties if the refractive index reduces with increasing the energy [24]. The normal dispersion ranges of type A are 0–1.51 eV and 3.13–5.58 eV, while those of type B are 2.00–2.73 eV and 4.97–11.34 eV. Type A shows abnormal dispersion properties with the

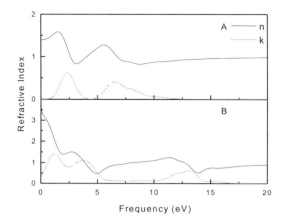

Figure 5. The refractive index of the two BCN$_2$.

energy ranges of abnormal dispersion properties 1.51–3.13 eV and 5.58–8.76 eV, whereas the abnormal dispersion ranges of type B are 0–2.00 eV, 2.73–4.97 eV and 8.76–13.96 eV. The peaks of the extinction coefficient $k(\omega)$ lead to a higher absorption coefficient, as shown in Figure 6(a).

Figure 6 shows the absorption coefficient, reflectivity and loss function of the two materials. When the incident light frequency is the same as the natural frequency of crystal atoms, it will lead to resonance absorption. The absorption coefficients of the two materials are shown in Figure 6(a). At the frequency of 6.67 eV and 13.18 eV, the absorption coefficients of the two materials reach the highest peaks of 43403.5 and 131785.9 cm^{-1}, respectively, which may arise from the transitions from B 2p to C 2p states. The optical absorption of type B is much stronger than that of type A. Furthermore, some peaks appear in the low-energy region, indicating that light can excite electronic transitions from the valance band to the conduction band in this region. As shown in Figure 6(b), the reflectivity reaches the maximum peak value of 0.08 at 2.27 eV for type A and 0.27 at 4.56 eV for type B, and the reflectivity of type B is much higher than that of type A. We also note that the reflectivity trends of the two structures are opposite in the energy region of 0–6.6 eV. The loss function is a crucial factor to describe the energy loss of a fast electron traversing the material. The peaks in the loss function spectra represent the characteristic associated with the plasma resonance and the corresponding frequency is called the plasma frequency [25]. As shown in Figure 6(c), both materials have two main loss peaks and the peaks for type A are at 2.97 eV and 7.20 eV that correspond to the steep reduction of $R(\omega)$, while the peaks for type B at 5.16 eV and 13.92 eV also correspond to the rapid decrease in the reflectivity.

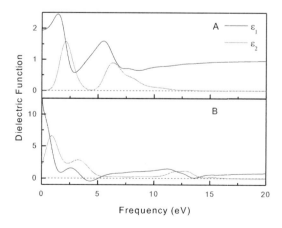

Figure 4. The dielectric function of the two BCN$_2$.

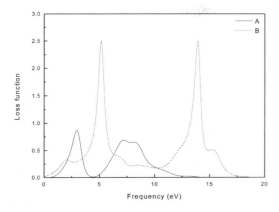

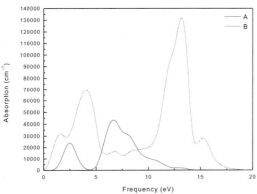

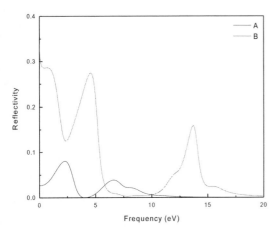

Figure 6. (a) Absorption coefficient, (b) reflectivity, (c) loss function of the two BCN_2.

4 CONCLUSION

Using the first-principles calculations based on the density functional theory, we systematically investigate the electronic structure and optical properties of two types of quasi-2D BCN_2. Our calculated results also show that both BCN_2 exhibit metallic properties, and the valence band maximum and the conduction band minimum are decided by the 2p states of B, C and N atoms. The optical properties are investigated by analyzing the dielectric function and other optical properties such as the refractive index, absorption coefficient, reflectivity and energy loss function. It is found that the structure with more number of B-N bonds is more stable than the other, whereas the structure with B-N, C-N, B-C and N-N bonds has much higher absorption coefficient, reflectivity and energy loss than another. In addition, the optical absorption may excite electronic transitions from the valance band to the conduction band in the low-energy region.

ACKNOWLEDGMENT

This research was financially supported by the Chongqing Scientific and Technological Projects (CSTC2013JCYJCYS5002), and by the Fundamental Research Funds for the Central Universities (XDJK2014C008), and the Chongqing Graduate Research and Innovation Projects (CYS14060).

REFERENCES

[1] Kawaguchi, M. & Kawashima, T. 1996. Syntheses and structures of new graphite-like materials of composition BCN (H) and BC3N (H), Chem. Mater 8: 1197–1201.
[2] Kim, D.H. & Byon, E. 2004. Characterization of ternary boron carbon nitride films synthesized by RF magnetron sputtering. Thin Solid Films 447: 192–196.
[3] Gago, R. & Jiménez, I. 2002. Growth and characterisation of boron–carbon–nitrogen coatings obtained by ion beam assisted evaporation. Vacuum 64: 199–204.
[4] Liu, A.Y. & Wentzcovitch, R.M. 1989. Atomic arrangement and electronic structure of BC2N, Phys. Rev. B 39: 1760.
[5] Nozaki, H. & Itoh, S. 1996. Lattice dynamics of BC2N. Phys. Rev. B 53: 14161–14170.
[6] Azevedo, S. & Paiva, R. De. 2006. Structural stability and electronic properties of carbon-boron nitride compounds. Europhys. Lett. 75: 126.
[7] Zhang, X. & Wang, Y. 2013. First-principles structural design of superhard materials. J. Chem. Phys. 138: 114101.
[8] Mazzoni, S.C.M. & Nunes, R.W. 2006. Electronic structure and energetics of BxCyNz layered structures. Phys. Rev. B 73: 073108.
[9] Sun, G. 2007. Chemical synthesis and characterization of B-C-N and C-N compounds. Qinhuangdao: Yanshan University.
[10] Raidongia, K. & Nag, A. 2010. BCN: A graphene analogue with remarkable adsorptive properties. Chem-Eur. J. 16: 149–157.

[11] Hernandez, E. & Goze, C. 1998. Elastic properties of C and BxCyNz composite nanotubes. Phys. Rev. Lett. 80: 4502.

[12] Sun, J. & Zhou, X.F. 2006. First-principles study of electronic structure and optical properties of heterodiamond BC2N, Phys. Rev. B 73: 045108.

[13] Hohenberg, P. & Kohn, W. 1964. Inhomogeneous Electron Gas. Phys. Rev. 136: 864–871.

[14] Ceperley, D.M. & Alder, B.J. 1980. Ground state of the electron gas by a stochastic method. Phys. Rev. Lett. 45: 566–569.

[15] Vanderbilt, D. 1990. Soft self-consistent pseudopotentials in a generalized eigenvalue formalism. Phys. Rev. B 41: 7892–7895.

[16] Monkhorst, H.J. & Pack, J.D. 1976. Special points for Brillouin-zone integrations. Phys. Rev. B 13: 5188–5192.

[17] Broyden, C.G. 1970. The convergence of a class of double-rank minimization algorithms 1. general considerations, IMA J. Appl. Math. 6: 76–90.

[18] Fletcher, R. 1970. A new approach to variable metric algorithms. Comput. J. 13: 317–322.

[19] Goldfarb, D. 1970. A family of variable-metric methods derived by variational means. Math. Comput. 24: 23–26.

[20] Shanno, D.F. 1970. Conditioning of quasi-Newton methods for function minimization. Math. Comput. 24: 647–656.

[21] Duan, M.Y. & Xu, M. 2007. First-principles study on the electronic structure and optical properties of ZnO doped with transition metal and N. Acta. Phys. Sin. 56: 5359–5365.

[22] Saha, S. & Sinha, T.P. 2000. Electronic structure, chemical bonding, and optical properties of paraelectric BaTiO3, Phys. Rev. B 62: 8828.

[23] Long, J.P. & Yang, L.J. 2013. First-principles calculations of structural, electronic, optical and elastic properties of LiEu2Si3. Solid State Sci. 20: 36–39.

[24] Yang, M.Z. & Chang, B.K. 2014. Research on electronic structure and optical properties of Mg doped Ga0.75Al0.25N. Opt. Mater. 36: 787–796.

[25] Sun, J. & Wang, H.T. 2005. Ab initio investigations of optical properties of the high-pressure phases of ZnO. Phys. Rev. B 71: 125132.

Material Science and Engineering – Chen (Ed.)
© *2016 Taylor & Francis Group, London, ISBN 978-1-138-02936-1*

Magnetic effect on electron spin with both hyperfine interaction and spin-orbit coupling

J.Q. Zhao, T. Wang, L.Y. Li, S.D. Zhuang, J.H. Mao & H.L. Wang
School of Science, Shandong Jianzhu University, Jinan, China

ABSTRACT: To analyze the magnetic effect on the electron spin state, we described the electron spin by a Hamiltonian matrix in a 12×12 uninteracted space, and calculated the time-dependent wave function by an eighth-order Runge-Kutta method. On this basis, the expectation values of spin angular momentum (S_z) were obtained, and the effects of hyperfine interaction, spin-orbit coupling and magnetic field were analyzed. With the increasing magnetic field, the electrons' spin period and flip probability decrease, and therefore S_z intends to the same as its initial value. The hyperfine interaction and spin-orbit coupling enhance the magnetic field-deduced spin flip. The enhancing effect of the spin-orbit coupling is much prominent than that of the hyperfine interaction. This analysis will be an valuable reference for deep understanding and developing of organic functional materials.

Keywords: spin angular momentum; hyperfine interaction; spin-orbit coupling; the eighth-order Runge-Kutta method; organic semiconductor

1 INTRODUCTION

The physical properties of materials depend on the electronic states of their constituent atoms. The electrons have both charge and spin. The magnetic effect of spins is significant in fundamental research. It may provide rich information of the material's structure, interactions and the resulting intrinsic connection of various kinds of physical properties. The Organic Semiconductor (OSC), for example, is an active emerging material with immense promise for innovative, convenient and high-performance electronics. With in-depth research on the performance and mechanism of such devices as Organic Light-Emitting Diode (OLED), organic solar cell and organic spintronic device, magnetic effects in OSCs have increasingly aroused researchers' interest (Xu 2006, Mermer 2005 & Hu 2009).

The magnetic effect is an intrinsic property universal in OSCs. It is usually characterized by Organic Magnetoresistance (OMAR). Much work has been done in exploring the OMAR mechanism (Hu 2009, Wu 2006, Sheng 2006, Desai 2007, Dong 2011, Harmon 2012, Li 2013, Zhao 2013).

Its origin, however, has not been known with certainty. Many proposed theoretical models rely on some degree of spin dynamics involving spin configuration, spin correlation and spin flip (Hu 2009, Sheng 2006, Zhao 2013). The dynamical spin is manipulated by the applied magnetic field combined with the local hyperfine interaction and spin-orbit coupling. The hyperfine interaction is an important

component in the organic spin dynamics (Sheng 2006, Harmon 2012, Nguyen 2007a, b, Bobbert 2010, Kersten 2011). The spin-orbit coupling is generally weak in OSCs. However, for OSCs containing heavy atoms, it is significant (Nguyen 2007b). The modification of either hyperfine interaction (Sheng 2006) or spin-orbit coupling (Wu 2006 & Wu 2007) can lead to a tuning of OMAR in OLEDs.

In the quantum theory, we can describe the electron spin by the time-dependent spin wave function. In principle, the wave function can be obtained by solving the Schrödinger equation. In practice, however, its analytic expression is difficult to be obtained for the Hamiltonian with both hyperfine interaction and spin-orbit coupling. Therefore, a lot of work discusses only the hyperfine interaction or spin orbit coupling effect (Zhao 2013, Wu 2007, Sheng 2007 & Rybicki 2010). Y. Sheng et al. (Sheng 2007) applied a perturbation theory to a four-dimensional subspace for calculating the spin angular momentum. In this work, we supply a physical description and numerical method to demonstrate the effects of the magnetic field combined with the hyperfine interaction and spin-orbit coupling on the spin angular momentum, aiming to present a complete quantitative solutions to the spin description in the spin-related studies.

2 MODEL

When a magnetic field $\boldsymbol{B} = B\boldsymbol{k}$ is applied to the OSC, we present the spin Hamiltonian of an electron including the hyperfine interaction and spin-orbit coupling as follows (Zhao 2013):

$$H = \omega_0[(1/2)L_z + S_z] + (a/\hbar)\mathbf{I} \cdot \mathbf{S} + (b/\hbar)\mathbf{L} \cdot \mathbf{S} \quad (1)$$

where $\omega_0 = 2\mu_B B/\hbar$; μ_B is the Bohr magneton; a and b are the hyperfine interaction and spin-orbit coupling constants; $\mathbf{I}$ is the nuclear spin of the host center; $\mathbf{L}$ and $\mathbf{S}$ are the orbital and spin momenta. Because of $[\mathbf{S}, \mathbf{I} \cdot \mathbf{S}] \neq 0$ and $[\mathbf{I}, \mathbf{I} \cdot \mathbf{S}] \neq 0$, $\mathbf{S}$ and $\mathbf{I}$ are unconserved. The spin-orbit coupling results in $[\mathbf{S}, \mathbf{L} \cdot \mathbf{S}] \neq 0$ and $[\mathbf{L}, \mathbf{L} \cdot \mathbf{S}] \neq 0$; therefore, $\mathbf{L}$ and $\mathbf{S}$ are no longer conserved, respectively, while the total angular momentum $\mathbf{J} = \mathbf{L} + \mathbf{S}$ is still conserved. Accordingly, S_z may change with time. We can calculate the expectation value of S_z with time by $S_z(t) = (\psi(t), S_z\psi(t))$, where $\psi(t)$ is the time-dependent wave function of electrons.

For simplicity, we assume $I_z = \hbar/2$ and $l = 1$. Introducing an equivalent hyperfine field B_{hyp} and a spin-orbit coupling field B_{so}, we have $a/(2\omega_0) = B_{hyp}/B$ for $I_z = \hbar/2$ and $b/\omega_0 = B_{so}/B$ for $L_z = \hbar$. Using the uninteracted basis vectors $|S_z L_z I_z\rangle$ ($i = 1, 2, \ldots, 12$), i.e. $\psi_1 = \uparrow\uparrow\uparrow$, $\psi_2 = \leftrightarrow\uparrow\uparrow$, $\psi_3 = \downarrow\uparrow\uparrow$, $\psi_4 = \uparrow\uparrow\uparrow$, $\psi_5 = \leftrightarrow\downarrow\uparrow$, $\psi_6 = \downarrow\downarrow\uparrow$, $\psi_7 = \uparrow\uparrow\downarrow$, $\psi_8 = \leftrightarrow\uparrow\downarrow$, $\psi_9 = \downarrow\uparrow\downarrow$, $\psi_{10} = \uparrow\downarrow\downarrow$, $\psi_{11} = \leftrightarrow\downarrow\downarrow$ and $\psi_{12} = \downarrow\downarrow\downarrow$, by a stepwise calculation of the matrix element $H_{ij} = (\psi_i, H\psi_j)$, we obtain the Hamiltonian in the 12×12 matrix:

The 12-dimensional uninteracted space could be divided into five subspaces: $\{j = 2: \psi_1\}$; $\{j = 1: \psi_{2,4,7}\}$; $\{j = 0: \psi_{3,5,8,10}\}$; $\{j = -1: \psi_{6,9,11}\}$; $\{j = -2: \psi_{12}\}$. The undiagonalized matrix implies that $\psi_{1,12}$ are eigenstates and, therefore, do not evolve with time other than through a trivial phase factor. However, $\psi_{2\ldots11}$ are mixed through the off-diagonal matrix elements and, therefore, will change with time.

We assume the time-dependent wave function $\psi(t) = \Sigma A_k(t)\psi_k$ ($k = 1, \ldots, 12$), or express it as a matrix $\psi(t) = (A_1(t), A_2(t), \ldots, A_{12}(t))^T$. For $A_m = RA_m + iI_m$ ($m = 1, \ldots, 12$), we get a system of 24 differential equations from the Schrödinger equation. Then, by the eighth-order Runge-Kutta method (Zhao 2011), the recurrence formula can be constructed for solving the system, and all the $A_m(t)$ ($m = 1, \ldots, 12$) can be obtained. The expectation value of S_z with time is given by

$$S_z(t) = \sum_{j,k=1}^{12} A_j(t)^* (S_z)_{jk} A_k(t)$$
$$= \frac{\hbar}{2}\left(\begin{array}{c} |A_1(t)|^2 + |A_2(t)|^2 + |A_3(t)|^2 - |A_4(t)|^2 - |A_5(t)|^2 \\ - |A_6(t)|^2 + |A_7(t)|^2 + |A_8(t)|^2 + |A_9(t)|^2 \\ - |A_{10}(t)|^2 - |A_{11}(t)|^2 - |A_{12}(t)|^2 \end{array} \right)$$

(3)

where $|A_m(t)|^2 = |RA_m(t)|^2 + |IA_m(t)|^2$ ($m = 1, 2, \ldots, 12$).

$$H = \hbar \begin{pmatrix} \omega_0 + \frac{a}{4} + \frac{b}{2} & 0 & 0 & 0 & 0 & 0 & 0 & 0 & 0 & 0 & 0 & 0 \\ 0 & \frac{\omega_0}{2} + \frac{a}{4} & 0 & \frac{\sqrt{2}b}{2} & 0 & 0 & 0 & 0 & 0 & 0 & 0 & 0 \\ 0 & 0 & \frac{a}{4} - \frac{b}{2} & 0 & \frac{\sqrt{2}b}{2} & 0 & 0 & 0 & 0 & 0 & 0 & 0 \\ 0 & \frac{\sqrt{2}b}{2} & 0 & -\frac{a}{4} - \frac{b}{2} & 0 & 0 & \frac{a}{2} & 0 & 0 & 0 & 0 & 0 \\ 0 & 0 & \frac{\sqrt{2}b}{2} & 0 & -\frac{\omega_0}{2} - \frac{a}{4} & 0 & 0 & \frac{a}{2} & 0 & 0 & 0 & 0 \\ 0 & 0 & 0 & 0 & 0 & -\omega_0 - \frac{a}{4} + \frac{b}{2} & 0 & 0 & \frac{a}{2} & 0 & 0 & 0 \\ 0 & 0 & 0 & \frac{a}{2} & 0 & 0 & \omega_0 - \frac{a}{4} + \frac{b}{2} & 0 & 0 & 0 & 0 & 0 \\ 0 & 0 & 0 & 0 & \frac{a}{2} & 0 & 0 & \frac{\omega_0}{2} - \frac{a}{4} & 0 & \frac{\sqrt{2}}{2}b & 0 & 0 \\ 0 & 0 & 0 & 0 & 0 & \frac{a}{2} & 0 & 0 & -\frac{a}{4} - \frac{b}{2} & 0 & \frac{\sqrt{2}}{2}b & 0 \\ 0 & 0 & 0 & 0 & 0 & 0 & 0 & \frac{\sqrt{2}b}{2} & 0 & \frac{a}{4} - \frac{b}{2} & 0 & 0 \\ 0 & 0 & 0 & 0 & 0 & 0 & 0 & 0 & \frac{\sqrt{2}b}{2} & 0 & -\frac{\omega_0}{2} + \frac{a}{4} & 0 \\ 0 & 0 & 0 & 0 & 0 & 0 & 0 & 0 & 0 & 0 & 0 & -\omega_0 + \frac{a}{4} + \frac{b}{2} \end{pmatrix}. \quad (2)$$

3 RESULTS AND DISCUSSION

First, we assign the hyperfine interaction constant a. If we estimate it according to the equivalent hyperfine field $B_{hyp} = 5$ mT, then we have $a = 1.75867 \times 10^3$ μs^{-1}. In the following calculation, we take that corresponds to the hyperfine field $B_{hyp} \approx 5\sim500$ mT. The spin-orbit coupling constant b is assigned around a within a certain range.

As referred above, the 12-dimensional uninteracted space could be divided into five subspaces. As an sample, we focus on the largest subspace $\{j = 0: \psi_{3,5,8,10}\}$. For an electron under state $\psi_3(S_z(0) = -\hbar/2)$, when we consider the effect of the magnetic field, the states $\psi_{3,5,8,10}$ will be mixed and, therefore, S_z will change with time. Figure 1 shows the calculated $S_z(t)$ under four different magnetic fields. The panel inset is the corresponding probability of electrons remaining in the initial ψ_3 state (P_3). At the initial moment $t = 0$, P_3 is 100% and S_z is $-\hbar/2$ for all the magnetic fields. Then, they change with time at different periods and amplitudes. With the increasing magnetic field, P_3 increases, and the period and spin flip decrease, and, therefore, S_z intends to the same as its initial value.

For the same subspace and initial ψ_3 state, we calculate the effect of the magnetic field on the average spin angular momentum, as shown in Figure 2. The panel inset shows the average probabilities of all the states in the subspace under $a = b = 5 \times 10^3$ μs^{-1}. It could be seen that the initial ψ_3 state is partially converted into $\psi_{3,5,8}$ states, and the magnetic field strengthens the spin flip with the increasing hyperfine interaction and spin-orbit coupling. For a certain hyperfine interaction, as shown in Figure 3, the increasing spin-orbit coupling enhances the magnetic field-deduced spin flip, and leads to the appearance of an S_z extremum in the low magnetic

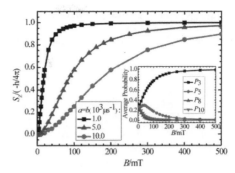

Figure 2. The effect of the magnetic field on the average spin angular momentum for different hyperfine interactions and spin-orbit couplings. The initial state is ψ_3. In the panel inset, $P_{3,5,8,10}$ are the average probability of states $\psi_{3,5,8,10}$ in the subspace $\{j = 0: \psi_{3,5,8,10}\}$ for $a = b = 5 \times 10^3$ μs^{-1}.

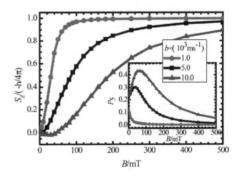

Figure 3. The effect of the magnetic field on the average spin angular momentum for a certain hyperfine interaction ($a = 5 \times 10^3$ μs^{-1}) and different spin-orbit couplings. The initial state is ψ_3. In the panel inset, the corresponding probability of state ψ_5 is illustrated as a sample of states in the subspace $\{j = 0: \psi_{3,5,8,10}\}$.

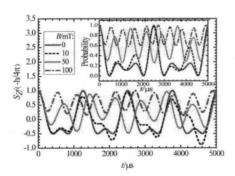

Figure 1. The calculated spin angular momentum changing over time under four different magnetic fields ($a = b = 5 \times 10^3$ μs^{-1}). The panel inset is the corresponding probability of the electronic state remaining in the initial ψ_3 state.

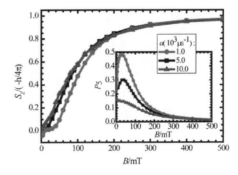

Figure 4. The effect of the magnetic field on the average spin angular momentum for a certain spin-orbit coupling ($b = 5 \times 10^3$ μs^{-1}) and different hyperfine interactions. The initial state is ψ_3. In the panel inset, the corresponding probability of state ψ_5 is illustrated as a sample of states in the subspace $\{j = 0: \psi_{3,5,8,10}\}$.

field. For a certain spin-orbit coupling, as shown in Figure 4, the increasing hyperfine interaction also induces a probability extremum. The average spin angular momentum, however, changes little especially in a high magnetic field.

4 CONCLUSIONS

In this paper, we consider both hyperfine interaction and spin-orbit coupling, deduce the Hamiltonian matrix in the 12×12 uninteracted space, and calculate the time-dependent wave function by the eighth-order Runge-Kutta method. Therefore, the expectation value of the spin angular momentum can be obtained for any combination of hyperfine interaction, spin-orbit coupling, magnetic field and time. As an sample, we present the results of the subspace $\{j = 0: \psi_{3,5,8,10}\}$ for the initial state ψ_3, and analyze the effect of the hyperfine interaction, spin-orbit coupling and magnetic field on the spin angular momentum. The other results can be similarly obtained by this method, even for any larger space composed of the nuclear spin, electron spin and orbital quantum numbers.

ACKNOWLEDGMENT

This research was financially supported by the National Natural Science Foundation of China (Grant No. 11204161).

REFERENCES

[1] Bobbert P A, Nguyen T D, Wagemans W, et al, 2010. Spin relaxation and magnetoresistance in disordered organic semiconductors, Synth Metals 160: 223–229.
[2] Desai P, Shakya P, Kreouzis T, et al, 2007. Magnetoresistance and efficiency measurements of Alq3-based OLEDs, Phys Rev B 75: 094423.
[3] Dong X F, Li X X and Xie S J, 2011. Theoretical investigation on organic magnetoresistance based on Zeeman interaction, Org Electron 12: 1835–1840.
[4] Harmon N J and Flatté M E, 2012. Semiclassical theory of magnetoresistance in positionally-disordered organic semiconductors, Phys Rev B 85: 075204.

[5] Hu B, Yan L and Shao M, 2009. Magnetic-field effects in organic semiconducting materials and devices, Adv Mater 21: 1500–1516.
[6] Kersten S P, Schellekens A J, Koopmans B, et al, 2011. Effect of hyperfine interactions on exciton formation in organic semiconductors, Synth. Metals 161: 613–616.
[7] Li S Z, Dong X F, Yi D, et al, 2013. Theoretical investigation on magnetic field effect in organic devices with asymmetrical molecules, Org Electron 14: 2216–2222.
[8] Mermer Ö, Veeraraghavan G, Francis T, et al, 2005. Large magnetoresistance in nonmagnetic π-conjugated semiconductor thin film devices, Phys Rev B 72: 205202.
[9] Nguyen T D, Sheng Y, Wohlgenannt M, et al, 2007a. On the role of hydrogen in organic magnetoresistance: A study of C60 devices, Synth Metals 157: 930–934.
[10] Nguyen T D, Sheng Y, Rybicki J, et al, 2007b. Magnetoresistance in π-conjugated organic sandwich devices with varying hyperfine and spin-orbit coupling strengths, and varying dopant concentrations, J Mater Chem 17: 1995–2001.
[11] Rybicki J, Nguyen T D, Sheng Y, et al, 2010. Spin-orbit coupling and spin relaxation rate in singly charged π-conjugated polymer chains, Synth. Metals 160: 280–284.
[12] Sheng Y, Nguyen T D, Veeraraghavan G, et al, 2006. Hyperfine interaction and magnetoresistance in organic semiconductors, Phys Rev B 74: 045213.
[13] Sheng Y, Nguyen T D, Veeraraghavan G, et al, 2007. Effect of spin-orbit coupling on magnetoresistance in organic semiconductors, Phys Rev B 75: 035202.
[14] Wu Y and Hu B, 2006. Metal electrode effects on spin-orbital coupling and magnetoresistance in organic semiconducting materials, Appl Phys Lett 89: 203510.
[15] Wu Y, Xu Z, Hu B, et al, 2007. Tuning magnetoresistance and magnetic-field-dependent electroluminescence through mixing a strong-spin-orbital-coupling molecule and a weak-spin-orbital-coupling polymer, Phys Rev B 75: 035214.
[16] Xu Z H, Wu Y and B Hu, 2006. Dissociation processes of singlet and triplet excitons in organic photovoltaic cells, Appl Phys Lett 89: 131116.
[17] Zhao J Q, Jia Z F, Qiao S Z, et al, 2011. Solving dynamical equations of polaron in PPV by Runge-Kutta method, Chinese J Comput Phys 28: 743–748.
[18] Zhao J Q, Wang T, Zhang M S, et al, 2013. Magnetic effect on hopping rate of electrons in organic semiconductors, Appl Phys Lett 103: 182104.

Material Science and Engineering – Chen (Ed.)
© *2016 Taylor & Francis Group, London, ISBN 978-1-138-02936-1*

Enhanced performance of HIZO Thin-Film Transistors by using a high-κ HfTaO gate dielectric

P. Xu & H.-J. Wang
Department of Electromachine Engineering, Jianghan University, Wuhan, P.R. China

ABSTRACT: Ta-doped HfO_2 (THO) films are fabricated by pulsed laser deposition in different O_2 content (0%, 10%, 25% and 50%) atmosphere. The structure and composition of the THO film were investigated using X-ray Photoelectron Spectroscopy (XPS). Then the high-κ THO film is applied to amorphous HfInZnO (α-HIZO) Thin Film Transistors (TFTs) as gate dielectric. The electrical characteristics of Metal–Insulator–Metal (MIM) capacitors and α-HIZO thin film transistors are then investigated. Electrical properties with capacitance equivalent thickness of 6.57 nm, and equivalent permittivity of 23.7 are obtained for an $Al/THO/n^+$-Si/Al MIM capacitor when THO films are deposited in the atmosphere of 25% O_2 content. Superior performance of THO/α-HIZO TFTs has also been achieved with a low threshold voltage of 0.75 V, a saturation mobility of 0.12 $cm^2V^{-1}s^{-1}$ and an on–off current ratio up to 2×10^5 (W/L = 200 μm /50 μm) at 2 V.

Keywords: HfInZnO; HfTaO; thin-film transistors

1 INTRODUCTION

In the past few years, Transparent Conducting Oxide (TCO)-based Thin-Film Transistors (TFTs) have attracted a great deal of interest for their widely application in Active Matrix Liquid Crystal Displays (AMLCD)[1] and Organic Light-Emitting Diode Displays (OLED)[2]. Among all kinds of TCOs, Zn-based amorphous oxide semiconductors such as ZnO[3], InZnO (IZO)[4], InGaZnO (IGZO)[5–7], HfInZnO (HIZO)[8–11] are nowadays drawing much attention because of their genuine characteristics such as high mobility, good uniformity, transparency in visible light, possibility of low-temperature process, and low cost.

Dielectric layer is a vital section of TFTs for the electrical properties of gate insulator and interface quality between active layer and dielectric layer, which affects the threshold voltage, subthreshold swing and mobility of TFTs greatly. Various high-κ materials such as ZrO_2[7], HfO_2[12] and Ta_2O_5[13, 14] have been selected to act as the gate dielectric of oxide TFTs to reduce the gate leakage current and increase the gate capacitance density. Some researchers have found that a small amount of Ta doped into HfO_2[15], Ti doped into Ta_2O_5[16] could enhance the dielectric properties of pure HfO_2, and Ta_2O_5.

However, Ta-doped HfO2 (THO) films on α-HIZO TFTs have been seldom reported up to now. In this work, the research examined the

improvements of the addition of Ta into the hafnium oxide (HfO2) dielectric films by Pulsed Laser Deposition (PLD) on n-type silicon wafer. Then inverted-staggered α-HIZO TFTs constructed with high-κ THO gate dielectric are fabricated and their electrical characteristics are investigated.

2 EXPERIMENT DETAILS

Heavily doped n-type silicon wafers were used as bottom gate in the experiment. The native oxide layer on the n^+ silicon substrate.

Surface was removed by using Radio Corporation of America (RCA) method followed by ultrasonic agitation in acetone, alcohol and deionized water sequentially. First, Ta-doped HfO_2 dielectric film (40 nm thick) was deposited onto Si substrate by pulsed laser deposition with a KrF (248 nm) laser using a Ta foil and HfO_2 target (99.9% purity). For comparison purposes, THO films with a nominal thickness of 40 nm were deposited under different $O_2/(Ar + O_2)$ ratio of 0%, 10%, 25% and 50% respectively. Then α-HIZO films whose atomic ratio Hf: In: Zn is 0.2: 1: 2 with a nominal thickness of 50 nm were in-situ deposited by the same method. To form the top S/D contact, a conventional photolithography followed by lift-off techniques was performed and Al was deposited on the patterned α-HIZO film and on the backside of the Si substrate as the contact electrodes, by using a RF sputtering metal Al target with an Ar flow ratio of 10 sccm. Consequently, the α-HIZO TFTs with a high-κ THO gate dielectric and Al S/D electrodes were achieved.

3 RESULTS AND DISCUSSION

To character dielectric property of THO films, Al/THO/n+-Si/Al capacitors and Al/HfO$_2$/ n+-Si/ Al capacitors were fabricated with the schematic structure shown in Figure 1a. Figure 1b shows capacitance (C) v.s. voltage character of the dielectric film at 1 MHz from where the accumulation capacitance per unit area (C_{ox}) is extracted. The Capacitance Equivalent Thickness (CET) is evaluated by $\varepsilon_{SiO2}\varepsilon_0/C_{ox}$, where ε_{SiO2} and ε_0 are the permittivities of SiO$_2$ and vacuum, respectively. The values of CET are 9.0, 8.6, 6.5 and 7.5 nm for the THO samples fabricated in 0%, 10%, 25% and 50% O$_2$ content atmosphere respectively. The κ values of the THO films are 17.3, 18.1, 23.7 and 20.9 as O$_2$ content increasing from 0% to 50%, calculated by $\varepsilon_{SiO2} \times (t_{ox}/CET)$, where t_{ox} is the value of physical gate dielectric thickness (40 nm). The CET and κ values are depicted in Figure 2. As can be seen from the figure, CET first decreases to 6.57 nm then increases to 7.47 nm with O$_2$ content increasing which are opposite with the change trends of

κ values. The highest κ value is 23.7 corresponding to the smallest CET value of 6.57 nm when O$_2$ content is 25 percent. In contrast, pure HfO$_2$ film showed a poor dielectric performance with CET of 11.2 nm and κ value of 13.9.

The structure and composition of the THO films according to the different O$_2$ content were investigated using X-ray Photoelectron Spectroscopy (XPS) as shown in Figure 3. Figure 3a–3h illustrates Hf 4f, Ta 4f, and Hf 5p spectra of the THO films for the as-deposited samples at various O$_2$ content from 0% to 50%. The binding energies of Hf 4f$_{7/2}$ and 4f$_{5/2}$ peaks at 16.8 eV and 18.4 eV are shown in Figure 2a–2d respectively, which are typical value of Hf^{4+} in HfO$_2$ and have no changes as the O$_2$ content increase. Figure 2e–2h show the Ta 4f spectra and Hf 5p spectra of the samples at different O$_2$ content. Hf 5p spectra has no changes as O$_2$ content changes while for Ta 4f spectra the situation is different. Ta 4f spectra have two peaks with lower binding energy of 4f$_{7/2}$ at 26.2 eV and higher binding energy of 4f$_{5/2}$ at 27.8 eV which first become strong then to weak with O$_2$ content increasing, which is in accordance with the law of capacitance density and the change trend of the dielectric constant changing in Figure 2. When O$_2$ content is 25%, the peak strength of Ta 4f spectra is the strongest.

Based on the analysis above, high κ THO films fabricated at 25 percent O$_2$ atmosphere has the best

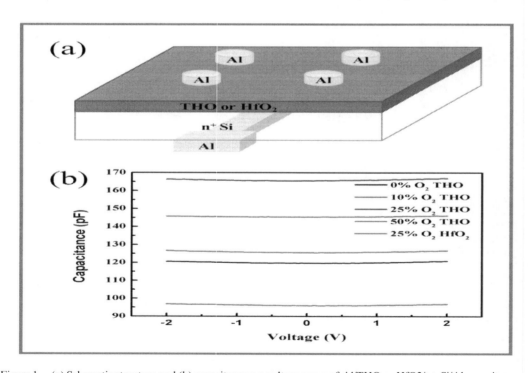

Figure 1. (a) Schematic structure and (b) capacitance v.s. voltage curve of Al/THO or HfO2/n+-Si/Al capacitors.

216

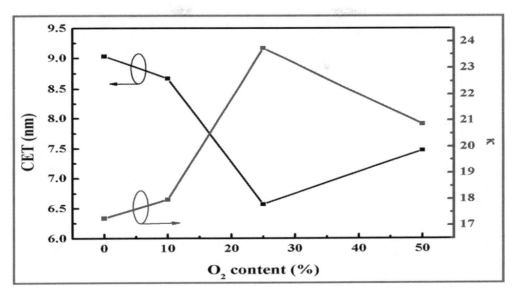

Figure 2. Electrical properties of the THO film deposited in different O_2 content atmosphere: Capacitance Equivalent Thickness (CET), equivalent permittivity.

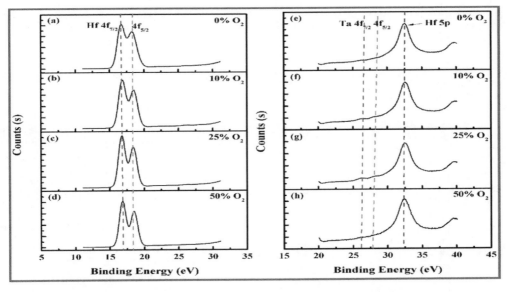

Figure 3. X-ray Photoelectron Spectroscopy (XPS) of the THO films: (a–d) Hf 4f double peaks, (e–h) Ta 4f double peaks and Hf 5f peaks.

dielectric properties and are proper to be used in thin film transistors. At the same time, HfO_2 TFTs are fabricated at the same.

Condition with THO TFTs for comparison. The device schematic diagram of the two different kinds of α-HIZO TFTs with bottom-gate and top-contact S/D electrodes is illustrated in Figure 4a.

The carrier concentration, resistivity, and hall mobility of the α-HIZO active channel are about $9.3 \times 10^{15} \, cm^{-3}$, $42 \, \Omega \cdot cm$ and $16 \, cm^2 \, V^{-1} \, s^{-1}$ measured by Hall measurement system, respectively. The output characteristics (drain-to-source current I_{DS} versus drain-to-source voltage V_{DS} at constant gate-to-source voltage V_{GS}) of the THO and HfO_2 TFTs

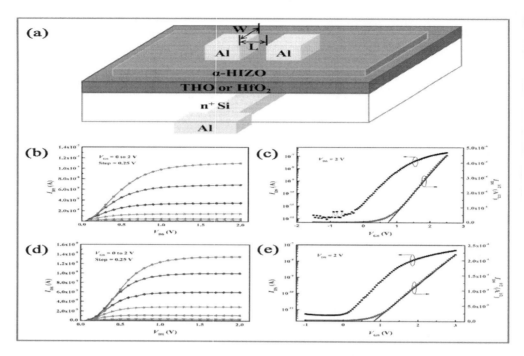

Figure 4. The schematic diagram (a) and operation characteristics of the α-HIZO TFT with W/L = 200 μm/50 μm: (b), (d) output characteristics of THO TFTs and HfO2 TFTs, (c), (e) transfer characteristics of THO TFTs and HfO2 TFTs respectively.

with W/L = 200 μm/50 μm are shown in Figure 4b and 4d respectively. All TFTs show n-type transistor behaviors. It can be seen from Fg. 4b that the saturation current for THO TFTs with Al S/D electrodes is 0.11 μA at bias of $V_{DS} = V_{GS} = 2$ V while the value reduces to 1.3×10^{-2} μA for HfO_2 TFTs in Figure 4d. The transfer characteristics (I_{DS} versus V_{GS} at fixed V_{DS}) of THO and HfO_2 TFTs are depicted in Figure 4c and 4e respectively. For the α-HIZO TFTs with THO dielectric, the ON-state current is about 2×10^{-7} A at a gate voltage of 2.5 V and a drain voltage of 2 V, whereas the OFF-state current is 1×10^{-12} A, corresponding to an ON-OFF current ratio of 2×10^5. For the α-HIZO TFTs with HfO_2 dielectric, the ON-state current is about 7×10^{-8} A at a gate voltage of 2.5 V and a drain-source voltage of 2 V, whereas the OFF-state current is $\sim 3.5 \times 10^{-12}$ A, corresponding to an ON-OFF current ratio of 2×10^4. The maximum OFF-state current ($\sim 10^{-12}$ A) is very important because the operating TFTs using in displays spends most of their time in the OFF-state. The small current is propitious to reduce the power consumption which is mainly attributed to the gate leakage current. The saturation mobility (μ_{sat}) is derived from the following expression:

$$I_{DS} = \left(\frac{\mu_{sat} C_{ox} W}{2L} \right) (V_{GS} - V_{TH})^2, \ V_{DS} > V_{GS} - V_{TH},$$

where W and L are the channel width and length of TFT devices, respectively, C_{ox} is the capacitance per unit area of dielectric layer, and V_{GS} and V_{DS} are the gate-source and drain-source voltages respectively, V_{TH} is determined as a horizontal-axis intercept of a linear fitting to the $I_{DS}^{1/2}$–V_{GS} plot. The saturation mobilities μ_{sat} of the α-HIZO TFT with THO and HfO_2 dielectric are calculated as 0.12 and 0.03 cm^{2}V^{-1}s^{-1}. As it is clearly shown in Figure 4c, the threshold voltage of the TFTs is 0.75 V. The small CET (6.57 nm) of gate dielectric leads to the small V_{TH} of the TaHfO/α-HIZO TFTs which indicate that the THO films fabricated at 25 percent O_2 atmosphere have a good control capability of gate voltage on channel conduction.

4 CONCLUSIONS

In conclusion, thin-film transistors with HIZO channel layer and Ta doped HfO_2 gate dielectric were fabricated. The optimized condition of pulsed laser depositing THO films is in 25% O_2

218

content atmosphere. The optimized THO films obtain capacitance equivalent thickness of 6.57 nm and equivalent permittivity of 23.7. It shows a great improvement in dielectric properties through a small amount Ta doped into pure HfO_2. The THO TFT device is achieved with a low threshold voltage of 0.75 V, a saturation mobility of 0.12 $cm^2V^{-1}s^{-1}$ and an on–off current ratio up to 2×10^5 (W/L = 200/50 μm) at 2 V.

ACKNOWLEDGMENTS

This work is supported by the Research Program of Wuhan Science and Technology Bureau (Project no. 20125049145-19).

REFERENCES

[1] T.S. Kim, et al. Current Applied Physics, 11(5) (2011) 1253.
[2] C. Chen, et al. Japanese Journal of Applied Physics 48/3 (2009) 03B025.
[3] J. Zhang, et al. Thin Solid Films 544 (2013) 281.
[4] H.S. Kim, et al. ACS applied materials & interfaces 4/10 (2012) 5416.
[5] X. Zou, et al. Microelectronics Reliability 50/7 (2010) 954.
[6] X. Zou, et al. Semiconductor Science and Technology 26/5 (2011) 055003.
[7] Jae Sang Lee, et al. IEEE Electron Device Letters 31/3 (2010) 225.
[8] H.-S. Choi, et al. Applied Physics Letters 99/18 (2011) 183502.
[9] E. Chong, et al. Applied Physic Letters 96/15 (2010) 152102.
[10] Y.W. Lee, et al. IEEE Electron Device Letters 33/6 (2012) 3.
[11] E. Chong, et al. Thin Solid Films 519/20 (2011) 6881.
[12] D. Wu, et al. Applied Physics Letters 96/12 (2010) 123118.
[13] C.-T. Lee, et al. Applied Physics Letters 103/8 (2013) 082104.
[14] R. Branquinho, et al. Journal of Display Technology 9/9 (2013) 6.
[15] T. Yu, et al. Vacuum 92 (2013) 58.
[16] C.H. Kao, et al. Surface and Coatings Technology 231 (2013) 512.

Material Science and Engineering – Chen (Ed.)
© 2016 Taylor & Francis Group, London, ISBN 978-1-138-02936-1

Cathodoluminescence and photoluminescence properties of Eu-doped $Ba_3Si_6O_{12}N_2$ green phosphors by microwave sintering

B. Han

Structural and Functional Integration of Ceramics Group, The Division of Functional Materials and Nanodevices, Ningbo, China
Institute of Materials Technology and Engineering, Chinese Academy of Sciences, Ningbo, China
Institute of Materials, Shanghai University, Shanghai, China

Y.F. Wang

Structural and Functional Integration of Ceramics Group, The Division of Functional Materials and Nanodevices, Ningbo, China
Institute of Materials Technology and Engineering, Chinese Academy of Sciences, Ningbo, China

Q. Liu

State Key Laboratory of High Performance Ceramics and Superfine Microstructure, Shanghai Institute of Ceramics, Chinese Academy of Sciences, Shanghai, China

Q. Huang

Structural and Functional Integration of Ceramics Group, The Division of Functional Materials and Nanodevices, Ningbo, China
Institute of Materials Technology and Engineering, Chinese Academy of Sciences, Ningbo, China

ABSTRACT: Green oxynitride $Ba_3Si_6O_{12}N_2$:Eu^{2+} phosphors were prepared by the microwave-assisted sintering method. Their photoluminescence properties and morphology were investigated. The particles exhibited a broad green emission centered at 527 nm with a shoulder at a higher wavelength under 428 nm excitation, and also showed a single intense broad band around 523 nm under low-energy electron beam (e-beam) excitation. Simultaneously, it was observed that particles at different temperatures exhibited different light intensities, according to the Cathodoluminescence (CL) in a Scanning Electron Microscope (SEM). These luminescence variations may originate from a variation in the composition. Moreover, a suitable excitation range makes it match well with the emission of near-UV LEDs or blue LEDs. The proposed method is expected to be potentially applicable to other oxynitride/nitride phosphors.

Keywords: cathodoluminescence; oxynitride; microwave-assisted sintering; phosphors

1 INTRODUCTION

Inorganic luminescent materials have been used in the applications of lighting, displays, and imaging, such as fluorescent tubes, white Light-Emitting Diodes (LEDs), cathode tube display, and Field Emission Display (FED) [1]. White LEDs are considered the next generation of solid-state lighting systems because of their excellent properties, such as low power consumption, high efficiency, long lifetime and little pollution, which made with the blue GaN-pumped yellow YAG:Ce^{3+} phosphor. Although this method is simple to get white light, some problems still exist: changing of emitting color with input power, low color rendering index (Ra) and rapid thermal quenching of phosphors. Therefore, for high demand applications, novel

host materials are needed [2]. Moreover, LEDs have already entered the market as backlights for Liquid Crystal Displays (LCD), replacing fluorescent lamps. Although LCDs have a great market at present, FEDs have applications in some specific areas, such as special military devices, besides the general flat display applications, owing to their improved performance and lower power consumption compared with the actual devices, such as mercury gas-discharge fluorescence lighting or plasma displays. The key component for FEDs is a phosphor that emits a strong luminescence under electron beam irradiation [3,4]. The phosphors play a key role in these performances, and their quantum efficiency and stability over time need to be improved [5]. The novel oxynitride and nitride phosphors, such as Ca(Y)-α-sialon:Eu^{2+}(Yb^{2+}), β-sialon:Eu^{2+},

221

AlN:Eu^{2+}, γ-AlON:Eu^{2+}, Mg^{2+}, CaAlSiN$_3$:Eu^{2+}, have attracted considerable attention due to their longer wavelength excitation, smaller thermal quenching, higher chemical and radiation stability than the conventional oxide or sulfide phosphors [2,6–13].

To achieve a large color gamut, highly efficient yellow, green and red phosphors are required for white LEDs, such as aluminate, silicate, and oxynitride phosphors. The red CaAlSiN$_3$:Eu^{2+} phosphors are a promising candidate, but the green phosphors still confront some difficulties. Sr$_2$SiO$_4$:Eu^{2+} phosphors have a strong thermal quenching, whereas the synthesis of β-SiAlON:Eu^{2+} phosphors requires the severe conditions such as high temperature and high pressure. The M-Si-O-N (M = Ca, Sr, Ba) system has attracted as a new phosphor system due to its easy fabrication, good thermal stability, and high luminescence efficiency [14]. Eu^{2+}-doped Ba$_3$Si$_6$O$_{12}$N$_2$ phosphor has been synthesized by a solid-state reaction method; however, the phase of pure Ba$_3$Si$_6$O$_{12}$N$_2$ is hardly obtained, due to the low chemical reactivity of Si$_3$N$_4$.

The microwave sintering is a relatively new technique. The advantages of high-temperature microwave processing have been demonstrated in fields such as sintering and joining of ceramic materials, inorganic synthesis, and development of composite materials [15], which has attracted the attention of many researchers due to its advantages over the conventional sintering, such as very rapid heating rate, decreased sintering temperature, improved physical and mechanical properties, high purity, and lower environmental hazards [16–20]. The microwave heating methods in the high-temperature domain are currently under active development, with prospects of industrial use in the near future. For the Al$_2$O$_3$ and Si$_3$N$_4$ systems, the microwave heating has been observed to accelerate the diffusional processes, such as sintering and grain growth [17,19]. However, due to the host-lattice and occupation site dependencies, it is necessary to identify and understand the local luminescence properties of rare-earth-doped oxynitride particles [3]. The low-energy Cathodoluminescence (CL) technique has decisive advantages for this purpose, and one can also map the luminescence distribution both laterally and along the sample depth that can provide precious information on the growth mechanisms. Finally, the combination of CL with other electron beam (e-beam) techniques allows to determine the origin of the luminescence [21]. Consequently, CL can reveal most of the luminescence processes in the materials. Beside the optical properties from CL, structural information can be obtained from other technologies such as surface morphology from Scanning Electron Microscopy (SEM) and Energy-Dispersive X-Ray Spectroscopy (EDS). However, few reports have existed on such applications of

e-beam microscopy on rare-earth-doped oxynitride phosphors [3,22–24]. In this study, we make use of the microwave-assisted sintering method for the effective synthesis of Eu^{2+}-doped Ba$_3$Si$_6$O$_{12}$N$_2$ green phosphors under a flowing N$_2$/H$_2$ gas. Their local luminescence properties of Eu-doped Ba$_3$Si$_6$O$_{12}$N$_2$ by CL and PL properties were investigated.

2 EXPERIMENTAL PROCEDURE

2.1 *Preparations*

The Ba$_3$Si$_6$O$_{12}$N$_2$:Eu^{2+} phosphors were synthesized by a microwave synthesis method. The starting materials BaCO$_3$, Eu$_2$O$_3$, SiO$_2$ (Aladdin Chemistry Co. Ltd., Shanghai, China), and Si$_3$N$_4$ (Ube Industries Ltd., Tokyo, Japan) were mixed in the required proportions and calcined at 1100–1250°C for 4 h in a flowing gas mixture (15%H$_2$/N$_2$). A microwave furnace (HAMiLab-V3, Synotherm Corporation, China) was used with a continuously variable power of 2.45 GHz microwaves up to 2.85 kW. The composition was generally defined by the formula: Ba$_3$Si$_6$O$_{12}$N$_2$:Eu^{2+} (Si$_3$N$_4$: SiO$_2$ = 0.75:3.75).

2.2 *Characterization*

The phase was analyzed by an X-Ray Diffractometer (XRD, Bruker AXS D8 Advance, Karlsruhe, German) using CuKα1 radiation. The PL spectra were measured at room temperature using a fluorescence spectrophotometer (F-4600, Hitachi Ltd, Tokyo, Japan). Energy-Dispersive X-Ray Spectroscopy (EDS) measurements were performed in a field emission scanning electron microscope (Quanta 250FEG, FEI, USA). CL measurements were carried out using a SEM (Quanta 400 FEG, FEI, USA) equipped with a CL system (GatanMonoCL3+). All the measurements were made at room temperature.

3 RESULTS AND DISCUSSION

3.1 *Phases*

Figure 1 shows the X-ray diffraction patterns of Ba$_3$Si$_6$O$_{12}$N$_2$:Eu^{2+} powders synthesized at 1100–1250°C with a 0.2 europium content. Generally speaking, temperature is considered as one of the most important factors determining the thermodynamic activity and final products. Moreover, different soaking temperatures have to be attempted according to different starting materials, catalysts, and reactors. The powders synthesized at temperature 1100°C contain many impurity phases of α-Si$_3$N$_4$, orthosilicate phases BaSi$_2$O$_5$, Ba$_4$Si$_6$O$_{10}$ and Ba$_5$Si$_8$O$_{21}$, but somewhat a Ba$_3$Si$_6$O$_{12}$N$_2$ phase

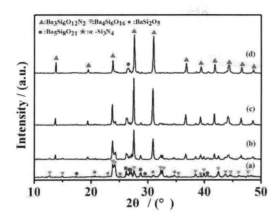

Figure 1. X-ray diffraction patterns of $Ba_3Si_6O_{12}N_2$:Eu^{2+} phosphors calcined at 1100–1250°C, (a) 1100°C, (b)1150°C, (c) 1200°C, (d) 1250°C.

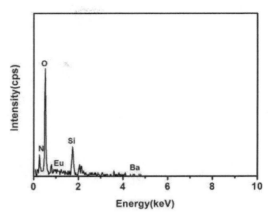

Figure 2. EDS spectra of $Ba_3Si_6O_{12}N_2$:Eu^{2+} phosphors calcined at 1250°C.

product has appeared. Therefore, it is possible that barium carbonate and silicon dioxide have dispersed uniformly, forming a solid solution with the structure of silicon nitride and europium oxide compounds. When the temperature is raised, the reaction product is stable with the increasing temperature and the crystallinity continues to improve, reaching the maximum value at 1250°C. The impurity phases gradually disappear with an increasing temperature, and most of the high intensity peaks match well with the reported data by Mikami et al. [14], and a few orthosilicate phases $BaSi_2O_5$ still remain. This may be due to the fact that strong microwave electric fields induce a nonlinear driving force (named ponderomotive force) for (ionic) mass transport near surfaces and structural interfaces (e.g. grain boundaries) in ceramic materials [25–27]. This is in accordance with the fact that a higher temperature benefits the crystallization, whereas a very high temperature can induce a mass of molten phases and hard agglomerations, which are greatly adverse for luminescent properties. The elemental analysis of the powder synthesized at 1250°C by the energy-dispersive spectrometer shows only the presence of Eu, Ba, Si, N, and O (Figure 2).

3.2 Photoluminescence properties of the samples

The samples show similar photoluminescence excitation and emission spectra of the samples, as shown in Figure 3. The powders show prominent luminescent properties, and the intensity keeps a systematic increase along with the increasing temperature. Each phosphor shows a broad excitation spectrum from 280 to 500 nm and a green emission band centered at about 527 nm, which is a characteristic $4f^65d^1 \rightarrow 4f^7$ transition of Eu^{2+}. This emission should

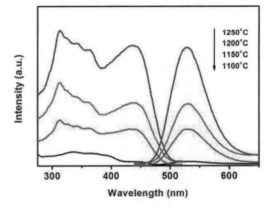

Figure 3. Photoluminescence spectra of the samples at different temperatures (λex = 428 nm, λem = 527 nm).

originate from Eu^{2+} in the $Ba_3Si_6O_{12}N_2$ host but not orthosilicate phases, which makes no contribution to the luminescence [28]. Sharp lines assigned to Eu^{3+} transitions have not been detected, indicating that there are no Eu^{3+} ions or no detection in the crystal lattice. Powders obtained at lower temperatures show a weak emission band due to the existence of impurity and bad crystallization of $Ba_3Si_6O_{12}N_2$. As shown in Figure 1, the impurity phases gradually disappear and the crystallization of $Ba_3Si_6O_{12}N_2$ gets better with an increasing temperature, resulting in a stronger emission intensity. The phosphor fired at 1250°C shows the best luminescence whose full width at half maximum of the band is below 70 nm.

3.3 Diffuse reflection spectra of the samples

The diffuse reflection spectra of the $Ba_3Si_6O_{12}N_2$:Eu^{2+} phosphors were recorded from

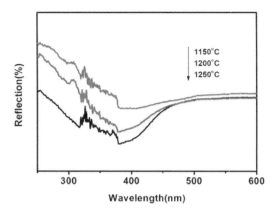

Figure 4. Diffuse reflection spectra of the samples.

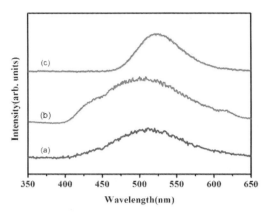

Figure 6. Cathodoluminescence emission spectra of the samples at different temperatures: (a) 1150°C, (b) 1200°C, (c) 1250°C.

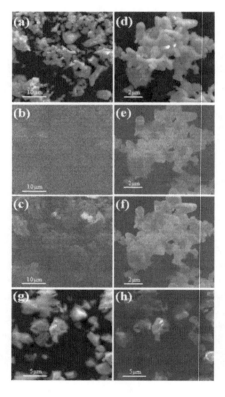

Figure 5. SEM (a) 1150°C, (d) 1200°C, (g) 1250°C and CL images at (b) 445 nm, (c) 511 nm for (a); (e) 447 nm (f) 507 nm for (d), and (h) 523 nm for (g).

1150 to 1250°C, as shown in Figure 4. With the increasing temperature, the Eu^{2+} absorption band becomes slightly intense with the different temperature. According the report [14], the computed optical band gap of $Ba_3Si_6O_{12}N_2$ is 4.63 eV. The powders show a strong drop in the reflection in the UV range below 300 nm, corresponding to

the valence-to-conduction band transitions of the host lattice. For the Eu^{2+}-doped samples, strong absorption bands are presented from 300 to 500 nm, which are assigned to the $4f^7 \rightarrow 4f^6 5d$ transition of Eu^{2+} ions. The absorption intensity of Eu^{2+} markedly increases over the range of 300–500 nm for $Ba_3Si_6O_{12}N_2:Eu^{2+}$, similar to the case of the PL spectra of the phosphors.

3.4 Cathodoluminescence properties of the samples

Figure 5 shows the SEM and CL images of the samples. The $Ba_3Si_6O_{12}N_2:Eu^{2+}$ particles are relatively irregular (see Figure 5a, 5d and 5g). The particles show some charging, but it does not noticeably affect the luminescence results. At 1150°C, the 445 nm emission is uniformly distributed on the particles regardless of their shapes (see Figure 5b). However, the 445 nm image consists of several faintly bright patches dispersed among the particles (Figure 5c). Interestingly, at 1200°C, the particle surface looks brighter on the 507 nm image, and grayer on the 447 nm image (Figure 5e and 5f). While the grains show much stronger 523 nm emission (Figure 5h) at 1250°C. This may suggest that there is a compositional variation between the surface and the inner part of the particles. The CL spectra are shown in Figure 6.

4 CONCLUSIONS

In summary, Eu^{2+}-doped $Ba_3Si_6O_{12}N_2:Eu^{2+}$ green phosphors were synthesized by the microwave sintering method. The particles exhibited a broad green emission centered at 527 nm with a shoulder at a higher wavelength under 428 nm excitation. Furthermore, it showed a single intense broad band around 523 nm under low-energy e-beam excitation for the sample sintered at 1250°C for 4 h. These luminescence variations may originate from a variation in the composition. Moreover, it is shown that the emission properties of the phosphors can be effectively studied by SEM-CL. A suitable excitation range makes it match well with the emission of near-UV LEDs or blue LEDs.

ACKNOWLEDGMENTS

This work was supported by the Zhejiang Provincial Natural Science Foundation of China (R12E020005, LQ14E020007), the Ningbo Natural Science Foundation (2013A610027), the Opening Project of State Key Laboratory of High Performance Ceramics and Superfine Microstructure (SKL201307SIC), and the Research Fund for the Postdoctoral Advanced Program of Zhejiang Province (No. BSH1301022). We would further like to thank Mr Gao Xiaodong (Suzhou Institute of Nano-Tech and Nano-Bionics, Chinese Academy of Sciences) for assisting in the CL measurements.

REFERENCES

[1] Shang MM, Geng DL, Yang DM, Kang XJ, Zhang Y, Lin J. 2013. Luminescence and energy transfer properties of $Ca_2Ba_3(PO_4)_3Cl$ and $Ca_2Ba_3(PO_4)_3Cl:A$ ($A = Eu^{2+}/Ce^{3+}/Dy^{3+}/Tb^{3+}$) under UV and low-voltage electron beam excitation. *Inorg Chem* 52:3102–12.

[2] Yang LX, Xu X, Hao LY, Wang YF, Yin LJ, et al. 2011. Optimization mechanism of $CaSi_2O_2N_2:Eu^{2+}$ phosphor by La^{3+} ion doping. *J Phys D Appl Phys* 44:355403.

[3] Dierre B, Takeda T, Sekiguchi T, Suehiro T, Takahashi K, et al. 2013. Local analysis of Eu^{2+} emission in $CaAlSiN_3$. *Sci Technol Adv Mat* 14:064201.

[4] Itoh S, Tanaka M, Tonegawa T. 2004. Development of field emission displays. *J Vac Sci Technol B* 22:1362–6.

[5] Swart HC, Terblans JJ, Coetsee E, Ntwaeaborwa OM, Dhlamini MS, et al. 2007. Review on electron stimulated surface chemical reaction mechanism for phosphor degradation. *J Vac Sci Technol A* 25:917–21.

[6] Xie RJ, Hirosaki N, Sakuma K, Yamamoto Y, Mitomo M. 2004. Eu^{2+}-doped Ca-alpha-SiAlON: A yellow phosphor for white light-emitting diodes. *Appl Phys Lett* 84:5404–6.

[7] Suehiro T, Onuma H, Hirosaki N, Xie RJ, Sato T, Miyamoto A. 2010. Powder synthesis of Y-alpha-SiAlON and its potential as a phosphor host. *J Phys Chem C* 114:1337–42.

[8] Xu X, Tang JY, Nishimura T, Hao LY. 2011. Synthesis of Ca-alpha-SiAlON phosphors by a mechanochemical activation route. *Acta Mater* 59:1570–6.

[9] Hirosaki N, Xie RJ, Kimoto K, Sekiguchi T, Yamamoto Y, et al. 2005. Characterization and properties of green-emitting beta-SiAlON:Eu^{2+} powder phosphors for white light-emitting diodes. *Appl Phys Lett* 86: 211905.

[10] Inoue K, Hirosaki N, Xie RJ, Takeda T. 2009. Highly efficient and thermally stable blue-emitting AlN:Eu^{2+}phosphor for ultraviolet white light-emitting diodes. *J Phys Chem C* 113:9392–7.

[11] Yin LJ, Xu X, Hao LY, Xie WJ, Wang YF, et al. 2009. Synthesis and photoluminescence of Eu^{2+}-Mg^{2+} co-doped gamma-AlON phosphors. *Mater Lett* 63:1511–3.

[12] Uheda K, Hirosaki N, Yamamoto Y, Naito A, Nakajima T, Yamamoto H. 2006. Luminescence properties of a red phosphor, $CaAlSiN_3:Eu^{2+}$, for white light-emitting diodes. *Electrochem Solid St* 9:H22-H5.

[13] Li YQ, Delsing ACA, de With G, Hintzen HT. 2005. Luminescence properties of Eu^{2+}-activated alkaline-earth silicon-oxynitride $MSi_2O_{2-delta}N_{2+2/3delta}$($M = Ca$, Sr, Ba): A promising class of novel LED conversion phosphors. *Chem Mater* 17:3242–8.

[14] Mikami, M., Shimooka, S., Uheda, K., Imura, H., Kijima, N. 2009. New green phosphor $Ba_3Si_6O_{12}N_2$:Eu for white LED: Crystal structure and optical properties, *Key Eng Mater* 403:11–14.

[15] Rybakov KI, Olevsky EA, Krikun EV. 2013. Microwave sintering: Fundamentals and modeling. *J Am Ceram Soc* 96:1003–20.

[16] Chockalingam S, Earl DA. 2010. Microwave sintering of Si_3N_4 with $LiYO_2$ and ZrO_2 as sintering additives. *Mater Design* 31:1559–62.

[17] Oghbaei M, Mirzaee O. 2010. Microwave versus conventional sintering: A review of fundamentals, advantages and applications. *J Alloy Compd* 494:175–89.

[18] Liu LH, Zhou XB, Xie RJ, Huang Q. 2013. Facile synthesis of Ca-alpha-SiAlON:Eu^{2+} phosphor by the microwave sintering method and its photoluminescence properties. *Chinese Sci Bull* 58:708–12.

[19] Brosnan KH, Messing GL, Agrawal DK. 2003. Microwave sintering of alumina at 2.45 GHz. *J Am Ceram Soc* 86:1307–12.

[20] Tiegs, TN., Kiggans, JO., Kimrey, HD. 1991. Microwave sintering of silicon nitride, *Ceram. Eng. Sci. Proc.* 12:1981–1992.

[21] Dierre B, Yuan XL, Sekiguchi T. 2010. Low-energy cathodoluminescence microscopy for the characterization of nanostructures. *Sci Technol Adv Mat* 11:043001.

[22] Liu TC, Kominami H, Greer HF, Zhou WZ, Nakanishi Y, Liu RS. 2012. Blue emission by interstitial site occupation of Ce^{3+} in AlN. *Chem Mater* 24:3486–92.

[23] Xu FF, Sourty E, Zeng XH, Zhang LL, Gan L, et al. 2012. Atomic-scaled investigation of structure-dependent luminescence in Sialon:Ce phosphors. *Appl Phys Lett* 101:161904.

[24] Dierre B, Zhang XM, Fukata N, Sekiguchi T, Suehiro T, et al. 2013. Growth temperature influence on the luminescence of Eu, Si-codoped AlN phosphors. *ECS J Solid State Sc* 2:R126-R30.

[25] Booske JH, Cooper RF, Freeman SA, Rybakov KI, Semenov VE. 1998. Microwave ponderomotive forces in solid-state ionic plasmas. *Phys Plasmas* 5:1664–70.

[26] Booske JH, Cooper RF, Freeman SA. 1997. Microwave enhanced reaction kinetics in ceramics. *Mater Res Innov* 1:77–84.

[27] Wang YF, Liu LH, Xie RJ, Huang Q. 2013. Microwave assisted sintering of thermally stable $BaMgAl_{10}O_{17}:Eu^{2+}$ phosphors. *ECS J Solid State Sc* 2:R196-R200.

[28] Li WY, Xie RJ, Zhou TL, et al. 2014. Synthesis of the phase pure $Ba_3Si_6O_{12}N_2:Eu^{2+}$ green phosphor and its application in high color rendition white LEDs. *Dalton T* 43: 6132–6138.

Material Science and Engineering – Chen (Ed.)
© 2016 Taylor & Francis Group, London, ISBN 978-1-138-02936-1

Influence of Mn doping on dielectric loss of La/Ca Co-doped $BaTiO_3$ ceramic with dielectric-temperature stability

D.Y. Lu
Research Center for Materials Science and Engineering, Jilin Institute of Chemical Technology, Jilin, China

Q.Y. Zhang
College of Sciences, Jilin Institute of Chemical Technology, Jilin, China

T.T. Liu
Research Center for Materials Science and Engineering, Jilin Institute of Chemical Technology, Jilin, China
College of Chemistry, Jilin University, Changchun, China

Y.Y. Peng
Research Center for Materials Science and Engineering, Jilin Institute of Chemical Technology, Jilin, China
College of Chemistry, Jilin Institute of Chemical Technology, Jilin, China

X.Y. Yu & Q. Cai
Research Center for Materials Science and Engineering, Jilin Institute of Chemical Technology, Jilin, China
College of Chemistry, Jilin University, Changchun, China

X.Y. Sun
Research Center for Materials Science and Engineering, Jilin Institute of Chemical Technology, Jilin, China

ABSTRACT: In order to lower the dielectric loss (tan δ), the influence of Mn doping ($x = 0.005, 0.010, 0.015$) on structure and dielectric properties of the $(Ba_{0.97}La_{0.03})$ $(Ti_{0.985}Ca_{0.015})$ O_3 ceramic (LC) with an X7R specification was studied using XRD, Raman spectroscopy, SEM, EPR and dielectric measurements. All of the samples have a tetragonal perovskite structure and a most fine-grained microstructure ($GS = 0.6$ μm) is obtained for the sample with $x = 0.015$. The Mn^{2+} content in LC doped with 0.5% Mn is most, but this sample becomes a semiconductor. When $x = 1\%$ and 1.5%, the tan δ of LC can be markedly reduced to be lower than 0.02 and the dielectric permittivity can be improved. However, Mn doping in LC destroys the X7R stability of LC.

Keywords: perovskites; dielectric properties; defect complexes; site occupation

1 INTRODUCTION

Temperature-stable $BaTiO_3$-based ceramics with higher dielectric permittivity generally have a core-shell structure, in which a fine grain has a ferroelectric $BaTiO_3$ core, a paramagnetic Nb-Co-rich coating surface layer (shell), and a transition region between the two formers [1,2]. However, the control of preparation process is delicate and slight variations in process have an unsetting effect on temperature stability and permittivity of materials. Recently, we adopted a conventional ceramic processing technique to have discovered a novel X7R-type dielectric ceramic $(Ba_{1-x}La_x)(Ti_{1-x/2}Ca_{x/2})$ O_3 ($x = 0.03$) (LC) (Note: X7R specification:

$|(\varepsilon' - \varepsilon'_{RT})/\varepsilon'_{RT}| \leq 15\%$ in a range of −55 to 125 °C), in which La^{3+} and Ca^{2+} ions can be incorporated completely into Ba and Ti sites, respectively [3]. This important finding is therefore a revolutionary breakthrough owing to a simplification in preparation process and the advances in compositional design.

Although the X7R-type LC ceramic showed a higher permittivity ($\varepsilon'_{RT} = 2170$), its dielectric loss was relatively higher (tan $\delta < 0.04$) [3]. It is well known that Mn ions in $BaTiO_3$ generally act as an acceptor owing to its high ability to conduct electron trapping, which contributes to the improvement in ceramic resistance [4,5]. In order to lower the dielectric loss of LC, in this work, Mn ions was incorporated into the BLTC lattice. The influence of Mn doping on structure and dielectric properties of LC was studied.

2 EXPERIMENTAL

The initial materials were reagent-grade $BaCO_3$, La_2O_3, TiO_2, $CaCO_3$, and MnO_2. La, Ca and Mn co-doped $BaTiO_3$ ceramics (abbreviated LCM) were prepared using the same mixed oxides method described elsewhere [3] according to the formula $(Ba_{1-y}La_y)(Ti_{1-y/2-x}Ca_{y/2}Mn_x)O_3$ ($y = 0.03$, $x = 0.005$, 0.010, 0.015) (LCM). The final sintering condition was 1400 °C for 12 h. The sample with $x = 0$, with a formula of $(Ba_{0.97}La_{0.03})(Ti_{0.985-x}Ca_{0.015})O_3$ (abbreviated LC), is provided by our previous work [3].

Powder X-Ray Diffraction (XRD) measurements were made at room temperature using a DX-2700 X-ray diffractometer (Dandong Haoyuan Inc.). XRD data were collected between $20° \leq 2\theta \leq 85°$ in steps of 0.02°. Crystal structures were determined by MS Modeling (Accelry Inc.) and Cu $K\alpha_1$ radiation ($\lambda = 1.540562$ Å). Temperature dependences of the dielectric permittivity and dielectric loss were measured at 1 kHz from –75 to 200 °C at a heating rate of 2 °C/min with a weak 1 kHz ac electric field using an RCL meter (Fluke PM6306, USA). Raman spectra of ceramic powders were measured using a LabRAM XploRA Raman spectrometer (Horiba Jobin Yvon, French), with a 532 nm laser. The microstructure was observed using an EVOMA 10 Scanning Electron Microscope (SEM) (Zeiss, Germany) operated at 15 keV. The surfaces of the ceramic samples were polished and then thermally etched for SEM observations. Electron Paramagnetic Resonance (EPR) spectra were measured using an A300 electron-spin resonance spectrometer system (Bruker BioSpin GMBH, Germany) at X band frequency (9.85 GHz). All EPR spectra were obtained for 30 mg ceramic powders.

3 RESULTS AND DISCUSSION

The pictures of LCM ceramics are shown in Figure 1. The sample gradually becomes black with increasing Mn content (x).

Powder XRD patterns of LC [3] and LCM ceramics are shown in Figure 2. All of the samples have a single-phase perovskite structure. The diffraction peaks in the vicinity of 45° and 56° are Gaussian-fitted, as shown in Figure 3. The ceramic with $x = 0$ (LC) has a pseudo-cubic structure [3], characteristic of broad and symmetric (200) and (211) peaks. The tetragonality in LCM can be observed from the separate (002)/(200) characteristic peaks and (112)/(211) peaks.

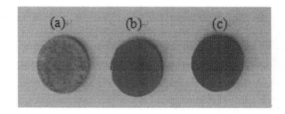

Figure 1. Pictures of $(Ba_{0.97}La_{0.03})(Ti_{0.985-x}Ca_{0.015}Mn_x)O_3$ ceramics: (a) $x = 0.005$, (b) 0.010, (c) 0.015.

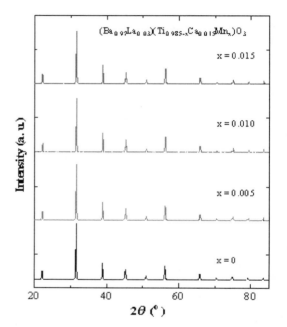

Figure 2. Powder XRD patterns of LC and LCM ceramics.

The unit cell volume (V_0) as a function of x for LCM is shown in Figure 4, where V_0 of the tetragonal $BaTiO_3$ (64.41 Å^3, from JCPDS Cards: No. 5-626) is shown in this figure for comparison. Table 1 lists ionic radii versus Coordinate Number (CN) [6]. The smaller V_0 of all the samples compared with $BaTiO_3$ indicates the formation of $La_{Ba}^{\bullet}$—$Ca_{Ti}^{''}$—$La_{Ba}^{\bullet}$ defect complexes in LC and LCM. It is well known that Mn ions occupy Ti sites in $BaTiO_3$ and may show valence-state change in the forms of Mn^{4+}, Mn^{3+} and Mn^{2+}; the average valence state of Mn ions in $BaTiO_3$ is +3.3~+3.4 for 0.2%~2% Mn-doped $BaTiO_3$ sintered at 1000 °C for 10 h in air [7]. An interesting phenomenon is that V_0 of LC with $x = 0$ is less than that of LCM with $x = 0.05$ but greater than those of two LCM samples with $x = 0.01$ and 0.015. On the basis of an ionic size comparison ($r(Mn^{2+}) > r(Ti^{4+})$, $r(Mn^{3+})$ and $r(Mn^{4+}) < r(Ti^{4+})$ [6]),

228

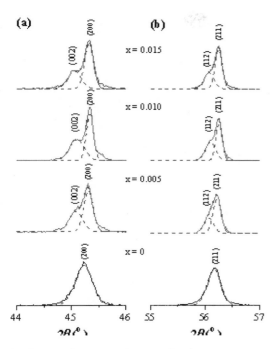

(a) **(b)**

x = 0.015 (211)
(002) (200) (112)

x = 0.010
(002) (200) (112) (211)

x = 0.005
(002) (200) (112) (211)

x = 0
(200) (211)

44 45 46   55 56 57
2θ(°)   2θ(°)

Figure 3.   Enlarged and Gaussian-fitted (002) and (200), (112) and (211) peaks in Figure1.

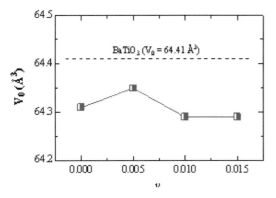

Figure 4.   The unit cell volume (V0) as a function of x for LCM ceramics.

Table 1.   Ionic radii versus Coordinate Number (CN).

Ion	CN	r (Å)
Ba^{2+}	12	1.61
Ti^{4+}	6	0.605
La^{3+}	12	1.36
Ca^{2+}	12	1.34
Ca^{2+}	6	1.00
Mn^{4+}	6	0.53
Mn^{3+}	6	0.58
Mn^{2+}	6	0.67

common optical modes of $BaTiO_3$—A_1 (TO_2), A_1 (TO_3), $B_1 + E(TO+LO)$, and $A_1(LO_3) + E(LO_3)$, peaking at 246, 517, 302, and 720 cm^{-1}, respectively [8]. The sharp 302 cm^{-1} band among these modes implies existence of the tetragonal transformations in LC. and LCM.

A peak at 832–837 cm^{-1} originates from the Raman charge effect [9], which indicates the internal deformation of the BO6 octahedrons caused by the charge difference of dopants. It was reported that the Raman charge effect will decrease when the content of B-site doping ions with smaller size compared with Ti4+ increases [9]. The intensity of this peak decreases with increasing x (Fig. 6), revealing that the number of Mn2+ ions in LCM decreases and Mn ions are in the mixed-valence forms of Mn4+, Mn3+ and Mn2+. This conclusion is in good agreement with the XRD results.

A new band at 674 cm^{-1}, not existing in LC, appears in LCM. This implies that the introduction of the third type of doping ions Mn in the LC lattice causes the additional deformation of the BO_6 octahedrons. The effect of Mn ions on $La_{Ba}^{\bullet}$—$Ca_{Ti}^{''}$—$La_{Ba}^{\bullet}$ defect complexes is responsible for appearance of this band.

EPR spectra of LCM are shown in Figure 7. A strong sextet signal was observed. All kinds of ions excluding Mn in LCM have no EPR response. Non-Kramer Mn^{3+} (3d^4) is ESR-inactive. Mn^{2+} (3d^5) and Mn^{4+} (3d^3), as Kramer ions, should be ESR-active and show the sextet Hyperfine Structure (HFS) of ^{55}Mn ($I = 5/2$) [10]. In order to determine the origin of this sextet signal, the HFS splitting constant (A) was calculated as $A = 89$ G according to the sextet signals observed in LCM. Thus, this strong sextet signal is assigned as Mn^{2+}, rather than Mn^{4+} [10], because the A value associated with Mn^{4+} is smaller with respect to that of Mn^{2+}. The $x = 0.005$ sample exhibits the strongest sextet signal with respect to other two samples, the Mn^{2+} content is inferred to be most among LCM samples. This result is in good agreement with the XRD and Raman results.

Dielectric-temperature curves of LC [3] and LCM are shown in Figure 8. The dielectric loss

it is inferred that for the $x = 0.05$ sample much more Mn ions were reduced into Mn^{2+}.

SEM images of LC [3] and LCM are shown in Figure 5. LC exhibits an intermediate grain size of 1.6 µm [3]. With continuous incorporations of Mn ions in LC, the grain growth of LCM is significantly depressed. The sample with $x = 0.015$ exhibits a fine-grained microstructure, with an average grain size of GS = 0.6 µm.

Raman spectra of LC [3] and LCM are shown in Figure 6. All of the samples exhibit the four

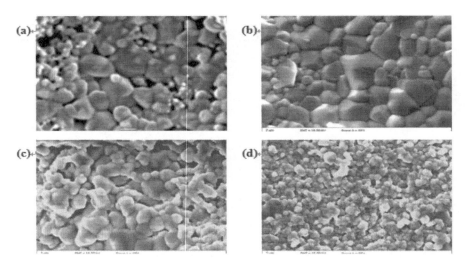

Figure 5.   SEM images of LCM ceramics with (a) $x = 0$, (b) $x = 0.005$, (c) $x = 0.010$, (d) $x = 0.015$.

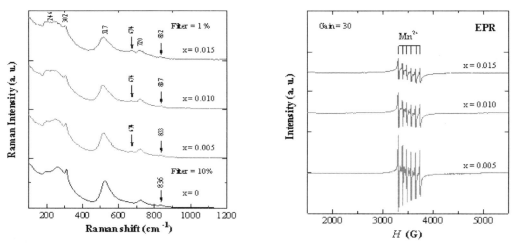

Figure 6.   Room temperature measured Raman spectra of LCM ceramics.

Figure 7.   EPR spectra of LCM ceramics.

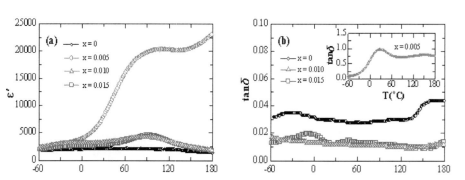

Figure 8.   Temperature dependence of (a) the dielectric permittivity (ε') and (b) the dielectric loss (tan δ) for LCM ceramics.

(tan δ) of LC can be markedly reduced from < 0.04 to < 0.02 due to 1% and 1.5% Mn doping in LC, showing a high ability to conduct electron trapping for Mn dopants. Unexpectedly, the $x = 0.005$ sample exhibits a semiconducting behavior with a very high tan δ of ~1 at room temperature.

LC exhibits X7R temperature stability, with a room-temperature permittivity of $\varepsilon'RT = 2170$. When $x = 0.005$, an abnormal high dielectric behavior implies that the electron conduction occurs in this sample; the observation for dielectric loss also verifies the emergence of electron conduction. The reason is unclear. When $x \geq 0.01$, Mn doping in LC can improve permittivity, but destroys the X7R stability of LC. The two broad peaks at −20 and 90 °C for $x = 0.010$ and 0.015 respond to orthorhombic-tetragonal and tetragonal-cubic phase transitions, respectively.

4 CONCLUSIONS

In order to lower the dielectric loss (tan δ) of the (Ba0.97La0.03) (Ti0.985Ca0.015) O3 ceramic (LC) with an X7R specification, (Ba0.97La0.03) (Ti0.985−xCa0.015Mnx) O3 (x = 0.005, 0.010, 0.015) (LCM) were prepared using the mixed oxides method and the influence of Mn doping on structure and dielectric properties of LC was studied. All of the LCM samples have a tetragonal perovskite structure and they are gradually refined to a fine-grained microstructure with increasing x. The introduction of the third type of doping ions Mn in the LC lattice causes the additional deformation of the $BO6$ octahedrons, which is indicated by a new Raman band at 674 cm^{-1}. The XRD, EPR and Raman results indicate that the Mn2+ content in LC doped with 0.5% Mn is most among LCM samples, but this leads to a semiconducting behavior. When x = 1% and 1.5%, the tan δ of LC can be markedly reduced to be lower than 0.02 and the dielectric permittivity can be improved. However, Mn doping in LC destroys the X7R stability of LC.

ACKNOWLEDGEMENTS

This work was financially supported by the National Natural Science Foundation of China (21271084) and Project of Jilin Provincial Science and Technology Department (20121825).

REFERENCES

[1] R.Z. Chen, X.H. Wang, L.T. Li, Z.L. Gui, Effects of Nb/Co ratio on the dielectric properties of $BaTiO_3$-based X7R ceramics, Mater. Sci. Eng. B 99 (2003) 298–301.

[2] H.I. Hsiang, L.T. Mei, Y.J. Chun, Dielectric properties and Microstructure of Nb-Co do-doped $BaTiO_3$-$(Bi_{0.5}Na_{0.5})TiO_3$ ceramics, J. Am. Ceram. Soc. 92 (2009) 2768–2771.

[3] D.Y. Lu, Y. Yue, X.Y. Sun, Novel X7R $BaTiO_3$ ceramics co-doped with La^{3+} and Ca^{2+} ions, J. Alloys Compd. 586 (2014) 136–141.

[4] H.T. Langhammer, T. Müller, K.H. Felgner, H.P. Abicht, Crystal structure and related properties of manganese-doped barium titanate ceramics, J. Am. Ceram. Soc. 83 (2000) 605–611.

[5] Jeong J., Han Y-H. Electrical properties of acceptor doped $BaTiO_3$. J. Electronceram. 13 (2004) 549–553.

[6] R.D. Shannon, Revised effective ionic radii and systematic studies of interatomic distances in halides and chalcogenides, Acta Cryst. A 32 (1976) 751–767.

[7] F. Ren, S. Ishida, Chemical states, roles and interrelations of additives in $BaTiO_3$ ceramics, J. Ceram. Soc. Jpn. 103 (1995) 759–766.

[8] D.Y. Lu, X.Y. Sun, M. Toda, A novel high-k 'Y5V' barium titanate ceramics co-doped with lanthanum and cerium, J. Phys. Chem. Solids 68 (2007) 650–664.

[9] D.D. Han, D.Y. Lu, X.Y. Sun, Structural evolution and dielectric properties of $(Ba_{1-x}Nd_x)(Ti_{1-y}Fe_y)O_3$ ceramics, J. Alloys Compd. 576 (2013) 24–29.

[10] D.Y. Lu, Q.L. Liu, T. Ogata, X.Y. Sun, X.F. Wang, Tetragonal phase stabilization caused by Pr ions in $Ba(Ti_{0.99}Mn_{0.01})O_3$ with mixed phases, Jpn. J. Appl. Phys. 50 (2011) 035806.

Material Science and Engineering – Chen (Ed.)

Electronic structure and power factor of Na$_x$CoO$_2$ from first-principles calculation

P.X. Lu

College of Materials Science and Engineering, Henan University of Technology, Zhengzhou, China

X.M. Wang

College of Information Science and Engineering, Henan University of Technology, Zhengzhou, China

ABSTRACT: In order to investigate the relationship between the electronic structure and the power factor of Na$_x$CoO$_2$ (x = 0.3, 0.5 and 1.0), the first-principles calculation was conducted by using density functional theory and semi-classical Boltzmann theory in this work. Our results suggest that with decreasing Na content a transition from semiconductor to semimetal is observed. Na$_{0.3}$CoO$_2$ possesses a higher electrical conductivity at 1000 K due to its increased density of states near Fermi energy level. However, an optimal Seebeck coefficient at 1000 K is obtained in Na$_{0.5}$CoO$_2$ because of its broadened band gap near Fermi energy level. Consequently, a maximum power factor is realized in Na$_{0.5}$CoO$_2$. So our work provides a complete understanding on the relationship between the electronic structure and the thermoelectric power factor of Na$_x$CoO$_2$.

Keywords: electronic structure; thermoelectric properties; Na$_x$CoO$_2$; first-principles calculation

1 INTRODUCTION

Thermoelectric materials, which can be used as solid-state Peltier cooler or power generator from waste heat, will play an important role in solving global fossil energy crisis and minimizing environmental pollution (L.E. 2008). A material with an excellent performance needs a high electrical conductivity, a large Seebeck coefficient and a low thermal conductivity simultaneously (G.J. 2008). Recently, sodium cobalt oxide Na$_x$CoO$_2$ has been recognized as encouraging thermoelectric materials due to their lower cost, nonpoisonous, simple synthesis procedure and the anti-oxidization in oxidized atmosphere for long time (K. 2003. & M. 2002). Many efforts in improving the thermoelectric properties of Na$_x$CoO$_2$ were made on either improving the Seebeck coefficient, or increasing the electrical conductivity or decreasing the thermal conductivity. The experimental results have indicated that Na$_x$CoO$_2$ possesses a large Seebeck coefficient, a small electrical conductivity and meanwhile a high thermal conductivity (T. et al. 2006). Meanwhile, theoretical calculations were also conducted widely to reveal the mechanism of the electron and phonon transport in Na$_x$CoO$_2$. The lattice thermoelectric properties of Na$_x$CoO$_2$ vary with its stoichiometry, structural defects, temperature and nanosheet geometry (D.O. et al. 2014). Its thermal conductivity could be suppressed significantly by an Einstein-like rattling mode at low energy (D.J. et al. 2013). Moreover, atoms doping could lead to a metal-insulator transition at low temperature in Na$_{0.8}$Co$_{1-x}$Sm$_x$O$_2$ (J. et al. 2012). The formation energies of defects and the resultant changes in electronic structure of Na$_x$CoO$_2$ were evaluated by using first-principles calculation (M. et al. 2010). In addition, the thermoelectric properties of the Na$_x$CoO$_2$ were also investigated by using simple lattice dynamical model or Boltzmann theory (P.K. et al. 2005 & S. et al. 2010). To the best of our knowledge, however, there have been no reports demonstrating the relationship among the electronic structure and the thermoelectric properties of Na$_x$CoO$_2$ up to now. Therefore, our work is aimed to provide a complete understanding on the relationship between the electronic structure and the thermoelectric properties of Na$_x$CoO$_2$ by using first-principles calculation.

2 CALCULATION METHODS

Na$_x$CoO$_2$ has a rhombohedral symmetry (R-3 mH, Group No. 166), its lattice parameter and the atoms site can be obtained through the Inorganic Crystal Structure Data (ICSD, findit 2008 software). Calculation was performed using CASTEP

(Cambridge Serial Total Energy Package), a first-principle pseudopotential method based on the Density-Functional Theory (DFT) (M.D. et al. 2002). The used pseudopotential was the Norm-conserving pseudopotential. As for the approximation of the exchange correlation term of the DFT, the Generalized Gradient Approximation (GGA) of Perdew was adopted with Perdew-Burke-Ernzerh of parameters. The cutoff energy of atomic wave functions was set at 830 eV. Sampling of the irreducible wedge of Brillouin zone was performed with a regular Monkhorst-Pack grid of special k-points, which was $12 \times 12 \times 4$. All atomic positions in the models had been relaxed according to the total energy and force using BFGS scheme, based on the cell optimization criterion (RMs force of 0.01 eV/nm, displacement of 0.00005 nm). The calculation of total energy and electronic structure was followed by cell optimization with SCF tolerance of 5.0×10^{-6} eV. The Partial Density of States (PDOS) for electrons was obtained by broadening the discrete energy levels using a smearing function of 0.05 eV Full-Width at Half-Maximum (FWHM) on a grid of k-points generated by the Monkhorst–Pack scheme (H.J. 1976.).

The effective mass of carriers m* is obtained by differentiating the energy at Fermi energy level twice with respect to the wave vector of the high symmetric points in Brillouin zone:

$$m^* = \hbar^2 / \left(\frac{\partial^2 E}{\partial k^2} \right) \qquad (1)$$

where $\hbar$ is the Plank reduced constant, E is the energy band near Fermi energy level, k is the wave vector of the high symmetric points in Brillouin zone. The Fermi energy E_F^0 is the highest energy level occupied by electrons at 0 K, and is obtained from Eq. 2:

$$E_F^0 = \frac{\hbar^2 k_0^2}{2m^*} = \frac{k_0^2}{2} \frac{\partial^2 E}{\partial k^2} \qquad (2)$$

where $\hbar$ is the Plank reduced constant, k_0 is the radius of the spherical surface with a energy $E = E_F^0$ at k space, m^* is the effective mass of carriers, E is the energy band and k is the wave vector in Brioullin zone. The temperature dependence of the Fermi energy is calculated by Eq. 3:

$$E_F = E_F^0 \left[1 - \frac{\pi^2}{12} \left(\frac{k_B T}{E_F^0} \right)^2 \right] \qquad (3)$$

where E_F^0 is the Fermi energy at 0 K, T is the absolute temperature and k_B is the Boltzmann constant.

The Fermi-Dirac distribution function $\left(-\frac{\partial f}{\partial E} \right)$ is written as Eq. 4:

$$\left(-\frac{\partial f}{\partial E} \right) = \frac{1}{k_B T} \frac{1}{\left(e^{(E-E_F)/k_B T} + 1 \right)} \cdot \frac{1}{\left(e^{-(E-E_F)/k_B T} + 1 \right)} \qquad (4)$$

where k_B is the Boltzmann constant, E is the energy band, E_F is the Fermi energy and T is the absolute temperature. The group velocity of carriers $v_e(E)$ is calculated by differentiating the energy with respect to the wave vector in Brioullin zone (C. et al. 2005). Consequently, the electrical conductivity and the Seebeck coefficient are defined as Eq. 5 and Eq. 6, respectively (Y. et al. 2010):

$$\sigma = \frac{e^2}{3} \int dE \left(-\frac{\partial f}{\partial E} \right) g(E) \tau(E) v^2(E) \qquad (5)$$

$$\alpha = \frac{e}{3T\sigma} \int dE \left(-\frac{\partial f}{\partial E} \right) g(E) \tau(E) v^2(E)(E - E_F) \qquad (6)$$

where e is the electron charge, $\left(-\frac{\partial f}{\partial E} \right)$ is the Fermi distribution function, $g(E)$ is the Density of States (DOS), $v_e(E)$ is the group velocity of electrons, E and E_F are the energy band and the Fermi energy, respectively. $\tau(E)$ is the relaxation time, which can be simplified as an energy-independent constant equal to 6.5×10^{-25} s using the experimental electrical conductivity. Therefore, the power factor is determined by Eq. 7:

$$P = \alpha^2 \sigma \qquad (7)$$

where α is the Seebeck coefficient, σ is the electrical conductivity.

3 RESULTS AND DISCUSSION

The band structure of Na_xCoO_2 (x = 0.3, 0.5, 1.0) is shown in Figures 1a, 1b and 1c, respectively. It can be seen that $Na_{1.0}CoO_2$ and $Na_{0.5}CoO_2$ possess a typical semiconductor behavior with an indirect energy gap of about 0.40 eV and 1.04 eV near the Fermi energy level, respectively. However, $Na_{0.3}CoO_2$ exhibits a typical semimetal characteristic since several energy bands exist across the Fermi energy level. Moreover, in comparison with $Na_{1.0}CoO_2$, an obvious anisotropy of the band structure for $Na_{0.3}CoO_2$ and $Na_{0.5}CoO_2$ can be observed because the band curves along the HK direction are very flat while very steep along the KG direction. In addition, the much more extrema of the band structure existing in $Na_{0.3}CoO_2$ and $Na_{0.5}CoO_2$ will be beneficial to increasing their

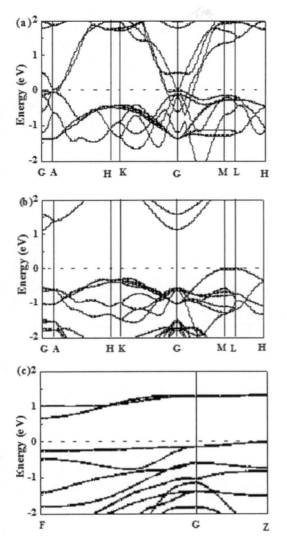

Figure 1. Band structure of Na_xCoO_2: (a) $Na_{0.3}CoO_2$; (b) $Na_{0.5}CoO_2$ and (c) $Na_{1.0}CoO_2$.

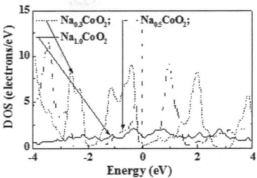

Figure 2. DOS of Na_xCoO_2.

The Density of States (DOS) of Na_xCoO_2 ($x = 0.3$, 0.5, 1.0) is presented in Figure 2. In general, the electrical conductivity is mainly determined by the concentration of carriers near the Fermi energy level (J.M. 1972). The small DOS of 2.15 electrons/eV for $Na_{1.0}CoO_2$ may be a main reason for its small electrical conductivity. However, the increased DOS of 9.0 electrons/eV and 4.0 electrons/eV for $Na_{0.3}CoO_2$ and $Na_{0.5}CoO_2$ is beneficial to enhancing their electrical conductivity. Pseudogap energy, an energy gap between the nearest two peaks near Fermi energy level in the DOS curve, can reflect the covalent bonding strength between the parent atoms. The pseudogap energy of $Na_{1.0}CoO_2$ bulk is merely 1.00 eV, which indicates the relatively weak covalent bonding would result in a low phonon vibration frequency and thus a small phonon thermal conductivity according to the following formula:

$$\omega = \sqrt{\frac{k}{m}} \qquad (9)$$

where ω is the phonon vibration frequency, k is the force constant, and m is the effective mass. Similarly, the increased pseudogaps for $Na_{0.3}CoO_2$ and $Na_{0.5}CoO_2$ means a relatively strong covalent bonding between the parent atoms and thus an increased phonon thermal conductivity.

The electrical conductivity for Na_xCoO_2 ($x = 0.3$, 0.5, 1.0) is presented in Figure 3. With decreasing Na content the electrical conductivity at 1000 K increases from 21000 $\Omega^{-1} \cdot m^{-1}$ to 91000 $\Omega^{-1} \cdot m^{-1}$ and 141000 $\Omega^{-1} \cdot m^{-1}$ respectively, which is well consistent with the experimental data (S. et al. 2006). The difference in electrical conductivity is mainly resulted from the difference of their density of states, mobility of carriers as well as band gap near Fermi energy level.

The Seebeck coefficient for Na_xCoO_2 ($x = 0.3$, 0.5, 1.0) is presented in Figure 4. The negative

electrical conductivity. It should also be noted that the effective mass of carriers in $Na_{0.3}CoO_2$ and $Na_{0.5}CoO_2$ (m^*/m_e equal to 3.0 and 3.7, respectively) is larger than that in $Na_{1.0}CoO_2$ (m^*/m_e equal to 1.8) since the width of the band structures at the bottom of conduction band is widened, which is helpful for enhancement of their Seebeck coefficients according to the following formula:

$$E_{gap} = 2eT\alpha_{max} \qquad (8)$$

where, E_{gap} is the band gap energy, e is the electron charge, T is the absolute temperature and α_{max} is the maximum Seebeck coefficient (H.J. et al. 1999).

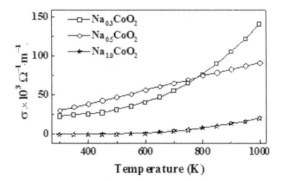

Figure 3. Electrical conductivity of Na_xCoO_2.

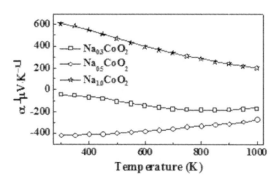

Figure 4. Seebeck coefficient of Na_xCoO_2.

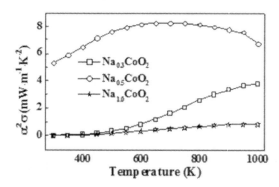

Figure 5. Power factor of Na_xCoO_2.

sign of the Seebeck coefficients for $Na_{0.3}CoO_2$ and $Na_{0.5}CoO_2$ indicates that the major carriers are electrons in the whole temperature, but the main carriers in $Na_{1.0}CoO_2$ are holes. With decreasing Na content the Seebeck coefficient at 1000 K varies from -165 $\mu V \cdot K^{-1}$ to -271 $\mu V \cdot K^{-1}$ and 200 $\mu V \cdot K^{-1}$, which is close to the experimental result (J. et al. 1983). Consequently, a maximum

Seebeck coefficient can be obtained in $Na_{0.5}CoO_2$, which is ascribed to its broadened band gap, the increased effective mass of carriers and the small density of states near Fermi energy level.

Figure 5 shows the power factor of Na_xCoO_2 ($x = 0.3, 0.5, 1.0$). It can be seen that with decreasing Na content the power factor of Na_xCoO_2 at 1000 K increases from 3.86 $mW \cdot m^{-1} \cdot K^{-2}$ to 6.73 $mW \cdot m^{-1} \cdot K^{-2}$ while then decreases to 0.85 $mW \cdot m^{-1} \cdot K^{-2}$. So, an optimal power factor of 6.73 $mW \cdot m^{-1} \cdot K^{-2}$ is achieved in $Na_{0.5}CoO_2$ for its increased electrical conductivity and Seebeck coefficient.

4 CONCLUSIONS

In conclusions, the electronic structure and the power factor of Na_xCoO_2 ($x = 0.3, 0.5, 1.0$) have been calculated successfully by using density functional theory and Boltzmann transport theory. With decreasing Na content the DOS at the Fermi energy level increased and a transition from semiconductor to semimetal is observed, which consequently leads to an increased electrical conductivity continuously. However, $Na_{0.5}CoO_2$ has an optimal Seebeck coefficient due to its broadened band gap near Fermi energy level. Therefore, a maximum power factor is realized in $Na_{0.5}CoO_2$.

ACKNOWLEDGEMENT

This project is supported financially by Science Foundation of Henan University of Technology (Grant Nos 2011BS056 and 11JCYJ12).

REFERENCES

[1] Bell, L.E. 2008. Cooling, heating, generating power, and recovering waste heat with thermoelectric systems. *Sci.* 321: 1457–1461.
[2] Snyder, G.J. & Toberer, E.S. 2008. Complex thermoelectric materials. *Nat. Mater.* 7: 105–114.
[3] Takada, K. Sakurai, H. Takayama-Muromachi, E. Izumi, F. Dilanian, R.A. & Sasaki, T. 2003. Superconductivity in two-dimensional CoO_2 layers. *Nat. (Lond.)* 422: 53–55.
[4] Ito, M. Nagira, T. Oda, Y. Katsuyama, S. Majima, K. & Nagai, H. 2002. Effect of partial substitution of 3d transition metals for Co on the thermoelectric properties of $Na_xCo_2O_4$. *Mater. Trans.* 43: 601–604.
[5] Seetawan, T. Amornkitbamrung, V. Burinprakhon, T. Maensiri, S. Tongbai, P. Kurosaki, K. Muta, H. Uno, M. & Yamanaka, S. 2006. Effect of sintering temperature on the thermoelectric properties of $Na_xCo_2O_4$. *J. Alloys Compds.* 416: 291–295.

[6] Demchenko, D.O. & Ameen, D.B. 2014. Lattice thermal conductivity in bulk and nanosheet Na_xCoO_2. *Computational Materials Science* 82: 219–225.

[7] Voneshen, D.J. Refson, K. Borissenko, E. Krisch, M. Bosak, A. Piovano, A. Cemal, E. Enderle, M. Gutmann, M.J. & Hoesch, M. 2013. Suppression of thermal conductivity by rattling modes in thermoelectric sodium cobaltate. *Nature materials* 12(11): 1028–1032.

[8] Sun, J. & Guo, Z.P. 2012. The metal-insulator transition of $Na_{0.8}Co_{1-x}Sm_xO_2$. *Intergrated Ferroelectrics* 137: 120–125.

[9] Yoshiya, M. Okabayashi, T. Tada, M. & Fisher, C.A.J. 2010. A first-principles study of the role of Na vacancies in the thermoelectricity of Na_xCoO_2. *J. Electr. Mater.* 39(9): 1681–1686.

[10] Jha, P.K. Troper, A. da Cunha Lima, I.C. Talatic, M. & Sanyald, S.P. 2005. Phonon properties of intrinsic insulating phase of the cobalt oxide superconductor $NaCoO_2$. *Physica B* 366: 153–161.

[11] Tosawat, S. Athorn, V.U. Prasarn, C. Chanchana, T. & Vittaya, A. 2010. Evaluating Seebeck coefficient of Na_xCoO_2 from molecular orbital calculations. *Computational Materials Science* 49: S225–S230.

[12] Segall, M.D. Lindan, P.J.D. Probert, M.J. Pickard, C.J. Hasnip, P.J. Clark, S.J. & Payne, M.C. 2002. First-principles simulation: ideas, illustrations and the CASTEP code. *J. Phys.: Cond. Matt.* 14: 2717–2744.

[13] Monkhorst, H.J. & Pack, J.D. 1976. Special points for Brillouin-zone integrations. *Phys. Rev. B* 13: 5188–5192.

[14] Stiewe, C. Bertini, L. Toprak, M. Christensen, M. Platzek, D. Williams, S. Gatti, C. Müller, E. Iversen, B.B. Muhammed, M. & Rowe, M. 2005. Nanostructured $Co_{1-x}Ni_x(Sb_{1-y}Tey)_3$ skutterudites: Theoretical modeling, synthesis and thermoelectric properties. *J. Appl. Phys.* 97: 044317.

[15] Kono, Y. Ohya, N. Taguchi, T. Suekuni, K. Takabatake, T. Yamamoto, S. & Akai, K. 2010. First-principles study of type-I and type-VIII $Ba_8Ga_{16}Sn_{30}$ clathrates. *J. Appl. Phys.* 107: 123720.

[16] Goldsmid, H.J. & Sharp, J.W. 1999. Estimation of the thermal band gap of a semiconductor from seebeck measurements. *J. Electron. Mater.* 28(7): 869–872.

[17] Ziman, J.M. 1972. *Principles of the Theory of Solid.* Cambridge University Press, Cambridge.

[18] Tosawat, S. Vittaya, A. Thanusit, B. Santi, M. Prasit, T. Ken, K. Hiroaki, M. Masayoshi, U. & Shinsuke, Y. 2006. Effect of sintering temperature on the thermoelectric properties of $Na_xCo_2O_4$. *J. Alloys Compds.* 416: 291–295.

[19] Molenda, J. Delmas, C. & Hagenmuller, P. 1983. Electronic and electrochemical properties of Na_xCoO_{2-y} cathode. *Solid State Ionics* (9–10): 431–435.

Material Science and Engineering – Chen (Ed.)
© 2016 Taylor & Francis Group, London, ISBN 978-1-138-02936-1

Electrodeposition of nickel under the conditions of applied magnetic field and no forced-convection

X.D. Zhang & P.M. Ming

School of Mechanical and Power Engineering, Henan Polytechnic University, Jiaozuo, Henan, China

ABSTRACT: The morphologic characteristics and microhardness of the nickel layers electrodeposited from an ammonium sulfonate bath under the conditions of applied magnetic field (0.82 T) and no forced-convection have been investigated. Nickel layers featuring good surface qualities with few defects and a smooth appearance and high microhardness varing from 342 HV to 395 HV were obtained at relatively high current densities, ranging from 3 to 11 A/dm². This is attributed to the positive roles coming from MHD and/or micro-MHD effects during electrodeposition. The findings indicated that the applicable nickel layers can be prepared by electrodeposition at fairly high current densities with MHD-driven natural convections alone.

Keywords: MHD; electrodeposition; morphology; microhardness

1 INTRODUCTION

Several effects such as Magnetohydrodynamic (MHD) effect and magnetomechanical effect may take place when an external magnetic field is applied to an electrochemical electrodeposition system. The MHD effect will be induced when a magnetic field is perpendicular to the current density applied to an electrochemical cell, all the charged species that move in the electrolyte will experience a Lorentz force, which is perpendicular to the current density and the magnetic field. This force will induce convective movements in the solution that reduce the diffusion layer thickness (magnetohydrodynamic or MHD effect). Numerous studies show that this MHD effect will induce convective movements in the solution, reduce the diffusion layer thickness, improve the deposition rate, influence the behaviors of electrochemical reaction and also the quality of deposition layer (Krause et al. 2007) (Koza et al. 2008) (Koza et al. 2010) (Leventis & Dass 2005) (Bodea et al. 1999). Therefore the MHD effect has been attracted by many researchers. Generally, the magnetomechanical effect includes magnetic field gradient effect and concentration gradient of charged particles in electrolyte. They can emerge when the uneven distribution of its own strength or the different concentrations of charged particles of different nature respectively happen. These two effects may influence the electrochemical reaction to some extend. The MHD effect has been considered as the most important factor that affects the electrodeposition of metals and thus it has been studied intensively.

Ebadi et al. (2011) have studied the influence of strong magnetic field on corrosion resistance and microstructure of electrodeposited Ni coatings in a Watts bath containing thiourea. Bund & Ispas (2005) reported that an increase of the surface roughness of nickel layer plated from an Amino-sulphonate type bath with the additive of sodium dodecyl sulfate or sodium 2-ethylhexyl sulphate occurred if the electrodeposition process was not applied to an external forced agitation although it was imposed to the magnetic field (0–800 mT). However, without exerting external forced convection, Devos et al. (1998) prepared a grain-refined coating in a Watts bath containing organic inhibitors of 2-butyne-1,4-diol and explained that the related mechanisms, that is, the refining effect results mainly from the increase in the concentration of inhibiting species near the cathode surface caused by the improved MHD-driven mass-transport, rather than directly from MHD effect. Unlike the above researchers, Krause et al. (2004) electroplated a smooth Co coating with less pinhole defects under the magnetic field and interpreted that the decisive contribution to this result was the MHD effect. Ispas et al. (2009) studied the influences of the combined effects of the MHD and natural convection on the electrochemical behaviors (such as nucleation and growth) and processes of Ni electrodeposition in magnetic field (0–0.7 T). They found that finer grains, more uniform layer, and higher current density applied and plating rate can be obtained when the direction of natural convection is the same as that of magneto-hydrodynamic convection.

Nevertheless, the influencing mechanisms of the MHD effect on the metal electrodeposition process including morphological characteristics and performances of the deposited coatings have not been sufficiently exploited yet. Furthermore, the reported results pertaining to MHD assisted electrodeposition of metals are, to some extent, contradictory possibly due to the different process conditions. On the other hand, some cases of metal electrodeposition in which the MHD effect is dominant, such as nickel electrodeposition in an ammonium sulfonate bath without an external forced stirring have not been studied so far. Therefore, this paper focuses on investigating the morphologic characteristics and properties of the nickel layers produced from an ammonium sulfonate and additive-free bath under the conditions of pure natural convections which may include gradient-driven convection and MHD-driven convection.

2 MATERIALS AND METHODS

A stainless steel (SUS 316) sheet (16 mm × 32 mm × 1 mm) was used as the cathodic substrate and a sulfur-containing nickel plate (99.99%) as the anode. The compositions in this study were: nickel sulphamate 450 g/l, nickel chloride 8 g/l and boric acid 40 g/l. All chemicals involved are p.a. grades and the electrolyte was prepared using bi-distilled water. The magnetic field generated by permanent magnets was parallel to the cathode surface with magnetic flux density, B, of 0.82 T which is almost homogenous. The applied current densities, I, were 3–19 A/dm^2.

Surface morphologies of the deposits were, respectively, examined by a measuring tool microscope (STM6-LM, Japan) and a scanning electron microscope (JSM-6390 LV, Japan). Microhardness tests were performed using an intelligent digital microhardness tester (THV-1MD, China) with a load of 490 mN exerted for 10 s. The microhardness of deposit is the average of seven measurements in different places.

3 RESULTS AND DISCUSSION

3.1 Morphological characteristic

Figure 1 shows the surface morphologies of the nickel deposits electroplated at different current densities. It can be seen that relatively satisfactory surface morphologies of nickel with a favorable smoothness and few observable defects such as pinhole and nodular can be achieved in a fairly wide range of current densities of 3–11 A/dm^2 if the magnetic field is applied to the electroplating cell,

(a) I=3 A/dm^2, B=0 T   (b) I=5 A/dm^2, B=0 T

(c) I=3 A/dm^2, B=0.82 T   (d) I=5 A/dm^2, B=0.82 T

(e) I=7 A/dm^2, B=0.82 T   (f) I=9 A/dm^2, B=0.82 T

(g) I=11 A/dm^2, B=0.82 T   (h) I=13 A/dm^2, B=0.82 T

(i) I=15 A/dm^2, B=0.82 T

Figure 1. Surface morphologies of Ni electrodeposits prepared at different current densities (Optical graphs, X50).

although they tend to deteriorate when the applied current densities exceed 13 A/dm^2. Conversely, in the absence of a magnetic field, the nickel deposits showed quite a few defects and a rough surface even though they were produced at a much low current density of 3 A/dm^2. This is possibly attributed to the positive roles resulting from MHD and/or micro-MHD effects: enhancing mass transportation in the vicinity of the cathode surface and thus reducing diffusion layer, which lead to a big plating rate and a quick removal of by-products such as hydrogen. The findings also indicated that the applicable electrodeposition of nickel can be conducted at fairly high current densities only with MHD-driven natural convections.

Observations from SEM images of electroplated nickel layers (shown in Fig. 2) further showed that the grains of nickel plated in a magnetic field were much finer and exhibits a more uniform distribution than those generated traditionally. This may result from the grain refining mechanisms which come from micro-MHD effect (Ispas et al. 2009).

3.2 *Microhardness*

The variations in the hardnesses of deposited nickel layers prepared at different current densities and with or without magnetic field were illustrated in Figure 3. With the current density increasing

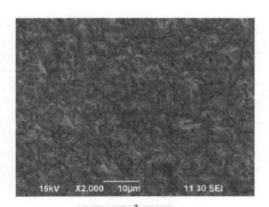

(a) I=5 A/dm^2, B=0 T

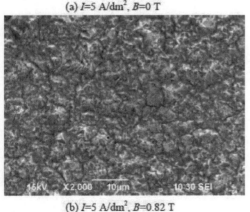

(b) I=5 A/dm^2, B=0.82 T

Figure 2. SEM images of nickel layers electrodeposited with or without a magnetic field.

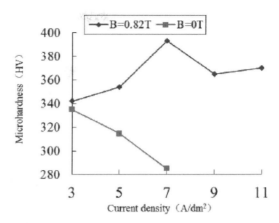

Figure 3. Relationship of the microhardness of the deposited nickel layers and current densities.

from 3 A/dm^2 to 11 A/dm^2, the hardnesses of the nickel layers electroplated under the magnetic field increase to a peak value (about 390 HV, 7 A/dm^2) and then slowly lower. This may contribute mainly to the change of MHD effect. At relatively low current densities of 3–7 A/dm^2, MHD effect can provide the electrochemical reactions with a fairly good mass transportation condition and more and more compact nickel layers featuring fine grains can be obtained, showing an increasingly higher microhardness, while when the applied current densities were greater than 7 A/dm^2, the natural convection driven by MHD effect cannot offer the required mass transfer conditions which massive particles including ions and hydrogen products are needed to be transferred timely and, in this situation, many coarse grains may form and a variety of defects such pinholes may exist in the metal layer, thus giving increasingly smaller hardness. In contrast, the microhardnesses of deposited layers produced without magnetic field were much lower than those created with MHD effect and decrease rapidly with increasing of the applied current densities.

4 CONCLUSIONS

The morphological characteristic and microhardness of nickel deposits produced from an ammonium sulfonate bath system with an imposition of magnetic field parallel to the cathode surface have been investigated in the absence of external forced convection and additives. Surface morphologies feature a favorable smoothness and few observable defects and the grains are finer. Relatively higher microhardnesses can be obtained. The results indicated that, applicable electrodeposited nickel coatings can be achieved only with the actions of the MHD effects including macro and micro MHD effect.

ACKNOWLEDGMENTS

This work was supported financially by National Natural Science Foundation of China (No. 50805077).

REFERENCES

[1] Krause, A., et al. 2007. Magnetic field induced micro-convective phenomena inside the diffusion layer during the electrodeposition of Co, Ni and Cu. *Electrochimica acta* 52(22): 6338–6345.

[2] Koza, J.A., et al. 2008. The effect of magnetic fields on the electrodeposition of CoFe alloys. *Electrochimica Acta* 53(16): 5344–5353.

[3] Koza, J.A., et al. 2010. Electrocrystallisation of CoFe alloys under the influence of external homogeneous magnetic fields—Properties of deposited thin films. *Electrochimica Acta* 55(3): 819–831.

[4] Leventis, N. & Dass, A. 2005. Demonstration of the elusive concentration-gradient paramagnetic force. *Journal of the American Chemical Society* 127(14): 4988–4989.

[5] Bodea, S., et al. 1999. Electrochemical growth of iron arborescences under in-plane magnetic field: Morphology symmetry breaking. *Physical review letters* 83(13): 2612.

[6] Ebadi, M., Basirun W.J. & Alias, Y. 2011. Influence of Magnetic Field on Corrosion Resistance and Microstructure of Electrodeposited Ni Coatings. *Advanced Materials Research* 264: 1383–1388.

[7] Bund, A. & Ispas, A. 2005. Influence of a static magnetic field on nickel electrodeposition studied using an electrochemical quartz crystal microbalance, atomic force microscopy and vibrating sample magnetometry. *Journal of Electroanalytical Chemistry* 575(2): 221–228.

[8] Devos, O., et al. 1998. Magnetic field effects on nickel electrodeposition. *Journal of The Electrochemical Society* 145(2): 401–405.

[9] Krause, A., et al. 2004. The effect of magnetic fields on the electrodeposition of cobalt. *Electrochimica Acta* 49(24): 4127–4134.

[10] Ispas, A., et al. 2009. Nucleation and growth of thin nickel layers under the influence of a magnetic field. *Journal of Electroanalytical Chemistry* 626(1): 174–182.

Material Science and Engineering – Chen (Ed.)
© 2016 Taylor & Francis Group, London, ISBN 978-1-138-02936-1

Synthesis and characterization of core-shell ion-imprinted polymers by the surface imprinting technology for selective separation lead ions from the aqueous solution

G.X. Zhong & W.L. Guo
School of Biology and Chemical Engineering, Jiangsu University of Science and Technology, Zhenjiang, China

Y. Xu
School of Chemistry and Chemical Engineering, Jiangsu University, Zhenjiang, China

L.L. Zhang & Y. Chen
School of Biology and Chemical Engineering, Jiangsu University of Science and Technology, Zhenjiang, China

ABSTRACT: This paper describes the preparation of core-shell ion-imprinted polymer nanoparticles by the surface imprinting technology for the selective removal of lead ions from the aqueous solution. The material was characterized by Fourier transform infrared spectroscopy and scanning electron microscopy. The adsorption by the material was studied by batch experiments with respect to the effects of the pH value and adsorption isotherms. The maximum sorption retention capacity of Pb (II) ions on the imprinted polymer was 92.08 mg g^{-1}. The prepared P-IIP was also proved to be applied to selectively remove Pb (II) from the solution. The result suggested that the sorbent could be used as an excellent adsorbent for the efficient removal of Pb (II) from the aqueous solution.

Keywords: ion-imprinted polymers; lead; selective adsorption

1 INTRODUCTION

Nowadays, the pollution by heavy metals deriving from various environmental sources has attracted much more attention. Lead is one of the most hazardous elements for human health, because it can cause an adverse effect on the metabolic processes of humans, and it has been proved to be a carcinogenic agent [1]. Consequently, the development of reliable methods for the removal of lead in water samples is of particular significance. Recently, novel materials called Ion Imprinted Polymers (IIP) have attracted much attention as a highly selective sorbents for the removal of metal ion [2]. Besides, the IIP show very interesting characteristics such as high selectivity, low cost, high surface area, durability and reusability [3]. However, the IIP prepared by the traditional technique will result in incomplete template removal, small binding capacity, and slow mass transfer. Fortunately, this problem can be resolved by surface imprinting, in which the imprinted templates are situated at the surface or in the proximity of the material's surface [4].

Silica nanoparticles are widely used as support materials due to their high surface-to-volume ratio and heat resistance [5]. Furthermore, the core-shell nano-composites have emerged as one of the hot research directions in recent years, which will further improve the functionality level of the silicon-based materials.

In this paper, we report the efficient synthesis of core-shell structural ion-imprinted polymeric nanoparticles for the fast and selective removal of Pb (II) from the aqueous solution. The polymeric nanoparticles have been characterized by Fourier Transform Infrared Spectroscopy (FT-IR) and Scanning Electron Microscopy (SEM). The adsorption behavior of Pb (II)-IIP for Pb (II) has been studied in batch experiments. Therefore, this ion-imprinted polymer is an efficient solid phase for the extraction of lead ions in complex matrices.

2 EXPERIMENTAL PROCEDURE

2.1 *Apparatus and chemicals*

Spectrometric measurements were carried out with a TAS-986 Flame Atomic Adsorption Spectrometer (FAAS) (Beijing, China). A pHS-3C digital pH meter (LIDA Instrument Factory, Shanghai, China) was used for the pH value adjustments.

Tetraethyl Orthosilicate (TEOS) was purchased from Guoyao Chemical Reagents Corp, Shanghai. In addition, the Acryl amide (AM) and the N, N'-methylenebisacrylamide (BIS) were obtained from Aladdin-reagent Co., Ltd (Shanghai, China). A standard stock solution of Pb (II) (1.0 g/L) was prepared by dissolving Pb (NO$_3$)$_2$. Doubly Distilled Water (DDW) was used for all dilutions.

2.2 Preparation of Pb (II)-IIP

2.2.1 Synthesis and modification of silica colloid nanoparticles

Monodisperse and uniform silica nanoparticles were synthesized according to the Stöber method [6]. Then, silica nanoparticles were chemically modified using MPS to introduce vinyl groups. Finally, the prepared MPS-modified silica nanoparticles were dried under vacuum at room temperature for further use.

2.2.2 Preparation of the core-shell structural Pb (II)-IIP

First, in a 100 mL quartz round bottomed flask, Pb (NO$_3$)$_2$·6H$_2$O (0.1 mmol), acryl amide (0.6 mmol), MPS-modified silica nanoparticles (50 mg) and BIS (1 mmol) were dispersed into 20 mL of methanol aqueous solution under heating and magnetic stirring. After purging the mixture with nitrogen, AIBN (5 mg) was added into this suspension. The resultant mixture was stirred at 65°C for 12 h under nitrogen. The products were separated by filtration and washed with the aqueous solution of methanol several times. Then, the obtained solid was treated with 2 mol L^{-1} HCl to completely leach the coordinated and non-coordinated Pb (II). Subsequently, the particles were washed with distilled water until neutral and then dried to constant weight under vacuum at 50°C. Finally, the core-shell structural Pb (II)-IIP can be obtained. By comparison, NIP was also prepared as a blank in parallel but without the addition of Pb (II).

2.3 Adsorption batch experiments

The adsorption of Pb (II) from the aqueous solution was conducted in batch experiments. 0.01 g of Pb (II)-IIP was added into a 25 mL colorimetric tube containing a certain amount of Pb (II) at a certain pH range. Then, the mixture was shaken vigorously for a certain moment. After centrifugation, the concentration of Pb (II) in the residue solution was determined by FAAS. The absorption capacity q_e (mg g^{-1}) and removal efficiency ($R\%$) at equilibrium were calculated as follows:

$$q_e = \frac{(C_0 - C_e)V}{W} \qquad (1)$$

$$R = \frac{C_0 - C_e}{C_0} \times 100\% \qquad (2)$$

where C_0 (mg L^{-1}) and C_e (mg L^{-1}) are the concentrations of Pb (II) at initial and equilibrium, respectively; and V (mL) and W (g) are the volume of solution and the mass of the adsorbent, respectively.

3 RESULTS AND DISCUSSION

3.1 Characterization and analysis

3.1.1 FT-IR results

FT-IR spectroscopy was employed to characterize the structural change in the different samples in the synthetic of the adsorbent. The FT-IR spectra of the pure silica are shown in Figure 2. Compared with the two spectra of pure silica and Pb(II)-IIP, the strong absorption peak at 1661 cm^{-1} and 1530 cm^{-1} is the result of the stretching vibration absorption of the amide carbonyl (amide I band) and in-plane bending vibration absorption of the N-H bond of the amide group (amide II band). The above phenomena suggest that core-shell Pb (II)-IIP is successfully prepared.

3.1.2 SEM

The morphology of pure silica and Pb (II)-IIP was analyzed by SEM. As can be seen from Figure 2a, the SEM image reveals that the silica particles are spherical in shape and almost uniform in size. However, it is obvious that the polymer layer is found on the surface of the particle, as shown in Figure 2b, which suggested that the core-shell Pb (II)-IIP can be successfully prepared by the surface imprinting technology.

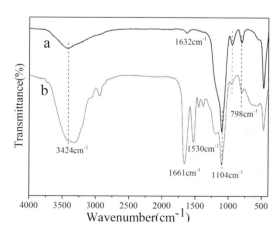

Figure 1. FT-IR of (a) pure silica and (b) Pb (II)-IIP.

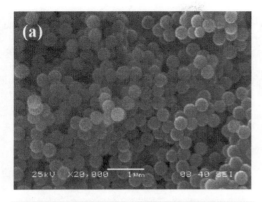

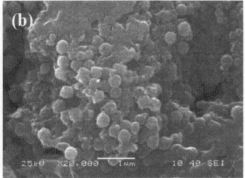

Figure 2.   SEM images of (a) pure silica and (b) Pb (II)-IIP.

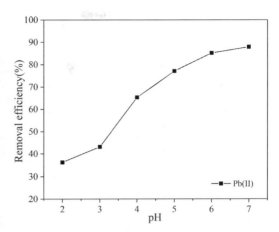

Figure 3.   Effect of pH on the removal efficiency of Pb (II)-IIP.

3.2   Adsorption experiments

3.2.1   Effect of pH

The acidity of the solution is an important factor affecting the adsorption process. The pH not only influences the surface charge and the protonation degree of the adsorbent, but also affects the speciation of the sorbate. As shown in Fig. 3, the removal efficiency of Pb (II) onto Pb(II)-IIP increased significantly from 2.0 to 7.0, then it remained constant with a further increase in the pH until the value was 6.0. At a higher value of pH, precipitation of the metal hydroxide is expected [7]. Therefore, pH 6.0 was selected as the suitable acidity for subsequent work.

3.2.2   Adsorption kinetics study

The adsorption rate of Pb (II) plays an important role in defining the efficiency of sorption, and the kinetic curves at different adsorbents are shown in Figure 5. As shown in Figure 5, the adsorption capacity increased sharply during the first 40 min, and then they tended to reach an equilibrium about 90 min. In addition, the adsorption rate rapidly increased at the original stage, which was attributed to the sufficient available active sites on the surface of the adsorbent. Furthermore, these sites

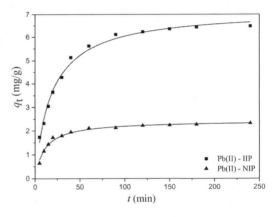

Figure 4.   Effect of time on the adsorption at different adsorbents.

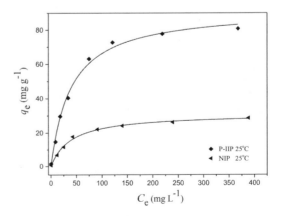

Figure 5.   Adsorption isotherms for the adsorption of Pb (II) onto Pb (II)-IIP and NIP.

245

were gradually occupied by Pb (II), leading to the decrease in the rate in the later period. Compared with Pb (II)-NIP, as shown in Figure 4, Pb (II)-IIP has a better performance in the adsorption capacity with the increasing adsorption time. Among them, the adsorption of Pb (II) onto Pb (II)-NIP quickly reached equilibrium. The phenomenon was due to, compared with imprinted polymers, the less available active sites on the surface of Pb (II)-NIP. The kinetic model was applied for fitting with the data, and the result indicated that the pseudo-second-order kinetic model was suitable to describe the adsorption process.

3.2.3 *Adsorption isotherms*

The adsorption capacity of Pb (II)-IIP and NIP was carried out with bath experiments. The Langmuir model was well fitted the data. As can be seen in Figure 4, the adsorption capacity of Pb (II)-IIP and NIP for Pb (II) was 92.08 and 31.10 mg g^{-1}, respectively. The results indicated that the prepared Pb (II)-IIP had a satisfactory adsorption capacity for Pb (II). However, the NIP is poor in the adsorption capacity for Pb (II).

3.2.4 *Selectivity study*

The imprinted polymers were also characterized by a uniform distribution of chelating sites. In ion-imprinted polymers, the cavities created after the removal of template were complementary to the imprint ion in size and coordination geometries. Competitive adsorption of Pb (II)/Co (II), Pb(II)/Zn(II), Pb(II)/Cu(II) Pb(II)/Ni(II) from their mixtures was investigated by using Pb(II)-IIP and NIP, respectively. The P-IIP exhibits an excellent adsorption selectivity for Pb (II) in the presence of competitive metal ions. However, the NIP is poor. The results indicated that the core-shell Pb (II)-IIP can selectively remove Pb (II) from the aqueous solution.

4 CONCLUSIONS

In the present work, the core-shell Pb (II)-imprinted polymer was prepared by the surface imprinting technique based on silica particles. The effects of various parameters on extraction efficiency were investigated, and the amount of Pb (II) was determined by FAAS. The adsorption behavior was carried out with Pb (II)-IIP in batch experiments. The prepared ion-imprinted polymer particles have an increased selectivity toward the Pb (II) ion over a range of competing metal ions with the same charge and similar ionic radius. The results indicated that the prepared core-shell Pb (II)-IIP could be used as a promising candidate for the selective removal of Pb (II) from the aqueous media.

ACKNOWLEDGMENTS

The corresponding author of this paper is Wenlu Guo.

REFERENCES

[1] Gama, E.M., da Silva Lima, A., & Lemos, V.A. Preconcentration system for cadmium and lead determination in environmental samples using polyurethane foam/Me-BTANC. *Journal of hazardous materials*, 136(3), pp. 757–762. (2006).

[2] Rao, T.P., Daniel, S., & Gladis, J.M. Tailored materials for preconcentration or separation of metals by ion-imprinted polymers for solid-phase extraction (IIP-SPE). *TrAC Trends in Analytical Chemistry*, 23(1), pp. 28–35. (2004).

[3] Liu, R., Guan, G., Wang, S., & Zhang, Z. Core-shell nanostructured molecular imprinting fluorescent chemosensor for selective detection of atrazine herbicide. *Analyst*, 136(1), pp. 184–190. (2011).

[4] Chen, L., Xu, S., & Li, J. Recent advances in molecular imprinting technology: current status, challenges and highlighted applications. *Chemical Society Reviews*, 40(5), pp. 2922–2942. (2011).

[5] Gao, D., Zhang, Z., Wu, M., Xie, C., Guan, G., & Wang, D. A surface functional monomer-directing strategy for highly dense imprinting of TNT at surface of silica nanoparticles. *Journal of the American Chemical Society*, 129(25), pp. 7859–7866. (2007).

[6] Stöber, W., Fink, A., & Bohn, E. Controlled growth of monodisperse silica spheres in the micron size range. *Journal of colloid and interface science*, 26(1), pp. 62–69. (1968).

[7] Behbahani, M., Bagheri, A., Taghizadeh, M., Salarian, M., Sadeghi, O., Adlnasab, L., & Jalali, K. Synthesis and characterisation of nano structure lead (II) ion-imprinted polymer as a new sorbent for selective extraction and preconcentration of ultra trace amounts of lead ions from vegetables, rice, and fish samples. *Food chemistry*, 138(2), pp. 2050–2056. (2013).

Material Science and Engineering – Chen (Ed.)
© 2016 Taylor & Francis Group, London, ISBN 978-1-138-02936-1

Low electrolyte dyeing technology with reactive dyes

X.P. Zeng

Wuhan Polytechnic, Wuhan, Hubei, China

ABSTRACT: The aim of the present study was to graft and modify cellulose fibers by synthesizing iod compounds. Consequently, cellulose fibers swelled sufficiently to increase the number of channels for reactive dyes entering the fiber, the affinity between the fiber and the dye is improved, and reactive dyes can be combined with fiber entirely. Thus, the fixation rate of the reactive dye increases and the inorganic salt is eliminated; meanwhile, the bath ratio is reduced.

Keywords: ionic pair; graft modification; low electrolyte; dyeing

1   THE TECHNICAL PRINCIPLE

In the knit industry, reactive dyes had already become the most important class of dyes. However, the color fixation rate of reactive dyes is low. About 20% to 30% of the dye exists in the waste water, thus the dyeing costs and burden on waste water treatment are increasing. How to improve the color fixation rate of reactive dyes, reduce waste water emission and improve the reuse rate of waste water has aroused wide concern of people.

The low electrolyte reactive dyeing process for cotton knitted fabric is to graft and modify cellulose fibers by synthesizing new compounds. Consequently, cellulose fibers swell sufficiently to increase the number of channels for reactive dyes entering the fiber, the affinity between the fiber and the dye is improved, and reactive dyes can be combined with fiber entirely. Thus, the fixation rate of the reactive dye increases from 70~80% to 92% or more. Therefore, the inorganic salt eliminates to achieve the purpose of low electrolyte dyeing. The solubilization of reactive dyes is played by the compound, the aggregation of high concentration dyes under a small bath ratio is reduced, and the activity of the dye levelness is improved. When the low green organic alternative alkali and aerosol dyeing techniques are applied to the low electrolyte dyeing process, the amount of the alternative alkali 1/8 to 1/10 of soda, and the bath ratio of aerosol dyeing can be controlled at 1:3–4.

2   TECHNICAL RESEARCHES

This paper studies the effects of different ions on the solution concentration for the color fixation rate. To obtain the optimal solution and graft modification effect for cellulose fiber by ions, the research focuses on the effect of different ions in the solution for the color fixation rate and the change in the color fixation rate between before and after the modification of the reactive dye.

Cotton fabrics is dipped into the solution with different amounts of ionic solution at room temperature for 60 min as a treatment. Then, Argazol Red BF-3B solution is used at concentrations of 0.5% and 2.5% to dye the fabrics before and after the treatment, respectively. The ionic pair concentrations were 10, 20, 30, 40, 50 (g/L). Then, the color fixation rate was compared by testing the exterior strength and absorbance of the dye solution. The effect of different concentrations of ions on the cotton fabric's dyeing color fixation rate is presented in Table 1.

Table 1.   Effect of different concentrations of ions on the cotton fabric's dyeing color fixation rate.

Amount of dye	Ion pair (g/L)	Without treatment	10	20	30	40	50
0.5%	Exterior strength %	Standard sample	105	113	113.6	114.2	114.7
	Color fixation rate %	80	87	91	92	93	93.1
2.5%	Exterior strength %	Standard sample	98.0	115.0	118.3	120.3	120.6
	Color fixation rate %	75	74	88	89	90.2	90.5

3 TECHNICAL APPLICATIONS

3.1 The leaching of ionic pairs in the solution

Ionic pair solution is one of the most active cross-linker compounds that have an affinity of the dyes. This compound can modify and swell the fiber. The fiber is sufficiently modified and swollen by the treatment with the cross-linking agent. In addition, multiple reactive gene groups are lead into the fiber to achieve the purpose of low electrolyte dyeing. The procedure of the ionic pair solution treatment is to leach the fabric into the solution after scouring and bleaching through the normal pre-treatment.

Ionic pair solution treatment:

Equipment: German THEN Aerosol dyeing machine;
Liquor ratio: 1:4
Machine parameters: the machine speed is 140 m/min; the power of the fan is 60%; the operation cycle is 3 min/ cycle.

Formulation:

Light and bright color ionic pair solution formulation

Ionic pair solution   2%
NaOH (caustic soda)   2%

A. Medium and dark color ionic pair solution formulation
Ionic pair solution   4%
NaOH (caustic soda)   2%

Process flow: the process flow of low electrolyte dyeing that leaches in the ionic pair solution is shown in Figure 1.

3.2 Low electrolyte dyeing new technology contrasts with conventional salt dyeing

The combinations of Remazol Red RGB/yellow RGB/blue RGB were selected. We performed

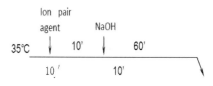

Figure 1.

Table 2. Comparison of the color fixation rate, process time consuming, process water consuming, color fastness and dyeing waste water testing.

Parameter				Conventional salt dyeing	Low electrolyte dyeing new technology	Comparison
Liquor ratio				1:12	1:4	Liquor ratio is low in the new technology process, only 1:4
Color fixation rate (%)				74	90	The new technology can save 16% of the dye
Sodium sulfate (g/L)				60	no	A lot of sodium sulfate can be saved by the new technology
Process time consuming (hour)				9.70	7.92	1.78 hours can be saved by the new technology
Process water consuming (ton)				140	99.2	40.8 tons water can be saved by the new technology
Color fastness	GB standard	Soap washing (rank)		4	4–5	Low electrolyte dyeing can improve the rank of color fastness on
		Water washing (rank)		4	4–5	soap washing, water washing, wet friction fastness 0.5 more
		Sweat stains (rank)	Acid	4–5	4–5	
			Alkali	4–5	4–5	
		Friction (rank)	Dry	4–5	4–5	
			Wet	3	3–4	
Waste water	Conductivity (μs/cm)			50000	2360	The pollution indicators in the low electrolyte dyeing new technology waste water pollution indicators are all significantly reduced. The conductivity decreased by 95%. The chroma is down by 70%. COD is dropped by 50%
	Chroma (times)			280	84	
	COD (mg/L)			530	265	
	pH			10.93	10.90	

gray-blue by low electrolyte reactive dyeing new technology under each small liquor ratio process and conventional salt dyeing process.

The comparison of the color fixation rate, process time consuming, process water consuming, color fastness and dyeing waste water testing is presented in Table 2.

4 CONCLUSIONS

By ionic pair compound acting on cellulose fibers for modifying and swelling, the affinity between the cellulose fibers and the dyes is improved. Then, a new knitted fabric salt-free dyeing technology with reactive dyes is developed. During the new technology process, the electrolyte, such as anhydrous sodium sulfate, is not added to anhydrous sodium sulfate.

The low dyeing is combined with the alternative alkali dyeing technology and aerosol dyeing. The dyeing process is salt free with a low liquor ratio. The alternative alkali electrolyte is used for color fixing during the process of fixing. The aerosol dyeing is with less water consumption and a variety of techniques is sensibly used in conjunction. Thus, the amount of electrolyte in waste water is greatly reduced.

Because the electrolyte in water is reduced, the burden of reverse osmosis water is relieved and the service life of the membrane is extended. The low cost of printing, dyeing and waste water reusing is achieved.

As a result of the increasing dyes' color fixation rate, the production cost is reduced, and production efficiency is improved. A total of 2 million tons of heavy smeary water is reduced per year. The difficulty in completing the treatment is reduced.

The amount of the electrolyte is greatly reduced. This creates favorable conditions for dyeing waste water reuse. The dyeing waste water reuse rate is 50%. So, the goal of energy saving and clean production is attained, which has significant economic and social benefits.

REFERENCES

[1] Liu Jiangjian. The Control of Low Bath Ratio Dyeing of Cloth in Rope. [J] Dyeing & Finishing; 2008,34(9): 24–25.
[2] Standard Committee of the IEEE Electromagnetic Compatibility Society. IEEE standard method for measuring the effectiveness of electromagnetic shielding enclosures, December, 1997.
[3] Liao Xuanting. Xu hua. Study on Presharpen Dyeing Technology of Reactive Dyes of Cotton knit Fabric. Textile Department, Wuhan Technology College, 2006(5):40–42.
[4] Zhang Hui. Zhang Hua. Zhang Limin. Effect of boiling and bleaching On the dyeing properties of hemp fabric [J] Process in Textile Science and technology, 2008(3):67–69.
[5] Song Xinyuan; ShenYuru. The Development of Reactive Dyes and Their Dyeing in Recent Years. [J] Dyeing & Finishing; 2002, 28(1): 45–49.

Material Science and Engineering – Chen (Ed.)
© 2016 Taylor & Francis Group, London, ISBN 978-1-138-02936-1

Photon-assistant transport properties of coupled Quantum Dots

Y.R. Ma, Y.J. Liu, R. Niu & Y.R. Huang
College of Automation Engineering, Beijing Polytechnic, Beijing, China

ABSTRACT: Based on the calculations of the electronic structures of coupled double and triple Quantum Dots (QDs), we study their transport properties under AC electric fields. With the help of the Floquet theorem, dependence of the average currents of these systems on electronic structures is investigated. It is found that a two-level structure has no significant impact on the transport properties. For both symmetric and asymmetric configurations, there are interferential Fano resonances in parallel Double Quantum Dots (DQDs) and photon-assistant Breit–Wigner resonance in serial DQDs. However, a Λ-type energy level of Triple Quantum Dots (TQDs) has a great impact on the transport properties: there is a symmetric Breit–Wigner resonance for asymmetric configuration due to photon-assistant tunneling; and there is an asymmetric Fano resonance for symmetric configuration due to the help of a 'trapping dark state' (coherent trapping) of electrons. We provide useful ways to design novel optoelectronic nanodevices.

Keywords: coupled quantum dots; photon-assistant transport; Fano resonance; Breit–Wigner resonance

1 INTRODUCTION

In recent years, the transport properties of semiconductor nanostructures have been paid much attention and studied widely. Great interest has recently focused on the quantum coherence and interference, leading to a lot of entertaining phenomena, such as Aharonov-Bohm (AB) oscillations [1], Kondo effects [2,3], coherent trapping [4,5], and Fano resonances [6–8]. In recent work, the double and triple QDs have attracted much attention due to their rich electronic structures and interesting physical phenomena. Recent experiments have shown that coherent coupling in DQDs embedded in an Aharonov-Bohm interferometer [9–12] and high-quality TQDs can be artificially produced [13,14]. A large number of relevant studies have been widely conducted on these systems [15–19]. In spite of many studies on multiple QDs, relatively little attention has been paid to the photon-assistant transport in coupled QDs driven by a time-period electric field. TQDs driven by the time-period field with the relevant three-level structure may have some interesting interference phenomena, such as coherent trapping, such as those we have known very well in quantum optics. In this paper, we study the photon-assistant transport in multiple coupled QDs (Figure 1), and pay special attention to the consequences of these wonderful interference phenomena.

In previous work on multiple coupled QDs, the quantum energy levels, which depend on the

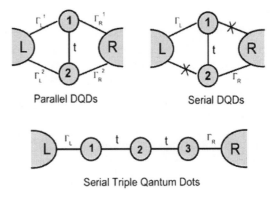

Figure 1. Schematic diagrams of the coupled QD systems.

structures of the systems, are usually assumed as constant parameters [20,21], so that the nature of the quantum properties of the systems cannot be presented quantitatively. Therefore, it is necessary to reveal the quantum behaviors of coupled QDs based on more practical models.

In our approach, we design coupled DQDs with a two-level structure and serial TQDs with a Λ-type three-level structure using a two-dimensional confining model in the effective mass frame. Based on the obtained level structure and with the help of the Floquet theory [22], we study the transport properties of these systems. It is found that Fano resonance will occur when the DQDs are coupled parallelly, and Breit–Wigner resonance [23] will occur when the DQDs are coupled serially for both symmetric and asymmetric cases. However, a Λ-type three-level serial TQD has great different transport properties: the photon-assistant tunneling leads to the symmetric Breit–Wigner resonance when the system is asymmetric (between left and right QDs). When the system is in a symmetric configuration (between left and right QDs), the formation of a 'trapping dark state' results in the interesting asymmetric Fano resonance under a resonant condition (photon energy equals the energy difference between the left/right quantum dot and the middle dot). Our research indicates that transport properties in a coupled QD system are very sensitive to the number of QDs.

2 MODEL AND ELECTRONIC STRUCTURE

We start our study by analyzing the electronic structures of serial TQDs as an example when the following two-dimensional potential is used:

$$V(x,y) = 0.5m^* \min\left\{\omega_{lx}^2(x+d)^2 + \omega_{ly}^2 y^2,\right.$$
$$\left.\omega_{mx}^2 x^2 + \omega_{my}^2 y^2, \omega_{rx}^2(x-d)^2 + \omega_{ry}^2 y^2\right\} \quad (1)$$

where d is the interdot distance; m^* ($m^* = 0.067m_e$ for GaAs QDs) is the effective mass of the electron; and $\omega_{lx(y)}$, $\omega_{mx(y)}$ and $\omega_{rx(y)}$ are confining trap frequencies of the left, middle and right dots in the x(y) direction, respectively. Here, we investigate the levels of the serial TQDs by varying $\hbar\omega_{rx} = \hbar\omega_{ry} = \hbar\omega_r$ with constant parameters of $\hbar\omega_{lx} = \hbar\omega_{ly} = \hbar\omega_l = 3.00\ meV$, $\hbar\omega_{mx} = \hbar\omega_{my} = \hbar\omega_m = 3.52\ meV$ and $d = 77.70\ nm$. For such parameters, the tunneling energy between the left and middle dot is about 51 μeV. A Λ-type three-level structure is obtained through solving the generalized eigenvalue of this system (see Figure 2).

The corresponding eigenwavefunctions are shown in the inset of Figure 2. As the value of $\hbar\omega_r$ increases, the level of the right dot increases. When $\hbar\omega_r = \hbar\omega_l = 3.00\ meV$, where the levels of the left and right dots are equal, the states $|0\rangle$ and $|1\rangle$ become a pair of delocalized bonding and antibonding states, while the other state $|2\rangle$ is localized in the middle quantum dot. The corresponding eigenwavefunctions are shown in the inset of Figure 2. When $\hbar\omega_r$ is far from 3.00 meV, the three levels are nearly the energies of the ground states of the left, middle and right dots, respectively, and the eigenstates are all localized in each quantum dot, as shown from their wavefunctions in the inset of Figure 2. We study the energy level structure of the coupled DQDs system using the same method

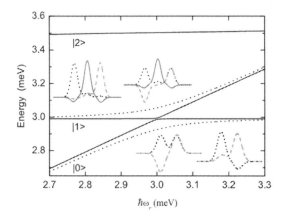

Figure 2. The three solid lines represent the lowest three eigenenergy levels as a function of $\hbar\omega_l$ for TQDs, and the two dotted lines represent the lowest two eigenenergy levels for DQDs. Insets: eigenwavefunctions for DQDs and TQDs at $\hbar\omega_l = 3.00\ meV$ and 2.70 meV, respectively. The solid lines represent state $|2\rangle$. The dotted lines represent state $|1\rangle$. The dashed lines represent state $|0\rangle$.

through changing $\hbar\omega_r$, where $\hbar\omega_l = 3.00\ meV$ and $d = 77.70\ nm$. In Figure 2, the lowest two eigenenergy levels of the coupled DQDs system are plotted as a function of $\hbar\omega_r$. The corresponding eigenwavefunctions are also shown in the inset of Figure 2.

3 PHOTON-ASSISTANT TRANSPORT PROPERTIES

Based on the results of the electronic structure, the transport properties of the time-dependent system can be studied with the help of the Floquet theorem. It can be shown that the average current of the system can be written as follows:

$$I = \frac{e}{\hbar} \sum_{K=-\infty}^{+\infty} \int d\varepsilon \left\{ T_{LR}^{(k)}(\varepsilon) f_R(\varepsilon) - T_{RL}^{(k)}(\varepsilon) f_L(\varepsilon) \right\} \quad (2)$$

where

$$T_{LR}^{(k)}(\varepsilon) = \Gamma_L \Gamma_R |\langle 1|G^{(K)}(\varepsilon)|N\rangle|^2 \quad (3)$$
$$T_{RL}^{(k)}(\varepsilon) = \Gamma_R \Gamma_L |\langle N|G^{(K)}(\varepsilon)|1\rangle|^2 \quad (4)$$

denote the transmission probabilities for electrons with the initial energy ε and the final energy $\varepsilon + k\hbar\Omega$ from the right lead and from the left lead, respectively. $f_l(\varepsilon) = (1 + \exp[(\varepsilon - \mu_l)/k_B T])^{-1}$ denotes the Fermi function. According to the Floquet theorem for a time-periodic Hamiltonian, there exists a complete set of solutions for the Schrödinger equation, which have the form $e^{-(i\varepsilon_\beta + \gamma_\beta)t}|\mu_\beta(t)\rangle$, where Floquet states $|\mu_\beta(t)\rangle$ are time-periodic functions, and ε_β γ_β ($\beta = 0,1,2$ for TQDs) are the real and imaginary parts of quasi-energies, respectively. The Fourier coefficients of the retarded Green function can be written as follows:

$$G^{(K)}(\varepsilon) = \sum_{\beta,k'} \frac{\left|u_{\beta,k'+k}\right>\left<u_{\beta,k'}^\dagger\right|}{\varepsilon - (\varepsilon_\beta + k'\hbar\Omega - i\gamma\beta)} \qquad (5)$$

where $\left|\mu_{\beta,k}\right>$ are the Fourier coefficient of the Floquet state $\left|\mu_{\beta,k}\right>$. These equations are also applied to the general serially coupled QDs system.

3.1 Transport properties of coupled DQDs

In our numerical calculations, we set the energy independent dots-lead hopping rate $\Gamma_L = \Gamma_R = 9\,\mu eV$, and assume that $k_B T = 0$. The external voltage $\mu_L - \mu_R = 15\,\mu V$ and the AC field magnitude $A = 1.54$ V/cm. First, to study the transport properties of the asymmetric system, we set different confining potentials on the left and right QDs. The average currents I as a function of the driving frequency for the serially and parallelly coupled DQDs are shown in Fig. 3(a) and 3(b) with $\hbar\omega_l = 3.00$ meV and $\hbar\omega_r = 3.10$ meV. We find that the current curve has a symmetric Breit–Wigner line type when the two QDs are coupled serially. However, the current curve has an asymmetric Fano line type when the two QDs are coupled parallelly. The Breit–Wigner line type suggests that there is a photon-assistant resonance, i.e. $\hbar\Omega = \sqrt{(E_1 - E_2)^2 + 4t_c^2}$. The Fano line type is due to the interference between photon-correct bonding and antibonding states. Then, we consider the case that the system is

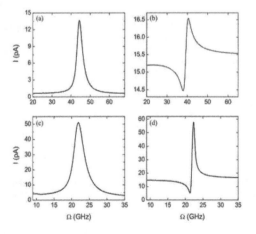

Figure 3. Average current as a function of the driving frequency Ω for serially (a) and parallelly (b) coupled DQDs. Parameters: $\hbar\omega_r = 3.10$ meV, $k_B T = 0$. (c), (d) are the same as (a), (b), respectively, except that $\hbar\omega_r = 3.00$ meV.

symmetric by applying the same confining potential to the two side dots, i.e. $\hbar\omega_l = \hbar\omega_r = 3.00$ meV. The current curves are shown in Figure 3(c) and 3(d). Analogously, we obtain symmetric Breit–Wigner resonance in serial DQDs and asymmetric Fano resonance in parallel DQDs. From the above discussion, we draw the following conclusion: an AC electric field will induce the photon-assistant Fano resonance for both symmetric and asymmetric parallel DQDs; however, it will be for neither symmetric nor asymmetric serial DQDs. It means that the energy level structure of the DQDs has no significant effect on the transport properties.

3.2 Transport properties of serially coupled TQDs

As it can be seen from Section 3.1, there is no Fano effect in serial DQDs due to the lack of obvious interference channels. We might suppose that the Fano effect will also not appear in serial TQDs. However, the transport properties have large differences between the serial TQDs and DQDs. It is the main result of this section.

Based on the result of the electronic structure, we study the transport properties through the system with a Λ-type three-level structure (Figure 2) under the action of an AC driving field with a frequency Ω. In the numerical calculation, the parameters are the same as DQDs, except that the external voltage $\mu_L - \mu_R = 120\,\mu V$. We first study the transport properties of an asymmetric system by applying different confining potentials to the left and right dots. The average current I as a function of the driving frequency Ω is shown in Figure 4(a) with $\hbar\omega_l = 3.10$ meV, $\hbar\omega_r = 3.00$ meV. A Breit–Wigner line shape appears in the vicinity of $\Omega = 129.48$ GHz, suggesting that there is a resonance for electrons in the system in this case (i.e. $\hbar\Omega = E_2 - E_1$). Figure 4(b) represents the time-dependent occupation probabilities for an electron in the left, middle and right dots. Here, we performed our calculation in a closed system (i.e. without the interaction with the leads), and used the initial condition $p_L = 1$ and the resonant condition $\Omega = 129.48$ GHz. It is clear that there is a photon-assistant mixing between the left and middle dots. This photon-assistant charge transfer leads to the occurrence of the current resonance phenomenon. Then, we consider the symmetric case by applying the same confining potential to the two side dots. The average current I as a function of the driving frequency Ω with $\hbar\omega_r = \hbar\omega_l = 3.00$ meV is shown in Figure 4(c). We find that the current curve has an asymmetric Fano line shape around $\Omega = 130.57$ GHz, and the current amplitude is much larger than that in the asymmetric case. In order to understand the intriguing phenomenon, we made a similar calculation of the time evolution of the occupation probabilities for an electron in the left,

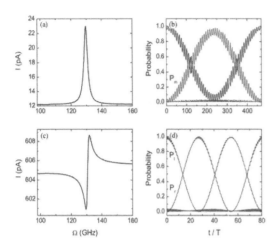

Figure 4. (a) Average current for serial TQDs. (b) Time-dependent occupation probabilities for an electron in the left, middle and right dots, where $\hbar\omega_r = 3.10\ meV$. (c), (d) are the same as (a), (b), respectively, except that $\hbar\omega_r = 3.00\ meV$.

middle and right dots, as shown in Figure 4(d). We find that the electron mainly occupies in the left and right dots, and the occupation probability for the middle dot is very small, indicating that the middle dot mediates the super-exchange interaction between the left and right dots. The states $|0\rangle$ and $|1\rangle$ behave like a 'trapping dark state' in an atomic system. Although state $|2\rangle$ has a very small occupation probability in the left and right dots, it plays an important role in the occurrence of the Fano-type resonance.

4 CONCLUSION

In this paper, we studied the transport properties of coupled DQDs and serially coupled TQDs under the action of an AC electronic field. Using a two-dimensional confining potential model in the effective mass frame, the two-level and Λ-type three-level structures are obtained through solving the generalized eigenvalue of the system. Based on the level structure and the Floquet theory, we investigate the dependence of the average current on the electronic structure of the system. It is found that the two-level structure has no significant impact on the transport properties of coupled DQDs. For symmetric and asymmetric configurations, there are Fano resonances in parallel DQDs and Breit–Wigner resonance in serial DQDs. However, it is a different story for serially coupled TQDs: a Λ-type level structure has an important impact on the transport properties. It is found that when the system is asymmetric, the symmetric Breit–Wigner

resonance appears due to photon-assistant tunneling. When the system is symmetric, the interesting asymmetric Fano resonance occurs under the resonant condition. In this case, quantum interference results in the formation of 'trapping dark states'. Our work presents the unique quantum interference features of the multiple QDs system, and is useful for the design of novel nanodevices. Briefly, for an N QDs system, the transport properties of N = 1 are different from N = 2, and N = 2 are also different from N = 3, so we cannot predict the transport properties of N > 3.

REFERENCES

[1] Y. Ji, M. Heiblum, D. Sptinzak, and Hadas Shtrikman, Science, 2000, 290: 779–783.
[2] D.Goldhaber-Gordon, J. Göres, M.A. Kastner, Hadas Shtrikman, D. Mahalu, and U. Meirav, Phys. Rev. Lett., 1998, 81: 5225–5228.
[3] B.R. Bulka, and P. Stefanski, Phys. Rev. Lett., 2001, 86: 5128–5133.
[4] T. Brandes, and F. Renzoni, Phys. Rev. Lett., 2000, 85: 4148–4151.
[5] W.D. Chu, S. Duan, and J. Zhu, Appl. Phys. Lett., 2007, 90: 222102-1-3.
[6] J. Göres, D. Goldhaber-Gordon, S. Heemeyer, M.A. Kastner, Hadas Shtrikman, D. Mahalu, and U. Meirav, Phys. Rev. B, 2000, 62: 2188–2194.
[7] A. Clerk, X. Waintal, and P.W. Brouwer, Phys. Rev. Lett., 2001, 86: 4636–4639.
[8] C. Johnson, C.M. Marcus, M.P. Hanson, and A.C. Gossard, Phys. Rev. Lett., 2004, 93: 106803-1-4.
[9] W. Holleitner, C.R. Decker, H. Qin, K. Eberl, and R.H. Blick, Phys. Rev. Lett., 2001, 87: 256802-1-4.
[10] W. Holleitner, R.H. Blick, A.K. Hüttel, K. Eberl, and J.P. Kotthaus, Science, 2002, 297: 70–72.
[11] W. Holleitner, R.H. Blick, and K. Eberl, Appl. Phys. Lett., 2003, 82: 1887–1889.
[12] J.C. Chen, A.M. Chang, and M.R. Melloch, Phys. Rev. Lett., 2004, 92:, 176801-1-4.
[13] J. Kim, D.V. Melnikov, J.P. Leburton, D.G. Austing, and S. Tarucha, Phys. Rev. B, 2006, 74: 035307-1-8.
[14] D. Schröer, A.D. Greentree, L. Gaudreau, K. Eberl, L.C.L. Hollenberg, J.P. Kotthaus, and S. Ludwig, Phys. Rev. B, 2007, 76: 075306-1-11.
[15] D.S. Saraga, and D. Loss, Phys. Rev. Lett., 2003, 90: 166803-1-4.
[16] Z. Jiang, Q. Sun, and Y. Wang, Phys. Rev. B, 2005, 72: 045332-1-7.
[17] R. Žitko, J. Bonča, A. Ramšak, and T. Rejec, Phys. Rev. B, 2006, 73: 153307-1-4.
[18] R. Žitko, and J. Bonča, Phys. Rev. Lett., 2007, 98: 047203-1-4.
[19] Wei-zhong Wang, Phys. Rev. B, 2007, 76: 115114-1-6.
[20] S. Kohler, J. Lehmann, M. Strass, and P. Hänggi, Adv. Solid State Phys., 2004, 44: 157–167.
[21] J. Lehmann, S. Camalet, S. Kohler, and P. Hänggi, Chem. Phys. Lett., 2003, 368: 282–288.
[22] S. Kohler, J. Lehmann, and P. Hänggi, Phys. Rep., 2005, 406: 379–443.
[23] G. Breit, and E. Wigner, Phys. Rev., 1936, 49: 519–531.

Material Science and Engineering – Chen (Ed.)
© 2016 Taylor & Francis Group, London, ISBN 978-1-138-02936-1

Research on a new oxide-scale detection technique based on magnetic induction

J.W. Chen, Z. Jiang & H. Mu
Guizhou Electric Power Test Research Institute, Guiyang, China

H.D. Liu, Q. Wan & B. Yang
School of Power and Mechanical Engineering, Wuhan University, Wuhan, China

ABSTRACT: Oxidation of the inner surface of the stainless steel pipe by high-temperature steam is a severe problem in the super-heater and reheater tube in a super-critical unit. Too much exfoliation oxide scale will choke the pipe, and can cause even pipe bursting. To reduce accident and decrease losses, it is important to detect the oxide scale that is formed in a stainless steel pipe in the boiler of the power plant. Based on the technology of magnetic induction, a method to detect the oxide scale was presented. A set of instrument for oxide-scale detection was designed and then performances of this instrument were carefully tested. The results showed that the method could be accurate to measure the amount of oxide scale in the pipe at the level of a gram, and could be possible to be applied to the oxide-scale detection in a stainless steel pipe in the boiler of the power plant.

Keywords: oxide scale; magnetic-induction technique; pipe bursting

1 INTRODUCTION

With much large-capacity and high-parameter generator units coming into service, oxidation problems in the inner surface of the boiler pipe of the power plant have become increasingly serious [1]. When the power generator suspends production for overhaul, a large amount of inner oxide scales would exfoliate due to its different expansion coefficients with the austenitic material. These oxide scales would deposit at the bends of vertical pipes of the boiler, resulting in the block of the pipes. Furthermore, the blocked high-temperature vapor would increase the temperature of the pipe to a higher point, which would lead to the failure of the super-heater or reheater [2–4]. Therefore, it is essential to detect the oxide scale for the safe operation of the power generator during the repairing process.

So far, the usual detective technology of the deposited oxide scale is the radiographic inspection based on the X-ray structure. However, drawbacks of this method are obvious: (I) the expense would increase the cost of the operation of the power generator; (II) it is difficult to measure the small amount of the deposited oxide scale due to the low sensitivity of this method; (III) the dangerous source of the X-ray is harmful to the detectors; (IV) the long detective process will increase the maintenance period, which has a bad economic impact on the power plant [5–6].

Therefore, efforts need to be made to develop a fast and effective method to exactly measure the amount of the deposited oxide scale. In this paper, we design a new magnetic detector based on the different magnetic properties between the austenitic heat-resistance steel and its oxide scale. In addition, experiments were designed to test the performances of this detector.

2 THEORY AND DESIGN

2.1 Basic theory

There is an obvious different magnetic property between the austenitic heat-resistance steel and its oxide scale. The former shows no obvious response to the outer magnetic field, exhibiting the paramagnetic behaviors. In contrast, the later exhibits much more strong response to the magnetic field, indicating its soft magnetic property. Therefore, when the austenitic heat-resistance steel pipe with oxide scales was put into a magnetic field, the oxide scales would be magnetized to be attracted by a magnetic force. That force could be measured by a magneto sensitive element, which can export an electric signal that would characterize the amount of oxide scales.

2.2 *Experimental design*

2.2.1 *Design of the detector*

In order to measure the amount of oxide scales in the pipe, the force between the magneto sensitive element and the oxide scales should be first measured. The opponents needed are as follows: a steady magnetic source, a magneto sensitive element, and a sensor measuring the force. Figure 1 shows the schematic diagram of detector for the austenitic heat-resistance steel oxide scales. When the detector gets closer to the austenite steel pipe with the oxide scales, the permanent magnet (2) would magnetize the oxide scales and could be attracted by the oxide scales. The force on the permanent magnet (2) will cause the bend of the cantilever (5), which will lead to the strain of the strain gages (6) and an electric signal could be exported by the wire (8) due to the change in electric resistance in the strain gages. As a result, the amount of the oxide scales is proportional to the output signal, which could be an effective method to characterize the amount of oxide scales in the pipe.

2.2.2 *Performance of the detector*

Figure 2 shows the whole system of the detector: the DC power supplies the steady DC source to the detector, then the detector collects the signal from the oxide scales, and the output signal could be measured by the digital multi-meter.

The experimental procedures are as follows:

1. Effect of the position of detector on the output signal. The detector was put on the upside and the downside of the oxide scale, as shown in Figure 3(a) and (b), respectively. The amount of the oxide scales varied from 0 to 20 g, and the output signals were collected when every one gram of the oxide scales was added to the pipe.
2. Effects of the size of the permanent magnet on the output signal. In this process, the output signal was collected when the size of the permanent magnet was chosen as $\Phi20 \times 15$ mm and $\Phi20 \times 10$ mm, respectively. The amount of the oxide scales increased from 0 to 50 g.
3. Effect of the distances between the detector and oxide scales on the output signal. The output signal was collected when the distances between

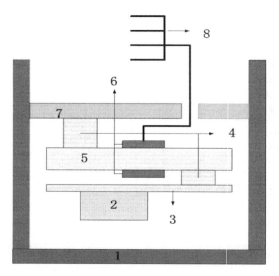

Figure 1. The schematic diagram of the detector for the austenitic heat-resistance steel oxide scales: 1-stainless steel shell, 2-permanent magnet, 3-holder, 4-connecting bolt, 5-cantilever, 6-strain gage, 7-cover with hold, 8-wires for input and output.

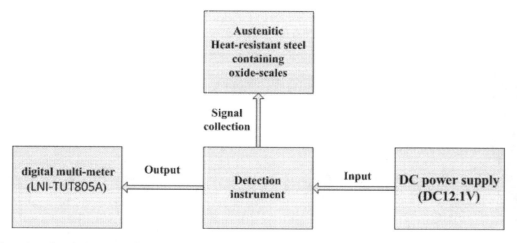

Figure 2. The whole system of the detector.

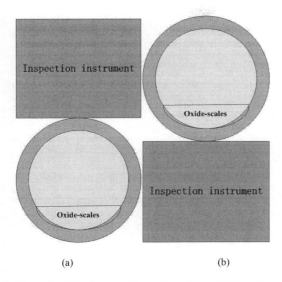

(a)   (b)

Figure 3.   The detector (a) on the upside and (b) on the downside.

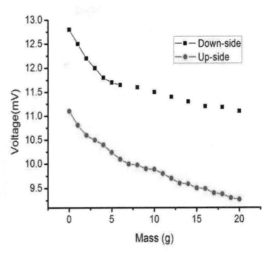

Figure 4.   The output voltage changed with the increase in the oxide scales at different positions.

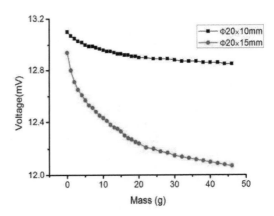

Figure 5.   Effects of the size of the permanent magnet on the output voltage.

the detector and the oxide scales varied from 4 to 9 cm, while the amount of the oxide scale increased from 0 to 50 g for every distance.

3   RESULTS AND DISCUSSION

3.1   *The position of the detector on the output signal*

Figure 4 shows the output voltage changes with the increase in oxide scales at different positions. The voltages decreased with a higher mass of the oxide scales at the upside and downside of the pipe. Furthermore, when the amount of the oxide scales was lower than 5 g, the downside position showed much fast changes when compared with the upside position. Moreover, in the range of 5–20 g, an obvious change in the output voltages could also be found with the increase in the amount of oxide scales. These results indicated that the down-position was beneficial for the detection of oxide scales.

3.2   *Size of the permanent magnet on the output signal*

Figure 5 shows the output voltages that vary with the increase in the amount of oxide scales at different sizes of the permanent magnets. Much more obvious change could be observed from the curve with a larger permanent magnet, which was mainly due to the stronger magnetic field provided by the larger magnet. Moreover, the detector with a larger magnet was more sensitive to the smaller amount of the oxide scales. Therefore, in order to

improve the sensitivity of the detector, a larger size of the permanent magnet should be considered as a choice for the detector.

3.3   *Distances between the detector and oxide scale on the output signal*

Figure 6 shows the output voltage changes with the increase in the oxide scales at different distances between the detector and the oxide scales. Lower output voltages could also be observed as the amount of oxide scales increased from 0 to 50 g. Moreover, the output voltage showed a slower decreasing trend with a higher mass of the oxide scales, as the distance between the detector and the oxide scales increased from 4 to 9 cm. As a result,

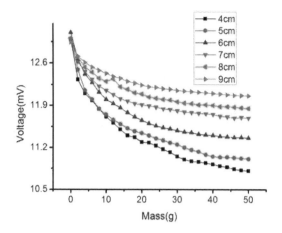

Figure 6. Effects of the distance between the oxide scales and the detector on the output voltage.

this method is more beneficial for the detection of the pipe with a thinner wall.

4 CONCLUSIONS

A new oxide-scale detection technique based on magnetic induction was put forward to detect the oxide scales in the austenitic heat-resistance steel pipe. A detector was designed to detect the oxide scales in the pipe. Furthermore, the effects of the position of the detector, the size of the permanent magnet, as well as the distance between the detector and the oxide scales on the output voltage were systematically investigated. The results showed that this method could detect the amount of oxide scales at a 1-gram level. Moreover, the downside position, the larger permanent magnet, and the shorter distance between the detector and oxide scales could improve the sensitivity of the detector. These results indicated that this method could be used to detect the oxide scales in the bend of the heat-resistance pipe of the boiler in the power plant.

REFERENCES

[1] Rehn, I.M. (1981). Corrosion problems in coal-fired boiler superheater and reheater tubes: steam-side oxidation and exfoliation. Development of a chromate-conversion treatment. *Final report (No. EPRI-CS-1812). Foster Wheeler Development Corp., Livingston, NJ (USA).*

[2] Otsuka, N., & Fujikawa, H. (1991). Scaling of austenitic stainless steels and nickel-base alloys in high-temperature steam at 973 K. *Corrosion*, 47(4), 240–248.

[3] Wright, I.G., & Dooley, R.B. (2010). A review of the oxidation behaviour of structural alloys in steam. *International Materials Reviews*, 55(3), 129–167.

[4] Dooley, R.B., Wright, I.G., & Tortorelli, P. (2007). Program on technology innovation: oxide growth and exfoliation on alloys exposed to steam. *EPRI Report, Palo Alto, CA.*

[5] Viswanathan, R. (1985). Dissimilar metal weld and boiler creep damage evaluation for plant life extension. *Journal of Pressure Vessel Technology*, 107(3), 218–225.

[6] Schilke, P.W., Foster, A.D., & Pepe, J.J. (1991). Advanced gas turbine materials and coatings. *General Electric Company.*

Nanomaterials

Material Science and Engineering – Chen (Ed.)
© 2016 Taylor & Francis Group, London, ISBN 978-1-138-02936-1

Effective properties prediction of Carbon Nanotubes reinforced nanomaterials

Y. Sha & C.H. Duan
CAE, Beijing University of Chemical Technology, Beijing, China

ABSTRACT: Based on micromechanics of composite materials, a Representative Volume Element (RVE) model applied displacement boundary conditions was established in Abaqus. The effective elastic properties of the nano material were obtained by the finite element method. In order to prove the accuracy of this method, the RVE model considering the influence of the interphase between carbon nanotubes and matrix was established as an example in this paper, which represents the square arrangement of carbon nanotubes reinforced nanomaterials. Compared with the other methods, the feasibility of the method can be proved.

Keywords: CNTs; RVE; finite element analysis; nanomaterials

1 INTRODUCTION

Carbon Nanotubes (CNTs) have been widely concerned because of its good physical and chemical properties. With a small amount of CNTs added in matrix, the performance of composite materials can be improved obviously. According to the study of CNTs in recent years, CNTs' enhancement effect is mainly dependent on the interphase between the matrix and CNTs. However, the properties of CNT reinforced nanomaterials have not been fully excavated because of the various restrictions.

At present, the research of nanomaterial properties mainly focuses on three aspects: the molecular dynamics method, the numerical method and the continuum mechanics method. The molecular dynamics method is restricted to a very short time and small scales, and it is difficult to predict the macroscopic properties of nanomaterials. Numerical method is limited to the fixed form. Continuum mechanics method can be considered to solve the problem of nanomaterials properties prediction.

X.L. Chen, Y.J. Liu [1,2] evaluated the effective mechanical properties of CNT reinforced nanomaterials using a 3-D cylindrical and square representative volume element model based on continuum mechanics. And formulas to extract the effective material constants from solution for the RVE are derived. G.I. Giannopoulos [3] considered the atomistic microstructure of single-walled CNTs, the 3-D structure of CNTs was established via a discrete spring-based finite element method, while appropriate joint elements were used to simulate the interphase between the reinforcement and the matrix material. B.S. Sindu [4] predicted the tensile stiffness of single-walled carbon nanotubes in different atomistic structure by finite element method. Carbon nanotubes random distribution model was set up to prove that when the carbon nanotubes were closely spaced, the area between carbon nanotubes would be a high stress concentration area, and cracks would be caused. K.I. Tserpes [5] established the short multi-walled carbon nanotubes RVE model and analyzed the various factors' effects on the effective properties of CNT nanomaterials, considering the interphase stiffness, thickness, CNT aspect ratio, material property, volume fraction and so on.

In homogeneous materials, the stress and strain relationship can be defined at any point. But it cannot be defined in composite materials because of its heterogeneity. In order to obtain the equivalent material properties of nanomaterials, a representative volume element which is a smallest repetitive element of the nanomaterials is studied. It can stand for the global mechanical performance.

The effective performance prediction of CNT reinforced nanomaterials are mostly based on a uniform arrangement of CNTs. Different from the original load test methods like uniaxial tension, lateral expansion and axial torsion, a new method based on micromechanics of composite materials was used in this paper. And a representative volume element model applied displacement boundary conditions was established in Abaqus. The nanomaterial effective elastic properties were obtained by the finite element method. In order to prove the accuracy of this method, the RVE model considering the influence of the interphase between carbon nanotubes and matrix was established as an example in this paper, which represents the square arrangement of carbon nanotubes reinforced nanomaterials. Compared with the original methods, the feasibility of the method can be proved. It provides a feasible method to predict the effective performance of CNT reinforced nanomaterials. This method also provides a way to solve the problem of the properties prediction of CNTs aggregation.

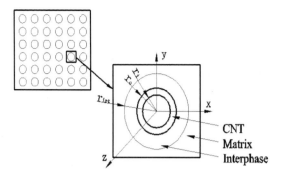

Figure 1. RVE model.

2 RVE MODEL

In this paper, CNTs were assumed to be squarely arranged in matrix. The RVE model containing carbon nanotubes, interphase and matrix was established in Abaqus to calculate the effective performance of nanomaterials. The RVE model is shown in Figure 1.

The displacement boundary conditions of RVE model was applied by the Eqs. (1):

$$u_i = \overline{\varepsilon_{ij}} x_j \tag{1}$$

Effective modulus of the nanomaterial can be defined as:

$$<\sigma_{ij}> = \overline{C_{ijkl}} <\varepsilon_{kl}> \tag{2}$$

For isotropic along CNT, the stress and strain relationship can be expressed as (1 aligns with the carbon nanotube length direction):

$$
\begin{bmatrix} <\sigma_{11}> \\ <\sigma_{22}> \\ <\sigma_{33}> \\ <\sigma_{23}> \\ <\sigma_{12}> \\ <\sigma_{13}> \end{bmatrix} =
\begin{bmatrix}
C_{1111} & C_{1122} & C_{1122} & & & \\
C_{1122} & C_{2222} & C_{2233} & & & \\
C_{1122} & C_{2233} & C_{2222} & & & \\
& & & C_{2222}-C_{2233} & & \\
& & & & C_{1212} & \\
& & & & & C_{1212}
\end{bmatrix}
$$
$$
\times
\begin{bmatrix} <\varepsilon_{11}> \\ <\varepsilon_{22}> \\ <\varepsilon_{33}> \\ <\varepsilon_{23}> \\ <\varepsilon_{12}> \\ <\varepsilon_{13}> \end{bmatrix} \tag{3}
$$

Among them, the stiffness matrix is made up of effective elastic constants:

$$E_{11} = C_{1111} - 2C_{1122}^2/(C_{2222}+C_{2233})$$

$$E_{22} = E_{33} = [C_{1111}(C_{2222}+C_{2233})-2C_{1122}^2]$$

$$\times (C_{2222}-C_{2233})/(C_{1111}C_{2222}-C_{1122}^2)$$

$$\mu_{12} = \mu_{13} = C_{1122}/(C_{2222}+C_{2233}) \tag{4}$$

$$\mu_{23} = [C_{1111}C_{2233}-C_{1122}^2]/(C_{1111}C_{2222}-C_{1122}^2)$$

$$G_{12} = G_{13} = C_{1212}$$

$$G_{23} = (C_{2222}-C_{2233})/2$$

Under the boundary conditions, Eqs. (5) can be proved as:

$$<\varepsilon_{ij}> = \frac{1}{V}\int_V \varepsilon_{ij}dV = \frac{1}{2V}\int_{\partial V}(u_i n_j + u_j n_i)dS$$

$$= \frac{1}{2V}\int_{\partial V}(\overline{\varepsilon_{ik}}x_k n_j + \overline{\varepsilon_{jk}}x_k n_i)dS$$

$$= \frac{1}{2V}\int_V(\overline{\varepsilon_{ik}}\delta_{kj} + \overline{\varepsilon_{jk}}\delta_{ki})dV = \overline{\varepsilon_{ij}} \tag{5}$$

Unit strain $\overline{\varepsilon_{11}} = 1$ was applied, then $<\sigma_{11}> = C_{1111}$ was available. In this equation, $<\sigma_{11}>$ can be obtained according to the volume average strain concept:

$$<\sigma_{ij}> = \frac{1}{V}\int_V \sigma_{ij}dV \tag{6}$$

C_{1111} can be obtained. In the same way, the other parameters in C_{ijkl} can be obtained.

3 THE EXAMPLE ANALYSIS

The dimensions of the RVE model are function of CNT volume fraction. The RVE model geometry sizes are shown in Table 1 respectively. One of the finite models is shown in Figure 2.

Table 1. The dimensions of model with different CNT volume fraction [6].

Volume fraction	x = y (nm)	z (nm)	r_i (nm)	r_o (nm)	r_{int} (nm)
1%	11.521	2.880	0.315	0.650	1.404
2%	8.147	2.037	0.315	0.650	1.404
3%	6.652	1.663	0.315	0.650	1.404
4%	5.760	1.440	0.315	0.650	1.404
5%	5.152	1.288	0.315	0.650	1.404

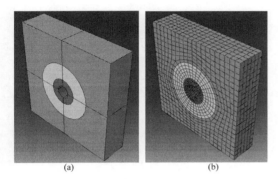

(a)   (b)

Figure 2. The finite element model.

Table 2. Related properties [6].

Material property	CNT	Matrix	Interphase
Young's modulus (GPa)	1054	2.026	16.1
Poisson's ratio	0.25	0.4	0.4

In this study, Carbon nanotubes, matrix and interphase were assumed as isotropic materials. The elastic properties of carbon nanotubes, matrix and interphase with all commonly-used values are displayed in Table 2.

4 THE RESULTS DISCUSSION

The material effective properties with different CNT volume fractions calculated by the finite element method and those from Zhong Hu [6] method are plotted in Figure 3.

Figure 3 shows that longitudinal Young's modulus (shown in Fig. 3(a)), longitudinal Poisson's ratio (shown in Fig. 3(d)(f)) calculated by the two methods agreed perfectly. In comparison, the transversal related parameters did not agree as well as the longitudinal related parameters. For transversal Young's modulus (shown in Fig. 3(b))

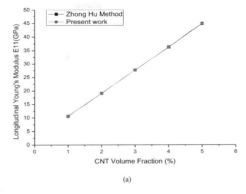

(a)

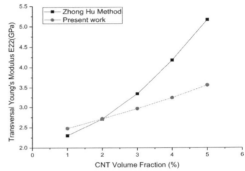

(b)

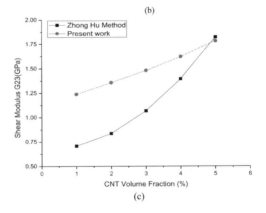

(c)

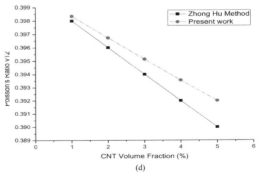

(d)

Figure 3. The results of two methods.

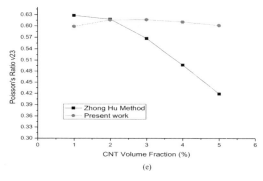

(e)

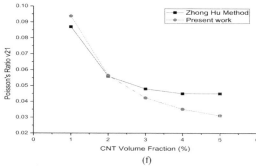

(f)

Figure 3.   (*Continued*)

and shear modulus (shown in Fig. 3(c)), the values from the two methods have the same variation tendency. For transversal Poisson's ratio, there is no obviously correlation between the results and the volume fraction of CNT in this method, but Zhong Hu method results decreases with increasing the CNT volume fraction.

5   CONCLUSION

1. In this paper, compared with Zhong Hu method, the feasibility of the method can be proved. It provides a feasible method to predict the effective performance of CNT reinforced nanomaterials.
2. CNT aggregation occurs during the process of preparation. CNT aggregation model can be established in this method, which provides a way to solve the problem of the properties prediction of CNT aggregation.

REFERENCES

[1] Y.J. Liu and X.L. Chen. 2003. Mechanics of Materials Vol. 35: p. 69.
[2] Y.J. Liu and X.L. Chen. 2004. Computational Materials Science Vol. 29: p. 1.
[3] G.I. Giannopoulos, S.K. Georgantzinos and N.K. Anifantis. 2010. Composites: Part B Vol. 41: p. 594.
[4] B.S. Sindu, Saptarshi Sasmal and Smitha Gopinath. 2014. Construction and Building Materials Vol. 50: p. 317.
[5] K.I. Tserpes and A. Chanteli. 2013. Composite Structures Vol. 99: p. 366.
[6] Zhong Hu, Md. Ragib Hasan Arefin, Xingzhong Yan and Qi Hua Fan. 2014. Composites: Part B Vol. 56: p. 100.

Material Science and Engineering – Chen (Ed.)
© *2016 Taylor & Francis Group, London, ISBN 978-1-138-02936-1*

Ion-implanted Au nanoparticles for encoding information in light

C. Torres-Torres
ESIME ZAC, Instituto Politécnico Nacional, México, D.F., México

R. Rangel-Rojo
Depto. de Óptica, Centro de Investigación Científica y de Educación Superior de Ensenada, Ensenada, B.C., México

R. Torres-Martínez
CICATA-Q, Instituto Politécnico Nacional, Santiago de Querétaro, Querétaro, México

J.C. Cheang-Wong, A. Crespo-Sosa, L. Rodríguez-Fernández & A. Oliver
Instituto de Física, Universidad Nacional Autónoma de México, D.F., México

ABSTRACT: Measurements of the nanosecond third order nonlinear optical response exhibited by optically irradiated Au nanoparticles are presented. The nanocomposites were nucleated by an ion-implantation method followed by a thermal annealing. Optical storage based on degenerated two-wave mixing recording is presented. The nonlinear refractive index of a high-purity silica matrix containing Au nanoparticles was modified by nanosecond ablation. Self-diffraction signals allowed us to recover vectorial information associated to the encoding process.

Keywords: optical materials; optical Kerr effect; two-wave mixing; optical storage-recording materials

1 INTRODUCTION

High-speed memory systems usually store one digital bit in each unitary cell; a single byte could be independently employed to write or read information. Different examples of outstanding memory designs based on advanced materials have been reported (Vontobel et al. 2009). Particularly, low-dimensional structures have attracted the attention of many researchers in regard to their unique optical response (Jeong et al. 2010). Quantum functions related to memories and logic operations have been proposed by using nanophotonic devices (Bussières et al. 2013). Moreover, extraordinary all-optical tasks have been envisioned taking into account the powerful and ultrafast nonlinear optical properties in nanomaterials (Rigneault et al. 2006). The optical characteristics that concern to metallic nanoparticles are strongly dependent on their Surface Plasmon Resonance; and then, nonlinear optical features could be also influenced by tuning Surface Plasmon Resonance excitations. It is well known that the Surface Plasmon Resonance effects in metallic nanoparticles can be tailored by changing the nanoparticle shape, size and density (Gonella & Mazzoldi 2000). Furthermore, considering that the surrounding of metallic nanoparticles is a notable contribution to their optical nonlinearities (Stepanov 2011), ablation processes seem to be a useful tool to selectively modify the nonlinear optical response features of nanocomposites.

On the other hand, optical two-wave mixing phenomena are polarization dependent. The superposition of high irradiance optical waves can generate an induced birefringence or a permanent optical damage that ought to be dependent on polarization too. Thus, the generation of an interference grating should be capable to encrypt vectorial information by optical ablation in a single cell sensitive to polarization. With this motivation, in this work are presented the resulting optical Kerr nonlinearities exhibited by Au Nanocomposites (NCs) exposed to third order nonlinear optical interactions and ablation effects. The samples were prepared by an ion-implantation processing route. Nanosecond measurements were performed by a two-wave mixing technique with self-diffraction at a near-resonance 532 nm wavelength. Potential applications for developing nanodevices with optical memory functions can be contemplated.

2 EXPERIMENT

2.1 *Preparation of the nanocomposites*

The preparation of the sample starts with high-purity silica glass plates ($16 \times 16 \times 1$ mm^3) with OH content less than 1 part per million (ppm), and impurity content less than 20 ppm, which are implanted at room temperature with a fluence of around 2.8×10^{16} ions/cm^2. The ion-implantation was carried out by using 2 MeV Au^{2+} ions. The implantation process was followed by a thermal annealing treatment that allows promoting the nanoparticle formation in an oxidizing atmosphere (air) at a temperature of 1100°C for 1 hr (Oliver et al. 2002). The concentration depth profile distribution of the metals, and the ion fluences, were determined by Rutherford Backscattering Spectrometry (RBS) measurements using a 2.54 MeV ^{4}He$^+$ beam. Ion implantation and RBS analysis were both performed at Instituto de Física of the Universidad Nacional Autónoma de México, with the 3 MV Tandem accelerator Pelletron (NEC 9SDH-2). The linear absorption spectrum of the sample was measured with a Varian UV-VIS spectrophotometer.

2.2 Nonlinear optical studies

The near-resonant optical nonlinearities of the NCs were explored using a vectorial self-diffraction technique (Torres-Torres et al. 2009). Figure 1 illustrates the scheme of the experimental setup for the two-wave mixing configuration. Initially, the system was calibrated by using a standard reference material Carbon Disulfide (CS$_2$). The CS$_s$ sample was contained in a quartz cuvette with 1 mm width. The magnitude of the third-order nonlinear susceptibility of CS$_2$ is $|\chi^{(3)}| = 1.9 \times 10^{-12}$ esu (Boyd 1992). The self-diffracted and transmitted optical signals were measured at 532 nm with a well defined linear polarization from a Nd-YAG laser system Surelite-II, featuring a 4 ns pulse duration and 30 mJ maximum pulse energy. A half wave

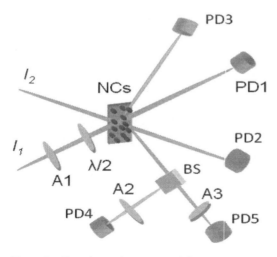

Figure 1. Experimental two-wave mixing setup.

plate, $\lambda/2$, together with a polarizer, A1, is used for controlling the plane of polarization of the probe beam. Pump and probe beams, with an irradiance relation of approximately 1:1, and parallel linear polarizations, were focused on the sample with a spot size of 1 mm. BS represents a beam-splitter. Two analyzers, A2–3, were employed to acquire the irradiance data associated with the orthogonal components of polarization of the self-diffracted beam. PD1–4 are photodetectors.

The polarization of the pump beam, named I_2, was fixed during the experiment, while the polarization of the probe beam, named I_1, was rotated in an interval of 180° with data captured each 10°. The estimation of the nonlinear optical parameters was carried out by means of a mathematical description of the electric fields in the interaction. We consider the expressions for representing the amplitudes of the transmitted and self-diffracted waves as (Torres-Torres 2009):

$$E_{1\pm}(z) = \left[E_{1\pm}J_0\left(\Psi_\pm^{(1)}\right) + \left(iE_{2\pm} - iE_{3\pm}\right)J_1\left(\Psi_\pm^{(1)}\right) + \ldots \right.$$
$$\left. - E_{4\pm}J_2\left(\Psi_\pm^{(1)}\right) \right] \exp\left(-i\Psi_\pm^{(0)} - \frac{\alpha(I)z}{2}\right)$$

(1)

$$E_{2\pm}(z) = \left[E_{2\pm}J_0\left(\Psi_\pm^{(1)}\right) + \left(iE_{4\pm} - iE_{1\pm}\right)J_1\left(\Psi_\pm^{(1)}\right) + \ldots \right.$$
$$\left. - E_{3\pm}J_2\left(\Psi_\pm^{(1)}\right) \right] \exp\left(-i\Psi_\pm^{(0)} - \frac{\alpha(I)z}{2}\right)$$

(2)

$$E_{3\pm}(z) = \left[E_{3\pm}J_0\left(\Psi_\pm^{(1)}\right) + iE_{1\pm}J_1\left(\Psi_\pm^{(1)}\right) + \ldots \right.$$
$$\left. - E_{2\pm}J_2\left(\Psi_\pm^{(1)}\right) - iE_{4\pm}J_3\left(\Psi_\pm^{(1)}\right) \right]$$
$$\exp\left(-i\Psi_\pm^{(0)} - \frac{\alpha(I)z}{2}\right)$$

(3)

$$E_{4\pm}(z) = \left[E_{4\pm}J_0\left(\Psi_\pm^{(1)}\right) - iE_{2\pm}J_1\left(\Psi_\pm^{(1)}\right) + \ldots \right.$$
$$\left. - E_{1\pm}J_2\left(\Psi_\pm^{(1)}\right) + iE_{3\pm}J_3\left(\Psi_\pm^{(1)}\right) \right]$$
$$\exp\left(-i\Psi_\pm^{(0)} - \frac{\alpha(I)z}{2}\right)$$

(4)

where $E_{1\pm}(z)$ and $E_{2\pm}(z)$ are the complex amplitudes of the circular components of the transmitted waves beams, $E_{3\pm}(z)$ and $E_{4\pm}(z)$ are the amplitudes of the self-diffracted waves; z is the length of the propagation of the wave through the nonlinear media and I corresponds to the total optical irradiance in interaction. The linear and nonlinear absorption coefficients are represented by α_o and β, respectively, with the absorption coefficient dependent on irradiance as $\alpha(I) = \alpha_o + \beta I$. The Bessel function of order m is represented by $J_m(\Psi_\pm^{(m)})$, and,

$$\Psi_\pm^{(0)} = \frac{4\pi^2 z}{n_0 \lambda}\left[\left(A + \frac{n_0\beta}{2\pi}\right)\sum_{j=1}^{4}\left|E_{j\pm}\right|^2 + \ldots \right.$$
$$\left. + \left(A + B + \frac{n_0\beta}{2\pi}\right)\sum_{j=1}^{4}\left|E_{j\mp}\right|^2\right] \qquad (5)$$

$$\Psi_\pm^{(1)} = \frac{4\pi^2 z}{n_0 \lambda}\left[\left(A + \frac{n_0\beta}{2\pi}\right)\sum_{j=1}^{3}\sum_{k=2}^{4}E_{j\pm}E_{k\pm}^* + \ldots \right.$$
$$\left. + \left(A + B + \frac{n_0\beta}{2\pi}\right)\sum_{j=1}^{3}\sum_{k=2}^{4}E_{j\mp}E_{k\mp}^*\right] \qquad (6)$$

are the phase increments.

The nonlinear refractive index, n_2, and the nonlinear absorption coefficient, β, can be related to $\chi^{(3)}$ (esu) by (Boyd 1992),

$$\chi^{(3)} = 2n_0^2\varepsilon_0 c_0 n_2 + i\frac{n_0^2\varepsilon_0 c_0^2}{\omega}\beta, \qquad (7)$$

where ε_0 represents the permittivity in the vacuum, c_0 the speed of the light, and ω the optical frequency. The magnitude of the third-order nonlinear optical susceptibility, $\chi^{(3)}$, can be expressed as,

$$\left|\chi^{(3)}\right| = \sqrt{\left(\mathrm{Re}\,\chi^{(3)}\right)^2 + \left(\mathrm{Im}\,\chi^{(3)}\right)^2}. \qquad (8)$$

2.3 Two-wave mixing recording

Experiments related to modulation of nonlinear refractive index by multi-wave mixing interactions are sensitive to the vectorial nature of light. The superposition of two beams in a nonlinear optical media is able to generate an interference irradiance grating capable to modulate the refractive index. When two coherent beams with the same wavelength interact with a material, and they present a high optical irradiance, an induced birefringence can be promoted; this effect is known as the optical Kerr effect. However, when absorption and nonlinear optical absorption in a sample are important and take place, a two-wave interaction produces optical damage of the material resulting from ablation mechanisms. So, encryption of vectorial information can be achieved by a two-wave mixing recording. We employed the experimental setup schematized by Figure 1 in order to record single cells with dependence on polarization in the studied ion-implanted NCs.

The reading of the recorded cells was also carried out by the transmittance of the second harmonic coming from our Nd:YAg laser system.

3 RESULTS AND DISCUSSION

Figure 2 depicts the optical absorption spectrum of the NCs studied. The Surface Plasmon Resonance absorption band emerging from the presence of Au nanoparticles can be clearly seen around 520 nm. This peak in the plot can be considered as an evidence of the presence of Au nanoparticles incorporated in the silica matrix by the ion-implantation technique (Torres-Torres et al. 2007).

Two-wave mixing experiments were performed for recording the sample by optical ablation. In order to guarantee a permanent ablation effect by the incident pulses, the linear optical absorption of the NCs was monitored after systematically increasing the irradiance of single shots emitted by the laser system; when a reduction in the optical absorption was detected, it was considered that an induced grating was stored. The ablated zone of the NCs contains information related to the ablation process. The transmittance of a single-beam through the grating originates a self-diffraction phenomenon that can be measured in order to identify the geometric angle of the interacting beams employed for recording the sample. Figure 3 schematizes the rotation of the sample to exemplify the information of the geometric grating encrypted by two-wave mixing ablation.

The observation of the self-diffraction effect resulting from the two-wave mixing experiment described by Figure 1 allowed us measure the optical Kerr effect of the sample. In this work is compared the modification of the nonlinear refractive index of the NCs before and after the ablation process. The vectorial self-diffraction method allows the estimation of third order optical parameters that describe β and n_2. Figure 4 shows the experimental results with marks and numerical fitting as solid lines. Each point represents the

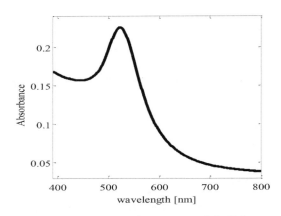

Figure 2. Optical absorption spectrum of the NCs.

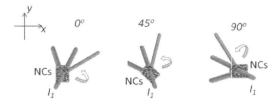

Figure 3. Scheme of the optical self-diffraction effect stored by two-wave mixing. Sample and self-diffraction signal oriented at (a) 0°, (b) 45°, (c) 90°.

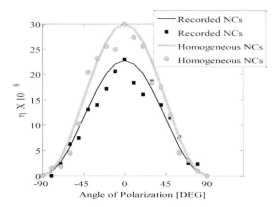

Figure 4. Self-diffraction efficiency η vs Angle of polarization of the incident beams.

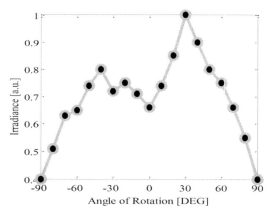

Figure 5. Self-diffraction irradiance as a function of the angle of rotation of a sample recorded.

average of 10 measurements detected under equivalent laboratory conditions. The self-diffraction efficiency, η, corresponds to the rate between self-diffracted and transmitted irradiances.

Our experiments indicate that an ablation process close to the threshold can generate a reduction in the average nonlinear refractive index exhibited by the NCs from $n_2 = -2 \times 10^{-15}$ m^2/W, to $n_2 = -1.6 \times 10^{-15}$ m^2/W. However, it is worth noting that a quite homogeneous third order optical response is exhibited by the sample; but a remarkable difference in the self-diffraction effect can be obtained in the recorded sample with a well-defined grating. Figure 5 describes the irradiance of the self-diffracted signal as a function of the orientation of the sample.

From Figure 5 can be deduced, that the permanent ablated grating can selectively enhance the self-diffraction phenomena in regards to the vectorial information stored as a nonlinear refractive index or as a permanent optical ablation effect. According to our experimental results, it can be stated that this process for tailoring the third order optical nonlinearities can be a good candidate for enabling memory functions recorded in ion-implanted materials.

4 CONCLUSIONS

Within this work was analyzed the third order nonlinear optical response exhibited by Au NCS featuring a spatially modulated distribution. A vectorial two-wave mixing experiment was conducted to evaluate the nonlinear refractive index and the ablation process for encoding information. Regarding the self-diffraction effect and optical transmittance obtained in the recorded sample, the possibility for developing potential applications in ultrafast optical storage can be considered. The optical Kerr effect related to the studied NCs, seems to promise potential applications for proposing near-resonant all-optical functions.

ACKNOWLEDGMENTS

The authors kindly acknowledge the financial support from UNAM, CICESE, IPN, and CONACyT.

REFERENCES

[1] Boyd, R.W. (Academic Press) 1992. *Nonlinear Optics*, San Diego.
[2] Bussières, F.; Sangouard, N.; Afzelius, M.; de Riedmatten, H.; Simon, C.; Tittel W. 2013. Prospective applications of optical quantum memories. *J. Mod. Optic* 60(18) 16:1519–1537.
[3] Gonella, F. & Mazzoldi (Academic Press) 2000. *Handbook of nanostructured materials and nanotechnology.* San Diego.
[4] Jeong, H.Y.; Kim, J.Y.; Kim, J.W.; Hwang, J.O.; Kim, Ji-Eun; Lee, J.Y.; Yoon, T.H.; Cho, B.J.; Kim, S.O.; Ruoff, R.S.; Choi, Sung-Yool 2010. Graphene Oxide Thin Films for Flexible Nonvolatile Memory Applications. *Nano Lett.*10(11):4381–4386.

[5] Oliver, A.; Cheang-Wong, J.C.; Roiz, J.; Rodríguez-Fernández, L.; Hernández, J.M.; Crespo-Sosa, A.; Muñoz, E. 2002. Metallic nanoparticle formation in ion-implanted silica after annealing in reducing or oxidizing atmospheres. *Nucl. Instr. and Meth. B* 191: 333–336.

[6] Rigneault, H.; Lourtiouz, J.-M.; Delalande, C.; Leven, A. (ISTE Ltd) 2006. *Nanophotonics*. New Port Beach.

[7] Stepanov, A.L. 2011. Nonlinear optical properties of implanted metal nanoparticles in various transparent matrixes: a review. *Rev. Adv. Mater. Sci.* 27:115–145.

[8] Torres-Torres, C.; Khomenko, A.V.; Cheang-Wong, J.C.; Crespo-Sosa, A.; Rodríguez-Fernández, L.; Oliver, A. 2007. Absorptive and refractive nonlinearities by four wave mixing in Au nanoparticles in ion-implanted silica. *Opt. Express* 15:9248–9253.

[9] Torres-Torres, C.; Trejo-Valdez, M.; Sobral, H.; Santiago-Jacinto, P.; Reyes-Esqueda, J.A. 2009. Stimulated emission and optical third order nonlinearity in Li-doped ZnO nanorods. *J. Phys. Chem. C* 113:13515–13521.

[10] Vontobel, P.O.; Robinett, W.; Kuekes, P.J.; Stewart, D.R.; Straznicky, J.; Williams, R.S. 2009. Writing to and reading from a nano-scale crossbar memory based on memristors. *Nanotechnology* 20:425204.

269

Material Science and Engineering – Chen (Ed.)
© 2016 Taylor & Francis Group, London, ISBN 978-1-138-02936-1

Cooperative buckling of parallel nanowires on elastomeric substrates

Y.L. Chen, L.L. Zhu, Y.L. Liu & X. Chen
SV Laboratory, School of Aerospace, International Center for Applied Mechanics, Xi'an Jiaotong University, Xi'an, China

ABSTRACT: The cooperative buckling behaviors of parallel nanowires on an elastomeric substrate are investigated based on a finite element framework. Under compression of the elastomeric substrate, an especial helical buckling mode of nanowires and buckling evolution with effective compressive strain are observed. It is revealed that typical buckling characteristics of two parallel nanowires strongly depend on the distance between them. When the distance is smaller than a critical value $h_s/2$ (half of a representative buckling spacing of a single nanowire on an elastomeric substrate), with decreasing of the separation distance, the buckling spacing, displacement amplitudes and the ratio between the in-/out-of-plane amplitudes all increase significantly. For parallel nanowire assembles, when the nanowires are closely arrayed, the buckling spacing notably increases with the number of nanowires. This study contributes to controllable buckling of stiff elements on a soft substrate and may shed some light on the design and optimization of flexible electronics.

Keywords: helical buckling; nanowire assemblies; finite element simulation; flexible electronics

1 INTRODUCTION

Buckling of stiff elements on elastomeric substrates has been widely studied (Audoly & Boudaoud 2008, Kim & Rogers 2009, Song et al. 2008, Xu et al. 2015, Yin et al. 2009) and significant practical applications, such as precision metrology (Stafford et al. 2004, Wilder et al. 2006), the stretchable electronic circuit (Song et al. 2009), the electronic eye camera (Ko et al. 2008), conformable skin sensors (Rogers et al. 2010, Someya et al. 2004), and flexible displays (Chen et al. 2002), have been found. As one of the most promising candidates for stretchable electronics, nanowire (NW) is usually fragile and stiff. To realize elasticity and flexibility, a particular configuration (e.g., the spiral line) is requisite. Fortunately, such a configuration is achievable through compression of a soft substrate with nanowires attached on its surface (Xu et al. 2010). A helical buckling mode makes extremely large deformation possible, even for intrinsically fragile materials (e.g., silicon nanowires), because the disastrous stress concentration in nanowires is alleviated by the helical configuration. Mechanical instability and buckling govern the helical formation process (Xu et al. 2010). As the substrate is compressed, nanowires buckle in the helical mode to reach minimization of strain energy and meanwhile the compressive strain of the nanowire is replaced by relatively small bending strain of the slender structure.

Previously, the buckling pattern of one single nanowire was studied (Xiao et al. 2008, Xu et al. 2010), while the simple fact is that, most of the electronic devices' functionality and performance rely on the cooperative behavior of a set of units. To study the buckling of nanowire assemblies on an elastomeric substrate, here we conduct a number of Finite Element Method (FEM) simulations to study the cooperative buckling behavior of a set of parallel nanowires attached to a soft substrate.

We start this work with two parallel nanowires tied to a soft substrate with varying distances between them. The influence of the separation distance on the buckling characteristics and helical evolution processes will be studied in detail. Furthermore, a new model of a set of parallel nanowires with fixed distance between them on a soft substrate is considered. With the number of nanowires varying, we are able to study the effect of the number of them on the overall buckling behavior. Results obtained here may shed useful insight on controllable buckling of stiff elements on a soft substrate, the design of flexible electronics as well as 3D complex nano-structure fabrications.

2 MODEL AND METHODS

The FEM model is shown in Figure 1. The nanowires with a length of 20 μm and a diameter of 30 nm are tied to the surface of the soft substrate

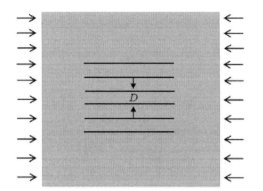

Figure 1. The FEA model (top view).

(40 μm × 40 μm × 5 μm). The density of the substrate and the nanowire is set to be 0.97×10^3 kg/m^3 and 2.4×10^3 kg/m^3, respectively; the Young's modulus and Poisson's ratio 6 MPa and 0.475, and 200 GPa and 0.3, respectively. Throughout all the simulations, a typical effective compression strain of 34% is applied to the substrate in the direction parallel to the nanowires.

The distance between nanowires, denoted as D, varies from $2 h_S$ to $h_S/16$, where h_S represents the initial buckling spacing of a single nanowire on a soft substrate with the magnitude of ~1.49 μm, calculated from theoretical results (Xiao et al. 2010). The buckling pattern of nanowires is investigated numerically using commercial FEM software ABAQUS where NWs consist of beam elements B31 and the substrate is described by three dimensional continuum element (C3D8R). A mesh convergence study is carried out to ensure appropriate mesh density and quality.

To verify the theoretical results about buckling spacing, we first investigate a model consisting of a single nanowire tied to a soft substrate. Then instead of studying just a single nanowire, two parallel nanowires are introduced to the substrate with varying distances between them so as to study the cooperative behavior between them with the typical buckling characteristics of nanowires (buckling spacing, buckling amplitude and the ratio between the in-/out-of-plane amplitudes) considered. Finally, a number of nanowires are considered and the influence of the number of them on the cooperative buckling behavior is investigated.

3 RESULTS AND DISCUSSION

3.1 Cooperative helical buckling of two parallel nanowires on a soft substrate

The model with only one nanowire (Fig. 2a) is first considered. Upon compression of the substrate, the

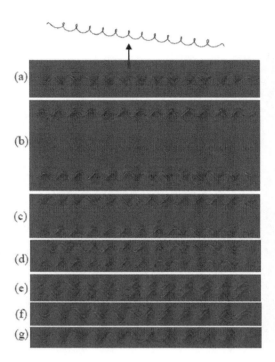

Figure 2. Helical buckling configurations of (a) a single nanowire, and two parallel nanowires with varying distances between them: (b) $2 h_S$, (c) h_S, (d) $h_S/2$, (e) $h_S/4$, (f) $h_S/8$, (g) $h_S/16$.

especial helical buckling pattern of the nanowire is observed which is consistent with experimental results (Xu et al. 2010). The effective buckling spacing h_S is ~1.50 μm calculated from simulation results, and according to the theoretical calculations, this value is 1.49 μm (Xiao et al. 2010). The good agreement between the simulation and theoretical results suggests the validity and effectiveness of the current model.

The helical buckling deformation of two parallel nanowires with different distances is shown in Figure 2. When the distance exceeds half of the typical buckling spacing of a single nanowire $h_S/2$, the deformation of the two parallel nanowires seems not to sense the existence of each other and the buckling configuration is almost the same to that of the single nanowire. With the decrease of separation ($D = h_S/2$, $h_S/4$, $h_S/8$ or $h_S/16$), the interaction between them becomes notable. For example, the buckling spacing of the nanowires increases significantly with the decrease of the distance between them. When the distance is large (e.g., $D = h_S$), there exist 12 intact thread pitches; however, when the distance decreases to $h_S/16$, the number decreases to 10. The parallelization of nanowires can be equivalent to the increasing of the structure stiffness, which may account for the increase of the buckling spacing with the decrease of the distance between nanowires, since a previous study indicates that the buckling spacing actually increases with the Young's modulus of nanowires (Xiao et al. 2010).

The helical buckling spacing of nanowires is given as a function of the distance between them (Fig. 3a) and both the spacing and distance are normalized by h_S. It seems that there is a turn point at $h_S/2$ in this curve. When $D < h_S/2$, a steep rise of the buckling spacing with narrowing of the nanowires is observed and the buckling spacing increases by ~20% when the distance drops from $h_S/2$ to $h_S/16$. As the distance exceeds $h_S/2$, the spacing converges slowly to the magnitude of the buckling spacing of a single nanowire, namely 1.49 μm (the dash line). Therefore, we can conclude that when the nanowires' separation distance is larger than $h_S/2$, the interaction between them can be ignored, while if the nanowires are sufficiently adjacent (smaller than $h_S/2$), the interaction between them will play a remarkable role in helical buckling behavior.

Figure 3b shows the buckling amplitudes within the surface (the in-plane amplitude, black squares) and perpendicular to the surface (the out-of-plane amplitude, red circles) of the substrate, where the effective compressive strain (ε_{com}, the compressive displacement divided by the initial length of the substrate) is 34%. Quite similar to the buckling

spacing, both the in-plane and out-of-plane components decrease steeply when the distance between nanowires is smaller than $h_S/2$; thereafter, the amplitudes decrease much slower and converge to the magnitude of a single nanowire. The analogical change can also be observed on the curve of the amplitude ratio of in-plane component to the out-of-plane component.

Another prominent characteristic of the buckling nanowires is the evolution of the amplitude ratio of in-plane component to the out-of-plane component as the substrate is compressed. As shown in Figure 4, the section of the helical buckling configuration of the nanowire in the first stage of compression is close to an ellipse with the amplitude ratio of in-plane component to the out-of-plane component much larger than 1), and when the substrate undergoes further compression, the figure tends to become a circle (the ratio converges to 1). This evolution path of the buckling nanowire's section agrees well with experimental results (Xiao et al. 2010).

3.2 Cooperative helical buckling of a number of nanowires on a soft substrate

In the above section, we focus on the buckling characteristics of two parallel nanowires on the compressed soft substrate. In this section, with the separation distance fixed, the number of nanowires is increased from 2 to 3, 4 and 8, and Figure 5 shows the cooperative buckling configuration of 8 NWs. The influence of the number of nanowires on the helical buckling behavior is investigated. As the most important buckling parameter described above, the buckling spacing is also a focus here (Fig. 6). When the distance between NWs is large (h_S, for example), the influence of the number of NWs is ignorable; for models with narrow

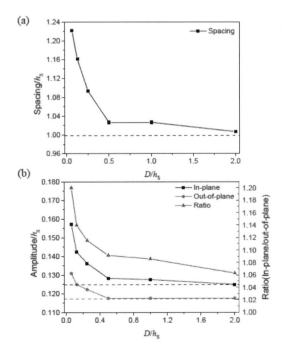

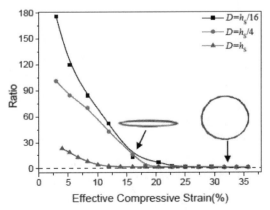

Figure 3. (a) Initial buckling spacing, (b) amplitudes and their amplitude ratio versus the normalized distance.

Figure 4.   The ratio of buckling amplitudes versus the effective compressive strain.

Figure 5. The cooperative buckling configuration of eight NWs.

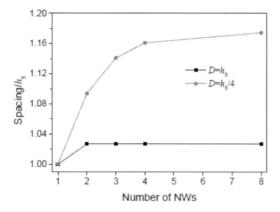

Figure 6. The normalized buckling spacing versus the number of parallel NWs.

separation distance (e.g. $h_s/4$), the buckling spacing increases notably whilst more NWs exist, but it is also noted that the increase speed gets much slower when the number is large enough.

4 CONCLUSION

Instead of just focusing on one single nanowire buckling patterns on a soft substrate, this work focuses on the cooperative helical buckling behavior of a set of nanowires, which is essential to the functionality and performance of electronic devices that often involve a large assembly of different flexible electronics.

Results of a single nanowire's buckling on a soft compressed substrate agree well with both theoretical and experimental results. We first focus on the buckling behavior of two parallel nanowires on a soft substrate. Essential parameters to describe the buckling configurations are investigated, as the separation distance between the nanowires is varied. It is found that the separation distance at half of a representative buckling spacing of a single nanowire ($h_s/2$) is a turning point for the cooperative behavior. Upon compression of the substrate, the buckling spacing, the in-plane and out-of-plane buckling amplitudes and the ratio between

them drop steeply as the separation increases from $h_s/16$ to $h_s/2$; thereafter, the amplitudes of them decrease much slower and converge to the results of a single nanowire. When the separation distance is large enough ($>h_s/2$), nanowires seem not to sense the existence of each other and the buckling pattern of each nanowire is close to that of an isolated one. Another characteristic influenced by the separation distances is the evolution of the ratio of the in-plane buckling amplitude to that of the out-of-plane buckling with the effective compression strain of the substrate. Though overall, all simulations show that this ratio is first very large and then decreases and converges to 1 (the final section of the helical buckling nanowire tends to be a circle) which is consistent with experiments, a small separation distance will lead to a much higher ratio during the initial stage of compression. For the buckling of nanowire assemblies, the buckling characteristics are also influenced by the number of nanowires when the distance between nanowires is small enough. For example, when the separation distance is $h_s/4$, increasing the number of NWs may leads to a ~18% increase of the buckling spacing.

The mechanical principles and patterns about nanowires' cooperative buckling behavior shown here may have a contribution to controllable buckling of stiff elements on a soft substrate and provide useful insight for the design and optimization of flexible electronics as well as three-dimensional micro-fabrications.

ACKNOWLEDGEMENTS

The authors acknowledge the support from the National Natural Science Foundation of China (11302163, 11172231 and 11372241), ARPA-E (DE-AR0000396) and AFOSR (FA9550-12-1-0159).

REFERENCES

[1] Audoly, B. & Boudaoud, A. 2008. Buckling of a stiff film bound to a compliant substrate—Part I. *Journal of the Mechanics and Physics of Solids* 56.2401–21.
[2] Chen, Z. et al. 2002. The fracture of brittle thin films on compliant substrates in flexible displays. *Engineering Fracture Mechanics* 69.597–603.
[3] Kim, D.H. & Rogers, J.A. 2009. Bend, buckle, and fold: mechanical engineering with nanomembranes. *ACS nano* 3.498–501.
[4] Ko, H.C. et al. 2008. A hemispherical electronic eye camera based on compressible silicon optoelectronics. *Nature* 454.748–53.
[5] Rogers, J.A. et al 2010. Materials and mechanics for stretchable electronics. *Science* 327.1603–7.

[6] Someya, T. et al. 2004. A large-area, flexible pressure sensor matrix with organic field-effect transistors for artificial skin applications. *Proc Natl Acad Sci U S A* 101.9966–70.

[7] Song, J. et al. 2009. Mechanics of noncoplanar mesh design for stretchable electronic circuits. *Journal of Applied Physics* 105.123516.

[8] Song, J. et al. 2008. An analytical study of two-dimensional buckling of thin films on compliant substrates. *Journal of Applied Physics* 103.014303.

[9] Stafford, C.M. et al. 2004. A buckling-based metrology for measuring the elastic moduli of polymeric thin films. *Nat Mater* 3.545–50.

[10] Wilder, E.A. et al. 2006. Measuring the modulus of soft polymer networks via a buckling-based metrology. *Macromolecules* 39.4138–43.

[11] Xiao, J. et al. 2008. Mechanics of buckled carbon nanotubes on elastomeric substrates. *Journal of Applied Physics* 104.033543.

[12] Xiao, J. et al. 2010. Mechanics of nanowire/nanotube in-surface buckling on elastomeric substrates. *Nanotechnology* 21.85708.

[13] Xu, F. et al. 2010. Controlled 3D buckling of silicon nanowires for stretchable electronics. *ACS nano* 5.672–78.

[14] Xu, S. et al. 2015. Assembly of micro/nanomaterials into complex, three-dimensional architectures by compressive buckling. *Science* 347.154–59.

[15] Yin, J. et al. 2009. Anisotropic buckling patterns in spheroidal film/substrate systems and their implications in some natural and biological systems. *Journal of the Mechanics and Physics of Solids* 57.1470–84.

Material Science and Engineering – Chen (Ed.)
© 2016 Taylor & Francis Group, London, ISBN 978-1-138-02936-1

Improvement of the field emission properties of post-annealed ZnO nanorod arrays by chemical bath deposition

Y.H. Huang, H.M. He, C.H. Lai, Y.M. Lee, W.S. Chen & C.H. Hsu
Department of Electronic Engineering, National United University, Miao-Li, Taiwan

J.S. Lin
Department of Mechanical Engineering, National United University, Miao-Li, Taiwan

ABSTRACT: ZnO nanorod arrays were prepared by the low-temperature chemical bath deposition method combined with traditional annealing and Rapid Thermal Annealing (RTA) for the investigation of structural and the field emission properties. For the condition of traditional annealing, the ZnO samples were post-annealed at 400°C for 1 h, while the RTA was carried out at 400°C for 1 min. X-Ray Diffraction (XRD) and Photoluminescence (PL) were further employed to investigate the structural and oxygen defect properties. Furthermore, we found that RTA effectively improves the electrical properties and field emission performance as well. The RTA-treated ZnO nanorod arrays exhibit a low turn-on field (0.091 V/um) and a high-field enhancement factor ($\beta = 1.1 \times 106$) due to the enhanced crystal quality and the oxygen defect reduction.

Keywords: ZnO nanorod arrays; structural properties; thermal annealing; RTA; field emission

1 INTRODUCTION

Zinc Oxide (ZnO) is a well-known oxide semiconductor with a wide band gap (about 3.3 eV), low electron affinity (2.1 eV) and a natural n-type electrical conductivity [1]. These unique electrical properties continue to create great interest for the ZnO applied in field emission displays. ZnO film was prepared by RF magnetron sputtering with hydrogen plasma treatment to reduce the work function. The field emission of the ZnO film was enhanced with a turn-on field of 13.6 V/μm and a threshold field of 24.7 V/μm [2]. One-dimensional (1-D) ZnO nanostructures have been proposed for the field emission devices due to their large geometric field enhancement and high aspect ratios [3–6]. Lee et al. [7] prepared ZnO nanorods grown on a ZnO seed layer (by RF sputtering), and performed a post-deposition annealing using a thermal furnace under O_2 and H_2/N_2 ambient conditions, respectively. It was found that the oxygen defects and OH groups on the ZnO surface were reduced, resulting in the increase of the UV-to-visible emission ratio. Lian et al. [8] further prepared ZnO nanostructures grown on vertically aligned Carbon Nanotubes (CNT) using thermal CVD. Their ZnO-coated CNTs had a lower threshold field (3.1 V/μm) when compared with the value (~5.0 V/μm) of pristine CNT.

In this study, we aim to synthesize ZnO nanorod arrays with a high aspect ratio by the low-temperature Chemical Bath Deposition (CBD) method, and to investigate the effect of traditional furnace annealing and Rapid Thermal Annealing (RTA) on the structural and emission properties of the ZnO nanorod arrays.

2 EXPERIMENTAL PROCEDURE

One-dimensional ZnO nanorods were grown on Indium Tin Oxide (ITO)-coated glass (AUO Co., Ltd) with a sheet resistance of 7 Ω/cm by chemi-Cal Bath Deposition (CBD). The ITO glasses were initially cleaned with wet chemical cleaning [9]. First, the deposition of the ZnO seeding layer is performed by dip coating, followed by the growth of ZnO nanorods by CBD (95°C, 1.5 h) in a ZnO aqueous solution. All the aqueous solutions were prepared using distilled water. For the deposition of the seeding layer, the ITO substrates were dipped in a 500 ml aqueous solution containing zinc acetate-2-hydrate ([$Zn(CH_3COO)_2 \cdot 2H_2O$], 99% purity) mixed with the hexamethylenetetramine ($C_6H_{12} N_4$, HMTA, 99.5% purity). The above process was repeated twice, and then dried in an oven at 100°C for 10 min to grow a ZnO seeding layer. The thickness of the seeded layer is around 170 nm. Then, the seeded substrates were immersed into the same zinc acetate solution with a reaction temperature kept at 95°C for 1.5 h to grow ZnO nanorods. After the CBD growth, all the samples received thermal annealing at 400°C for 1h or rapid thermal annealing at 400°C for 1 min. The morphology of the resulting samples was characterized by Field Emission Scanning Electron Microscopy (FE-SEM), and the crystalline structure of the samples was characterized by the X-Ray Diffraction (XRD) measurement. The defect properties were studied using a Photoluminescence (PL) spectroscopy (mini PL system, Pneun Japan) by a deep UV laser of 248 nm excitation source at room temperature. The field emission properties of the samples were measured by Electrochemical Workstation (Jiehan-5000), the ZnO nanorods sample as a cathode, and the ITO glass substrate as an anode. The distance between the cathode and the anode was 10 μm, and the measured emission area in this experiment was 0.5×0.5 cm^2.

3 RESULTS AND DISCUSSION

SEM was used to investigate the nanostructure of ZnO nanorods. Figure 1 shows the surface SEM micrographs of ZnO nanorod arrays. It can be seen that the average diameter is ~55 nm by traditional thermal annealing, and the average diameter is ~94 nm by RTA, respectively. After each annealing, ZnO nanorods are uniformly distributed across the examined area. Also, the high aspect ratio of 18.6~21.3 is obtained for CBD-derived ZnO.

The X-Ray Diffraction pattern (XRD) of the synthesized ZnO nanorod arrays is shown in Figure 2. The curves in the figure represent the XRD of ZnO nanorods by traditional thermal annealing and RTA in the range of 20°–80°. All of the samples show sharp (100), (002) and (101) peaks. This means that the ZnO nanorod array is a hexagonal wurtzite structure and highly preferentially oriented along the c-axis. It is worth mentioning that the RTA treatment improves the crystal quality and the nanorod alignment.

The PL emission spectra of ZnO nanorods are shown in Figure 3. The curves in the figure represent PL of ZnO nanorods by traditional thermal annealing and RTA. In general, the PL spectrum of ZnO is composed of ultraviolet emission from free excitons and a visible blue-green band related to the deep-level defect emission. The peak centered at 370 nm is the UV emission, while the broad peak from 450 to 550 nm is the visible emission. The intensity of the UV emission increases and the intensity of the visible emission decreases

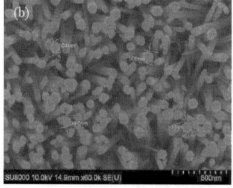

Figure 1. SEM micrographs of (a) ZnO nanorod arrays by traditional thermal annealing and (b) ZnO nanorod arrays by RTA.

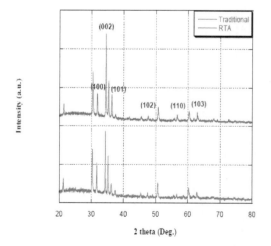

Figure 2. XRD pattern of ZnO nanorod annealing by traditional thermal annealing and RTA. The peaks at 21.3°, 30.3°, 35.2° and 60.4° were originated from the ITO substrate.

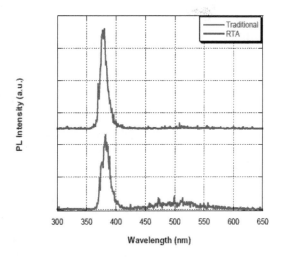

Figure 3.   Photoluminescence spectra of ZnO nanorods by traditional thermal annealing and RTA.

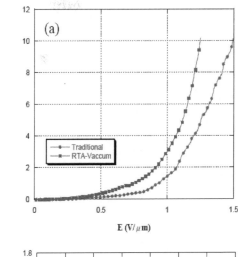

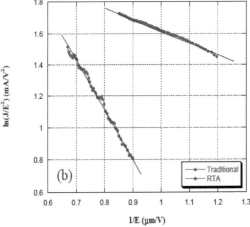

Figure 4.   (a) J-E plot and (b) F-N plot of ZnO nanorods by traditional thermal annealing and RTA.

after the RTA, indicating that the oxygen defects are effectively reduced.

According to the Fowler–Nordheim (FN) theory, we can explain the relationship between the emission current density (J) and applied electric field (E) by the Fowler–Nordheim (F-N) equations [10] as follows:

$$J = A\left(\frac{\beta^2 E^2}{\phi}\right)\exp\left(\frac{-B\phi^{\frac{3}{2}}}{\beta E}\right) \qquad (1)$$

where J is the emission current density (mA/cm^2); E is the applied electric field; A (1.54×10^{-6} (eV.V^{-2})) and B (6.38×10^9 (V.m$^{-1}\cdot$eV$^{3/2}$)) are constant; φ is the work function with a value of ~5.3 eV for ZnO; and β is the field enhancement factor.

According to Equation (1), the plot of ln(J/E^2) vs. E^{-1} (the F-N plot) should be a straight line. The slopes can be calculated from $S = -\frac{Bd\phi^{3/2}}{\beta}$ in the F-N plots and used to estimate the β value. The turn-on field (E$_{On}$), which is defined as the applied field, needs to draw an emission current of 10 (μA/cm^2). As it can be seen in the J-E plot (Figure 4(a)), the turn-on field of ZnO nanorods by RTA is lower than ZnO nanorods by traditional thermal annealing, while the ZnO nanorods by RTA has a high-field enhancement factor. As a result, the ZnO nanorods by RTA show good field emission properties with a lower turn-on field and a higher-field enhancement factor. Figure 4(b) shows that the β value for ZnO nanorods by traditional thermal annealing and RTA is 2.77×10^5 and 1.1×10^6, respectively.

4   CONCLUSIONS

In summary, vertically well-aligned single crystal ZnO nanorod arrays were prepared on the ITO substrates by chemical bath deposition. The SEM micrographs of ZnO nanorods can be seen that the average diameter is ~72 nm by traditional thermal annealing, and the average diameter is ~94 nm by RTA. Thus, the high aspect ratio of 18.6~21.3 is achieved. For ZnO receiving the traditional annealing, the turn-on field is 0.165 V/μm at the emission current density of 10 μA/cm^2. The field enhancement factor β is 2.77×10^5. However, RTA further reduces the turn-on field (0.091 V/μm) and improves the field enhancement factor (β~1.1 $\times$ 10^6). The improvement of the field emission of the ZnO nanorod arrays is mainly attributed to the enhanced crystal quality and the reduced oxygen defects. The results indicate that the CBD-derived ZnO nanorod arrays combined with RTA at 400°C for 1 min can be an effective conducting cathode for field emission applications.

REFERENCES

[1] L.G. Ma, X.Q. Ai, X.L. Huang, S. Ma, Superlattices and Microstructures 50, 703 (2011).

[2] J.B. You, X.W. Zhang, P.F. Cai, J.J. Dong, Y. Gao, Z.G. Yin, N.F. Chen, R.Z. Wang, H. Yan, Applied Physics Letters 94, 262105 (2009).

[3] B. Weintraub, S. Chang, S. Singamaneni, W.H. Han, Y.J. Choi, J. Bae, M. Kirkham, V.V. Tsukruk, Y.L. Deng, Nanotechnology 19, 435302 (2008).

[4] C.J. Park, D.K. Choi, J. Yoo, G.C. Yi, C.J. Lee, Applied Physics Letters 90, 083107 (2007).

[5] C. Li, K. Hou, W. Lei, X.B. Zhang, B.P. Wang, X.W. Sun, Applied Physics Letters 91, 163502 (2007).

[6] N.S. Xu, S. Ejaz, Materials Science and Engineering R 48, 47 (2005).

[7] J. Lee, J. Chung, S. Lim, Physica E 42, 2143 (2010).

[8] H.B. Lian, J.H. Cai, K.Y. Lee, Vacuum 84, 534 (2010).

[9] Y.M. Lee, C.H. Hsu, H.W. Chen, Applied Surface Science 255, 4658 (2009).

[10] J. Tong, L. Li, N.J. Chu, H.X. Jin, Q. Tang, Q. Lu, L.N. Sun, D.F. Jin, H.L. Ge, X.Q. Wang, Physica E 40, 3166 (2008).

Material Science and Engineering – Chen (Ed.)
© *2016 Taylor & Francis Group, London, ISBN 978-1-138-02936-1*

Synthesis and structural characterization of a nanotube potassium titanate anode material for lithium-ion batteries

X.L. Ma & Y.J. Zhao

College of Life Science and Chemistry, Wuhan Donghu University, Wuhan, Hubei, China

ABSTRACT: Nanotube potassium titanate was synthesized by using a hydrothermal method. The structure, morphology and electrochemical properties were characterized by powder X-Ray Diffraction (XRD) and Transmission Electron Microscopy (TEM). The impact of the reaction temperature, time and other factors on the formation of nanotubes were discussed. The potassium titanate nanotubes obtained were used as anode electrode materials for rechargeable lithium-ion batteries. The first discharge capacities were 175 mAh/g for potassium titanate at the current density of 100 mA/g at ambient temperatures. The specific capacities were stabilized at around 80 mAh/g after 20 cycles.

Keywords: nanostructured materials; potassium titanate

1 INTRODUCTION

With their excellent properties of high specific energy, high working voltage, long cycle life, Li-ion batteries have received much attention [1–3]. With high capacity and power, lithium-ion batteries have been considered to be one of the heat energy conversion and storage systems [4–6]. Rechargeable lithium-ion batteries have been considered as the next generation of power sources for electric vehicles, hybrid electric vehicles, and plug-in hybrid electric vehicles. An intensive research has been devoted to search for an alterative anode material [7–10]. Nanostructured materials exhibit good electrochemical properties of performance. Zhang et al. showed that the nanocomposite of Fe_3O_4 hold a high specific capacity of more than 900 mAh/g after 50 cycles [11–13].

Nowadays, nanostructured materials have attracted enormous attention for the potential applications in nanodevices, nanobiology, and nanocatalysis [14–15]. There has been no report about using one-dimensional nanostructured potassium titanate as the anode material for rechargeable Li-ion batteries.

Here, we synthesized the K2Ti8O17 nanotube by using a hydrothermal method at different temperatures and time periods, and assembled the as-synthesized nanomaterials into cells as the working electrode. The results showed that the sizes of the K2Ti8O17 nanotube could be well controlled. When used as the anode material for lithium-ion batteries, the discharge capacity of the first cycle could reach as high as 175 mAh g⁻¹ and, finally, stabilize at about 80 mAh g⁻¹ after 20 cycles.

The electrochemical test showed that the K2Ti8O17 nanotube hold an excellent electrochemical performance and long-term cyclability. These results showed that this material could be one of the candidates for the negative materials.

2 EXPERIMENTAL PROCEDURE

2.1 Synthesis

The K2Ti8O17 nanotube was synthesized by using a hydrothermal method. KOH was dissolved in water to give a 10 mol/L precursor. In a typical synthesis, the TiO_2 powder was added to distilled water and stirred for 30 min. Then, KOH was added and stirred. The mixture was stirred until a homogeneous solution was formed, then sealed in a 40 ml teflonlined stainless steel autoclave, and heated at different temperatures and time periods to obtain the products.

2.2 Characterization

The phase purity of the products was examined by powder X-Ray diffraction on a Bruker D8 Advance X-ray diffractometer using Cu Kα radiation ($\lambda = 1.54056$ Å).

The crystal size and morphology of the products were examined with a transmission electron microscope (TEM, JEM-2010FEF, Japan; SEM, QUANTA 200, Holland).

Electrochemical measurements were carried out using two-electrode cells with the lithium metal as the counter electrode. The working electrode was fabricated by compressing a mixture of the sodium titanate composite/acetylene black/Polyvinylidene Fluoride (PVdF) with a weight ratio of 65:30:5. The weight of the active materials varied between 2.0 and 3.0 mg. The electrolyte was a 1 M LiPF6 in a 1:1 mixture of Ethylene Carbonate (EC)/Diethyl Carbonate (DEC), and the separator was Celgard 2500. The cell was assembled in a glove box filled with a high-purity argon gas. The galvanostatic charge/discharge experiment was performed between 1.0 and 3.0 V at a current density of 100 mA/g, with each experiment repeated at least 5 times.

3 RESULTS AND DISCUSSION

3.1 *XRD*

The X-Ray Diffraction (XRD) results for the K2Ti8O17 nanotube are shown in Figure 1. The diffraction peaks are in good agreement with the standard values for the K2Ti8O17 nanotube (JCPDS No. 41–1100), which shows the high-purity of the as-synthesized samples.

Figure 2 shows that all the products synthesized at 180°C are nanocrystalline since all the XRD patterns have wide peaks with low intensities.

3.2 *TEM*

Figure 3 shows the Transmission Electron Microscope (TEM) images of the K2Ti8O17 nanotube, which are synthesized at 120°C, 160°C and 180°C. The materials consist of a fine nanotube with the average diameter size of 8 nm. The measurement is in agreement with the crystallite size deduced from the XRD analysis.

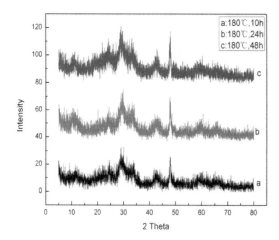

Figure 2. XRD patterns of the as-synthesized K2Ti8O17 heated at different time periods.

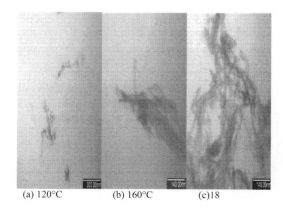

(a) 120°C   (b) 160°C   (c)18

Figure 3. TEM images of the as-synthesized Na2Ti3O7 heated at different temperatures.

Figure 4 shows that the reaction time is an important parameter for the synthesis of the K2Ti8O17 nanotube: the longer the reaction, the better the K2Ti8O17 nanotube.

3.3 *SEM*

Figure 5 shows the Scanning Electron Microscope (SEM) image of the as-synthesized K2Ti8O17 nanotube. The nanoparticles are of morphological nature with very small sizes. The measurement is in agreement with the crystallite size deduced from the XRD analysis.

3.4 *Electrochemical test*

Figure 6 shows that the discharge-charge cycling of the K2Ti8O17 nanotube was carried out in

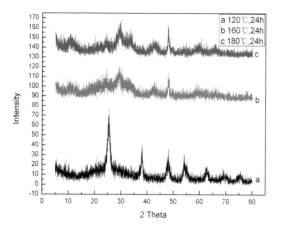

Figure 1. XRD patterns of the as-synthesized K2Ti8O17 heated at different temperatures.

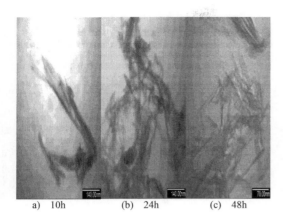

a) 10h   (b) 24h   (c) 48h

Figure 4.   TEM images of the as-synthesized K2Ti8O17 heated at different time periods.

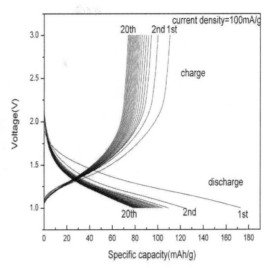

Figure 6.   The charge-discharge capacities of the as-synthesized K2Ti8O17 nanotube for the first 20 cycles.

Figure 5.   SEM patterns of the as-synthesized K2Ti8O17 nanotube.

the voltage windows 1.0 and 3.0 V (vs. Li+/Li) at a current density of 100 mA/g at room temperature. The sample delivers a discharge capacity of 175 mAh·g^{-1}. From Figure 5, we can conclude that the discharge specific capacity of the sample can stabilize at about 80 mAh·g^{-1} after 20 cycles.

4   CONCLUSIONS

In this paper, the K2Ti8O17 nanotube is synthesized by using a hydrothermal method. The materials consist of a fine nanotube with the average diameter size of 8 nm. The nanostructure sample used as the anode material for the Li-ion battery can deliver a better discharge capacity of 175 mAh·g^{-1}. The K2Ti8O17 nanotube shows an excellent electrochemical performance and long-term cyclability.

ACKNOWLEDGMENT

This study was supported by the Youth Foundation of Wuhan Donghu University and the Natural Science Foundation of Hubei Province of China (Grant No. 2014CKC526).

REFERENCES

[1]   Zhang S., Deng C., Fu B.L., Yang S.Y. and Ma L. *Electrochimica Acta*, 2010, 55, 8482–8489.

[2] Z.L. Jian, L. Zhao, H. Pan, Y.S. Hu, H. Li, W. Chen and L.Q. Chen, *Electrochem. Commun.*, 2012, 14, 86.

[3] B.L. Ellis, W.R.M. Makahnouk, Y. Makimura, K. Toghill and L.F. Nazar, *Nature Mater.*, 2007, 6, 749.

[4] Wang B.L. et al., *Chemical Physics Letters*, 406 (2005), 95–100.

[5] Wang X., Zhuang J., Peng Q. and Li Y.D., *Nature*, 437, 121–124.

[6] Kasuga T. and Hiramatsu M. et al. *Langmuir*, 1998, 14, 3160–3163.

[7] Kong X.Y. et al. *Science*, 2004, 303, 1348–1351.

[8] Chen Q. and Du G.H. et al. *Acta Cryst.* 2002, B58, 587–593.

[9] Sun X.M. and Li Y.D., *Chem. Eur. J.*, 2003, 9, 2229–2238.

[10] C.-Y. Chen, K. Matsumoto, T. Nohira, and R. Hagiwara, *Electrochem. Commun.*, 2014, 45, 63.

[11] Xiang J.Y., Tu J.P., Zhang L., Wang X.L., Zhou Y., Qiao Y.Q. and Lu Y. *Journal of Power Sources*, 2010, 195: 8331.

[12] Zhang B., Zheng J., Yang Z. *Ionics*, 2011, 17: 859.

[13] Guo X., Zhong B., Liu H., Song Y., Wen J. and Tang Y. *Transactions of Nonferrous Metals Society of China*, 2011, 21: 1761.

[14] Cheng F.Y., Zhao J.Z. and Song W. *Inorganic Chemistry*. 2006, 45, 2038–2044.

[15] Lou X.M., Wu X.Z. and Zhang Y.X., *Electrochemistry Communications*, 2009, 11, 1696–1699.

Material Science and Engineering – Chen (Ed.)
© 2016 Taylor & Francis Group, London, ISBN 978-1-138-02936-1

The fabrication of the InP nanoporous structure and the optical property of sulfur-passivated InP

S.Z. Niu, D. Fang, J.L. Tang, H.M. Jia, X. Gao, R.X. Li, X. Fang & Z.P. Wei
Changchun University of Science and Technology, Changchun, China

F. Fang
Nanchang University, Nanchang, China

ABSTRACT: In this paper, the Indium Phosphide (InP) nanoporous structure was successfully fabricated by the electrochemical anodization reaction in HCl. To this end, the Sulfur (S) passivation process was applied to the samples. The detail morphology and PL spectrum of the samples were characterized by Scanning Electron Microscopy (SEM) and Photoluminescence (PL), respectively. After the samples were passivated by the Ammonium Sulfide $((NH_4)_2S)$ solution for 10 minutes, the peak of PL for the samples caused by the band edge transition increased. The Energy Dispersive Spectroscopy (EDS) determined that the surface of the samples had Sulfur (S) elements after they were passivated by the Ammonium Sulfide $((NH_4)_2S)$ solution.

Keywords: anodization; InP; nanoporous; passivation; photoluminescence

1 INTRODUCTION

Group III-V compound semiconductor has a high carrier mobility and high efficiency of light emission characteristics; thus, it has been widely used in the fields of high-speed electronic devices, micro- and optoelectronic devices. Uniform nano-hole arrays forming on a semiconductor substrate can be widely used in the applications of photonic and electronic devices. Forming nanostructures of group III-V compound semiconductors traditional approaches utilizing expensive Molecular Beam Epitaxy (MBE) or Metal-Organic Vapor Phase Epitaxy (MOVPE) equipment have been used. High cost is a limitation of these traditional approaches. Similar to the porosity of Si structures, there is considerable interest in applying the electrochemical formation of porosity in InP, which is at low temperature, and the process is simple and of low cost. The formation of porous structures has been achieved on InP, Si, GaAs, CdSe, and GaN. The mechanisms and structural properties have been investigated on these group III-V materials in order to improve their electrical and optical properties.

The high density of surface states or interface states affects the use of group III-V materials. In order to eliminate the effect of the surface states for InP, Sulfur (S) treatment has been utilized. Recent research has shown that treatment with Sulfur (S) can reduce the surface Fermi-level pinning, interface states density in metal–insulator–semiconductor,

and the surface recombination velocity. After being passivated by $(NH_4)_2S$, the surface defects (V_p), surface states and non-radiative recombination velocity decreased, and then the photoluminescence intensity increased.

In this study, a straight InP nanoporous structure was obtained by the electrochemical anodic reaction in a HCl solution. A series of samples under different substrate potentials and etching times have been fabricated. The surface morphology was characterized by Scanning Electron Microscopy (SEM). The photoluminescence for the nanoporous InP, which was passivated by $(NH_4)_2S$, was rarely studied. In this paper, the nanoporous InP was passivated by the $(NH_4)_2S$ solution, and the photoluminescence properties were characterized by PL mapping measurements at room temperature.

2 EXPERIMENTAL PROCEDURE

In this paper, crystalline (100)-oriented substrates of n-type InP of thickness 500 μm (prior to anodic etching) were used. Before etching, the sample was purified by the following steps: the sample was soaked in acetone for 60 seconds to remove the organic contamination on the surface, then soaked in alcohol for 60 seconds to remove the acetone residual on the sample, and then rinsed in De-Ionized (DI) water and finally blown dry with compressed N_2 gas.

The anodization was carried out in the electro-chemical cell as has been described elsewhere. The anodic etching process was carried out in an electro-lyte solution (7.5% HCl aqueous solution) at room temperature. We chose the substrate potential V_s as 5 V and 7 V. The anodization time was 10 seconds, 30 seconds, 60 seconds for 5 V and 10 seconds, 30 seconds, 60 seconds for 7 V, respectively. The substrate potential was supplied in the DC mode throughout the experiments. The Sulfur (S) passiva-tion process was also carried out. The sample was cleaned by a standard process before the passiva-tion experiment, and then soaked in an Ammonium Sulfide $((NH_4)_2S)$ solution for 10 minutes at 60°C. Scanning Electron Microscope (SEM, JEOL-6010LA) measurement was used to characterize the surface morphology and structure for the porous samples by plan-view and cross-sectional scanning. The optical properties of the anodic etched samples and passivated samples were characterized by the PL mapping measurement at room temperature. For PL mapping, the wavelength of the laser light source is 532 nm, and the power is 150 mW.

3 RESULTS AND DISCUSSION

The surface morphology of the anodization pro-duced nanoporous InP, as shown in Figure 1. The plan-view SEM image clearly shows that the nano-porous InP is etched on the InP substrate after the electrochemical anodic reaction in the HCl elec-trolyte. It is obvious that there are regular straight parallel pore arrays, as shown in Figure 1 (d), (e) and (f), while in Figure 1 (a) and (b), there are branches around the wall of the porous structure. This agrees with the explanation in the reference: lower overpotentials cannot restrict the diffusion of the holes, and the holes can diffuse relatively free within their short diffusion length and attack weak bonds on the pore walls, which results in the formation of branches. Higher overpotentials restrict the motion of the holes, and the holes can only reach the pore tip and thus a fairly regular porous structure without branches was formed. Figure 1 (e) and (f) also shows branches, which may relate to the long etching time. The diameter of the pores shown in Figure 1 (d), (e) and (f) was about 200~500 nm, and the longer etching time corresponded to a larger pore diameter. The lat-eral thickness of the walls for the InP nanoporous structure becomes thinner as the pore formation proceeds, as shown in Figure 1 (d) and (f). A longer etching time increases the probability for weak bonds on pore walls; thus, the resulting pores have a thinner wall and a larger diameter.

Cross-sectional SEM images of the porous sample formed at the substrate potential $V_s = 7$ V

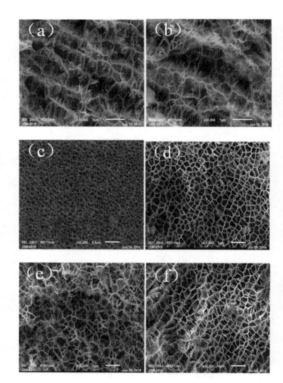

Figure 1. Plan-view SEM images of the sample anodi-zation at $V_s = 5$ V for (a) t = 10 s, (b) t = 40 s, (c) t = 60 s; and $V_s = 7$ V for (d) t = 10 s, (e) t = 30 s, (f) t = 60 s.

are shown in Figure 2. As shown in Figure 2, the anodization produced a straight InP nanoporous structure, penetrating from the surface into the bulk and forming on the InP substrate. The total thickness of the pores increases with the etching time. It is apparent that the thickness shown in Figure 2 (b) is about 2-fold compared with that shown in Figure 2 (a) and Figure 2 (c).

Figure 3 shows the result measured by EDS for the sample at $V_s = 7$ V, t = 10 s. The result shows that Sulfur (S) elements can be detected from the surface of the sample, which indicates that the sul-fur layer is formed on the InP nanoporous struc-ture and the wall surface of the pores.

PL measurements were also carried out for all of the pores formed in HCl electrolytes by the electrochemical anodizing reaction. As shown in Figure 4, the PL intensity of the bulk InP (remov-ing the surface oxide layer by 5% HF solution for 1 minute) is apparently higher than that of porous InP samples, and the PL line of porous samples are slightly blue shifted with respect to that of the bulk InP. The blue shift of the PL may be due to the larger average size of porous nanostructures (~20 nm), which corresponds to the weak quantum

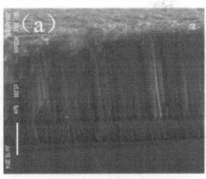

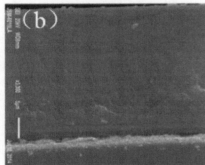

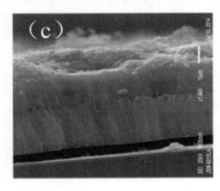

Figure 2. Cross-sectional SEM images of the sample anodization at $V_s = 7$ V for (a) t = 10s, (b) t = 30s, (c) t = 60s.

effect. We can also observe that the blue shift of the sample under 5 V is smaller than 7 V, as shown in Figure 4. This is because the pore diameter of the sample under the potential of 5 V is larger than 7 V. In addition, the FWHM of the sample under the potential of 5 V is apparently large, which may relate to the defect of V_P induced during the etching process.

Figure 5 and Figure 6 show the etching time dependent on the room temperature PL spectra on the potential voltage of 5 V and 7 V and their corresponding PL spectra after sulfur passivation. As shown in the figure, PL intensity decreases with the increase in the etching time. The long

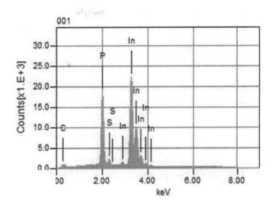

Figure 3. EDS image of nanoporous InP passivated by the $(NH_4)_2S$ solution at $V_s = 7$ V, t = 10s.

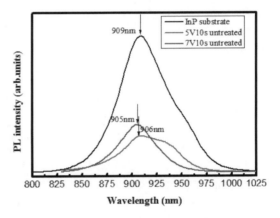

Figure 4. Room-temperature PL spectra of the InP substrate and the anodization produced InP nanoporous structure at $V_s = 7$ V, t = 10 s, respectively.

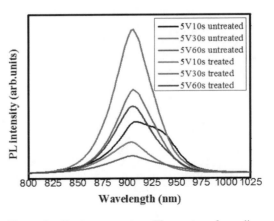

Figure 5. Room-temperature PL spectra of anodization produced InP nanoporous structure at $V_s = 5$ V, t = 10 s, t = 30 s, t = 60 s and the corresponding PL spectra for the sample passivated by the $(NH_4)_2S$ solution.

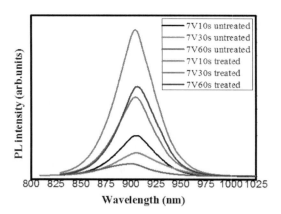

Figure 6. Room-temperature PL spectra of anodization produced InP nanoporous structure at $V_s = 7$ V, t = 10 s, t = 30 s, t = 60 s and the corresponding PL spectra for the sample passivated by the $(NH_4)_2S$ solution.

etching time results in a larger diameter and wall area of the pores, which increases the probability to induced high density of surface states and surface non-radiative recombination. Sulfur passivation for InP, which can reduce the surface defects and decrease the density of surface states, increases the intensity of PL. Here, porous InP was passivated by the $(NH_4)_2S$ solution (8% of concentration). The PL intensity of all the samples shown in Figure 5 and Figure 6 was increased after sulfur passivation, which indicates that a stable sulfur layer with In-S and P-S bonds is formed on the InP surface.

4 SUMMARY

By using electrochemical etching methods, a straight nanoporous structure on the InP (1 0 0) substrate has been successfully fabricated in a 7.5% HCl electrolyte. The pore diameter and pore layer thickness depend on the anodizing overpotentials and the etching time. The pores have a large diameter but not uniform at the 5 V overpotential, and the pore diameter is smaller but uniform at 7 V. The slight blue shift of the InP nanoporous structure compared with the bulk InP is due to the pore structure that causes a quantum effect and to the large diameter of the pores having a weak quantum-size effect. The long etching time may induce more surface defect, and after sulfur passivation, the PL intensity increases, which indicates that the surface defects and the density of surface states have been reduced.

ACKNOWLEDGMENT

This work was supported by the National Natural Science Foundation of China (61076039, 61204065, 61205193, 61307045, 61404009, 61474010), the Research Fund for the Doctoral Program of Higher Education of China (20112216120005), the Developing Project of Science and Technology of Jilin Province (20140520107JH, 20140204025GX), and the National Key Laboratory of High Power Semiconductor Lasers Foundation (No. 9140C310101120C031115, 9140C310104110C3101, 9140C310102130C31107, 9140C31010240C310004).

REFERENCES

[1] Bsiesy A, Vial J C, Gaspard F, et al. 1991. Photoluminescence of high porosity and of electrochemically oxidized porous silicon layers. Surface science 254: 195–200.
[2] Hamamatsu A, Kaneshiro C, Fujikura H, et al. 1999. Formation of <001>-aligned nano-scale pores on (001) n-InP surfaces by photoelectrochemical anodization in HCl. Journal of Electroanalytical Chemistry 473: 223–229.
[3] Han I K, Kim E K, Lee J I, et al. 1997. Stability of sulfur-treated InP surface studied by photoluminescence and x-ray photoelectron spectroscopy. Journal of applied physics 81: 6986–6991.
[4] Jung M, Lee S, Tae Byun Y, et al. 2008. Characteristics and fabrication of nanohole array on InP semiconductor substrate using nanoporous alumina. Microelectronics Journal 39: 526–528.
[5] Langa S, Carstensen J, Christophersen M, et al. 2001. Observation of crossing pores in anodically etched n-GaAs. Applied Physics Letters 78: 1074–1076.
[6] Liu A. 2001. Microstructure and photoluminescence spectra of porous InP. Nanotechnology 12: L1–L3.
[7] Martens K, Wang W, De Keersmaecker K, et al. 2007. Impact of weak Fermi-level pinning on the correct interpretation of III-V MOS CV and GV characteristics. Microelectronic Engineering 84: 2146–2149.
[8] Mynbaeva M, Titkov A, Kryganovskii A, et al. 2000. Structural characterization and strain relaxation in porous GaN layers. Applied Physics Letters 76: 1113–1115.
[9] O'Dwyer C, Buckley D N, Sutton D, et al. 2006. Anodic formation and characterization of nanoporous InP in aqueous KOH electrolytes. Journal of The Electrochemical Society 153: G1039-G1046.
[10] Sato T, Fujino T, Hasegawa H. 2006. Self-assembled formation of uniform InP nanopore arrays by electrochemical anodization in HCl based electrolyte. Applied Surface Science 252: 5457–5461.
[11] Sato T, Mizohata A. 2008. Photoelectrochemical etching and removal of the irregular top layer formed on InP porous nanostructures. Electrochemical and Solid-State Letters 11: H111-H113.

[12] Shikata S, Hayashi H. 1991. Photoluminescence studies on over-passivations of $(NH_4)_2S_x$-treated GaAs. Journal of applied physics 70: 3721–3725.

[13] Smith R L, Collins S D. 1992. Porous silicon formation mechanisms. Journal of Applied Physics 71: R1-R22.

[14] Tian S, Wei Z, Li Y, et al. 2014. Surface state and optical property of sulfur passivated InP. Materials Science in Semiconductor Processing 17: 33–37.

[15] Tiginyanu I M, Ursaki V V, Monaico E, et al. 2007. Pore etching in III-V and II-VI semiconductor compounds in neutral electrolyte. Electrochemical and Solid-State Letters 10: D127-D129.

[16] Wang Y, Yang X, He T C, et al. 2013. Near resonant and nonresonant third-order optical nonlinearities of colloidal InP/ZnS quantum dots. Applied Physics Letters 102: 021917.

[17] Yablonovitch E, Sandroff C J, Bhat R, et al. 1987. Nearly ideal electronic properties of sulfide coated GaAs surfaces. Applied physics letters 51: 439–441.

[18] Yang X, Zhao D, Leck K S, et al. 2012. Full Visible Range Covering InP/ZnS Nanocrystals with High Photometric Performance and Their Application to White Quantum Dot Light-Emitting Diodes. Advanced Materials 24: 4180–4185.

[19] Zhao H, Shahrjerdi D, Zhu F, et al. 2008. Inversion-Type InP MOSFETs with EOT of 21 Å Using Atomic Layer Deposited Al2O3 Gate Dielectric. Electrochemical and solid-state letters 11: H233–H235.

Material Science and Engineering – Chen (Ed.)
© 2016 Taylor & Francis Group, London, ISBN 978-1-138-02936-1

Photoluminescence analysis of ZnO microrods grown by the hydrothermal method

H.Q. Zhang, Y. Jin & L.Z. Hu
School of Physics and Optoelectronic Technology, Dalian University of Technology, Dalian, China

ABSTRACT: ZnO microrods were prepared by using a simple hydrothermal method at a temperature of 180°C on a Si substrate. The sample was characterized by Scanning Electron Microscopy (SEM), X-ray Diffraction (XRD), and Photoluminescence (PL). XRD data showed that the sample was a wurtzite ZnO structure. The SEM analysis revealed that ZnO microrods grew vertically on the substrate. The two emission peaks, 3.307 eV and 3.245 eV, were found in the room-temperature photoluminescence. The low-temperature photoluminescence analysis showed that the 3.245 eV emission maybe has the same origin with the 3.313 eV emission at 20 K.

Keywords: ZnO microrods; Hydrothermal; Temperature-dependent photoluminescence

1 INTRODUCTION

Since crystalline ZnO has a direct gap of 3.37 eV at room temperature, it is expected to be applied as light emitting sources for near- to ultraviolet regions (Look 2001). In addition, the free exciton binding energy is about 60 meV (Look 2001), significantly larger than the thermal energy (~26 meV) corresponding to the room temperature, allowing, in principle, the room temperature operation of exciton-based light emitting devices. Understanding the role of defects and the carrier recombination processes in ZnO structures is essential for the development of optoelectronic devices (Urgessa et al. 2012). PL spectroscopy in the band edge emission region is one of the most versatile techniques to study the defects in detail (the term "defect" can include both foreign impurities and native defects) in semiconductors (Watanabe et al. 2005). There have been many PL studies of ZnO structures (Urgessa et al. 2012, Xu et al. 2012, Chakraborty et al. 2013). Despite many studies, the exact mechanisms of the NBE PL in ZnO structures are still a subject of considerable debate. In this work, ZnO microrods were synthesized by using the hydrothermal method using zinc acetate dehydrate ($Zn(OOCCH_3)_2 \cdot 2H_2O$) and hexamethylenetetramine ($C_6H_{12}N_4$) as precursors, and a comprehensive investigation of the PL of ZnO microrods is presented.

2 EXPERIMENTAL PROCEDURE

Vertically aligned microrods of ZnO were prepared by a two-step process consisting of the deposition of ZnO nanoparticles by spin coating a solution of zinc acetate dihydrate and PVP in ethanol onto a pre-cleaned Si(111) substrate as a seed layer, followed by an immersion of the seeded substrate into a 50 ml solution containing equimolar (0.03M) aqueous solution of $Zn(OOCCH_3)_2 \cdot 2H_2O$ (99.9%) and $C_6H_{12}N_4$ (99.9%) kept in a Teflon-lined stainless steel autoclave for the growth of ZnO microrods. The detailed procedure for the seed particle formation and microrod hydrothermal growth has been described in Ren et al. (2013).

A field emission scanning electron microscope (FESEM, NOVA NanoSEM 450) was used to investigate the morphologies of the microrods, and the crystalline structure of the wires was analyzed using an X-ray powder diffractometer (SHIMADZU XRD-6100 Lab), using a CuKα line (0.154 nm). PL measurements, in the temperature range from room temperature to 50 K, were carried out using a HeCd laser (325 nm) as the excitation source.

3 RESULTS AND DISCUSSION

Figure 1 shows the FESEM image of the ZnO microstructures. The ZnO microstructures exhibit a rod-like morphology. From Figure 1, it can be confirmed that the diameter of ZnO microrods is larger than 1 μm, and the ZnO microrods basically grew vertically on the Si substrate with the hexagonal shapes and the well-faceted surfaces.

The XRD study was performed to evaluate the crystal orientation of the grown ZnO microrods. The 2θ-θ measurement result is shown in Figure 2.

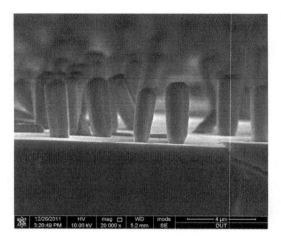

Figure 1. FESEM image of ZnO microrods grown on the (111)-Si substrate.

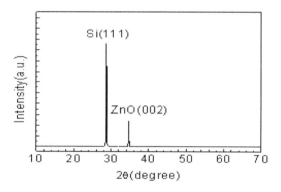

Figure 2. 2θ-θ XRD pattern obtained from ZnO microrods grown on the (111)-Si substrate.

Besides the diffraction from the Si substrate, the 34.64° peak is identified as the (002) diffraction of hexagonal wurtzite ZnO, suggesting that the ZnO microrods grew with the (002) orientation along the c-axis perpendicular to the substrate. In addition, the (002) peak at $2\theta_{002} = 34.64°$ has a full width at half-maximum (FWHM) of 0.17°, implying good crystal quality of ZnO microrods.

Room-temperature PL spectrum of the fabricated ZnO microrods is shown in Figure 3. There are two peaks appearing at 3.245 eV (382 nm) and 3.307 eV (375 nm), respectively. The 3.307 eV peak is attributed to the room-temperature free exciton (FX)-related NBE emission of ZnO (Watanabe et al. 2005, Xu et al. 2012). The exact mechanisms of the 3.245 eV peak are still a subject of considerable debate. The 3.245 eV peak may be attributed to the strong exciton phonon coupling (1 LO)

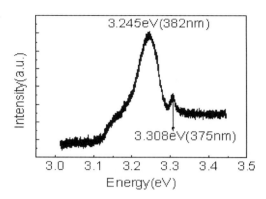

Figure 3. Room-temperature PL spectra of ZnO microrods.

(Chakraborty et al. 2013, Tainoff et at. 2010). The phonon replica's peak was observed to be more asymmetric on its high-energy side (Zhao et al. 2003, Wagner et al. 2007, Segall et al. 1968). From Figure 3, we can see that the 3.245 eV peak does not exhibit the characteristic asymmetric shape. So, this peak cannot be accounted by the phonon replica of the free exciton. In some articles (Zhao et al. 2004, Suzuki et al. 2010, Dai et al. 2014), the 3.245 eV peak is also attributed to the radiative recombination of an exciton-exciton collision process when the excitation intensity is high enough. During our PL measurement, the excitation intensity is about 1 W/cm^2, which is not enough to excite out a large number of excitons to produce the exciton-exciton collision scattering. So, where does the 3.245 eV emission band originate from?

In order to get more insight into the origin of the 3.245 eV emission, we study the temperature-dependent PL spectra. Figure 4 shows the temperature-dependent PL spectra from 50 K to 300 K. Each PL spectrum is normalized at the peak intensity. At 50 K, the main peak of the PL spectrum is 3.365 eV, which is commonly assigned to hydrogen-related ionized donor-bound excitons (3.365 eV, D$^+$X) (Meye et al. 2004). As expected, with the decreasing measurement temperature, the FX emission quenches gradually and the ionized donor-bound exciton emission emerges (D$^+$X, this emission position is 3.365 eV at 50 K) (Meye et al. 2004). As the measurement temperature decreases, the exciton will be bound to become the bound exciton (Watanabe et al. 2005). Using the room temperature value of free exciton peak position of 3.307 eV and the parameters (U, S, V and θ) from the literature (Hamby et al. 2003) together, the temperature dependence of FX emission energy in the range of 300–50 K was fit according to a model by Manoogian and Woolley:

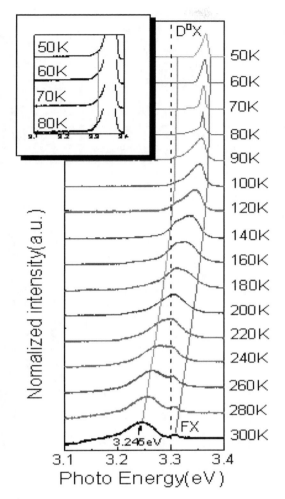

Figure 4. The temperature-dependent PL spectra of ZnO microrods from 300 K to 50 K. Each PL spectrum is normalized at the peak intensity (FX: the free exciton; D+X: the ionized donor-bound exciton). The insert is the magnified figure of 80–50 K PL spectra.

$$E_{FX}(T) = E_{FX}(0) + UT^S$$
$$+ V\theta \left[\coth\left(\frac{\theta}{2T}\right) - 1 \right] \qquad (1)$$

where E_{FX} is the energy of FX emission energy. The fit plot is shown in Figure 4. Figure 4 shows that the temperature dependence of the 3.245 eV emission energy is similar to the temperature dependence of the FX emission, which is the characteristic of the bound exciton emission (Klingshirn 1997). So, we propose the hypothesis that the 3.245 eV emission originates from the bound exciton emission. The binding energy of the bound exciton (E_b) is (Klingshirn 1997)

$$E_b = E_{FX} - E_{ex,b} \qquad (2)$$

where $E_{ex,b}$ is the emission energy of the bound exciton, and $E_b = 62$ meV calculated from the room temperature E_{FX} and $E_{ex,b}$. So, the temperature dependence of this bound exciton emission energy can be calculated from 300 K to 50 K. The calculated result is shown in Figure 4. Comparing the measured PL spectra and the plot, the 3.245 eV emission is present until 50 K (shown by the magnified insert figure of Figure 4 from 80 K to 50 K), and the emission energy is consistent with the calculated value at different temperatures. The calculated emission energy is 3.313 eV at 20 K. The 3.313 eV emission can be assigned to the exciton bound to the neutral acceptor (A^0X) (Watanabe et al. 2005, Meye et al. 2004) or the exciton bound to the structural defect (Meye et al. 2004, Wagner et al. 2011). The analysis shows that the 3.245 eV emission at room temperature and 3.313 eV emission at 20 K maybe have the same origin.

4 CONCLUSIONS

We performed a study of the PL spectra at room temperature and low temperature using the sample of ZnO microrods. At room temperature, the two emission transitions are present in the PL spectroscopy. The 3.307 eV emission originates from the free exciton transition. The variable-temperature PL spectroscopy shows that the 3.245 eV emission is present at all measurement temperatures and corresponds to the 3.313 eV emission at 20 K.

ACKNOWLEDGMENT

This work was supported by the Fundamental Research Funds for the Central Universities (DUT14 LK35).

REFERENCES

[1] Zhao, D.X. Liu, Y.C. Shen, D.Z. Lu, Y.M. Zhang, J.Y. & Fan, X.W. 2004. a thermally activated exciton-exciton collision process in ZnO microrods. Chin. Phys. Lett. 21(8):1640–1643.
[2] Meye, B.K. Alves, H. Hofmann, D.M. Kriegseis, W. Forster, D. Bertram, F. Christen, J. Hoffmann, A. Straßburg, M. Dworzak, M. Haboeck, U. & Rodina, A.V. 2004. Bound exciton and donor–acceptor pair recombinations in ZnO. Phys. Stat. Sol. (b) 241(2):231–260.

[3] Wagner, M.R. Callsen, G. Reparaz, J.S. Schulze, J.-H. Kirste, R. Cobet, M. Ostapenko, I.A. Rodt, S. Nenstiel, C. Kaiser, M. & Hoffmann, A. 2011. Bound excitons in ZnO: Structural defect complexes versus shallow impurity centers. Physical Review B 84:035313.

[4] Tainoff, D. Masenelli, B. Mélinon, P. Belsky, A. Ledoux, G. Amans, D. & Dujardin, C. 2010. Competition between exciton-phonon interaction and defects states in the 3.31 eV band in ZnO. Pysical Review B 81:115304.

[5] Suzuki, K. Inoguchi, M. Fujita, K. Murai, S. Tanaka, K. Tanaka, N. Ando, A. & Takagi, H. 2010. High-density excitation effect on photoluminescence in ZnO nanoparticles. Journal of Applied Physics 107:124311.

[6] Chakraborty, S. & Kumbhakar, P. 2013. Observation of exciton–phonon coupling and enhanced photoluminescence emission in ZnO nanotwins synthesized by a simple wet chemical approach. Materials Letters 100: 40–43.

[7] Zhao, Q.X. Willander, M. Morjan, R.E. Hu, Q.-H. & Campbell E.E.B. 2003. Optical recombination of ZnO nanowires grown on sapphire and Si substrates. Appl. Phys. Lett. 83:165–167.

[8] Segall B. & Mahan, G.D. 1968. Phonon-Assisted Recombination of Free Excitons in Compound Semiconductors. Phys. Rev. 171:935–948.

[9] Xu, N. Cui, Y. Hu, Z.G. Yu, W.L Sun, J. Xu, N. & Wu, J.D. 2012. Photolumicescence and low-threshold lasing of ZnO nanorod arrays. Optics Express 20(14):14857–14863.

[10] Watanabe, M. Sakai, M. Shibata, H. Tampo, H. Fons, P. Iwata, K. Yamada, A. Matsubara, K. Sakurai, K. Ishizuka, S. Niki, S. Nakahara, K. & H. Takasu. 2005. Photoluminescence characterization of excitonic centers in ZnO epitaxial films. Applied Physics Letters 86: 221907.

[11] Urgessa, Z.N. Oluwafemi, O.S. Dangbegnon, J.K. & Botha, J.R. 2012. Photoluminescence study of aligned ZnO nanorods grown using chemical bath deposition. Physica B 407:1546–1549.

[12] Look, D.C. 2001. Recent advances in ZnO materials and devices. Mater. Sci. Eng. B 80:383–387.

[13] Wagner, M.R. Zimmer, P. Hoffmann, A. & Thompsen, C. 2007. Resonant Raman scattering at exciton intermediate states in ZnO. Phys. Status Solidi (RRL)1: 169–171.

[14] Klingshirn, C.F. Semiconductor optics, Springer-verlag Berlin Heidelberg, 1997.

[15] Hamby, D.W. Lucca, D.A. Klopfstein, M.J. & Cantwell, G. 2003. Temperature dependent exciton photoluminescence of bulk ZnO. Journal of Applied Physics 93:3214.

[16] Ren, X.M. Zhang, H.Q. Hu, L.Z. Ji, J.Y. Li, Y. Liu, J.L. Liang, H.W Luo, Y.M. & Bian, J.M. 2013. The Effect of Growth Time on the Morphology of ZnO Nanorods by Hydrothermal Method. Advanced Materials Research 622–623:855–859.

[17] Dai, J. Xu, C.X. Zhu, G.Y. Lin, Y. & Shi, Z.L. 2014. Ultraviolet micro photoluminescence resonance and lasing action in a single ZnO micro-tetrapod. Physica B 442:70–73.

Composite and polymer materials

Material Science and Engineering – Chen (Ed.)
© 2016 Taylor & Francis Group, London, ISBN 978-1-138-02936-1

Study on the preparation and characteristics of HDPE/dialdehyde starch composites

X.Q. Liao

College of Materials Science and Engineering, Chongqing University of Technology, Banan, Chongqing, China

J. Zhu

Chongqing Key Laboratory of Environmental Materials and Remediation Technologies, Chongqing University of Arts and Sciences, Yongchuan, Chongqing, China

ABSTRACT: HDPE/dialdehyde starch composites were prepared by using a single-screw extruder, and the properties of composites, such as impact strength, hardness, melt index and the hydrophobic properties, were investigated by the Izod impact test machine, the melt index apparatus and the multi-functional plastic hardness instrument. The experimental data showed that with the increase in the dialdehyde starch content, the hydrophobic and flow properties of composites greatly increased, while the impact strength and the hardness decreased.

Keywords: environmental-friendly material; high-density polyethylene; dialdehyde starch; blending modification

1 INTRODUCTION

Industries have been working to decrease the dependence on products based on petroleum due to their threat and destruction to the environment. The tremendous increase in the production and use of plastics in our daily life has resulted in the generation of huge plastic wastes, which would be buried or burned. Additionally, it leads to investigating the environmental-friendly sustainable materials to replace the existing ones. High-Density Polyethylene (HDPE) is a kind of widely used polymer material that facilitates our life, especially for its use in the fast-food business and food packaging industry. While the stability of HDPE offers its excellent property, it also brings a disadvantage as waste after use. These wastes from polyethylene products usually are burned or buried into landfill. Neither of these methods is environmental friendly.

Recently, many scientists have been trying to give polyethylene products the property of degradation via chemical or physical methods. Starch is a kind of natural hydrophilic carbohydrate material, which is renewable, biodegradable and inexpensive, and it is one of the most important polysaccharide polymers used to develop biodegradable materials [1]. The modified starch-based polymers represent the extensively studied biodegradable polymers [1–5]. For example, Ren et al. prepared a high-content starch/HDPE composite with the surface of starch modified by the aluminum coupling agent

and stearic acid. The result showed that the properties of composites can be improved while the starch was modified by the aluminum coupling agent and stearic acid [2]. Ma et al. studied that starch, as an inert filler, was added to the Polyvinyl Chloride (PVC) formulation, and these composites were added to the rigid polyurethane plastics to produce plastics. The results indicated that the biodegradable starch-filled PVC plastics can prevent the environment from the pollution of waste plastics [3]. Teng et al. prepared the starch-polyethylene biodegradable plastics film and investigated the physical and mechanical properties of the starch-polyethylene biodegradable plastics film. It was shown that the degradable rate of the testing material was more than 20% in 20 or 30 days, and the water-absorbed rate and the permeating rate of this material were higher than the general plastics PP and CPP [4].

This paper mainly investigated the influence of dialdehyde starch on the hydrophobic property, flowability, hardness and impact strength of HDPE. We tried to find a proper addition amount of dialdehyde starch, which could balance the negative influence on the properties of HDPE.

2 EXPERIMENTAL PROCEDURE

2.1 *Materials*

HDPE (5000S) was purchased from Daqing petrochemical Company. Dialdehyde starch was obtained from Taian Jinshan Modified Starch Co. Ltd and was dried at 70°C for 48 h under vacuum before use. Glycerol was offered by ChengDu KeLong Chemical Co. Ltd and was used without any further purification. Other solvents with AP grade were purchased from Bodi Chemical Factory (Tianjin, China) and used without further purification.

2.2 The preparation of HDPE/dialdehyde starch composites

A typical procedure to prepare HDPE/dialdehyde starch composites was as follows. First, dialdehyde starch (10 g) was completely mixed with glycerol (2% W/W). The blended chips containing the mixtures were fabricated with HDPE (190 g) by a single-screw extrusion machine (SJ-25, Nanjing Giant Machinery Co. Ltd), with three temperature zones of 110°C, 120°C and 125°C, respectively, and a speed of 30 rpm. Then, the products were compressed into the plate at 140°C by a vulcanizing machine (XLB-400*400*2/0.25MN, Xincheng Yiming Rubber Machinery Co, Ltd Qing dao, China) under the pressure of 5 MPa. Finally, all specimens were stored at room temperature in vacuum before the analysis. The compositions of HDPE/dialdehyde starch composites used in the experiments are listed in Table 1.

2.3 The investigation of the properties of composites

Hydrophobic property test was carried out according to GB1034-70: first, the granules of each sample (14 g) were dipped in water at 20°C for 24h. Then, the external water was removed with filter paper and weighed the specimens (M1). Finally, wet samples were dried at 60°C in vacuum to constant weight and the mass of each sample (M2) was recorded. The values of water absorption can be estimated by Equation (1):

Table 1. The compositions of HDPE/dialdehyde starch composites.

Sample number	Glycerol (w/w)	Dialdehyde starch (g)	HDPE (g)	Content of dialdehyde starch (%)
1	2%	0	200	0
2	2%	10	190	5
3	2%	20	180	10
4	2%	30	170	15
5	2%	40	160	20
6	2%	50	150	25
7	2%	60	140	30
8	2%	70	130	35

Water absorption (%) = $(M1 - M2)/M2 \times 100\%$

$$(1)$$

Melt index test was investigated by GB/T3682-2000. The melt index test was tested by a melt flow rate meter (ZZKTZ400, ShengTai Machinery Equipment Co. Ltd Suzhou Branch, China). The test conditions were 145°C, 5 kg and 30 s. The values of the melt index test can be estimated by Equation (2):

MFR $(\theta, mnom) = tref \times m/t$
$= 600 \times m/t(g/10 \text{ min})$ (2)

where θ is the testing temperature (°C); mnom is the nominal load (Kg); tref is the reference time (10 min); m is the average quality of cut (g); and t is the time interval of cut (s).

Izod impact test was performed according to GB/T1843–2008 and investigated by a machine (SM-8215, Mark Technology Co. Ltd FoShan Branch in GuangDong, China). First, the impact section was marked on the testing specimen, and the thickness and width of the spline were measured, then the measured spline was fixed to keep the impact pendulum from falling automatically. Finally, the corresponding number was referred to the value of thrust energy of the spline.

The hardness test was investigated by a multi-functional plastic ball indentation hardness tester (QYS-96, Intelligent Instrument Equipment Co. Ltd, Changchun, China) according to GB3398-2008. The values of the hardness test can be estimated by Equation (3):

H = $0.21F/[0.25D \times (h - 0.04)]$ (3)

where H is the hardness of the ball indentation test (N/mm^2); F is the testing force (N); D is the ball diameter (mm); and h is the depth of indentation after the fixed frame deformation (mm).

3 RESULTS AND DISCUSSION

3.1 Hydrophobic property of HDPE/dialdehyde starch composites

Figure 1 shows the water absorption of HDPE/dialdehyde starch composites. It can be observed that the water absorption hardly changes when the content of dialdehyde starch is less than 10%, but it increases sharply when the content of dialdehyde starch is more than 10%. This is because HDPE is a hydrophobic material and dialdehyde starch attracts water. The water absorption of HDPE/dialdehyde starch composites depends on the content of dialdehyde starch. The addition of dialdehyde starch into HDPE helps the water mol-

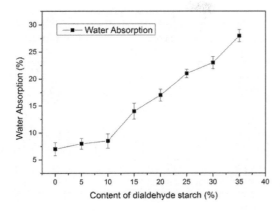

Figure 1. Water absorption of HDPE/dialdehyde starch composites.

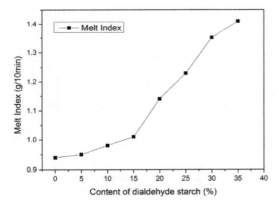

Figure 2. Melt index of HDPE/dialdehyde starch composites.

ecule to permeate inside of HDPE and promote its degradation.

3.2 Melt index of HDPE/dialdehyde starch composites

Figure 2 shows the melt index of HDPE/dialdehyde starch composites. When the content of dialdehyde starch is less than 15%, the melt index increases slowly with the content of dialdehyde starch increasing. However, when the content of dialdehyde starch is more than 15%, the melt index increases sharply. This means that the addition of dialdehyde starch improves the flowability of composites. Compared with starch, the crystalline degree of dialdehyde starch is lower and its flowability is higher than starch. Dialdehyde starch could act as a kind of plasticizer. When its content

is low, the plasticizing effect is not obvious, which explains why the melt index increases slowly when the content of dialdehyde starch is less than 15%.

3.3 Impact strength of HDPE/dialdehyde starch composites

Figure 3 shows the Izod impact of HDPE/dialdehyde starch composites. When the content of dialdehyde starch is less than 15%, the impact strength decreases slowly with the content of dialdehyde starch increasing. However, the Izod strength decreases sharply when the content of dialdehyde starch is more than 15%. This is due to the poor compatibility between the two parameters. A small content of dialdehyde starch does not influence the continuous phase of HDPE, but it does when the content of dialdehyde starch is dispersed into HDPE.

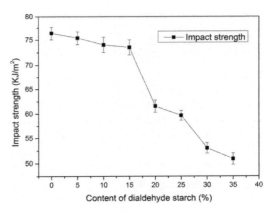

Figure3. Impact strength of HDPE/dialdehyde starch composites.

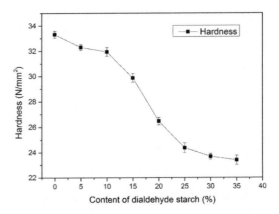

Figure 4. Hardness of HDPE/dialdehyde starch composites.

3.4 *Hardness of HDPE/dialdehyde starch composites*

The hardness of HDPE/dialdehyde starch composites is shown in Figure 4. It is obvious that the hardness decreases only to a small extent when the content of dialdehyde starch is less than 10%. However, it decreases sharply when the content of dialdehyde starch is more than 10%. So, it is possible that a small content of dialdehyde starch offers a proper degradation ability without a significant decrease in hardness.

4 CONCLUSION

HDPE/dialdehyde starch composites were prepared by using a single-screw extruder, and their properties of composites were investigated. The results indicated that a small amount of dialdehyde starch increases their water absorption and flowability gently, and decreases their impact strength and hardness gently. However, when the content of dialdehyde starch is above 10%, these influences become sharp and obvious.

ACKNOWLEDGMENT

This study was financially supported by (a) the Natural Science Foundation of Chongqing Municipal Science and Technology Commission, China (Grant No. CSTC2013 JCYJA50029), (b) the Local Fund for Chongqing University of Arts and Sciences (Grant No. R2012CH08), and (c) the Science and Technology Project Affiliated to the Education Department of Chongqing Municipality (KJ1401122).

REFERENCES

[1] V. Alves, N. Costa, L. Hilliou, et al, Design of biodegradable composite films for food packaging, Desalination, 199(2006), 331–333.
[2] I.M. Martins, S.P. Magina, L. Oliveira, et al, New biocomposites based on thermoplastic starch and bacterial cellulose, Composites Science and Technology, 69(2009), 2163–2168.
[3] N. Cañigueral, F. Vilaseca, J. Méndez, et al, Behavior of biocomposite materials from flax strands and starch-based biopolymer, Chemical Engineering Science, 64(2009), 2651–2658.
[4] F. Vilaseca, J. Mendez, A. Pelach, M. Llop, N. Canigueral, J. Girones, X. Turon and P. Mutje, "Composite materials derived from biodegradable starch polymer and jute strands," Process Biochemistry. (2007), 329–334.
[5] Y.Z. Wan, H. Luo, F. He, H. Liang, Y. Huang and X. Li, "Mechanical, moisture absorption, and biodegradation behaviours of bacterial cellulose fibre-reinforced starch biocomposites," Composites Science and Technology, 69(2009), 1212–1217.

Material Science and Engineering – Chen (Ed.)
© 2016 Taylor & Francis Group, London, ISBN 978-1-138-02936-1

Effect of silicon on the microstructure and properties of Si$_p$/ZA40 composite materials

N.S. Xie, H. Liu & B.Z. Hu

School of Materials Science and Engineering, Shaanxi University of Technology, Hanzhong, Shaanxi, China

ABSTRACT: In order to fabricate an *in situ* composite of high-aluminum zinc-based alloys, obtain fine grain and good interfacial bonding of reinforcement and matrix of composite materials, Si$_p$/ZA40 composites with different contents of silicon particles were produced using the *in situ* synthetic method by adding a certain content of the silicon element. The effect of silicon on the microstructure and mechanical properties of high-aluminum zinc-based alloys with different silicon contents was studied by metallurgical microscopy, Scanning Electron Microscopy (SEM), tensile test and hardness test in this work. The results show that a sort of alloy with a fine silicon phase distributed uniformly can be obtained under an appropriate chemical composition. The comparison capability between high-aluminum zinc-based alloys and silicon-modified high-aluminum zinc-based alloys shows that the hardness of silicon-modified high-aluminum zinc-based alloys increases. However, the addition of the silicon element decreases the tensile strength and elongation of the test alloys. The tensile fracture of high-aluminum zinc-based alloys, which contains 6wt% silicon, shows brittle rupture.

Keywords: high-aluminum zinc-based alloys; Si$_p$/ZA40 composite materials; microstructure; mechanical properties

1 INTRODUCTION

High-aluminum zinc-based alloy is an important cast zinc/aluminum alloy. It has excellent mechanical properties, foundry properties, machining properties and particularly wearable properties [1]. Furthermore, the raw material is of low cost and is non contaminated, and has a wide range of sources. Due to all of these advantages, this kind of alloy has quickly grown up in the domestic and overseas as new type of alloy [2]. Copper is the main strengthening element of the zinc/aluminum alloy with a certain degree of solid solution; a white strip phase will precipitate when copper content exceeds the solid solution degree [3]. Aluminum is the main strengthening element of high-aluminum zinc-based alloys with a certain degree of solid solution as well. With the increase in aluminum content, the dendrite of the newborn α' phase increases, the eutectoid volume reduces obviously, α dendrites grows coarsening in the alloy, and the friction coefficient increases [4]. Silicon is a main element of the zinc-based alloy. It can improve the casting properties, improve the wear resistance of the zinc-based alloy, increase the silicon content suitably and refine the alloy matrix, improve the density, and reduce the tendency of shrinkage [5]. In this paper, taking the ZA40 alloy without adding alloy elements as the reference, the effects of silicon on the metal-lurgical structure and mechanical properties of the ZA40 alloy were investigated through examining the change in mechanical properties after the silicon element was added to the base alloy [6]. Using the zinc alloy has an important significance on saving copper. High-aluminum zinc-based alloys have been applied as a new kind of non-ferrous materials in recent decades. The squeeze casting process can reduce macro-segregation, refining the composite materials' grain, decrease the matrix air holes and shrinkage cavities, and prevent the internal defect of composite casting parts.

2 EXPERIMENTAL MATERIALS AND METHODS

2.1 *Experimental materials*

The raw materials used were industrial pure zinc, industrial pure aluminum, and industrial pure magnesium. The aluminum-copper intermediate alloy was the Al-30wt%Cu alloy. The aluminum-silicon intermediate alloy was the Al-30wt%Si alloy, and the composition of industrial pure zinc is listed in Table 1. The composition of industrial pure aluminum is listed in Table 2. The base alloy of Si$_p$/ZA40 composites was the ZA40 high-aluminum zinc-based alloy, and the chemical composition of the base alloy ZA40 was Zn-40%Al–2.0%Cn-0.5%Mg.

Table 1. The chemical compositions of industrial pure zinc.

Element	Zn	Fe	Si	Cu	Zn	Ti
Content (wt%)	Margin	0.02	0.02	0.01	0.002	0.002

Table 2. The chemical compositions of industrial pure aluminium.

Element	Al	Pb	Cd	Fe	Sn	Cu
Content (wt%)	Margin	0.001	0.002	0.005	0.005	0.002

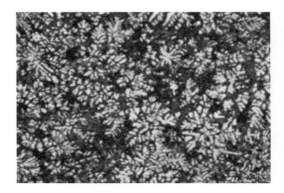

Figure 2. The microstructure of the ZA40 alloy.

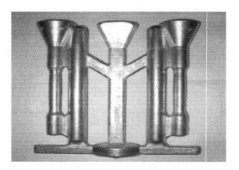

Figure 1. The casting tensile specimen of Sip/ZA40 composites.

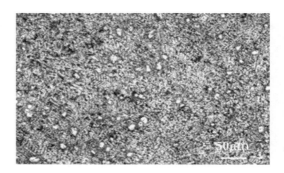

Figure 3. The microstructure of the Sip/ZA40 composite.

2.2 *Experimental methods*

The Si_p/ZA40 composites were fabricated by using melting method. The aluminum-silicon alloy was the silicon carrier of Si_p/ZA40 composites. The Si_p/ZA40 composites were fabricated using the gravity casting technique, with sodium salt as the modifier. The casting tensile specimen of composites is shown in Figure 1. The microstructure of Si_p/ZA40 composites was analyzed by the EPIPHOT-300U inverted microscope model. The fracture characteristics of Si_p/ZA40 composites were analyzed by JSM-6390 LV scanning electron microscopy. The tensile properties of the composites were tested with invariable strain rates on the CMT5105 material test system. Hardness was tested by the HB-3000B hardness tester.

3 EXPERIMENTAL RESULTS AND DISCUSSION

3.1 *The effect of silicon on the microstructure of Si_p/ZA40 composites*

The microstructure of the ZA40 high-aluminum zinc-based alloy is shown in Figure 2. As it can be seen from Figure 2, the microstructure of the based alloy consists of the dendrite α' phase, the eutectoid $\alpha+\eta$ phase, the η phase, the Mn phase and the eutectic volume, and as the newborn α' phase increased, α dendrite grew coarsening in the alloy. Because the aluminum content was high, the number of the α' phase increased and the eutectoid $(\alpha+\eta)$ phase decreased. The microstructure of Si_p/ZA40 composites is shown in Figure 3. As shown in Figure 3, a sort of alloy with a fine silicon phase distributed uniformly can be obtained under an appropriate chemical composition, and there are not only eutectoid silicon phases but also primary silicon phases with the increase in silicon content in the alloy, and the eutectoid volume phase precipitates little and in small size. It has microstructure characteristics of metal matrix composites reinforced with particles. So, the *in situ* Si_p/ZA40 composite has high toughness and high-temperature performance because of its fine grain and good interfacial bonding, and due to its high specific strength, high specific modulus, and high wear resistance.

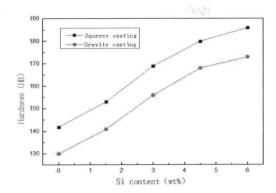

Figure 4. The effect of silicon on the hardness of Sip/ZA40 composites.

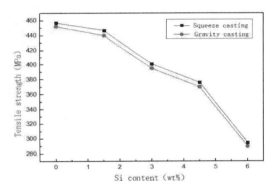

Figure 5. The effect of silicon on the tensile strength of Sip/ZA40 composites.

3.2 *Effect of silicon on the properties of $Si_p/ZA40$ composites*

The effect of silicon on the hardness of $Si_p/ZA40$ composites is shown in Figure 4. As it can be seen from Figure 4, the hardness of different silicon samples containing the zinc-based alloy was investigated, the quantity of silicon phase particles increases with the increase in silicon content in the alloy, and the hardness of $Si_p/ZA40$ composites increases, but the impact toughness of the alloy decreases. This is because the dissolved silicon element in the ZA40 matrix is very small, so in the alloy substrate, silicon exists in a granular form. Silicon particles of $Si_p/ZA40$ composites have high hardness, so the hardness of $Si_p/ZA40$ composites is higher than that of the ZA40 alloy. The effect of silicon on the tensile strength of $Si_p/ZA40$ composites is shown in Figure 5. As shown in Figure 5, the tensile strength of $Si_p/ZA40$ composites decreases with the increasing silicon content, and silicon particles are distributed uniformly. This is because

the silicon particles of $Si_p/ZA40$ composites have high hardness, and it is not easy to buffer stress, and so in the process of drawing, silicon particles easily separate from the ZA40 matrix alloy. From Figure 4 and Figure 5, it can be seen that the influence of the squeeze casting process on mechanical properties of $Si_p/ZA40$ composites is obvious, and the tensile strength and hardness of $Si_p/ZA40$ composites by squeeze casting are 1.15 times and 1.85 times that of $Si_p/ZA40$ composites by gravity casting, respectively. The tensile fracture of high-aluminum zinc-based alloys, which contains 6wt% silicon, shows brittle rupture.

3.3 *Effect of silicon on the wear resistance of $Si_p/ZA40$ composites*

The wear characterization of $Si_p/ZA40$ composite materials was investigated by the MG-2000 type friction and wear testing machine at high speed and high temperature. SM-6390 LV scanning electron microscopy was used to analyze the wear surface of the composite materials. The surface morphological characteristics of the wear surface of $Si_p/ZA40$ composite materials are shown in Figure 6. The results of wear resistance showed that with the increase in the silicon contents, the wear rate decreased and the wear resistance significantly improved. The wear resistance of $Si_p/ZA40$ composites increased by nearly twice than that of the ZA40 alloy. This is because not only the $Si_p/ZA40$ composites have an eutectoid volume phase, but also silicon particles that are very hard can play a load-bearing role and hinder the movement of the plastic deformation of matrix dislocations. The reason why squeeze casting composite is better than others is that the defects are difficult to appear, and silicon particles on the surface of the matrix support the surface of the matrix and keep it from further

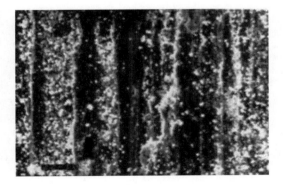

Figure 6. The surface morphology characteristics of the wear surface of the Sip/ZA40 composite material.

wear. On the other hand, the output of Si_p/ZA40 composites products in the ordinary state tends to be difficult to avoid the defects; however, its organization or performance and on the material for further processing is also a matter of question, which must consider the production. From Figure 6, it can be seen that the wear mechanism of Si_p/ZA40 composite materials is mainly abrasive wear and delaminate wear. This study not only extends the theoretical scope of the study of the particle-reinforced ZA40 matrix composite, but also applies the practical significance of the ZA40 matrix composites instead of ZA40 and other alloys. Therefore, the deep and systemic research on Si_p/ZA40 composite materials has a promoting effect on the fabrication and application of the *in situ* composite of high-aluminum zinc-based alloys.

4 SUMMARY

1. The microstructure of the ZA40 high-aluminum zinc-based alloy consists of the dendrite α' phase, the eutectoid $\alpha+\eta$ phase, the η phase, the Mn phase and the eutectic volume.
2. The *in situ* Si_p/ZA40 composite has a fine grain and good interfacial bonding, and the fine silicon phase distributed uniformly can be obtained under an appropriate chemical composition.
3. The quantity of silicon phase particles increases with the increase in silicon content in the alloy, and the hardness of Si_p/ZA40 composites increases.

4. The tensile strength of Si_p/ZA40 composites decreases with the increasing silicon content, and silicon particles are distributed uniformly.
5. The tensile strength and hardness of Si_p/ZA40 composites by squeeze casting are 1.15 times and 1.85 times that of composites by gravity casting, respectively.
6. The tensile fracture of the high-aluminum zinc-based alloys, which contains 6wt% silicon, shows brittle rupture.

REFERENCES

[1] Xie Niansuo. & Wang, & Yan. & Wu, Lizhi. 2010. Research Application and Prospects of High Zinc-based alloy. *Hot Working Technology*, 39(14): 50–53.
[2] Zhihui Zhang. & Zhiwei Fu. & Guozhi Ruan. 2011. Effect of Al-Si alloy powder on properties of corundum-mullite composites. *Rare Metals*, 30(1): 511–514.
[3] Wang Yan. & Xie Nian-suo. & Li Chun-yu. 2010. Recent Development in Preparing Particles Reinforced Zinc base Composites by In situ Crystallization Technique. *Foundry Technology*, 31(5): 656–659.
[4] Xie Nian-suo. & Wang Yan. 2011. Study on Wear Properties of Sip/ZA40 Composites. *Foundry Technology*, 32(6): 834–836.
[5] K. Al-Helal, & I.C. Stone, & Z. Fan. 2012. Simultaneous Primary Si Refinement and Eutectic Modification in Hypereutectic Al–Si Alloy. *Transactions of the Indian Institute of Metals*, 65(6): 663–667.
[6] Sharma S.C., & Girish B.M., & Somashekar D.R., et al. 1999. Sliding wear behavior of zircon particles reinforced ZA27 alloy composite materials. *Wear*, 224(1): 89–94.

Material Science and Engineering – Chen (Ed.)
© 2016 Taylor & Francis Group, London, ISBN 978-1-138-02936-1

Preparation and mechanical properties of PP/morphology of $CaCO_3$ composites

X.Q. Liao & J. Zhu

Chongqing Key Laboratory of Environmental Materials and Remediation Technologies,
Chongqing University of Arts and Sciences, Yongchuan, Chongqing, China

ABSTRACT: In this study, calcium carbonate nanoparticles with different shapes (rhombohedral and spherical) were used to modify isotactic polypropylene (PP). The properties of composites were investigated by SEM, Izod impact test machine and the multi-functional plastic hardness instrument. The results showed that the morphology of calcium carbonate played an important role in the toughening effect on PP matrix.

Keywords: $CaCO_3$ nanoparticles; mechanical properties

1 INTRODUCTION

Isotactic polypropylene (PP) is a kind of commercial polymers with excellent all-round properties, which were widely used in many areas because of its good processing performance and low price. However, the extensive application of PP is still limited by its high shrink age, low modulus, low temperature impact strength and relatively poor impact resistance [1,2]. Therefore, the toughening research of PP is widely concerned. So far, the main toughening methods include elastomer toughening [2], toughening modification of rigid body [3] and nucleating agent modification [4]. Especially the modification of inorganic particles was widely investigated because of its good stiffness and the economic benefit. In recent years, with the development of ultra-fine $CaCO_3$ technology, nano-$CaCO_3$ had been meticulously researched [5,6]. But, the morphology of $CaCO_3$ impact on the performance of the PP was ignored.

Previously, the crystallization properties of PP/$CaCO_3$ composites were investigated, and the results indicated that the morphology of calcium carbonate played a major role in the crystallization behavior of PP, and the RCC promotes more beta crystal formation, the articles will be presented in another paper. In this study, the mechanical properties of PP/$CaCO_3$ composites with RCC and SCC were studied.

2 EXPERIMENTAL SECTION

2.1 Materials

The isotactic polypropylene (T30 s) was offered by China National Petroleum Corporation Lanzhou petrochemical company and its melt flow index is 2.6 g/10 min. The Rhombohedra Calcium Carbonate (RCC) was provided by Nano Materials Technology Co, Ltd, (Ruicheng Shanxi China). The Spherical Calcium Carbonate (SCC) was made by our lab according to the procedure reported by Volodkin et al [7]. All other chemicals and solvents used were of analytical grade.

2.2 Preparation of PP/$CaCO_3$ composites

The mixture of PP and $CaCO_3$ with the different weight ratio were fed into the Torque Rheometer (RM-200A, Harpo Harbin Electric Technology), with three temperature zones of 185 °C, 190 °C and 185 °C, respectively and a torque speed of 60 rpm. Then the products were compressed into the plate at 190 °C by a vulcanizing machine (XLB-400*400*2/0.25MN, Qing dao xincheng yiming Rubber Machinery Co, Ltd.) under the pressure of 5 MPa. Finally, all specimens were stored at room temperature before the analysis.

2.3 Characterization

The morphology of nano particles and their dispersion in PET were examined by Scanning Electron Microscopy (SEM) (Philips XL-3, FEI, Oregon, USA) with an accelerating voltage of 20 kV. The samples were sputtered with a thin layer of gold before testing.

The tensile testing was performed on a universal test machine (CMT4104, Shenzhen SANS Testing Machine Co., China) at a crosshead speed of 50 mm/min. The bending test was carried out with the same universal test machine at a rate of 5 mm/min. The un-notched impact examinations were executed on a pendulum impact testing machine (ZBC2000, Shenzhen SANS Testing Machine Co., China). At least five samples were conducted for each property test and the averaged results were reported.

3 RESULTS AND DISCUSSION

3.1 Scanning electron microscope images analysis

Figure 1 shows the morphology of the selective $CaCO_3$ particles and their composites with PP. From Figure 1(a) and (d), it is clearly that RCC nanoparticles exhibited bigger crystal size (~900 nm) and regular rhombohedral shape composed of the laminated calcite. SCC nanoparticles had dimensions ranging from 400~600 nm and shown sphere-like appearance. Images of Figure 1 (b), (d), (e) and (f) show that both RCC and SCC particles were uniformly dispersed in PP and they had the good compatibility with PP matrix.

3.2 Mechanical properties

In order to investigate the effect of $CaCO_3$ particles content on the mechanical properties of PP, a series of the mechanical testing were carried out, including the tensile testing, the bending testing and the un-notched impact examinations. Table 1 shows that the variations of tensile strength, elongation at break, bending strength and un-notched impact strength of PP/$CaCO_3$ composites as a function of the different amount of $CaCO_3$

particles. From Table 1, it can be seen that both the tensile strength and the bending strength of composites almost progressively decreased with the increment of $CaCO_3$ particles in the range from 2 to 4 wt%. When it was beyond 8 wt%, the values of the tensile strength and the bending strength increased monotonically. On the other hand, compared with the change of the tensile strength and the bending strength, the elongation at break and the un-notched impact strength of PP/$CaCO_3$ composites exhibited the opposite trend with the increase of $CaCO_3$ content. For example, when the addition of $CaCO_3$ was 4%, both the elongation at break and the un-notched impact strength arrived the maximum, whereas, decreased greatly above 4%. These data indicated that the addition of $CaCO_3$ with the different shapes can ameliorate the mechanical properties of PP. As we know, the calcium carbonate acts as a rigid particle, which could have a toughening and enhancing effect on PP composites. In general, α crystal form dominates in PP, as Figure 1 shown. The size of α spherule is large and has a clear boundary between spherules [8]. And it could produce defects during stretching, so the toughness of PP composites was poor. However, with the addition of $CaCO_3$ particles, the β crystal increased gradually in PP matrix, which shows the excellent impact resistance and toughness [9]. So the ductility of PP composites increases. As far as the SCC was concerned, it was smaller than RCC so that there was a larger surface area with PP matrix (seen in Fig. 1). The SCC would

Figure 1. The morphology of the selective $CaCO_3$ particles and their dispersion in PP: (a) RCC, (b) (c) 4% RCC/$CaCO_3$, (d) SCC, (e) (f) 4% SCC/$CaCO_3$.

Table 1. Mechanical properties of PP/SCC.

Sample	Tensile strength (MPa)	Elongation at break (%)	Bending strength (MPa)	Un-notched impact strength (KJ/m^2)
Pure PP	41.09 ± 0.51	26.23 ± 4.13	36.37 ± 1.73	40.83 ± 0.11
2%PP/RCC	39.13 ± 1.32	50.47 ± 17.98	37.76 ± 1.53	38.27 ± 3.35
4%PP/RCC	34.34 ± 2.07	84.69 ± 10.63	36.76 ± 1.46	42.60 ± 2.90
8%PP/RCC	37.23 ± 1.63	19.89 ± 4.49	40.59 ± 3.75	29.61 ± 2.14
12%PP/RCC	38.22 ± 1.11	15.18 ± 1.12	49.11 ± 3.00	28.58 ± 3.86
2%PP/SCC	34.23 ± 1.24	75.27 ± 1.28	42.61 ± 2.82	39.18 ± 2.13
4%PP/SCC	32.01 ± 2.93	171.93 ± 27.10	31.97 ± 2.33	51.68 ± 1.89
8%PP/SCC	35.50 ± 2.04	20.14 ± 3.91	45.36 ± 5.35	32.05 ± 0.78
12%PP/SCC	36.75 ± 0.71	18.22 ± 5.07	47.47 ± 4.37	30.75 ± 1.37

easily produce some stress concentrations in PP matrix. When the mechanical testing was carried out, the interface between the PP matrix and the particles would consume large amounts of energy by deboning forming holes [10]. So the elongation at break of PP/SCC composites was better than that of PP/RCC composites.

4 CONCLUSION

In this study, calcium carbonate nanoparticles with different shapes (rhombohedral and spherical) were used to modify isotactic PP. The results showed that the morphology of calcium carbonate played an important role in the toughening effect on PP matrix. The un-notched impact strength and elongation at break of the composite increase first and then decrease gradually with increasing calcium carbonate content, Show good toughness, especially add the SCC.

ACKNOWLEDGMENTS

This study was financially supported by (a) the natural science foundation of Chongqing Municipal Science and Technology Commission, China (Grant No. CSTC2013 JCYJA50029), (b) the Local Fund for Chongqing University of Arts and Sciences (Grant No. R2012CH08), (c) the Opening-project Fund for Chongqing Key Laboratory of Micro/Nano-Materials Engineering and Technology (Grant No. KFJJ1209) and the graduate research project for Chongqing University of Arts and Sciences (Grant No. M2013ME04).

REFERENCES

[1] Zaman Hu, Hun PD, Khan RA, Yoon K-B, 2012. Comparison of effect of surface-modified micro-/nano-mineral fillers filling in the polypropylene matrix, Journal of Thermoplastic Composites, 26(8):1100–1103.
[2] Grein C. Toughness of neat, 2005. Rubber modified and filled β-nucleated polypropylene: from fundamentals to applications, Intrinsic Molecular Mobility and Toughness of Polymers II: Springer, 188:43–104.
[3] Shelesh-Nezhad K, Orang H, Motallebi M, 2013. Crystallization, shrinkage and mechanical characteristics of polypropylene/CaCO$_3$ nanocomposites, Journal of Thermoplastic Compositesvol. 26(4):544–554.
[4] Xu L, Xu K, Chen D, Zheng Q, Liu F, Chen M, 2009. Thermal behavior of isotactic polypropylene in different content of β-nucleating agent, Journal of thermal analysis and calorimetry, 96(3):733–740.
[5] Lin Y, Chen H, Chan C-M, Wu J, 2011. Effects of coating amount and particle concentration on the impact toughness of polypropylene/CaCO$_3$ nanocomposites, European Polymer Journal, 47(3):294–304.
[6] Dmitry V. Volodkin, NataliaI. Larionova and Gleb B. Sukhorukov, 2004. Protein Encapsulation via Porous CaCO$_3$ Microparticles Templatin, Biomacromolecules, 5(5):1962–1972.
[7] H. Cheraghi; F.A. Ghasemi; G. Payganeh, 2013. Morphology and Mechanical Properties of PP/LLDPE Blends and Ternary PP/LLDPE/Nano-CaCO$_3$ Composites, Strength of Materials, 45(6):730–738.
[8] Li L, Dou Q, 2011. Effect of malonic acid treatment on crystal structure, melting behavior, morphology, and mechanical properties of isotactic polypropylene/nano-CaCO$_3$ composites, Journal of Macromolecular Science, Part B, 50(5):831–845.
[9] Qiang Fu and Guiheng Wang, 1993. Effect of morphology on brittle-ductile transition of HPDE/CaCO$_3$ blends, Journal of Applied Polymer Science, 49(11):1985–1988.

Material Science and Engineering – Chen (Ed.)
© 2016 Taylor & Francis Group, London, ISBN 978-1-138-02936-1

Dynamic Mechanical Analysis of PP composites modified by $CaCO_3$ with different morphologies

S. Yang & J. Luo
College of Electrical Engineering, Chongqing University, Chongqing, China

H.B. Chen
Bishan Power Supply Branch, Chongqing City Power Company, Chongqing, China

ABSTRACT: In this study, different shapes (rhombohedral and spherical) of calcium were used to modify isotactic Polypropylene (PP), and the properties of composites were investigated by XRD and DMA. The results showed that the morphology of calcium carbonate has important effects on the dynamic mechanical properties of PP. With the increase in $CaCO_3$ content, the storage modulus and dissipation factor (tanδ) first increased and then gradually decreased.

Keywords: $CaCO_3$ nanoparticles; Dynamic Mechanical Analysis

1 INTRODUCTION

Isotactic Polypropylene (PP) is a kind of commercial polymers with excellent all-round properties, which has been widely used in many areas because of its good processing performance and low price. However, polypropylene composites tend to lack sufficient toughness, low-temperature impact strength and relatively poor impact resistance [1, 2], so its extensive application is still limited. Therefore, the toughening research of PP is widely concerned. Especially, the modification of inorganic particles has been widely investigated because of its good stiffness and economic benefit. In recent years, with the development of ultra-fine $CaCO_3$ technology, nano-$CaCO_3$ has been meticulously researched [3, 4], but the morphology of the calcium carbonate effect on the properties of PP is ignored.

DMA is an important method to detect viscoelastic polymer materials and the response of a given material to an oscillatory deformation as a function of the temperature. In addition, it also allows the prediction of composite mechanical behavior under real-life conditions. In this study, the dynamic mechanical properties and the crystallization properties of PP/$CaCO_3$ composites were studied by DMA and XRD.

2 EXPERIMENTAL SECTION

2.1 Materials

Isotactic polypropylene (T30 s) was supplied by China National Petroleum Corporation Lanzhou Petrochemical Company, whose melt flow index was 2.6 g/10 min. Rhombohedral Calcium Carbonate (RCC) was provided by Nano Materials Technology Co., Ltd (Ruicheng Shanxi China). Spherical Calcium Carbonate (SCC) was prepared in our laboratory according to the procedure reported by Volodkin et al [5]. All other chemicals and solvents used were of analytical grade.

2.2 Preparation of PP/CaCO3 composites

The mixture of PP and $CaCO_3$ at different weight ratios was fed into the Torque Rheometer (RM-200 A; Harpo Harbin Electric Technology), with three temperature zones of 185°C, 190°C and 185°C, respectively, and a torque speed of 60 rpm. Then, the products were compressed into the plate at 190°C by a vulcanizing machine (XLB-400 * 400 * 2/0.25 MN, Qing dao xincheng yiming Rubber Machinery Co, Ltd) under the pressure of 5 MPa. Finally, all specimens were stored at room temperature before the analysis.

2.3 Characterization

Wide-angle X-Ray Diffraction (WXRD) analyses were performed on an XRD-6000 diffractometer (Shimadzu, Japan) with an X-ray generator of 3 kW, graphite monochromatic, and Cu Kα radiation (wavelength = 1.5406 Å), and operated at 40 kV and 20 mA. The samples were scanned at room temperature from 10° to 50° at a scanning rate of 2°/min.

The Dynamic Mechanical Analysis (DMA) was conducted on a DMA Q800 (TA Instruments,

USA) in the tension mode using rectangular samples. The samples were subjected to a cyclic tensile strain with an amplitude of 0.2% at a frequency of 1 Hz and a static force of 0.1 N. The temperature range was from 0°C to160°C, with a heating rate of 3°C /min.

3 RESULTS AND DISCUSSION

3.1 *The XRD analysis*

Figure 1 shows the XRD patterns of pure PP and PP/CaCO$_3$ composites. For the spectrum of pure PP, the most intense reflections presented at 2θ of 14.0°, 16.8°, 18.6°, 21.2° and 21.8° corresponded to (110), (040), (130), (111) and (131) lattice planes of the α-monoclinic crystal in pure PP, and the weaker characteristic peak presented at 2θ = 16.0° was ascribed to the (300) plane of β-PP. During the addition of CaCO$_3$ particles, it can be seen that all PP/CaCO$_3$ composites had the obvious characteristic diffraction peaks at 2θ = 16.0°, besides those reflections belonging to the α crystal in pure PP. This phenomenon indicated that CaCO$_3$ particles act as the nucleating agent, and the addition of CaCO$_3$ particles does cause the crystal form trans-

formation of the PP matrix in nanocomposites. With the increase in CaCO$_3$ particles, the characteristic diffraction peak of β-PP decreased gradually. It was interesting that when the content of RCC particles was 12%, the characteristic reflection of the β crystal almost could not be observed in composites. These results meant that the morphologies of CaCO$_3$ particles would have an important effect on the formation of the β crystal in the PP matrix.

3.2 *Dynamic mechanical analysis*

3.2.1 *Storage modulus*

Figure 2 shows the effect of different CaCO$_3$ contents on the storage modulus of PP/CaCO$_3$ composites at a frequency of 1 Hz. The corresponding parameters are given in Table 1. From Figure 2, it was found that with the addition of CaCO$_3$ particles, the storage modulus of PP composites was higher than that of pure PP in the whole range from 0 to 160°C. At the same time, from the data of the selected samples (see Table 1), it was found that, first, the storage modulus of PP/CaCO$_3$ composites increased slightly, and then decreased as the content of CaCO$_3$ particles increased further. When the content of calcium carbonate was 4%, both PP/RCC and PP/SCC composites had the

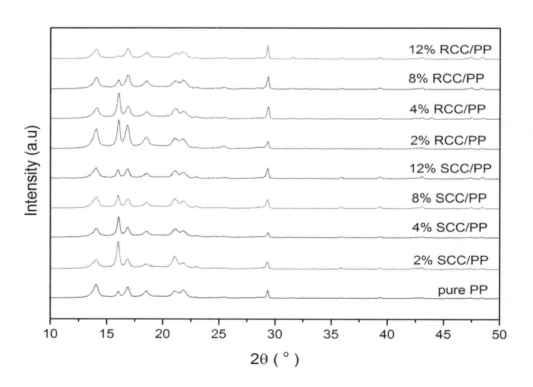

Figure 1. The XRD analysis of PP/CaCO$_3$ composites with different CaCO$_3$ contents.

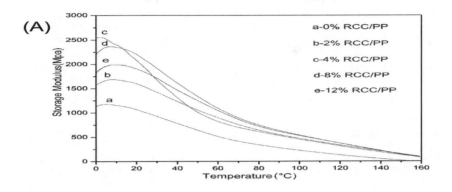

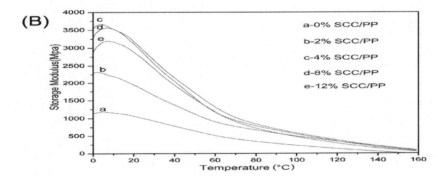

Figure 2. Storage modulus as a function of temperature of PP/CaCO$_3$ composites: (A) PP/RCC composites, (B) PP/SCC composites.

Table 1. Storage modulus and glass transition temperature of the PP/SCC composite.

Sample	Storage modulus (GPa)				Glass transition temperature
	10 (°C)	50 (°C)	90 (°C)	130 (°C)	T_g^α (°C)
Pure PP	1.15	0.64	0.28	0.08	2.17
2%PP/RCC	1.68	1.05	0.55	0.27	3.93
4%PP/RCC	2.40	1.06	0.53	0.25	7.45
8%PP/RCC	2.35	1.34	0.64	0.29	8.80
12%PP/RCC	1.99	1.25	0.62	0.30	8.90
2%PP/SCC	2.19	1.12	0.58	0.27	2.98
4%PP/SCC	3.53	1.56	0.66	0.26	3.39
8%PP/SCC	3.50	1.71	0.70	0.29	6.78
12%PP/SCC	3.17	1.54	0.59	0.21	6.91

Tgα was obtained from the value of the dissipation factor (tanδ) curves.

highest storage modulus at about 10°C, respectively. It was observed that the increment of the storage modulus of PP/SCC composites was more obvious than the counterpart of PP/RCC composites at the same CaCO$_3$ content. These results indicated that both RCC and SCC can significantly improve the rigidity of the composites. It was well known that when the polymer is in the glassy phase, its fluidity is poor because of the hindrance of the chain-segment motion in the polymer matrix. Moreover, the addition of inorganic particles to the polymer matrix can also reduce the liquidity of polymers and lead to the higher modulus for polymers [6, 7]. So, in the low-temperature stage, the storage modulus of PP composites increased. However, the superfluous $CaCO_3$ particles could easily induce the agglomeration in the PP matrix, so that the storage modulus of composites decreased when the content of $CaCO_3$ particles was above 4%. On the contrary, at a high temperature, the storage modulus of composites was dominated by the intrinsic matrix modulus of the PP matrix [4]. So, it can be clearly seen that the modulus of $PP/CaCO_3$ composites was not very obvious at a high temperature. Compared with RCC particles, the storage modulus of PP/SCC composites was higher than that of PP/RCC composites because of the larger surface area of SCC particles in the PP matrix.

3.2.2 *Dissipation factor (tanδ)*

Figure 3 shows the dissipation factor (tanδ) curves for $PP/CaCO_3$ composites. The glass transition temperatures (T_g) corresponding to $PP/CaCO_3$ composites with different $CaCO_3$ contents are listed in Table 1. From Figure 3, it can be seen that the tanδ of $PP/CaCO_3$ composites almost progressively increased with the increment of calcium carbonate particles. When the amount of $CaCO_3$ was beyond 4%, the tanδ value decreased and then appeared to plateau. From the data (see Table 2), we found that the glass transition temperature (T_g) gradually increased with the addition of the rigid $CaCO_3$ particles for all specimens. These results can be explained by the fact that because the dissipation factor indicates the ratio of loss modulus to the polymer storage modulus, the value of tanδ shows the movement of PP molecular chains. That is, the higher the tanδ was, the greater the friction between the filler and PP matrix was. In general, polypropylene exhibits three molecular relaxations corresponding to the different crystalline forms [2]. The tanδ peak belonging to the β-relaxation was clearly visible from Figure 3. This indicated that the addition of calcium carbonate could not only be helpful to the crystal transformation in the PP matrix from form α to β form, but also improve the toughness of PP. This is consistent with the previous analysis of loss modulus.

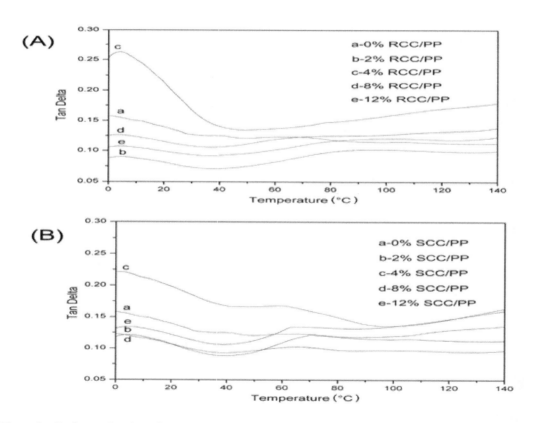

Figure 3. Tanδ as a function of temperature for $PP/CaCO_3$ composites: (A) PP/RCC composites, (B) PP/SCC composites.

4 CONCLUSION

In this study, calcium carbonate nanoparticles with different shapes (rhombohedral and spherical) were used to modify isotactic PP. The results showed that the morphology of calcium carbonate played an important role in the toughening effect on the PP matrix. With the increase in calcium carbonate content, the storage modulus and dissipation factor (tanδ) first increased and then gradually decreased, when the addition of 4% calcium carbonate resulted in a better comprehensive performance.

REFERENCES

[1] Zaman H.U., Hun P.D., Khan R.A., and Yoon K.-B. "Comparison of effect of surface-modified micro-/nano-mineral fillers filling in the polypropylene matrix". Journal of Thermoplastic Composites, vol. 26, no. 8, pp. 1100–1103, January, 2012.

[2] Grein C. "Toughness of neat, rubber modified and filled β-nucleated polypropylene: from fundamentals to applications," Intrinsic Molecular Mobility and Toughness of Polymers II: Springer, vol. 188, pp. 43–104, September, 2005.

[3] Lin Y., Chen H., Chan C.-M., and Wu J. "Effects of coating amount and particle concentration on the impact toughness of polypropylene/CaCO₃ nanocomposites", European Polymer Journal, vol. 47, no. 3, pp. 294–304, 2011.

[4] Dmitry V., Natalia I., and Gleb B., "Protein Encapsulation via Porous CaCO₃ Microparticles Templating", Biomacromolecules, vol. 5, no. 5, pp. 1962–1972, August, 2004.

[5] H. Cheraghi, F.A. Ghasemi, and G. Payganeh. "Morphology and Mechanical Properties of PP/LLDPE Blends and Ternary PP/LLDPE/Nano-CaCO₃ Composites", Strength of Materials vol. 45, no. 6, pp. 730–738, December, 2013.

[6] S. Karamipour, H. Ebadi-Dehaghani, D. Ashouri, and S. Mousavian, "Effect of nano-CaCO₃ on rheological and dynamic mechanical properties of polypropylene: Experiments and models", Polymer Testing, vol. 30, no. 1, pp. 110–117, February, 2011.

[7] Selvin Thomas P., Thomas S., and Bandyopadhyay S., et al. "Polystyrene/calcium phosphate nanocomposites: Dynamic mechanical and differential scanning calorimetric studies", Composites Science and Technology, vol. 68, no. 15–16, pp. 3220–3229, December, 2008.

Material Science and Engineering – Chen (Ed.)
© 2016 Taylor & Francis Group, London, ISBN 978-1-138-02936-1

Improving the toughness properties of CFRP composites by Aramid Nonwoven Fabric

W.M. Zhao & X.S. Yi

National Laboratory of Advanced Composites Materials, Beijing Institute of Aeronautical Materials, Beijing, China

ABSTRACT: Aramid nonwoven fabric was interlaminated into a Carbon Fiber-Reinforced Polymer (CFRP) composite panel by using the resin transfer molding process because of its excellent mechanical and thermal performance. The *"ex situ"* toughened aramid-RTM composites were demonstrated to have mechanical properties. The interlaminate fracture toughness showed a significant increase in G_{IC} of 549.9 J/m^2 to 1071 J/m^2, and an increase in G_{IIC} up to 58.1%, respectively. Meanwhile, the CAI value increased by 17%. The toughening mechanism was found that the aramid nonwoven fabrics still kept the original structure in the interlayer, and formed a macroscopic bicontinuous structure with the resin matrix, resulting in a remarkable toughening effect.

Keywords: RTM; aramid nonwoven fabric (ANF); interlaminar toughened; compression strength after impact (CAI)

1 INTRODUCTION

Carbon Fiber-Reinforced Polymer (CFRP) composites have been well widely used in aerospace, automotive, sports and civil engineering because of their high strength-to-weight ratio, rigidity, and large/complex manufacturing advantages. Considering the laminate feature of CFRP composites, efforts on how to improve the interlaminar toughness have been made over the past years, such as rubber, thermoplastics or nanoparticles, or phase changing toughened polymer matrices [1]. However, this may bring difficulty for certain manufacturing processes, such as Resin Transfer Molding (RTM), because the viscosity and rheology will be changed. In contrast to these *"in situ"* toughening philosophies, the *"ex situ"* concept was proposed by LAC/BIAM [2]. Studies [3–8] showed that the *"ex situ"* toughening technique could effectively resolve the *"in situ"* dilemma with excellent mechanical performance. Aramid is a promising material with outstanding mechanical and thermal properties. Aramid Nonwoven Fabric (ANF) has been mass produced in the industry to make filter or flame-resistant applications.

In this paper, aramid nonwoven fabric-toughened CFRP composites were prepared by the RTM process. The key mechanical properties, such as G_{IC}, G_{IIC} and CAI values, together with the microstructure were compared with the control material. The toughening mechanism is also discussed.

2 EXPERIMENTAL SECTION

2.1 *Materials and CFRP composite preparation*

The National Key Laboratory of Advanced Composites (LAC)/BIAM produced aerospace-grade epoxy (EP-3266), which is used as the resin system. The toughening layer of aramid nonwoven fabrics has an area density of 27.5 g/m^2. The TGA showed that the thermal stable temperature can be as high as 440°C, and the fiber diameter is 15 µm, as observed from the SEM microstructure shown in Figure 1.

The carbon fiber was laminated in a closed RTM mold, and the toughening layers of ANF were inserted between every two layer of carbon fibers. Then, the 3266 resin was injected into the mold under a pressure of 0.1 MPa at 40°C–45°C until the preform was impregnated. The curing procedure of the composites is shown in Figure 2. The composite laminate was cooled down in the oven to 40°C. To evaluate whether there is a defect inside the cured composite, C-scan was utilized and no voids or other defects were found inside. Thereafter, the composite laminate was cut to represent G_{IC}, G_{IIC}, and CAI test specimens.

2.2 *The G_{IC} test set-up*

The G_{IC} test was carried out according to the HB 7402–1996 standard. The dimension of the test specimens was 180 mm (0°) × 25 mm (90°). The test machine used was INSTRON 8803, and the

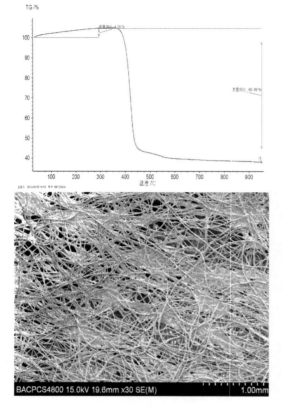

Figure 1. TGA curve and the SEM image of aramid nonwoven fabric (ANF).

set-up mode is shown in Figure 3. The interlaminar toughness G_{IC} can be calculated by the following equation:

$$G_{IC} = \frac{mp\delta_e}{2aw} \quad (1)$$

where m is the factor; p is the critical load to propagate the crack (N); δ_e is the displacement (mm); a is the crack length; and w is the sample width (mm).

2.3 G_{IIC} test set-up

The test was conducted according to the HB 7402–1996 standard, and the dimension of test the specimens was 140 mm (0°) × 25 mm (90°). The test machine used was INSTRON 8803. The test set-up is shown in Figure 4. The interlaminar shear toughness can be calculated by the following equation:

$$G_{IIC} = \frac{9P\delta a^2}{2W(2L^3 + 3a^3)} \quad (2)$$

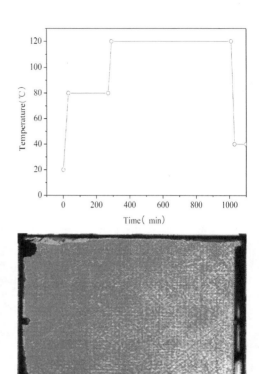

Figure 2. The RTM composites' curing procedure, 240 mm × 240 mm, and the C-scan result.

where G_{IIC} is the shear toughness (kJ/m²); P is the critical load (N); δ is the displacement (mm); a is the crack length (mm); w is the specimen width (mm); 2L is the length between two points (mm).

2.4 Compression strength after impact (CAI) test

The impact damage resistance of composites was characterized by the CAI test according to the ASTM D7137 /D7137–05. The size of the test specimens was 150 mm × 100 mm × 5.2 mm

Figure 3. The G_{IC} test set-up in INSTRON 8803.

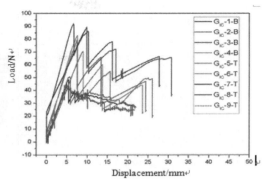

Sample	G_{IC} / J/m^2	Peak load/N
Control panel	549.89	56.93
ANF toughened	1071.26	98.08

Figure 5. G_{IC} curve and the average values compared between the blank and toughened composites.

Figure 4. G_{IIC} test set-up in INSTRON 8803.

rectangular laminates with quasi-isotropic plies of $[45°/0°/-45°/90°]_{nS}$. In addition, the CAI test was under the low-velocity impact energy of 6.67 J/mm. After the impact, the damage area was evaluated by the ultrasonic C-scan. Then, the residual strength of specimens (CAI values) was obtained after compression. Each CAI date reported here was an average of four effective tests.

3 RESULTS AND DISCUSSIONS

3.1 G_{IC} test results

The G_{IC} test results are shown in Figure 5. The average G_{IC} value of the ANF-toughened composite was 1071.26 J/m^2 (G_{IC}-5~9-T), and the blank panel (control) was 549.87 J/m^2 (G_{IC}-1~4-B), increasing by 94.8%. Additionally, the critical crack load increased from 56.93 N to 98.08 N. The SEM image of the cracked surface, shown in Figure 6, indicated that the aramid fibers still kept the original structure in the interlayer, and the resin infiltrated into the nonwoven structure. Thereafter, the aramid fibers were finally broken at a high force.

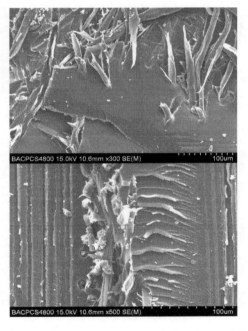

Figure 6. SEM images of the cracked surface and the toughened layer between carbon fibers.

3.2 The G_{IIC} test results

The G_{IIC} results are shown in Figure 7. The maximum loading of the ANF-toughened composite was 1408 N, while the control panel was 1082 N. The G_{IIC} increased by 30%. When taking the bending displacement and crack length into account, the G_{IIC} increased by 58.1%.

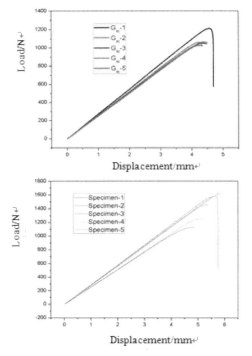

Sample	Displacement /mm	Peak Load /N	G_{IIC} value /J/m^2
Control panel	4.466	1082	1268.0
ANF toughened	5.418	1408	2005.3

Figure 7. Displacement-load curves of the blank sample and the toughened sample, together with the peak load and displacement.

Table 1. Impact damage area, compression strength after impact (CAI) of composite laminates toughened by thermoplastic nonwoven fabric.

System	C-scan after impact	Pmax/kN	CAI/Mpa
Control panel		79.3	206
ANF toughened		106.3	241

3.3 Compression strength after impact properties

CAI is an important criterion for an anisotropic carbon fiber-reinforced plastic, and widely used to evaluate the toughness of composites. In this study, the damage areas after the impact and CAI values of the toughened and blank composites are

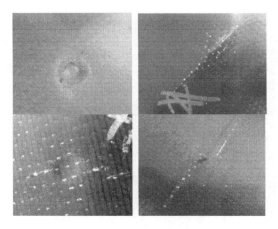

Figure 8. Photograph of the CAI impact test for the control panel (above 2 photos) and the ANP-toughened panel (below 2 photos).

summarized in Table 1. The peak load of crash was 79.3 kN with the CAI of 206 MPa for the blank composite, and the peak load of crash was 106.3 kN with a CAI value of 241 MPa for the toughened composite. The improvement of the peak load was achieved by 34% with the CAI value of 17%. Besides, the damage area dropped from 962 mm^2 to 490 mm^2. The damage area on the other side of the panel also demonstrated the limited damaged area.

The cross-section morphology of the toughened composite is shown in Figure 6. The aramid nonwoven fabric still maintained the original structure in the interlayer, and the matrix was uniformly distributed. However, a macroscopic bicontinuous structure was formed by a reaction-induced mechanism between the ANF and thermoset matrix resin in the interlaminar toughness of composites. As a result, at a low-speed impact, this macroscopic bicontinuous structure could absorb the impact energy by the elastic deformation of aramid fiber. Consequently, the interlaminar crack could be effectively prevented and the damage area diminished, and hence the compression strength after impact can be remarkably increased.

4 CONCLUSIONS

The interlaminar toughness of composites has been increased remarkably after interlaminar toughened by ANF. G_{IC} was increased by 94.8%, G_{IIC} increased by 58.1% and CAI increased by 17%.

The ANF still maintained the original structure in the interlayer, and the matrix distributed through it with a good interface. This kind of structure can be regarded as a macroscopic bicontinuous structure consisting of ANF and resin as two different continuous phases. This kind of structure demonstrated an obvious toughness effect.

REFERENCES

[1] Yi, X.S. 2006. Research and Development of Advanced Composite Materials Technology. *National Defense Press.*

[2] Yi, X.S. & An, X. A method to significantly improve the composite toughness. *Chinese Patent No.* ZL011009810.

[3] Thouless, M. D. & Du, J. 2000. Mechanics of Toughening Brittle Polymers [J]. *ACS Symposium Series* 70–85.

[4] Long, W. & Xu Y.H. 2004. Preliminary study on resin transfer molding of highly-toughened graphite laminates by ex-situ method. *Journal of Materials Science* 39:2263–2266.

[5] Cheng, Q. F. & Fang, Z. P. 2008. "Ex Situ" Concept for Toughening the RTMable BMI Matrix Composites, Part I: Improving the Interlaminar Fracture Toughness. *Journal of Applied Polymer Science* 109:625–1634.

[6] Cheng, Q. F. & Fang, Z.P. 2008. Ex-situ Concept for Toughening the RTMable BMI Matrix Composites. II. Improving the Compression After Impact [J]. *Journal of Applied Polymer Science*, 108:2211–2217.

[7] Cheng Q. F. & Xu Y.H. 2009. Morphological and Spatial Effects on Toughness and Impact Damage Resistance of PAEK-toughened BMI and Graphite Fiber Composite Laminates [J]. *Chinese Journal of Aeronautics*, 22:87–96.

[8] Cheng, Q. & Yi, X.S. 2006. Improvement of the Impact Damage Resistance of BMI/Graphite Laminates by the Ex-situ Method. *High Performance Polymers* 18:907–917.

[9] Zhang, P. & Yi X.S. 2011. Improving the Compression Strength after Impact of RTM Composites by Interlaminar Thermoplastic Nonwoven Fabric. *American Journal of Engineering and Technology Research* 11(12):40–43.

Material Science and Engineering – Chen (Ed.)
© 2016 Taylor & Francis Group, London, ISBN 978-1-138-02936-1

Anisotropic thermal expansion behaviors of $Al_{18}B_4O_{33}$ whisker-reinforced aluminum matrix composites

P.T. Zhao
China Electronics Technology Group Corporation No. 38 Research Institute, Hefei, P.R. China

C.Q. Liu
School of Materials Science and Engineering, Dalian Jiaotong University, Dalian, P.R. China

S.C. Xu
Key Laboratory of Functional Materials Physics and Chemistry of the Ministry of Education, Jilin Normal University, Siping, P.R. China

Z.M. Du, L.D. Wang & W.D. Fei
School of Materials Science and Engineering, Harbin Institute of Technology, Harbin, P.R. China

ABSTRACT: The thermal expansion behaviors of $Al_{18}B_4O_{33}$ whisker-reinforced aluminum matrix composites prepared by squeeze casting were studied. The results show that the composites exhibit a significant anisotropy in the Coefficient of Thermal Expansion (CTE) attributed to the preferred orientation of whiskers. A whisker distribution function model was established based on the thermoelasticity theory. The calculation results according to the model are highly consistent with the test results within a low temperature range.

Keywords: Composite materials; Thermal expansion; X-ray diffraction topography; Thermoelasticity

1 INTRODUCTION

Whisker—or short fiber-reinforced Metal Matrix Composites (MMCs) have been investigated extensively for their excellent properties, such as high strength, Young's modulus, and low CTE [1–3]. Low CTE is very important for the applications that require a high-dimensional stability [4]. A large number of experimental and theoretical investigations have been carried out on the CTE prediction and design of the MMCs [4–8]. Some models have been established and widely used for the prediction and analysis of the CTEs of the MMCs. Turner [9] and Kerner models [10] are effective for the MMCs reinforced by particles, and the Schapery model [11] for the continuously reinforced MMCs. In addition, the models based on Eshelby's equivalent inclusion theory [12] and the mean stress and strain field approaches [13, 14] have been used for the CTE analysis of MMCs.

Previous studies on the CTEs of MMCs have mainly focused on two types of Discontinuously Reinforced Aluminum Matrix Composites (DRAMCs). One is the composite with a unidirectional aliment whisker or short fiber, which can be obtained by extrusion. The other is the composite with a random distribution of whisker or short fiber, which can be obtained by stir casting and powder metallurgy. However, whiskers or short fibers may not be accurately unidirectional or random distributions in the composites, especially the compressed or rolled composites with a low deformation degree [15, 16]. For the as-cast ABOw/Al (ABOw, aluminum borate whisker) composite, it has been found that the whiskers tend to be perpendicular to the squeeze casting direction [16]. In this case, the CTE theories based on the whisker unidirectional or random distributions in the composites are difficult to be used for the analysis of the CTEs of the composites. However, the whisker or short fiber distributions can cause the anisotropy of the CTE of composites [14, 17–22]. Rather surprisingly, there have been no intensive theoretical studies on the relationship between the CTE and the whisker distribution.

In the present study, the anisotropic CTEs of ABOw/Al composites fabricated by squeeze casting were studied on the basis of X-Ray Diffraction (XRD) characterization [23] of whisker preferential orientation and Eshelby's model.

2 EXPERIMENTAL SECTION

The ABOw/Al composite with about 25vol% ABOw was fabricated by the squeeze casting method. The original dimensions of ABOw were 0.5–1.5μm in diameter and 10–30μm in length. The 6061 aluminum (0.4–0.8% Si, 0.7% Fe, 0.8–1.2% Mg, 0.15–0.4% Cu, 0.15% Mn, 0.04–0.35% Cr, 0.25% Zn, balance Al) was used as the matrix. The temperatures of molten aluminum and die were 800°C and 560°C for the squeeze casting, respectively. The microstructure of the composite was observed on a Hitachi S-3000-type Scanning Electron Microscope (SEM). XRD analysis of the distribution of ABOw in the composite was carried out on a Philips X'Pert X-ray diffractometer with CuKα radiation. Thermal expansion experiments were performed on a Netzsch DIL 402C dilatometer with a heating rate of 5°C/min and a temperature range of 100–400°C. The thermal expansion specimens with a dimension of 5 mm × 20 mm were cut from the composite billet.

3 WHISKER ORIENTATION DISTRIBUTION AND CTE MODEL

3.1 *Whisker orientation distribution*

Figure 1 shows the SEM photographs of the whisker orientation distribution in the composite. The homogeneous distribution of the whisker is shown in Figure 1(a) and (b). It can also be found that the longitudinal directions of the whiskers are dominant in the plane perpendicular to the Squeeze Casting Direction (SCD), as shown in Figure 1(a), while the transverse sections of the whiskers are dominant in the plane parallel to SCD, as shown in Figure 1(b). The above results suggest that the whiskers exhibit an orientational alignment with

the longitudinal axes perpendicular to the SCD, which can be defined as an in-plane texture. The texture of ABOw may result from the whisker rotation under the press during whisker preform molding and composite fabrication by squeeze casting.

As a single crystal, the in-plane texture of ABOw can be quantitatively analyzed by the XRD method. For the ABOw, <001> is the growth direction and (001) is the transverse plane [24]. The intensity of the (002) diffraction peak of ABOw will be the highest if the longitudinal directions is perpendicular to the SCD, because of the perpendicular relationship between the scattering vector (the difference in wave vectors between the incidence and diffraction X-rays) and the SCD. As shown in Figure 2(a), the specimen was cut from the composite ingot for the ω-scan XRD measurement [23], where ω is the X-ray incidence angle. The measurement was carried out with the fixed $2\theta_{002}$ value of 31.5° for the (002) diffraction peak of ABOw and the ω value ranging from 2° to $2\theta_{002}$. The ω-scan XRD curve is shown in Figure 2(b). It can be seen that the ω-scan curve of the (002) diffraction peak is symmetric about the diffraction angle corresponding to the (002) plane of ABOw, suggesting a preferred alignment with the whisker longitudinal axis perpendicular to the SCD. In the present study, the Lorentzian function is used to fit the curve and calculate the whisker distribution. As a result, the distribution function of the ABO whisker in the composite can be expressed as follows:

$$f(\psi) = \cfrac{1}{\left(\psi - \cfrac{\pi}{2}\right)^2 + \cfrac{w^2}{4}} \tag{1}$$

where $f(\psi)$ is the orientation density of the whisker longitudinal axis; $\psi = \theta - \omega$ (rad) is the angle between the whisker longitudinal axis and the

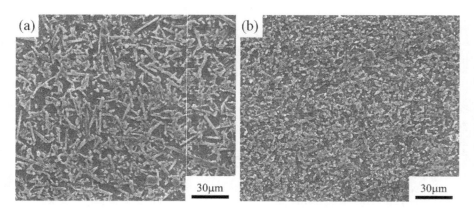

Figure 1.   SEM photographs of the composite perpendicular (a) and parallel to the SCD (b).

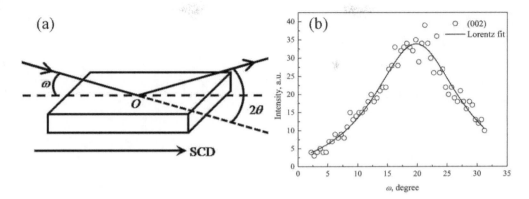

Figure 2. Schematic diagrams of the ω-scan XRD mode (a) and the ω-scan curve of (002) ABOw and Lorentz fit curve (b).

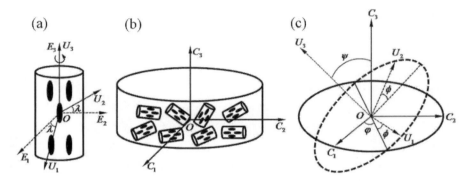

Figure 3. Schematic diagrams of (a) the composite unit containing Eshelby's ellipsoids, (b) the distribution of the composite unit in the composite, and (c) the coordinate relation between the composite and the composite unit.

Figure 5. Measured and calculated CTEs curves of the composite, and also the CTE of 6061 Al.

SCD; and $w = 0.32$ rad is the width of the ω-scan curve of the (002) plane of ABOw.

3.2 CTE model

In the present study, some simplifications are used to analyze the CTE of the composites. First, it is assumed that the whiskers are ellipsoid with an average aspect ratio, as indicated in Eshelby's theory [25].

Second, we assume that the composite is composed of many composite units, and a unit has a small amount of unidirectional whiskers. The CTE of the composite unit can be evaluated according to Eshelby's theory [25]. The composite unit is shown in Figure 3(a). Generally, the inclusion in the CTE model of Eshelby's theory is isotropic. In the present study, Eshelby's theory is extended to treat the composite unit with anisotropic inclusions because of the low anisotropic CTEs of ABOw. We assume that all whiskers in the $O\text{-}E_1E_2E_3$ coor-

dinate have the same orientation, and the CTEs (α_E) of a composite unit can be calculated using Eshelby's theory [25]:

$$\alpha_E = \alpha_m - f_w\{(C_m - C_w)[S - f_w(S - I)] - C_m\}^{-1} \times C_w(\alpha_w - \alpha_m) \quad (2)$$

where f is the whisker volume fraction; C is the stiffness tensor; S is Eshelby's tensor; I is the identity matrix; and the subscripts c, m and w refer to the composite, matrix and whisker, respectively.

Because the orientations of the whiskers are symmetric about the E_3 direction [see Figure 3(a)], the CTE of the composite unit in the $O\text{-}U_1U_2U_3$ coordinate has a rotational symmetry about the U_3 axis, so the average CTE of the composite unit can be obtained as follows:

$$\alpha_U = \bar{\alpha}_E = \frac{1}{2\pi}\int_0^{2\pi} \alpha_E U_{ij} d\lambda \quad (3)$$

323

$$\mathbf{U}_{ij} = \begin{bmatrix} \cos\lambda & \sin\lambda & 0 \\ -\sin\lambda & \cos\lambda & 0 \\ 0 & 0 & 1 \end{bmatrix} \qquad (4)$$

where λ is the angle between the E_1 axis and the U_1 axis. The $\boldsymbol{\alpha}_U$ can be obtained as follows:

$$\boldsymbol{\alpha}_U = \begin{bmatrix} \dfrac{\alpha_{E1}+\alpha_{E2}}{2} & 0 & 0 \\ 0 & \dfrac{\alpha_{E1}+\alpha_{E2}}{2} & 0 \\ 0 & 0 & \alpha_{E3} \end{bmatrix} \qquad (5)$$

Finally, the distributions of E_3 or U_3 directions of the composite unit are the same as those of the whisker longitudinal direction, as shown in Equation (1) and Figure 3(b).

The relationships between the composite units and the composite in the coordinate $O\text{-}C_1C_2C_3$ are shown in Figure 3(c), and the composite unit orientation in the coordinate $O\text{-}C_1C_2C_3$ can be expressed by the Euler transformation ($\mathbf{N}_{hk}$) as follows:

$$\mathbf{N}_{hk} = \begin{bmatrix} \cos\varphi\cos\phi - \sin\varphi\cos\psi\sin\phi & \cos\varphi\sin\phi - \sin\varphi\cos\psi\sin\phi & \sin\varphi\sin\psi \\ -\sin\varphi\cos\phi - \cos\varphi\cos\psi\sin\phi & -\sin\varphi\sin\phi - \cos\varphi\cos\psi\cos\phi & \cos\varphi\sin\psi \\ \sin\psi\sin\phi & -\sin\psi\cos\phi & \cos\psi \end{bmatrix} \qquad (6)$$

where ϕ, φ and ψ are the Euler angles, as shown in Figure 3(c) [5]. In the composite coordinate $O\text{-}C_1C_2C_3$, the CTE tensor ($\boldsymbol{\alpha}_{U\text{-}C}$) of the composite unit with the Euler angles of φ, ϕ and ψ is given in Equation (7). The composite CTEs ($\boldsymbol{\alpha}_C$) can be considered as the average value of all composite units with the orientation distribution of Equation (1), which are given in Equations (8)–(10).

The CTEs of the composite can be obtained by averaging $\boldsymbol{\alpha}_{U\text{-}C}$ as follows:

$$\boldsymbol{\alpha}_{C1} = \frac{\int_0^\infty \boldsymbol{\alpha}_{U-C} f(\psi)d\psi}{\int_0^\infty f(\psi)d\psi} = \begin{bmatrix} \alpha_{C1} & 0 & 0 \\ 0 & \alpha_{C1} & 0 \\ 0 & 0 & \alpha_{C3} \end{bmatrix} \qquad (8)$$

Thus, the CTE components of the composite can be obtained as follows:

$$\alpha_{C1} = \frac{(\alpha_{E1}+\alpha_{E2})}{4} \frac{\int_0^\infty (1+\cos^2\psi)f(\psi)d\psi}{\int_0^\infty f(\psi)d\psi}$$
$$+ \frac{\alpha_{E3}}{2} \frac{\int_0^\infty \sin^2\psi f(\psi)d\psi}{\int_0^\infty f(\psi)d\psi} \qquad (9)$$

$$\alpha_{C3} = \frac{(\alpha_{E1}+\alpha_{E2})}{2} \frac{\int_0^\infty \sin^2\psi f(\psi)d\psi}{\int_0^\infty f(\psi)d\psi}$$
$$+ \alpha_{E3} \frac{\int_0^\infty \cos^2\psi f(\psi)d\psi}{\int_0^\infty f(\psi)d\psi} \qquad (10)$$

4 RESULTS AND DISCUSSION

The measured CTEs of the composite are shown in Figure 4. The CTE of 6061 Al is also shown for calculation and comparison. It is found that the CTE curves of the composite are much lower than that of 6061 Al over the entire temperature range, suggesting a reducing effect of ABOw on the CTE

$$\boldsymbol{\alpha}_{U-C} = \boldsymbol{\alpha}_U \mathbf{N}_{hk} = \begin{bmatrix} \alpha_{U1}\left(\dfrac{1+\cos^2\psi}{2}\right) + \alpha_{U3}\dfrac{\sin^2\psi}{2} & 0 & 0 \\ 0 & \alpha_{U1}\left(\dfrac{1+\cos^2\psi}{2}\right) + \alpha_{U3}\dfrac{\sin^2\psi}{2} & 0 \\ 0 & 0 & \alpha_{U1}\sin^2\psi + \alpha_{U3}\cos^2\psi \end{bmatrix} \qquad (7)$$

Table 1. Non-zero components of Eshelby's tensor for the composite.

Component	$S_{11} = S_{22}$	$S_{12} = S_{21}$	$S_{13} = S_{23}$	$S_{31} = S_{32}$	S_{33}	$S_{44} = S_{55}$	S_{66}
Value	0.6475	−0.0478	0.0961	−0.003	0.0211	0.2475	0.3469

of the composite. Furthermore, the CTE of the composite perpendicular to the SCD is smaller than that parallel to the SCD, resulting from the preferred orientation of the whiskers. The difference between the CTEs parallel and perpendicular to the SCD increases with the increasing temperature above 220°C.

The CTE model given in Equations (9) and (10) is applied to calculate the CTEs of the composite. Before the calculation, first, αE should be calculated. The physical properties of the ABOw and matrix used herein are obtained from References [5, 26], which are considered as temperature independent. The assumption of temperature independence of the physical properties of the matrix can only be used within the low temperature range. It is obvious that the stiffness coefficient will decrease with increasing temperature, which can cause additional deviations between the calculated and measured CTE values at the high temperature. The measured CTE of 6061 Al, as shown in Figure 5, is used for the calculation of αE. The non-zero components of Eshelby's tensor (S) for prolate spheroid are calculated using the method of Brown et al. [14], as listed in Table 1. The non-zero components of the stiffness tensor for the ABO whisker are calculated using the data given in Reference [26]. The non-zero components of the stiffness tensor for 6061 Al used herein are obtained from Reference [5].

The calculated CTEs of the composite are shown in Figure 4. It can be found that the calculation results agree well with the measurement results at a low temperature range, especially, the difference between the calculation and the experimental results is less than 5% for the CTE perpendicular to the SCD in the temperature range of 100~300°C. However, the difference between the calculation and experimental results is large at a higher temperature.

It is well known that the thermal mismatch stress (TMS), resulting from the CTE difference between the reinforcement and the matrix, significantly affects the thermal expansion of the MMCs [7, 25, 27]. The TMS will be relaxed through the elastic and plastic deformation of the matrix on heating. The plastic relaxation of TMS is dominant owing to the lower yield strength of the aluminum matrix at the elevated temperature. Generally, the TMS in the matrix is tensile at the low temperature and compressive at the elevated temperature [7, 27]. So, the TMS in the matrix will transform a tensile stress into a compressive stress accompanied with the plastic relaxation of TMS owing to the lower yield strength of the matrix, when the temperature is raised above a certain temperature. As a result, the plastic relaxation of compressive TMS causes the decrease in CTE perpendicular to the SCD. Taking into account the Poisson effect, the tensile strain in the matrix can be induced parallel to the SCD, resulting in the increasing CTE along the direction at the higher temperature, as shown in Figure 4. In addition, the variations in stiffness coefficients at the high temperature can also cause the difference between the calculated and experimental CTEs.

Because the whiskers tend to be perpendicular to the SCD, the above results suggest that the plastic relaxation of TMS along the longitudinal direction of the whisker is easier than that along the radial direction of the whisker. When the plastic relaxation of the TMS takes place, the CTEs of the composite cannot be predicted based on the elastic model, so the measured CTEs of the composite are quite different from those calculated by the model provided in the present study.

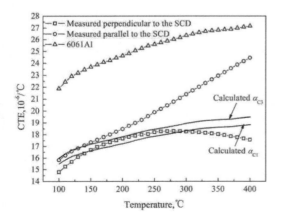

Figure 4. Measured and calculated CTEs curves of the composite, and also the CTE.

5 CONCLUSIONS

1. The whiskers in the composite fabricated by the squeeze casting method exhibit a preferred orientation with the longitudinal axis perpendicular to the squeeze casting direction. The in-

plane texture or the orientation distribution of the whisker can be quantitatively characterized by using the ω-scan XRD method.

2. The CTE of the composite is anisotropic, resulting from the preferred orientation of the whiskers in the composite.

3. A CTE model of the whisker distribution function based on Eshelby's theory is established and consistent with the measured results within a low temperature range. The deviation of the calculated CTEs from the experimental data is caused by the plastic relaxation of TMS in the matrix.

ACKNOWLEDGMENT

This work was supported by the National Basic Research Program of China (No. 2011CB612200), the National Natural Science Foundation of China (No. 50875059, No. 51301075, No. 51174064), the National High Tech Project (No. 2008 AA03 A239), and the China Postdoctoral Science Foundation (No. 20070420023).

REFERENCES

[1] V. Tvergaard. Acta Mater. 38 (1990) 185–194.

[2] D.B. Miracle. Compos. Sci. Technol. 65 (2005) 2526–2540.

[3] R.J. Arsenault, L. Wang, C.R. Feng. Acta Mater. 39 (1991) 47–57.

[4] Y.D. Huang, N. Hort, K.U. Kainer, Compos. Part A 35 (2004) 249–263.

[5] M.L. Dunn, H. Ledbetter, Z. Li, Metall. Mater. Trans. A 30A (1999) 203–212.

[6] L.S. Shi, H.B. Liu, B. Wang, Mater. Sci. Technol. 15 (1997) 5–7.

[7] M. Hu, W.D. Fei, C.K. Yao. Scripta Mater. 46 (2002) 563–567.

[8] L.D. Wang, W.D. Fei, M. Hu, L.S. Jiang, C.K. Yao, Mater. Lett. 53 (2002) 22–24.

[9] P.S. Turner, J. Res. Nat. Bu. Stan. 37 (1946) 239–250.

[10] E.H. Kerner. Proc. Phys. Soc. B 69 (1956) 808–813.

[11] R.A. Schapery. J. Compos. Mater. 2 (1968) 380–384.

[12] J.D. Eshelby. Proc. Roy. Soc. A 241 (1957) 376–396.

[13] T. Mori, K. Tanaka. Acta Mettall. 21 (1973) 571–574.

[14] L.M. Brown, D.R. Clarke. Acta Metall. 23 (1975) 821–830.

[15] W.D. Fei, W.Z. Li, C.K. Yao. J. Mater. Sci. 37 (2002) 211–215.

[16] H.Y. Yue. Ph.D. thesis, Harbin Institute of Technology, Harbin, (2009).

[17] Y. Takao, M. Taya. J. Compos. Mater. 21 (1987) 140–156.

[18] T. Ohnuki, Y. Tomota. Scripta Mater. 34 (1996) 713–720.

[19] H.E. Nassini, M. Moreno, C.G. Oliver. J. Mater. Sci. 36 (2001) 2759–2772.

[20] Y.D. Huang, N. Hort, H. Dieringa, K.U. Kainer. Compos. Sci. Technol. 65 (2005) 137–147.

[21] A. Rudajevová, S. Kúdela Jr.,S. Kúdela, P. Lukáč. Scripta Mater. 53 (2005) 1417–1420.

[22] X. Luo, Y. Yang, C. Liu, T. Xu, M. Yuan, B. Huang. Scripta Mater. 58 (2008) 401–404.

[23] W.D. Fei, C.Q. Liu, M.H. Ding, W.L. Li, L.D. Wang. Rev. Sci. Instrum. 80 (2009) 093903.

[24] L.J. Yao, H. Fukunaga. Scripta Mater. 36 (1997) 1267–1271.

[25] T.W. Clyne, P.J. Withers, An Introduction to Metal Matrix Composites, first. Cambridge University Press, Cambridge, England; 1993.

[26] K. Suganuma, G. Sasaki, T. Fujita, N. Suzuki. ICCM/8, Honolulu, HI, United States; 1991.

[27] W.D. Fei, L.D. Wang. Mater. Chem. Phys. 85 (2004) 450–457.

Material Science and Engineering – Chen (Ed.)
© 2016 Taylor & Francis Group, London, ISBN 978-1-138-02936-1

Effect of SnO_2 coating content on the microstructure of $Al_{18}B_4O_{33}$w/6061Al composite extruded at 350 °C

P.T. Zhao
China Electronics Technology Group Corporation No. 38 Research Institute, Hefei, P.R. China

C.Q. Liu
School of Materials Science and Engineering, Dalian Jiaotong University, Dalian, P.R. China

S.C. Xu
Key Laboratory of Functional Materials Physics and Chemistry of the Ministry of Education, Jilin Normal University, Siping, P.R. China

Z.M. Du, L.D. Wang & W.D. Fei
School of Materials Science and Engineering, Harbin Institute of Technology, Harbin, P.R. China

ABSTRACT: SnO_2-coated $Al_{18}B_4O_{33}$ whisker reinforced 6061Al matrix composites were fabricated using squeeze casting method, and extruded at 350 °C. The texture and microstructure of the extruded composites were studied by ω-scan X-ray diffraction, SEM and TEM techniques. The results indicate that both <111> and <100> fiber textures coexist in the matrix of all extruded composites, and <111> texture is stronger than <100> texture. The dispersion degrees of textures increase with the SnO_2 coating content.

Keywords: $Al_{18}B_4O_{33}$ whisker; aluminum matrix composite; extrusion texture

1 INTRODUCTION

$Al_{18}B_4O_{33}$ whiskers reinforced aluminum matrix (ABOw/Al) composites are highly regarded as a potential material for broad applications in aerospace and automotive industry [1, 2], due to its remarkable properties, such as higher Young's modulus and strength [3], better wear resistance [1–4], and excellent mechanical property at elevated temperatures [5, 6]. It not only shows much better performance than traditional monolithic metallic materials, but also far exceeds the SiCw/Al composites under economic consideration [7]. However, the development of ABOw/Al composites is hindered by their poor deformation ability. Such drawback usually requires extrusion of the composite at very high temperatures, which makes production complex and expensive.

Recently, the ABOw/Al composites were extruded at relatively low temperatures, by introducing the interphase with low melting point into the interface of the composites in our group [8, 9]. The effect of this interphase on the low temperature extrusion has been illustrated in our previous study [10]. It is well known that the content of this interphase is one of the key parameters for the material optimum design [8, 9]. However, the effect of this interphase content on the low temperature extrusion deformation of ABOw/Al composites is still not illustrated intuitively hitherto. To solve this problem, we propose a throughout and precise way to elucidate this mechanism by investigating the formation and evolution of textures which are closely related to the deformation of extruded materials [11]. Similar investigations are reported in extensive literature [12–14], but, to our best of knowledge, no research effort is made to explore the texture evolution of ABOw/Al composites containing various content of low melting point interphase.

In the present work, 6061Al matrix composites reinforced with SnO_2-coated $Al_{18}B_4O_{33}$ whisker (ABOw/SnO/6061Al) were fabricated and extruded at 350 °C. Different quantities of low melting point interphase emerged within these materials, and thus its impact on the microstructure and texture formation was carefully explored. Moreover, the effect of low melting point interphase content on the tensile properties of the composite was discussed.

2 EXPERIMENTS

The reinforcement used for the 6061 aluminum alloy matrix was ABOw with a diameter of 0.5–1.5 μm and a length of 10–30 μm. With a reported chemical method, SnO_2 was coated on the surface of whiskers prior to their preform process at 1000 °C for 1 h [15]. The composites containing about 20% ABOw were fabricated by squeeze casting method at a casting temperature of 800 °C. The mass ratios of SnO_2 and composite were set as 1.1%, 0.73%, 0.55% and 0, respectively. The corresponding abbreviations are C1.1, C0.73, C0.55 and C00 respectively.

The extrusion billets machined from the as-cast composite ingots were 39 mm in diameter and 20 mm in length. The composites were extruded at 350 °C with a extrusion ratio of about 20:1 (area ratio) and a die angle of 45°. The extrusion die and composite billets were heated to 350 °C, together with the lubricant made of graphite powders and high temperature oil. The extrusion process was carried out on a 2000 kN pressure machine with a extrusion rate of 18 mm/s.

The morphologies of SnO_2 coatings and microstructures of the composites were observed by Scanning Electron Microscope (SEM) on an S-3000 SEM and Transmission Electronic Microscope (TEM) on a Tecnai G^2 F30 TEM. X-Ray Diffraction (XRD) analysis was carried out on a Philips X'Pert x-ray diffractometer with Cu K radiation. The textures of extruded composites were examined by a ω-scan XRD method [16]. All specimens for SEM and ω-scan XRD characterization were grinded and polished, and then etched in 5% HF solution.

3 RESULTS AND DISCUSSIONS

The discrepancy in XRD patterns of as-cast C00 and C1.1 composites indicates the presence of Sn in the C1.1 composite (Fig. 1). Figure 2(a) presents the TEM image of the as-cast C1.1 composite. Some fine particles are found to disperse randomly at and near the whisker/aluminum interface. The particle A in Figure 2(a) is mainly composed of O, Mg, Al and Sn elements, as characterized by Energy Dispersive Spectrometer (EDS) and element quantitative analysis (see Fig. 2(b)). Based on these results, it could be concluded that Sn particles in the C1.1 composite exist at the whisker/aluminum interface, which is understood that Sn rarely forms a solid solution with aluminum according to Al-Sn phase diagram [17]. The same result has also been obtained in the previous study [10].

The effect of SnO_2 coating on the whiskers distribution of extruded composites could be

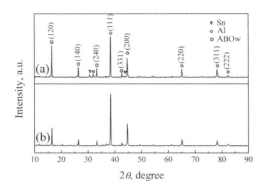

Figure 1. XRD patterns of as-cast composites: (a) C1.1 and (b) C00.

apparently noticed from Figure 3. Along the longitudinal direction, the average length of whiskers in C1.1 composite is larger than that in C00 composite (Fig. 3(a, b)), indicating that the Sn effectively protects the integrality of whiskers during extrusion. At an extrusion temperature of 350 °C higher than the melting point of Sn (231.9 °C), it becomes a liquid at the interface of composites, and thus relaxes the stress concentration at the interface and facilitates the interface sliding and whisker rotation. Along the transverse direction, the distribution of ABOw is found to be more homogeneous in C1.1 composite than in C00 composite, as shown in Figure 3(c, d). Borrego et al. [13] claimed the cluster phenomenon of SiC whiskers in the SiCw/6061Al composite extruded at 359 °C. In our study, the homogeneity of whisker distribution in extruded composites can be greatly improved due to the origination of Sn particles on the whisker surface. It is also worth noting here that the whiskers tend to parallel to the extrusion direction and the whisker fractures are universal in the extruded composites, which agrees well with previous investigations [8, 9, 13]. Figure 4 shows ω-scan XRD curves of extruded composites. According to the physical meaning of ω-scan XRD curve which is illustrated in the previous study [10, 16], it can be confirmed that <111> and <100> textures of matrix, and <001> texture for ABOw were formed in the extruded composites. Besides, the peaks of ω-scan XRD curves for matrix become more dispersive as the SnO_2 coating is increased. It can be also found that the intensity of ω-scan XRD of $\{200\}_{Al}$ is lower than that of $\{111\}_{Al}$ for a given SnO_2 content, which means that the $<100>_{Al}$ texture is weaker than $<111>_{Al}$ texture. Figure 5 exhibits the variations of Gaussian Width (GW) of ω-scan XRD curves with the SnO_2 coating content. Obviously, the GWs of $\{111\}_{Al}$, $\{200\}_{Al}$ and $\{004\}_{ABOw}$ depends greatly

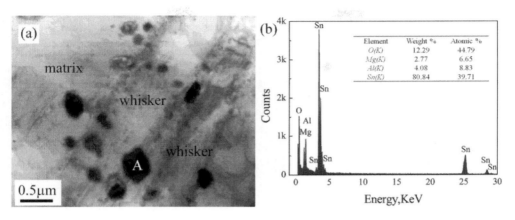

Figure 2. (a) TEM image of as-cast C1.1 composite and (b) EDS for A region in (a).

on the SnO_2 coating content. Now that GW values are increased under higher dispersion degree of the texture [10, 16], it is comprehensible that more SnO_2 coatings cause the GWs of $\{111\}_{Al}$ and $\{200\}_{Al}$ to arise and enhance the dispersion of $<111>$ and $<100>$ textures correspondingly. The other finding is the GWs of $\{200\}_{Al}$ reflection larger than those of $\{111\}_{Al}$ reflection, indicating higher dispersion degree of $<100>$ texture for all extruded composites.

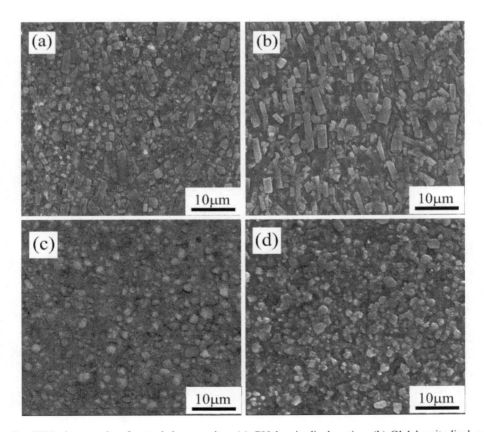

Figure 3. SEM photographs of extruded composites: (a) C00 longitudinal section, (b) C1.1 longitudinal section, (c) C00 transverse section and (d) C1.1 transverse section.

329

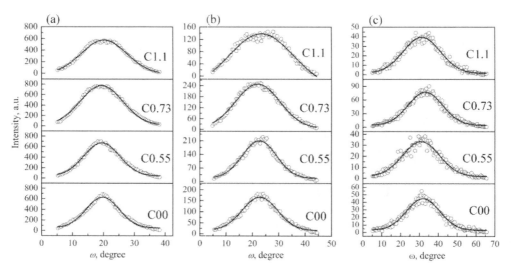

Figure 4. ω-scan XRD data (open circle) and Gaussian fit curves (black solid line) for (a) $\{111\}_{Al}$, (b) $\{200\}_{Al}$ and (c) $\{004\}_{ABOw}$ reflections.

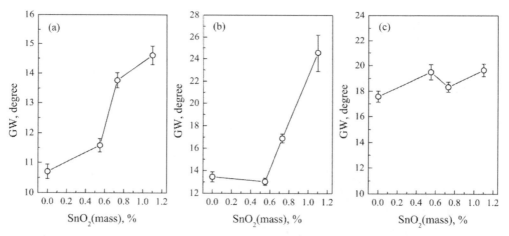

Figure 5. Variations of the GW dependency on the ω-scan XRD data for (a) $\{111\}_{Al}$, (b) $\{200\}_{Al}$ and (c) $\{004\}_{ABOw}$ the SnO_2 coating contents.

Generally, the matrix deformation and interface sliding construct the deformation of composites at elevated temperatures [14], which could also affect the textures formation and evolution. The <100> and <111> textures in the extruded face center cubic metals are determined by recrystallization and the dislocation slip, respectively [11, 14, 18]. In the present work, the interphase of Sn rarely dissolves into the matrix and thus has no influence on the stacking fault energy of matrix. At this point, the variation of the dispersion degree of texture with the SnO_2 coating content is considered as the result of interfacial sliding and interface stimulated nucleation for recrystallization. The liquid Sn effectively reduces the interface stress and strength during extrusion, favors interface sliding and weakens recrystallization and grain growth at the interface [11]. As the SnO_2 coating is increased, more Sn originates accordingly in the composites, in order to further help interface sliding for higher dispersion degree of <111> texture, and to weaken the interface recrystalization for higher dispersion degree of the <100> texture as well.

The non-linear dependence of $\{004\}_{ABOw}$ GWs on SnO_2 coating content means an involvement of more complex mechanism in ABOw rotation during hot extrusion process. As discussed above, the melted interphase Sn has lubricant effect for helping the interface sliding and ABOw rotation. Nevertheless, the whisker may fracture at initial stage of extrusion process to form shorter whiskers that rotate readily even at small amount of SnO_2 coatings. Therefore, the GWs of $\{004\}_{ABOw}$ depends on the competition between the short-size effect of whiskers and lubricant effect of liquid Sn.

4 CONCLUSIONS

1. SnO_2 coating was coated on ABO whiskers for ABOw/6061Al composites. The coating induces origination of Sn particles at the interface that can protect the integrity of ABOw whisker in extrusion.
2. For all of the composites, <111> and <100> fiber textures are developed during hot extrusion, and the former is generally sharper than the latter. However, both the textures are undermined with increasing amount of SnO_2 coating.

ACKNOWLEDGEMENT

The work was supported by the National Basic Research Program of China (No. 2011CB612200), the National Natural Science Foundation of China (No. 50875059) (No. 51301075) (No. 51174064), the National High Tech Project (No. 2008AA03A239) and the China Postdoctoral Science Foundation (20070420023).

REFERENCES

[1] S.V. Prasad, and R. Asthana, Tribol. Lett. 17 (2004) 445–453.
[2] D.B. Miracle, Materials Park, OH: ASM International, (2001) 1043–1049.
[3] K. Suganuma, T. Fujita, N. Suzuki and K. Niihara, J. Mater. Sci. Lett. 9 (1990) 635–663.
[4] J.P. Tu, Y.Z. Yang, Comp. Sci. Technol. 60 (2000) 1801–1809.
[5] S.J. Zhua, T. Iizuka, Comp. Sci. Technol. 63 (2003) 265–271.
[6] H.Y. Yue, L.D. Wang, W.D. Fei, Mater. Sci. Eng. A 486 (2008) 409–412.
[7] W.D. Fei, Y.B. Li, Mater. Sci. Eng. A 379 (2004) 27–32.
[8] Z.J. Li, W.D. Fei, H.Y. Yue, L.D. Wang, Comp. Sci. Technol. 67 (2007) 963–973.
[9] P.T. Zhao, Z.M. Du, L.D. Wang, W.D. Fei, Adv. Mater. Res. 299–300 (2011) 692–695.
[10] P.T. Zhao, L.D. Wang, Z.M. Du, S.C. Xu, P.P. Jin, W.D. Fei, Compos. Part A 43 (2012) 183–188.
[11] E.A. Calnan, Acta Metall. 2 (1954) 865–874.
[12] C.G. Jiao, Z.K. Yao, and Y.F. Han, Chin.J.Met.Sci. Technol. 8 (1992) 25–29.
[13] A. Borrego, R. Fernández, M.C. Cristina, J. Ibáñez, G. González-Doncel, Comp. Sci. Technol. 62 (2002) 731–742.
[14] W.D. Fei, W.G. Li, C.K. Yao, J. Mater. Sci. 37 (2002) 211–215.
[15] J. Hu, X.F. Wang, S.W. Tang, Comp. Sci. Technol. 68 (2008) 2297–2299.
[16] W.D. Fei, C.Q. Liu, M.H. Ding, W.L. Li, L.D. Wang, Rev. Sci. Instrum. 80, 093903 (2009) 1–6.
[17] L.F. Mondolfo, London: ButterWorths/Boston Press; (1976). 384.
[18] C.J. McHargue, L.K. Jetter, J.C. Ogle, Trans. Met. Soc. AIME. 215 (1959) 831–837.

Material Science and Engineering – Chen (Ed.)

Effect of an enclosed cage structure on the polymer chain characteristics of Ope-POSS/PU hybrid composites

R. Pan & L.L. Wang

Chemistry and Material Science College, Sichuan Normal University, Chengdu City, P.R. China

Y. Liu

Key Laboratory of Special Waste Water Treatment, Sichuan Province Higher Education System, Chengdu City, P.R. China

ABSTRACT: Based on the experimental results, a molecular simulation approach was applied in elucidating the polymer chain characteristics of Ope-POSS/PU hybrid composites with octa(propylglycidyl ether) Polyhedral Oligomeric Silsesquioxane (Ope-POSS) containing eight glycidyl groups at various concentrations. The hybrid composite models were constructed and characterized by pair correlation functions g(r) and simulated X-ray plots. The results indicate that with the Ope-POSS concentration increasing up to 15 wt% in hybrid composites, due to the humping enclosed cage structure, the number of contacts between neighboring chains decreased and distinct assembled clusters were observed, which indicated the micro-phase separation in composites.

Keywords: POSS; polyurethane; molecular dynamics; molecular mechanics

1 INTRODUCTION

In previous research, Ope-POSS/PU hybrid composites with different Ope-POSS concentrations have been characterized by FT-IR, DSC and TGA techniques [1]. The experimental results indicate that the thermal property of polyurethane can be tailored by Ope-POSS incorporation. However, due to the complex units and multiple interactions between different parts in hybrid composites, the mechanism of how the covalent incorporation of Ope-POSS into the PU backbone modifies the polymer chain characteristics and thus leads to the improvement of hybrid composite is deserved to be elucidated.

2 SIMULATION PROCEDURES

Accelrys Amorphous Cell module and COMPASS force field in Materials Studio software were adopted in all simulation processes in our research, which has been used successfully for the simulation of polymer nanocomposites containing POSS [2]. As periodic boundary conditions were imposed, an initial low density was used to construct the bulk cubic structures of random hybrid copolymers. For each sample, 10 initial configurations were optimized by the molecular dynamics technique under the NPT (constant particle numbers, pressure and temperature) conditions at 4 Gpa with a minimization

involving 30000 steps to relax and equilibrate. After this minimization procedure, the density fluctuation of each system was less than 0.05 g/cm³ under a given condition. Since these optimized configurations might not be in a local energy minimum state, an annealing procedure from 623 K to 273 K was applied on the above optimized configurations by conducting the velocity Verlet algorithm in NVT dynamics to reduce the possible potential energy. Finally, configurations with the highest energy were rejected, and 5 configurations for each sample were selected for further analysis of the composites' characteristics. The weight fractions of Ope-POSS in composites and sample codes are listed in Table 1. The molecular structures of the Ope-POSS /PU hybrid composite are shown in Figure 1 [3].

Table 1. Characteristics of Ope-POSS/PU hybrid composites.

Sample code	Ope-POSS: TDI: PPG (mole ratio)	Ope-POSS (wt%)	Initial density (g/cm³)	Final density (g/cm³)
0PU	0:0.6:0.3	0	1.32	1.28
5PU	1:12:0.6	5	1.24	1.20
10PU	1:0.6:0.3	10	1.21	1.25
15PU	3:12:0.6	15	1.19	1.17
20PU	2:0.6:0.3	20	1.17	1.21

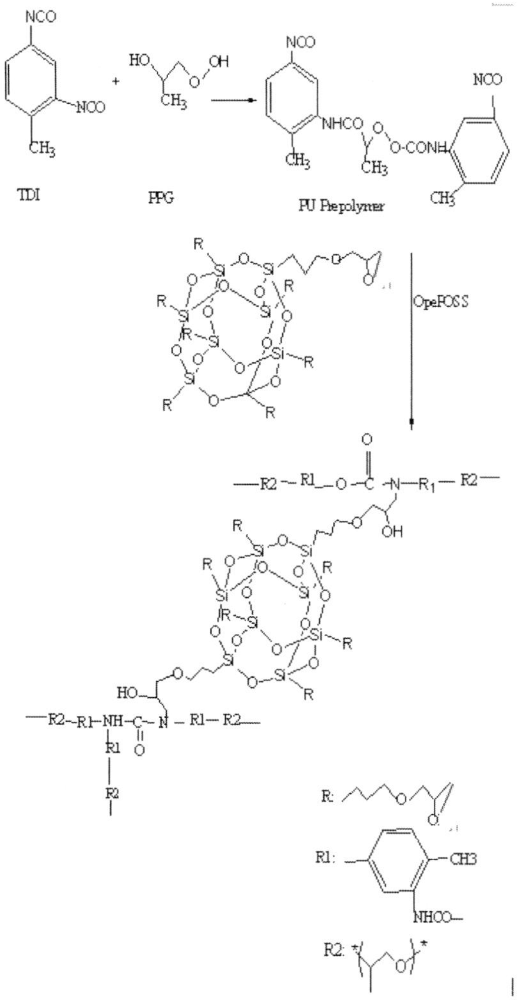

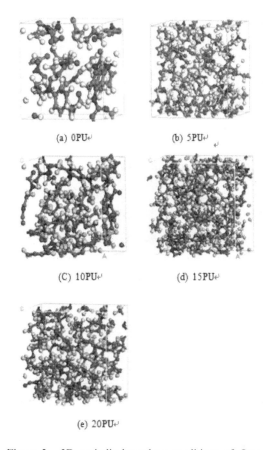

(a) 0PU

(b) 5PU

(C) 10PU

(d) 15PU

(e) 20PU

Figure 2. 3D periodic boundary conditions of Ope-POSS/PU hybrid composites: (a) 0PU; (b) 5PU; (c) 10PU; (d) 15PU; (e) 20PU.

Figure 1. Chemical structures of Ope-POSS/PU hybrid composites.

3 3D BOUNDARY CONDITION STRUCTURE CONSTRUCTION

Recently, it has been reported that the structure and energy of the POSS/PU hybrid composites were successfully simulated by using the COMPASS force field [4]. A model structure was generated through several cycles of molecular mechanics and molecular dynamics energy minimization. After the above minimization procedure, the density fluctuation of each system was less than 0.05 g/cm^3 under a given condition, indicating that the structure generated was fully relaxed and in the equilibrium state, which can be confirmed by energy optimization.

4 RADIAL DISTRIBUTION FUNCTION ANALYSIS

In statistical mechanics, the radial distribution function (pair correlation function) g(r) is a measure of the probability of finding a pair of atoms (α,β) that is separated by a radial distance $r_{\alpha\beta}$, which is expected for a complete random distribution. Thus, details of polymer chain packing can be estimated by defining the atoms in the polymer main chain as given reference particles in the inter-molecular pair correlation function, respectively [5]. Figure 3 shows the inter-molecular pair correlation function based on all atoms in the main chain. With the increase in the Ope-POSS concentration, the number of contacts between the main chains is decreased. It can be concluded that at any given distance, the number of contacts between the neighboring chains is decreased due to the presence of Ope-POSS with a humping enclosed cage structure, which makes the neighboring chains depart from each other and extend their distance.

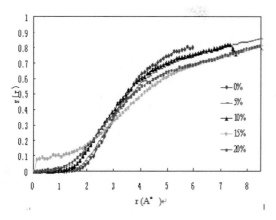

Figure 3. Pair correlation functions g(r) of 0PU, 5PU, 10PU, 15PU and 20PU hybrid composites.

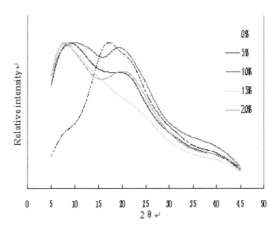

Figure 4. Simulated WAXS patterns of Ope-POSS/PU composites.

5 SIMULATED WAXS PATTERNS ANALYSIS

Simulated WAXS patterns of each sample are shown in Figure 4. In the 0PU sample, one broad peak ranges from approximately 18° to 22° corresponding to a Bragg distance of 4.4 A°, which can be attributed to the amorphous polyurethane phase part and hard-soft interactions in analogous to conventional PU systems. The same trend can be observed in all Ope-POSS-treated composites. Some residual crystallinity peaks appear in the 7.5°–10° range and the intensity of these peaks increases sharply with the increasing Ope-POSS concentration. In the 20PU sample, the halo at 7.3° (d = 12.03 A°) coupled with a strong increase in intensity can be observed. The shape and intensity of this peak elucidate a strong tendency to form a new distinct phase, which can be attributed to the inter-molecular interaction of

Ope-POSS cages, and is supposed to form the self-assembled clusters in composites, which have been already observed in the reported literature [6]. Moreover, for PU composites with a high concentration of Ope-POSS, the amorphous broadened halo at about 20° is decreased and sharpened, which indicates that the amorphous region becomes denser with increasing Ope-POSS concentration.

6 CONCLUSION

In this study, molecular simulation was applied to study the effect of the Ope-POSS cage structure on the chain packing characteristics of Ope-POSS/PU hybrid composites at the molecular level. As the result shows, with the concentration increasing in hybrid composites, Ope-POSS, as a rigid core linked to the polymer chains, apparently departs from the polymer chain distance and tends to form a distinct cluster, which leads to the micro-phase separation in hybrid composites.

ACKNOWLEDGMENTS

This work was supported by the Sichuan Education Office Foundation (Project No. 15ZA0040), China and the Key Laboratory of Special Waste Water Treatment in the Sichuan Province Higher Education System, China.

REFERENCES

[1] Yonghong Liu & Yong Ni et al. 2006. Polyurethane networks modified with octa(propylglycidyl ether) polyhedral oligoeric silsesquioxane, *Macromol. Chem. Phys.* 207:1842–1851.
[2] Jianqing Zhao et al. 2008. Polyhedral oligomeric silsesquioxane (POSS)-Modified thermoplastic and thermosetting nanocomposites: A review', *Polym. Polym. Compos.* 16(8): 483–500.
[3] Zheng L. & Coughlin E.B. et al. 2001. X-ray Characterizations of Polyethylene Polyhedral Oligomeric Silsesquioxane Copolymers. *Macromol.*35: 2375–2379.
[4] Lingling Wang & Rui Pan. 2015. Influence of trisilanolphenyl POSS on structure and thermal properties of polyurethane hybrid composites: a molecular simulation approach. *Acta Polymerica Sinica.* doi: 10.11777/j.issn1000–3304.2015.14231.
[5] YinYani & Monica H. Lamm 2009. Molecular dynamics simulation of mixed matrix nanocomposites containing polyimide and polyhedral oligomeric silsesquioxane (POSS). *Polym.* 50: 1324–1332.
[6] Rui Pan & Robert Shanks et al. 2014. Trisilanolisobutyl POSS/polyurethane hybrid composites: preparation, WAXS and thermal properties. Polymer Bulletin. 71:2453–2464.

Material Science and Engineering – Chen (Ed.)
© 2016 Taylor & Francis Group, London, ISBN 978-1-138-02936-1

Experimental and numerical study of single-sided scarf-repaired composite honeycomb sandwich structures

Z.W. Deng, Y. Yan & H.B. Luo
School of Aeronautic Science and Engineering, Beihang University, Beijing, China

ABSTRACT: This paper describes our current research on the compression behaviors of composite honeycomb sandwich structures. The compression performance of three types of specimens, namely intact panels, damaged panels and repaired panels, was tested individually. The test results demonstrate that repair methods can recover the compressive strength of the specimens effectively, and the strength of the repaired panel could restore around 80%. The ultimate compressive load of the three types of honeycomb sandwich panel was analyzed, and then the effects of repair parameters and damage variables are discussed. Through the Abaqus software platform, the compression tests of the composite sandwich panel were simulated. The simulation ultimate loads and the test results were in good agreement, verifying the correctness of the finite element model. The numerical method provided the theoretical basis for the analysis of the mechanical properties of the honeycomb sandwich structure.

Keywords: composite; honeycomb sandwich structure; compressive strength; scarf repair

1 INTRODUCTION

Composite materials have many excellent properties such as high specific strength, high specific stiffness, good fatigue performance, and anti-corrosion, so the use of advanced composite materials is increasingly common. A modern advanced aircraft requires the structural weight to be as light as possible, which means the skin material must be thinner, satisfying the strength and rigidity requirements. Because the composite honeycomb sandwich structure has the ability to reduce the structure weight, reduce the stress concentration, improve the fatigue life, improve the surface quality and maintain the steady state of shape, its application on a modern advanced aircraft increases constantly. The honeycomb panel in the process of manufacture and use is prone to damage and defect, so the repair issues of the honeycomb panel are worthy of further study.

Sung-Hoon Ahn and George S Springer performed many experiments to evaluate the fiber-reinforced composite material repair technology, mainly through the tensile strength to measure the effect of repairing. Data were generated with the following parameters having been varied: type of material of the damaged laminate, type of repair materials, scarf angle and number of external piles in scarf repair, length and number of repair piles in uniform and stepped lab repair, test temperature, and humidity. Since then, many scholars have carried out a large number of experimental studies.

Alex B. Harman used an analytical method to optimize the profile of the scarf joint between dissimilar modulus adherends, so that adhesive stresses are approximately uniform along the joint. The optimized scarf repair is expected to enhance the joint strength and reduce the amount of material removal. Besides, the finite element method is used to verify the validity of the optimization method and evaluate the results of the repair.

Ulrich Hansen studied the buckling strain of the debonded region shown to be the key parameter in assessing the structural integrity of a sandwich component in compression. In order to simplify the calculation, he proposed a one-dimensional spring model to simulate the honeycomb core.

This research was based on the compression performance of the composite honeycomb sandwich structure before and after the repair. Three types of specimens, namely intact panel, damaged panel and repaired panel, were manufactured to conduct a compressed destructive experiment. The different damage variables and repair parameters were established to explore the effects of different factors on compression performance. Through the Abaqus software platform, the composite sandwich panel compression tests were simulated, and the experimental results and the simulation results were shown to be in good agreement.

2 EXPERIMENTAL PROCEDURE

2.1 *Test equipment and fixtures*

The testing machine model was WDW-200E, as shown in Figure 1. Specimens were compressed to

Figure 1. Test equipment and fixtures.

failure by this machine to test their compression strength. The fixtures were arranged in parallel to the compression direction on both sides of the specimen in order to avoid the specimens' failure caused by panel buckling in compression.

2.2 Specimens

A total of 44 specimens were tested in the study. They belonged to three categories: 2 intact panels, 12 damaged panels, and 30 repaired panels.

All specimens were composite honeycomb sandwich panels. The skin of the sandwich panel was fabricated with T300/QY8911. The thickness of each unidirectional prepreg ply was 0.12 mm. The ply sequence of the sandwich panel was [45/90/-45/90/0] s/Core/[45/90/-45/90/0]s. The honeycomb core material was Nomex NRH-3-64 with a thickness of 15 mm. The panel skin and the honeycomb core were adhesive by glue J116A after the co-curing. All specimens were 150 mm long and 100 mm wide.

Intact panel means that the panel was undamaged apparently. Damaged panel had an initial damage on one side of the panel, which was a round opening in the center of one skin. The opening sizes were 20 mm and 30 mm for 6 specimens, respectively. Repaired panel was a scarf-repaired specimen, which had an initial damage before repair. The repair ply stacking sequence corresponded to the stacking sequence of the original skin structure. Besides, there was a 0° additional repair ply applied to the repair region, and its diameter was larger than most outside repair ply of 25 mm. The repair glue film was J116B with a thickness of 0.12 mm, its diameter was the same as the diameter of the additional repair ply. In order to explore the effect of the repair taper ratio on the compression strength, specimens adopted three repair parameters, namely 1/10, 1/15 and 1/20, respectively.

2.3 Test results

Table 1 presents the compression test results, including the average ultimate load, coefficient of variation, recovery rate and the number of test pieces. In Table 1, Intact represents the intact panel, the first letter D stands for the damaged panel, R stands for the repaired panel, the first number represents the damage diameter, and second figures represent the taper ratio. As listed in the table, the coefficients of variation for all specimens were under 10%. For this reason, the dispersion of the test results was within an acceptable range, and the data were sufficiently reliable for use in engineering applications. Damaged panels showed the worst capacity to sustain the compression load with its lowest ultimate loads due to the high stress concentrations at the edges of the damaged areas. The experimental results from the damaged panel indicated that the larger the damage diameter, the lower the compression strength.

After repair, the compression strength of the repaired panel could restore about 80% that of the intact specimens, which means that the scarf repair method can obviously increase the compressive properties of the damaged composite honeycomb sandwich panel.

For the honeycomb panel at different repair taper ratios, there was no obvious changing trend in compression strength, which indicated that the compression performance of the repaired panel was not significantly influenced when the taper ratio varied from 10 to 20 degrees.

Figure 2 shows the failure process of the repaired panel from the beginning of compression to the panel completely destroyed. The failure process could be divided into three stages. In the first stage, when the compression load increased to an initial damage threshold value, delamination occurred in the repaired region, and then the ply buckling occurred, which was observed as an additional repair ply uplift. In the second stage, with

Table 1. Variables and test results of the specimens.

Specimen	Ultimate load (KN)	Coefficient of variation (%)	Strength recovery (%)	Specimen numbers
Intact	80.01	1.03	100	2
D-20	54.38	6.35	67.97	6
D-30	51.69	3.15	64.61	6
R-20-10	66.52	4.62	83.15	5
R-20-15	66.07	2.88	82.58	5
R-20-20	65.46	5.97	81.82	5
R-30-10	63.09	8.15	78.86	5
R-30-15	62.55	4.87	78.18	5
R-30-20	64.27	6.51	80.33	5

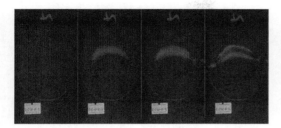

Figure 2. Failure process of the repaired panel.

the load increasing further, the delaminated region expanded continually and bulging became bigger. In the third stage, the delaminated area gradually expanded too close to the edge of the repair ply, and then the collapse crack suddenly appeared beside the repair region along the bulging lateral extension direction, and the repaired panel was instantly completely destroyed.

3 SIMULATION

3.1 Finite element model

The explicit finite element analysis software Abaqus/Explict was used to simulate the compression performance of the repaired composite honeycomb sandwich structure. The mesh of the finite element model is shown in Figure 3. In order to accurately simulate the stress distribution of the composite skin, each ply was established as a layer of three-dimensional 8-node hexahedron element C3D8R. Because the load and deformation of the honeycomb core were small in the compression process, to reduce the computation quantity, the honeycomb core was simplified as a homogeneous anisotropic material, using the three-dimensional 8-node hexahedron element C3D8R. A sharp slope appeared in the repair region at the taper ratio, simulated by the three-dimensional 6-node wedge element C3D6R. In order to simulate the delaminate failure of the layered composite plate, cohesive elements COH3D8 were established between each ply of the original plate and the patch. The thickness of the cohesive element ply was 0.005 mm. The cohesive element was also used to simulate the role of adhesion between the patch and the original plate.

3.2 Material properties

The properties of single-layer composite materials used in the finite element model are outlined in Table 2. From Table 2, it can be seen that E represents the elastic modulus of the material, G represents the shear modulus of the material, μ represents Poisson's ratio of the material, XT,

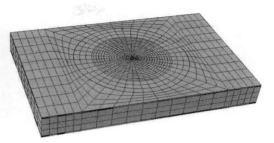

Figure 3. Mesh of the finite element model.

Table 2. Properties of the composite material T300/QY8911.

E_{11}/MPa	E_{22}/MPa	E_{33}/MPa	G_{12}/MPa	G_{13}/MPa
135000.0	8800.0	8800.0	4500.0	4500.0
G_{23}/MPa	μ_{12}	μ_{13}	μ_{23}	
3000.0	0.33	0.33	0.33	
X_T/MPa	X_C/MPa	Y_T/MPa	Y_C/MPa	
1548.0	1226.0	55.5	218.0	
S_{12}/MPa	S_{13}/MPa	S_{23}/MPa		
89.9	89.9	89.9		

Table 3. Properties of the honeycomb core and adhesive film J116B.

Core	E_T/MPa	G_L/MPa	G_W/MPa	
	229.2	65.1	35.0	
J116B	E/MPa	G/MPa	$t_n = t_s = t_t$/MPa	$G_n = G_s = G_t$/ (N*mm^{-1})
	5000	5000	35	1.45

XC, YT, YC, respectively, represent the tensile strength and compressive strength of the material along the fiber direction and perpendicular to the fiber direction, and S represents the shear strength of the material.

The properties of the honeycomb core and adhesive film are outlined in Table 3. In this paper, the honeycomb core was considered as carrying only in three directions, and the failure of the honeycomb core was ignored. ET, GL, GW, respectively, represent the elastic modulus, longitudinal shear modulus and transverse shear modulus of the honeycomb core. E and G, respectively, represent the elastic modulus and shear modulus of the adhesive film, ti (i = n, s, t), respectively, present the normal strength and shear strength along two directions, and Gi (i = n, s, t) represents the fracture energy in three directions.

3.3 Failure criteria and damage evolution

Hashin criteria was used to predict the failure and damage evolution of the composite skin, considering four kinds of initial damage.

Fiber tension failure:

$$F_f^t = \left(\frac{\sigma_{11}}{X_t}\right)^2 \tag{1}$$

Fiber compression failure:

$$F_f^c = \left(\frac{\sigma_{11}}{X_c}\right)^2 \tag{2}$$

Matrix tension failure:

$$F_m^t = \left(\frac{\sigma_{22}}{Y_t}\right)^2 + \left(\frac{\tau_{12}}{S}\right)^2 \tag{3}$$

Matrix compression failure:

$$F_m^c = \left(\frac{\sigma_{22}}{2S}\right)^2 + \left[\left(\frac{Y_c}{2S}\right)^2 - 1\right]\frac{\sigma_{22}}{Y_c} + \left(\frac{\tau_{12}}{S}\right)^2 \tag{4}$$

Once the panel failure, damage evolution model changed according to the equation (5) below:

$$
\begin{Bmatrix} \sigma_{11} \\ \sigma_{22} \\ \tau_{12} \end{Bmatrix}
= \frac{1}{D}
\begin{bmatrix}
(1-d_f)E_1 & (1-d_f)(1-d_m)\nu_{21}E_1 & 0 \\
(1-d_f)(1-d_m)\nu_{12}E_2 & (1-d_m)E_2 & 0 \\
0 & 0 & (1-d_s)GD
\end{bmatrix}
\times
\begin{Bmatrix} \varepsilon_{11} \\ \varepsilon_{22} \\ \gamma_{12} \end{Bmatrix}
\tag{5}
$$

In Equation (5), $D = 1 - (1-d_f)(1-d_m)\nu_{12}\nu_{21}$. The damage parameters d_f, d_m and d_s were derived from the failure parameters d_f^t, d_f^c, d_m^t and d_m^c by Equations (6), (7) and (8) as follows:

$$d_f = \begin{cases} d_f^t & if\ \sigma_{11} \geq 0, \\ d_f^c & if\ \sigma_{11} \geq 0, \end{cases} \tag{6}$$

$$d_m = \begin{cases} d_m^t & if\ \sigma_{22} \geq 0, \\ d_m^c & if\ \sigma_{22} \geq 0, \end{cases} \tag{7}$$

$$d_s = 1 - (1-d_f^t)(1-d_f^c)(1-d_m^t)(1-d_m^c) \tag{8}$$

The failure of the honeycomb core was ignored.

4 CONTRAST SIMULATION AND EXPERIMENT

4.1 Failure mode

In the compression process, first, the stress of the cohesive element in the repair region increased to the maximum stress. Delamination occurred as the cohesive element failure began. For further increasing compression load, more number of cohesive elements failed, causing the buckling of repair plies. The load on repair plies transferred to the original plate, and then the original plate suddenly destroyed as the pressure reached the ultimate load.

The finite element method was able to simulate all the experimentally observed phenomena accurately. The specimens' picture shown in Figure 4 was in the first stage (left) and the third stage (right) of the compression process of the repaired panel. The simulation and the experimental results were in good agreement.

4.2 Ultimate load

The comparison of the ultimate load between the experimental values and simulation values is shown in Figure 5, which indicated that the

Figure 4. Comparison between the experimental and simulation results.

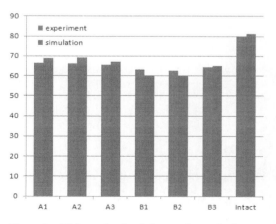

Figure 5. Ultimate loads of the specimens in the test and simulation.

simulation loads were in good agreement with the test results. The error did not exceed 10%, so the correctness of the finite element model was verified.

5 CONCLUSIONS

After repair, the compression strength of the repaired panel could restore about 80% that of intact specimens, which meant that the scarf repair method could obviously increase the compressive properties of the damaged composite honeycomb sandwich panel. The compression performance of the repaired panel was not significantly influenced when the taper ratio changed between 10 and 20 degrees.

The failure process of the repaired composite honeycomb sandwich panel under the compression load could be divided into three stages: delamination and buckling of repair plies, bulging expanding and completely destroyed.

The finite element method was able to simulate all the experimentally phenomena accurately, including initial damage, damage evolution and failure pattern. The simulation ultimate loads and the test results were in good agreement, verifying the correctness of the finite element model.

REFERENCES

[1] Ahn, S., & Springer, G. S. 1998. Repair of composite laminates-i: test results. *Journal of Composite Materials, 32, 11*, 1036–1074.

[2] Chen, B.Y., Tay, T.E., Baiz, P.M., & Pinho, S.T. 2013. Numerical analysis of size effects on open-hole tensile composite laminates. *Composites Part A: Applied Science and Manufacturing, 47, 2*, 52–62.

[3] D.S.G. Campilho, R., & F.S.F. de Moura & J.J.M.S. Domingues, M. 2012. Stress and failure analyses of scarf repaired cfrp laminates using a cohesive damage model. *Journal of Adhesion Science and Technology, volume 21, 9*, 855–870.

[4] Hansen, U. 1998. Compression behavior of FRP sandwich specimens with interface debonds. *Journal of Composite Materials, 32, 4*, 335–360.

[5] Harman, A.B., & Wang, C.H. 2006. Improved design methods for scarf repairs to highly strained composite aircraft structure. *Composite Structures, 75, 1–4*, 132–144.

[6] Hashin, Z. 1980. Failure criteria for unidirectional fiber composites. *Transactions of the ASME. Journal of Applied Mechanics, 47, 2*, 329–334.

[7] H. Wang, C., & Venugopal & L. Peng, V. 2014. Stepped flush repairs for primary composite structures. *The Journal of Adhesion, 91, 1*, 95–112.

[8] Soutis, C., & Z. Hu, F. 2012. Failure analysis of scarf-patch-repaired carbon fiber/epoxy laminates under compression. *AIAA Journal, 38, 4*, 737–740.

[9] Vadakke, V., Carlsson, L.A., Vadakke, V., & Carlsson, L.A. 2004. Experimental investigation of compression failure of sandwich specimens with face/core debond. *Composites Part B: Engineering, 35*, 583–590.

[10] Wang, C.H., & Gunnion, A.J. 2008. On the design methodology of scarf repairs to composite laminates. Composites Science and Technology, *68, 1*, 35–46.

[11] Xiaoquan, C., Baig, Y., & Renwei, H. 2013. Study of tensile failure mechanisms in scarf repaired CFRP laminates. *International Journal of Adhesion and Adhesives, 41, 1*, 177–185.

Material Science and Engineering – Chen (Ed.)
© 2016 Taylor & Francis Group, London, ISBN 978-1-138-02936-1

Differences in influences of fumed silica and a kind of "cage-like silica" on properties of polydimethylsiloxane rubber based composites at low temperatures

Z.Q. Dong, Y.L. Zhu, M.P. He & J.Y. Pei
College of Chemical and Material Engineering, Quzhou University, Quzhou, China

ABSTRACT: In this study, fumed silica and a kind of "cage-like silica" octaphenylsilsesquioxane (POSS) were separately incorporated into polydimethylsiloxane (PDMS), and then properties of the resulting composites were characterized carefully. Silica is dispersed uniformly and amorphously in PDMS. And because of the large volume of silica, it forms a network in PDMS matrix. The formed silica network inhibits the mobility of PDMS during crystallization and glass transition which lowered crystallinity (Xc) and enhanced glass transition temperature (Tg). However, the situation is quite different for the composite PDMS/POSS. In PDMS/POSS, POSS itself forms crystal domains which provide templates for PDMS chain segments to become ordered and crystalline. In the process of glass transition, the mobility of PDMS is confined by POSS crystals, thus Tg of the composite is significantly elevated.

Keywords: polydimethylsiloxane; silica; polyhedral oligomeric silsesquioxanes; crystallization; dynamic mechanical properties

1 INTRODUCTION

Polydimethylsiloxane (PDMS) rubber, composed of—$Si(CH_3)_2O$—structure, is the most important member in the family of polysiloxanes. Since the oil resources are drying up, PDMS which comes from the non-oil chemical route is considered as the foremost candidate for new materials. PDMS is endowed with many outstanding properties, say ultraviolet resistance, weatherability, electric insulativity, air permeability, physical inertia, high and low temperature resistance and so on (Jovanovic, Govedarica, Dvornic, & Popovic, 1998; Zheng, Tan, Dai, Lv, Zhang, & Xie, 2012). Nevertheless, it's mechanical strength and thermal resistance is far from perfect, so it should be modified before practical use (Pradhan, Srivastava, Ananthakrishnan, & Saxena, 2011).

During the past decades, silica is the most widely used filler for PDMS (Salazar-Hernández, Alquiza, & Salgado, 2010). Silica is alarmingly widespread in nature, with amorphous and crystalline morphology. The most efficient silica in modifying PDMS is fumed silica, whose structure is shown in Figure 1. In silica molecule, each silicon atom is bonded with four oxygen atom and each oxygen atom is bonded with two silicon atom. So the molecular formula of SiO_2 is the simplified one which cannot accurately reflect the structure of silica. As early as 1980s, Sun et al.

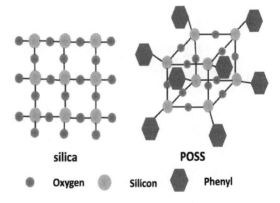

silica   POSS

● Oxygen   ● Silicon   ⬢ Phenyl

Figure 1.   Schematic drawings of silica and POSS.

found that mechanical properties of fumed silica/PDMS composite could be prominently improved (Sun, & Mark, 1989). Later, Pissis et al. discovered that crystallinity of PDMS was depressed by the addition of silica (Fragiadakis, Pissis, & Bokobza, 2005). Moreover, in the characterization of dielectric relaxation spectroscopy, two α relaxations of PDMS were detected. The two relaxations were ascribed to unrestricted PDMS chains and PDMS chains which were blocked by silica, respectively.

In recent years, a kind of "cage-like silica"—octaphenylsilsesquioxane (POSS) has been attracted the attention of researchers all over the world. POSS is characterized by the formula (RSiO1.5)n in which n can be 6, 8, 10, 12, etc, and it could be regarded as the smallest silica particle for the size of POSS cage is about 1.5 nm (Huang, Xie, Jiang, Wang, & Yin, 2009). The most universally applied POSS is (RSiO1.5)8 and the structure of octapheyl substituted one is illustrated in Figure 1.

Because of the organic groups in the outer space, POSS is compatible with most polymers, such as polyethylene (Zheng, Waddon, Farris, & Coughlin, 2002), polypropylene (Fina, Tabuani, Frache, & Camino, 2005), polybutadiene (Liao, Zhang, Fan, Wang, & Jin, 2011), polyvinylchloride (Soong, Cohen, Boyce, & Mulliken, 2006), polystyrene (Vielhauer, Lutz, & Reiter, 2013), poly(acrylate) (Escude, & Chen, 2009), polyester (Gao, Li, & Kong, 2011), polyamide (Lin, Shao, & Wang, 2013), polyimide (Dasgupta, Sen, & Banerjee, 2010), epoxy (Xin, Ma, Chen, Song, & Qu, 2013), etc. In addition, POSS is also a kind of excellent filler in the modification of PDMS. Pan et al. (Pan, Mark, & Schaefer, 2003) found through chemical bonding of POSS with PDMS network, the mechanical properties of the resulting composite could be significantly improved. Meng et al. (Meng, Wei, Liu, Zhang, Nishi, & Ito, 2013) reported that when POSS was below certain content, POSS could distribute evenly in PDMS and the thermal stability of POSS/PDMS composite could be enhanced, whereas in case more POSS was incorporated, it tended to aggregate and the thermal stability of the resulting composite decreased.

In this article, silica and POSS were separately composited with PDMS through blending and curing. By means of Differential Scanning Calorimetry (DSC) and Dynamic Mechanical Analysis (DMA), the influences of silica and POSS on crystallization behaviors, melting behaviors and dynamic mechanical properties of PDMS were studied. Afterwards, the influence mechanisms were explored through Scanning Electron Microscopy (SEM) and X-Ray Diffraction (XRD).

2 EXPERIMENTAL

2.1 Materials

Both pretreated fumed silica (abbreviated as silica) and PDMS (600,000 g/mol) were purchased from Zhonglan Chenguang Chemical Research Institute. Octaphenylsilsesquioxane (abbreviated as POSS) was provided by Aladdin. 2,5-Dimethyl-2,5-di(tert-butylperoxy)hexane (DBPH), the peroxide vulcanizing agent, was obtained from Dongyang chemical, Haian, China.

2.2 Preparation of PDMS/silica and PDMS/POSS composites

PDMS was firstly blended with 5wt% silica and 5wt% POSS respectively in an open two-roll mill, and then predetermined amount of DBPH was added. Afterwards, the mixture obtained was placed to plate vulcanizing press at 150 °C and a pressure of 10 MPa for 10 min, followed by post-curing at 180 °C for 4 hours.

3 RESULTS AND DISCUSSION

3.1 Crystallization and melting behaviors of the composites measured by differential scanning calorimetry

Figure 2 illustrates the crystallization curves of PDMS and the composites. It is obvious that silica and POSS impose opposite effects on the crystallization of PDMS where, as shown in Table 1, crystallization temperature (T_c) of PDMS + silica is as low as −75.06 °C, but that of PDMS + POSS is elevated to −70.90 °C. What is well-known is that crystallization is a heat-release process which mainly happens during cooling, so the higher T_c means the facilitation of crystallization.

Further information can be found in the DSC melting curves which are displayed in Figure 3. Currently, the degree of perfection of crystal is

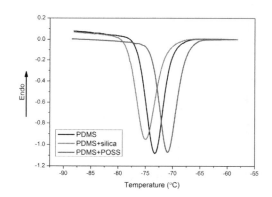

Figure 2. DSC crystallization curves of the samples.

Table 1. Crystallization and melting parameters of the three samples.

Sample	T_c (°C)	T_m (°C)	X_c (%)
PDMS	−73.28	−45.13	67.75
PDMS + silica	−75.06	−45.91	65.80
PDMS + POSS	−70.90	−44.61	67.80

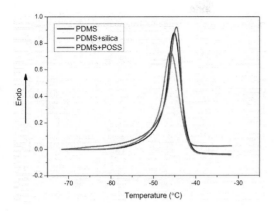

Figure 3. DSC melting curves of the samples.

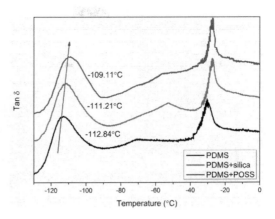

Figure 4. Changes of tan δ as a function of temperature.

weighed by crystallinity (Xc) of crystal. Xc of crystal can be determined from the equation:

$$Xc = \Delta Hf / \Delta Hf0 \times 100\%$$

where ΔHf is the heat of fusion of the sample, and $\Delta Hf0$ is the heat of fusion of a perfectly (100%) crystalline polymer. As a semi crystalline polymer, PDMS normally can't achieve the crystallinity of 100%. So Xc of the samples could be calculated according to the conclusion given by Mirta, I.A. that the value of $\Delta Hf0$ for PDMS rubber is 37.4 J/g. The calculation results, as well as melting temperatures (Tm) of the samples are listed in Table 1.

The order of Tm for the three samples is consistent with that of Tc. This is because higher Tc leads to higher mobility of PDMS chain segments, thus PDMS chain segments are more readily to arrange orderly into crystal lattice. As regard to Xc of the samples, Xc of PDMS + silica is largely decreased, while that of PDMS + POSS remains nearly the same as that of neat PDMS. The results demonstrate that the decreased Tm of PDMS + silica is caused by the lowered Xc of the composite, whereas the increased Tm of PDMS + POSS is not caused by the enhanced Xc but by the more intact PDMS crystals.

3.2 Dynamic mechanical analyses of the composites

DMA was also conducted to bridge macro behaviors with micro structure. Among all the parameters obtained from DMA, the loss factor (tan δ) is the most convincing one. As shown in Figure 4, the tan δ curves for all the three samples show two peaks. Associating the curves with DSC melting curves (Fig. 3), peaks at higher temperatures could be attributed to the melting of polysiloxane. Nevertheless, as measured by different means, the

test results are not able to be compared with each other.

Evidently, peaks at lower temperatures are assigned to glass transition of polysiloxane chain segments. With the addition of silica, glass transition temperature (Tg) can be elevated by 1.63°C (from −112.84°C to −111.21°C), while the addition of POSS could dramatically enhance Tg of the composite (3.73°C). The mechanism of modifications in dynamic mechanical properties may be deduced from the micro structures of the composites. So SEM and XRD characterizations were also conducted to uncover the internal relationships between aggregational structures of fillers and properties of the composites.

3.3 Dispersion state of silica and POSS in PDMS matrix studied by scanning electron microscope

SEM images in Figure 5 show the dispersion state of silica and POSS in PDMS. In the composites, silica is dispersed uniformly in PDMS matrix without the existence of large-size aggregation. Nevertheless, although POSS is also dispersed evenly in the matrix, its aggregations with diameters in the range 0.5–3 μm can be obviously seen in the photograph. In addition, XRD was applied to provide further information on the aggregational state of silica and POSS.

3.4 Mechanism of influences of silica and POSS on PDMS

Crystallization always begins with the formation of crystal nucleus which could provide templates for the chain segments. However, the nucleating ability is based on the crystalline structure of crystal nucleus.

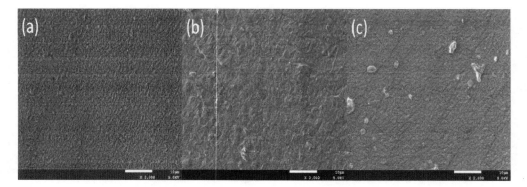

Figure 5. SEM images of PDMS (a), PDMS + silica (b) and PDMS + POSS (c).

In PDMS + silica, silica is dispersed so uniformly that no ordered silica crystal is shaped, resulting in that the crystallization of PDMS is hardly induced by silica. Furthermore, as silica has a significantly low density (25–50 g/L), merely 5wt% silica accounts for extremely large volume. The result is that silica in PDMS matrix forms its own network which prevents PDMS chain segments from orderly arranging, leading to enormously decreased Xc. During glass transition, the formed silica network also hinders the mobility of chain segments, so Tg of PDMS + silica is elevated.

4 CONCLUSIONS

The composites PDMS/silica and PDMS/POSS were prepared by open two-roll mill blending and vulcanizing. DSC measurements show that Tc and Tm increase with the addition of POSS, whereas Tc and Xc decrease substantially when incorporated with silica. The reasons lie in the different micro structures of silica and POSS in PDMS matrix. Silica is amorphously dispersed in PDMS and forms a network which keeps PDMS chain segments from orderly arranging during crystallization, whereas POSS forms crystal domains which could act as templates for PDMS chain segment to become ordered. DMA results indicate that Tgs of both PDMS/silica and PDMS/POSS are elevated. Wherein, the mobility of PDMS in PDMS/silica is confined by the silica network, while that in PDMS/POSS is inhibited by POSS crystals. Silica network, while that in PDMS/POSS is inhibited by POSS crystals.

ACKNOWLEDGEMENT

This research was financially supported by the Quzhou Science and Technology Project (2014045) and Quzhou University talent training project (BSYJ201411).

REFERENCES

[1] Dasgupta, B., Sen, S.K., & Banerjee, S. (2010). Aminoethylaminopropylisobutyl POSS-Polyimide nanocomposite membranes and their gas transport properties. Mater Sci Eng: B. 168, 30–35.
[2] Escude, N.C., & Chen, E.Y.X. (2009). Stereoregular methacrylate-poss hybrid polymers: syntheses and nanostructured assemblies. Chem Mater. 21, 5743–5753.
[3] Fina, A., Tabuani, D., Frache, A., & Camino, G. (2005). Polypropylene–polyhedral oligomeric silsesquioxanes(POSS)nanocomposites. Polymer. 46, 7855–7866.
[4] Gao, J., Li, S., & Kong, D. (2011). Reaction kinetics and physical properties of unsaturated polyester modified with methylacyloxylpropyl-POSS. J Polym Res. 18, 621–626.
[5] Fragiadakis, D., Pissis, P., & Bokobza, L., (2005). Glass transition and molecular dynamics in poly(dimethylsiloxane)/silica nanocomposites. Polymer. 46, 6001–6008.
[6] Huang, X.Y., Xie, L.Y., Jiang, P.K., Wang, G. L., & Yin, Y. (2009). Morphology studies and ac electrical property of low density polyethylene/octavinyl polyhedral oligomeric silsesquioxane composite dielectrics. Eur Polym J.45, 2172–2183.
[7] Jovanovic, J.D., Govedarica, M.N., Dvornic, P.R., & Popovic, I.G. (1998). The thermogravimetric analysis of some polysiloxanes. Polym Degrad Stab. 61, 87–93.
[8] Liao, M., Zhang, X., Fan, C., Wang, L., & Jin, M. (2011). Preparation and characterization of polybutadiene/allylisobutyl polyhedral oligomeric silsesquioxane nanocomposites by anionic polymerization. J Appl Polym Sci. 120, 2800–2808.
[9] Lin, X.X., Shao, Q., & Wang, Y.M. (2013) Morphology studies and ac electrical property of low density polyethylene/octavinyl polyhedral oligomeric. Adv Mater Res 710, 45–50.
[10] Meng, Y., Wei, Z., Liu, L. Zhang, L., Nishi, T. & Ito, K. (2013). Significantly improving the thermal stability and dispersion morphology of polyhedral oligomeric silsesquioxane/polysiloxane composites by in-situ grafting reaction. Polymer, 54, 3055–3064.

[11] Pan, G., Mark, J.E., & Schaefer, D.W. (2003). Synthesis and characterization of fillers of controlled structure based on polyhedral oligomeric silsesquioxane cages and their use in reinforcing siloxane elastomers. J Polym Sci Part B: Polym Phys. 41, 3314–3323.

[12] Pradhan, B., Srivastava, S.K., Ananthakrishnan, R., & Saxena, A. (2011). Preparation and characterization of exfoliated layered double hydroxide/silicone rubber nanocomposites. J Appl Polym Sci.119, 343–351.

[13] Salazar-Hernández, C., Alquiza, M.J.P., & Salgado, P. (2010). TEOS–colloidal silica–PDMS-OH hybrid formulation used for stone consolidation. Appl Organomet Chem. 24, 481–488.

[14] Soong, S.Y., Cohen, R.E., Boyce, M.C., & Mulliken A.D. (2006). Rate-Dependent deformation behavior of POSS-Filled and plasticized poly (vinyl chloride). Macromolecules. 39, 2900–2908.

[15] Sun, C.C., & Mark, J.E. (1989). Comparisons among the reinforcing effects provided by various silica-based fillers in a siloxane elastomer. Polymer. 30, 104–106.

[16] Vielhauer, M. Lutz, P.J., & Reiter, G. (2013). Direct arylation polycondensation for the synthesis of bithiophene-based alternating copolymers. J Polym Sci Part A: Polym Chem. 51, 947–952.

[17] Xin, C., Ma, X., Chen, F., Song, C. & Qu, X. (2013). Synthesis of EP-POSS mixture and the properties of EP-POSS/epoxy, SiO2/epoxy, and SiO2/EP-POSS/epoxy nanocomposite. J Appl Polym Sci. 130, 810–819.

[18] Zheng, L., Waddon, A.J., Farris, R.J., & Coughlin, E.B. (2002). X-ray characterizations of polyethylene polyhedral oligomeric silsesquioxane copolymers. Macromolecules. 35, 2375–2379.

[19] Zheng, Y., Tan, Y.X., Dai, L.N., Lv, Z., Zhang, X.Z., & Xie, Z.M. (2012) Synthesis, characterization, and thermal properties of new polysiloxanes containing 1,3-bis(silyl)-2,4-dimethyl-2,4- diphenylcyclodisilatelzane. Polym Degrad Stab. 97, 2449–2459.

Material Science and Engineering – Chen (Ed.)
© 2016 Taylor & Francis Group, London, ISBN 978-1-138-02936-1

Influence of Vacuum Ultraviolet radiation on the mass loss of polymer materials

Y. Wang
Lanzhou Hongrui Ht Mechanical and Electrical Equipment Co. Ltd., Lanzhou, China

X. Guo, X.R. Wang, S.S. Yang & X.J. Wang
*Science and Technology on Vacuum Technology and Physics Laboratory, Lanzhou Institute
of Space Technology Physics, Lanzhou, China*

ABSTRACT: Outgassing of polymer materials induced by Vacuum Ultraviolet (VUV) can cause molecular contamination. Based on the diffusion theory and radiation chemistry, simulation of the mass loss of polymer materials induced by VUV radiation was researched. The mathematical function between the mass of volatile components in the material and irradiation time was obtained using the variable separation approach. Mass loss for the Polyimide (PI) or Polyethylene Terephthalate (PET) film obtained from experiments and parameters in the model was calculated through curve fitting in order to validate the applicability of the model. The results show that the model of mass loss is applicable to the PI or PET film, and can be used for the prediction on the mass loss of polymer materials induced by VUV irradiation.

Keywords: vacuum ultraviolet (VUV); polymer materials; mass loss; molecular contamination; diffusion theory

1 INTRODUCTION

Outgassing of polymer materials induced by space irradiation can cause molecular contamination. (Laikhtman et al., 2009) The presence of contamination on thermal control surfaces will alter absorptance/emittance ratios and change the thermal balance, while contamination on solar arrays will decrease the power output. Contamination on optical instruments will decrease the signal throughput and can scatter the signal beyond the diffraction design, thus further decreasing the performance (Rampini et al., 2004). Nevertheless, the contamination on the sensitive surfaces of optical systems and cells of solar batteries is a complex physical and chemical process, which contains three principal courses, including outgassing of volatile components from materials, transporting and deposition on the sensitive surfaces. The effects of molecular contamination not only lie on polymer materials bestowed on the spacecraft, but also relate to the dimension of vehicles and the location of sensitive surfaces, and are also influenced by the temperature and irradiation. (Bertrand, 1995) The simulation on the effect of contamination is based on a series of physical models. However, most existing models do not take into account the irradiation, so that they can only forecast the

contamination effects to a certain extent, which leads to differences in the orbit (Hall, 2000; Wang, 1989, 1994; Khassanchine, 2004). Consequently, elaboration of prediction models describing the outgassing processes of polymer materials is a present-day problem to predict the sensitive surface contamination.

The high-energy portion of the ultraviolet spectrum containing wavelengths approximately below 200 nm is generally referred to as Vacuum Ultraviolet (VUV) radiation. VUV radiation, which can be strongly absorbed on the polymer material, will decompose the film material and increase mass loss induced by these materials, hence causing contaminations and ultraviolet-enhanced contaminations, thereby influencing the optical, chemical and physical performances of sensitive surfaces of optical systems and cells of solar batteries (Brinza et al., 1991; Keith, 2007; Dever, 2005). Studies on the performance of polymer materials caused by VUV radiation have mainly focused on that of common materials, such as polyimides and silicone rubbers, whereas simulations or predictions of mass loss induced by radiation are scarcely researched. In this work, in order to acquire in-depth knowledge of the outgassing kinetics characteristics of a set of polymer film materials, the model of mass loss of polymeric materials induced by VUV radiation is studied based on the physical and chemical mechanisms between ultraviolet radiation and materials. Experiments of VUV radiation on Polyimide (PI) and Polyethylene Terephthalate (PET) are also carried out, which validate the applicability of the model.

2 SIMULATION OF MASS LOSS

Several principal postulates applied in the model of outgassing at thermal-vacuum action are utilized in order to describe mathematically the influence of VUV on the physical and chemical processes that occur in the material and on its surface. The change in the concentration of volatile components in the material applied to a hermetically enclosed substrate is stipulated by the following processes: desorption from the surface at the material-vacuum boundary; photodecomposition reactions of the material; and diffusion resulting from the aforementioned processes.

This model is based on the following assumptions:

1. Thickness of the polymer film is significantly less when compared with other linear dimensions, so that it can be treated as a one-dimensional massive plate.
2. Temperature of the materials is fixed.
3. Coefficients of diffusion desorption, and VUV decomposition only depend on time.
4. Some components of the outgassing process can be produced as a result of the destruction of other components; desorption is not considered in the model, that is, molecules diffused to the surface all volatilize to vacuum.
5. Volatile components in the materials are involved only in the first-order reaction.
6. Outgassing occurs through the material-vacuum boundary only.
7. Coefficients used in the model are considered as effective coefficients.

The process of VUV radiation on film materials is shown in Figure 1. Under these assumptions, the change in the concentration of volatile components $C(x, t)$ in the film material being in vacuum under exposure to VUV radiation can be described by the following partial differential equations:

$$\begin{cases} \dfrac{\partial C}{\partial t} = D\dfrac{\partial^2 C}{\partial x^2} - kC + Ae^{-\alpha(l-x)} & 0 < x < l, t > 0 \\[2mm] C(0,t) = C(l,t) = 0 & t > 0 \\[2mm] C(x,0) = C_0 & 0 \le x \le 1 \end{cases} \quad (1)$$

where l is the thickness of the material; k is the coefficient of photodecomposition reactions;

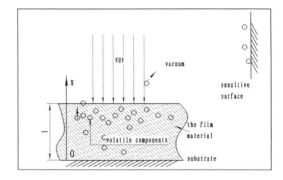

Figure 1. Outgassing induced by VUV radiation for polymer materials.

$S(x, t) = Aexp[-\alpha(l - x)]$ is the source function of volatile components in the material, in which A is the parameter that depends on the composition of the material and the UV source; α is the effective coefficient of linear reduction of VUV radiation; C_0 is the initial concentration of volatile components.

Based on the partial differential equations theory, Function (1) can be calculated by using the variable separation approach:

$$C = \sum_{n=1}^{\infty} \left\{ \frac{4C_0}{n\pi} e^{-\left(\frac{n\pi}{l}\right)^2 Dt} + \frac{2Al^2}{[(\alpha l)^2 + (n\pi)^2]Dn\pi} \right.$$
$$\left. \times \left[1 - e^{-\left(\frac{n\pi}{l}\right)^2 Dt} \right] + \frac{4C_0}{n\pi} e^{-\left(k + \frac{Dn^2\pi^2}{l^2}\right)t} \right\} \sin\frac{n\pi}{l}x$$

$$n = 1, 3, 5 \qquad (2)$$

The outgassing rate of volatile components from the unit of surface can be determined by Fick's second law:

$$F(x,t) = -\frac{D}{l} \sum_{n=1}^{\infty} \left\{ 4C_0 e^{-\left(\frac{n\pi}{l}\right)^2 Dt} + \frac{2Al^2}{(\alpha l)^2 + (n\pi)^2} \right.$$
$$\left. \times \left[1 - e^{-\left(\frac{n\pi}{l}\right)^2 Dt} \right] + 4C_0 e^{-\left(k + \frac{Dn^2\pi^2}{l^2}\right)t} \right\} \cos\frac{n\pi}{l}x$$

$$n = 1, 3, 5 \qquad (3)$$

For the side of the one-dimensional massive plate, the outgassing rate is given by

$$F(l,t) = \frac{4C_0 D}{l} \sum_{n=1}^{\infty} \left[e^{-\left(\frac{n\pi}{l}\right)^2 Dt} + e^{-\left(k + \frac{Dn^2\pi^2}{l^2}\right)t} \right]$$
$$- \sum_{n=1}^{\infty} \frac{2Al}{(\alpha l)^2 + (n\pi)^2} e^{-\left(\frac{n\pi}{l}\right)^2 Dt} + \sum_{n=1}^{\infty} \frac{2Al}{(\alpha l)^2 + (n\pi)^2}$$

$$n = 1, 3, 5 \qquad (4)$$

Then, the mass loss of the film $W\,(T,\,t)$ is obtained by integrating $F(l,\,t)$ from 0 to t:

$$W(T,t) = \frac{l^2}{\pi^2 D}\left[\frac{4DC_0}{l} - b\right]\left[1 - e^{-\left(\frac{\pi}{l}\right)^2 Dt}\right]$$
$$+ \frac{4DC_0 l}{kl^2 + \pi^2 D}\left[1 - e^{\left[-k-\left(\frac{\pi}{l}\right)^2 D\right]t}\right] + bt \qquad (5)$$

where $b = \sum_{n=1}^{\infty} \frac{2Al}{(\alpha l)^2 + (n\pi)^2}$.

3 MATHEMATICAL FIT AND VALIDATION OF THE MASS LOSS MODEL

3.1 *Mathematical fit of the mass loss model*

To determine the mass loss model for the special film material, experiments of PI and PET irradiated by VUV are carried out in the VUV Radiation Simulation Experimental Equipment, respectively. The thickness of PI is 20 μm, while that of PET is 12.5 μm. Additionally, the Total Mass Loss (TML) is tested in the Contamination Condensation Effect Equipment (CCEE) (Wang, 2002). The outgassing data determined in the experiment are listed in Table 1 and Figure 2.

As shown in Figure 2, the mass loss caused by VUV radiation for PI or PET increases sharply at the initial time, whereas it enhances mildly later.

A mathematical curve fit of the change in mass loss with time is performed for each mass measurement, in order to extrapolate the mass loss at times longer than the experimental ones.

Utilizing the experimental data, coefficients in (5) are fitted. The result is summarized in Table 2, and comparisons between the fit line and the experimental data are shown in Figure 3.

The fitted coefficients in the mass loss equation of PI or PET are listed in Table 2. When introducing the coefficients listed in Table 2 to formula (5), the mathematical formula of mass loss caused by VUV radiation on PI or PET can be obtained, which is demonstrated in (6) and (7), respectively:

Table 1. Outgassing data caused by VUV irradiation.

Irradiation time ESH	PI TML (g/cm²)	PET TML (g/cm²)
500	4.24E-6	8.74E-7
1000	9.47E-6	1.03E-6
2200	1.66E-5	1.19E-6
3000	1.92E-5	1.26E-6
4400	2.23E-5	1.28E-6
5000	2.31E-5	1.29E-6

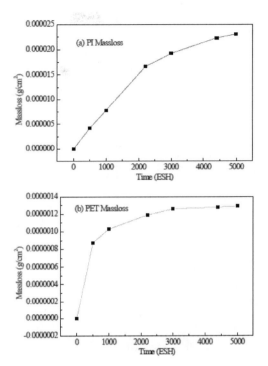

Figure 2. Mass loss of PI or PET induced by VUV irradiation.

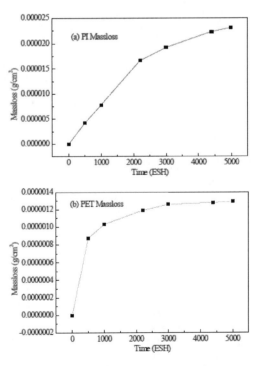

Figure 3. Comparison of (a) PI or (b) PET mass loss between the fitting line and the experimental data.

Table 2. Fit coefficients in the mass loss equations of PI or PET.

Coefficients	C_0 g/cm^3	k 1/h	D cm^2/h	b g/(cm$^2 \cdot$ h)	R^2
PI fit data	1.6E-2	2.2E-6	1.7E-10	1.0E-15	0.99
PET fit data	1.3E-3	7.0E-11	3.2E-10	7.9E-12	0.98

$$
\begin{aligned}
W(T,t) &= \frac{l^2}{\pi^2 D}\left[\frac{4DC_0}{l}-b\right]\left[1-e^{-\left(\frac{\pi}{l}\right)^2 Dt}\right] \\
&\quad + \frac{4DC_0 l}{kl^2+\pi^2 D}\left[1-e^{[-k-(\frac{\pi}{l})^2 D]t}\right]+bt \\
&= 2.66\times10^{-5}\times(1-\exp(-4.3\times10^{-4}t)) \\
&\quad + 2.5\times10^{-15}\times(1-\exp(-4.6\times10^{6}t)) \\
&\quad + 1\times10^{-10}t
\end{aligned}
\tag{6}
$$

$$
\begin{aligned}
W(T,t) &= \frac{l^2}{\pi^2 D}\left[\frac{4DC_0}{l}-b\right]\left[1-e^{-\left(\frac{\pi}{l}\right)^2 Dt}\right] \\
&\quad + \frac{4DC_0 l}{kl^2+\pi^2 D}\left[1-e^{[-k-(\frac{\pi}{l})^2 D]t}\right]+bt \\
&= -2.74\times10^{-9}\times(1-\exp(2.26\times10^{-2}t)) \\
&\quad + 6.03\times10^{-7}\times(1-\exp(-2.26\times10^{-2}t)) \\
&\quad + 7.9\times10^{-12}t
\end{aligned}
\tag{7}
$$

3.2 Validation of the mass loss model

To validate the application of (6) and (7) for a longer irradiation time, the irradiation time is extended to 200 h (i.e. 20000ESH). For the sake of comparison, the fitting line and the experimental data are shown in Figure 4.

The simulation and experiment data are compared in Table 3. From this table, it can be found that deviation factors between the simulations and the experiments are less than 2.5%, which testify that the simulation is valid for the mass loss of PI or PET irradiated by VUV. The reasons of such a difference in the results are under investigation. The main contributing factor is believed to be attributable to the different test methodology. However, the present results are considered to be conservative for mass loss predictions.

4 CONCLUSIONS

Based on the chemical and physical processes predicted for VUV radiation on the polymer film material, mathematical models are proposed for the description of the change in the mass loss of materials and irradiation time exposed to the space environment during a long operation period.

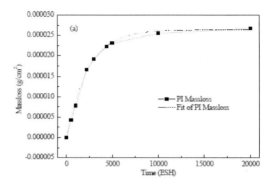

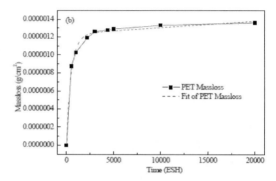

Figure 4. Comparison of (a) PI or (b) PET mass loss between the extrapolative line and the experimental data.

Table 3. Comparison of mass loss between the simulations and the experiments.

Irradiation time (ESH)	10000	20000
PI mass loss experiment data (g · cm^{-2})	2.55E-5	2.67E-5
PI mass loss simulation data (g · cm^{-2})	2.61E-5	2.64E-5
Deviation factors (%)	2.35	−1.12
PET mass loss experiment data (g · cm^{-2})	1.33E-6	1.36E-6
PET mass loss simulation data (g · cm^{-2})	1.30E-6	1.38E-6
Deviation factors (%)	−2.26	1.47

The mathematical function about mass loss and irradiation time is acquired by analyzing and solving the model. The mass change of the PI or PET film is studied using the VUV radiation equipment and the contamination condensation effect equipment. Coefficients in the model are calculated if values for the mass of the sample and irradiation time are introduced into the mass loss function. The results obtained from the mass loss mathematical model for PI or PET are in good agreement with that of long radiation experiments. They show that the model of mass loss is valid for the PI or PET film material, and it can be used for the prediction on the mass loss of polymer materials induced by VUV radiation.

REFERENCES

[1] Brinza D E, Stiegman A E, Staszak P R, et al. 1991. Vacuum Ultraviolet (VUV) Radiation Induced Degradation of Fluorinated Ethylene Propylene (FEP) Teflon Aboard the Long Duration Exposure Facility (LDEF). LDEF-69 Months in Space-First Post-Retrieval Symposium, 817–829.

[2] Bertrand W T. 1995. Effects of Spacecraft Material Outgassing on Optical System in the Vacuum Ultraviolet. ADA296935.

[3] Dever, Joyce A, Yan L. 2005. Vacuum Ultraviolet Radiation Effects on DC93-500 Silicone Film Studied. NASA, 20050217221.

[4] Hall D F, Arnold G S, Simpson T R, et al. 2000. Progress on Spacecraft Contamination Model Development. SPIE 4096:138–156.

[5] Keith C, Albyn. 2007. Outgassing Measurements Combined with Vacuum Ultraviolet Illumination of the Deposited Materials. Journal of Spacecraft and Rochets, 44(2):102–108.

[6] Khassanchine R H, Grigorevskiy A V, Galygin A N. 2004. Simulation of Outgassing Processes in Spacecraft CoatingsInduced by Thermal Vacuum Influence [J]. Journal of Spacecraft and Rochets, 41(3):384–388.

[7] Laikhtman A, Gouzman, I, Verker R, et al. 2009. Contamination Produced by Vacuum Outgassing of Kapton Acrylic Adhesive Tape. Journal of Spacecraft and Rochets, 46(2):236–240.

[8] Rampini R, Grizzaffi L, Lobascio C. 2003. Outgassing Kinetics Testing of Spacecraft Materials. Materials Sicence & Engineering Technology, 34(4): 359–364.

[9] Wang X R. 1989. Mathematical Analysis and Experimental Verification on Spacecraft Materials Mass Loss Process. Chinese Space Science and Technology, 3:8–20.

[10] Wang X R, Fan C Z. 1994. A Diffusion Model upon Spacecraft Materials Acceleration Outgassing Process. Journal of Astronautics, (1):55–59.

[11] Wang X R, Ma W J. 2002. A Simulation Equipment Used for Determining the Outgassing Contamination Condensation Effect on Cryogenic Sensitive Surface in Space. Journal of Astronautics, 23(3):68–71.

Bio and medical materials

Material Science and Engineering – Chen (Ed.)
© *2016 Taylor & Francis Group, London, ISBN 978-1-138-02936-1*

The progress of vegetable oil-based lubricants

H. Wen, Y.G. Shi, G.Z. Fan, P.F. Xu & Y.X. Jiang
Department of Training, Logistical Engineering University, Chongqing, P.R. China

B. Su
Oil Application Group, Being Oil Institute, Beijing, P.R. China

H.F. Gong
Vehicle Engineering College, Chongqing University of Technology, Chongqing, P.R. China

ABSTRACT: Vegetable oils are promising substitutes for mineral lubricants for their excellent lubricity, biodegradability and viscosity-temperature characteristics; however, because of thermo-oxidative unstability and lower viscosity, they could not be utilized widely. Here, the compositions and properties of the vegetable oils used as biodegradable lubricants are discussed systematically, and the modification methods of vegetable oils, such as adding additives, using addition reactions and chemical modification, are reviewed thoroughly.

Keywords: vegetable oil, biodegradable, lubricants and chemical modification

1 INTRODUCTION

Generally, vegetable oil-based lubricants, with a poor thermal-oxidation stability and lower viscosity, however, have some excellent properties as lubricants, such as high-viscosity index, high lubricity, low volatility, and, especially, both low toxicity and high biodegradability. In 1970s, Europe began to study vegetable oil-based lubricants and had legislated strict regulations to promote their development and application. The past studies in the USA have shown the potential of vegetable oil-based lubricants instead of mineral oil-based products for metal-cutting fluids and transmission fluids. Recently, Nankai University, Shanghai Jiaotong University and Chinese Petroleum University have also conducted research on vegetable oil-based lubricants and their derivatives. Currently, a range of vegetable oil-based products have been developed and commercially available from various suppliers, and the demand for vegetable oil-based lubricants is increasing strongly.

as lubricating oils have a long history. D. Johnson [5] in Colorado State University combined rapeseed oil, sunflower oil, soybean oil and castor oil together as engine oil, resulting in a reduced harmful substance of 20%~30%. M. Shahabuddin [6] reported a bio-lubricant that consists of 10 wt% to 50 wt% Jatropha oil and SAE 40. Vegetable oil containing 5% graphite and 2% serpentine was used as drilling fluids with excellent anti-wear and antifriction properties at a high or low temperature [7]. Bhawna Khemchandani [8] studied the tribological properties, thermo-oxidative stability of the mixture of safflower oil, synthetic ester and 0.1 wt% to 0.6 wt% antioxidant. F. Murilo [9] showed that the vegetable oil had a poor anti-oxidation stability, but a high biodegradability at 700 kPa, 140°C compared with mineral oils and synthetic esters. Obviously, vegetable oils can directly be used as a lubricant. However, the major concerns with vegetable oils are their poor low-temperature, low-viscosity and poor-oxidative properties.

2 VEGETABLE OILS

Fatty acid glyceride esters are the major components of vegetable oils and also a little free fatty acids, phospholipids, wax, tocopherol, and VE. The kinds of fatty acids determine the performance of vegetable oils [3–4]. Vegetable oils, such as palm oil, soybean oil and castor oil, directly used

3 MODIFICATION OF VEGETABLE OILS

3.1 *Additives*

Two kinds of additives are added into the oils for enhancing their performances. The one takes its action by molecular deformation, adsorption and solubility, such as viscosity index improver, oiliness agent, and pour point depressant. The other one takes its function in terms of the chemical reactions occurring at the contact interface, such as extreme pressure anti-wear agent, antioxidant, and dispersant. At present, more research is focused on the extreme pressure and anti-wear agents and antioxidants [10].

3.1.1 Extreme pressure and anti-wear agents

Extreme pressure and anti-wear agents, which contain specific ingredients, namely S, P, Cl, B or their salts, react with the metal surface to form a reaction film with a lower shear strength than the metal, playing an extreme pressure and anti-wear effect. Song Jiantong [11] studied the effect of sulfurized isobutylene T321, sulfo ammonium phosphate T307 and two benzyl two sulfide T322 on the chemical and friction properties of rapeseed oil, and showed that these additives can improve the anti-wear and extreme pressure properties of vegetable oil-based lubricants and reduce their friction. Weimin Li [12] used Natural Garlic Oils (NGO) as a high-performance environmental-friendly extreme pressure additive for lubricating oils. The four ball test results revealed that the 1 wt% NGO in the base fluids could significantly improve the weld point from approximately 1236 N to 8000 N or higher. L.A. Quinchia [13] used an Ethylene–Vinyl Acetate copolymer (EVA) as the viscosity modifier for vegetable oils.

3.1.2 Antioxidants

Vegetable oils are oxidized according to the allyl-radical oxidation mechanism. Antioxidants can be generally divided into three categories, namely chain termination agent, peroxide decomposer and metal inhibitor. Metal inhibitor mainly prevents the metal catalytic oxidation reaction, which forms a stable complex with metal ions, such as Fe, Cu and Mn, and can produce a synergistic effect when used with other antioxidants. C.G. Tsanaktsidis [14] synthesized a biodegradable thermal oxidation stability agent TPA (Thermal Polyaspartate Anion), and its good thermal oxidation stability was shown in rapeseed oil used for diesel engine oil. Sevim Z. Erhan [15] investigated the relationship between ZDDC and soybean oil's oxidation stability. The results showed that there is an obvious synergistic effect between ZDDC and anti-wear agent in soybean oil. The antioxidant can effectively improve most of the vegetable oil's oxidation stability when 0.1%~0.2% are added. Weimin Li, L.A. Quinchia and C. G. Tsanaktsidis et al. conducted some pioneer work in the special additive for vegetable oils. However, the conventional additives also present a negative effect on biodegradability. In order to improve the oxidation stability, low-temperature fluidity and lubricity, it is very necessary to design biodegradable additives for vegetable oils.

3.2 Chemical modification

The double bond in vegetable oils is vulnerable to oxygen attack, resulting in poor oxidation stability. If the double bond could be saturated by chemical methods, the vegetable oil's properties would be enhanced radically.

3.2.1 Addition

Some functional groups can be linked, such as alkane, hydroxyl or other functional elements such as S, P, Cl, N and B. Copper, nickel or its alloys could be used as the catalyst. Hu Zhimeng [16] synthesized hydroxyl vegetable oils by the addition reaction, and their tribological properties were investigated. The results showed that the preferable extreme pressure and anti-wear properties of the modified vegetable oils were achieved. Lou A.T. Honary [17] developed a biodegradable additive for soybean oil-based multigrade hydraulic oils. Ravasio [18] found that the catalyst Cu/SiO_2 prepared by "chemical adsorption and hydrolysis" presents a good selectivity for vegetable oil's hydrogenation, and the methyl ester of the hydrogenated vegetable oil gives an excellent oxidation stability.

3.2.2 Epoxidation

Peracetic acid (1:0.5~2) with H_2O_2 was used as the epoxidation agent. Zeolite, ion exchange resin, $NaCO_3$, $(NH_4)_2SO_4$ or $SnCl_2$ were used as the catalyst. Adhvaryu [19] studied the Epoxy Soybean Oil (ESBO), Soybean Oil (SBO) and High Oleic Soybean Oil (HOSBO) by means of NMR. The ball-on-disk experiment showed that ESBO presented better lubricating properties and oxidative stability. Yan Weili [20] studied $Al_2Si_2O_5 \cdot nH_2O$ modified by the quaternary ammonium functional group and phosphotungstic acid as catalysts for the epoxidation of soybean oil.

R. D. Kulkarni [24] studied the thermal oxidative stability and low-temperature fluidity of mustard oil, and pointed out that the base oil prepared with the ring-open by 2-ethylhexyl alcohol had a higher viscosity index and a good low-temperature fluidity.

3.2.3 Esterification

Li Qinghua [25] found that acetylated rapeseed oil produced less sludge in the rotary oxygen bomb test, and the viscosity-temperature properties and biodegradability were also improved when mixed with mineral oil. G. Gorla [26] prepared the epoxidized karanja fatty acid methyl, butyl, 2-methyl-1-propyl, and 2-ethylhexyl esters from the renewable nonedible source karanja oils. Zhou Shirou [27] gained rapeseed acid-Neopentyl-Glycol (NPG)-Sebacic Acid (SA) ester, which had excellent lubrication performances. Generally, the esterification products have a poor low-temperature performance

Table 1. The common methods for the vegetable oil's modification [19–26].

	Addition reaction	Epoxidation reaction	Esterification reaction
Catalyst	Simple metal, alloy Anti-acidosis	H_2SO_4, 120 ion exchange resin, Na_2CO_3, $(NH_4)_2SO_4$ or $SnCl_2$	H_2SO_4, solid acid, metal and metallic oxide
Process conditions	Three-phase system, strict condition, 200°C, 2 MPa, remove phospholipid and protein, H_2SO_4, Cl^-	Strict catalytic agent, inertness condition, removing residual catalyst Environmental pollution	Mild condition, without pretreatment, homogeneous reaction, removing residual catalyst and post-processing difficultly
Development direction	Optimizing catalyzer, improving catalytic activity, designing functional group molecule	Optimizing catalyzer, improving catalytic activity, reducing environmental pollution Ring opening of the epoxy group is beneficial to the molecular design	Molecular design with addition reaction and epoxidation, nanometer catalyst

unless mixed with synthetic ester, PAO or mineral oil. Sevim Z. Erhan [16], and Cecilia Orellana Akerman [28] conducted some research on the PAO and synthetic ester as the diluent, respectively. Table 1 outlines the advantages and disadvantages of the modification methods.

3.3 Biological modification

Biological modification is a method that utilizes the genetic or biological gene technology for increasing saturated components and reducing the unsaturated component in the vegetable oils for gaining the expected properties. Kaab [27] reported one kind of the sunflower seed oil, in which the oleic acid content can be as high as 90% and stearic acid content ranged only between 1.0% and 1.5%. Grant Ian [28] gained rapeseed with high oleic acid content. Cole Glenns [29] obtained sunflower seed oil with 65% content of the oleic acid. Compared with other techniques, the biological modification is environmental friendly, and promising for promoting a new way of the transgenic plant oil's deep process.

4 SUMMARY

With the progress of science and technology, an increasing number of vegetable oil-based lubricants will be applied for meeting the needs of sustainable development of environments. Recently, studies have mainly focused on overcoming the inherent weaknesses of vegetable oil-based lubricants by the varieties of the modification, such as modern biological technique for obtaining the desired vegetable oil, molecular design for gaining expected vegetable oil and developing environmental-friendly additives.

Also, it is very imperative to constitute the standards of vegetable oil-based lubricants for normalizing their production and management.

ACKNOWLEDGMENTS

This research was supported by NSFC 21206204.

REFERENCES

[1] Jin Zhiliang, Xiong Jing, Wang Yu-min. The Biodegradable Green lubricating oil [J]. Environmental Protection and Security, 2006(4):86–88.

[2] Wang Lingyun, Yang Li-ting etc. Research progress of green polymers based on vegetable oil [J]. Applied Chemical Industry, 2009(5):724.

[3] Han Hengwen, Liu Xuebin. Research Progress of modification methods of vegetable oils as environmentally friendly lubricant basestocks [J]. Lubricating Oil, 2008, 23(6): 6.

[4] Huang Fenglin, Huang Yong, Zhu Shan. Research and Application of the Biodegradable Green lubricating oil [J]. Petrochemical Technology & Application, 2009 (11): 567.

[5] Bai Yang, Zhan Lin-cong etc. Progress of research on green lubricating oils based on vegetable oils [J]. Journal of Wuhan Polytechnic University. 2009(6):50

[6] M. Shahabuddin H.H. Masjuki, M.A. Kalam. Experimental investigation into tribological characteristics of bio-lubricant formulated from Jatropha oil [J]. SciVerse ScienceDirect/Procedia Engineering, 2013(56):597–606.

[7] Wang Wanjie. LI Changsheng. Study on the preparation and Tribological Properties of Drilling Fluid Vegetable Oil Lubricant [J]. Lubrication Engineering. 2010(1):29–32.

[8] Bhawna Khemchandani, A.K. Jaiswal etc. Mixture of safflower oil and synthetic ester as a base stock for biodegradable lubricants. Lubrication Science (2013) DOI: 10.1002/ls.

[9] F. Murilo T. Luna, Breno S. Rocha etc. Assessment of biodegradability and oxidation stability of mineral, vegetable and synthetic oil samples. [J]. Industrial Crops and Products, 33(2011):579–583.

[10] Xu Xiaohong. The industrial Application of Rapeseed Oil [J]. Grain science and technology in Si Chuan, 1996, 4(52): 32.

[11] Song Jiantong, LV Jiang-yi, Zhu Chunhong. Effect of sulfur and phosphorus additives on antiwear properties of vegetable oil [J]. Journal of Southwest Petroleum University (Science & Technology Edition), 2010, 8(4):160–162.

[12] Weimin Li1, Cheng Jiang1, Mianran Chao. Nature Garlic Oils as High Performance Environmental Friendly Extreme Pressure Additive in Lubricating Oils [J]. ACS Sustainable Chemistry & Engineering 2014.2.17.

[13] L.A. Quinchia, M.A. Delgado*, C. Valencia, J.M. Franco, C. Gallegos. Viscosity modification of different vegetable oils with EVA copolymer for lubricant applications [J]. Industrial Crops and Products, 32 (2010): 607–612.

[14] Constantinos G. tsanaktsidis, Stavros G. Christidis etc. A novel for improving the physicochemical properties of diesel oil and jet fuel using polyaspartate polymer additives [J]. Fuel, 104(2013):155–162.

[15] Sevim Z. Erhan et al. Oxidation and low temperature stability of vegetable oil-based lubricants [J]. Industrial Crops and Products, 24 (2006):292–299.

[16] Hu Zhimeng. Hydroxy vegetable oil fatty acid synthesis and lubricity [J]. Chemical science and technology, 2001, 9(3): 1–4.

[17] Lou A T Honary. Soybean Based Hydraulic Fluid [P]. US:5972855, 1999.

[18] Ravasio, Zaccheria F. Environmental friendly lubricants through selective hydrogenation of rapeseed oil over supported copper catalysts [J]. Applied Catalysis, 2002, 233:1–6.

[19] Adhvaryu, S Z Erhan. Epoxidized Soybean Oil as a Potential Source of High—temperature Lubricants [J]. Industrial Crops and Products, 2002,15(3):247–254.

[20] Liwei Yan. The Study of Soybean Oil Epoxidated by Heterogeneous Catalyst [D]. School of Chemical Engineering and Energy. May 2011.

[21] Alejandrina Campanellaa, Eduardo Rustoyb, Alicia Baldessarib etc. Lubricants from chemically modified vegetable oils [J]. Bioresource Technology, 2010(101): 245–254.

[22] Wang Bin. Application and research on lubrication of Two new environmental friendly lubricants [D]. Shang Hai University. Shang Hai, 2005, 8, Doctor's Thesis.

[23] Arumugam S, Sriram G. et.al. Synthesis, Chemical Modification and Tribological Evaluation of Plant oil as Bio-Degradable Low Temperature Lubricant [J]. Procedia Engineering, 2012(38):1508–1517.

[24] Ravindra D. Kulkarnia, Priya S. Deshpande etc. Epoxidation of mustard oil and ring opening with 2-ethylhexanol for bio-lubricants with enhanced thermo-oxidative and cold flow characteristics. Industrial. Crops And Products, 49 (2013): 586–592.

[25] Cecilia Orellana Akerman, Yasser Gaber. et al. Clean synthesis of biolubricants for low temperature applications using heterogeneous catalysts [J]. Journal of Molecular Catalysis B: Enzymatic, 2011 (72):263–269.

[26] Brajendra K Sharma, A Adhvaryu, Zengshe Liu, et al. Chemical Modification of Vegetable Oils for Lubricant Applications [J]. JAOCS, 2006, 83(2):129–136.

[27] Kaab. H. Proceeding of the final conference on CTVO-NET-Chemical-Technical utilization of vegetable oils [J]. Bonn June, 2000(50):20–21.

[28] Grant Ian, Charne David G. et al. Brassicanapus vegetable oil wherein the levels of oleic, alpha-linolenic,and saturated fatty acids are endogenously formed and are simultaneously provided in a typical highly beneficial distribution via genetic control [P]. US:6011164A, 2000, 1–4.

[29] Cole Glenns S. Hazebroek Jan P. Endogenous vegetable oil derived from helianthus annus seeds wherein the levels of palmitic acid and oleic acid are provided in an atypical combination [P]. US:587227A, 2000, 2–16.

Material Science and Engineering – Chen (Ed.)
© 2016 Taylor & Francis Group, London, ISBN 978-1-138-02936-1

Numerical simulation and analysis of heat and mass transfer process in hot air drying of agricultural products

M. Wang
SINOPEC Shijiazhuang Refining and Chemical Branch, Shijiazhuang, China

T. Lin
China Nuclear Power Engineering Co. Ltd., Hebei Branch, Shijiazhuang, China

ABSTRACT: In order to study the heat and mass transfer process in convection drying of agricultural product, two-dimensional mathematical model was established, and the validation of model was verified by experimental data. The contrastive analysis using average dry basis moisture content showed that the errors between simulated results and experimental data were less than 5%. The simulation results showed that: dry basis moisture content gradually decreases while the temperature of material increased under the action of hot air; the temperature and dry basis moisture content showed a gradient inside the agricultural product during hot drying process.

Keywords: convection drying; heat and mass transfer; numerical simulation

1 INTRODUCTION

The hot drying process of agricultural products is an operation consisting of evacuating the solvent (water or otherwise) that existed in void (pore) space. This process has an advantage of storing, transporting and secondary processing to food products. It has been judged as a complex phenomenon in which heat and mass transfer simultaneously occurred.

Investigations on heat and mass transfer mechanism is important to enhance in understanding the overall drying process. Simulation of temperature and moisture distribution in drying process has been studied by many authors in recent years [1–4]. They did some experimental research on the drying characteristics of agricultural products such as carrots, peppers, and the relationship of drying time and moisture content was fitted with polynomial curve, however, they didn't study of heat and mass distribution in the process of hot drying. Within this context, the objective of the present work is to establish mathematical model to analyze the heat and mass transfer processes in hot air drying of agricultural product. And comparison studies with experimental data from drying of potatoes will make to validate the mathematical model.

2 MATHEMATICAL MODEL

The physical model (i.e. the calculation zone) was shown in Figure 1, the following assumptions were made in order to establish mathematical model of two-dimensional unsteady heat and mass transfer: 1) The thermal physical properties remain the same in the process of hot drying; 2) Ignore the contraction deformation in the process of hot drying; 3) Without heat source in the interior of the material; 4) The temperature of the hot air stays the same; 5) Ignore the thermal radiation in the drying process.

2.1 Control equation

Heat transfer control equation [5]:

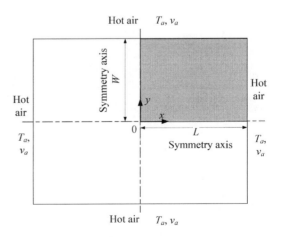

Figure 1. Two-dimensional physical model.

$$\frac{\partial(\rho C_p T)}{\partial t} = \kappa \left(\frac{\partial^2 T}{\partial x^2} + \frac{\partial^2 T}{\partial y^2} \right)$$

$$+ D_{eff}(\rho C_p) \left(\frac{\partial X}{\partial x} \frac{\partial T}{\partial x} + \frac{\partial X}{\partial y} \frac{\partial T}{\partial y} \right) \quad (1)$$

where ρ is bulk density, kg/m^3; C_p is heat capacity, J/kg/K; T is temperature, K; t is time, s; κ is thermal conductivity, W m/K; D_{eff} is diffusion coefficient, m^2/s; X is dry basis moisture content, kg/kg d.b.

Mass transfer control equation [5]:

$$\frac{\partial X}{\partial t} = D_{eff} \left(\frac{\partial^2 X}{\partial x^2} + \frac{\partial^2 X}{\partial y^2} \right) \quad (2)$$

2.2 Initial conditions and boundary conditions

The initial conditions:
When $t = 0$,

$$X = X_0 \text{ and } T = T_0 \quad (3)$$

The boundary conditions:
at $0 \leq x \leq L$, $y = 0$ and $x = 0$, $x = 0$, $0 \leq y \leq W$ as shown in Figure 1:

$$\frac{\partial X}{\partial x} = 0, \frac{\partial X}{\partial y} = 0;$$

$$\frac{\partial T}{\partial y} = 0, \frac{\partial T}{\partial x} = 0 \quad (4)$$

at $x = L$, $0 \leq y \leq W$ as shown in Figure 1:

$$\dot{m}_s = -\rho D_{eff} \frac{\partial X}{\partial x}$$

$$h(T_a - T_{surf}) = (\dot{m}_s L_v(T_{surf})) + \kappa \frac{\partial T}{\partial x} \quad (5)$$

at $0 \leq x \leq L$, $y = W$ as shown in Figure 1:

$$\dot{m}_s = -\rho D_{eff} \frac{\partial X}{\partial x}$$

$$h(T_a - T_{surf}) = (\dot{m}_s L_v(T_{surf})) + \kappa \frac{\partial T}{\partial y} \quad (6)$$

where $\dot{m}_s$ is surface rate of moisture vaporization, kg/m^2/s.
and

$$\dot{m}_s = \frac{h_m M_v}{R} \left(\frac{P_{v,sat \, at \, T}}{T} - \frac{RH \times P_{v,sat \, at \, T_a}}{T_a} \right) \quad (7)$$

where h_m is mass transfer coefficient, m/s; M_v is molecular weight of water, kg/kmol; R is universal gas constant, J/kmol/K; RH is relative humidity.

3 MODEL VALIDATION

The explicit finite difference method was used for the time term and the central difference scheme was used for the space term in the numerical resolution of the model. The simulation steps were as follows: 1) input initial drying conditions, the geometry model and the number of grid division; 2) calculate the physical parameters of material; 3) calculate boundary and internal nodes difference equation in the time dependent heat and mass parameters at that time t; 4) solve the differential equation, calculate dry basis moisture content and temperature distribution at each point; 5) at that time t+1, return to step 3)~5), until the difference of moisture content between calculated result and setted value less 0.001, then output simulation results, the calculation ends.

Due to the difficulty of direct measurement of moisture content inside the material, average dry basis moisture content was adopted to valid the mathematical model in this paper. Experiment process was as followed: fresh potatoes used in experiment were brought from local supermarket. Before drying, potatoes were hand peeled and cut into rectangle-shaped slices (dimensions in mm: 45 × 20 × 10). The slices were blanched in water at 95°C during 5 min to prevent enzymatic browning and immediately cooled in water at 20°C during few minutes to remove excess heat. Finally, the water remaining at the surface was removed with a filter paper. Two of the long sides of the samples was covered with a thin aluminum film in order to make it waterproof. The dry mass of the product was determined after each experiment by vacuum oven drying method at 80 °C for 24 h.

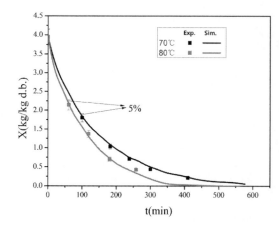

Figure 2. Dry basis moisture content trendy over time for experiment and simulation.

Verification experiment conditions: air temperatures were 70 °C and 80 °C respectively, air velocity was 1 m/s, initial dry basis moisture content was 3.95 kg/kg d. b., environment temperature was 25 °C, relative humidity was 85%.

Figure 2 shows a comparison between numerical result and experimental data for the mean moisture content. As shown, the deviation between calculated results and experimental values for all four cases was less than 5%. Therefore, the mathematical model this article was a reliable theoretical model for describing agricultural material drying process.

4 RESULTS AND DISCUSSION

The results discussed in this section were obtained under the numerical calculation conditions of va = 1 m/s, Ta = 343.15 K.

The temperature distribution at different instants of time in the drying process of potato were shown in Figure 3. It can be seen that when the drying process began, the temperature of material surface rose quickly under the action of heat transfer between material and hot air, as shown in Figure 3 a) and b); as the drying process going, the distribution of temperature inside material showed a gradient under the action of heat conduction, and the diffusion distribution of temperature was low near the centre and high near the surface, as shown in Figure 3 b) and c); after a time of heat transfer and heat conduction the temperature inside material rose to the close lever of the temperature of hot air, as shown in Figure 3 d).

Figure 4 showed the distribution of dry basis moisture content at various time points during the hot drying process of potato. It can be seen that dry basis moisture content showed a gradient inside the the material; the distribution of dry basis moisture content was high near the centre and low near the surface in general.

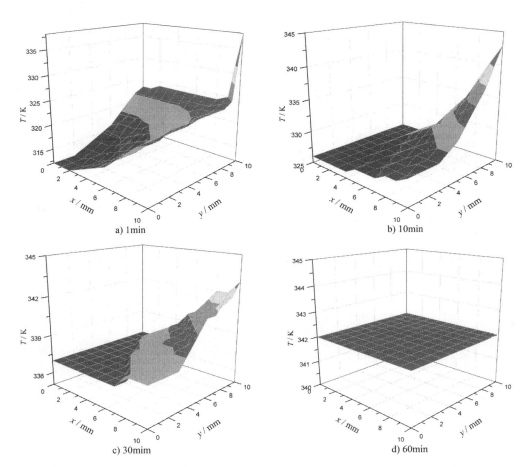

a) 1min

b) 10min

c) 30mim

d) 60min

Figure 3.   Temperature distribution for different time.

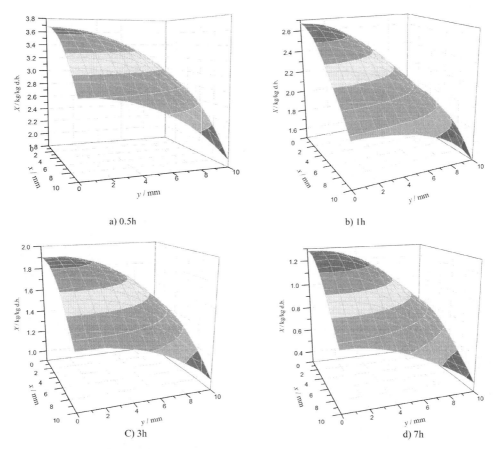

a) 0.5h b) 1h C) 3h d) 7h

Figure 4. Dry basis moisture content distribution for different time.

5 CONCLUSION

1. The mathematical model presented in this article can be used to analyze the distribution of temperature and dry basis moisture content inside the agricultural product.
2. The distribution of temperature and dry basis moisture content inside the agricultural product showed a gradient inside the agricultural product during hot drying process, and the diffusion distribution of temperature was low near the centre and high near the surface; the distribution of dry basis moisture content was high near the centre and low near the surface in general.

REFERENCES

[1] Huang Yanbin, Zheng you, Chen Haiqiao, Li ying, Chen Hourong. Hot-air drying characteristic and mathematical model of lemon. Science and Technology of Food Industry, 2012 33(14): p. 169–172 + 191.

[2] Jiang Gan, Chen Manqi, Liang Jiexin, et al. Drying characteristics and mathematical model of hoy air drying of carrot [J]. Guangdong Agricultural Sciences, 2013, 40(22): 106–110.

[3] Li jing, Xiao Xia PuXiaoLu, Huang Yanbin, Chen Hourong. Characteristics and Mathematical Model of Hot-Air Drying for Purple Sweet Potato. Food science, 2012 33(15): p. 90–94.

[4] Zhang jianjun, hai-xia wang, Ma Yongchang Zheng Yan. Experimental research on hot-air drying properties of capsicum. Transactions of the CSAE, 2008 24(3): p. 298–301.

[5] Yu, C.M. Numerical Analysis of Heat and Mass Transfer for Porous Materials (A Theory of Drying). Beijing: Tsinghua University Press, 2011.

Material Science and Engineering – Chen (Ed.)
© 2016 Taylor & Francis Group, London, ISBN 978-1-138-02936-1

The establishment of HPLC analysis method of 4-chloro-4'-fluorobutyrophenone

Y. Zhang, S.C. Zhao, H.Y. Qu, H. Zhao, H.X. Du, J.X. Yang, S.H. Mu & C. Zheng
School of Chemical and Pharmaceutical Engineering, Hebei University of Science and Technology, Shijiazhuang, Hebei, China

ABSTRACT: 4-Chloro-4'-fluorobutyrophenone is always used as an important intermediate in the pharmaceutical industry to synthesize dipfluzine, haloperidol and droperidol. In this paper, a stability-indicating and accurate high performance liquid chromatography method for analyzing 4-chloro-4'-fluorobutyrophenone was developed and validated. Chromatographic analysis was performed on a Waters E2698 liquid chromatograph equipped with Thermo C18 column (250×4.6 mm, 5 μm) and PDA detector (full wavelength scanning and extracting chromatograms under 243 nm) using a mobile phase composed of methanol and water (70/30, v/v) with flow rate of 1.0 mL·min^{-1} and injection volume of 20 μL. Under the established condition, a linear response was obtained for concentration range of 0.05–0.2 mg·mL^{-1} ($R^2 = 0.9994$). In addition, the accuracy was acquired as 101.06% of average recovery with RSD = 1.85%, the precision and repeatability were also achieved with RSD = 1.69% and 1.89% respectively. It was shown that the method will be adequate for routine analysis of 4-chloro-4'-fluorobutyrophenone.

Keywords: 4-Chloro-4'-fluorobutyrophenone; HPLC; stability; analytical method; methodology validation

1 INTRODUCTION

4-Chloro-4'-fluorobutyrophenone can be used as a very crucial intermediate in the pharmaceutical field for the synthesis of many drugs such as haloperidol, droperidol and dipfluzine. So it is very necessary to establish a completely stable, accurate and validated method for the analysis of 4-chloro-4'-fluorobutyrophenone.

4-Chloro-4'-fluorobutyrophenone can be synthesized by Friedel-Crafts acylation reaction from 4-chlorobutyryl chloride and fluorobenzene. The content determination methods for 4-chloro-4'-fluorobutyrophenone have been reported in literature including polarimetric method and HPLC method. Polarimetric method is easy to use but always not accurate enough for content determination and the reported HPLC method can not be used as a routine detection because it has no methodology validation. In this paper, an accurate and stability-indicating high performance liquid chromatography method was developed and validated for analyzing 4-chloro-4'-fluorobutyrophenone.

2 EXPERIMENTAL SECTION

2.1 Instrumentation and materials

A liquid chromatography system (Waters E2698) which was equipped with a quaternary pump, an automatic injector, a vacuum degasser, a Thermo C18 column (250×4.6 mm, 5 μm) and a PDA detector (Waters 2998) was used to perform the analysis of 4-chloro-4'-fluorobutyrophenone. The mobile phase was methanol (Thermo Fisher Scientific, chromatographically pure) and water (HPLC grade). For weighing the materials, an analytical balance (Mettler Toledo AL104) was used. The standard sample of 4-chloro-4'-fluorobutyrophenone with a purity of 99.38% were synthesized according to the literature and purified by column chromatography in our lab. The samples with a small amount of impurities were synthesized with the same method in our lab. Hydrochloric acid, sodium hydroxide and 30% hydrogen peroxide solution used were all of analytically pure.

2.2 Chromatographic conditions

An isocratic mobile phase which consisted of methanol and water (70/30, v/v) was used. Prior to HPLC analysis, the mobile phase was vacuum-filtered through a 0.45 μm membrane and degassed by ultrasonication for 15 min. The temperature of the column was maintained at 45°C. The flow rate of mobile phase was 1.0 mL·min^{-1} and the injection volume was 20 μL. The Photo Diode Array (PDA) detector scanned on full wavelength and the chromatogram at 243 nm was extracted when the chromatograms and data were processed.

2.3 Preparation of standard and sample solutions

About 0.01 g standard sample of 4-chloro-4′-fluorobutyrophenone was weighed accurately and diluted with methanol to 100 mL in a volumetric flask to obtain the standard solution with the concentration of 0.1 mg·mL^{-1}. The sample solution of 4-chloro-4′-fluorobutyrophenone which was synthesized in our lab was prepared with the same method.

2.4 System suitability

Under the chromatographic conditions, the standard solution and the sample solution were analyzed. As shown in the obtained chromatograms, the retention time (Rt) of 4-chloro-4′-fluorobutyrophenone was 5.3 min, the number of theoretical plates (N) was 6890, separating degree (R) of the main peak and adjacent impurity peak was greater than 1.5, symmetry of the main peak was well and the tailing factor was 1.05. Therefore, the chromatographic conditions were appropriate for the analysis of 4-chloro-4′-fluorobutyrophenone.

2.5 Linearity

Linearity is the degree of the linear relationship between test results (or measured response signal) and the test items which are variable. In this research, linearity was determined by the peak areas of 4-chloro-4′-fluorobutyrophenone measured by chromatograms and the concentration of 4-chloro-4′-fluorobutyrophenone in the solutions. 0.1047 g standard sample was weighed precisely and diluted with methanol to yield seven solutions with concentrations of 0.05205 mg·mL^{-1}, 0.08328 mg·mL^{-1}, 0.1041 mg·mL^{-1}, 0.12492 mg·mL^{-1}, 0.16656 mg·mL^{-1} and 0.2082 mg·mL^{-1} respectively. Each solution was detected twice. A linear regression equation (Y = 66489055.876 X + 307732.8186) was obtained according to the measured average peak area of each concentration. The R^2 obtained was 0.9994, which meant that the linearity was acceptable for the analysis within the concentration range evaluated. The data obtained is shown in Table 1 and the linear regression figure is shown in Figure 1.

2.6 Specificity

Specificity refers the characteristics that the method can correctly determine the analyte when other components such as impurities, degradation products or excipients may exist. In this experiment, samples were treated with acid, alkali or oxide for a certain time respectively and then analyzed. From the chromatographs and the data (Table 2), we find that some degradation products form but these do not interfere

Table 1. Results for linearity.

Number	Concentration (mg·mL^{-1})	Average peak area (mAU)
1	0.05205	3892867
2	0.08328	5802641
3	0.10410	7133776
4	0.12492	8548691
5	0.16656	11411005
6	0.20820	14200141

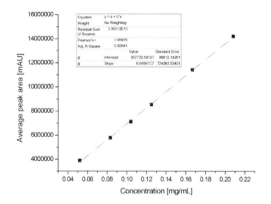

Figure 1. The linear regression figure.

Table 2. Results for specificity.

Reagent	Time (min)	Average peak area (mAU)	Undegraded analyte (%)
0.1 mol·L^{-1} HCl	15	8632748	97.7
	30	8040335	90.7
	45	7363296	82.8
	60	6792577	76.1
0.01 mol·L^{-1} NaOH	5	9368551	87.3
	10	8027894	74.4
	15	7186752	66.3
	30	6023440	55.1
30% H$_2$O$_2$	30	9965425	98.0
	60	9533276	93.6
	120	8968810	87.9
	360	7829773	76.3

with the analysis of 4-chloro-4′-fluorobutyrophenone. As shown in Table 2, the degradation of 4-chloro-4′-fluorobutyrophenone in 0.1 mol·L^{-1} HCl solution can be reached to 9.3% after 30 min and the damage extent increases with the extension of time. In the solution of 0.01 mol·L^{-1} NaOH, degradation of 4-chloro-4′-fluorobutyrophenone is more easy with 12.7% for 5 min. Besides, 12.1% of 4-chloro-4′-fluorobutyrophenone is damaged in 30% H$_2$O$_2$ for 120 min and degradation increases with the time

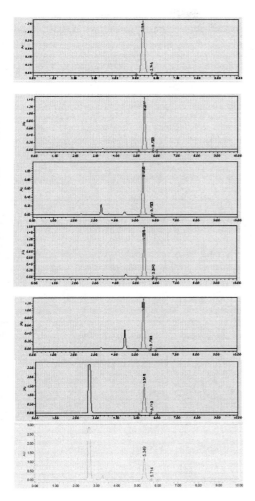

Figure 2. The first picture is liquid chromatogram of original sample; the second and the third liquid chromatograms are for samples in 0.1 mol·L⁻¹ HCl solution for 30 min and 6 h respectively; the next two liquid chromatograms are for samples in 0.01 mol·L⁻¹ NaOH solution for 5 min and 10 min respectively; and the last two liquid chromatograms are for sample in 30% H₂O₂ solution for 1 h and 6 h respectively.

extension. Some of the liquid chromatograms are shown in Figure 2.

2.7 Accuracy

Accuracy is determined by the recovery test. Known amounts of standard samples corresponding to 80%, 100% and 120% concentration of 4-chloro-4′-fluorobutyrophenone of the test samples were added to the test sample solutions and each sample solution was prepared in triplicate. 17.5 mg sample was weighed and dissolved with methanol to 10 mL, 0.5 mL of the above solution was taken

out with a pipette to a 10 mL volumetric flask. A certain amount of 0.135 mg·mL⁻¹standard solution was added into the above volumetric flask and diluted with methanol to 10 mL. Through measuring and calculating, the results are shown in Table 3. As seen from Table 3, the mean percentage of recovery is 101.06% and the RSD is 1.85% which indicate that the accuracy of the method is satisfied.

2.8 Precision

In our experiment, 0.0559 g standard sample was weighed and diluted with methanol to 50 mL, 1 mL of the above solution was taken out with a pipette to metred volume to 10 mL. The solution was detected for 6 times continuously and the peak area is shown in Table 4. As seen from Table 4, the satisfied RSD is obtained.

2.9 Repeatability

In this process, different concentrations of sample solutions (n = 6) were prepared and analyzed under the conditions. About 21 mg samples were diluted with methanol to 100 mL and detected by HPLC respectively. The contents of 4-chloro-4′- fluorobutyrophenone in the samples were calculated basing on the linear regression equation (Table 5) and the results show that the RSD is 1.89% which indicates that the repeatability of the method is satisfied.

Table 3. Results for accuracy.

Sample weighed (mg)	Standard solution added (mL)	Recovery (%)	Average recovery (%)	RSD (%)
17.4	3	102.57		
17.5	3	99.41		
17.8	3	99.03		
18.5	4	98.40		
17.9	4	103.64	101.06	1.85
17.6	4	100.34		
17.2	5	101.13		
17.3	5	102.00		
17.6	5	102.98		

Table 4. Results for precision.

Number	Peak area (mAU)	RSD (%)
1	7606854	
2	7289829	
3	7419367	
4	7544070	1.69
5	7401049	
6	7311051	

Table 5. Results for repeatability.

Sample weighed (mg)	Peak area (mAU)	Content (%)	RSD (%)
20.3	7834930	55.77	
20.8	7783372	54.05	
21.6	7884755	52.76	1.89
21.5	7981576	53.68	
21.0	7762684	53.39	
21.7	8044518	53.62	

Table 6. Results for stability of solution.

Sample weighed (mg)	Time (h)	Peak area (mAU)	RSD (%)
	0	11946894	
	2	11751017	
	4	11841011	
18.4	6	11801367	0.74
	8	11928379	
	10	11966127	

2.10 Stability of solution

Stability of solution refers the change of the composition with time extension under the condition of room temperature, from which we can determine the available time of the sample solution. In our experiment, 18.4 mg sample was weighted and diluted with methanol to 100 mL and the sample solution was stayed at room temperature for 0, 1, 2, 4, 8, 10 h respectively before injected to analyze. Detecting result is shown in Table 6. Seen from Table 6, the obtained RSD is 0.74% which illustrates that the solution is stable within 10 h.

2.11 Detection Limit (DL) and Quantification Limit (QL)

DL is the minimum amount of analyte that could be detected but not necessarily quantified as an exact value. The DL reflects whether the method has sufficient sensitivity or not and it is defined as the concentration of the analyte in the sample solution with the fixed signal-to-noise ratio (S/N = 3). The detected result shows that DL of 4-chloro-4'-fluorobutyrophenone is 131 ng·mL^{-1}.

QL is the smallest possible quantity of analyte that can be quantified and it is defined as the concentration of the analyte in the sample solution with the fixed signal-to-noise ratio (S/N = 10). The chromatograms for DL and QL determination of 4-chloro-4'-fluorobutyrophenone are shown in Figure 3. The QL of 4-chloro-4'-fluorobutyrophenone in this experiment is 350 ng·mL^{-1}.

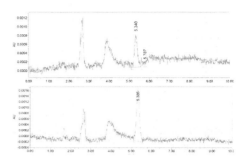

Figure 3. The first picture is liquid chromatogram of DL; the second picture is liquid chromatogram of QL.

3 SUMMARY

A HPLC method for analysis of 4-chloro-4'-fluorobutyrophenone was established and validated with linear for concentrations from 0.05 to 0.2 mg·mL^{-1}. The results demonstrated that the method is selective, sensitive, precise and can be used as a good option for routine analysis of 4-chloro-4'-fluorobutyrophenone since there was no official one.

ACKNOWLEDGEMENTS

We express our gratitude for Hebei Research Center of Pharmaceutical and Chemical Engineering, Pharmaceutical Molecular Chemistry Key Laboratory of Ministry Technology and Hebei Province Key Laboratory of Molecular Chemistry for Drug.

REFERENCES

[1] Gao, Y. 2012. Synthesis of 4-chloro-4'-fluorobutyrophenone and 1-phenyl-2- pyrrolidone. Nanjing University of Science and Technology.
[2] Gobetti, C. Pereira, R.L. Mendez, A.S.L. & Garcia, C.V. 2014. Determination of the new antiplatelet agent ticagrelor in tablets by stability-indicating HPLC method. Current Pharmaceutical Analysis 10(4): 279–283.
[3] Mondal, P. & Rani, S.S. 2014. Novel stability indicating validated RP-HPLC method for simultaneous quantification of artemether and lumefantrine in bulk and tablet. Current Pharmaceutical Analysis 10(4): 271–278.
[4] Silva, A.C. Lopes, C.M. Fonseca, J. Soares, M.E. Santos, D. Souto, E.B. & Ferreira, D. 2012. Risperidone Release from Solid Lipid Nanoparticles (SLN): Validated HPLC method and modelling kinetic profile. Current Pharmaceutical Analysis 8(4): 307–316.
[5] Wang, Y.L. Chen, Z.M. Bao, C.H. & Zhang, Y.J. 2003. Preparation methods of dipfluzine and application for the treatment of disease of heart head blood-vessel. Chinese Patent CN1,654,461.
[6] Zhao, G. Ma, F.L. & Hang, Z. 1994. Preparation of 4-chloro-4'-fluorobutyrophenone. Shandong Chemistry Industry 01: 12–14.

Material Science and Engineering – Chen (Ed.)
© 2016 Taylor & Francis Group, London, ISBN 978-1-138-02936-1

Study on the optimization of the fermentation process by lactic acid bacteria in jelly fish

H. Wang, S.S. Chen & Z.Z. Shi
College of Food Science and Technology, Shanghai Ocean University, Shanghai, P.R. China

ABSTRACT: In order to enhance the sensory and preservation effect of desalted jellyfish, dynamic changes in the fermentation process in jellyfish caused by *Lactobacillus* were studied in different conditions, such as fermentation time, direct *Lactobacillus* fermentation inoculation amount, fermentation temperature and inoculation proportion. The evaluation indices were pH value, total acid content and sensory evaluation. Optimal fermentation conditions obtained through the dynamic changes and response surface analysis methods were as follows: the fermentation time is 35 hours, direct *Lactobacillus* fermentation inoculation amount is 6%, fermentation temperature is 37°C, and the inoculation proportion between *Lactobacillus plantarum* and *Lactobacillus brevis* is 2:1. Under the optimum condition, the sensory evaluation value of fermented jellyfish is 23.58.

Keywords: direct *Lactobacillus* fermentation; desalted jellyfish; response surface analysis

1 INTRODUCTION

Jellyfish is kind of favorite seafood of people. It is rich in nutrition [1]. It has been regarded as a kind of high-quality diet aquatic products [2]. China is a big country in aquaculture, and full of wealthy jellyfish resources. With the development of the use of marine resource for people's food, much attention has been paid to jellyfish by domestic and foreign counterparts [3]. The main products are instant jellyfish and salted jellyfish. Salted jellyfish can be preserved conveniently and kept for much longer, but desalination must be proceeded before eating. The instant jellyfish is convenient for eating but cannot be preserved for a long time, and the preservation conditions are harsh [4–10]. Many studies have reported on the instant jellyfish in China. However, making use of the anticorrosion mechanism of *Lactobacillus*, acidic material, metabolic product and dominant bacterial community [11–12], inoculating *Lactobacillus* into desalted jellyfish for fermentation to enhance the flavor of desalted jellyfish and preservation have not been reported. Therefore, this research optimized the fermentation process of inoculating *Lactobacillus* into jellyfish.

2 MATERIALS AND METHODS

2.1 *Materials and instruments*

2.1.1 *Raw materials*
Materials such as salted jellyfish, salt, and glucose were purchased from the local market.

2.1.2 *Strains*
The strains used in the study were purchased from the China Center of Industrial Culture Collection (CICC). *Lactobacillus plantarum* and *Lactobacillus brevis* were used in the study.

2.1.3 *Instruments*
BCM-1000A biological clean bench, f-120 pH meter, AL204 electronic analytical balance, thermostatic incubator, mixer, and Bioscreen automatic growth curve analyzer were used in the study.

2.2 *Experimental design*

The salted jellyfish was soaked in fresh water for 24 hours in order to remove the salt. The tools employed should be disinfected by 75% ethanol or boiling water, and the jellyfish should be cut into pieces with the shape of $2.0 \times 2.0\,cm^2$. Before fermentation, the samples should be scalded in $60 \pm 2°C$ water for 30 ± 2 seconds in order to kill some residual microorganisms, remove part of the water and keep tissues close. In addition, the fermentation seed liquid was prepared. The preserved strains were inoculated in a 10 mL liquid MRS medium and activated at a 37°C incubator for 18–24 hours, and then transferred to a 100 mL liquid medium with the ultimate number of viable lactic acid bacteria reaching up to 10^8 CFU/mL. The ratio of material to solvent was 5 to 2. 6% glucose and 3% salt was added to the bacterial suspension, and then added the chopped jellyfish and kept them completely submerged in the water. The fermentation time and fermentation temperature were determined by the single-factor experiment and response surface analysis, and then terminated fermentation and stored at a low temperature.

2.2.1 *The dynamic changes in rapid fermentation by lactic acid bacteria (single-factor experiment)*

Different fermentation temperatures, fermentation times, *Lactobacillus* inoculation quantities, and inoculation proportions were used for the single-factor experiment in the process described above. The measurement index was the total acid content and pH. All experiments were repeated three times, and the average was used as the results for analysis, as presented in Table 1.

2.2.2 *Response surface analysis*

The response surface analysis was carried out according to the results of the single-factor experiment, used the response surface Box-Benhken central composite design [13], and regarded the sensory evaluation value as the response value. Each factor was tested at three levels. All the experiments were repeated three times, and the average of these three experimental results was used as the final results, as summarized in Table 2.

2.2.3 *Verification experiments and sensory evaluation*

The jellyfish was fermented by direct *Lactobacillus* fermentation in accordance with the best condition

Table 1. Design of the single-factor experiment.

Factors	Level				
	1	2	3	4	5
Fermentation time/h	15	20	25	30	35
Fermentation temperature/°C	27	32	37	42	47
Inoculation quantity/%	0	2	4	6	8
Inoculation proportion (pla:bre)	1:2	1:1	2:1	2:3	3:2

Table 2. Design of the response surface analysis.

Code	Factor	Level		
		−1	0	+1
A	Fermentation time/h	30	35	40
B	Inoculation quantity/%	4	6	8
C	Inoculation proportion (pla:bre)	1:1	2:1	1:2

of the response surface analysis. The total acid content was tested and the sensory evaluation was carried out on the jellyfish.

2.2.4 *Indicators and methods of measurement*

a. The total acid content of the zymotic fluid was measured by using the titration method to understand the indirect fermentation effect of *Lactobacillus* [14].
b. The pH value was measured by a pH meter.
c. The texture, taste, odor and color of the fermented jellyfish with different treatments were used as sensory quality assessment indices. It is scored by eight professionals assessment team on a 10-point scale: 8–10 are categorized as very good, 5–7 as good, and 1–4 as bad. Then, the average of the three repeated results were calculated [15].

3 RESULTS AND DISCUSSION

3.1 *Single-factor experiment*

3.1.1 *Dynamic changes over time of pH value and total acid content in the direct Lactobacillus fermentation liquor*

The inoculation amount used in the experiment was 8%, and the inoculated proportion between *Lactobacillus plantarum* and *Lactobacillus brevis* was 2:1. The fermentation temperature was set at 37°C to analysis the relationship between the fermentation time and the changes in the total acid content and the pH value. The results are showed in Figure 1 and Figure 2.

As shown in Figure 1, with the extension of the fermentation time, the pH value of the fermentation liquor dropped from 5.0 to 4.2 at 35 hour and maintained steadily after 35 hour. The pH value of natural fermentation changed more gently than the inoculated ones. It revealed that *Lactobacillus*

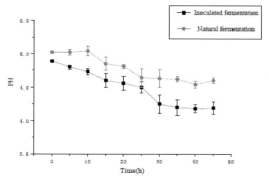

Figure 1. The effects of fermentation time on the pH value change in desalted jellyfish fermentation.

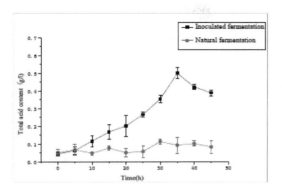

Figure 2. The effects of fermentation time on the total acid content change in desalted jellyfish fermentation.

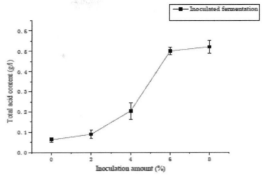

Figure 3. The effects of *Lactobacillus* inoculation amount on the total acid content of desalted jellyfish fermentation.

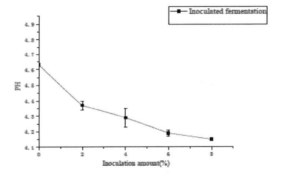

Figure 4. The effects of *Lactobacillus* inoculation amount on the pH value change in desalted jellyfish fermentation.

produced acid and made the pH lower after the inoculation of lactic acid bacteria. As shown in Figure 2, with the extension of fermentation time, the total acid content of the inoculated ones increased first and then decreased. The growth vitality of *Lactobacillus* increased gradually and constantly using glucose to produce lactic acid with the increase in the total acid content in the first 35 hours. After 35 hours, the vitality of *Lactobacillus* dropped, which was the same for the total acid content. Combined with the sensory evaluation analysis, the time at which *Lactobacillus* was inoculated for jellyfish fermentation was decided as 35 hours for the response surface analysis.

3.1.2 Dynamic changes in total acid content and pH under the condition of different Lactobacillus inoculation amounts

Lactobacillus inoculation quantity was selected, respectively, at 0%, 2%, 4%, 6%, and 8% for the experiment to test the relationship between different amounts of inoculum for jellyfish fermentation by *Lactobacillus* and changes in total acid content and pH. The experiment conditions were as follows: the fermentation time was 35 hours, the inoculation ratio was 2:1, and the fermentation temperature was 37°C. The experimental results are shown in Figure 3 and Figure 4.

As it can be seen from Figure 3, the total acid content increased gradually with the increase in inoculation quantity. The total acid content increased rapidly when the inoculation amount rose from 0% to 6%, and the total acid content began to increase slowly when the amount rose from 6% to 8%. As shown in Figure 4, with the increase in *Lactobacillus* inoculation quantity, the pH value decreased from 4.63 to 4.15. In combination with the sensory evaluation, the jellyfish tasted too sour and was not suitable for consumption when the inoculation amount was at 8%. The

inoculation amount at 6% was appropriate and the fermentation effect was best. Based on the above results, a 6% inoculation amount was used for the response surface analysis.

3.1.3 Dynamic changes in total acid content and pH at different fermentation temperatures

In the experiment, the fermentation time was fixed at 35 hours, the *Lactobacillus* inoculation amount was 6%, the inoculation ratio was 2:1, and the fermentation temperature was set at five levels, namely 27°C, 32°C, 37°C, 42°C, and 47°C. The experimental results are shown in Figure 5 and Figure 6.

Lactobacillus was inoculated into the desalted jellyfish, and the vitality and fermentation ability of *Lactobacillus* need an appropriate fermentation temperature. As shown in Figure 5 and Figure 6, when the fermentation temperature was below 37°C, the pH of the desalted jellyfish showed a trend of decline with the increase in temperature. Otherwise, the growth of *Lactobacillus* was restrained when the fermentation temperature was

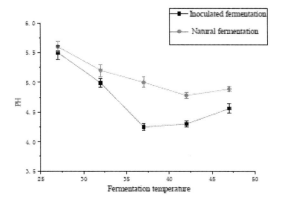

Figure 5. The effects of fermentation temperature on the pH value change in desalted jellyfish fermentation.

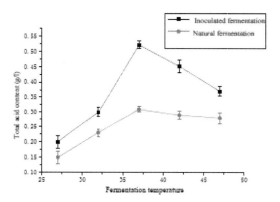

Figure 6. The effects of fermentation temperature on the total acid content of desalted jellyfish fermentation.

above 37°C and the pH value began to increase. The total acid content had the same changing trend as the pH value. The total acid content increased with the rise of temperature, but the total acid content began to decrease when the fermentation temperature was above 37°C. There was no such changing trend in the control sample. Therefore, the fermentation temperature of the jellyfish was 37°C.

3.1.4 The effects of the inoculation proportion between Lactobacillus plantarum and Lactobacillus brevis on the change in total acid content and pH value

The inoculation proportion between *Lactobacillus plantarum* and *Lactobacillus* used for the experiment was 1:2, 1:1, 2:1, 2:3, and 3:2 to test the effects of different inoculation proportions on the change in total acid content and pH value. The fermentation time was 35 hours, the fermentation temperature was 37°C, and the inoculation amount was 6%. The results are shown in Figure 7.

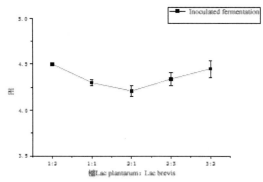

Figure 7. The effects of inoculation proportion on the pH value change in desalted jellyfish fermentation.

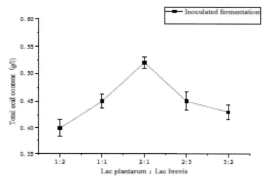

Figure 8. The effects of inoculation proportion on the total acid content of desalted jellyfish fermentation.

As shown in Figure 7 and Figure 8, different inoculation proportions caused different changes in total acid content and pH value. When the inoculation proportion was 2:1, the pH value of the desalted jellyfish was about 4.25, lower than the other proportion, and the total acid content was higher than the other. Meanwhile, in combination with the sensory evaluation, when the inoculation proportion was 2:1, the taste was appropriate, and the flavor and smell were not unpleasant.

3.2 Determination of the direct fermentation process by the response surface analysis

3.2.1 Response surface analysis and variance analysis

According to the above experimental results, the fermentation time, the *Lactobacillus* inoculation amount, and the *Lactobacillus* inoculation proportion were chosen as significant factors influencing the total acid content and the pH value. The response surface Box-Benhken central composite design was applied to the analysis. Each factor was designed at three levels. The response surface

Table 3. Experimental design and results of the central composite design.

Number	X_1-Fermentation time/h	X_2-Inoculation amount/%	X_3-Inoculation proportion	Sensory evaluation
1	40.00	8.00	0.50	16
2	35.00	6.00	0.50	21
3	35.00	6.00	0.50	22
4	30.00	4.00	0.50	15
5	35.00	4.00	0.67	18
6	30.00	6.00	0.33	20
7	35.00	6.00	0.50	25
8	30.00	8.00	0.50	17
9	40.00	6.00	0.33	19
10	35.00	4.00	0.33	14
11	35.00	8.00	0.67	18
12	40.00	6.00	0.67	24
13	35.00	8.00	0.33	18
14	30.00	6.00	0.67	23
15	35.0	6.00	0.50	20
16	40.00	4.00	0.50	13
17	35.00	6.00	0.50	21

Table 4. Analysis of variance (AVOVA) of the regression model.

Source of mean square	Sum of squares	Degree of freedom	Mean square	F	P Pr > F
Model	158.19	3	52.73	22.29	<0.0001
B	10.13	1	10.13	4.28	0.0590
C	18.00	1	18.00	7.61	0.0163
B^2	130.07	1	130.07	54.99	<0.0001
Residual	30.75	13	2.37		
Lack of fit	15.95	9	1.77	0.48	0.8350
Pure error	14.80	4	3.70		
Cor total	188.94	16			

experimental designs and results are shown in Table 3.

Multivariate regression and analysis of variance were carried out by using the software Design Experiment 8.0.6 in the data of Table 1. A Quadratic multinomial regression equation was obtained (formula 2–1) as follows:

$$Y = -35.99510 + 17.18750X_2 + 8.82353X_3 - 1.38542X^2 \tag{2-1}$$

The sensory evaluation (Y) of the directly fermented jellyfish was the dependent variable, and the fermentation time-X1, inoculation amount-X2 and inoculation proportion—X3 were the independent variables. The analysis of variance of this regression equation is presented in Table 4.

As it can be shown from Table 4, the variance F value of the regression equation was 22.2 and the P value (<0.0001) was highly significant. This indicated that the model was very significant and meaningful. The p value of lack of fit = 0.8350 > 0.05, which indicated that the fitting of this model was good, and the residual may be caused by random error. The model and equation can be used to analysis and predict the fermentation effect of the desalted jellyfish.

The results of the analysis of variance conducted to test the significance of the coefficient of the regression model revealed that B (inoculation amount) and C (inoculation proportion) were significant, and B^2 was affected very significantly. Therefore, every affecting factor for *Lactobacillus* fermentation in the desalted jellyfish was not a simple linear relationship. Factors were sequenced by their effects on the results in the selecting level range: inoculation proportion > inoculation amounts > fermentation time.

After optimization by Design Expert 8.0.6, the desalted jellyfish fermentation process parameters were obtained. The best combination of the fermentation time, inoculation amount and inoculation proportion was as follows: the fermentation time was 35 hours, the inoculation amount was 6.20%, and the inoculation proportion was 0.67. Combined with the actual operation, the best parameters were determined as follows: the fermentation time was 35 hours, the inoculation amount was 6%, and the inoculation proportion was 0.67. The predicted value of the sensory evaluation of the jellyfish was 23. 22; meanwhile, the result of the verification test was found to be 23.58.

3.2.2 Response surface analysis and optimization

As shown in Figure 9, when C (inoculation) was constant, the higher sensory evaluation results of the fermented jellyfish were at the positive level of A (fermentation time) and B (inoculation amount) in the range of the selected conditions. The change in the two factors can cause the changes in the sensory evaluation.

As shown in Figure 10, when B (inoculation amount) was constant, the higher sensory evaluation results of the fermented jellyfish were at the positive level of A (fermentation time) and C (inoculation) in the range of the selected conditions. The change in the two factors can cause the changes in the sensory evaluation.

As shown in Figure 11, when A (fermentation time) was constant, the higher sensory evaluation results of the fermented jellyfish were at the positive level of B (inoculation amount) and C (inoculation) in the range of the selected conditions. The change in the two factors can cause the changes in the sensory evaluation.

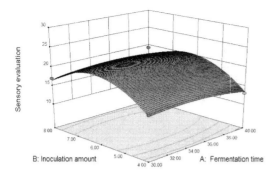

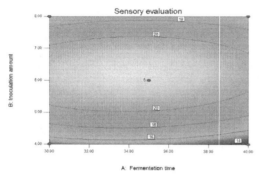

Figure 9.   The response surface analysis of A and B.

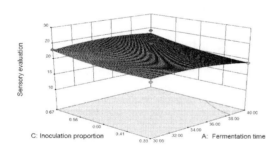

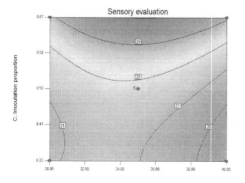

Figure 10.   The response surface analysis of A and C.

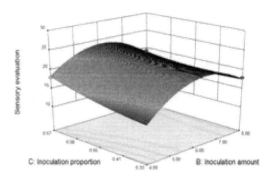

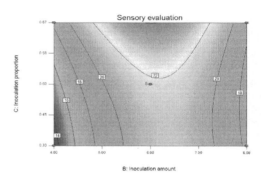

Figure 11.   The response surface analysis of B and C.

3.2.3   Verification of the experiment results

The experiment was verified at the stable point with three replications to measure the effects of *Lactobacillus* fermentation in the jellyfish. The sensory evaluation was 23.58, which was close to the predicted value.

4   CONCLUSIONS

1. The dynamic changes in *Lactobacillus* fermentation in desalted jellyfish with different conditions, such as fermentation temperature, fermentation time and inoculation proportion, were studied in this research. For determining the factors, 37°C was chosen.
2. According to the variation pattern, three significant factors influencing the fermentation of jellyfish by *Lactobacillus* were chosen to be used to design the response surface experiment. These factors were *Lactobacillus* inoculation amount, fermentation time and *Lactobacillus* inoculation proportion. The control conditions of direct *Lactobacillus* fermentation in desalted jellyfish were confirmed. The fermentation time was 35 hours, the *Lactobacillus* inoculation amount was 6%, and the inoculation pro-

portion between *Lactobacillus plantarum* and *Lactobacillus brevis* was 2:1.

3. Measuring the fermentation effect of *Lactobacillus* on the desalted jellyfish in the verification experiment, the sensory evaluation value of the fermented jellyfish was 23.58, which was close to the predicted one. Therefore, the process parameters obtained by the response surface analysis method are accurate and reliable, and have the practical value.

REFERENCES

[1] Wang C.Y. Jellyfish canned production [J]. Food Science, 1993(8):76–78.

[2] Xia S.Y. Aquatic food processing [M]. Beijing: Chemical Industry Press, 2008:228.

[3] Hsieh Y P, Rudloe J. Potential of utilizing jellyfish as food in Western countries [J]. Trends in Food Science & Technology, 1994, 5(7):225–229.

[4] Shan J.H. Effect of packaging and acetum treatment on the quality of desalted jellyfish in storage [J]. Science and Technology of Food Industry, 2011 (08):364.

[5] Li J R, Hsieh Y P. Traditional Chinese food technology and cuisine [J]. Asia Pacific J Clin Nutr, 2004, 13(2):147–155.

[6] Hsieh Y P, Leong F M, Rudloe J. Jellyfish as food [J]. Hydrobiologia, 2001, 451:11–17.

[7] Peter Fratzl. Collagen [M]. USA: Springer Science, 2008:10.

[8] Peck M W. Clostridium botulinum and the safety of refrigerated processed foods of extended durability [J]. Trends in Food Science & Technology, 1997, 8:186–192.

[9] Silva F V, Gibbs P A. Non-proteolytic Clostridium botulinum spores in low acid cold distributed foods and design of pasteurization processes [J]. Trends in Food Science & Technology, 2010, 21:95–105.

[10] Rajkovic A, Smigic N, Devlieghere F. Contemporary strategies in combating microbial contamination in food chain [J]. International Journal of Food Microbiology, 2010, 141:S29–S42.

[11] Pilar Calo-Ma ta, Samuel Arlindo, Karola Boehme, et al. Current Applications and Future Trends of Lactic Acid Bacteria and their Bacteriocins for the Biopreservation of Aquatic Food Product. Food Bioprocess Technol, (2008) 1:43–63.

[12] Tiejun Li, Aiyun Li. Study on the progress of Lactobacillus antibacterial mechanism [J]. Microbiology, 2002, 29(5):82–85.

[13] Rui Yang, Wei Zhang. Influence on microbial strains of fermentation conditions on kimchi fermentation process [J]. Food and Fermentation Industries, 2005(3):90–92.

[14] Montgomery. Experimental design and analysis [M]. Beijing: China Statistics Press, 1998:591–626.

[15] Hua Yi, S.C. Liu. Effect on the finished product quality and nitrite content of pickled pepper artificially inoculated fermentation [J]. Hunan Agricultural Sciences, 2007 (3):144–146.

Material Science and Engineering – Chen (Ed.)
© *2016 Taylor & Francis Group, London, ISBN 978-1-138-02936-1*

Experimental study on the biomechanical behavior of the expanding ring for treating glaucoma

Y.B. Cong & L.Y. Wang
School of Mechanical Science and Engineering, Jilin University, Changchun, China

H.Q. Yu
Department of Ophthalmology, First Hospital, Jilin University, Changchun, China

ABSTRACT: Glaucoma Drainage Devices (GDDs) are well accepted in the therapy of refractory glaucoma. The expanding ring that is implanted in the anterior chamber is used to expand the closed or narrow anterior chamber angle. The primary aim of this study is to experimentally evaluate the biomechanical properties of the expanding ring made of hydrophobic acrylic materials. The Shimadzu Autograph AG-X Plus universal testing machine was used for the compression test, stress relaxation test and creep deformation test. The results showed that there was a nonlinear constitutive relationship between load and deformation, linear-elastic, super-elastic and viscous-elastic behavior that occurred in the loading and unloading processes under different load rates. The load had an influence on the results of the stress relaxation and creep deformation, and the stiffness of the expanding ring was very sensitive to the loading rate. The expanding ring has biomechanical stability and meets the needs of design for treating glaucoma. Using biomechanical testing techniques, the mechanical properties of the expanding ring can evaluate the efficacy and provide more valuable data to relevant research.

Keywords: glaucoma drainage devices; hydrophobic acrylic; expanding ring; viscoelastic properties; stress relaxation; creep deformation

1 INTRODUCTION

Glaucoma is a major cause of blindness worldwide, which is characterized by optic nerve damage, leading ultimately to irreversible blindness. There will be approximately 80 million people with glaucoma and more than 20 million people in China by 2020 (Quigley and Broman, 2006). Primary Angle-Closure Glaucoma (PACG) is the most common type in Asia. It is a disease of ocular anatomy that is related to pupillary-block and angle-crowding mechanisms of the anterior chamber angle (Tarongoy et al., 2009). The implant surgery is a continuously evolving field in glaucoma treatment. Glaucoma drainage implants have a history reaching back over more than a century. A significant progress was achieved in the 1970s with Molteno's concept of a tube–plate system, with the plate being placed away from the limbus region. Ahmed and Baerveldt glaucoma drainage implants were introduced clinically in the 1990s, and are now both offered commercially, with several modified versions (Dietlein et al., 2008). The secretion of aqueous humor (AH) and the regulation of its outflow are physiologically important processes for the normal function of the eye. In the healthy eye, the flow of AH against resistance generates an average Intraocular Pressure (IOP) of approximately 15 mmHg (Ethier et al., 2004). IOP is a measurement of the fluid pressure inside the anterior chamber of the eye. AH exits primarily through the Trabecular Meshwork (TM) and eventually drains into the venous blood. The basic concept that the impairment in aqueous humor outflow results in the elevation of the IOP is a central tenet of glaucoma pathology and treatment (Goel et al., 2010). Therapy of glaucoma focuses on IOP reduction by means of medical therapy, laser treatment and filtering surgery. In case of refractory glaucoma, drainage devices are increasingly accepted as an alternative to conventional filtering surgery (Siewert et al., 2012).

Angle closure is a disorder of ocular anatomy characterized by the closure of the drainage angle by the appositional or synechia approximation of the iris against the trabecular meshwork. The final common result in related disorders is an elevation of the IOP, due to the blocking of the aqueous humor outflow from the eye, followed by the development of glaucomatous optic neuropathy (Tarongoy et al.,

2009). The expanding ring for treating glaucoma was made according to the patent (Hongquan, 2007), which is designed to implant in the anterior chamber angle to make a close or narrow anterior chamber distraction. The function of the anterior chamber would be to recover and the aqueous humor drains normally. With the development of a variety of foldable Intraocular Lens (IOLs) biomaterials, several new materials are currently available for the fabrication of the IOL. According to the design requirements, the expanding ring must be made of silicone, hydrogel, and soft acrylic. Hydrophobic acrylic has proved its efficiency in the clinical setting, which does not permanently change its optical and mechanical properties as a result of folding and compression during injection, preserves excellent biocompatibility, and provides sufficient refractive stability (Chehade and Elder, 1997). Several studies on the biomechanical properties of foldable intraocular lens have been reported to assess the performance of different IOL designs and materials. Lane et al. (Lane et al., 2004) used industry-standard biomechanical testing techniques to compare and find the superior biomechanical characteristics in various IOL designs. Bozukova et al. (Bozukova et al., 2013) assessed the biomechanical properties of 11 commercially available IOLs to compare their compressibility, deformability, injectability, and optical quality before and after injection to help surgeons make proper IOL selections. Xuhui et al. (Xuhui et al., 2010) analyzed the biomechanical properties of adult brachial plexus by observing tensile mechanical properties, stress relaxation and creep deformation by comparing the brachial plexus in normal human cadavers and brachial plexus from simulated brachial plexus injury anastomosis samples.

The purpose of this study was to experimentally evaluate the biomechanical and viscoelastic properties of the expanding ring of hydrophobic acrylic materials, which was made according to the patent (Hongquan, 2007). Because the expanding ring is made from a polymeric material, a certain amount of stress relaxation and creep deformation can take place when the expanding ring is compressed. As the biomechanical performance of the expanding ring can directly affect the clinical outcomes, these parameters must play an important role in the design and application of the GDDs. Given the lack of published data on the biomechanical characteristics of the expanding ring, we attempted to provide an outline of its biomechanical behavior. We conducted this experimental study to assess the adhesive force and to evaluate its possible role in preventing the migration of cornea epithelial cells. The results of this study would be helpful in assessing the treatment effect in vivo and guiding the next animal study.

2 MATERIALS AND METHODS

2.1 Materials and equipment

The expanding ring comprised the upper and lower ring, whose outer diameter was 10 ± 0.05 mm and 11 ± 0.05 mm, and the inside diameter was 9 ± 0.05 mm and 8.6 ± 0.05 mm respectively. The total height was 1.5 mm. There were six holes with the diameter of 0.4 mm in the upper parts from which AH would flow to the anterior chamber angle (Figure 1). The expanding ring was made up of hydrophobic acrylic, which has been extensively used in the IOLs.

All tests were performed at the Mechanics Experimental Center, Jilin University, China from January to February 2015 with the autograph AG-X Plus universal testing machine, Shimadzu, Japan.

2.2 Compression test

Compression test was conducted at load rates of 0.03 mm/min, 0.1 mm/min, 0.3 mm/min, and 0.6 mm/min, respectively. According to the different load rates, a total of 24 samples were divided into 4 groups. There were 6 samples at every load rate. The maximum of the compressive load was set to be 100 N. When the load reached the maximum value, the compression stopped, and then the samples were unloaded immediately. The value of the force and displacement was obtained by a computer following the samples.

2.3 Stress relaxation test

Stress relaxation was referred to stress decay under a constant strain, and the stress reached equilibrium after a certain time. According to the different load conditions, a total of 24 samples were divided into 4 groups. There were 6 samples in every load condition. The load rate exerted on the samples

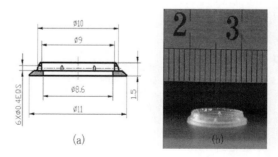

Figure 1. (a) Two-dimensional CAD model of the expanding ring (dimensions in millimeters); (b) photograph of the expanding ring.

was 0.3 mm/min. The initial force of 0.01 N was identical in each group. When the force reached to 2 N, 4 N, 6 N and 8 N, respectively, the displacement remained constant. Then, the force decreased over time. The experimental duration was 7200 seconds, and 975 time points were collected. Force relaxation data and curve were recorded at set time by a computer.

2.4 *Creep deformation test*

Creep deformation was referred to strain changes over time under a constant stress. Stress relaxation test and creep deformation test were the important methods to study the viscoelastic properties of biological materials. The number of the samples, load rate and initial condition were similar to the stress relaxation test. When the force reached to 2 N, 4 N, 6 N and 8 N, respectively, the force remained constant. Then, the displacement increased over time. The experimental duration was 7200 seconds, and 975 time points were collected. Displacement creep data and curves were recorded at set time by a computer.

The data in all tests were analyzed using Matlab 2011 software, which was expressed as Mean ± SD.

3 RESULTS AND DISCUSSION

3.1 *Comparison of compression test results under different load rates*

Load-deformation curves of sample loading and unloading under different load rates are illustrated in Figure 2. As can be observed in Figure 2, there was a nonlinear constitutive relationship between load and deformation of the expanding ring.

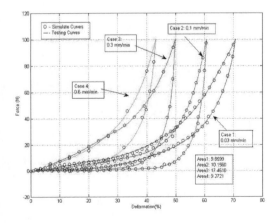

Figure 2. Load-deformation curves of sample loading and unloading under different load rates.

According to the experimental data, the analytical expression of the constitutive relationship was simulated between load and deformation by using the cubic polynomial for cases with four different load rates. The constitutive equation can be expressed as follows:

$$F(x) = kx + \alpha x^2 + \beta x^3 \qquad (1)$$

Comparison between the simulated data and the experimental data is shown in Figure 2. For the different load rates, the parameters k, α, and β had different values. The values of these parameters are provided in Table 1.

For case 1 and case 4, the constitutive relationship can be expressed by a cubic polynomial with different coefficients under the loading process. Comparing the magnitude of the coefficients of linear term, square and cubic terms, it can be found that the weak linear-elastic and super-elastic of the material basically play a dominated role in the condition that the loading is less than about 40 N, but the strong linear-elastic and viscous-elastic dominated the mechanical behavior after loading is over 40 N under the loading process (Figure 2 and Table 1). However, for case 2 or case 3, the constitutive relationship needs to be divided

Table 1. The data fitting parameters of the load-deformation curves under different load rates.

	Loading			
	k	α	β	F(N)
Case 1	0.6094	−0.0235	0.0005	0 < F < 100
0.03 mm/min				
Case 2	0.3573	−0.0136	0.0004	0 < F ≤ 30
0.1 mm/min	12.1020	−0.4936	0.0053	30 < F < 100
Case 3	0.4754	0.0305	−0.0004	0 < F ≤ 30
0.3 mm/min	2.6702	−0.1157	0.002	30 < F < 100
Case 4	1.0907	−0.0514	0.0018	0 < F < 100
0.6 mm/min				

	Unloading			
	k	α	β	F(N)
Case 1	0.1467	−0.074	0.0002	0 < F ≤ 40
0.03 mm/min	22.8612	−0.7404	0.0062	40 < F < 100
Case 2	0.0248	−0.0014	0.0000	0 < F ≤ 3.6
0.1 mm/min	9.8747	−0.4169	0.0044	3.6 < F ≤ 39
	315.283	−0.9214	0.0948	39 < F < 100
Case 3	−0.0047	−0.0021	0.0001	0 < F ≤ 3.6
0.3 mm/min	6.8328	−0.3748	0.0052	3.6 < F ≤ 9
	277.048	−11.78	0.1256	39 < F < 100
Case 4	0.3425	−0.0376	0.0014	0 < F ≤ 39
0.6 mm/min	76.5600	−3.9435	0.0515	40 < F < 100

into two cubic polynomials during the whole loading process. The former is dominated by the weak linear-elastic and super-elastic property of the material when the load is less than about 30 N, the latter is dominated by the strong linear-elastic and viscous-elastic when the load is larger than 30 N (Figure 2 and Table 1).

Comparing these 4 cases, it can be seen that the mechanical behavior of the expanding ring is very complicated due to the dynamic constitutive relationship with respect to the different loading rates. However, the whole varying trends of them are similar. The strongest linear-elastic occurred in case 4 and the weakest linear-elastic occurred in case 2 when loading was less than about 30 N. However, when loading was larger than about 30 N, the strongest linear-elastic occurred in case 2, the next was case 3, and the weakest linear-elastic occurred in case 1. The square term played a super-elastic role and the cubic term played a viscous-elastic role in all cases except one situation, which was the loading process of case 3 when the load was less than 30 N. In this situation, the square term plays the viscous-elastic and cubic term plays the super-elastic roles (Figure 2 and Table 1). When the loading was larger than about 60 N and reaching to 100 N, the maximum of deformation was inversely proportional to the loading rate, and the maximum deformation were, respectively, about 70% (case 1), about 60% (case 2), about 50% (case 3) and about 43% (case 4).

The mechanical behavior of the expanding ring under the unloading process seemed to be more complicated than the loading situation. For case 1 or case 4, the constitutive equation needs to be expressed by two piecewise cubic functions, but for case 2 or case 3, the constitutive equation needs to be expressed by three piecewise cubic functions. The unloading process was similar to a "jump down" process when the load reduced from 100 N to about 40 N; the strong linear-elastic and viscous-elastic played the dominated roles in this situation. The strongest linear-elastic occurred in case 2 and the most strong viscous-elastic occurred in case3.

For case 2 and case 3, the coefficients of the linear-elastic were larger than 1 order of the magnitude of the coefficient of the square term and larger than 2 orders of the magnitude of the coefficient of the cubic term when the loading reduced under about 40 N but was above 4 N. At the same time, the super-elastic played the dominated role when the load reduced to lower than 5 N. In other words, the strong linear-elastic and the viscous-elastic dominated the deformation when the deformation from their maximum value restored into 10%, the rest of the deformation restored was dominated by the super-elastic term. For case 1, the weak linear-elastic and super-elastic dominated the behavior

when the load reduced to lower than about 40 N. For case 4, the super-elastic played a dominated role when the load reduced to below 20 N.

According to the 4 cases examined, the dissipation energy within a cycle period of the loading and unloading processes was calculated by using the experimental data. The maximum of dissipation energy was about $17.4610 \times 1.5/100 = 0.26$ N·mm that occurred in case 3, and the minimum of the dissipation energy was about $9.2721 \times 1.5/100 = 0.14$ N·mm that occurred in case 4. The dissipation energy of the rest of two cases was approximated to the minimum value. It implied that the inner dissipation energy of the expanding ring was more complicated with respect to the different load rates.

From Figure 2, it can be seen that the maximum stress should occur in the situation when the viscous-elastic dominated the mechanical behavior. The faster the load rate was, the faster the maximum stress reached. The stress increasing rate was much slower than the strain increasing rate when the load rate was equal or less than 0.1 mm/min under the load that was less than about 30 N (case 1 and case 2). However, when the load increased above 40 N, the stress sharply increased with respect to the strain for all cases. The stress increasing rate was equal to the strain increasing rate when the load rate was equal to or larger than 0.3 mm/min under the loading that was less than about 30 N (case 3 and case 4). When the loading was larger than 30 N, there was more significant effect of the load rate on the constitutive relationship of stress and strain. Interestingly, there were two obvious characteristics observed from Figure 2 when the load was less than 30 N: 1) the loading curves of case 1 and case 2 were similar and the loading curves of case 3 and case 4 were similar (Figure 2); 2) however, these two group of curves were very different from each other, and the gradient of the second group of curves was larger than that of the first group of curves. It implied that the stiffness of the expanding ring was outstandingly increased as the loading rate increased from 0.1 mm/min to 0.3 mm/min. Therefore, there must be existing a critical region between 0.1 mm/min and 0.3 mm/min of the loading rate within this critical region. The stiffness of the expanding ring was very sensitive to the loading rate.

3.2 *Comparison of stress relaxation test results*

Stress relaxation curves of the samples under different load conditions are shown in Figure 3. It can be found that in Figure 3, in the F = 2 N group, the initial load was 2 N, and the load at 1000 seconds was decreased to 0.48 N. In the other three groups, the initial load was 4 N, 6 N, 8 N, and the load at 1000 seconds was decreased to 1.06 N, 1.28 N, 2.51 N

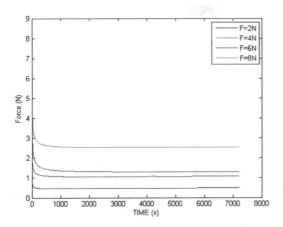

Figure 3. Stress relaxation curves of the samples under different load conditions.

Table 2. The data fitting parameters of the stress relaxation curves.

	k_1	α	k_2	β
F = 2 N	0.8183	−0.07614	0.4901	−1.808e-5
F = 4 N	1.519	−0.04231	1.159	−3.118e-5
F = 6 N	2.55	−0.03001	1.546	−5.753e-5
F = 8 N	2.669	−0.03208	2.734	−2.667e-5

respectively. In all groups, the load was dramatically decreased, which fell to nearly 25% of the initial load in a very short time range, and then kept the relative stability. As stress relaxation occurred, the amount of pressure that the expanding ring exerted on the cornea and the iris would be decreased. It was beneficial to reduce the injury risk.

As can be seen from Figure 3, the stress relaxation curve was a regular curve, which would be carried out on the data fitting with an exponential relationship. The fitting formula was proposed as equation (2):

$$X(t) = k_1 \times e^{\alpha t} + k_2 \times e^{\beta t} \qquad (2)$$

The experimental data were put into Equation (2). The related parameters are listed in Table 2.

3.3 Comparison of creep deformation test results

Creep deformation curves of the samples under different load conditions are shown in Figure 4. It can be found from Figure 4, in the F = 2 N group, the initial deformation was 4.88%, and deformation at 7200 seconds was increased to 9.33%. In the other three groups, the initial deformation

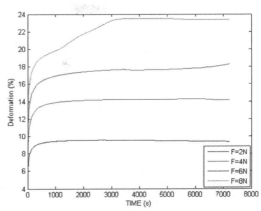

Figure 4. Creep deformation curves of the samples under different load conditions.

Table 3. The data fitting parameters of the creep deformation curves.

	k_1	α	k_2	β
F = 2 N	9.351	5.255e-6	−3.526	−.01109
F = 4 N	13.73	9.206e-6	−5.23	−.00918
F = 6 N	16.75	1.436e-5	−5.909	−.00799
F = 8 N	19.65	4.318e-5	−6.697	−.00719

was 7.36%, 9.55%, 11.24%, and deformation at 7200 seconds was increased to 14.15%, 18.28%, 23.37%, respectively. It can be found that with the load increased, the quantity of deformation would be increased simultaneously. The maximum deformation occurred in the F = 8 N group, and it was obvious that the deformation took more time than in the other three groups. It may show that the expanding ring was difficult to be stable under large load conditions. Considering the physiological environment of the eyeball, the load exerted on the expanding ring would be relatively small. The expanding ring would maintain the deformation and meet the needs of the design.

Similarly with the stress relaxation curves, the creep deformation curves would be carried out on the data fitting with an exponential relationship. The fitting formula was still proposed as Equation (2). The related parameters are listed in Table 3.

4 CONCLUSIONS

The expanding ring, which based on the patent (Hongquan, 2007), was made of hydrophobic acrylic for mechanical IOP regulation and, in particular, for the prevention of ocular hypotension.

Although valid data on the clinical application of the expanding ring are not available as yet, profound knowledge about its theoretical concept and practical management is necessary for its design and optimization. As mentioned previously, the aim of this study was to experimentally assess the biomechanical properties of the expanding ring of hydrophobic acrylic materials. According to the compression properties and viscoelastic properties of the expanding ring, it contributed to reveal the major problems and revise the design. To achieve this goal, we constructed the compression test, stress relaxation test and creep deformation test with the Shimadzu Autograph AG-X Plus universal testing machine. Our key finding was that 1) the nonlinear constitutive relationship between load and deformation of the expanding ring, linear-elastic, superelastic and viscous-elastic behavior occurred in the loading and unloading processes under different load rates. The stiffness of the expanding ring was very sensitive to the loading rate. 2) The load had an influence on the results of the stress relaxation and creep deformation; with the load increased, the quantity of stress relaxation and creep deformation both increased. The change in both stress relaxation and deformation was relatively large at the initial time, and then reached to equilibrium in 7200s. 3) The expanding ring has an excellent compression viscoelasticity, which is very important for the research on the implantation *in vivo*.

In conclusion, the results of this study demonstrate the comprehensive biomechanical behavior of the expanding ring for treating glaucoma through a series of biomechanical laboratory tests, and assess the performance of the expanding ring for treating glaucoma. It can reflect the complex mechanical competency of the expanding ring, and provide relevant subjects in predicting the expanding ring's treatment *in vivo*. The material of the expanding ring may be others, such as silicone, hydrogel, and NiTi shape memory alloys, based on their mechanical properties. This experimental study has the potential to contribute to the understanding of the structural analysis and optimization of the expanding ring, and would serve as a basis for further animal experiment and evaluations of different implantations of glaucoma drainage devices.

REFERENCES

[1] Bozukova, D., Pagnoulle, C. & Jérôme, C. (2013) Biomechanical and optical properties of 2 new hydrophobic platforms for intraocular lenses. Journal of Cataract & Refractive Surgery, 39, 1404–1414.

[2] Chehade, M. & Elder, M.J. (1997) Intraocular lens materials and styles: a review. Australian and New Zealand journal of ophthalmology, 25, 255–263.

[3] Dietlein, T.S., Jordan, J., Lueke, C. & Krieglstein, G.K. (2008) Modern concepts in antiglaucomatous implant surgery. 246, 1653–1664.

[4] Ethier, C.R., Johnson, M. & Ruberti, J. (2004) Ocular biomechanics and biotransport. 6, 249–273.

[5] Goel, M., Picciani, R.G., Lee, R.K. & Bhattacharya, S.K. (2010) Aqueous humor dynamics: a review. 4, 52.

[6] Hongquan, Y. (2007) Expanding ring for treating glaucoma. China patent application CN 1887245 A. January 3, 2007.

[7] Lane, S.S., Burgi, P., Milios, G.S., Orchowski, M.W., Vaughan, M. & Schwarte, E. (2004) Comparison of the biomechanical behavior of foldable intraocular lenses. Journal of Cataract & Refractive Surgery, 30, 2397–2402.

[8] Quigley, H.A. & Broman, A.T. (2006) The number of people with glaucoma worldwide in 2010 and 2020. 90, 262–267.

[9] Siewert, S., Roock, A., Schmidt, W., Löbler, M., Hinze, U., Chichkov, B., Guthoff, R., Sternberg, K. & Schmitz, K.P. (2012) Development of a novel valved drug-eluting glaucoma implant for safe and durable reduction of intraocular pressure. 57, 1.

[10] Tarongoy, P., Ho, C.L. & Walton, D.S. (2009) Angle-closure glaucoma: the role of the lens in the pathogenesis, prevention, and treatment. 54, 211–225.

[11] Xuhui, Hou, Xinying, Li, Songbai, Yang, Jian, Yin, Hongshun & MA (2010) Effects of brachial plexus injury anastomosis simulation on biomechanical properties of adult brachial plexus. Neural Regen Research, 5, 471–475.

Functional and ceramic materials

Material Science and Engineering – Chen (Ed.)
© 2016 Taylor & Francis Group, London, ISBN 978-1-138-02936-1

Visible-light refraction going from negative to positive in Au/TiO$_2$/Au multilayered structures

Y. Zang
College of Electronic Science and Engineering, Nanjing University of Posts and Telecommunications, Nanjing, P.R. China

M. Sun
The 28th Research Institute of China Electronic Technology Group Corporation, Nanjing, P.R. China

J.W. Chen
College of Electronic Science and Engineering, Nanjing University of Posts and Telecommunications, Nanjing, P.R. China
National Key Laboratory of Electromagnetic Environmental Effects and Optoelectric Engineering, Nanjing, P.R. China

ABSTRACT: Based on the generalized reflection and refraction theory, visible-light refraction going from negative to positive in Au-TiO$_2$-Au multilayered structures is predicted, being similar to the refraction behaviors occurring at the Au-air interface. In addition, the effects of the symmetry of the multilayered structures on refraction behaviors are addressed. This work suggests a common physics mechanism for the refraction phenomena occurring both at the noble metal-air interface and in metal-dielectric-metal multilayered structures.

Keywords: refraction; multilayered structure; interface

1 INTRODUCTION

Recently, all-angle negative refraction of Transverse Magnetic (TM) polarized wave and positive refraction of Transverse Electric (TE) polarized wave in Ag/TiO$_2$/Ag multilayered structures have been demonstrated experimentally [1]. On the other hand, it has been reported that visible-light refraction at the Ag-air interface is negative [2], and visible-light refraction at the Au-air interface may go from negative to positive with the increasing frequency of the incident light [3]. The origin of light refraction phenomena at the metal/air interface has not been fully understood yet [2, 3].

In principle, refraction phenomena occurring both in Ag-TiO$_2$-Ag multilayered structures and at the metal/air interface are associated with the reflection and refraction of light at the metal/dielectric interface, thus a common physics mechanism for the above-mentioned refraction phenomena may be suggested based on the reflection and refraction theory [4].

In this work, we explore the origin of the refraction behaviors of light at the Au/air interface, and then predict the possible refraction properties of light in Au/TiO$_2$/Au multilayered structures. This work may be useful for understanding the refraction behaviors of light occurring at the metal/dielectric interface and in multilayered structures.

The remainder of this paper is organized as follows. In Section 2, the generalized reflection and the refraction theory and calculation model are briefly introduced. In Section 3, numerical results and discussions are presented. Finally, some conclusions are drawn in Section 4.

2 THEORY AND CALCULATION MODEL

For simplicity, we consider the case of an obliquely incident TM-polarized Harmonic Homogeneous Plane Wave (HHPW) traveling through a lossy interface. It is known that a lossy homogeneous isotropic medium can be represented by a complex scalar relative permittivity $\tilde{\varepsilon} = |\tilde{\varepsilon}| \exp(j\alpha_\varepsilon)$ and permeability $\tilde{\mu} = |\tilde{\mu}| \exp(j\alpha_\mu)$, respectively. (In this paper, the complex-valued parameters are marked with "~".) Apparently, in a lossy medium and at a lossy interface, the variation in electromagnetic wave parameters of $\tilde{E}(t)$, $\tilde{D}(t)$, $\tilde{H}(t)$ and $\tilde{B}(t)$

is usually non-synchronous. Thus, the real-valued boundary conditions with time terms are adopted [4,5] as follows:

$$\text{Re}(\tilde{H}_i(t)) + \text{Re}(\tilde{H}_r(t)) = \text{Re}(\tilde{H}_t(t)), \tag{1}$$

$$\begin{aligned} a_i \, \text{Re}(\tilde{H}_i(t))\cos\theta_i &- a_r \, \text{Re}(\tilde{H}_r(t))\cos\theta_r \\ &= a_t \, \text{Re}(\tilde{H}_t(t))\cos\theta_t, \end{aligned} \tag{2}$$

$$\begin{aligned} b_i \, \text{Re}(\tilde{H}_i(t))\sin\theta_i &+ b_r \, \text{Re}(\tilde{H}_r(t))\sin\theta_r \\ &= b_t \, \text{Re}(\tilde{H}_t(t))\sin\theta_t. \end{aligned} \tag{3}$$

where $a_\varsigma = \text{Re}(\tilde{\eta}_\xi) - \text{Im}(\tilde{\eta}_\xi)\,\text{Im}(\tilde{H}_\varsigma)/\text{Re}(\tilde{H}_\varsigma)$ (since $\text{Re}(\tilde{E}_\varsigma(t)) = \text{Re}(\tilde{\eta}_\xi \tilde{H}_\varsigma(t)) \equiv a_\varsigma \, \text{Re}(\tilde{H}_\varsigma(t))$, $\tilde{\eta}_\xi = \sqrt{\tilde{\mu}_\xi \mu_0/\tilde{\varepsilon}_\xi \varepsilon_0}$ is the wave impedance); $b_\varsigma = \text{Re}(\tilde{k}_\xi) - \text{Im}(\tilde{k}_\xi)$ $\text{Im}(\tilde{H}_\varsigma)/\text{Re}(\tilde{H}_\varsigma)$ (since $\text{Re}(\tilde{D}_\varsigma(t)) = \text{Re}(\tilde{k}_\xi \tilde{H}_\varsigma(t)) \equiv b_\varsigma$ $\text{Re}(\tilde{H}_\varsigma(t))$, $\tilde{k}_\xi = \frac{\omega}{c}\sqrt{\tilde{\mu}_\xi \tilde{\varepsilon}_\xi}$ is the wave vector), with $\varsigma = i, r, t$ refers to the incident, reflected and transmitted waves; and $\xi = 1, 2$ indicates the two media between the interfaces, respectively. According to Equations (1) and (3), we have

$$b_i \sin\theta_i = b_r \sin\theta_r = b_t \sin\theta_t. \tag{4}$$

At the lossless interface, Equation (4) is just the phase matching condition [6]. The generalized Snell's law, including the effects of energy losses, is obtained from Equation (4) as follows:

$$\theta_r = \theta_i, \tag{5a}$$

$$\begin{aligned} \sin\theta_t &= b_i \sin\theta_i / b_t \\ &= \sin\theta_i \frac{\text{Re}(\tilde{k}_1)}{\text{Re}(\tilde{k}_2)} \frac{1 + tg(\alpha_{k_1})tg(\omega t)}{1 + tg(\alpha_{k_2})tg(\omega t)}. \end{aligned} \tag{5b}$$

Furthermore, solving Equations (1) and (2) gives the formulas for the transmission and reflection coefficients as follows:

$$T_{||} \equiv \frac{\text{Re}(\tilde{H}_t(t))}{\text{Re}(\tilde{H}_i(t))} = \frac{a_i \cos\theta_i + a_r \cos\theta_r}{a_r \cos\theta_r + a_t \cos\theta_t}, \tag{6a}$$

$$\Gamma_{||} \equiv \frac{\text{Re}(\tilde{H}_r(t))}{\text{Re}(\tilde{H}_i(t))} = \frac{a_i \cos\theta_i - a_t \cos\theta_t}{a_r \cos\theta_r + a_t \cos\theta_t}. \tag{6b}$$

Causally, the Time-Averaged Poynting Vector (TAPV) of the incident wave propagates toward the interface, while the TAPVs of the reflected and transmitted waves propagate away from the interface [5]. It is stressed that the direction of the Time-Dependent Poynting Vector (TDPV) of wave in lossy medium may alter with the change in time; thus, the direction of the TDPV of the incident wave may become to propagate away from the interface, whereas the direction of the TDPV of either reflected wave or transmitted wave may propagate toward the interface. To obtain the generalized for-

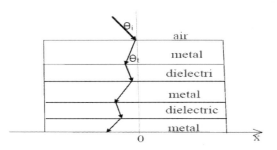

Figure 1. The adopted model and the coordinate axis x. The transmitted wave is taken as a result of the wave that refracts at an interface and then propagates through a layer of a homogeneous film step by step.

mulas of the reflection and refraction of the wave at an interface formed by two types of arbitrarily homogeneous isotropic media, all possible direction relationships between the electric and magnetic parameters of the incident, reflected and transmitted waves must be considered [5].

Analogously, the generalized formulas of the reflection and refraction of an obliquely incident TE polarized HHPW traveling through a lossy interface may be obtained [4,5].

In numerical simulations, the adopted model and the coordinate axis x are shown in Figure 1. At the interface, the time-dependent refracted angle θ_t, the fields of $H_\varsigma(t)$ and $E_\varsigma(t)$ are calculated by using the above-mentioned formulas, the propagation of the wave in the layer of the homogeneous film is considered by the term of $e^{-\text{Im}(\tilde{k})\bullet\tilde{r}}$ (where r is the propagation distant of the wave in the film). When the wave transmits through the final interface, the TDPV of the transmitted wave is given as

$$\vec{S}_\varsigma(x,t) \equiv \vec{E}_\varsigma(t) \times \vec{H}_\varsigma(t). \tag{7}$$

Furthermore, the TAPV may be obtained as follows:

$$< \vec{S}_\varsigma(x) > \equiv \sum_{t=0}^{T} \vec{S}_\varsigma(x,t)\Delta t / T_{period}. \tag{8}$$

3 NUMERICAL RESULTS AND DISCUSSION

First, we consider the cases of the obliquely incident TM-polarized light traveling through the Au/air interface. The incident angle is arbitrarily chosen as 20°, the free-space wavelength of the incident light as $\lambda = 532$ and 473 nm, and thus permittivity of Au is taken as $\tilde{\varepsilon}_{Au} = -3.93 + j1.74$ and $-1.27 + j4.54$, respectively [3]. The calculated TAPV versus refracted angle are shown in Figure 2. It is found that the TAPV of the transmitted wave distributes within a range of the refracted angle. Sometimes, a weak peak may occur in the TAPV curve except for the main peak (see Figure 2(a)), which is related to the negative TDPV of the incident wave [5]. We define the position of the main peak as the energy flow refracted angle. Apparently, the main peak of the TAPV curve for the light with $\lambda = 532$ corresponds to the negative refracted angle. When $\lambda = 473$ nm, the peak of the TAPV curve relates to the positive refracted angle. These results are in agreement with the observed results reported in Ref. [3].

Then, we consider the cases of the obliquely incident TM-polarized light traveling through a MDM (metal, Au; dielectric, TiO$_2$) multilayered structure. The structure formed by three vertically stacked MDMDM unit cells, shown in Figure 1(a) in Ref. [1], is adopted. Analogously, the individual layer thicknesses of the MDMDM layers are taken as 33, 28, 30, 28 and 33 nm, respectively. The incident angle is arbitrarily chosen as 20°, the free-space wavelength of the incident light as $\lambda = 532$ and 473 nm, respectively, the TiO$_2$ permittivity is approximately taken as $\bar{\varepsilon}_{TiO_2} = (4.3 + j0.2)^2 = 18.4 + j1.7$ [7]. The TAPV of the transmitted wave versus x coordination is calculated and presented in Figure 3. It is found that the TAPV of the transmitted wave distributes in a range along the x axis. The peak of the TAPV curve for the light with $\lambda = 532$ is located on the negative side of the x coordinate. In contrast, the main peak of the TAPV curve for the light with $\lambda = 473$ nm is located on the positive side of the x coordinate. According to the peak position x_{peak}, the refracted angle is given as $\theta_{peak} = \arctan(x_{peak}/h_{thickness})$, where $h_{thickness}$ refers to the thickness of the film. Furthermore, the power refraction index [1] is defined as $n_S = \sin(\theta_i)/\sin(\theta_{peak})$. Apparently, the n_S of the MDM multilayered structure may go from

negative to positive when the wavelength of light decreases. From the results shown in Figure 2, we conclude that the refraction behaviors of light in Au/TiO$_2$/Au multilayered structures mainly relate to the refraction properties of light at the Au/TiO$_2$ interface.

Finally, we pay some attention on the effects of the symmetry of the waveguide modes, which is demonstrated to be crucial to the realization of negative refraction in the meta-material [8]. However, according to our model (see Figure 1), the symmetry of the waveguide modes seems not to be crucial to the realization of negative refraction. The numerical examples, shown in Figure 4,

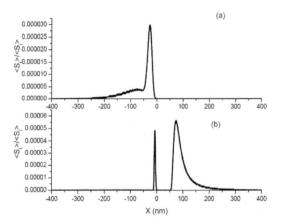

Figure 3. For the cases of an obliquely incident light traveling through a MDM multilayered structure with an incident angle of 20°, the calculated TAPV versus x coordinate, the free-space wavelength is chosen as (a) 532 nm and (b) 473 nm, respectively.

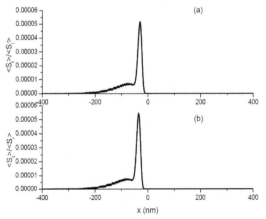

Figure 4. The calculated TAPV versus x coordinate for the obliquely incident TM-polarized light traveling through MDM multilayered structures without symmetry. The incident angle is chosen as 20° and the free-space wavelength of the light is 532 nm.

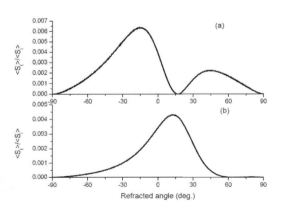

Figure 2. For the cases of an obliquely incident TM-polarized light traveling from Au into the air with an incident angle of 200, the calculated TAPV versus refracted angle, the free-space wavelength is chosen as (a) 532 nm and (b) 473 nm, respectively.

demonstrate clearly that the symmetry of the waveguide modes is not necessary, so the design and preparation of the multilayered structure may be simplified.

4 CONCLUSIONS

In summary, based on the generalized formulas of reflection and refraction, we demonstrate that visible-light refraction may go from negative to positive both at the Au/air interface and in Au/TiO$_2$/Au multilayered structures. Our work suggests a common basis for the negative refraction induced by the noble metal-air interface, single-layered noble metal film and MDM multilayered structures, providing a convenient approach to optimize the device design and address the issue on making a perfect lens.

REFERENCES

[1] Xu, T. Agrawal1, A. Abashin, M. Chau, K.J. & Lezec, H.J. 2013 All-angle negative refraction and active flat lensing of ultraviolet light. Nature 497: 470–474.

[2] Wu, Y.H. Gu, W. Chen, Y.R. Dai, Z.H. Zhou, W.X. Zheng, Y.X. & Chen, L.Y. 2008 Negative refraction at the pure Ag/air interface observed in the visible Drude region. Appl. Phys. Lett. 93: 071910.

[3] Wu, Y.H. Gu, W. Chen, Y.R. Li, X.F. Zhu, X.F. Zhou, P. Li, J. Zheng, Y.X. & Chen, L.Y. 2008 Experimental observation of light refraction going from negative to positive in the visible region at the pure air/Au interface. Phys. Rev. B 77: 035134.

[4] Chen, J.W. & Lu, H.X. 2011 Generalized laws of reflection and refraction derived from real valued boundary conditions. Opt. Comm. 284: 3802–07.

[5] Sun, N. Mao, Y.M. Chen, J.W. Zang, Y. Wang, W. Tao, Z.K. & Xie, G.Z. 2014 Effects of non-synchronized variations of electric and magnetic properties on transmitted waves at lossy interface. J. Quant. Spectr. Rad. Trans. 138: 50–59.

[6] Landau L.D. & Lifschitz, E.M. 1984 Electrodynamics of Continuous Media, Pergamon Press, Oxford.

[7] EI-Raheem, M.M.A. & AI-Baradi, A.M. 2013 Optical properties of as-deposited TiO$_2$ thin films prepared by DC sputtering technique. International J. Phys. Sciences 8:1570–78.

[8] Verhagen, E. Waele, R. Kuipers, L. & Polman, A. 2010 Three-dimensional negative index of refraction at optical frequency by coupling plasmonic waveguides. Phys. Rev. Lett. 105: 223901.

Material Science and Engineering – Chen (Ed.)
© 2016 Taylor & Francis Group, London, ISBN 978-1-138-02936-1

Influence of the pore types of mesoporous SiO$_2$ on material removal during CMP by molecular dynamics simulation

Y. Wang, R.L. Chen, H. Lei & R.R. Jiang
Nano-science and Nano-technology Research Center, Shanghai University, Shanghai, China

X.C. Lu
State Key Laboratory of Tribology, Tsinghua University, Beijing, China

ABSTRACT: The influence of the pore types of clusters on the material removal process was analyzed on the basis of the combinational effects of the plough of the cluster, the adhesion between the cluster and the substrate, and the permeation of substrate atoms. The results show that the number of removed atoms by the hollow cluster is smaller than that by the mesoporous cluster with through pores or blind pores due to the absence of the permeation effect. Besides, the number of removed atoms by the mesoporous cluster with blind pores is more than that by the mesoporous cluster with through pores due to the presence of micro-concavities, which can obtain a better combination of the aforementioned three effects. We propose an ideal model for abrasive during the Chemical Mechanical Polishing (CMP) process that can realize a higher material removed rate and lower damage.

Keywords: material removal mechanism; mesoporous silica; pore type; molecular dynamics simulation; chemical mechanical polishing

1 INTRODUCTION

The collision process between the cluster and the machined surface has been widely used in the field of precision surface machining, such as the ion cluster bombardment process under the vacuum condition (Aoki et al. 2011), Abrasive Water Jet (AWJ) cutting (Stoić et al. 2014, Orbanic & Junkar 2008, Chen et al. 1998), Fluid Jet Polishing (FJP) (Fähnle et al. 1998), Elastic Emission Machining (EEM) (Mori et al. 1990, Kim 2002) and Chemical Mechanical Polishing (CMP) (Ali et al. 1994, Levert & Korach 2009) under the assistance of water. However, so far, clusters in these applications have been mostly solid clusters.

Compared with solid clusters, mesoporous clusters have long been studied as absorbers, catalysts and molecular sieves in the field of biology and chemistry (O'Connor et al. 2006, Vinu et al. 2003, Corma 1997, Zhu et al. 2003, Vallet-Regi et al. 2001, Giri et al. 2005, Szegedi et al. 2014). However, in recent years, they have been applied in the field of microelectronics engineering, CMP for instance.

CMP is the only technology to provide global planarization in the manufacturing of Integrated Circuit (IC) (Ali et al. 1994). A typical commercial CMP process functions as follows (Levert & Korach 2009): a workpiece, crystal silicon wafer

for instance, is held in position and loaded against a polishing pad that is adhered to a rotating platen. The platen that provides mechanical input will energize abrasives in the slurry. While the chemically active abrasive particles in the slurry provide the chemical activity for CMP (Fig. 1). A very fine Material Removal (MR) is accomplished at the nanoscale level as a consequence of a combined mechanical and chemical activity. Therefore, paying attention to abrasive particles in the slurry is of great importance. As the feature size of IC approaches to 14 nm or below, a new polishing technology becomes more critical to meet the

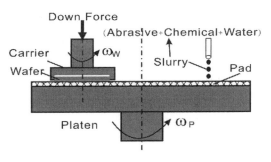

Figure 1. Schematic of the Chemical Mechanical Polishing (CMP).

demand of manufacturing. Therefore, in recent years, mesoporous abrasives have been applied to the CMP technology by some researchers.

A series of experiments performed by Lei et al. proved that the slurry containing mesoporous silica (Liu et al. 2010), mesoporous alumina (Lei et al. 2012b) or copper-incorporated mesoporous alumina abrasives (Lei et al. 2012a) could obtain much larger Material Removal Rate (MRR) and much lower roughness than that containing solid ones under the same polishing conditions. However, since the experiment conditions are limited, the correlations between the polishing performance and the pore structure of mesoporous abrasives are difficult to be acknowledged through the experiments. Moreover, the ideal structure of abrasives for CMP, which is favorable to realize the higher material removed rate and lower roughness, has not been understood clearly. These have limited the applications of new types of abrasives in the CMP process.

Therefore, studying the interaction between mesoporous abrasives with various pore types and the surface of the workpiece through Molecular Dynamics (MD) simulations has an important significance to develop new types of abrasives and understand the material removal mechanism during the CMP process. At present, the MD models used to investigate the physical essence of the CMP process are mostly the abrasive wear model and the abrasive impact model.

On the basis of the abrasive wear model, Si et al. (2011, 2012) studied the effects of abrasive rolling on material removal and surface finish in the CMP process. The influence of the ratio of radii of the particle and asperity on material removal was studied by Chagarov et al. (2003). Agrawal et al. (2010) reported that the material removal rate constant is linearly dependent on both the density and velocity of abrasives based on a phenomenological model. Zhang et al. (2013) studied the influence of moisture and slider surface roughness on the friction and wear behavior.

On the basis of the abrasive impact model, Jame et al. (2013) made a material removal mechanism map that could capture the effects of the impact velocity and abrasive grain size on the occurrence and transition between plasticity-dominated and fracture-dominated behaviors. A large-scale classical MD simulation of interaction among multiple nanoparticles and the solid surface was carried out by Han et al. (2011) to investigate the physical essence of surface planarization. Chen et al. (2010, 2011) studied the surface damage and dynamic phase transformation of crystalline silicon under the dry and wet impact of the large silica cluster by MD simulation.

However, these simulations reported were performed by the cutting or impact process of solid clusters on the substrate, not by mesoporous clusters. To our knowledge, at present, there is a

small amount of research on mesoporous clusters impacting on a substrate by MD simulation besides our previous study (Chen et al. 2013). In our previous study, by changing the diameter of the pores of the cluster, it was shown that the mesoporous cluster could obtain much larger material removal rate and much lower roughness than the solid one under the same impact conditions.

In this study, a further study on the influence of the pore type of the mesoporous cluster on the material removal process was performed by MD simulations, in order to obtain a more optimized structure for the mesoporous cluster. On the basis of the simulation results, the ideal model of the abrasive for CMP was proposed. This is instructive in the development of new types of abrasives during the CMP process.

2 SIMULATION METHODOLOGY

The impacting model of the cluster and the impacted target is shown in Figure 2. MD simulation methods have been described in more detail in our previous work (Chen et al. 2008, 2010, 2011, 2013). A Stillinger-Weber-like potential (Watanabe et al. 1999) models the inner atomic interactions of the silica cluster or the silicon substrate, and the interaction between the silica cluster and the silicon substrate. For the silicon substrate, the Stillinger-Weber-like potential is the same as the Stillinger-Weber potential. In addition, for the short distance between the atoms, high strains can occur and even the ensemble can break down. So, in this study, the modified potential function of $M(r)$ was adopted when the distance between the atoms is smaller, as given in Equation (1):

$$M(r) = A \times r^2 + B \times r + C \qquad (1)$$

Besides, the coefficients of A, B, C were solved by Equation (2):

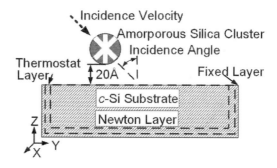

Figure 2. The schematic diagram of a cluster impacting the crystal silicon substrate for a molecular dynamics simulation.

$$\begin{cases} M(r)\,|_{r=r_0} = U(r)\,|_{r=r_0} \\ M(r)'\,|_{r=r_0} = U(r)'\,|_{r=r_0} \\ M(r)'\,|_{r=0} = 10 \times M(r)'\,|_{r=r_0} \end{cases} \qquad (2)$$

where r is the distance between two atoms. In addition, the variable r_0 was selected on the basis of some tentative calculations.

The Si (001) substrate contains 499200 silicon atoms within a space of 217.2Å × 325.8Å × 141.18Å (i.e. 40 × 60 × 26 unit cells). The outer layer of the substrate, except the top surface, was fixed in space and owned a thickness of 5 Å. In addition, a thermostat layer with a thickness of 10 Å was used to simulate the heat conduction in a reasonable way.

There are three types of amorphous silica clusters in our simulations. First, a solid cluster with a diameter of about 89 Å and 24000 atoms was prepared by quenching melted beta-cristobalite. On the basis of the solid cluster, we prepared a mesoporous cluster with three through pores, which is called the MT-Cluster. The central axes of the pores intersect at the mass center of the cluster, as shown in Figure 3a. On the basis of MT-Cluster, the hollow cluster and the mesoporous cluster with blind pores (MB-Cluster) were prepared. In the hollow cluster, both ends of the pore are closed up by a thickness of 10 Å, as shown in Figure 3b. On the contrary, in the MB-Cluster, the middle part of the pore is stuffed with a length of 46Å, so there appears six holes on the surface of the cluster (Figure 3c). The pore diameter of the hollow cluster decreases in order that the number of

Table 1. Information about the pore diameter, the length of pores in each cluster and the number of atoms comprised the cluster.

Type of cluster	Pore diameter/Å	Length of pore/Å	Number of atoms
Hollow cluster	17	69	21240
MT-cluster	20	89	19362
MB-cluster	20	22	21240

atoms in the hollow cluster is equal to that in the MB-Cluster. Table 1 presents the pore diameter, the pore length of these clusters and the number of atoms contained in the cluster.

In addition, at the very beginning of the impact, clusters are located 20Å above the silicon substrates. The simulation system was initiated with a temperature of 293 K. After a relaxation of 7000 fs with a time step of 1fs, the initial impact velocities are 4313 m/s, while the incidence angles of the clusters are 0°, 45° and 75°. The impact process would continue over 10,000 fs. During the simulation, the temperature of the thermostat atoms was kept at the ambient temperature by the Gauss–Hoover method.

3 RESULTS AND DISCUSSION

3.1 Simulation results

As shown in Figures 4–6, the damage of the substrate resulting from the collision of MB-Cluster is more serious than that impacted by the hollow cluster or MT-Cluster under the same conditions. Additionally, there is no obvious difference in the damage between the MT-Cluster and the hollow cluster. This can also be verified by the variance curves of the coordination number (CN) of silicon atoms of the substrate during the impact process, as shown in Figure 7.

More importantly, it can also be found that the number of atoms removed by the MB-Cluster is much greater than that by the MT-Cluster. Meanwhile, compared with the MT-Cluster, the hollow cluster had a significant decrease in the number of removed atoms, as shown in Figure 8.

3.2 Results analyses

The wear mechanisms in the material removal process during CMP are mostly abrasive wear and adhesive wear (Corma 1997). Therefore, as it has been described in our previous work (Chen et al. 2013), we analyze the material removal process during the impact process according to the following

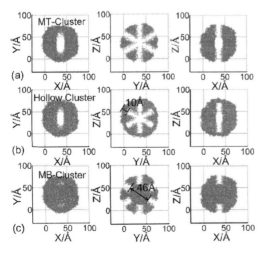

Figure 3. The cross-sectional view of mesoporous clusters in the X-Y plane, X-Z plane, and Y-Z plane, respectively. (a) MT-Cluster; (b) Hollow Cluster; (c) MB-Cluster.

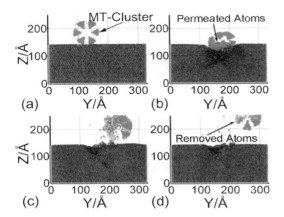

Figure 4. Side cross-section view of the impact zone at the different moments under 4313 m/s and 45°of the MT-Cluster. The red asterisk dots represent the silica cluster atoms. The magenta dots and blue dots represent five- and three-fold silicon atoms, respectively. Cyan dots represent the silicon atoms with a coor-dination number under three. The rest of the atoms represent the four-fold silicon atoms. (a) 600 fs; (b) 2450 fs; (c) 7150 fs; (d) 10000 fs.

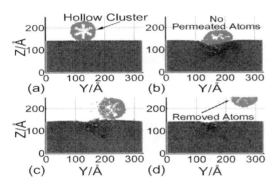

Figure 5. Side cross-section view of the impact zone at the different moments under 4313 m/s and 45°of the hollow cluster. The description of the colored atoms is the same as that given in Figure 4. (a) 600 fs; (b) 2500 fs; (c) 6400 fs; (d) 10000 fs.

factors: the plough effect based on the abrasive wear ($N_{R\text{-Plough}}$), the adhesion effect based on the adhesive wear ($N_{R\text{-Adhesion}}$), and the permeation effect based on the porous structure ($N_{R\text{-Permeation}}$). Namely, the number of atoms removed from the surface by an impacted cluster per unit time, N_R, is given as

$$N_R = N_{R\text{-}Plough} + N_{R\text{-}Adhesion} + N_{R\text{-}Permeation} \quad (3)$$

When one cluster moves through a distance $L(t)$ along a surface, it will sweep out a volume, which is described by $N_{R\text{-Plough}}$ given by (Chen et al. 2013, Ahmadi & Xia, 2001)

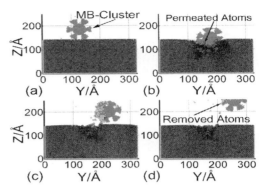

Figure 6. Side cross-section view of the impact zone at the different moments under 4313 m/s and 45°of the MB-Cluster. The description of the colored atoms is the same as that given in Figure 4. (a) 600 fs; (b) 2450 fs; (c) 6650 fs; (d) 10000 fs.

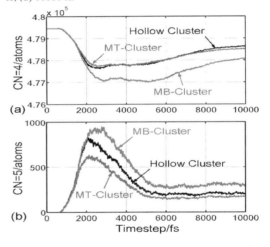

Figure 7. Variance curves of the coordination number (CN) of silicon atoms from the substrate under 4313 m/s and 45° of the hollow cluster, MT-Cluster and MB-Cluster, respectively. For a more thorough study on the variance of CN during the impact, refer to Chen et al. (2010). (a) CN = 4; (b) CN = 5.

$$N_{R\text{-}Plough} \approx c_1 \int_0^t \delta(t) b(t) U_{slide}(t) dt \quad (4)$$

Meanwhile, the volume removed resulting from the adhesion effect between the cluster and the surface is given by (Chen et al. 2013, Ahmadi & Xia, 2001)

$$N_{R\text{-}Adhesion} = N_{R\text{-}Adhesion-slide} + N_{R\text{-}Adhesion-roll}$$
$$\approx c_2 \int_0^t \pi a(t) b(t) U_{slide}(t) dt + c_3 \int_0^t a(t) b(t) U_{roll}(t) dt$$
$$(5)$$

where c_1, c_2 and c_3 are coefficients. As shown in Figure 9, $\delta(t)$ is the penetration depth of the

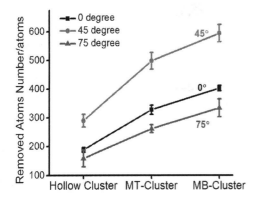

Figure 8. . Number of removed atoms from the surface of the silicon substrate after the impact of the hollow cluster, MT-Cluster and MB-Cluster under the same conditions, respectively.

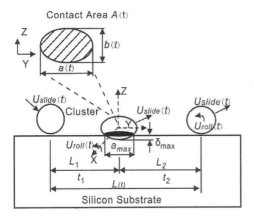

Figure 9. The schematic diagrams of collision process. $L(t)$ is the distance of cluster moving after the contact moment between the cluster and the substrate. Δt_1 and ΔL_1 are the contact time and contact distance from the contact moment between the cluster and substrate to the moment when the cluster penetrated up to the deepest position, respectively. Δt_2 and ΔL_2 are the contact time and contact moment when the cluster detached from the silicon substrate, respectively. a_{max}, b_{max} and δ_{max} are the contact radii and penetration depth between the cluster and the substrate at the moment of the deepest penetration position of the cluster. $A(t)$ is the area of the projected ellipse of contact between the cluster and the impacted substrate. $U_{slide}(t)$ and $U_{roll}(t)$ re the relative sliding speed and rolling speed between the cluster and impact surface, respectively.

cluster into the impact silicon surface. $A(t)$ or $b(t)$ are the radii of the projected ellipse of the contact between the cluster and the impact surface. $U_{slide}(t)$ and $U_{roll}(t)$ are the relative sliding speed and the rolling speed between the cluster and the impact

surface, respectively. In addition, the distance $L(t)$ is given by

$$L(t) = \int_0^t U_{slide}(t)dt \qquad (6)$$

Because the mass of MT-Cluster is smaller than that of the hollow cluster or MB-Cluster, its projectile energy will be lower than that of the hollow cluster or MB-Cluster. This will result in the decrease in the impact load of the MT-Cluster. So, $\delta(t)$ of the MT-Cluster is smaller than that of the hollow cluster or MB-Cluster for the decrease in the impact load (Table 2). Therefore, the damage of silicon surface impacted by the MT-Cluster will be weaker than that impacted by the MB-Cluster (see Figure 7).

Meanwhile, the elastic module and the hardness of the hollow cluster are less than those of the MB-Cluster because of the hollow structure of the cluster. So, $b(t)$ or the contact area ($a(t) \times b(t)$) of the hollow cluster is a little larger than that of the MT-Cluster or MB-Cluster, as presented in Table 2. It also will result in the decrease of $\delta(t)$ of the hollow cluster relative to the MB-Cluster. Therefore, the damage of the silicon surface impacted by the hollow cluster will be weaker than that impacted by the MB-Cluster, although both of their numbers of comprised atoms are equal. In addition, the damage of the hollow cluster is similar to that of the MT-Cluster (Figure 7).

In addition, as presented in Table 2, there is no obvious difference in the $L(t)$ among the MT-Cluster, MB-Cluster and hollow cluster. Therefore, the $N_{R\text{-}Plough}$ of the MT-Cluster or hollow cluster will be less than that of the MB-Cluster. Furthermore, the $N_{R\text{-}Adhesion}$ of the MT-Cluster or MB-Cluster will be less than that of the hollow cluster.

As shown in Figure 4 and Figure 6, some atoms located on the surface of the silicon substrate will permeate and diffuse into the pores of the mesoporous cluster in the impact loading stage. Eventually, a few atoms diffused into the mesoporous cluster will escape from the surface of the silicon substrate together with the flying cluster. So, the permeation effect plays an important role in the increase of removed atoms by the mesoporous cluster. However, the permeation phenomenon will not occur during the impact process of the hollow cluster (see Figure 5). So, the $N_{R\text{-}Permeation}$ of the MT-Cluster and MB-Cluster will be more than that of the hollow cluster.

In summary, because of the decrease in $\delta(t)$, the $N_{R\text{-}Plough}$ of the hollow cluster or MT-Cluster was less than that of the MB-Cluster. However, the $N_{R\text{-}Adhesion}$ of the hollow cluster was more than that of the MT-Cluster or MB-Cluster owing to the increase in the contact area. In addition, the

Table 2. Data for the contact areas between the cluster and the substrate during the impact of the hollow cluster, MT-Cluster and MB-Cluster with 4313m/s and 45°. The description of all variables is the same as given in Figure 9.

	Hollow cluster	MT-cluster	MB-cluster
Number of atoms	21240	19362	21240
Δt_1 (fs)	1850	1850	1900
ΔL_1 (Å)	46.6	44.8	44.2
δ_{max} (Å)	12.4	11.22	15.91
b_{max} (Å)	83.7	80.9	81.6
$b_{max} \times \delta_{max}$ (Å²)	1037.88	907.70	1298.26
a_{max} (Å)	68.9	68.7	69.2
$b_{max} \times a_{max}$ (Å²)	5766.93	5557.83	5646.72
Δt_2 (fs)	3900	4700	4200
ΔL_2 (Å)	39.3	38.3	39.5

$N_{R\text{-Permeation}}$ of the MT-Cluster or MB-Cluster was more than that of the hollow cluster since there are no pores on the surface of the latter. Therefore, the MT-Cluster and MB-Cluster removed more atoms than the hollow cluster mostly due to the permeation effect of the porous structure on the surface of the cluster. This also verifies again that the pore structure plays an important role in the material removal process during the impact process.

As previously mentioned, the $N_{R\text{-Plough}}$ of the MB-Cluster would be larger than that of the MT-Cluster due to the increase in $\delta(t)$. Furthermore, the outer layer of the former may be softer than that of the latter due to the micro-concavities on the surface. This would result in the increase of the contact area ($a(t) \times b(t)$) between the cluster and the impacted surface (see Table 2). This would be helpful for the increase in $N_{R\text{-Adhesion}}$. Meanwhile, though the length of the pores of the MB-Cluster is shorter than that of the MT-Cluster, the effective number of pores and their pore length in the contact area are almost the same (Figures 4 and 6). So, there is no obvious difference in $N_{R\text{-Permeation}}$ between the former and the latter. Therefore, the number of removed atoms by the MB-Cluster will be more than that by the MT-Cluster (Figure 7).

3.3 Discussion

Therefore, the above-mentioned results indicate that an ideal model for the abrasive (see Figure 10) during the CMP process should satisfy the following two requirements:

1. The inner layer of the abrasive is relatively harder, which could prevent the excess reduction of $N_{R\text{-Plough}}$. Meanwhile, the outer layer of the abrasive is relatively softer, which is helpful for the increase in $N_{R\text{-Adhesion}}$.

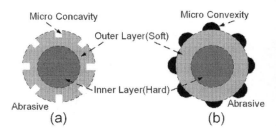

Figure 10. The ideal model for the abrasive during the chemical mechanical polishing (CMP) process.

2. The surface of the abrasive should be rough rather than smooth. Besides, this rough surface is studded with micro-concavities or micro-convexities. These concavities or convexities will be helpful for the increase in $N_{R\text{-Permeation}}$. Meanwhile, they will also result in a further increase of the contact area between the abrasive and the surface of the wafer, which will be beneficial to the increase in $N_{R\text{-Adhesion}}$.

Therefore, changing the microstructure on the surface of particles may be a good way to improve the MRR. In other words, an ideal abrasive for CMP should realize the optimal balance between the plough effect, the adhesion effect and the permeation effect.

4 CONCLUSIONS

The influence of the pore type of mesoporous cluster on material removal and surface quality during the impact process was studied by molecular dynamics simulation. An ideal model of the abrasives for CMP was also proposed. Thus, the following conclusions can be drawn:

On the one hand, owing to the decrease in the penetration depth, the plough effect of the hollow cluster and the mesoporous cluster with through pores will be weakened. Thus, the number of atoms removed by these clusters will be less than that removed by the mesoporous cluster with blind pores. One the other hand, the contact area between the cluster and the impacted substrate will increase as the hardness and the elastic module of the cluster decrease. So, the number of removed atoms due to the adhesion effect caused by the hollow cluster will be more than that caused by the mesoporous cluster. In addition, compared with the hollow cluster, the permeation effect of the pores of the mesoporous cluster contributes much to the number of removed atoms.

Therefore, with the combination of the plough effect, the adhesion effect and the permeation effect, the number of removed atoms by the hollow cluster will be obviously less than that by the mesoporous cluster with through pores or blind pores under the same impact conditions.

Meanwhile, compared with the mesoporous cluster with through pores, the effective contact area of the mesoporous cluster with blind pores will increase due to the presence of micro-concavities. Also, the cluster with blind pores will penetrate much deeper than the cluster with through pores for the larger projectile energy. In other words, it is easier for the cluster with blind pores to achieve the best combination of the three impact effects, which will help the cluster realize a higher material removal rate than the cluster with through pores under the same conditions. However, the damage of the substrate surface caused by the mesoporous cluster with blind pores will be a little more serious than that caused by the other two types of clusters.

In other words, the ideal model for the abrasive during the CMP process could be a mesoporous abrasive with a rough surface studded with micro-concavities or micro-convexities whose outer layer is softer and the inner layer is harder. This kind of ideal model will be beneficial to realize the higher material removed rate and lower damage of the machined surface, which is the final aim of CMP. These findings are instructive in the optimization of experimental parameters and the development of new techniques during the CMP process.

ACKNOWLEDGMENT

This work was supported by the NSFC (No. 51375291, No. 51475279, No. 91323302), the Innovation Program of Shanghai Municipal Education Commission (13YZ004), and the Research Fund for the Doctoral Program of Higher Education of China (No. 20123108110016).

REFERENCES

[1] Agrawal, P.M. & Narulkar, R. & Bukkapatnam, S. et al. 2010. A phenomenological model of polishing of silicon with diamond abrasive. *Tribology International* 43(1–2): 100–107.

[2] Ahmadi, G. & Xia, X. 2001. A Model for Mechanical Wear and Abrasive Particle Adhesion during the Chemical Mechanical Polishing Process. *Journal of the Electrochemical Society* 148(3): G99–G109.

[3] Ali, I. & Roy, S. & Shinn, G. 1994. Chemical-mechanical polishing of interlayer dielectric: A review. *Solid State Technology* 37(10): 63–70.

[4] Aoki, T. & Seki, T. & Matsuo, J. 2011. Molecular dynamics simulations of large fluorine cluster impact on silicon with supersonic velocity. *Nuclear Instruments & Methods in Physics Research Section B-beam Interactions with Materials and Atoms* 269(14): 1582–1585.

[5] Chagarov, E. & Adams, J.B. 2003. Molecular dynamics simulations of mechanical deformation of amorphous silicon dioxide during chemical-mechanical polishing. *Journal of Applied Physics* 94(6): 3853–3861.

[6] Chen, L. & Siores, E. & Wong, W.C.K. 1998. Optimising abrasive waterjet cutting of ceramic materials. *Journal of Material Processing Technology* 74(1–3): 251–254.

[7] Chen, R.L. & Jiang, R.R. & Lei, H. et al. 2013. Material removal mechanism during porous silica cluster impact on crystal silicon substrate studied by molecular dynamics simulation. *Applied Surface Science* 264: 148–156.

[8] Chen, R.L. & Liang, M. & Luo, J.B. et al. 2011. Comparison of surface damage under the dry and wet impact: Molecular dynamics simulation. *Applied Surface Science* 258(5): 1756–1761.

[9] Chen, R.L. & Luo, J.B. & Guo, D. et al. 2008. Extrusion formation mechanism on silicon surface under the silica cluster impact studied by molecular dynamics simulation. *Journal of Applied Physics* 104: 104907.

[10] Chen, R.L. & Luo, J.B. & Guo, D. et al. 2010. Dynamics phase transformation of crystalline silicon under the dry and wet impact studied by molecular dynamics simulation. *Journal of Applied Physics* 108(7): 073521.

[11] Corma, A. 1997. From microporous to mesoporous molecular sieve materials and their use in catalysis. *Chemical Reviews* 97(5): 2373–2419.

[12] Fähnle, O.W. & Brug, H. & Frankena, H.J. 1998. Fluid jet polishing of optical surfaces. *Applied Optics* 37(28): 6771–6773.

[13] Kim, J.D. 2002. Motion analysis of powder particles in EEM using cylindrical polyurethane wheel. *International Journal of Machine Tools & Manufacture.* 42(1): 21–28.

[14] Levert, J.A. & Korach, C.S. 2009. CMP Friction as a function of slurry silica nanoparticle concentration and Diameter. *Tribology Transactions* 52(2): 256–261.

[15] Mori, Y. & Yamauchi K. & Endo K. et al. 1990. Evaluation of elastic emission machined surfaces by scanning tunneling microscopy. *Journal of Vacuum Science and Technology A.* 8(1): 621–624.

[16] Giri, S. & Trewyn, B.G. & Stellmaker, M.P. et al. 2005. Stimuli-responsive controlled-release delivery system based on mesoporous silica nanorods capped with magnetic nanoparticles. *Angewandte Chemie-International Edition* 44(32): 5038–5044.

[17] Han, X.S. & Gan, Y.X. 2011. Analysis the complex interaction among flexible nanoparticles and materials surface in the mechanical polishing process. *Applied Surface Science* 257(8): 3363–3373.

[18] James, S. & Sundaram, M.M. 2013. A molecular dynamics study of the effect of impact velocity, particle size and angle of impact of abrasive grain in the Vibration Assisted Nano Impact-machining by Loose Abrasives. *Wear* 303(1–2): 510–518.

[19] Lei, H. & Jiang, L. & Chen, R.L. 2012. Preparation of copper-incorporated mesoporous alumina abrasive and its CMP behavior on hard disk substrate. *Powder Technology* 219:99–104.

[20] Lei, H. & Wu, X. & Chen, R.L. 2012. Preparation of porous alumina abrasives and their chemical mechanical polishing behavior. *Thin Solid Film* 520(7): 2868–2872.

[21] Liu, P. & Lei, H. & Chen, R.L. 2010. Polishing properties of porous silica abrasive on hard disk substrate CMP. *International Journal of Abrasive Technology* 3(3): 228–237.

[22] O'Connor, A.J. & Hokura, A. & Kisler J.M. et al. 2006. Amino acid adsorption onto mesoporous silica molecular sieves. *Separation and Purification Technology* 48(2): 197–201.

[23] Orbanic, H. & Junkar, M. 2008. Analysis of striation formation mechanism in abrasive water jet cutting. *Wear* 265(5–6): 821–830.

[24] Si, L.N. & Guo, D. & Luo, J.B. et al. 2011. Abrasive rolling effects on material removal and surface finish in chemical mechanical polishing analyzed by molecular dynamics simulation. *Journal of Applied Physics* 109(8): 084335.

[25] Si, L.N. & Guo, D. & Luo, J.B. et al. 2012. Planarization process of single crystalline silicon asperity under abrasive rolling effect studied by molecular dynamics simulation. *Applied Physics A: Materials Science & Processing* 109(1): 119–126.

[26] Stoić, A. & Duspara, M. & Kosec B. et al. 2014. The influence of mixing water and abrasives on the quality of machined surface. *Metalurgija* 53(2): 239–242.

[27] Szegedi, Á. & Popova, M. & Valyon, J. et al. 2014. Comparison of silver nanoparticles confined in nanoporous silica prepared by chemical synthesis and by ultra-short pulsed laser ablation in liquid. *Applied Physics A-Materials Science & Processing* 117(1): 55–62.

[28] Vallet-Regi, M. & Rámila, A. & del Real, R.P. et al. 2001. A new property of MCM-41: Drug delivery system. *Chemistry of Materials* 13(2): 308–311.

[29] Vinu, A. & Streb, C. & Murugesan V. et al. 2003. Adsorption of cytochrome c on new mesoporous carbon molecular sieves. *Journal of Physical Chemistry B* 107(33): 8297–8299.

[30] Watanabe, T. & Fujiwara, H. & Noguchi, H. et al. 1999. Novel Interatomic Potential Energy Function for Si, O Mixed Systems. *Japanese Journal of Applied Physics Part 2-Letters* 38(4 A): L366–L369.

[31] Zhang, L.C. & Tang, C. 2013. Friction and wear of diamond–silicon nano-systems: Effect of moisture and surface roughness. *Wear* 302(1–2): 929–936.

[32] Zhu, J. & Kónya, Z. & Puntes, V.F. et al. 2003. Encapsulation of metal (Au, Ag, Pt) nanoparticles into the mesoporous SBA-15 structure. *Langmuir* 19(10): 4396–4401.

Material Science and Engineering – Chen (Ed.)
© 2016 Taylor & Francis Group, London, ISBN 978-1-138-02936-1

Study on the tribological characteristics of Si_3N_4-hBN ceramic materials sliding against cast iron

W. Chen, X. Ai, D.Q. Gao & Z.L. Lv
Shaanxi University of Science & Technology, Xi'an, Shaanxi, China

ABSTRACT: The tribological behaviors of Si_3N_4-hBN composites sliding on cast iron under the dry friction condition were studied on a MMU-5G-type pin-on-disk wear tester. The materials and worn surfaces were characterized by means of Scanning Electron Microscopy (SEM) and X-Ray Diffraction (XRD), and the worn surfaces were analyzed using X-Ray Diffraction Spectroscopy (EDS). The results show that the addition of hBN to Si_3N_4 cannot effectively improve the tribological characteristics of Si_3N_4-hBN/cast iron sliding pairs. There is no friction in the chemical reaction film, and a smooth worn surface is formed. The analysis indicates that the abrasive wear and high friction coefficients of Si_3N_4-hBN/QY600 sliding pairs under the dry friction condition result from cracking and fracture, and the graphite in QT600 spalls off during the wear test. The occurrence of severe adhesive wear on the worn surface of Si_3N_4-hBN/Cr20 sliding pairs under the dry friction condition is observed. The debris is oxidized and adheres to the worn surface, limiting the solid lubricant to play its role in reducing the friction, and hindering the formation of the tribo-chemical reaction film. At the same time, the adhesion layer peeling off is responsible for the increase in the worn surface roughness, so that the friction coefficients are at a higher level.

Keywords: composite ceramics; nodular cast iron QT600; white cast iron Cr20; friction coefficient

1   INTRODUCTION

It is well known that silicon nitride ceramics are widely used in many engineering fields, such as aerospace, key parts of engine, and high-speed cutting tools, because of their excellent comprehensive performance (e.g. high hardness, high strength, high stiffness, lower density, excellent chemistry stability and fine mechanical properties at high temperatures) (F. Wang et al. 2007, R.G. Wang et al. 2002, M. Liu et al. 2004, D.Q. Wei et al. 2006, W. Chen et al. 2012 & E. Carrasquero et al. 2005). However, studies have shown that silicon nitride ceramics with high friction coefficients (generally 0.7~1.0) under the dry friction condition are vulnerable when a serious wear occurs, which limits their application in practical engineering. At present, on account of some shortcomings of traditional ways of lubrication (X.Z. Zhao et al. 1996), researchers have paid more attention to the study on the realization of self-lubricating performance of ceramic materials in order to decrease their friction and wear coefficients. The way that was often adopted to improve the tribological properties of ceramic materials is to add a solid lubricant to realize the self-lubricating performance.

J.M. Carrapichan et al. (2002) investigated the tribological behavior of self-mated Si_3N_4/BN composite ceramics. Their results revealed that the addition of BN could decrease the friction coefficients of silicon nitride composite ceramics, and the wear coefficients were less than 10^{-5} mm³/(N·m); however, it was sharply increased when the amount of added BN is more than 10%. Under this experimental condition, the tribo-chemical reaction film was not formed, which may be attributed to the mild condition that is not enough to cause the tribo-chemical reaction between the composite ceramic and the ambient environment. W. Chen et al. (2012) investigated the tribological characteristics of Si_3N_4-hBN/Fe-B alloy pairs. When Si_3N_4-hBN sliding against the Fe-B alloy under the dry friction condition, the wear coefficients of the Si_3N_4-hBN pin specimen were increased with the increase in hBN content, while that of the Fe-B alloy disk specimen decreased, but they were all higher than 10^{-5} mm³/(N·m). The friction coefficients of the same pairs with water lubrication reduced obviously because the debris was flowed away by water in order to avoid abrasive wear; meanwhile, the worn surface of the pin and disk was smooth due to chemical polishing of the Si_3N_4-hBN pin specimen surface. So, the friction coefficients and wear coefficients were, respectively, below 0.1 and 10^{-6} mm³/(N·m). In the study by W. Chen et al. (2009), the function of water was in accordance with the above-mentioned experiment, in which the worn surface became smooth because of the tribo-chemical reactions between Si_3N_4 and hBN. Another study (W. Chen et al. 2010) reported that the formation of the tribo-chemical film (containing B_2O_3, SiO_2 and Fe_2O_3), which protects and lubricates the worn surface as well as reduces the friction and wear coefficients, was observed when Si_3N_4-hBN composite ceramics was mated with austenitic stainless steel 1Cr18 Ni9Ti under the dry friction condition.

In summary, in terms of the present status of the research, different tribological properties were obtained when Si_3N_4 and its composite ceramics were mated with various materials under different environmental and parameter conditions. There were differences in the starting point and the purpose of different researchers: the data obtained from the same test that was carried out many times on the same tester under the same conditions were also different (W. Chen et al. 2010 & 2013, C.C Liu et al. 2004, Z.X. Xiang et al. 2000, Z.X. Xiang et al. 2006, Y.M. Gao et al. 2001, Y.Y. Huang et al. 2010, Y.M. Gao et al. 1996, 1997 & 1999). So, this paper has chosen the Si_3N_4-hBN/cast iron pairs as the research object to carry out the dry friction experiment. Combining with the friction surface analysis and morphology observation, the friction coefficients were measured to study the function of the solid lubricant hBN during the friction process, and to investigate the formation situation of the tribo-chemical reaction film and the tribological performance of friction pairs. It can provide the experimental basis for further study on the wear mechanism of Si_3N_4 composite ceramics, and enrich their tribological theory.

2 EXPERIMENTAL SECTION

2.1 The preparation of the specimen

2.1.1 The preparation of Si_3N_4 and Si_3N_4-hBN pin specimens

We took Si_3N_4 powder (purity > 99.99%, α phases > 94%, average particle size 0.3 µm) as the substrate, hBN (purity > 99.99%, average particle size 0.3µm) as the solid lubricant, and Al_2O_3 and Y_2O_3 (purity > 99.5%, average particle size 1µm, total volume fraction 10%) as sintering additives. The specimens of Si_3N_4-hBN composed of 0, 5, 10, 20, 30 vol.% hBN were sintered by hot-pressing under 30 MPa pressure at 1800°C for 30 min of dwell time. The XRD analysis result and the microstructure of sintered Si_3N_4 20%hBN composite show that it composed of elongated β-Si_3N_4 and layered hBN.

Si_3N_4-hBN composites were cut into rectangular-type pins in the size of 5 mm × 5 mm × 20 mm and 5 mm × 5 mm × 10 mm; the former was used for performance testing and the latter was used for friction and wear tests. Density and porosity, Vickers hardness, bending strength and fracture toughness of the specimens were measured, respectively, by using the Archimedes drainage method, the Vickers hardness tester, the three-point bending method and the indentation method. The results are summarized in Table 1. It revealed that the physical and mechanical properties of Si_3N_4-hBN composites exhibit a downward trend with the increasing content of hBN.

2.1.2 Mating plate

The disk, as one of the friction pairs, was machined from the pearlitic nodular cast iron QT600 and the abrasion resistant white cast iron Cr20, whose microstructures are shown in Figure 1, of diameter 44 mm and thickness of 6 mm. Figure 1a shows the microstructure of QT600, where the white part is ferrite and the black part is graphite. Figure 1b shows the microstructure of Cr20, composed of eutectic carbide (Cr,Fe) 7C3, austenite and its transformation products. The main components of QT600 and Cr20 are given in Table 2.

The pins and disks were finished by grinding to achieve a surface roughness that was 0.8 µm or less. Then, the specimens were tested after acetone ultrasonic cleaning.

2.2 Experimental program

Friction and wear experiments were carried out on a pin-on-disk device in which an upper disk contacted a bottom pin under a normal load (10 N) without lubrication. The pin specimen was fixed, and the disk specimen rotated at different linear

Table 1. Physical and mechanical properties of Si_3N_4-hBN specimens.

hBN amount (vol.%)	0 (SN0)	5 (SN5)	10 (SN10)	20 (SN20)	30 (SN30)
Density (g/cm³)	3.31	3.17	3.10	2.97	2.94
Porosity (%)	0.84	0.90	0.91	1.04	1.05
Bending strength (MPa)	812	758	613	541	465
Vickers hardness (GPa)	19.9	19.6	15.3	9.3	6.7
Fracture toughness (MPa·m$^{1/2}$)	8.58	8.01	7.14	5.97	5.50

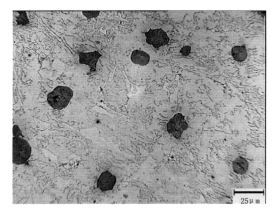

(a) Microstructure of QT600

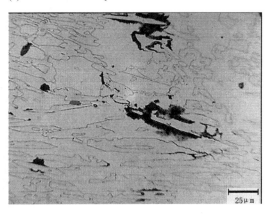

(b) Microstructure of Cr20

Figure 1. Microstructures of mating plates.

Table 2. Main components of QT600 and Cr20 (wt%).

Elements	QT600	KmTBCr20
C	3.3–3.6	3.0–3.2
Si	2.51–2.98	0.5–0.8
Cr	–	18–22
Mn	0.30	0.6–0.9
P	<0.05	–
S	<0.02	–
Mg	0.041	–

speeds (0.65 m/s and 1.31 m/s). The sliding distance was 850 m.

The friction coefficients were automatically recorded by the tester and output by its data processing software. The morphologies of the worn surface of the pairs were examined by SEM. The elements' constitution of the worn surface was analyzed by EDS. The specimens' phase constitution was determined by XRD.

3 RESULTS AND DISCUSSION

3.1 Friction and wear properties of Si_3N_4-hBN/QT600 and Si_3N_4-hBN/Cr20 pairs

Figure 2a shows the friction coefficients of the Si_3N_4-hBN/QT600 pairs sliding under the dry friction condition at different linear speeds. It can be seen from the figure that the friction coefficients of the pairs, which were higher than 1.0 at 0.65 m/s, were greater than 0.8 when the speed was 1.31 m/s. The friction coefficients of the pairs at both linear speeds were higher than 1.4 when the content of hBN was up to 30%. The above analysis indicated that the addition of hBN to Si_3N_4 cannot effectively reduce the friction.

Figure 2b shows the friction coefficients of the Si_3N_4-hBN/Cr20 pairs sliding under the dry fric-

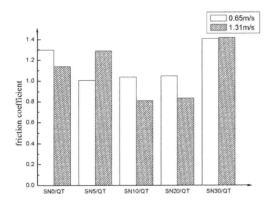

(a)Si_3N_4-hBN/QT600 pairs

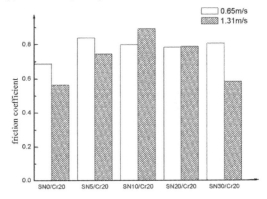

(b) Si_3N_4-hBN/Cr20 pairs

Figure 2. Friction coefficients of the pairs under the dry friction condition at different linear speeds.

tion condition at different linear speeds. From this figure, it could be easily found that the change in the friction coefficients of the pairs with the increase in hBN content was stable, and all the values were higher than 0.68 at the speed of 0.65 m/s. With the increasing content of hBN at the speed of 1.31 m/s, first, the friction coefficients exhibited an upward tendency to 10% content of hBN, and then showed a downward tendency. The maximum value was about 0.9 when the SN10 pin was mated with the Cr20 disk. So, it can be concluded that the addition of hBN to Si_3N_4 also cannot effectively decrease the friction coefficients.

3.2 Wear mechanism analysis

3.2.1 Wear mechanism of Si_3N_4-hBN/QT600 pairs

The worn surface morphology of the QT600 disk mated with the SN30 pin at 1.31 m/s is illustrated in Figure 3. As shown in the figure, the friction surface was rough, a small number of furrow-like abrasions existed, a lot of peeling-off pits were generated and wear debris were embedded into

those pits. At the initial friction stage, there was collision and friction between asperities the on Si_3N_4-hBN surface and the spheroidal graphite of QT600. According to the relevant literature (Y.M. Gao et al. 1995), when the friction test of the pairs was carried out in air, spalling-off pits were formed on the surface of the cast iron plate because of the brittle fracture of graphite in the spheroidal graphite cast iron QT600, which easily occurred due to the increase in the friction interface temperature, leading to the embrittlement of the graphite. Meanwhile, a lot of debris were generated and resulted in the abrasive wear of the pairs.

Figure 4 shows the worn surface morphology of the SN30 pin. It can be seen from the figure that the surface of the pin was uneven, and the existence of a large number of wear debris was observed. Abrasive wear of the pairs prevented the formation of the tribo-chemical reaction film; therefore, the presence of the film on the QT600 and SN30 surface could not be observed, and the friction coefficients were higher.

The SEM image and the EDS analysis results of the smooth area on the QT600 disk worn surface were obtained, as shown in Figure 5 and Figure 6. The black areas in the figure are nodular graphites in QT600, which is mainly composed of graphite (C). Furrow-like abrasion (e.g. point 2) contained more content of Fe and less content of C and Si, which was consistent with the composition of QT600. Besides, a certain amount of O was also detected. The reason is that the material cannot contact closely with the mating plate after the furrow-like abrasion is formed, so the oxidation phenomenon occurs under the effect of high friction temperature. Moreover, the smooth area of the worn surface (e.g. point 3) always has a sliding friction with the counterpart, so there is a little amount of O and the components are consistent with QT600.

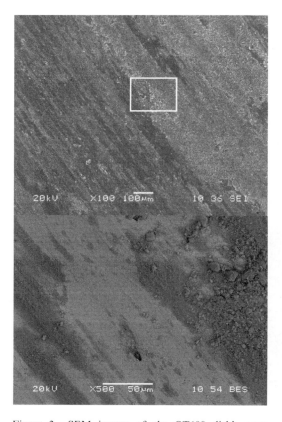

Figure 3. SEM images of the QT600 disk's worn surface.

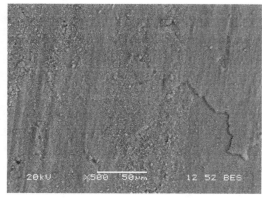

Figure 4. SEM image of the Cr20 pin's worn surface.

400

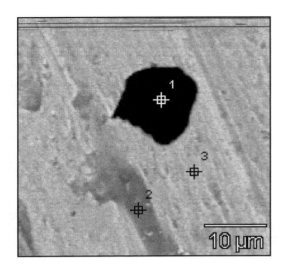

Figure 5.   SEM image of the smooth area on the QT600 disk's worn surface.

Figure 7 shows the SEM image and the EDS line scanning analysis results of the rough area on the QT600 disk's worn surface. As can be seen, C distributes uniformly along the vertical direction of the furrow-like wear, and more content of O is relatively present at the furrow and the edge of the furrow due to the oxidation phenomenon with air that occurs at a high friction temperature. The distribution of Si is uniform on the whole, but increases sharply at the edge of the furrow because the debris generated by Si_3N_4-hBN is embedded into the furrow. A small amount of Fe was detected

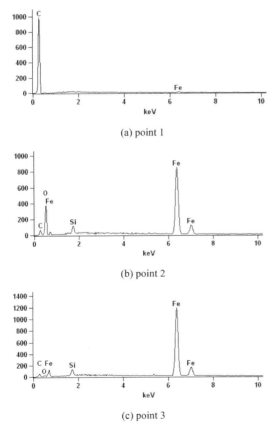

(a) point 1

(b) point 2

(c) point 3

Figure 6.   EDS analysis of the smooth area on QT600 disk's worn surface.

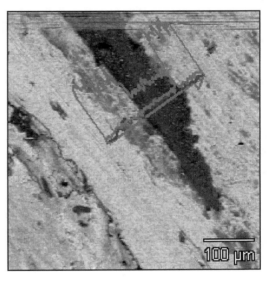

(a) SEM image

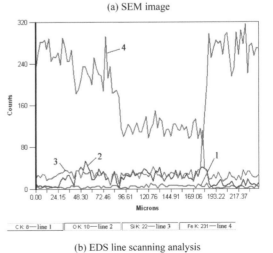

(b) EDS line scanning analysis

Figure 7.   The analysis results of the rough area on the QT600 disk's worn surface.

401

at the furrow, while the amount at the other position was large. This phenomenon resulted from the wear of the material.

3.2.2 *Wear mechanism of Si₃N₄-hBN/Cr20 pairs*

The worn surface morphology of the Cr20 disk mated with the SN30 pin at 1.31 m/s is illustrated in Figure 8. It can be clearly seen from the figure, there is serious adhesive wear on the Cr20 worn surface and a lot of spalling-off pits caused by brittle fracture. A large number of wear debris adhered to these pits.

Figure 9 shows the EDS line scanning analysis results of the Cr20 disk's worn surface. At the early friction stage, the collision and friction occurs between the asperities on the surface of Si_3N_4-hBN and Cr20. Owing to the high brittleness, the fracture occurs on the surface of white cast iron easily and the peeling-off pits are generated, so a large number of wear debris are generated during the friction process and embedded into those pits, and then oxidized at a high temperature and adhere to the surface of the Cr20 disk. Therefore, the amount of Fe is sharply decreased and the amount of Si and O is significantly increased at the posi-

(a)SEM image

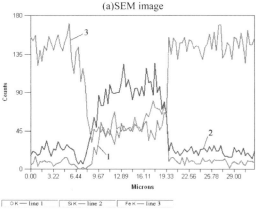

(b) EDS line scanning analysis

Figure 9. The analysis results of the Cr20 disk's worn surface.

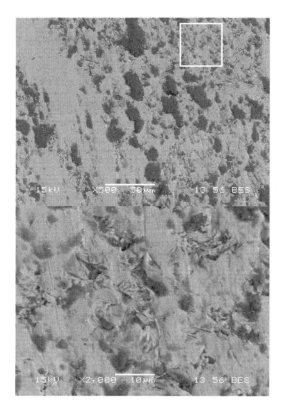

Figure 8. SEM images of the Cr20 disk's worn surface.

tion where the peeling-off pits are located. Meanwhile, the adhesion of the oxidized wear debris to the worn surface has limited the solid lubricant component hBN to play its role in reducing friction and lubricating the pairs, hindering the formation of the tribo-chemical reaction film. When the friction of the pair continues, the adhesion of the Cr20 disk and Si_3N_4-hBN pin occurs under a high temperature and pressure. A portion of the material on the Cr20 disk surface will be taken away by the Si_3N_4-hBN pin when the relative slide continues, which leads to the wear of the disk. At the same time, the adhesion layer on the Cr20 disk flakes increases the roughness of the worn surface, and keeps the friction coefficients at a high level.

4 CONCLUSIONS

1. Collectively, when the Si₃N₄-hBN pins slide against the cast iron plates under the dry friction condition, the solid lubricating phase hBN fails to play an effective anti-friction effect. Meanwhile, the smooth surface of both the ceramics and the metals cannot be obtained, and there is no presence of the tribo-chemical reaction film, and the friction coefficients are at a high level.

2. When the Si₃N₄-hBN pins are mated with the QT600 disks under the dry friction condition, a large amount of wear debris are generated. The reason for the generation of the debris is that the brittle fracture of graphites in QT600 occurs and then results in the peeling off from the surface during the friction process.

3. When the Si₃N₄-hBN pins are mated with the Cr20 disk under the dry friction condition, a serious adhesion wear on the friction surface occurs, leading to the generation of a lot of wear debris. Meanwhile, the adhesion layer that is peeled off increases the roughness of the worn surface.

ACKNOWLEDGMENTS

The authors would like to thank the Natural Science Foundation of China (No. 51405278), the Special Scientific Research Plan Project of Education Department of Shaanxi Province (14 JK1082), and the Scientific Research Starting Foundation for Introducing Doctor of Shaanxi University of Science & Technology (BJ11–01).

REFERENCES

[1] C.C. Liu, J.L. Huang 2004. Tribological characteristics of Si₃N₄-based composites in unlubricated sliding against steel ball. Materials Science and Engineering A 384: 299–307.

[2] D.Q. Wei, Q.C. Meng, D.C. Jia 2006. Mechanical and tribological properties of hot pressed h-BN/Si₃N₄ ceramic composites. Ceram. Int. 32: 549–554.

[3] E. Carrasquero, A. Bellosi, M.H. Staia 2005. Characterization and wear behavior of modified silicon nitride. Int. J. Refract. Met. Hard Mater 23: 391–397.

[4] F. Wang, Z.K. Fan, Y.Y. Sun 2007. Study on Al₂O₃/hBN self-lubricating ceramic composite. Aerospace Material & Technology, (1): 64–67.

[5] J.M. Carrapichano, J.R. Gomes, R.F. Silva 2002. Tribological behavior of Si₃N₄-BN ceramic materials for dry sliding applications. Wear 253: 1070–1076.

[6] M. Liu, Z.L. Lu, Y.Z. Zhang, et al. 2004. Friction and wear behaviors of ceramic and ceramic composite materials under dry friction. Hot Working Technology 47(10): 42–45.

[7] R.G. Wang, W. Pan, M. Jiang, J. Chen, Y.M. Luo 2002. Investigation of the physical and mechanical properties of hot-pressed machinable Si₃N₄/h-BN composites and FGM. Mater. Sci. Eng. B 90: 261–268.

[8] W. Chen, Y.M. Gao, F.L. Ju, Yong Wang 2009. Tribo-chemical behavior of Si₃N₄-hBN ceramic materials with water lubrication. Journal of Xi'an Jiaotong University, 2009, 43(9): 75–80.

[9] W. Chen, Y.M. Gao, C. Chen 2010. Tribological behavior of Si₃N₄-hBN ceramic materials against stainless steel under the dry friction condition. Tribology 30(3): 243–249.

[10] W. Chen, Y.M. Gao, C. Chen, J.D. Xing 2010. Tribological characteristics of Si₃N₄-hBN ceramic materials sliding against stainless steel without lubrication. Wear 269: 241–248.

[11] W. Chen, Y.M. Gao, L. Chen, H.Q. Li, C. Chen 2012. Study on the tribological characteristics of Si₃N₄ –hBN ceramic materials sliding against Fe-B alloy. Lubrication engineering 37(3): 29–36.

[12] W. Chen, Y.M. Gao, J.Z. Chen 2013. Tribological behavior of Si₃N₄-hBN ceramic materials sliding against Fe-B alloy without lubrication. Applied Mechanics and Materials Vols. 268–270: 32–36.

[13] X.Z. Zhao, J.J. Liu, B.L. Zhu, H.Z. Miao, Z.B. Luo 1996. Tribological properties of Si₃N₄/45 steel pairs under lubricated and unlubricated conditions. Journal of the Chinese ceramic society 24(5): 515–522.

[14] Y.M. Gao, L. Fang, J.Y. Su 1995. Air and water lubricated sliding wear of Si₃N₄ against cast irons. Journal of Xi'an Jiaotong University 29(2): 73–78, 84.

[15] Y.M. Gao, L. Fang, J.Y. Su, Z.G. Xie 1996. Formation and composition of surface film for Si₃N₄-gray iron sliding pairs lubricated with distilled water. Journal of the Chinese ceramic society 24(5): 523–530.

[16] Y.M. Gao, L. Fang, J.Y. Su 1997. The effect of tribofilm formation on the tribological characteristics of ceramic-cast iron sliding pairs. Wear 210: 1–7.

[17] Y.M. Gao, E.Z. Wang, L. Fang, J.Y. Su 1999. Observation on the process of tribofilm formation for Si₃N₄-white iron pair lubricated with distilled water. Mechanical Science and Technology 18(3): 478–480.

[18] Y.M. Gao, X.J. Ma, L. Fang, X.F. Zhang, Jun-Yi Su 2001. The effect of inhibitors on the corrosion and tribological characteristics of gray iron sliding against Si₃N₄ under lubricated conditions. Wear 248: 1–6.

[19] Y.Y. Huang, Y. Chen, M. Gao, Y.Z. Zhang 2010. Friction and wear characteristics of cast iron under high speed dry sliding condition. Material & Heat Treatment 39(10): 33–36.

[20] Z.X. Xiang, S.C. Chen, Z. Wang, Y. Gao 2000. Study on the tribological behavior of Si₃N₄ ceramic against chilled cast iron. Tribology 20(3): 183–185.

[21] Z.X. Xiang, G. Dong, B. Lin, Z.G. Shen 2006. Tribological characteristics of silicon nitride-chilled cast iron with oil-less lubrication. Journal of Materials Engineering (4): 24–27, 32.

Material Science and Engineering – Chen (Ed.)
© 2016 Taylor & Francis Group, London, ISBN 978-1-138-02936-1

Analysis of packer rubber aging properties

J.N. Wang & J. You
Petroleum Production Engineering Research Institute of Huabei Oilfield Company, China

J. Wang
First Exploitation Factory of Huabei Oilfield Company, Hebei, China

L. Wang
Exploration and Development Research Institute of Huabei Oilfield Company, China

L. Liu & Q. Liu
Petroleum Production Engineering Research Institute of Huabei Oilfield Company, China

C. Zeng
Data Center of HuaBei Oilfield Company, China

W. Meng
First Exploitation Factory of Huabei Oilfield Company, Hebei, China

L. Zhang
Petroleum Production Engineering Research Institute of Huabei Oilfield Company, China

G.L. Zhu
The Second Exploitation Factory of Huabei Oilfield Company, China

ABSTRACT: Based on the real environment surrounding packer rubber and seal elements, material thermal-oxidative aging test and Akron abrasion experiments are conducted by using the laboratory experimental apparatus, followed by the analysis of Shore hardness, compression set, tensile modulus test, microstructure morphology and aging abrasion properties before and after aging and wearing. The accelerated aging test for the samples is performed. The laboratory findings show the rubber properties with the variation of temperature and time. The evaluation results provide theoretical references for future improvement of packer rubber service life in an operating environment.

Keywords: shore hardness; compression set; stretch modulus; morphology; wear; aging test

1 INTRODUCTION

Packer is one of the important downhole tools used in oil drilling and production activities. The packer seals the annulus space and isolates producing zones by elastic sealing elements to control the production of fluid and to protect the casing. Thus, there are various functions of the packer; of which, its important role is to seal the elastic elements. In oil exploration and production activities, packers are mainly used for drilling, well testing, layered recovery, water shutoff, water injection and stimulation-related operations. It is the unique sealing property of packer that reliably ensures the smooth going of regular production and varying underground technological measures in oil and gas wells. Sealing failure of packer rubber greatly is detrimental to the normal production in terms of economy and security. Packer rubber is a superelastic material, and the sealing performance is the key to evaluate the packer operating performance. The complexity of the downhole operating environment and the nonlinearity of the packer rubber material bring great difficulties in the theory study on rubber sealing performance. Rubber failure is primarily due to a high downhole temperature and pressure, resulting in rubber aging and tearing. By using the laboratory experimental apparatus, the simulation test for the packer rubber material is conducted, followed by the analysis of performance parameter. The results provide a theoretical basis for future evaluation of packer rubber service life and optimization of rubber sealing performance.

2 THERMAL OXIDATIVE AGING TEST OF THE RUBBER MATERIAL

According to the national standard GB/T 3512–2001 "Vulcanized Rubber or Thermoplastic Rubber Hot Air Accelerated Aging and Heat Test" and the national industry standard HG/T 3087–2001 "Rapid Method for Determination of Static Sealing Rubber Parts Storage Life", 5 test temperature points, 90°C, 110°C, 130°C, 150°C and 170°C, respectively, were selected. The equipment ventilation frequency was set up to 4 times/hour. The rotary table speed was set to 6 rev/min. Also, the aging chamber ambient temperature was adjusted to the corresponding value, so that the temperature fluctuation was ± 0.5°C. The size of the cylindrical and slender strip hydrogenated nitrile rubber sample was 17.5 × 25 mm and 120 × 10 × 2 mm, respectively. When loading the cylindrical sample, spacing between every two samples in the clamping apparatus should not be less than 5 mm, while the slender strip sample was suspended. When the sample was placed in the aging chamber, the timing started. Every time, when the specified aging time came, the sample should be brought out and the measurements of

Figure 1. Diffusion-limitation oxidation.

the compression set, Shore A hardness and stretch modulus should be made in accordance with the relevant national standards such as GB1683–1981, GB/T 531–1999 and GB/T 528–1998.

2.1 Effect of thermal oxidative aging on packer rubber mechanical properties

1. *Shore A hardness*
Cylindrical rubber samples under 5 aging test temperature points were sampled at different aging times, respectively. The sampling found that circular bosses were shaped obviously on the top and bottom end faces of the aged cylindrical samples (Figure 1). Additionally, the diameter of the boss surface decreased somewhat as aging time extended. In view of this, Shore A hardness measurement of the boss surfaces and the outer edges of the cylindrical samples was performed after sampling. A total of 6 samples were measured at each sampling measure point. For each sample, measurement points should not be less than 3. The value was obtained by the arithmetic mean.

Figure 2 (left) shows that the cylindrical boss surface hardness trend at the 5 test temperature points at a 25% compression ratio changes slightly during the 18-day accelerated aging period. While Figure 2 (right) shows that all aging sample hardness trends at the 5 test temperature points increase evidently. The higher the temperature is, the faster the hardness increases. From the overall curve trace, the experimental data are generally in line with the time-temperature equivalence rules. As shown in Figure 2 (left), when the temperature reaches 170°C and aging time comes to the ninth day, the rubber material hardness is close to the maximum value and little change in hardness takes place subsequently. The rubber hardness represents the capability of resistance to deformation; in this case, it can be now seen that the elasticity of this rubber material gradually disappears. The reason for the material hardness increase is that the residual cured rubber composition in the curing process continues to make cross-linking under the action

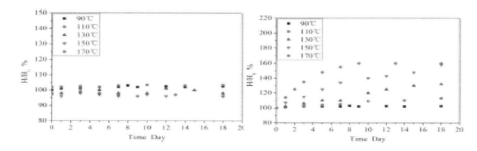

Figure 2. Change rate of boss surface hardness (left) and outer edge hardness (right) with time at a 25% compression ratio.

of free radicals, or a multiple disulfide bond is desulfated to generate more single bonds and disulfide bonds. In other words, the cross-linking reaction increases vulcanized rubber hardness.

However, the reason for the great difference in hardness changes of two parts of this rubber material is that the oxygen for the rubber thermal oxidative aging reaction is supplied by diffusion from exterior to interior. Therefore, if the diffusion is slow, sufficient oxygen supply to the interior of the rubber material would fail, resulting in reduced oxidation reaction rate. When the sample is thick, the oxygen diffused to the interior of the material is less, thereby resulting in slower oxidation rate of interior than exterior, and then producing diffusion-limited oxidation. For different types of rubber materials, due to the different oxygen diffusibility in the interior of different rubber materials, the critical penetration thickness of the material is somewhat different. For the phenomena shown in Figure 1, at different aging times, the diameter of the boss surface is different. It can be determined that a diffusion limitation oxidation effect occurs with this rubber sealing material. Because of the oxygen diffusibility limitation and compression effect on both the top and bottom end faces of the sample, the oxygen fails to spread to a deep level inside the cylinder, which causes abnormal internal oxidation, subsequently resulting in more severe oxidation and, therefore, more significant change in hardness in the outer edge than in the convex area.

2. Compression set

Compression test factors such as compression displacement and compression speed can be determined according to the tested sample performance. Prior to the test, the height and diameter should be measured at three points, respectively. The value is obtained by the arithmetic mean and accurate within 0.01 mm. The compressor is mounted on a tensile testing machine, and the load and dial gage zero of the tensile testing machine should be calibrated before the test. After the test machine starts, the compressor compresses the sample at a required falling rate. When the sample is compressed to the predetermined compression displacement, it should be shut down immediately. The compression deformation test is appraised. Then, the height of postaging cylindrical rubber samples is measured. Each preload clamp contains six cylindrical samples. According to the national industry standard HG/T3087–2001 "Rapid Method for Determination of Static Sealing Rubber Parts Storage Life", each clamp apparatus removed from the aging chamber should rest for one day in the standard test environment. Thereafter, the load will be removed, followed by one more day of rest. Subsequently, the height will be measured by the digital display slide caliper. The effective mean height can be determined

by calculating the mean value of the measurement data of the six test samples. The expression for the compression set is as follows:

$$\xi = \frac{H_0 - H_t}{H_o - H_1} \times 100\% \qquad (1)$$

where

H_0—original height of the rubber sample after physical relaxation;

H_t—height after resting for one more day after the load is removed, following resting for one day in the standard environment after aging for time t; and

H_1—clamp stop collar height.

The height of the cylindrical sample is measured after every sampling, and the compression set value can be found by expression (1). Figure 3 shows the change pattern of the HNBR material compression set with aging temperature and time.

At the same aging temperature, Figure 3 shows the longer the aging time is, the greater the compression permanent deformation is. At the same aging time, the higher the aging temperature is, the greater the compression permanent deformation is. As shown in Figure 3, the entire trend of the compression permanent deformation change is roughly in line with the time-temperature equivalence rule. When the aging time is 7 days, the compression permanent deformation exceeds 99% at 170°C of aging temperature, while the compression permanent deformation change will be gradually steady within the following aging time. Because the compression permanent deformation can also be a direct response to the elasticity of the material, it is considered that the material is completely ineffective when the aging time is 7 days and the aging temperature is 170°C.

3. Tensile modulus

When rubber products are used, they all suffer from certain external forces, which requires the rubber to possess physical and mechanical properties correspondingly, particularly the tensile performance of

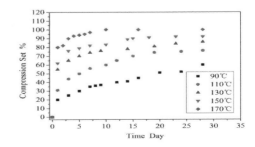

Figure 3. Curve for the HNBR material compression set alternation with time at five different temperatures.

the material. For quality inspection of the product, the determination of the composite formula of the design and process conditions and the comparison of aging resistance and anti-media performance of rubber, and tensile performance play a unique role. Therefore, the tensile property of rubber is very important.

After each sampling of the slender strip sample, its tensile modulus is measured. According to the real-time load-displacement curve recorded online, the data is processed, and then the elastic modulus can be obtained for each test material.

The expression for the elastic modulus is as follows:

$$E = \frac{(P_2 - P_1)}{S(\varepsilon_2 - \varepsilon_1)} \qquad (2)$$

where

S—sample sectional area (mm);

P_i—load at any point along the load-strain curve (N); and

ε_i—strain corresponding to point P_i on the load-strain curve ($\times 10^6$).

Five samples are measured at each aging time point and each aging temperature. The value is obtained by the arithmetic mean. Figure 4 shows the tensile modulus change pattern of the HNBR material with time at three aging temperatures. Prior to aging, the HNBR tensile modulus is approximately 2 Mpa. After a fourteen-day accelerated aging test, the tensile modulus at three aging temperatures was increased to a different extent. However, at the beginning of aging, the tensile modulus shows a declining trend, which further confirms the conclusion that the HNBR material presents primarily a main chain scission reaction in the early stage of aging. Figure 4 also shows that the higher the temperature is, the faster the tensile modulus of the HNBR material at three temperatures, which increases during the following aging time after the declining trend in the early stage of aging.

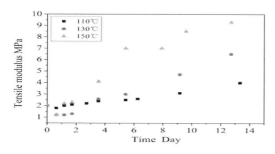

Figure 4.   Curve for the tensile modulus change in the HNBR material with time at three temperatures.

1.2   *Rubber microstructure morphological analysis*

Hot air aging test is the most commonly used aging test for rubber in air at a high temperature and atmosphere pressure, known as the thermal oxidative aging test. This test can be used to evaluate the heat resistance of rubber, protection performance of aging inhibitor and contamination of rubber ingredient, screen the formula, and determine the storage period.

Table 1 presents a comparison chart of the HNBR material slender strip sample surface topography before aging and after accelerated aging tests at three temperatures, which shows that sample surface oxidation gets more severe with the rise in the aging temperature. The microgram for eight-day aging at 170°C shows that non-uniform oxidation appears on the sample surface. The sample is produced by mixing natural rubber, HNBR and additives. However, the mixing process cannot guarantee the uniform distribution of the aging inhibitor in the material, resulting in inconsistent aging resistance of different regions of the material. Thus, after long-term aging at a high temperature, surface regions exhibit non-uniform aging. Table 2 presents a comparison chart of the HNBR material cylindrical sample top convex surface topography before aging and after accelerated aging tests at three temperatures. After accelerated aging at a high temperature, an obvious change does not take place around the top convex surface of the cylindrical sample, resulting from diffusion-limitation oxidation that causes indistinct aging of the sample surface with time. Now, the oxygen consumption rate of the rubber material is greater than the diffusion rate of oxygen into the interior of the material. Because the increase in oxidation degree would harden the material, further result-

Table 1.   HNBR material slender strip sample surface topography contrast before and after aging under optics microscopy.

Scale	Condition	
	200×	500×
Before aging		
8-Day aging at 90°C		
8-Day aging at 110°C		
8-Day aging at 150°C		

Table 2. HNBR material cylindrical sample top convex surface morphology contrast before and after aging under optics microscopy.

Scale	Condition	
	200×	500×
Before aging		
15-Day aging at 130°C		
15-Day aging at 150°C		
15-Day aging at 170°C		

ing in the decrease of oxygen penetration capability in the rubber material, the material surface area exposed to the air suffers from more oxygen than the interior of the material, leading to a more pronounced oxidation of the region with a higher oxygen concentration than that with a lower oxygen concentration. Therefore, the complete understanding is that oxidation non-uniformity may not be due to the decrease in penetration capacity caused by changes in material structure. Along with the change in time, even if the penetration rate remains constant, oxidation non-uniformity would gradually grow greater.

1.3 Summary

Aging test cabinet is used to carry out the accelerated aging of the cylindrical and slender strip sample of the HNBR rubber sealing material. After sampling, hardness, compression set and tensile modulus test of the sample as well as morphological and microstructure analysis are conducted. The thermal oxidative aging mechanism of HNBR materials under a static load is summarized as follows: the test results of hardness, compression set and tensile modulus demonstrate that hardness, compression set and tensile modulus increase with aging time extending and aging temperature rise under the circumstances of sufficient oxygen participating in the reaction, and the data is approximately in line with the time-temperature equivalence rule. Simultaneously, the results also show that there exists a cross-linking reaction accompanied by a fracturing reaction while HNBR material aging. At the early stage of aging, the fracturing reaction is dominant followed by the cross-linking reaction at the later stage.

2 WEAR TEST OF THE RUBBER MATERIAL

2.1 HNBR, NBR and FPM rubber performance comparison test

IIn the Akron abrasion test, under a certain load, the rotary circular sample at a constant speed contacts, at an angle, the grinding wheel for rolling friction. Since the sample lies at different levels from the grinding wheel axis, a different relative slip velocity wears the rubber. In this test, three different types of rubber are first tested, and Akron wear is compared for the evaluation of HNBR abrasion resistance. Three selected rubber types are all commonly used materials for packer rubber stator in the downhole environment such as high temperature, high pressure and oily medium.

Figure 5 shows the HNBR, NBR and FPM rubber Akron abrasion test. These three materials are commonly used for packer rubber. The expression for the sample wear volume V is as follows:

$$V = \frac{m_1 - m_2}{\rho} \tag{3}$$

where

V—sample wear volume of unit mileage (1.61 km), cm³;

m_1—sample mass after pre-grinding, g;

m_2—sample mass after the test, g; and

ρ—sample density, g/cm³.

The test number should be no less than two, and test data is obtained by the arithmetic mean with an allowable deviation of ±10%. According to the method specified in GB/533, the density of HNBR and the other two tested rubbers is determined, respectively.

At ambient temperature, the Akron wear volumes of the three materials are 0.062 cm³/1.61 km, 0.076 cm³/1.61 km and 0.034 cm³/1.61 km, which shows that the wear resistance of HNBR is better

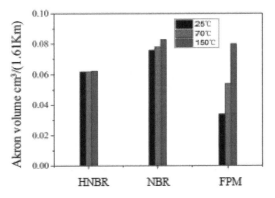

Figure 5. Hydrogenated nitrile, ordinary nitrile and fluorine rubber Akron abrasion test.

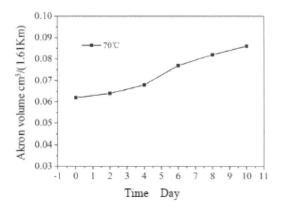

Figure 6. Effect of aging time on the HNBR Akron volume.

than that of NBR but worse than that of FPM rubber. However, when the temperature rises, HNBR wear resistance remains nearly unchanged, while FPM rubber wear resistance decreases dramatically. Hence, HNBR possesses a sound wear resistance not only at the low temperature but also at the high temperature.

2.2 *Wear resistance of the rubber material after aging*

Figure 6 shows the variation pattern of the HNBR Akron wear volume with aging time at 70°C. As shown in this figure, the Akron wear volume continuously increases with time extending, which indicates that material wear resistance gradually declines along with the increasing aging degree of HNBR. Moreover, it is also found that the change in the range of Akron wear volume is small at the early stage of aging, but larger after 5-day aging. Therefore, the decline in the range of rubber wear resistance will increasingly become larger with increasing aging.

3 CONCLUSION

In this paper, aging resistance and wear resistance of HNBR used for packer rubber are primarily studied. Aging resistance of rubber is studied mainly by the thermal oxidative accelerated aging method. After sampling, mechanical properties such as hardness, compression set and tensile modulus of the sample are tested. In addition, the sample microstructure morphology is observed. This study demonstrates that hardness, compression set and tensile modulus increase with extending aging time and increasing aging temperature under the circumstances of sufficient oxygen participating in the reaction, and the data is approximately in line with the time-temperature equivalence rule. HNBR wear

resistance is investigated by Akron abrasion experiments and compared with NBR and FPM rubber, which shows that its wear resistance is comparatively not very strong. The study on HNBR wear resistance at varying aging time finds that rubber wear resistance gradually declines with increasing aging time.

ACKNOWLEDGMENTS

The authors wish to thank the PETROCHINA major scientific and technological project "Key Technique Research and Application for Stable Yields of Eight Million Tons in Huabei Oilfield" (2014E-35) for its support.

REFERENCES

[1] N.E. Forost. An examination of the environment for aging of polymer [A]. *1997 IEE Annual Report—Conference on Electrical Insulation and Dielectric Phenomena, Minneapolis: Oct. 19–22, 1997. 354–357.*

[2] J.Y. Koo, I.T. Kim. An experiment investigation on the degradation characteristic of the outdoor silicone rubber due to sulfate and nitrate ions [A]. *1997 IEE Annual Report—Conference on Electrical Insulation and Dielectric Phenomena, Minneapolis: Oct. 19–22, 1997. 370–373.*

[3] S.J. Wang, J.P. Xiong, Yu Zuo. Study on Aging Mechanism of Rubbers [J]. *Synthetic Materials Aging and Application, No. 2, vol. 38, 2009, pp. 23–33.*

[4] Ying Zhang, Z.H. Lian, Lei Shi, C.F. Song, T.J. Lin, Analysis of Relationship between Load on Mechanical Packer and Lowering Hanging Load at the Wellhead, *China Petroleum Machinery, No. 1, vol. 17, 2003, pp. 28–31.*

[5] X.D. Liu, J.N. Xie, Z.X. Feng, Y.F. Xie. Research Progress on Accelerated Aging and Life Prediction Method for Rubber Material. *Synthetic Materials Aging and Application, No.1, vol. 43, 2014, pp.69–73.*

[6] Yan Xiao, B.R. Wei, M.P. Du, Accelerated Aging Test of Rubber and Calculation of its Storage Life. *Synthetic Materials Aging and Application, No. 1, vol. 36, 2014, pp. 40–43.*

[7] X.J. Zhang, X.L. Chang, S.X. Chen, M.J. Hua. Thermal Oxidation Aging Test and Life Assessment of Fluorine Rubber Sealing Materials., *Equipment Environmental Engineering, No. 4, vol. 9, 2012, pp. 35–38.*

[8] M.Z. Cheng, X.S. Hu, W.K. Mo. Research progress on thermal stability of RTV silicone rubber. *Silicone Material, No. 1, vol. 20, 2006, pp. 38–41.*

[9] Y.X. Zhou, Xu Zhang, Rui Liu, Fei Hou, Y.X. Zhang, S.Y. Gao. Study on Micromorphology of Electrical Trees in Silicon Rubber, High Voltage Engineering, No. 1, vol. 40, 2014, pp. 9–15.

[10] J.C. Zhao, X.L. Feng, Y.X. Cui, X.J. Liu. The Preliminary Research on the Interface Aging of Ethylene-propylene Rubber in Lye. *Materials Review, No. 23, vol. 28, 2014, pp. 300–306.*

[11] H.Y. Pan. Methods for Improving Styrene-butadiene Rubber's Aging Properties. *Guangzhou Chemical Industry, No. 22, vol. 41, 2013, pp. 28–29,48.*

Material Science and Engineering – Chen (Ed.)
© 2016 Taylor & Francis Group, London, ISBN 978-1-138-02936-1

Colorimetric analysis of green jadeite

D.J. Nie, Y. Zou & X.K. Zhu
Faculty of Materials Science and Engineering, Kunming University of Science and Technology, Kunming, Yunnan, China

ABSTRACT: According to the principle of colorimetry, reflection curves and the colorimetric parameters of light green and green jadeite are obtained by the USB2000+ fiber spectrometer. The analysis shows that the reflection peak of light green jadeite is located at 500 nm and is not sharp, and the reflectivity is 10%–20%. For green jadeite, the reflection peak is sharp and located near 530 nm, and the reflectivity is 10%–30%. For light green jadeite, the appearance of green is inhibited when b* is negative, and the appearance of green is produced when b* is positive. For green jadeite, the degree of bright is influenced by dominant wavelength, color purity, a*, and b*. Dominant wavelength reflects the dominant hue. With the value of a* being smaller and b* being larger, color purity is higher and the green color is brighter.

Keywords: jadeite; colorimetry; fiber optic spectrometer

1 INTRODUCTION

Colorimetry is a developing discipline since the early twentieth century, which is based on the physical optics, visual physiology, visual psychology, and psychophysics. The theory and technology of color measurement and the law of human color vision have been studied. Moreover, colorimetry provides a scientific basis and the unified standard for objective, quantitative description of colors (2007). The CIE colorimetric system is based on a lot of psychological physics experiment and research on visual rules of human eyes. This system uses the digital and instrument to represent and measure colors.

In the study of jadeite, one important topic is to evaluate the color of jade objectively and quantitatively, as color is the main factor affecting the value evaluation of jadeite, and green is the most important and most valuable color of jadeite (2009). Zhang, Zhang and Wang (2006) analyzed the characteristics, differences and relationships between the colorimetric method and the color matching method in evaluating the color of jadeite. Wang and Yuan (2007) thought that using the fiber optic spectrometer, the color of jadeite could be described objectively and quantitatively. Zhang and Guo (2011) studied the relationships between lightness, chroma, and hue angle for the green color of jadeite. Guo, Zhang and Mo (2010) proposed that the green quality evaluation of jade should be primarily based on the brightness, and then followed by the chroma and the hue angle.

According to colorimetry as the theoretical basis, fiber optics is used to collect diffuse reflection signals on the samples' surface to quantify the colors of samples. Reflection curves and colorimetric parameters of the samples could be obtained to judge the relationship between them and the color of samples.

2 EXPERIMENTAL SAMPLES AND EXPERIMENTAL CONDITIONS

2.1 *Experimental samples*

Experimental samples were divided into two categories: one category is colorless to pale green jadeite samples, labeled Q-1-Q-7; the other category is green jadeite samples, labeled S-1-S-5.

2.2 *Experimental conditions*

The experimental instrument was a USB2000+ fiber optic spectrometer produced by America Ocean Optics Company. The colorimetric parameters of the samples were obtained from the reflected signals by the integrating sphere. The experimental conditions were as follows: the wavelength range was 380–1000 nm, the light source was D65 in CIE standard illumination; the wavelength interval was 10 nm, the integration was 1600 μs; and the observer's field of view was 2° in the dark environment.

3 RESULTS AND DISCUSSION

3.1 *Colorimetric characteristics of light green samples*

Samples Q-1-Q-7 are light green jadeite jade, ranging from colorless to light green.

Figure 1 shows the reflection curves of light green samples. The reflection peaks of the samples are located at about 500 nm, while the absorption peaks near 438 nm. The reflection peaks of the samples are located at 500 nm, and are not sharp, indicating that the samples of jadeite jade are green but the chroma of green is not high. It shows absorption at 438 nm in the reflection curves, indicating that the light of purple is absorbed.

For sample Q-7, the reflection curve is smooth, the reflectivity is under 10%, and reflection peak is not obvious; that is, for each band of light, the reflectivity is almost the same. The reflectivity of sample Q-5 reaches up to 30% and its reflection curve is more obvious than Q-7. The reflectivity of the other five samples ranges between 10% and 20%. For these samples, the intensity of reflection peaks increases with the increasing reflectance values, showing that with the samples' color gradually deepening, the chroma becomes higher and higher.

Table 1 lists the colorimetric parameters of light green samples. The dominant wavelength of the samples is located at the 483 nm-510 nm band, and the change trend is consistent with the color green. Therefore, the dominant wavelength can reflect the dominant hue of the samples.

In colorimetry, y is chromaticity coordinates. In the system of CIE1976 L*a*b*, a* represents green when it is a negative number. Also, b* represents yellow when it is a negative number, while representing blue when it is a positive number. The numerical value of y increases gradually, indicating that the proportion of green monochrome is increasing. Observing the numerical value of a* and b*, b* increases when changing from negative to positive, and a* increases after a minimum. That is to say, when the miscellaneous tone of the samples changes from blue to yellow, though green monochrome decreases, the whole reflection of green is more obvious. Namely, the blue tone

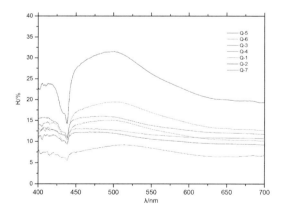

Figure 1. Reflection curves of light green samples.

Table 1. The colorimetric parameters of light green samples.

Sample no.	Tristimulus values			Chromaticity coordinate			Dominant wavelength (nm)	Color purity	L	a*	b*
	X	Y	Z	x	y	z					
Q-1	10.872	11.653	13.778	0.2995	0.3210	0.3795	483.784	0.0555	40.659	-1.5026	-2.7224
Q-2	9.812	10.634	12.815	0.2950	0.3197	0.3853	485.005	0.0726	38.958	-2.3274	-3.2590
Q-3	12.533	13.799	16.172	0.2949	0.3246	0.3805	488.417	0.0687	43.943	-3.8755	-2.5684
Q-4	11.324	12.827	14.682	0.2916	0.3303	0.3781	492.166	0.0763	42.502	-6.1305	-1.6934
Q-5	22.170	25.903	28.326	0.2902	0.3390	0.3708	496.888	0.0764	57.845	-10.9374	-0.1853
Q-6	14.386	16.622	17.911	0.2941	0.3398	0.3661	498.561	0.0621	47.781	-8.4469	0.3807
Q-7	6.949	8.072	8.236	0.2988	0.3471	0.3541	508.426	0.0461	34.132	-7.0060	1.8492

inhibits the appearance of green, and the yellow tone may contribute to the appearance of green. Also, the numerical value of color purity and a* are negatively correlated, as shown in Table 1. It shows that color purity increases with the proportion of green monochrome. With color purity being higher, the color of the samples is closer to the spectral color. This means the higher the saturation, the brighter the color.

3.2 Colorimetric characteristics of green samples

The samples S-1-S-5 are jadeite from light green to dark green, ranging from light green to dark green. S-2 is the brightest sample, S-1 and S-2 are light green, and S-4 and S-5 are dark green.

Figure 2 shows the reflection curves of green samples. The reflection peaks of the samples are located at about 530 nm and the two wide absorption bands are centered at 450 nm and 650 nm. The reflection peaks of the samples are located at 530 nm and are sharp, indicating that the samples of jade are green and the chroma of green is high. The two wide absorption bands are centered at 450 nm and 650 nm, indicating that the violet and red-orange light are absorbed. They increase after 690 nm with the maximum at 770 nm.

The reflectivity of S-4 and S-5 is 10%–20%, and that of S-3, S-2, and S-1 is 20%–30%. The reflection peak intensity increases with the increasing reflectance values, showing that the samples' color is gradually bright and the chroma becomes higher and higher.

Table 2 lists the colorimetric parameters of green samples. In X,Y and Z, X is close to Z, while Y is bigger than X and Z, indicating that the samples are green and the larger the difference, the brighter the color. The dominant wavelength of samples is located at the 520 nm-540 nm band with

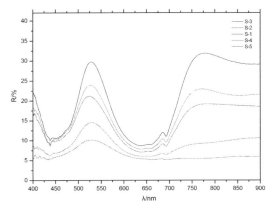

Figure 2.   Reflection curves of green samples.

Table 2.   The colorimetric parameters of green samples.

Sample no.	Tristimulus values			Chromaticity coordinate			Dominant wavelength (nm)	Color purity	L	a*	b*
	X	Y	Z	x	y	z					
S-1	10.820	15.276	12.329	0.2816	0.3976	0.3209	521.247	0.1363	46.010	−24.9545	10.1558
S-2	11.830	17.013	12.762	0.2843	0.4089	0.3067	528.295	0.1652	48.276	−27.4077	12.9437
S-3	13.691	20.588	13.732	0.2852	0.4288	0.2860	534.183	0.2185	52.496	−33.1347	17.7970
S-4	8.091	10.919	8.278	0.2965	0.4001	0.3034	537.725	0.1624	39.444	−19.0283	10.8654
S-5	6.510	8.228	6.743	0.3031	0.3830	0.3139	541.152	0.1291	34.454	−12.8929	7.8619

an increasing trend, and the change in the trend is consistent with the color green.

In colorimetry, y is chromaticity coordinates. In the system of CIE1976 L*a*b*, a* represents green when it is a negative number. Also, b* represents yellow when it is a negative number, while representing blue when it is a positive number. The numerical value of y increases gradually, indicating that the proportion of green monochrome is increasing. Observing the numerical value of a* and b*, b* increases when changing from negative to positive, and a* increases after a minimum. That is to say, when the miscellaneous tone of the samples changes from blue to yellow, though green monochrome decreases, the whole reflection of green is more obvious. Namely, the blue tone inhibits the appearance of green, and the yellow tone may contribute to the appearance of green. Also, the numerical value of color purity and a* are negatively correlated, as shown in Table 1. It shows that color purity increases with the proportion of green monochrome. With color purity being higher, the color of the samples is closer to the spectral color. This means the higher the saturation, the brighter the color.

4 CONCLUSIONS

Through the analysis of the reflection curves and the colorimetric parameters on light green and green samples, the following conclusions can be drawn:

The reflection peak at 500 nm is not sharp and the reflectivity is 10%–20%, which means that the jadeite is light green. The reflection peak at 530 nm is sharp and the reflectivity is 10%–30%, indicating that the jadeite is green.

For light green jadeite, the appearance of green is inhibited when b* is negative, and the appearance of green is produced when b* is positive. For green jadeite, the degree of bright is influenced by dominant wavelength, color purity, a*, and b*. Dominant wavelength reflects the dominant hue. With the value of a* being smaller and b* being larger, color purity is higher and the green color is brighter.

REFERENCES

[1] Guo, Y. & Zhang, J. & Mo, T. Quality evaluation of green jadeite jade's lightness based on CIE 1976 L*a*b* uniform color space. *Bulletin of The Chinese Ceramic Society*, 2010 (3): 560–566.
[2] Hu, W.J. & Tang, S.Q. & Zhu, Z.F. 2007. *Modern color science and application*. Beijing: Beijing Institute of Technology Press.
[3] Qiu, Z.L. & Li, L.P. & Chen, B.H. 2009. *Jewelry evaluation of system introduction*. Wuhan: China University of Geosciences Press.
[4] Wang, R. & Yuan, X.Q. Feasible research on colour of jadeite jade measured by colorimetry. *Journal of Gems and Gemmology*, 9–2 June 2007: 20–24.
[5] Zhang, H. & Zhang, B.L. & Wang, M.J. Application of method of colour measurement in color appraisement of jadeite jade. *Journal of Gems and Gemmology*, 8–3 September 2006: 16–20.
[6] Zhang, X.W. & Guo, Y. Distribution regularity of the green color of jadeite jade. *Acta petrologica et mineralogica*, 2011 (8): 23–26.

Material Science and Engineering – Chen (Ed.)
© 2016 Taylor & Francis Group, London, ISBN 978-1-138-02936-1

Synthesis of AlON powders by the hydrothermal method following by SPS calcination

Q.H. Deng
Jiangsu Yuncai Materials Co. Ltd., Changshu, Jiangsu, P.R. China
Division of Functional Materials and Nanodevices, Ningbo Institute of Materials Technology and Engineering,
Chinese Academy of Sciences, Ningbo, Zhejiang, P.R. China

Z.H. Ding, A. Yann & D.W. Shi
Division of Functional Materials and Nanodevices, Ningbo Institute of Materials Technology and Engineering,
Chinese Academy of Sciences, Ningbo, Zhejiang, P.R. China

Z.J. Liu
Jiangsu Yuncai Materials Co. Ltd., Changshu, Jiangsu, P.R. China

Q. Huang
Division of Functional Materials and Nanodevices, Ningbo Institute of Materials Technology and Engineering,
Chinese Academy of Sciences, Ningbo, Zhejiang, P.R. China

ABSTRACT: AlON powders have been successfully synthesized by a novel combination of the hydrothermal and carbothermal reduction nitridation methods. The effects of the reaction temperature and the initial theoretical C/Al mole ratio on the resultant phase compositions have been investigated. In addition, the reaction mechanism has been discussed. The hydrothermal method favored the homogeneous mixing of AlON precursors in solution, and resulted in low particle sizes. At SPS calcination temperature of 1680°C, lower than other reported synthesis temperatures, pure AlON powder could be synthesized.

Keywords: AlON; nanopowder; hydrothermal method; spark plasma sintering process; the carbothermal reduction nitridation method

1 INTRODUCTION

AlON is a cubic polycrystalline ceramic material, which composition is a solid solution of Al2O3 and AlN (Al23O27 N5 or 9 Al2O3·5 AlN) [1]. This material possesses excellent chemical and mechanical properties, thus it may find potential applications where high-performance structural ceramics are required [2–4]. In addition, fully dense AlON can achieve high transparency in optical range from 0.2 µm in the UV to 6.0 µm in the infrared[5]. Due to these properties, AlON ceramics may be widely used both in civilian and military areas such as optical windows, transparent armors, POS scan windows, etc [6–8].

However, the sintering process to achieve high transparency of polycristalline AlON ceramics remains a challenge. Indeed, very high sintering temperatures (up to 2000°C) are often required. A key solution to lower the sintering temperature may be the synthesis of ultrafine, high-purity and high-activity AlON powders.

Nowadays many methods have been developed for AlON powder synthesis. On one hand, solid-state reaction methods have been widely used for this powder synthesis [9–12]. For example, Qi and co-workers [9] used nanosized high-purity Al_2O_3 and AlN powders as raw materials, then they calcined the mixtures at 1750°C for 4 h to obtain the single phase AlON powder. Liu et al. [12] also reported the synthesis of aluminum oxynitride (AlON) powders by solid-state reaction, and studied in detail the effects of the calcination parameters i.e. temperature, heating rate and holding time on the AlON formation. Solid-state reaction methods generally lead to better properties such as purity, sinterability and stability against humidity. However, using AlN powders with high purity and fine particle size is not cost-effective. On the other hand, the carbothermal reduction and nitridation processing is an alternative method also used method to synthesize AlON powder [13–15]. For example, Liu et al. [13] have synthesized AlON powders by carbothermal reduction and nitridation processing with γ–Al_2O_3

and carbon black (C) as starting material. Ma et al.[14] have obtained aluminium oxynitride powders from three different precursors (Al_2O_3+C, Al(OH)3+C, NH4 Al(OH)2CO3+C). According to these authors, calcination of precursors at 1750°C resulted in single phase AlON powder. Although the carbothermal and nitridation processing seems cost-effective, it still faces with the issue of homogeneous mixing of starting precursors. To overcome the issue, Wang et al.[16] have initiated the Aluminothermic Reduction method. These authors observed that the temperature and atmosphere of AlON synthesis could change the eutectic composition of the resulting single phase AlON powder. They also obtained a single phase AlON powder after calcination at 1750°C. Recently, Jin et al.[17] have proposed a combination of Sol-Gel and carbothermal reduction nitridation method. The latter have used Sol-Gel method to synthesize Al_2O_3/C precursor. Then, a single phase AlON powder has been obtained after calcination at 1720°C.

However, to date there is no report on attempts to produce AlON powder by hydrothermal method followed by calcination through spark plasma sintering, thus enabling lower calcination temperatures and shorter processing times. In this study, pure AlON powders are for the first time prepared by hydrothermal method combined with spark plasma sintering process at a relatively low temperature (1680°C). The crystallographic phases in the calcined sample were analyzed by X-ray diffractometry. Additionally, the effects of the calcination parameters and the initial theoretical C/Al mole ratio on the AlON formation were analyzed.

2 EXPERIMENTAL PROCEDURE

AlON powders were prepared by hydrothermal method combined with spark plasma sintering process. In a typical procedure, 12 g (0.05 mmol) of aluminum chloride hexahydrate ($AlCl_3 \cdot 6H_2O$, 99%, Aladdin, China) were dissolved in 30 ml of deionized water with magnetic stirring to form a homogeneous solution. Then NaOH (4 g, 0.01 mol, Aladdin, China) was dissolved in 70 ml of deionized water, and the NaOH solution was added dropwise into the Aluminum chloride solution stirred until the precipitate vanished. Subsequently, glucose ($C_6H_{12}O_6$, 99%, Aladdin, China) was dissolved in the above solution. The amounts of added glucose were 1.5 g, 2 g and 2.5 g corresponding to the initial theoretical C/Al mole ratio of 1.0, 1.34, and 1.67 respectively. This colloidal solution was transferred to a Teflon-linked autoclave and maintained at 200°C for 12 h. When the autoclave was cooled down to room temperature,

the desired AlON precursor was collected, washed with deionized water four times, and finally dried at 80 °C. Then the precursor was calcined in a spark plasma sintering furnace at 1600°C–1680°C for 5 min under N_2 atmosphere, with a heating rate of 100°C/min. for the carbothermal reduction and nitridation method. Two modifications were made on the SPS graphite die for this powder treatment. One is a modification of the geometry of the two graphite punches such that the minimum load was supported by the graphite die itself instead of the loaded powder inside. The other modification is a throughout hole made in the center of the top graphite punch to allow a direct measurement of the powder temperature. Thus, a fast temperature ramp and an accurate temperature control could be fulfilled by using this die design.

The precursors synthesized by hydrothermal method were analyzed at room temperature using a fluorescent spectrophotometer (F-4600, Hitachi Ltd., Tokyo, Japan). XRD patterns of the powders before and after SPS were collected by powder X-ray diffraction (XRD, Bruker AXS D8 Discover, Germany), using Cu Kα radiation. Diffraction pattern were recorded using CuKa radiation from 10° to 90° 2θ in steps of 0.01° with a collection time of 10 s per step. The infrared spectra of the AlON precursors were recorded on a Nicolet MAGNA-IR 750 spectrometer over the range 4000–400 cm^{-1} using KBr pellets. The weight loss during pyrolysis of the precursor in argon was monitored by thermogravimetric analysis (TG, ZRY-1P, Germany) at a heating rate of 10°C/min. The morphology of the composite granules was characterized by scaning electron microscopy (SEM, Hitachi S-4800, Japan).

3 RESULTS AND DISCUSSION

Figure 1 shows the XRD pattern of AlON precursor prepared by hydrothermal method at 200°C for 12 h. All detected XRD diffraction peaks were labeled and were readily attributed to diaspore (JCPDS Card No. 49-0133) and Carbon (JCPDS Card No. 46-0943). The average crystallite size of diaspore (9 nm) was determined X-ray line broadening of the three strong peaks from using Warren-Averbach analysis [5].

The FT-IR spectrum of the precursor synthesized by hydrothermal method is shown in Figure 2. Similar curves are observed for the precursors with different initial theoretical C/Al mole ratios. The strong bands at 3293 and 3093 cm^{-1} can be assigned to the v_{as} (Al) O-H and v_s (Al) O-H stretching vibrations [18]. And the bands at 1070 and 1155 cm^{-1} are, respectively, attributed to the δs Al-O-H and δ as Al-O-H modes of diaspore.

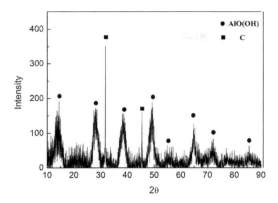

Figure 1. The XRD patterns of precursor synthesized by hydrothermal method.

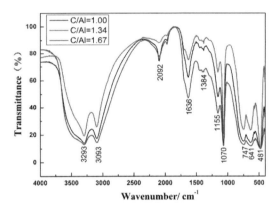

Figure 2. Infrared spectrum of the AlON precursor with different contents of initial C/Al mole ratio.

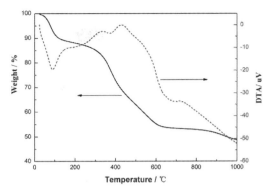

Figure 3. TG/DTG curves of the AlON precursor.

The torsional modes at 749 and 661 cm^{-1} of diaspore are also observed in the spectrum. The bands at 2092 cm^{-1} can be assigned to the combination bands. The weak band at 1636 cm^{-1} can be assigned to the stretching and bending modes of the adsorbed water. The result of the above FT-IR analysis is consistent with the XRD result, both substantiating that the crystalline phase of the as-prepared sample is diaspore. Moreover, the band at 1384 cm^{-1} is related the torsional vibrations of—CH2—, which is a direct evidence of the existence of the adsorbed surfactant.

Figure 3 shows the thermogravimetry (TG) and differential thermal analysis (DTA) curves of the precursor during its pyrolysis in N$_2$. The physically adsorbed water is desorbed at a temperature lower than 200°C. A sharp weight loss of about 33wt% is observed at a narrow temperature range from 200°C to 600°C. These losses can be attributed to the desorption of the chemically adsorbed water,

the dehydration of diaspore AlO(OH) to form Al$_2$O$_3$ and the removal of hydrocarbon species of the adsorbed surfactants[19]. Beyond 600°C up to 1000°C, the weight loss rate gradually decreased to a limit. These observations suggest that the calcination treatment of the precursor in argon at 600°C can thoroughly remove the water and surfactant, leading to the formation of Al$_2$O$_3$. Al$_2$O$_3$ and carbon are the precursors of AlON in the carbothermal reduction and nitridation processing at high temperature.

Excitation and emission spectra of the AlON precursor in water with different contents of initial C/Al mole ratio is shown in Figure 4. For all the samples, a broad emission spectrum with a peak around 445 nm is observed under the 350 nm excitation. A broad excitation spectrum is noticed from 210 nm to 410 nm. This phenomenon is correlated with the fluorescence spectra of carbon dots which were reported by Zhu[20]. No difference in the positions of excitation and emission peaks was observed among the three samples. A sharp increase in the emission intensity of the samples is observed with increasing initial C/Al molar ratio. Carbon quantum dots can reduce the reaction temperature due to the lower activation energy[21].

The XRD spectra of the samples calcined at1600°C, 1650°C and 1680°C at a heating rate of 100°C/min for 5 min are compared in Figure 5. From this Figure, it can be seen that the main crystallographic phase in calcined samples is the single-phase AlON with the following chemical formula Al$_{23}$O$_{27}$N$_5$. In the samples sintered at 1600°C and 1650°C, the presence of Al$_2$O$_3$ and AlN indicates that AlN has not reacted completely with Al$_2$O$_3$. When the calcination temperature is increased to 1680°C, only AlON phase existed. Therefore, pure AlON could be formed at 1680°C with this method. In comparison with previous investigations [11–13, 15–17], where mechanically mixed compounds of

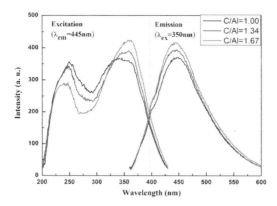

Figure 4. Excitation and emission spectra of the AlON precursor in water with different contents of initial C/Al mole ratio.

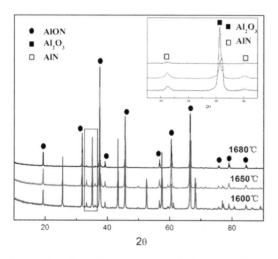

Figure 5. The XRD patterns synthesized at different sintering temperatures for 5 min.

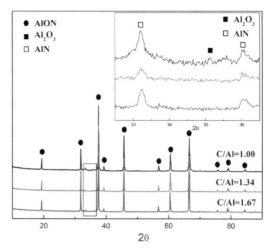

Figure 6. The XRD patterns synthesized at 1680°C for 5 min with different contents of initial C/Al mole ratio.

Figure 7. SEM image of AlON powder sintered by spark plasma sintering at 1680°C for 5 min.

Al_2O_3/C or Al_2O_3/AlN were used for AlON synthesis, the synthesis temperature of AlON in the present work is about 40°C–120°C lower.

Figure 6 shows the XRD spectra of the samples sintered at 1680°C at a heating rate of 100°C /min for 5 min with different contents of initial C/Al mole ratio. It can be seen that the main phase of all samples is pure AlON. When the initial C/Al mole ratios are 1 and 1.67, there still remain small weak peaks of AlN and Al_2O_3 in the pattern. Nevertheless, it is not expected that residual AlN and Al_2O_3 affect the sintering of AlON ceramics, because these residuals should be solubilized into the lattice of AlON under the high sintering temperature. Therefore, the AlON powder was synthesized by spark plasma sintering process with the initial

C/Al mole ratio of 1.34. The average crystallite size of diaspore was 68 nm as determined from X-ray line broadening of the three strong peaks using Warren-Averbach analysis [5].The SEM image of AlON powder calcined at 1680°C at heating rate of 100°C /min for 5 min with initial C/Al mole ratio of 1.34 is shown in Figure 7. A bimodal agglomerate particle size distribution can be observed in the range of 0.5 to 1 μm. The largest agglomerate particle size is seemingly below 1 μm. This fine particle size obtained from is presumably suitable for sintering.

4 SUMMARY

In conclusion, a combination of hydrothermal method and carbothermal reduction nitridation processing for preparing the AlON powders has been reported. AlON powders were prepared for the first time by hydrothermal method followed by spark plasma sintering process. It was found that pure AlON powder can be formed at 1680°C, at a heating rate of 100°C /min for 5 min. The AlON powder synthesis temperature is about 40°C–120°C lower than other reported methods. It has been shown that the initial C/Al mole ratio could affect the purity of AlON powder. For an initial theoretical C/Al mole ratio of 1.34, a pure single phase AlON powder can be synthesized. This simple method allows one to explore the novel properties of AlON powder and expand their practical applications.

ACKNOWLEDGEMENT

This work study was financially supported by 'Hundred Talents Program' of The Chinese Academy of Sciences (KJCX2-EW-H06), National Natural Science Foundation of China (Grant No. 51172248 and 91226202), Ningbo Municipal Natural Science Foundation (No.2014 A610006), State Key Laboratory of Porous Metal Materials (PMM-SKL-1-2013).

REFERENCES

[1] P. Tabary; Servant, C.; Alary, J.A.: submitted to Journal of Journal of the European Ceramic Society (2000).
[2] Z. Ning; Zhao, X.J.; Ru, H.Q.; Zhu, K.W.; Wang, X.Y.; Kan, H.M.: submitted to Journal of Rare Metal Materials and Engineering (2009).
[3] A. Shimpo; Ueki, M.; Naka, M.: submitted to Journal of Journal of the Ceramic Society of Japan (2002).
[4] N. Zhang; Zhao, X.J.; Ru, H.Q.Y.; Wang, X.Y.; Chen, D.L.: submitted to Journal of Ceramics International (2013).
[5] R.J. Matyi; Schwartz, L.H.; Butt, J.B.: submitted to Journal of Catalysis Reviews (1987).
[6] A. Krell; Hutzler, T.; Klimke, J.: submitted to Journal of Journal of the European Ceramic Society (2009).
[7] B. Paliwal; Ramesh, K.T.; McCauley, J.W.: submitted to Journal of Journal of the American Ceramic Society (2006).
[8] Y.Z. Wang; Lu, T.C.; Gong, L.; Qi, J.Q.; Wen, J.S.; Yu, J.A.; Pan, L.; Yu, Y.; Wei, N.A.: submitted to Journal of Journal of Physics D-Applied Physics (2010).
[9] J.Q. Qi; Wang, Y.Z.; Lu, T.C.; Yu, Y.; Pan, L.; Wei, N.A.; Wang, J.: submitted to Journal of Metallurgical and Materials Transactions a-Physical Metallurgy and Materials Science (2011).
[10] J.Q. Qi; Zhou, J.C.; Pang, W.; He, J.F.; Su, Y.Y.; Liao, Z.J.; Wu, D.X.; Lu, T.C.: submitted to Journal of Rare Metal Materials and Engineering (2007).
[11] W.Z. Sun; Chen, Y.H.; Wu, L.; Jiang, Y. In Powders and Grains 2013; Yu, A., Dong, K., Yang, R., Luding, S., Eds., 2013; Vol. 1542; pp 137–140.
[12] X.J. Liu; Li, H.L.; Huang, Z.R.; Wang, S.W.; Jiang, D.L.: submitted to Journal of Journal of Inorganic Materials (2009).
[13] X.J. Liu; Yuan, X.Y.; Zhang, F.; Huang, Z.R.; Wang, S.W.: submitted to Journal of Journal of Inorganic Materials (2010).
[14] F.Z. Ma; Lei, J.X.; Shi, Y.; Xie, J.J.; Lei, F. In Chinese Ceramics Communications Iii; Hu, J., Chen, N.C., Zhang, C., Eds., 2013; Vol. 624; pp 42–46.
[15] X.Y. Yuan; Liu, X.J.; Zhang, F.; Wang, S.W.: submitted to Journal of Journal of the American Ceramic Society (2010).
[16] Y.Z. Wang; Lu, T.C.; Yu, Y.; Qi, J.Q.; Wen, J.S.; Wang, H.P.; Xiao, L.; Yang, Z.L.; Yu, J.; Wen, Y.; Wei, N.: submitted to Journal of Rare Metal Materials and Engineering (2009).
[17] X.H. Jin; Gao, L.; Sun, J.; Liu, Y.Q.; Gui, L.H.: submitted to Journal of Journal of the American Ceramic Society (2012).
[18] H. Hongwei; Yi, X.; Qing, Y.; Qixun, G.; Chenrong, T.: submitted to Journal of Nanotechnology (2005).
[19] L. Qu; He, C.; Yang, Y.; He, Y.; Liu, Z.: submitted to Journal of Materials Letters (2005).
[20] H. Zhu; Wang, X.; Li, Y.; Wang, Z.; Yang, F.; Yang, X.: submitted to Journal of Chemical Communications (2009).
[21] R. Asatryan; Bozzelli, J.W.: submitted to Journal of Physical Chemistry Chemical Physics (2008).

Environmental materials

Material Science and Engineering – Chen (Ed.)
© *2016 Taylor & Francis Group, London, ISBN 978-1-138-02936-1*

Effective removal of copper (II) ions from the solution using the carbonization of Pomelo Peel as the adsorbent

J.H. Zhang, X. Ren, C.G. Zhao & J. Li
Department of Chemistry and Pharmacy, Zhuhai College of Jilin University, Zhuhai, Guangdong, P.R. China

ABSTRACT: Pomelo Peel is an abundant biomass waste, and has been shown to have the potential to serve as a resource of chemicals. The current work reports a low-cost way to deal with Pomelo Peel waste by the carbonization process. The carbonization obtained as the adsorbent was applied for the removal of copper (II) ions from the aqueous solution. The efficiency of such adsorbent in adsorbing the copper ion from the aqueous solution is systematically studied as a function of adsorbent dosage and pH. It was found that the adsorbent is both effective and economically viable.

Keywords: biosorption; copper ions; carbonization; pomelo peel

1 INTRODUCTION

The presence of toxic heavy metals in the environment through industrial waste disposal is currently an important environmental concern. The removal of heavy metal ions from drinking water is an essential process because they are able to accumulate in living tissues causing various diseases, not biodegradable and present an harmful effect on the ecosystem. Copper is one of the most toxic heavy metals to living organisms and one of the most widespread heavy metals in the environment [1, 2]. Copper is a very common substance that is widely used in many industries. The potential sources of copper ions in industrial effluents include metal cleaning, plating bath, paper board, mining, anti-fouling for paint and pigment, fertilizer, and wood pulp [3,4]. The World Health Organization has recommended that a maximum acceptable concentration of Cu^{2+} in drinking water is about 1.5 mg/L [5]. In order to solve heavy metal pollution in the environment, it is important to bring applicable solutions. Some in-place treatment technologies available for the removal of heavy metal ions from aqueous solutions are chemical precipitation, ion exchange, coagulation, and bioremediation and sorption/adsorption. Of all these techniques, adsorption at the solid substrate is preferred because of its high efficiency, easy handling, and cost effectiveness as well as the availability of different adsorbents [6–10].

The present study is conducted to evaluate the efficiency of a carbon adsorbent prepared from the direct carbonization of pomelo peel for the removal of copper ions from the aqueous solution, since it carries a large surface area and active sites to adsorb the metal ion under study.

2 MATERIALS AND METHODS

2.1 *Preparation of the adsorbent*

The adsorbent was prepared by the direct thermolysis of pomelo peel. A quartz boat containing the sample powder was placed in a tube furnace. The furnace was heated from room temperature to 600°C under nitrogen flow at the rate of 5°C/min, and then maintained for 240 min.

2.2 *Adsorption studies*

The stock solution of Cu^{2+} was prepared by dissolving $CuSO_4 \cdot 5H_2O$ (Tianjin Science and Technology Development Co.) in the deionized water at the concentration of 1 g/L. The experimental solutions were prepared by diluting the copper stock solution in accurate proportions at needed initial concentrations. The adsorption experiments were conducted in a 250 mL conical flask with 50 mL of standard solution. The solution was shaken and placed for some time. Subsequently, the filtrate was analyzed by an atomic absorption spectrophotometer (Shimadzu AA6300CModel). A Flame Atomic Absorption Spectrophotometer (FAAS) was used under optimal working conditions (Table 1) for the determination of metal ions in aqueous solutions before and after adsorption. Each experiment was performed in duplicate under identical conditions. The effect of pH on the adsorption of Cu^{2+} ions was studied over the pH range from pH 2.0 to pH 8.0 [11]. The adsorbent dosage mass added was 0.30 g and the initial concentration of Cu^{2+} was 20 mg/L. The pH of the Cu^{2+} solution was adjusted by using either 0.1M HCl or 0.1MNaOH. The effect of the

Table 1. FAAS operating conditions.

Parameters	Value
Wavelength (nm)	324.8
Lamp current (mA)	6.0
Spectral bandwidth (nm)	0.7
Height of the flask (mm)	7.0
Acetylene flow (L/min)	1.8
Air flow (L/min)	15.0

adsorbent dosage mass was conducted by different adsorbent dosage mass values ranging from 0.01 to 0.50 g with the Cu^{2+} solution (10 mg/L, pH 5.0) for 90 min. The adsorption capacity of the adsorbent at equilibrium was calculated using the equation:

$$q_e = \frac{(C_0 - C_e)V}{m} \qquad (1)$$

where C_0 and C_e (mg/L) are the initial and equilibrium concentrations of the metal ion solution, respectively. V (L) is the volume of the adsorbate in liters and m (g) is the amount of adsorbent in grams. The percentage adsorption was determined using the equation:

$$\text{Adsorption } \% = \frac{C_0 - C_e}{C_0} \times 100 \qquad (2)$$

2.3 Characterization of the adsorbent

The Thermogravimetric Analysis (TGA) was performed using a Shimadzu DTG-60 thermal analyzer system at the heating rate of 5°C· min⁻¹ at 900°C in the nitrogen atmosphere, and the nitrogen flow rate was 30 mL/min. The morphologies were characterized by SEM (FEI Quanta 400 Thermal FE Environment Scanning Electron Microscope). The specific surface area and pore structure of the carbon samples were determined by N_2 adsorption-desorption isotherms at 77 K (Quantachrome Autosorb-iQ) after being vacuum-dried at 150°C for 10 hours. The specific surface areas were calculated by a BET (Brunauer-Emmett-Teller) method. The cumulative pore volume and the pore size distribution were calculated by using a slit/cylindrical Nonlocal Density Functional Theory (NLDFT) model. The surface functional groups of the adsorbent before and after adsorption were analyzed by the Fourier Transform Infrared Spectrophotometer, Shimazdu IRPrestige-21. Potassium bromide disks were prepared by mixing 1 mg of the samples with 200 mg KBr, and the spectra were recorded from 4000 cm⁻¹ to 400 cm⁻¹.

3 RESULTS AND DISCUSSION

3.1 Characterizations

3.1.1 Thermogravimetric analysis

TGA was carried out in the interest of studying the thermal behavior of pomelo peel (see Figure 1). The experiment was performed under the N_2 atmosphere with a heating rate of 5°C/min. It shows that the major weight loss occurs at 380°C (74.5% weight loss), corresponding to the departure of absorbed water molecules. At the temperature higher than 550°C, pyrolysis almost completes. Then, the slow and continuous weight loss may be attributable to the decomposition of the organic functional groups. Therefore, 600°C was chosen for the ideal thermal decomposition temperature for pomelo peel.

3.1.2 Infrared spectral analysis

Figure 2 shows the FT-IR spectra of Pomelo Peel (PP), the carbonization (CPP) and the carbonization loaded by Cu^{2+} ions (CPP-Cu). In all spectra,

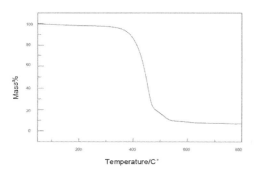

Figure 1. TGA curve of pomelo peel with the heating rate of 5°C/min under the N_2 atmosphere.

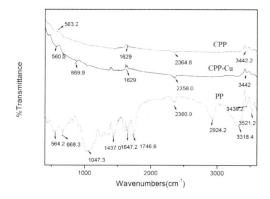

Figure 2. FTIR spectra of Pomelo Peel (PP), the carbonization (CPP) and the carbonization loaded by Cu^{2+} ions (CPP-Cu).

424

the broad peak in the region of 3500–3200 cm^{-1} is the characterization of –OH stretching in the alcohol, phenols and organic acids. Those peaks at 1629 cm^{-1} are related to the aromatic C = O bonds. When comparing between the CPP and CPP–Cu spectra, the intensity of the –OH stretching vibration at the wavenumber 3442 cm^{-1} decreases due to the adsorption of Cu^{2+} ions.

3.1.3 *SEM analysis*
The SEM images of PP and CPP are shown in Figure 3. These images show that the surfaces of PP and CPP are rough and porous. Such characteristics are suitable for the adsorption of metal ions.

3.1.4 *Surface area*
Biomass has been widely used as an adsorbent due to its microporous structures with high adsorption capacity and high surface area. The chemical nature and pore structure usually determine the sorption activity. The multipoint BET surface area analysis of pomelo peel is performed with Quantachrome Nova 2200e. Figure 4 shows the N$_2$ adsorption-desorption isotherm of the sam-

ple with a relative pressure (P/P$_0$) of 0.0–1.0. The surface area of the carbonization is found to be 9.43 m^2/g. It implies that the surface area of the carbonization is strongly related to the mesopore structure.

3.2 *Optimization of the conditions for adsorption*

3.2.1 *Effect of acidity on the adsorption efficiency of Cu (II)*
The pH of the solution is an important variable in the adsorption process. The influence of pH on the initial solution of Cu^{2+} was examined at a different pH ranging from 3.0 to 8.0. The results are shown in Figure 5. It shows that Cu^{2+} adsorption was constant at pH 2–3. However, the uptake of Cu^{2+} ions increased significantly from 43.0% to 82.0% at the pH value ranging from 3.0 to 5.0. A similar observation was reported for the adsorption of Cu^{2+} on biosorbent Uncaria gambir [12]. A low pH value of the Cu^{2+} solution contains a high concentration of H$^+$ ions, which may compete effectively with Cu^{2+} ions for exchangeable cations on the surface active sites of the carbonization. At pH 5.0, the Cu^{2+} solution consists of free Cu^{2+} ions that are mainly involved in the adsorption process, leading to an increase in the amount of Cu^{2+} adsorbed. From Figure 5, it can be observed that the adsorption percentage of Cu^{2+} increases significantly from pH 6.0 to 8.0. This phenomenon is mainly caused by the presence of three ion species in the solution, as suggested by Larous et al. [13] that the Cu^{2+} ions are present in a very small quantity, whereas Cu(OH)$^+$ and Cu(OH)$_2$ exist in large quantities. Therefore, at a higher pH (>6.0), Cu^{2+} ions are attached by the hydroxide ions to form Cu(OH)$_2$. So, the pH value of 5.0 was selected as the optimal pH condition.

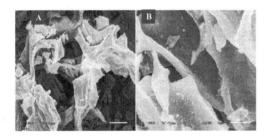

Figure 3.   Typical SEM images: A) pomelo peel, B) the carbonization of pomelo peel.

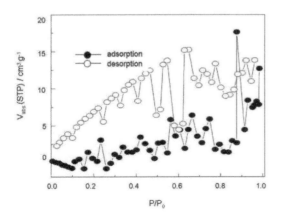

Figure 4.   N$_2$ adsorption-desorption isotherms of the carbonization of pomelo peel.

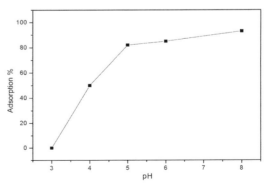

Figure 5.   Effect of pH on the sorption of Cu^{2+} onto the carbonization (initial concentration of solution is 10 mg/L, adsorbent dosage is 0.30 g, temperature is 20°C, and the contact time is 90 min).

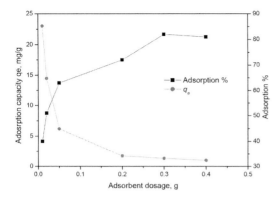

Figure 6. Effect of adsorbent dosage on the adsorption capacity and percentage adsorption of Cu^{2+} ions (initial pH of solution = 5.0, initial concentration of solution = 10 mg/L, temperature = 20°C, contact time = 90 min).

3.2.2 *Effect of dosage on the adsorption efficiency of Cu (II)*

The effect of the carbonization dosage on the adsorption of Cu^{2+} ions was studied, and the results are shown in Figure 6. Based on Figure 6, it can be observed that the adsorption of Cu^{2+} increases rapidly from 43.0% to 82.0% with an increase in the adsorbent dosage from 0.01 to 0.30 g. This is because by increasing the adsorbent dosage, the total number of adsorption sites available for the interaction of Cu^{2+} ions increases as well. A further increase in the carbonization dosage mass from 0.30 to 0.50 g seems not to affect the sorption in any wise. However, the adsorption capacity of Cu^{2+} is observed to be decreased with an increase of adsorbent dosage. When the adsorbent dosage mass increases from 0.01 g to 0.30 g, the loading capacity of the carbonization decreases from 23.0 mg/g to 1.3 mg/g. This phenomenon may be due to two reasons: the first is an increasing adsorbent amount at a constant Cu^{2+} ion concentration and volume, which may cause the unsaturation of the adsorption sites; the second is due to the particulate interaction such as aggregation resulting from a high adsorbent dosage. Moreover, the adsorbent surface and metal ions reach to an equilibrium point at which no further Cu^{2+} ion will be adsorbed [14]. The adsorbent dosage mass is fixed at 0.30 g for further adsorption studies.

4 CONCLUSION

This study revealed that the direct carbonization of pomelo peel can be used as an alternative adsorbent for the removal of heavy metal ions in industrial wastewater due to its higher efficiency of Cu^{2+} adsorption in the aqueous solution. The adsorp-

tion of Cu^{2+} onto the carbonization was affected by the pH and adsorbent dosage. It was found that the uptake percentage of Cu^{2+} was 82.0% when 0.30 g of adsorbent were placed with 50 mL of Cu^{2+} solution of 10 mg/L for 90 min at the pH value of 5.0.

REFERENCES

[1] M.W. Wan, C.C. Kan, B.D. Rogel, M.L.P. Dalida, Adsorption of copper (II) and lead(II) ions from aqueous solution on chitosan-coated sand, Carbohydr. Polym. 80 (2010) 891–899.

[2] I. Villaescusa, N. Fiol, M. Martínez, N. Mialles, J. Poch, J. Serarols, Removal of copper and nickel ions from aqueous solutions by grape stalks wastes, Water Res. 38 (2004) 992–1002.

[3] N. Li, R. Bai, Copper adsorption on chitosan–cellulose hydrogel beads: behaviors and mechanisms, Sep. Purif. Technol. 42 (2005) 237–247.

[4] C.S. Zhu, L.P. Wang, W. Chen, Removal of Cu (II) from aqueous solution by agricultural by product: peanut hull, J. Hazard. Mater. 168 (2009) 739–746.

[5] C.S. Rao, Environmental Pollution Control Engineering, Wiley Eastern, New Delhi, 1992.

[6] S. Arivoli, "Kinetic and Thermodynamic Studies on the Adsorption of Some Metal Ions and Dyes onto Low Cost Activated Carbons", PhD. Thesis, Gandhigram Rural University, Gandhigram, 2007.

[7] J. Paul Chen, Wu Shunnian Wu, and Kai Hau Chong, "Surface Modification of a Granular Activated Carbon by Citric Acid for Enhancement of Copper Adsorption", Carbon, 41 (2003), pp. 1979–1986.

[8] K. Vijayaraaghavan, J. Jegan, K. Palanivel, and M. Velan, "Batch and Column Removal of Copper from Aqueous Solution using a Brown Marine Alga", J. Chem. Engg., 106 (2005), pp. 177–184.

[9] G. Sekaran, K.A. Shanmugasundaram, M. Mariappan, and K.V. Raghavan, "Adsorption of Dyes by Buffing Dust of Leather Industry", Indian J. Chem Technol, 2 (1995), p. 311.

[10] K. Selvarani, "Studies on Low Cost Adsorbents for the Removal of Organic and Inorganics from Water", PhD. Thesis, Regional Engineering College, Thiruchirapalli, 2000.

[11] S. Larous, A.H. Meniai, M.B. Lehocine, Experimental study of the removal of copper from aqueous solutions by adsorption using sawdust, Desalination 185 (2005) 483–490.

[12] K.S. Tong, M. Jain Kassim, A. Azraa/Adsorption of copper ion from its aqueous solution by a novel biosorbent Uncaria gambir: Equilibrium, kinetics, and thermodynamic studies Chemical Engineering Journal 170 (2011) 145–153.

[13] S. Larous, A.H. Meniai, M.B. Lehocine, Experimental study of the removal of copper from aqueous solutions by adsorption using sawdust, Desalination 185(2005) 483–490.

[14] ÖGöK, A. Özcan, B. Erdem, A.S. Özcan, Prediction of the kinetics, equilibrium and thermodynamic parameters of adsorption of copper (II) ions onto 8-hydroxy quinoline immobilized bentonite, Colloids Surf. A: Physicochem. Eng. Aspects 279 (2006) 238–246.

Material Science and Engineering – Chen (Ed.)
© 2016 Taylor & Francis Group, London, ISBN 978-1-138-02936-1

Analysis of financial management environment in new materials enterprises

L. Yang

North China Electric Power University, Beijing, P.R. China

ABSTRACT: Financial management of enterprises is always living in a certain environment, but the environment restricts the enterprise's whole financial management activities of the economy. Therefore, understanding and learning the environment of enterprise's financial management is very important. This article begins from the related concepts of financial management environment, discusses the influence on the development of new materials enterprises, and puts forward the measures for improving the financial environment.

Keywords: new materials; enterprise; environment of the financial management influence; measures

1 THE CONTENT AND CLASSIFICATION OF THE ENTERPRISE'S FINANCIAL MANAGEMENT ENVIRONMENT

Robert Owen once stated that "the environment determines people's language, religion, culture, custom, ideology and behavior". In the modern society, the environment affects not only the individual, but also the enterprise; thus, the financial management environment plays a vital role in the survival and development of the enterprise.

Enterprise's financial management environment refers to the sum of internal and external conditions that affect financing, investment, production and other economic activity. There is no doubt that as to an enterprise, its establishment, growth, prosperity, decline and demise are all in the financial management environment. In today's competitive market economy, as an enterprise, fully understanding and learning to use environmental factors are the premise condition to succeed.

According to different standards, it can be classified into different financial management environments. For example, according to the source of influencing factors, the financial management environment can be divided into macroscopic environment and microcosmic financial management environment, namely external financial management environment and internal financial management environment. According to the way that influences the enterprise, it can be divided into direct and indirect financial management environments. Some others have divided it into soft and hard financial environment. These divisions all have their reasons and research significance. Specifically, the enterprise financial management mainly

includes the political environment, economic environment, social and cultural environment, technology environment and legal environment. Here, we mainly begin from these aspects to start our discussion.

1.1 *The political environment*

Political environment on the influence of enterprise's financial management activities should not be ignored. It includes the degree of social stability, the effectiveness of the various economic policies formulated by the government, as well as the efficiency of the government agencies. Peace and development are the themes of our era, peaceful and harmonious environment at home and abroad improves the continuity and stability of economic activities, and the government's policy support and efficient improvement provide a strong impetus for the enterprise.

1.2 *The economic environment*

Economic environment refers to the entire macro situation faced by enterprises in their production and business operation activities, including the enterprise's economic cycle, fiscal taxation, financial markets, and monetary banking. These are the most direct financial management environment factors, and enterprises should seek opportunity and speed up their own development.

1.3 *Social and cultural environment*

Social and culture environment is the sum of people's outlook on life, values and moral development, which are formed in the long term at a certain social environment, mainly including education, literature, art, customs, perception and attitudes. It becomes a kind of important intangible force that affects all aspects of social life. It is also an important part of the enterprise's financial management environment.

1.4 Science and technology environment

Science and technology is the first productive force. Modern enterprises should pay more attention to the key role of science and technology in their success or failure. The application of new materials is one of the main directions of development of science and technology in the 21st century, and is the deeper exploration for human in high-tech material fields.

1.5 Legal environment

Legal environment refers to the rules and regulations abided by the law when a company has an economic relationship with others, mainly including the enterprise organization law, taxation and financial rules and regulations. At present, governors have decided to promote the process of ruling the country by law in China; the legal environment becomes more and more important.

Nowadays, every environment influences and promotes each other, and enterprises should conform to the trend of the times and the environmental changes, adjust their development strategy and planning; only in this way can they have an impregnable position in the competition.

2 THE INFLUENCE OF FINANCIAL MANAGEMENT ENVIRONMENT IN NEW MATERIAL ENTERPRISES

"The new material industry" includes new materials and related products and technology, which belong to the technology-intensive industry. The added value of its products is high, the application range is wide, and its level of research becomes the standard of a country's scientific and technological progress. Generally speaking, the new material industry mainly includes the information materials industry, energy industry, and biological material industry. Next, we will take a look at the influence of the financial management environment in new material enterprises.

2.1 Development assistance to new material enterprises provided by the government

New material industry is the national strategic emerging industry. Its capital and technology are intensive, the development prospect is broad, all countries in the world attach great importance to fostering the new material industry, and it is true of our country as well. In order to promote the breakthrough of major key technology, the government has established special strategic emerging industry development funds and projects, such as "863" plan and the torch plan. At the same time, it has issued the relevant policies and regulations, encouraging financial institutions to intensify credit scale to the new material industry, so as to provide a positive political environment and legal environment for the development of new materials enterprises.

2.2 The progress of science and technology provides technical guarantee for the development of new materials enterprises

As mentioned earlier, the new material industry is a technology-intensive industry. In our opinion, as to the scientific progress and the development of the new material industry, they typically complement each other to some extent. On the one hand, there is no doubt that, recently, science and technology are developing rapidly, and the new knowledge, new technology, new products emerge endlessly, providing a favorable environment of science and technology for the new material industry development. On the other hand, the unprecedented development of the new material industry puts forward higher technical requirements, which, in turn, promotes the continuous innovation of science and technology.

2.3 The intensified competition in the market increases the risk of the development of new materials enterprises

In recent years, with the deepening of economic globalization and the development of market economy, the demand for new materials of emerging industries such as new energy and environmental protection increases rapidly, which enhances the heat of the domestic development of new materials, leading to the increasingly fierce competition. New materials enterprises not only need to face the domestic counterparts, but also have to meet the challenges from abroad, and foreign enterprises mostly have significant advantages in the aspects of capital and technology, which greatly increase the risk of the development of new materials enterprises.

2.4 The financing difficulty becomes the bottleneck of the development of new materials enterprises

At present, the development speed of new materials enterprises, especially small and medium-sized enterprises, is slow. This is often because of financing difficulties and the shortage of funds. Therefore, enterprises have great difficulty in attracting high-level professionals, and are unable to purchase advanced equipment to engage in productive activities and introduce foreign advanced technology. The above-mentioned reasons greatly limit the development of new materials enterprises in China.

2.5 Enhancement of national investment concept promotes the development of new materials enterprises

Since the reform and opening up, China has gradually transited from the planned economy to the market economy system, especially after joining the WTO, the concepts such as "market", "investment", have become more attractive in the majority of the people's hearts. Part of the nationals put their money out from the bank, in order to invest some industries and companies that they are optimistic about, thus achieving the objective of return. New materials enterprises benefit a lot in such a strong investment climate.

Financial management environment inevitably affects new materials enterprises in all aspects of the business, how to make full use of favorable factors, and avoid the adverse factors, which become an urgent problem that new materials enterprises have to face currently.

3 IMPROVING THE FINANCIAL MANAGEMENT ENVIRONMENT; PROMOTES THE HEALTHY DEVELOPMENT OF THE NEW MATERIALS ENTERPRISES

3.1 Deepening the understanding and research of the enterprise financial management environment

For a long time, some people think that the financial management environment is a vague concept, so it is unnecessary to discuss that this view is completely wrong. When the human society entered the 21st century, the importance of the financial management environment has reached unprecedented levels. If enterprises want to pursue their own long-term development, further research on the information of the financial management environment needs to be carried out, so as to make a suitable financial management concept and strategic blueprint. Moreover, the financial management environment is not immutable and frozen, and the enterprise should also constantly adapt to the changing financial environment, so as to broaden

their horizons, understand relevant information, and make reasonable financial decisions.

3.2 Adapting to changes in the external environment of financial management

External environment tends to be objective. It does not transfer with subjective will. However, this does not mean that the enterprise is incapable of action on the external environment. On the contrary, enterprises should take an active step to understand the external environmental changes, and enhance their ability to predict and innovate. It should also make full use of various preferential conditions provided by the external environment, and reduce the adverse effects on their own factors, to build a good external environment for the survival and development of enterprises.

3.3 Setting up a good internal financial management environment

Sometimes the enterprise's strength does not match their ambitions in improving external environment; however, enterprises are the real "leader" in establishing a good internal financial management environment. Enterprises should adopt feasible financial policies, optimize the structure of financial management, broaden the content of financial management, establish a high-level management team, and strengthen the guidance and training of managers and ordinary employees. Thus, it should make each part of the internal of the enterprise attend to their duties and play their proper roles.

3.4 Adhering to the principle of leading technology and giving priority to efficiency

Core competence is an inexhaustible motive force for new materials enterprises to succeed in an increasingly competitive market. Enterprises should always put the leading technology in the top-drawer position of the enterprise strategy. At the same time, the new materials industry should start specialized production. Instead of casting a net everywhere, it should specialize in one place, which will change into the advantage region, as a result, it will form a solid market.

In the process of improving the financial management environment in new materials enterprises, we should pay attention to seeking truth from facts, not only to do the best what we can do, but also within our capabilities. Managers should formulate the development strategy of the enterprise timely, make the enterprise to have breakthroughs in the key projects, and to achieve leapfrog development in the fierce competition.

4 CONCLUSION

Financial management environment plays a vital role in the development of new material enterprises. While improving the enterprise's financial environment is not achieved overnight, it is a systematic, overall project. The enterprise have to fully realize the importance of the financial management environment, and make a reasonable method to adapt and improve the constantly changing financial environment. Only in this way can they seek their own long period of stability in the fierce competition in today's society.

REFERENCES

[1] Liu Zhiyuan, Enterprise strategic management, 1997.
[2] Liu Yonghui, modern accounting to the twenty-first Century, 1996.

Material Science and Engineering – Chen (Ed.)
© *2016 Taylor & Francis Group, London, ISBN 978-1-138-02936-1*

Eco-friendly and cost-effective biodegradable films prepared via compounding polyester with thermoplastic starch

Y. Pan
Guangxi Key Laboratory of Petrochemical Resource Processing and Process Intensification Technology, School of Chemistry and Chemical Engineering, Guangxi University, Nanning, China

H. Wang & H. Xiao
Department of Chemical Engineering, University of New Brunswick, Fredericton, New Brunswick, Canada

ABSTRACT: Over the past decade, the biodegradable polymers, such as poly (butylene adipate-co-terephthalate) (PBAT), have been extensively used in various applications. However, the high cost of such materials has hindered their further application. In order to reduce the general cost of the material, Thermoplastic Starch (TPS) was melt blended with PBAT in this work to produce an eco-friendly biodegradable composite material with relatively low cost. The anhydride based compatibilizers as well as an inorganic filler, talc, were used to improve the compatibility between PBAT and TPS. The results showed that the mechanical properties and the thermal stabilities of the prepared composite films were improved in the presence of the compatibilizers. However, the inorganic filler, talc, generated a negative effect due to the different polarity of the unmodified outer layers. The scanning electronic images also confirmed the results from the micro-scale.

Keywords: thermoplastic starch; biodegradation; polyester; film

1 INTRODUCTION

As one of the most popular common materials, plastic has been widely used in various applications, such as packaging and mulching films, but it also generates negative effects with regard to its waste [Chiellini et al, 1996]. In order to minimize the environmental problems, researchers have developed a series of biodegradable polymers which are ultimately able to be converted into carbon dioxide, water, or other small molecules by the open environment and microorganisms [Miller et al. 2014; Jbilou et al. 2013; Witt et al. 2001]. Among all of the biodegradable plastics, aliphatic-aromatic polyester, poly (butylene adipate-co-terephthalate) (PBAT), has attracted great attention due to its similar processing conditions and mechanical properties compared to the low density polyethylene. As a linear random copolyester, PBAT is fully complied with the compostable plastic specification of ASTM D6400. However, the high cost has hindered the further application of PBAT.

Starch, a renewable bio-resource, has been considered as a naturally occurring polymer with a high potential to be blended with synthetic biodegradable polymers with affordable cost during the past decades [Schwach et al. 2004; Averous. 2004]. Usually, native starch is not able to be blended with synthetic polymers due to its relatively low thermal degradation temperature. Therefore, Thermoplastic Starch (TPS) is often prepared and applied in the polymer blends, where starch granules are plasticized using plasticizers (such as water or glycerol) under heating and shearing.

However, the incompatibility between the hydrophilic starch and the hydrophobic biodegradable polyester PBAT resulted in the poor mechanical properties of PBAT/TPS composites. Developing a blend with satisfied properties depends on the ability to control the interfacial tension, which is able to improve the stress transferring between the component phases.

In this work, two anhydride based compatibilizers and inorganic filler talc were used to improve the compatibility of the PBAT/TPS composite films via reactive extrusion. The mechanical properties, thermal stabilities and cross-section morphologies were investigated in order to evaluate the compatibility of the prepared composite materials.

2 EXPERIMENTAL

2.1 Materials

Poly(butylene adipate-co-butylene terephthalate), (PBAT, product of BASF, trade name: Ecoflex FBX 7011, Batch NO. 2010476283) and potato starch (food grade) were kindly provided by Al Pack Company (Moncton, New Brunswick, Canada). Glycerol, polyethylene-graft-maleic anhydride (PE-g-MAH, saponification value 32–36 mg KOH/g), poly(maleic anhydride-alt-1-octadecene) (PMAO, saponification value 310–315 mg KOH/g), and talc (particle size: 10 μm) were purchased from Sigma-Aldrich without any further purification.

The molecular structures of the chemicals are presented in Figure 1.

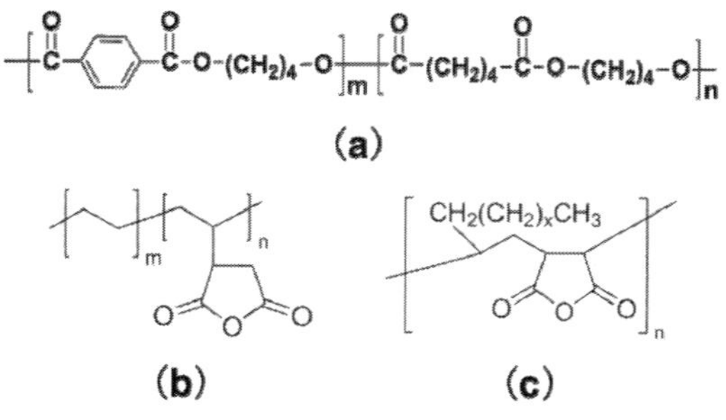

2.2 Preparation of Thermoplastic Starch (TPS) and PBAT/TPS composite films

The plasticization of starch was conducted in a ZSK 18 MEGAlab twin-screw extruder (screw outer diameter: 18 mm, ratio of length to diameter: 40:1, Coperion, USA). Potato starch, glycerol and deionized water were homogenized and extruded for the TPS pellets. The temperature profile used during extrusion was 85/110/135/155/170/160/150°C from feed throat to die with the melt temperature of 165°C. The screw speed was set at 150 rpm.

The preparation of PBAT/TPS pellets was to mix PBAT pellets with previously obtained TPS pellets and the additive (PE-g-MAH, PMAO and/or talc). Then, the mixtures were extruded using the same twin-screw extruder to produce the PBAT/TPS composite pellets. The temperature profile used for extrusion was 90/115/135/160/175/180/170°C from feed throat to die with the melt temperature of 175°C. The screw speed was set at 180 rpm. The recipes are shown in Table 1.

The films were produced using a Saturn Laboratory Blown Film Extrusion System (Future Design Inc., Mississauga, Canada). The single

Table 1. Recipes of PBAT/TPS composite films (unit: phr).

Entries	PBAT	TPS	PE-g-MAH	PMAO	Talc
PBAT	100		/	/	/
P1	80	20	/	/	/
P2	60	40	/	/	/
M1	80	20	3	/	/
M2	60	40	3	/	/
M3	60	40	3	/	3
M4	50	50	3	/	/
O1	80	20	/	3	/
O2	60	40	/	3	/
O3	60	40	/	3	3

screw outer diameter and the ration of length to diameter of the extrusion system were 25 mm and 30:1, respectively. The die diameter of the blown film system was 6.3 cm with the gap of 1 mm. The temperature profile used was 160/170/180/195/180/170°C from feed throat to die with the melt temperature of 185°C. The screw speed was set at 25 rpm. The thickness of the resultant films was 20–45 μm.

The mechanical properties were measured using a universal testing machine (Instron Model 4465, Norwood, USA) at room temperature (23°C). The tensile strength and elongation at break were recorded in the Machine Direction (MD) with a cross head speed at 500 mm/min, according to ASTM D882. At least ten dumbbell shaped samples (115.0 mm × 6.0 mm) were tested for each film, and the average thickness was recorded using a micrometer caliper prior to the test.

2.3 Thermal gravimetric analysis

Thermal Gravimetric Analysis (TGA) was performed using a thermogravimetric analyzer (model SDT Q600, TA instruments, USA). A sample of approximately 10 mg was heated from room temperature to 650°C at 20°C/min under nitrogen atmosphere. The Thermogravimetric (TG) data were recorded.

2.4 Scanning electron microscopy

A Scanning Electron Microscope (SEM) (Model JSM6400, JEOL, USA) was used to examine the surface morphologies before and after the soil burial test. The top surfaces of the samples were coated with a thin layer of carbon. An accelerated voltage of 10 KV and a vacuum pressure of approximately 10-6 Pa were used as the operation conditions.

Figure 1. Molecular structures of the chemicals: (a) PBAT; (b) PE-g-MAH and (c) PMAO.

3 RESULTS AND DISCUSSION

3.1 *Mechanical properties*

The tensile strength and elongation at break of the prepared PBAT/TPS films are shown in Table 2. The tensile strength of the pure PBAT was 27.4 MPa. As a typical hydrophilic biomaterial, starch contains numerous amount of hydroxyl groups, which are hardly compatible with the hydrophilic main chain of PBAT. The direct compounding of PBAT and TPS without any compatibilizer usually results in a severe phase separation or agglomeration [Mohanty et al. 2009]. Therefore, after starch was added into PBAT, the tensile strength of P1 and P2 dropped to 15.9 MPa and 18.3 MPa, respectively. It was interesting to notice that the strength of P2 was higher than that of P1, even though the TPS content of P2 was twice of P1. The reason could be that, less amount of TPS tended to aggregate and form large particles within the PBAT matrix during extrusion, which became the stress concentration points and further impaired the strength while performing mechanical test. On the other hand, TPS particles could coalesce and become a rather continues phase when the content of TPS was increased to 40phr. The less amount of TPS particles with smaller particle sizes had less negative impact on the strength of P2 than P1.

Maleic anhydride has been applied as a compatibilizer or coupling agent in a number of areas [Olivato et al. 2012]. The reactive anhydride group is able to generate covalent bonding among different functional groups, such as hydroxyl groups and amine groups. In this work, the anhydride containing PE-g-MAH and PMAO were used as the compatibilizers to improve the compatibility between PBAT and TPS. The tensile strength of M1 and O1 was higher than that of P1, though, in a relatively limited level. However, M2 was 17.5% higher than that of P2, which proved that the addition of PE-g-MAH

did generate linkages between PBAT and TPS, and further improved the interfacial adhesion between different phases. The strength of O2 was lower than that of P2, which was unexpected. The possible explanation could be that the compatibilizer was poorly dispersed during reactive extrusion due to its high molecular weight (Mn = 30,000~50,000). In addition, the side chain on PMAO might not be compatible with either PBAT or TPS.

The introduction of inorganic fillers to binary blends could decrease the viscosity of the constituent polymers and improve the compatibility by reducing the domain size of the high viscous component. Talc is one of the fillers usually used in polymer compounding system, which could provide unique advantages, such as low cost and high performance. In this work, M3 and O3 both contained the same amount of talc, 3phr. But it seemed that the addition of talc offered a negative effect on the mechanical properties as the strength of M3 was lower than that of M2. O3 was even worse, of which the tensile strength was only 14.2 MPa. Other research showed that a modification or chemical treatment on the inorganic fillers might be necessary prior to the extrusion since the polarity on the outer layer of the filler could be different among the other blending components. The direct melt blending of talc with the current polymer system could be this case, in which the unmodified talc was incompatible with PBAT and/or TPS, even in the presence of the compatibilizers, PE-g-MAH and PMAO. Although the modification might be able to improve the compatibility, the cost of the composite material will be affected.

A further addition of TPS from 40phr to 50phr was applied to the PE-g-MAH system, and the tensile strength of M4 was 11.2 MPa. As a comparison, both PBAT/TPS and PBAT/TPS/PMAO systems were unable to produce film samples as the low melt strength and phase separation resulted in an unstable bubble forming process, which confirmed the good compatibilization effect of PE-g-MAH.

Elongation at break exhibited a similar pattern to tensile strength, yet was less sensitive to the content of TPS compared to tensile strength.

3.2 *Thermogravimetric analysis*

The thermal stability was assessed by thermogravimetric analysis in Figure 2. Due to the existence of TPS, the TG thermograms exhibited a two-stage weight loss pattern (except the pure PBAT), in which the first one was mainly attributed to TPS and the second one resulted from PBAT. The thermal decomposition of starch started from 220–230°C, while PBAT was after 300°C.

Both M1 and O1 were above P1 in the TG thermograms, suggesting the compatibilizers, PE-g-MAH

Table 2. Mechanical properties of PBAT/TPS films.

Entries	Tensile strength (MPa)	Standard deviation (MPa)	Elongation at break (%)	Standard deviation (%)
PBAT	27.4	2.5	553	61
P1	15.9	1.4	460	47
P2	18.3	1.7	412	54
M1	16.3	0.8	465	48
M2	21.5	2.3	447	45
M3	18.6	1.9	423	39
M4	11.2	2.6	399	31
O1	16.1	2.2	478	57
O2	17.9	1.6	386	64
O3	14.2	1.7	379	45

and PMAO, did improve the thermal stability of PBAT/TPS system. M2 and O2 presented a similar pattern. The addition of talc within M3 and O3 has barely affected the thermal stability.

3.3 *SEM images*

The cross-section morphologies of the prepared PBAT/TPS films are shown in Figure 3.

The morphology of the pure PBAT film was quite smooth, while the other films exhibited vacant sites and embedded TPS or talc particles in the matrix. The size of the TPS particles within P1 was larger than that of P2, which confirmed that the less amount of TPS tended to generate agglomeration. After PE-g-MAH and PMAO was added as compatibilizers, the cross-section showed less starch particles, and meanwhile the size of the TPS particles decreased as well. The unmodified talc formed large aggregates, which impaired the matrix of the film and resulted in a poor tensile strength.

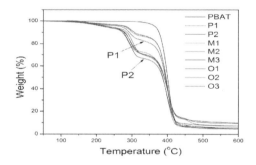

Figure 2.   TG thermograms of the film samples.

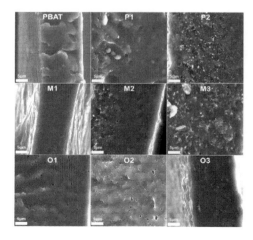

Figure 3.   SEM images of film cross-section.

4   CONCLUSIONS

The eco-friendly and cost-effective biodegradable PBAT/TPS films were successfully prepared via melt compounding in the presence of compatibilizers and inorganic filler. The compatibilizer PE-g-MAH was more efficient than PMAO in improving the compatibility between PBAT and TPS. The addition of compatibilizers also enhanced the thermal stability of the prepared films. However, the inorganic filler, talc, generated a negative effect on the compatibility since the polarity of the unmodified outer layer might be different from the other compounding components.

ACKNOWLEDGEMENTS

The authors would like to thank National Natural Science Foundation of China (No. 21466005), Guangxi Natural Science Foundation (No. 2014GXNSFAA118036) and the Dean Project of Guangxi Key Laboratory of Petrochemical Resource Processing and Process Intensification Technology (2014Z003) and NSERC & AIF (Canada).

REFERENCES

[1] Averous L. 2004. Biodegradable multiphase systems based on plasticized starch: A review. *J Macromol Sci Polym Rev.* C44(3): 231–74.
[2] Chiellini E., Solaro R. 1996. Biodegradable polymeric materials. *Adv Mater.* 8(4): 305–13.
[3] Jbilou F., Joly C., Galland S, Belard L., Desjardin V., Bayard R., 2013. Biodegradation study of plasticised corn flour/poly(butylene succinate-co-butylene adipate) blends. Polym Test. 32(8): 1565–75.
[4] Miller K.R., Soucek M.D. 2014, Degradation kinetics of photopolymerizable poly(lactic acid) films. *J Appl Polym Sci.* 131(13): 10–14.
[5] Mohanty S., Nayak S.K. 2009. Starch based biodegradable PBAT nanocomposites: Effect of starch modification on mechanical, thermal, morphological and biodegradability behavior. *Int J Plast Technol.* 13(2): 163–85.
[6] Olivato J.B., Grossmann M.V., Yamashita F., Eiras D., Pessan L.A. 2012. Citric acid and maleic anhydride as compatibilizers in starch/poly (butylenes adipate-co-terephthalate) blends by one-step reactive extrusion. *Carbohyd Polym* 87(4): 2614–8.
[7] Schwach E., Averous L. 2004. Starch-based biodegradable blends: morphology and interface properties. *Polym Int.* 53(12): 2115–24.
[8] Witt U., Einig T., Yamamoto M., Kleeberg I., Deckwer W.D., Muller R.J. 2001. Biodegradation of aliphatic-aromatic copolyesters: evaluation of the final biodegradability and ecotoxicological impact of degradation intermediates. Chemosphere. 44(2): 289–99.

Material Science and Engineering – Chen (Ed.)
© 2016 Taylor & Francis Group, London, ISBN 978-1-138-02936-1

Experimental study of direct flocculation for the oil removal of waste water in ship tanks

Z.M. Qin, J. Yu, J.H. Xie & B.H. Cai
Wuhan Second Ship Design and Research Institute, Wuhan, China

ABSTRACT: Direct flocculation for the oil removal of waste water in ship tanks is attractive and potential. While organic synthetic flocculants are widely used in waste water treatment, the density of the flocs generated from oily waste water is often not high enough, which hinders the flocculants' usage in tanks. So, a compound flocculant made from inorganic ferric chloride and organic polymeric flocculant is proposed for the treatment of oily waste water in ship tanks, in which ferric chloride is used to increase the density of flocs. Although the molecular weight of the organic flocculant is less than 6 million, the compound flocculant can effectively sink oil in the waste water with more than 95% removal. The optimized mass ratio of the organic flocculant to ferric chloride is 5:2. The compound flocculant can deal with oily waste water with oil concentrations up to 10^4 mg/L. The treatment of different kinds of oil is also investigated. Both vegetable oil and animal oil are investigated. Although the minimum flocculant needed for full flocculation is higher for animal oil, the flocculant needed for sinking is basically the same, higher than the quantity required for engine oil.

Keywords: flocculation; oily waste water; experiment

1 INTRODUCTION

Oil removal of waste water is important for ships, which is carried out conventionally using dedicated equipment such as oil water separators. As a chemical method, flocculation takes advantage of material rather than equipment and is easy to operate and less expensive, which makes it very potential.

Lee et al. gave a review of the application of coagulation and flocculation in waste water treatment. Mao et al. summarized the research status of the flocculating mechanism. Song studied the relationship by shooting moving floc particles during the dynamic flocculation process using high-definition digital cameras. Wang used the combined flocculation of ferric chloride and PAM to treat the oily waste water of a large circulating water system. Li et al. used the flocculation method to remove suspended solids and oil from the oily waste water of oil field. Tian utilized a flocculant composed of aluminum sulfate and PAC(poly-aluminum chloride) to treat the ship's bilge water with an oil concentration lower than 200 mg/L after composition and process optimization by the orthogonal test.

There is also a lot of oily waste water stored in ship tanks besides the bilge areas. The oily waste water in tanks needs to be treated with pertinence because it is different from the oily waste water in bilge areas. First, the bilge water coming from machine leakages contains engine oil, diesel oil, and hydraulic oil, while the tank waste water coming partially from kitchen contains animal and vegetable oil. The oil concentration may be higher than 200 mg/L. Second, the waste water in tanks is often emulsified due to the use of cleaning agents and stirring from transportation pumps. Finally, flocculation production will float on the top of the waste water if its density is lower, which will hinder the flocculant dosage afterwards.

While trying to solve the above problems, a kind of flocculant with high density production obtained through the compounding of organic and inorganic flocculants is proposed and experimented in this paper. First, the organic flocculant is optimized. Flocs generate quickly after the treatment, but does not sink in the waste water. Then, ferric chloride is used to compound with the organic flocculant, so that the flocs generated can get enough weight to sink. Finally, the compound flocculant with an optimized mass ratio is used for the treatment of different oils with different oil concentrations.

2 ORGANIC FLOCCULANT PREPARATION

2.1 Preparation procedure

Considering the presence of ions with negative charges, a cationic polymeric flocculant is selected for the treatment of oily waste water. The P(DMC-AM) chosen in this paper has advantages including the amount of charge control, uniformity of charge distribution, less dosage and sludge, fast reaction speed, and ease of microbial degradation.

The preparation procedure is as follows: AM (acrylamide) and DMC (methacryloyloxyethyltrimethy ammonium chloride) are dissolved in deionized water adequately, the combination of emulsifiers is put in the oil phase, the flask is stirred to dissolve and emulsify sufficiently, the water phase is added slowly into the oil phase, the temperature is raised, the pH value is adjusted, nitrogen is flushed for 10 minutes, the initiator solution is added, the temperature is adjusted for the reaction to cool down.

2.2 Process optimization

The process is optimized to get the maximum molecular weight.

2.2.1 The influence of the volume ratio of oil to water

During inverse emulsion polymerization, oil, as the continuous phase, contributes to the separation of individual droplet and as the media of polymerization heat emission. So, the volume ratio of oil to water influences not only the emulsification effect and the emulsion stability, but also the intrinsic viscosity of the polymerized production and the solid content of the polymerization system.

All emulsions are stable, as shown in Table 1. From the table, it can be seen that the largest molecular weight can be obtained when the volume ratio of oil to water is 1.3.

2.2.2 The influence of the monomer mass ratio

As shown in Table 2, the viscosity and cationic degree of the copolymer rises first and then falls with the addition of cationic monomer DMC. This occurs because the flocculation decreases and the flocs become small when the DMC quantity is too large. The largest molecular weight can be obtained when the mass ratio of DMC to AM is 6:4.

2.2.3 The influence of emulsifier concentration

As shown in Figure 1, the best emulsifier concentration is about 8%. The molecular weight will decrease if the concentration is above 8%.

Table 1. The effect of the volume ratio of oil to water.

Volume ratio	Intrinsic viscosity (dL/g)	Molecular weight ($\times 10^6$)
0.7	6.42	4.04
0.9	6.72	4.24
1.1	6.93	4.51
1.3	7.35	4.89
1.5	6.91	4.45

Table 2. The effect of the monomer mass ratio.

DMC: AM	Intrinsic viscosity (dL/g)	Cationic degree (%)
3:7	6.74	22.4
4:6	6.85	27.6
5:5	7.06	31.2
6:4	7.21	35.5
7:3	6.94	29.6

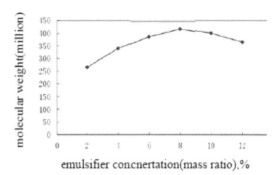

Figure 1. The effect of emulsifier concentration.

2.2.4 The influence of initiator quantity

As shown in Figure 2, the largest molecular weight can be obtained when the ratio of initiator mass to all monomers' mass is 0.5%.

2.2.5 The influence of reaction temperature

The polymerization temperature is chosen between 30°C and 65°C because the reaction temperature of the redox initiation system is relatively low. The initiator decomposes faster, and both the polymerization rate and the free radical concentration increases at a higher temperature. However, the rate of chain termination and chain transfer also increases, which will lower the viscosity and molecular weight of the product.

As shown in Table 3, the monomer conversion rate and production molecular weight rises first at a higher reaction temperature, reaches the climax at about 50°C, and then falls gradually. So, the best reaction temperature is controlled between 45°C and 55°C.

2.3 Waste water treatment

Oily waste water preparation. A sample of 5 g 32[#] engine oil, 2 g detergent and some water is added to a flask and stirred sufficiently for emulsification. The oil phase is added into 5 L water and stirred evenly. The mimic waste water is turbid. The overall oil concentration is about 1000 mg/L and has

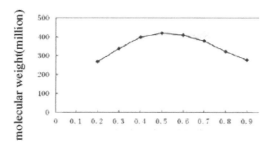

Figure 2. The effect of initiator quantity.

Table 3. The effect of reaction temperature.

Temperature (°C)	Molecular weight (×10⁶)	Conversion rate (%)
30	2.5	78.63
35	3	85.51
40	4.55	90.62
45	4.85	92.85
50	5	94.73
55	5.1	93.02
60	4.7	89.24
65	3.2	80.30

visible oil on the top. The measured pH and turbidity value is about 9 and 213 mg/L, respectively.

Flocculation of the waste water. A sample of 1000 mL oily waste water is added to a 2000 mL flask, and then the aforementioned 50 mg flocculent is added and stirred quickly (400 r/min) for several minutes. Flocs are generated quickly and float up in the waste water. The flocs are thick and dense, which look like snivel containing grains.

Combined flocculation of waste water. A sample of 1000 mL oily waste water is added to a 2000 mL flask, and then 50 mg inorganic flocculant PAC is added and stirred quickly (400 r/min), and the aforementioned 50 mg organic flocculant is added and the stirring is stopped. Flocs are generated quickly and float up in the waste water. The oil concentration and turbidity are lowered to 60.6 mg/L and 26.5 mg/L, respectively. The oil removal and turbidity removal are 93.9% and 87.5%, respectively.

3 COMPOUND FLOCCULANT PREPARATION

3.1 *Preparation procedure*

Ferric chloride is used for the inorganic weight gain of the aforementioned organic flocculent to increase the production density, so that it can be employed better in ship tanks.

The preparation procedure is as follows: ferric chloride is added to the aforementioned organic flocculent and stirred evenly, and then the temperature is raised to 70°C and reacted for two hours and cooled down to the room temperature.

3.2 *Optimization of mass ratio*

The oily waste water is prepared again with an oil concentration at about 5000 mg/L by increasing the engine oil to 25 g. The measured pH value is still about 9.

The compound flocculant is formed with a mass ratio of organic flocculant to ferric chloride of 10:2, 10:3 and 10:4, and then added to the oily waste water gradually and stirred quickly. The experimental results are as follows:

As shown in Table 4 and Figure 3, the sinking type production is obtained with the increased dosage of the compound flocculant. The production is yellow and composed of cigarette-ash-like fine particles with a density larger than the water. There is no suspended substance in the water after holding still if the sinking type production is generated. The dosage of the compound flocculant needed decreases gradually at a less mass ratio of the organic flocculant. The inorganic flocculant plays an important role.

However, the viscosity of the compound flocculant will increase with more number of inorganic floccu-

Table 4. Experimental results of different compound flocculants.

Mass ratio	Dosage (mg/L)	Separation time (min)	Production density	Suspended substance
10:2	1500	15	Float up	Little
10:2	2000	30	Float up	Some
10:2	2500	30	Sink	No
10:3	1500	15	Float up	No
10:3	2000	25	Sink	No
10:3	2500	15	Sink	No
10:4	1500	20	Sink	No
10:4	2000	15	Sink	No

Figure 3. Typical flocculation result.

lants. This will decrease the fluidity of the flocculant and cause inconvenience in the addition of the flocculant to the tanks. So, the mass ratio of 10:4 is selected to fulfill both the need of sinking and fluidity.

3.3 Waste water treatment

3.3.1 The influence of oil concentration

The compound flocculant is used to treat oily waste water with different oil concentrations. The oil concentration and turbidity before and after the treatment is given in Table 5.

Sinking type production can be formed quickly for oil concentrations of 800 mg/L and 10000 mg/L. The treatment of oily waste water with an oil concentration of 10^5 mg/L is also performed. There is a lot of oil on the top of the waste water. Sticky flocs are generated and float on top of the water after the addition of the flocculant. The flocs do not sink with more dosage of the flocculant, which means that the compound flocculant cannot deal with waste water with a very high oil concentration.

3.3.2 The influence of oil kinds

Both vegetable oil and animal oil are tested.

A sample of 1 g or 0.1 g vegetable oil is used to measure the cylinder. A 0.1 g emulsifier is added. Some water is added and stirred. Water is supplemented to the volume of 100 mL. Then, the oily waste water with oil concentrations of 1000 mg/L and 104 mg/L is obtained. Compared with the engine oil, it is clearer. The pH value is about 7.

The treatment results are given in Table 6. The production is composed of fine particles. There is no suspended substance in the water after holding still. The measurement results are given in Table 7.

A sample of 1 g or 0.1 g liquid animal oil is added to the flask. A 0.1 g emulsifier is added. Some 50°C hot water is added and stirred. Hot water is supplemented to the volume of 100 mL. Then, the oily waste water with oil concentrations of 1000 mg/L and 10^4 mg/L is obtained. Compared with the vegetable oil, the turbidity of waste water containing animal oil is obviously higher, which is similar to the engine oil. There is much oil on the top.

The treatment results are given in Table 8. The production is composed of fine particles. There is

Table 5. Test results before and after the treatment of oily wastewater containing engine oil.

Oil concentration			Turbidity		
Before (mg/L)	After (mg/L)	Removal (%)	Before (mg/L)	After (mg/L)	Removal (%)
800	32.5	95.9	196.4	7.8	87.5
10000	126.4	98.7	1678.5	80.6	96.8

Table 6. Experimental results of different vegetable oil concentrations.

Oil concentration (mg/L)	Dosage (mg/L)	Separation time (min)	Production density
10^4	500	5	Float up
10^4	1500	15	Float up & sink
10^4	2000	20	Float up & sink
10^4	2500	20	Sink
10^4	3000	18	Sink
10^3	500	5	Float up
10^3	1000	10	Float up & sink
10^3	1500	15	Sink
10^3	2000	15	Sink

Table 7. Test results before and after the treatment of oily wastewater containing vegetable oil.

Oil concentration			Turbidity		
Before (mg/L)	After (mg/L)	Removal (%)	Before (mg/L)	After (mg/L)	Removal (%)
1000	19.8	98.0	43.8	4.2	90.4
10000	45.6	99.5	342.5	5.4	98.4

Table 8. Experimental results of different animal oil concentrations.

Oil concentration (mg/L)	Dosage (mg/L)	Separation time (min)	Production density
10^4	500	15	Float up*
10^4	1000	5	Float up
10^4	2000	20	Float up & sink
10^4	2500	20	Sink
10^4	3000	15	Sink
10^3	500	5	Float up
10^3	1000	10	Float up
10^3	1500	15	Sink
10^3	2000	15	Sink

* Water is still turbid after treatment.

no suspended substance in the water after holding still, except the dosage of 500 mg/L. The measurement results are given in Table 9.

As shown, the animal oil is more difficult to flocculate because a 50 mg/L flocculant cannot make a full flocculation. However, the flocculant quantity needed for sinking is basically the same, which is about 2500 mg/L for the oil concentration of 10^4 mg/L. This is higher than the quantity required for engine oil, which is about 1500 mg/L.

Table 9. Test results before and after the treatment of oily wastewater containing animal oil.

Oil concentration			Turbidity		
Before (mg/L)	After (mg/L)	Removal (%)	Before (mg/L)	After (mg/L)	Removal (%)
1000	25.4	97.5	278.4	12.9	95.4
10000	64.3	99.4	1436.5	12.9	99.1

4 CONCLUSIONS

A compound flocculant made from ferric chloride and P (DMC-AM) is proposed for the treatment of oily waste water in ship tanks, in which ferric chloride is used to increase the density of flocs. The molecular weight of P (DMC-AM) can reach 5 million with a volume ratio of oil to water of 1.3, a mass ratio of DMC to AM of 6:4, an emulsifier concentration at 8%, a mass ratio of initiator at 0.5% and the reaction temperature at about 50°C. However, the density of the flocs generated is not high enough to sink, which hinders the flocculants' usage in tanks. The combined flocculation of PAC and P (DMC-AM) is tried, which also generates float-up flocs.

So, ferric chloride is used to gain weight. After compounding with ferric chloride, P (DMC-AM) can effectively sink oil in the waste water with more than 95% removal (mass ratio of organic flocculant to ferric chloride 5:2). The compound flocculant can deal with oily waste water with oil concentrations up to 10^4 mg/L. The treatment of vegetable oil and animal oil is also investigated. Although the minimum flocculant needed for full flocculation is higher for animal oil, the flocculant needed for sinking is basically the same, higher than the quantity required for engine oil.

With regard to the next step, the substitution of the stirring operation will be investigated to facilitate further the application of this flocculant in tanks.

REFERENCES

[1] Wang, C.Q. 2013. Study and practice of oil pollution treatment technology for large circulating water system. *Metallurgical power* 159(5): 61–63.
[2] Lee, C.S, et al. 2014. A review on application of flocculants in wastewater treatment. *Process safety and environmental protection* 92: 489–508.
[3] Song, J.J. 2009. Study on fractal growth characteristics of flocs in flocculation process in water treatment. *Master's Degree Disseration.* Harbin Institute of Technology.
[4] Tian, S.S, et al. 2007. Flocculation technology to dispose oily sewage in ship cabin. *Ship & Ocean Engineering* 36(3): 104–106.
[5] Li, Y.F, et al. 2005. Study on the treatment of wastewater containing heavy oil with demulsification and flocculation method. *Journal of Shenyang Architectural and Civil Engineering Institute* 21(6): 711–714.
[6] Mao, Y.L, et al. 2008. Advances in Flocculation Mechanisms and Research of Water-treatment Flocculants. *J. of HUST* 25(2): 78–82.

Building material and interior design

Material Science and Engineering – Chen (Ed.)
© 2016 Taylor & Francis Group, London, ISBN 978-1-138-02936-1

Analysis on the methods and changes of the interior space design of express inns

T.T. Yu
City College, Wenzhou University, Zhejiang, China

Y.P. Wang
Hangzhou Dianzi University, Zhejiang, China

X.F. Sun
Ouhai District Party Committee and Commission, Zhejiang, China

ABSTRACT: Since the introduction into China in the 1990s, express inns have spread from the coastal areas to the interior land in an astounding expansion rate. Its market share has been gradually expanding. However, in its development process, imitation from each other and the sole type have made the inn space become boring and have no individuality, which should be energetic and imaginative, thus losing their appeal for the guests. To have more room for the development, express inns must make a constant innovation in decoration, style selection, color matching, theme selection, site selection and other aspects.

Keywords: express inn; interior space design; design innovation; change

1   OVERVIEW OF EXPRESS INNS

Express inn, also known as a limited-service hotel, has a biggest feature of low price. Its service model is "b & b" (bed & breakfast). It first appeared in the 1950s in the USA. Currently, it has been a mature hotel form in European and American countries.

Express inn has a huge market potential, with advantages of low investment, high return, and short cycle. Since the introduction into China in the 1990s, express inns have spread from the coastal areas to the interior land in an astounding expansion rate. Nowadays, chain hotels, which are supported by powerful network platforms, are becoming mature. It has a large number of express inn brands, including 7DaysInn Group, Jinjiang Inn, and Home Chain. They have occupied an important position in the Chinese hotel market. However, in its development process, imitation from each other and the sole type have made the inn space become boring and have no individuality, which should be energetic and imaginative, thus losing their appeal for the guests. To have more room for the development, express inns must make a constant innovation on the original basis.

2   DISCUSSIONS ON THE DESIGN PRINCIPLES OF EXPRESS INNS

The main design purpose of express inns is primarily the rational and efficient use of space to create a warm and comfortable environment, and secondarily the formation of the inn's distinctive brand features. The ultimate goal is to use the design to attract customers to promote consumption, so that the merchant can get the maximum benefits.

2.1   *Decoration principles and methods*

Express inns are usually chain hotels. Because of the reasonable price, comfortable environment, and large competitiveness, the inns have made rapid and successful development, spreading over various large cities.

The first consideration of decoration is the decoration cost, and then is the preferences of local main consumer groups (in China, the main consumer groups are the middle-income young people). Finally, it should be adapted to local conditions and the inn's conditions. The so-called adaptation to the local condition is to fully consider the inn's location. In the design, in addition to ensure the style of the brand inns, it should add local elements to form a unified feature in the inns of the same brand as much as possible. For example, in the express inns of Suizhou, Hubei, it is a good choice to add some chimes cultural elements of Suizhou in the express inn design. It can increase tourists' curiosity and freshness, and also make the hotel have its unique style. The adaptation to the inn's condition is that even the inns of the same brand in the same city should consider the differences in various inns' sites, and it should make such differences as the inn's unique feature. For example, in the design of Jinjiang Inn of Fruit Lake, as it is near East Lake, in its design, it has an extensive use of partitions and wavy blue decorations. It is the case that adds the environmental elements and color in the design elements of express inns.

2.2 Rational use of style to create warm atmosphere

The hotel's lobby is the necessary pass for guests to enter and exit the hotel, which will leave a first impression to the guests. The hotel's lobby is the place where guests make check-in and check-out, and a transportation center to the guest rooms and other major public areas of the inn. It is the hub of the entire inn. The unique atmosphere of its design and style will directly affect the image of the hotel and the exertion of its own functions. As for city express inns, the lobby space should be an intensive functional layout within a limited area with full use of space, striving to achieve the "small but good", and delicate and rich effects. The lobby space area should be adapted to the size of the whole hotel and its actual function structure, rather than blindly pursuing for the largeness.

The design decoration style of hotel entrance can substantially be divided into garden, scaffolding and facade styles. While the general inns have small facades, large LOGO can be installed at the intersections, external walls and other striking places to attract and guide more guests. Also, it can take advantage of glass doors and windows to post huge advertising art paintings, and install neon facade to demonstrate the features of the hotel.

There are many design decoration styles of hotel lobby. Generally, express inns can choose a simple fashion style. Its characteristics are clean and bright open with flowing lines. For example, laying a large area of the mirror wall in the lobby makes the lobby look more spacious and dainty and exquisite. Also, the match of metal edge strips around the mirror makes the entire space have a rich and full modern feeling. The partial match fabric sofa and curved chandelier increase a soft and harmony morbidezza in the bold space, making the block surface have smooth lines.

As the fundamental of hotel's services, and also the hotel's main source of income, the design of the guest room becomes extremely important. Mentioning the hotel rooms, most people will come to mind of contemporary cabinet, large soft bed, paved floor carpet, and health and stylish bathroom sanitary wares, and other integral design elements in modern hotels under the soft light of the hotel that is in good order. There is no exception on this aspect for express inns. Moreover, express inns should spend more effort in ensuring the guests' sleeping comfort, and make the basic requirement of creating a sleeping and sensory comfort as a focus of the inn, and try to simplify the design style of those unnecessary extravagances with a grasp of the style standard of simplicity, convenience, and comfort.

2.3 Reasonable mix of colors to improve comfort feeling

For consumer groups, express inns' guest rooms are a private space for rest. The interior colors should make the customers feel relaxed and warm and create the pleasant atmosphere. Most hotel rooms are designed to adopt a unified standard color and a simple and unified style to reduce the cost greatly. However, there are very few hotels, for example boutique hotels have different styles in each room to attract different customers of preferences. The guest room design should maintain the comfort of the room, and also use colors to make them different under the forming of unified styles, which is economical and practical.

First, the color design of inn rooms should make the consumers feel cozy and comfortable. In the use of colors, it should use the colors that most people like. Although as for people who live in the inn, guest rooms are temporary private environment. With respect to consumer groups, inn business model is public space. In the selection of colors, especially that of the color tone, it should look for the popular color combinations that have similarity, which people like most, and that conform to the living space, in order to meet different needs of customers on the visual enjoyment, and allow customers to enjoy the spiritual joy and great value of the money. It is a good choice to select the warm colors as the basic tone to design guest rooms to make the room look gorgeous and noble.

Second, the color matching of inn guest rooms should give full consideration to the local climate and spatial orientation. In hot regions, cool colors should be the main color of the room, which can make the customer feel a little cool in the heat; contrarily, in the places of high latitudes, it is no doubt a very clever approach of using warm colors in guest rooms, for people will get the psychological temperature through vision, and even under the circumstances of no heating facilities, they can also feel the warmth. In the mid-latitude regions, the issue to be considered in the room is probably the direction that the room faces that whether it is north or south, or full of sunshine or no sunshine all year round. It determines the general orientation of cold and warm colors.

Third, if the hotel is in the region of the rich customs and culture, it must take the traditional local color into account. Because customers are mainly tourists from other places, they may mainly come for these customs. This is very suitable for China's national conditions. China has multiple ethnic nations, and every nation also has different cultural characteristics. The choice of color is among them. For example, people in the Tibetan region like painting on the interior walls with auspicious patterns. Some paint blue, green, and red ribbons to represent blue sky, sea and land. Therefore, such colors can be cleverly used in the hotel rooms.

Fourth, the sheets, curtains and other sewing items in the guest rooms are the best items to change the room atmosphere, which can change according to the fashion trends or seasons to bring fresher atmosphere to guests.

2.4 Making the theme prominent to create a new concept of hotel culture

Express inn design should take simplicity and comfort as the theme, and the reduction of price and the increase of comfort as the goal. The lobby is the hotel's facade. Guest rooms are an important part of the inns. Guest rooms of express inns should be simple, warm, and have a comfortable environment and low price, and be welcomed by the masses. However, now a lot of guest rooms and furnishings are too unified. The cabinet and room furnishings are too popular, which are prone to cause consumption fatigue to the consumers. Therefore, the further development of the inn should make breakthroughs on the original basis. The characteristic guest rooms in simple and low price can make changes in the color of the guest room, the shape of the cabinet, and the style design of the layout to attract consumers, improve the comfort of the consumption group, and ease consumption fatigue.

For example, there are seven floors in the inn, and in each floor, there are standard room, king size room and suite. Guest rooms should be mainly with low carbon and simple decor. Two of the floors are white and black colors. In the same floor, different guest rooms make some simple decorations on the basis of white and black colors. The color combinations can be added to make changes in the style and color of the guest rooms. However, generally, it should maintain the

unity of the style. Other floors are auxiliary floors. They also adopt simple decorations. For example, each guest room on the second floor can have the similar layout and ceiling. But the materials can choose some unusually used new materials like tawny glass to give customers the freshness. The fifth floor can adopt a mixing style, which makes the interior design have more new visual experience. And the sixth and seventh floors adopt a simplified Japanese style and European style. Such different styles according to the floors make the inn design more flexible and bouncing, which can be easy to attract consumption group and to increase competitiveness.

3 DESIGN INNOVATION OF EXPRESS INNS

3.1 Combination of reasonable matching and home decoration

Generally, the design of express inn's guest rooms is single with little color combination. To make breakthroughs on the original basis, it can add color combinations in the guest room design, making the room more humane. In each area of the room, and in different home decorations, the "utilization" of color can make the customer's mind active.

Bedroom color must make people feel relaxed and comfortable. So, it should choose low saturation and high bright colors. The choices of cold and warm colors are generally determined by the owner and the environment. Bedroom colors are often the reflections of owner's personality and lifestyle habits, which can be said that the concept of local color masters can be given full play. But hotel rooms are not designed for one person but for the masses. Therefore, the choice of colors in guest rooms should be some popular color matching. The ceiling walls should be mainly white. The floor carpet should be in light colors with few other color combinations. It should seek changes in the simplicity, making the space full of changes.

In the selection of furniture, on the one hand, it should combine the style in the guest room, grasping the main color to make matching. Also, it should consider the color feelings under lights.

In designing the home decoration, there are many styles to choose. How to choose the right style to apply to the decorated rooms is very important. For example, a simple European modern style is a simple form simplified from the European style. The combination of it into the guest room design can efficiently control the decoration design and also add new consumption elements, giving consumers simple but not easy feeling in order to attract consumers.

3.2 *Combination of reasonable matching and home decoration*

Low-carbon life is to minimize energy consumption during life, and reduce carbon emissions, especially carbon dioxide emissions, thus reducing the pollution of the atmosphere and slowing ecological degradation. It mainly starts from power saving, energy saving and recycling to change living details.

In the design of express inn, the simple design of the lobby and guest rooms is suitable for most young consumption group. At the same time, it can reduce the waste of certain materials that conform to the requirements of the low cost of express inns. Also, it is an important application of "low-carbon life" concept in the express inn design. The color combination, which is mainly white and supplemented by other colors, cannot be too single, but also relatively environmental protection and energy saving.

In addition to make it more "low-carbon", in the design implementation of express inns, the material selection can use recycling materials as much as possible. The materials of high energy consumption and high carbon, like clay solid bricks and aluminum windows and doors, should be avoided. In the design of the guest rooms, spotlight should

be avoided. Only in specific theme rooms can it be used to improve the environment warmth.

3.3 *Site diversity of express inns*

Express inns are mainly distributed in the city. While at the present stage, the development of our domestic express inns is trapped into a relatively slow stage. To break this impasse, it should not only make efforts in the design concept and interior decoration of inns, but also try to make breakthroughs in the site selection. In addition to develop in cities, express inns can share the markets of the original small hostels and farmhouses, and open chain inns in some tourist attractions to compete against the hotels in tourist attractions.

REFERENCES

[1] Zhengfeng Yue, Three Elements of Residential Interior Design [J]. Shanxi Architecture. 33(9) 225–226. (2007).

[2] Ji Wen, Spatial Development Pattern of Guangzhou Star Hotels [J]. Economic Geography. 26(3) 451–455. (2006).

[3] China Industry Advisory Network. 2011–2015 Market research and Development Planning Advisory Report of China's Economy Hotel Industry [S]. (2011).

[4] Kessels K, The Worst Hotel in the World, The Hans Brinker Budget Hotel Amsterdam [M]. Booth-Clibborn Editions. (2009) 112–117.

Material Science and Engineering – Chen (Ed.)
© 2016 Taylor & Francis Group, London, ISBN 978-1-138-02936-1

Properties of plant-based asphalt cement-stabilized fine-grained soil

X.B. Zhang & H.T. Sun
Dezhou Highway Engineering Corporation, Dezhou, Shandong, China

Q.B. Zhu, W.D. Cao & S.T. Liu
School of Civil Engineering, Shandong University, Jinan, Shandong, China

ABSTRACT: To find out whether the plant-based asphalt can improve the shrinkage performance of cement-stabilized fine-grained soil, the compaction test, the unconfined compressive strength test, the dry and temperature shrinkage tests of plant-based asphalt cement-stabilized fine-grained soil were conducted. The results show that the plant-based asphalt has a good effect on restraining the low temperature and dry shrinkage of cement-stabilized fine-grained soil, but has certain adverse effects on the growth of compressive strength.

Keywords: plant-based asphalt; cement-stabilized fine-grained soil; compressive strength; dry shrinkage; temperature shrinkage

1 INTRODUCTION

Cement-stabilized soil can be used as the base of second-class highway and below, or sub-base of all kinds of roads. Because of the evaporation and mixture hydration inside, the water in the mixture will reduce continually after mixing the cement-stabilized soil. The capillary effect and adsorption caused by decreasing water and changing temperature will cause the volume shrinkage of the material and shrinkage cracks of cement-stabilized soil further. With the increasing cement content the crack will be more significant.

Selecting the lower plasticity index soil, adding a certain amount of fiber and reducing the water content are the main solutions to cracking problems nowadays. These methods can improve the performance of cement-stabilized soil to a certain extent, but also bring many problems such as improvement of the construction cost and increase in the difficulty of construction.

Plant-based asphalt has the natural molecular polarity and alkaline, which might be good at easing the shrinkage crack of cement-stabilized soil. There has been barely related research on plant-based asphalt in China; most studies of other countries have focused on using it in HMA. This study conducted the dry shrinkage test and the temperature shrinkage test to determine the plant-based asphalt's modification effect on cement-stabilized soil.

2 MATERIALS AND PROPERTIES

2.1 Soil

The silty soil was from the periphery of Ji'nan, Shandong Province. Experiments of the sample showed that it was a low liquid limit silty soil. Most particles were powder with diameters ranging between 0.074 and 0.002 mm and their content were generally higher than 90%. The clay content was less than 10%, and the uniformity coefficient of soil was $Cu > 5$.

2.2 Plant-based asphalt

The related indices of plant-based asphalt are presented in Table 1.

Table 1. Reference values of plant-based asphalt's quality.

Items	Units	Reference values
Appearance	–	Dark brown
Relative density [20°C]	g/cm³	1.200–1.300
Kinematic viscosity [100°C]	mm²/s	220.00–250.00
Pour point	°C	≤28.00
Flash point [open]	°C	≥230.00
Volatility [163°C, 3 h]	%	≤1.00
Water content	%	≤1.00

2.3 Cement

R 42.5 ordinary Portland cement, which has 6%~15% active mixture and 6%~10% non-active materials, was used in this study. The water content of standard consistency was W/C = 0.34. The initial and final setting times met the standard.

3 DESIGN OF THE EXPERIMENT

3.1 The composition design of plant-based asphalt cement-stabilized soil

According to related research of cement-stabilized silty soil, the amount of cement accounts for 12% of soil's dry weight. Plant-based asphalt contains moisture and oil; therefore, it is necessary to estimate the water in plant-based asphalt percentage of total addition water.

Three doses of plant-based asphalt were chosen, namely 5%, 10% and 15% by total weight of cement. The plant-based asphalt was preheated to 60°C in the shape of fluid and mixed with normal temperature water, fully stirred it with a stirring rod and cooled it to room temperature. Using the VEKA instrument, the drop depth of the test cone was measured. The results are plotted as the diagram of the proportion of plant-based asphalt to replace the water and standard consistency in Figure 1.

According to Figure 1, the curve range satisfied the requirements of the specifications (drop depth of the test cone 33 mm~35 mm), corresponding to the horizontal coordinate numerical around 0.8. It means a unit mass of plant-based asphalt will replace the water whose weight equals 0.8 times of a unit plant-based asphalt's weight.

After determining the ratio, the coagulation time was measured. Therefore, the approximate content range of plant-based asphalt can be estimated. Four different adding amounts of plant-

Table 2. Setting times (W/C = 0.34).

Group	Initial setting time [min]	Final setting time [min]
Original cement slurry	253	364
5% plant-based asphalt cement	270	379
10% plant-based asphalt cement	288	390
15% plant-based asphalt cement	294	390
20% plant-based asphalt cement	436	>600

based asphalt were selected: 5%, 10%, 15%, and 20%. Cement slurry was mixed according to the specification and measured its initial setting times and final setting times, as presented in Table 2.

As shown in Table 2, while the addition amount of plant-based asphalt reached 20%, the initial setting time of the mixed slurry had already reached 7 h, which had exceeded the requirements of the specifications.

Plant-based asphalt contains oil that will hinder the hydration reaction of cement in the condensation process, resulting in the slow growth of early strength and prolonging the setting time. Based on former research, to promote the hydration of plant-based asphalt cement and improve its early strength, this study determined to use 5% plant-based asphalt and add 0.5 mol/L NaOH to provide the alkaline environment for the hydration reaction.

3.2 Testing program

This study conducted the compaction test to measure the maximum dry density and the optimum moisture content of the raw materials, and then performed the unconfined compression test, temperature shrinkage test, and dry shrinkage test to study the performance of plant-based asphalt cement-stabilized soil. There were two groups for the unconfined compression test. The first group was the original cement-stabilized soil and the second group was added 5% plant-based asphalt and changed water to 0.5 mol/L NaOH solution.

Temperature shrinkage test's specimens were small fine-grained soil beams whose size was 50 mm*50 mm*200 mm and made by the static compaction method, using 98% of standard compaction. It still needed two control groups: the first group was fine-grained soil; the second group was added 5% plant-based asphalt and changed water to 0.5 mol/L NaOH solution.

Dry shrinkage test's specimens' making method was the same as the temperature shrinkage test.

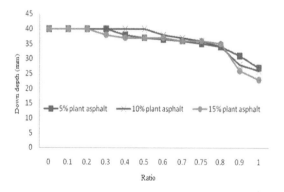

Figure 1. Relationship between the ratio of plant-based asphalt to replace water and standard consistency.

It needed two control groups: one was fine-grained soil and the other was added 5% plant-based asphalt and changed water to 0.5 mol/L NaOH solution.

4 TESTING RESULTS AND ANALYSIS

4.1 Compaction test of raw materials

Cement-stabilized materials should be compacted before the cement began to hydrate. All the test results are shown in Figure 2.

As shown in Figure 2 and based on the test results, the maximum dry density of the soil sample was 1.781 g/cm^3 and the corresponding optimal moisture content was about 13.2%.

4.2 Unconfined compression strength at different ages

By observing the two groups and measuring strength values of all specimens from different ages, the tests showed that with the age extending, specimens of the second group became smaller. It means that the interaction of cement, plant-based asphalt and NaOH inside the specimens produced tiny pores and then hindered the strength's growth.

Figure 3 shows the comparison of different ages' average unconfined compressive strength between cement-stabilized fine-grained soil and plant-based asphalt cement-stabilized fine-grained soil.

Figure 3 shows that the unconfined compressive strength of cement-stabilized fine-grained soil grew slowly with the prolonged age. 7 d strength had already met the standard requirements of expressway's sub-base. The strength of plant-based asphalt cement-stabilized fine-grained soil decreased with age, which also showed a rising trend, and the rate was greater than the cement-stabilized fine-grained soil, but its 7 d unconfined compressive strength was significantly lower, amounting to 25.2%. Strength differences between them gradually reduced with

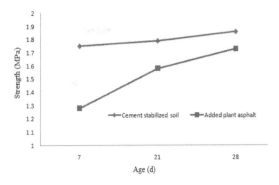

Figure 3. The unconfined compressive strength of two kinds of cement stabilized soil at different ages.

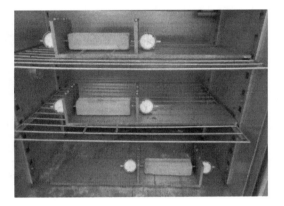

Figure 4. Temperature shrinkage test.

the extension of age, which fell by 13.4% at 21 d and finally reached 7.7% at 28 d. Therefore, the addition of NaOH and plant-based asphalt had some adverse effects on the early strength of cement-stabilized fine-grained soil, but had few effects on later strength and was conductive to speed up the growth rate of the later strength.

4.3 Temperature shrinkage test

The instrument of the temperature shrinkage test is shown in Figure 4.

Specimens of the two groups were observed and measured at all levels of temperature. The results of temperature shrinkage deformation were the sum of two micrometer's deformation divided by specimens' length, expressed as a percentage. By calculating the data of the temperature shrinkage test, all results were obtained, as shown in Figures 5 and 6.

The test results showed that in the temperature region of above 0°C, the temperature shrinkage strain and coefficient were small and had little

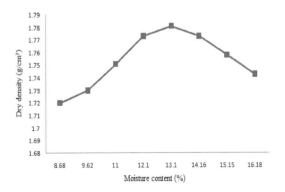

Figure 2. The compaction curve.

differences between two kinds of cement-stabilized fine-grained soil. It means that the shrinkage deformation of cement-stabilized fine-grained soil caused by the temperature change was not sensitive but under the higher temperature.

The temperature shrinkage deformation's most sensitive section of two kinds of cement-stabilized fine-grained soil was 0°C–10°C. It means that in this region, the temperature shrinkage deformation and its corresponding coefficient were the largest.

Compared with the original cement-stabilized fine-grained soil, under the same temperature conditions, the plant-based asphalt cement-stabilized fine-grained soil's temperature shrinkage deformation would decrease, especially the low-temperature shrinkage deformation, which was significantly improved. The temperature shrinkage coefficient decreased by 35%. It had a positive effect on improving the cement-stabilized fine-grained soil sub-base's low-temperature crack resistance performance.

4.4 Temperature shrinkage test

Dry shrinkage refers to the volume shrinkage of the semi-rigid base material caused by the change in its internal moisture content. Settling the specimen on the shrinkage device with both ends placed the dial gauge after maintenance, and the shrinkage deformation of specimens was measured through the numerical changes in the dial indicator.

The data were calculated and the specimens' dry shrinkage strain and dry shrinkage coefficient were obtained. Thus, the dry shrinkage strain–time curve and the dry shrinkage coefficient–time curve were generated, as shown in Figures 7 and 8.

Figures 7 and 8 show these following points:

1. Specimens' shrinkage strains of the two groups were all continuously increasing with extending time. The dry shrinkage strain and its growth of plant-based asphalt cement-stabilized fine-grained soil were both lower than cement-stabilized fine-grained soil at early 14 days, which means that the plant-based asphalt cement-sta-

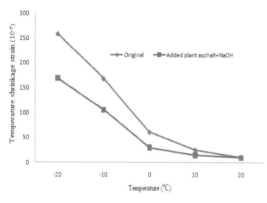

Figure 5. Temperature shrinkage strain—temperature curve.

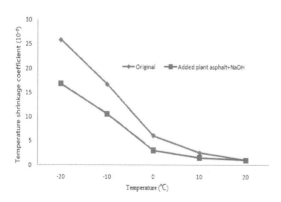

Figure 6. Temperature shrinkage coefficient—temperature curve.

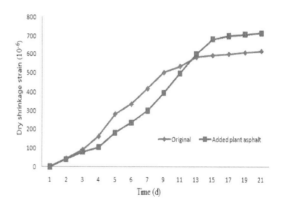

Figure 7. Dry shrinkage strain—time curve.

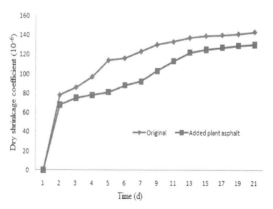

Figure 8. Dry shrinkage coefficient—time curve.

bilized fine-grained soil had a better early dry shrinkage performance.

2. The dry shrinkage coefficient and its growth of plant-based asphalt cement-stabilized fine-grained soil were both lower than cement-stabilized fine-grained soil, and plant-based asphalt cement-stabilized fine-grained soil's dry shrinkage coefficient had a significant decrease during 5 days to 15 days, and the maximum drop was about 20%; until the final stability, the total shrinkage coefficient was 139.95×10^{-6}, while cement-stabilized fine-grained soil's total shrinkage coefficient was 141.94×10^{-6} until it tended to stabilize at 21 days. Compared with the cement-stabilized fine-grained soil, the plant-based asphalt cement-stabilized fine-grained soil's total dry shrinkage coefficient reduced by 8.5%, which shows that the plant-based asphalt had a better inhibiting effect on the cement-stabilized fine-grained soil.

5 CONCLUSIONS

1. Plant-based asphalt had a baffling effect on the development of cement's hydration, hardening and strength in cement-stabilized soil, which makes the specimens hard to mold, thus the NaOH was needed to add to the water to provide a strong alkali environment for cement's hydration.

2. The original specimen's 7 d unconfined compressive strength was about 0.9 MPa, which grew slowly after 7 d and the 28 d's strength was about 1 MPa. The control group's specimen, which was added 5% plant-based asphalt and NaOH, had growing strength with the time extended at 28 d and tended to be stable after 28 d, and the 28 d's unconfined compressive strength was 0.87 MPa, decreased by 13% compared with the original group.

3. The control group with 5% plant-based asphalt and NaOH showed an improved effect on cement-stabilized soil's temperature shrinkage performance; especially in the low-temperature sensitive section below 0°C, the temperature shrinkage coefficient could be improved by 30% compared with the original group.

4. Specimen with 5% plant-based asphalt and NaOH had an improvement on its dry shrinkage performance. Although its dehydration rate and dry shrinkage strain were almost as large as the original group, its total dry shrinkage coefficient still decreased by 8.5%, which had a less improved effect than the temperature shrinkage coefficient.

REFERENCES

[1] China standard GB175-2007. 2007. Common Portland Cement. China Standards Press, 12:1~9.
[2] China standard GB/T 1346-2001. 2001. Test Methods for Water Requirement of Normal Consistency, Setting Time and Soundness of the Portland Cement. China Standards Press, 4:1~6.
[3] China standard JTG/E 30-2005. 2005. Test Methods of Cement and Concrete for Highway Engineering. China Communications Press, 7:76~79; 133~135.
[4] China standard JTJ 034-2000. 2000. Technical Specifications for Construction of Highway Roadbases. China Communications Press, 6:6~11.
[5] Dongjun Chen. 2014. The Mechanism and Key Technology Research of Plant-based asphalt Half Flexible Base. Shan Dong University. 4:43~75.
[6] Elham H. Fini, M. ASCE, Eric W. Kalberer et al. 2011. Chemical Characterization of Biobinder from Swine Manure: Sustainable Modifier for Asphalt Binder. *Journal of Materials in Civil Engineering.* American Society of Civil Engineers.
[7] Peng Zhang & Qingfu Li. 2010. Experimental Study on Shrinkage Properties of Cement-stabilized Macadam Reinforced with Polypropylene Fiber. *Journal of Reinforced Plastics and Composites.* Vol. 29, No. 12:1851~1860.
[8] Peng Zhang, Ph.D. Qingfu Li & Hua Wei, PhD. 2010. Investigation of Flexural Properties of Cement-Stabilized Macadam Reinforced with Polypropylene Fiber. *Journal of Materials in Civil Engineering,* 1282~1287.
[9] Weidong Cao, Xiaobo Zhang, Xinlong Qi, Haitao Sun, Shutang Liu. 2014. Advances in Bio-asphalt Research. *Petroleum Asphalt.* 5:1~4.
[10] Xiaohua Li, Dongqi Zheng, Wenrui Xu. 2004. Strength and Dry Shrinkage Crack of Cement Stabilized Soil. *China Western Science and Technology,* 5:40~41.
[11] Zhifeng Chen. 2009. Effect of Polypropylene Fiber on Shrinkage Properties of Cement-stabilized Macadam. *Modern Applied Science.* February, vol. 3, No. 2:71~74.

Material Science and Engineering – Chen (Ed.)
© 2016 Taylor & Francis Group, London, ISBN 978-1-138-02936-1

Influences of concrete mechanical properties when using zeolite powder as admixture

Q.W. Liu

Southwest Jiaotong University, Chengdu, Sichuan, China

ABSTRACT: Using zeolite powder as admixture is a common way to produce high-strength concrete. However, it remains unknown that what kind of influence will it cause when the concrete is in a hot and damp condition. So, we performed an experiment with two contrast experiments to determine how the concrete mixed with zeolite powder performs in compressive and tensile splitting properties under such condition. In these experiments, there were 36 concrete blocks cared in a normal and hot water tank. After the caring, we tested their mechanical properties, and found that the concrete mixed with zeolite powder cared in hot water had a lower performance than the blocks cared under the normal condition and those mixed with or without fly ash. This warns us to take into account the disadvantages of zeolite powder as admixture when producing high-strength concrete.

Keywords: zeolite powder; admixture; damp and hot condition; mechanical properties; concrete

1 INTRODUCTION

1.1 *Background*

There are various kinds of engineering in the world today and an increasing demand on the properties of concrete as well. So, concrete with traditional materials (e.g. cement, sand, water) can barely meet the structure or durability requirements under many harsh conditions. So, we combine the natural or artificial mineral material as concrete admixtures in the concrete mixture preparation in order to save cement, improve the concrete performance, and adjust the strength grade of concrete [1].

Concrete admixture can be divided into active mineral admixture and non-active mineral admixture. Commonly, the non-active mineral admixture and cement components cannot afford to the chemistry reaction, such as ground fine quartz sand, limestone, and slag. However, zeolite powder as a kind of active admixture can reach chemical reaction $Ca(OH)_2$, creating hydraulic cementing materials [2]. This kind of admixture is a good choice to replace cement and save water to improve the concrete performance. However, the influence under the damp and hot condition when using zeolite powder as admixture has not yet been defined. So, we perform an experiment to determine how the concrete performs.

1.2 *Zeolite powder*

Zeolite powder is milled from natural zeolite (Figure 1). Zeolite rock is a kind of calcining

aluminum silicate minerals after natural pozzolanic with some active silica and alumina. Zeolite powder has a large internal surface area and open structure. The fineness is less than 5% when 0.08 mm sieving [3]. Its average particle size is 5.0~6.5 microns and white in colour.

It contains mostly all the major elements and trace elements that the aquatic animal growth needs, all these elements exist in the ionic form that can be easily used by aquatic animals. Therefore, zeolite powder is most commonly used in aquaculture [4]. Zeolite rock series has more than 30 varieties, and ash zeolite and mordenite are commonly used as admixtures of concrete (Figure 2).

Figure 1. Zeolite powder.

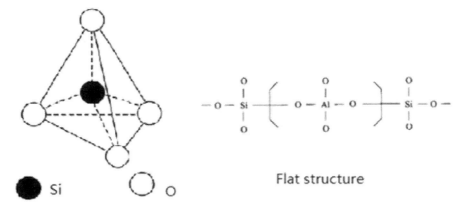

Figure 2.   Micro structure.

Using zeolite powder as concrete admixture has basically the following several advantages:

a. It can be the replacement of cement; with the cooperation of a water-reducing agent, it can reduce the dosage of the water and improve the strength of the concrete to make high-strength concrete.
b. Improving the workability of concrete and producing pumping concrete and fluid concrete [5].

2   THE EXPERIMENTAL DESIGN

2.1   *Preparing*

In order to distinguish the experiment from using 100% cement, the samples were mixed with common admixture fly ash. We set two groups of contrast experiment. In the first group, we use cement as the only gelled material. In the second group, we use 70% cement and 30% fly ash [4]. In the main experiments we use 70% cement and 30% zeolite powder. Each experiment is divided into two parts. One part for the compressive experiment and the other is for the tensile splitting experiment. In each part, we produce six blocks. Three of them will be cured under the normal condition. The other three will be cured in a high-temperature water tank, and the temperature is set as 70 degrees Celsius. Every block is made into one unified size of 100 mm*100 mm*100 mm, and cared for four weeks. The strength level of concrete is C30.

2.2   *The experimental group*

2.2.1   *Group1 (70%cement+30%zeolite powder)*
Compressive experiment: normal condition (ICN), high-temperature water tank (ICW).

Figure 3.   Blocks made and waiting for maintenance.

Tensile splitting experiment: normal condition (ISN), high-temperature water tank (ISW).

2.2.2   *Group2 (70%cement+30%fly ash)*
Compressive experiment: normal condition (IICN), high-temperature water tank (IICW).
Tensile splitting experiment: normal condition (IISN), high-temperature water tank (IISW).

2.2.3   *Group3 (100%cement)*
Compressive experiment: normal condition (IIICN), high-temperature water tank (IIICW).
Tensile splitting experiment: normal condition (IIISN), high-temperature water tank (IIISW).

Each group has 3 blocks for the compressive experiment, and another 3 blocks for splitting the compressive experiments. The blocks made is shown in Figure 3.

2.3   *The experimental data (MPa)*

The experimental data are provided in Table 1.

Table 1. Experimental data.

ICC	27.91	28.24	29.08
Average	28.41		
ICW	32.69	31.85	31.95
Average	32.46		
ISC	3.15	2.63	2.83
Average	2.87		
ISW	2.63	2.77	2.75
Average	2.72		
IICC	34.56	34.61	32.89
Average	34.02		
IICW	40.16	37.78	36.17
Average	38.04		
IISC	2.57	2.97	2.61
Average	2.72		
IISW	4.11	4.08	3.84
Average	4.01		
IIICC	37.99	37.89	38.48
Average	38.12		
IIICW	29.84	32.84	28.92
Average	30.53		
IIISC	5.19	4.35	5.95
Average	5.16		
IIISW	5.02	5.39	5.39
Average	5.27		

3 ANALYSIS AND CONCLUSION

As we can see from Table 1, both the compressive properties and tensile splitting properties of concrete mixed with zeolite or fly ash cured in the water tank are clearly higher than those cured under the normal condition. However, under the same curing condition, concrete mixed with zeolite shows a worse performance than that mixed with or without fly ash.

So, we can preliminarily reckon that though using zeolite powder as admixture can promote the workability of concrete, it leads to a decrease in compressive properties and tensile splitting properties when the intensity of concrete develops in a hot and damp environment. We should consider these losses when using such kind of concrete under these conditions. This warns us to take into account the disadvantages of zeolite powder as admixture when producing high-strength concrete.

REFERENCES

[1] Zeolite Powder and Its Use in High Strength Concrete (In Chinese)-Shujin Li, Keru Wu,—TU528.31-1003-1324(2004)01-0040-04.
[2] Use of Zeolite Powder as Auxiliary Cementitious Material (In Chinese)—Shoudong Li, Weihao Wu, Jianjun Cai, Yunfeng Luo.—1674-2133(2010)05-16-06.
[3] Information on http://baike.baidu.com.
[4] Use of Natural Zeolite in aquaculture (In Chinese)-Hongmei Gao, Mingxue Wang, 1003-1278(2005)01-0001-03.
[5] Produce and Usage of Zeolite Powder in Pumping Concrete (In Chinese)-Xiaoli Zhong, Zemin Zhu, 1002-3550(2005)07-0084-02 *Typography for references.*

Material Science and Engineering – Chen (Ed.)
© *2016 Taylor & Francis Group, London, ISBN 978-1-138-02936-1*

The effect of pH value on the measuring accuracy of the bundled chloride ion selective electrodes for concrete

Y. Jiang, Z. Chen, L.F. Yang & M. Zhou
Key Laboratory of Disaster Prevention and Structural Safety, School of Civil Engineering and Architecture, Guangxi University, Nanning, Guangxi, China

ABSTRACT: The potential method could measure the concentration of chloride ion in the solution quickly, but the measuring accuracy is affected by the pH levels of the solution. A bundled chloride ion selective electrode was developed, and the measuring accuracy of the electrode developed was analyzed in four different pH values which were relevant to concrete. The results show that the calibration curve of electrode is affected by the pH value of solution significantly, but the relationship between electrode potential and chloride concentration are still compliant with the Nernst equation. The selective electrode developed could provide high sensitivity and stability to the measurement of chloride concentration in concrete within a pH range of approximately 3.74 to 12.53.

Keywords: chloride ion; the ion selective electrode; pH value; the potential method; concrete

1 INTRODUCTION

The chloride ion within concrete is one of the most predominant factors on the durability destruction of concrete structures in marine environment. Numerous studies indicate that the diffused chloride leads to substantial reduction of the cross-section of the rebar. Subsequently, the concrete cover would deteriorate as cracking and spalling (Poulsen & Mejlbro 2010). The precise determination of chloride ion in concrete is very important to evaluate the durability of concrete structures. Thus a variety of standard methods are developed to quantify the chloride content in concrete, including some standard test methods (ASTM C1152 2004). The potential measurement method has been used in determining the concentration of chloride ion (Angst & Polder 2014). However, the calibration curve of electrode is affected by the pH value of solution significantly and the measuring accuracy of the electrodes decreases when the pH value of solution is out of the application range of the electrode. There is a wide range of pH values of the chloride solutions which are relative with concrete. For instance, the free chloride in concrete is extracted by distilled water in neutral environment (pH = 7.32); the total chloride in concrete is extracted by nitric acid solution in acidic environment (pH < 7); the inner pore solution of concrete prepared is an alkaline solution (pH > 11.50).

So it's necessary to invent a kind of chloride ion selective electrodes which have a wide application range of the pH values and can be embedded in concrete specimens. Meanwhile, the measuring accuracy of the electrodes in the solutions with different pH value should be analyzed for the application in different condition.

2 THE DEVELOPMENT OF BUNDLED CHLORIDE ION SELECTIVE ELECTRODE

In the potential method for measuring the concentration of chloride ion, the chloride ion selective electrode is used as an indicator electrode and the calomel electrode is used as a reference electrode.

When the electrode is placed in a solution containing chloride ions, an electrode potential develops across the sensing element, the magnitude of which depends on the chloride ion activity in the test solution. This potential is measured with a digital mV meter or an ISE meter, and follows the Nernst equation as follows:

$$E = E^{\ominus} + S \cdot \lg C_{Cl^-}; \qquad S = \frac{RT}{zF} \times 2.303 \qquad (1)$$

where E is the reduction potential at the temperature of interest; $E^{\ominus}$ is the standard reduction potential; R is the universal gas constant; C is the chloride ion activity in solution; T is the absolute temperature; z is the number of moles of electrons transferred; F is the Faraday constant, the number of coulombs per mole of electrons.

A plot of E vs $\lg C_{Cl^-}$ will produce a calibration curve, which will be linear over a range of chloride concentrations, the range depending on the response of the electrode under measurement conditions. According to the relationship between E and $\lg C_{Cl^-}$, the concentration of chloride ion, actually activity of chloride ion, can be calculated by the calibration curve.

According to the principle of potential measurement, the bundled chloride ion selective electrodes are subsequently developed as Figure 1. For measuring the concentration of chloride ion in concrete specimens, the bundled chloride ion selective electrodes are embedded in concrete specimens, and the salt bridge is served by the concrete pore solutions.

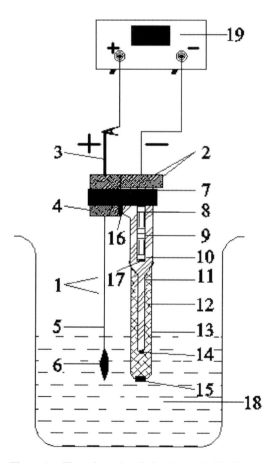

Figure 1. The schematic of the bundled chloride ion selective electrodes (1. The silver chloride electrode, 2. The Saturated Calomel Electrode (SCE), 3. Copper conductor, 4. PTFE, 5. Silver wire, 6. The layer of silver chloride (AgCl), 7. Polymeric tape, 8. Hg/Hg2Cl2, 9. 11 & 13. Glass tube, 10. The saturated KCl solution, 12. 0.1 M NaNO3, 14. 15 & 17, the porous material, 16. Rubber plug, 18. The NaCl solutions, 19. The device of potential measurement).

3 EXPERIMENT

The chloride concentration of a series of NaCl solutions with NaCl concentrations ranging from 0.0001 to 0.005 M and four different pH environments are tested by the bundled electrodes and commercial selective electrodes provided by NELD instrument company. The experimental temperature is controlled in the range of 20~25°C and the four environments with different pH value are as follows.

Neutral (pH = 7.32). In concrete, some chloride ions which are named the free chloride ion remain free in the pore solution. A common way to measuring the free chloride ion in concrete is to extract the free chloride ion from powder samples of concrete to solution by deionized water. In this condition, the extraction solution including the free chloride ion is neutral. In this experiment, All the NaCl solutions with different concentration in neutral condition are prepared with deionized water and the pH values of NaCl solution all are 7.32 by measurement.

Acidity (pH = 3.74). In concrete, some chloride ions which are named the bound chloride ion are bound to the hydration products and the total chloride ion include the free chloride ion and the bound chloride ion. A method of extraction of the total chloride ion is to extract the total chloride ion from powder samples of concrete to solution by 15% nitric acid (Chen et al. 2010a). In this condition, the extraction solution including the total chloride ion is an acid solution (Chen et al. 2010b). In this experiment, All the NaCl solutions with different concentration in acid condition are prepared with dilute nitric solution and the pH values of the NaCl solution are 3.74 by measurement.

Alkalinity (pH = 12.53) and Weak Alkalinity (pH = 10.45). For producing amount of $Ca(OH)_2$ in the process of hydration of cement, the pH value of inner pore solution of concrete almost reaches that of the saturated calcium hydroxide solution and is about 12.5 at room temperature condition (Du, et al., 2005). Ordinary iron and steel products are normally covered by a thin iron-oxide film that becomes impermeable and strongly adherent to the steel surface in the alkaline environment, thus making the steel passive to corrosion (Vladimír 2010). However, the passivity of the stable reinforcement can be fluctuated when the pH value of the concrete is in the range of 9.88 and 11.5. The stability of passive film would completely collapsed with the pH value under 9.88, or in the event that the Cl^-/ OH^- ratio in the pore solution at the iron matrix interface exceeds a threshold level of 0.61 (Angst et al. 2009). So the pH value of 10.45 which is the average of 11.5 and 9.88 is selected for experiment in weakly alkaline environment. In this experiment, two kinds of NaCl solution with different alkaline environment are prepared by Ca(OH)$_2$ solution, and the pH value of the NaCl solution are 10.45 and 12.53 respectively by adjusting the content of calcium hydroxide in solution.

4 RESULTS AND DISCUSSION

The Figure 2 to Figure 5 show that the electrode potentials measured by the bundled electrodes and a commercial selective electrodes for 4 kinds of solutions versus lgC_{Cl^-}. In the Figures, the circle data points (labeled "A" in Figures) denote the electrochemical potentials measured by the bundled electrodes, and the dotted line fitted by the circle data points denotes the calibration curve

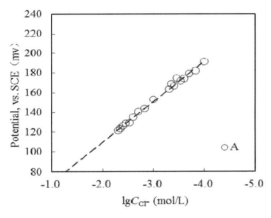

Figure 2.   The relationship between the potential and the concentration of chloride ion in acid environment (pH = 3.74).

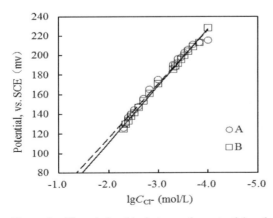

Figure 3.   The relationship between the potential and the concentration of chloride ion in neutral environment (pH = 7.32).

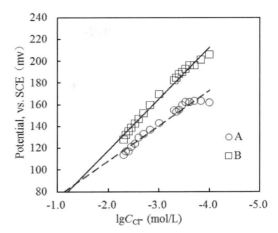

Figure 4.   The relationship between the potential and the concentration of chloride ion in weakly alkaline environment (pH = 10.45).

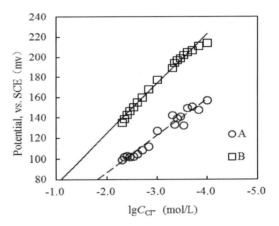

Figure 5.   The relationship between the potential and the concentration of chloride ion in alkaline environment (pH = 12.53).

for the bundled electrodes. The square data points (labeled "B" in Figures) denote the electrochemical potentials measured by the commercial selective electrodes, and the solid line fitted by the circle data points denotes the calibration curve for the commercial selective electrodes. The commercial selective electrodes aren't used in the measurement for acid solution (pH = 3.74), because the pH value of acid solution is out of the allowable measurement range.

The relationship between electrode potentials and concentration of chloride ion is studied by logarithm regression and correlation method, the equations of the calibration curves are deduced as Table 1.

Table 1. The fitted equations of the calibration curves for the bundled electrodes and a commercial selective electrodes in 4 kinds of solution.

Environment	Electrodes	Equations	R^2
Acidity	A	$E = -40.94 \lg C_{Cl^-} + 28.31$	0.99
	B	—	—
Neutral	A	$E = -55.20 \lg C_{Cl^-} + 5.09$	0.98
	B	$E = -58.00 \lg C_{Cl^-} - 5.04$	0.99
Weak alkalinity	A	$E = -32.53 \lg C_{Cl^-} + 43.01$	0.95
	B	$E = -47.61 \lg C_{Cl^-} + 22.69$	0.99
Alkalinity	A	$E = -36.00 \lg C_{Cl^-} + 14.58$	0.95
	B	$E = -49.06 \lg C_{Cl^-} + 26.74$	0.99

*The pH value of acid solution is out of the allowable measurement range of the commercial selective electrodes.

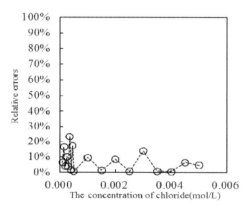

Figure 6. The relative errors between the concentration value calculated by the calibration curves of the bundled electrodes and true concentration value in acid environment (pH = 3.74).

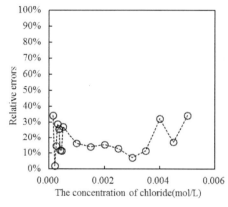

Figure 8. The relative errors between the concentration value calculated by the calibration curves of the bundled electrodes and true concentration value in weakly alkaline environment (pH = 10.45).

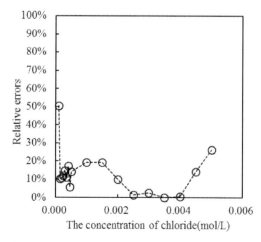

Figure 7. The relative errors between the concentration value calculated by the calibration curves of the bundled electrodes and true concentration value in neutral environment (pH = 7.32).

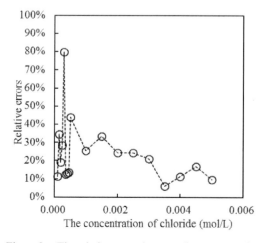

Figure 9. The relative errors between the concentration value calculated by the calibration curves of the bundled electrodes and true concentration value in alkaline environment (pH = 12.53).

A logarithmic relation between electrode potential measured by two kinds of electrodes and the concentration of chloride ion is presented in Figure 2 to Figure 5 for two kinds of electrodes and four kinds of solutions with different pH value. It is seen from Table 1 the equations of the calibration curves are affected by the pH value of solution significantly.

The relative errors between the concentration value calculated by the calibration curves of the bundled electrodes and true concentration value in different environment are shown in Figure 6 to Figure 9.

It is seen from the figures that the concentration values calculated by the calibration curves of the bundled electrodes developed just have a relative error less than 20% with the true concentration value in acid and neutral environment (pH = 3.74 & pH = 7.32), and a relative error less than 30% with the true concentration value in alkaline environment (pH = 10.45 & pH = 12.53). In generally, all the above can sufficiently demonstrate the suitability of the bundled chloride ion selective electrodes in the pH value range of 3.74–12.53.

5 CONCLUSIONS

A bundled chloride ion selective electrode is developed, and the measuring accuracy of the electrode developed is analyzed in four different pH values which are relevant to concrete. The research results show that:

1. The logarithmic relation between electrode potential measured by two kinds of electrodes and the concentration of chloride ion is obvious in four kinds of solutions with different pH value, but the equations of the calibration curve of the bundled chloride ion selective electrodes developed are different with those of the commercial electrodes in different environment.
2. The concentration values calculated by the calibration curves of the bundled electrodes developed just have a relative error less than 20% with the true concentration value in acid and neutral environment, and a relative error less than 30% in alkaline environment. This illustrates that the selective electrode developed could provide high

sensitivity and stability to the measurement of chloride concentration in concrete within a pH range of approximately 3.74 to 12.53.

ACKNOWLEDGEMENTS

This work was financially supported by the National Natural Science Foundation of China (NSFC, No: 51208120 & 51468004), the Doctoral Fund of Ministry of Education of China (20124501120005) and the Projects of Guangxi Natural Science Foundation (2014GXNSFAA118309 & 2012GXNSFEA053002).

REFERENCES

[1] Angst, U., Elsener, B., Larsen, C.K. & Vennesland, Ø. 2009. Critical chloride content in reinforced concrete—A review. *Cement and Concrete Research, 39: 1122–1138.*
[2] Angst, U.M. & Polder, P. 2014. Spatial variability of chloride in concrete within homogeneously exposed areas. *Cement and Concrete Research, 56: 40–51.*
[3] ASTM C1152. 2004. Standard Test Method for Acid-Soluble Chloride in Mortar and Concrete. *American Society for Testing and Materials.*
[4] Chen, Z., Yang, L.F., Gao, Q., Feng, Q.G, Jiang, Q.M., Zhou, M. & Hong B. 2010a. The Methods of Extraction of Chloride Ion from High-Performance Concrete. In Proceedings of the 7th International Symposium on Cement & Concrete/Proceedings of the 11th International Conference on Advance in Concrete Technology & Sustainable Development, *Foreign Languages Press: 1349–1352, Jinan, May 2010.*
[5] Chen, Z., Yang, L.F. & Zeng, J.C. 2010b. Determination method for chloride content in concrete extraction solution. *New Building Material: 73–76 (in Chinese).*
[6] Du, R., Huang, R. & Hu, R. 2005. Embeddable combination probe for in-situ measuring Cl⁻ and pH at the reinforcing steel/concrete interface. *Chinese Journal of Analytical Chemistry, 33 (1): 29–32.*
[7] Poulsen, E. & Mejlbro, L. 2010. Theory and Application in Diffusion of Chloride in Concrete. *CRC Press. Londan and New York.*
[8] Vladimír, P. 2000. Water extraction of chloride, hydroxide and other ions from hardened cement pastes. *Cement and Concrete Research, 30: 895–906.*

Material Science and Engineering – Chen (Ed.)
© *2016 Taylor & Francis Group, London, ISBN 978-1-138-02936-1*

The effects of graphite on the mechanical and thermal properties of cement pastes

C. Huang, Q. Wang & P. Song
School of Materials Science and Engineering, University of Jinan, Jinan, China

ABSTRACT: This paper studied the effects of different additive amounts of high-purity graphite on the mechanical and thermal properties of cement. By testing the water requirement of normal consistency and compressive strength, the effect law of graphite on the mechanical properties and absolutely dry bulk density of cement was analyzed; the effects of different additive amounts of graphite on the thermal properties and microstructure of cement were analyzed by using different measurements such as hot disk thermal constant analyzer and SEM. The results show that with the increase in graphite additive amounts, the water requirement of normal consistency increases, lattice defects such as holes and cracks of the sample increase, and the sample's absolutely dry bulk density and compressive strength both reduce. However, the addition of graphite significantly increases the sample's thermal properties. The comprehensive analysis show that the graphite's suitable additive amount is 8%~20%.

Keywords: graphite; mechanical properties; thermal conductivity coefficient; microstructure

1 INTRODUCTION

With the development of the global economic society, fossil fuel's consumption is constantly increasing, owing to the fossil fuel's non-renewability and serious environment pollution after burning, so the development and utilization of renewable clean energy is of great significance (Philippe Menanteau et al., 2003). Solar energy is a kind of high efficient, clean and renewable energy that never fails, has a broad application prospect, but at the same time solar energy is decentralized, indirect and random that makes it difficult to be utilized in a wider and broader range (Melissa A. Schilling and Melissa Esmundo, 2009). Most of the human's buildings are cement concrete structures and spread all over the world. If we can use concrete to absorb and store solar energy, this will make the application range of solar energy greatly expanded, and we will truly achieve the integration of solar energy utilization and buildings (Jonas Nässén and Fredrik Hedenus, 2012; Jan-Olof Dalenbäck, 1996).

Using concrete to absorb the solar energy is an innovative new energy technology. It is the first and also the most crucial step of this technology to prepare a cement-based absorber material whose mechanical properties conform to the requirements and whose thermal properties are excellent (Parameshwaran. R et al., 2012; Wang Fan et al., 2012). Xiao Jianzhuang studied the influence factors of concrete thermal parameters and made a theoretical calculation, indicating that aggregate mineralogical characteristics and thermodynamic characteristics are the crucial effect factors of concrete thermal conductivity coefficient (Xiao Jianzhuang et al., 2008), so it is of great significance to select the appropriate adding material as the thermal conductivity phase to improve the thermal properties of concrete.

Graphite is a kind of excellent thermal conductive material, and its thermal conductive property even surpasses many metal materials. Its chemical property is stable and has a good compatibility with cement. It can stably exist in the alkali environment of concrete, and also has a good light stability and can bear the high temperature caused by absorbing the solar energy (Yuan Huiwen et al., 2012). So, this experiment chose to add graphite into cement, studied the effect law and the action mechanism of graphite on the mechanical and thermal properties of cement, and eventually prepared the cement-based absorber material whose mechanical properties conformed to the requirements and whose thermal properties were excellent. Thus, it will lay a good material basis for the research of concrete thermal collection and the utilization of solar energy.

2 EXPERIMENTAL PROCEDURE

2.1 Materials

The P.O52.5R early strength ordinary Portland cement produced by Sunnsy Group was used to this experiment, its chemical compositions are given in Table 1; the T-299 type graphite powder produced by Qingdao was used to this experiment, its main parameters are given in Table 2.

2.2 *Methods*

In combination with the graphite's practical application in the building materials industry, graphite additive amounts, respectively, were 4%, 8%, 12%, 16%, and 20%. According to the proportion, the cement mixer was used to premix cement and graphite for 30 min until smooth, and then the mixture was added into water, and the water requirement of normal consistency of different graphite additive amounts were measured on the basis of GB/T1346–2001 "The Detection Methods of Water Requirement of Cement Normal Consistency, Setting Time and Stability", and the experiment was carried out under the water requirement of normal consistency. Molding samples were placed in a standard curing box (20 ± 1°C, relative humidity ≥ 90%) for maintenance, and were demolded after 1d, and water cured to the stipulated age (3d,7d,28d). After testing the compressive strength, the samples were tapped into small pieces and added in anhydrous ethanol to terminate hydration. Finally, they were baked to oven dry to carry out the SEM test. The test of absolutely dry bulk density referred to the American ASTMC188–2009 "The Standard Experiment Method of Hydraulic Cement Density".

The test of thermal properties adopted the Transient Plane Source method (TPS), because it can simultaneously measure the thermal conductivity coefficient, thermal diffusion coefficient and volumetric heat capacity (Ferriere. A et al., 2003). The experiment was conducted according to the German ISO22007–2 "The Standard Test Method of Thermal Conduction", and the hot disk thermal constant analyzer was used for measurement.

3 RESULTS AND DISCUSSION

3.1 *The analysis of water requirement of normal consistency, absolutely dry bulk density and compressive strength*

In combination with the graphite's practical application in the building materials industry, graphite additive amounts were, respectively, 4%, 8%, 12%, 16%, and 20%, and the water requirement of normal consistency of different graphite additive amounts was measured. The experiment was carried out under the water requirement of normal consistency, and then the samples' absolutely dry bulk density and compressive strength were tested. The specific experiment scheme and test results are presented in Table 3. In order to intuitively reflect the effect of different graphite additive amounts on the absolutely dry bulk density and compressive strength of cement, Figure 1 shows the absolutely dry bulk density and Figure 2 shows the compressive strength.

It can be seen from Table 3 that with the increase in graphite additive amounts, the water requirement of normal consistency increased gradually. When the graphite additive amount was 20 wt%, the water cement ratio increased by 0.295 to 0.327. This is because graphite particles are very small, have a certain water absorbability and are not as sleek as cement particles. The two grain compositions also have a big difference, so after addition of graphite, the water requirement of normal consistency increased. It can be seen from Figure 1 that with the increase in graphite additive amounts, the samples' absolutely dry bulk density gradually reduced. When the graphite additive amount was 20 wt%, the samples' absolutely dry bulk density reduced by 10.38% when compared with the blank samples. First, from the standpoint of material density, the graphite volume density was 826.5 kg/m^3, the cement stone volume density was 2059.4 kg/m^3, and the graphite volume density was far less than the cement stone volume density. Second, graphite particles were uniformly distributed in cement, the continuity of the cement stone structure was split, a number of holes and cracks were introduced, and the increase in the water requirement of normal consistency multiplied the samples' defects such as holes and cracks. So, the addition of graphite reduced the samples' absolutely dry bulk density.

Table 1. Chemical composition of ordinary Portland cement with early strength.

Compound	SiO_2	Fe_2O_3	CaO	Al_2O_3	MgO	f-CaO
Content (%)	19.68	2.89	67.23	5.29	3.46	0.76

Table 2. Major parameters of graphite.

Parameter	C content/%	Ash/%	Particle size range/μm	Specific surface area/ $(m^2 \cdot g^{-1})$	Volume density/ $(kg \cdot m^{-3})$	True density/ $(kg \cdot m^{-3})$	Thermal conductivity/ $[W \cdot (m \cdot K)^{-1}]$
Indicator	≥99.5	≤0.2	0.05~5	2.5	826.5	2108.9	135

Table 3. Effects of different additive amounts of graphite on the cement paste performances.

Graphite/ cement (wt%)	Water cement ratio	Compressive strength/MPa			Relative strength			Absolutely dry bulk density/ kg·m⁻³
		3d	7d	28d	3d	7d	28d	
0	0.295	49.20	78.00	85.50	1	1	1	2059.4
4	0.305	34.43	54.77	62.33	0.6998	0.7022	0.7013	2003.7
8	0.311	31.85	49.90	56.85	0.6473	0.6398	0.6424	1966.4
12	0.317	30.59	47.49	53.98	0.6219	0.6089	0.6113	1923.3
16	0.322	27.36	43.16	49.21	0.5562	0.5534	0.5572	1889.8
20	0.327	25.04	40.08	44.25	0.5089	0.5139	0.5210	1845.6

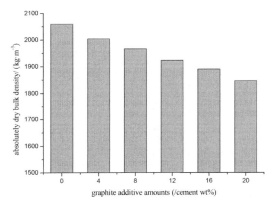

Figure 1. Effects of different additive amounts of graphite on the absolutely dry bulk density.

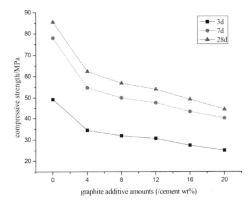

Figure 2. Effects of different additive amounts of graphite on the compressive strength.

It can be seen from Table 3 and Figure 2 that with the increase in graphite additive amounts, the samples' 3d,7d,28d compressive strength reduced significantly. When graphite additive amount was 20 wt%, the samples' 3d,7d,28d compressive strength, respectively, reduced by 49.11%, 48.61%, and 47.90% when compared with the blank samples; thus, it can be seen that when the graphite additive amount was certain, the reductions in compressive strength at all ages were consistent when compared with the blank samples. Graphite is a kind of inert material, which does not react with cement in order to introduce a number of "impurities" to the cement. These "impurities" uniformly filled between the cement hydration products and formed many interfaces with them, split the continuity of the cement stone structure, reduced the samples' compressive strength. At the same time, because only van der Waals force exists between graphite layers, they can easily dissociate along the interlayer and the graphite particle surface has a certain lubricity, so the external force would destroy the interfaces between graphite and cement hydration products,

and reduce the samples' compressive strength. In addition, with the increase in graphite additive amounts, the water requirement of normal consistency increased, defects such as holes and cracks multiplied, and also reduced the samples' compressive strength.

3.2 The analysis of thermal properties

The test of thermal properties adopted the Transient Plane Source (TPS) method, because it can simultaneously measure the thermal conductivity coefficient, thermal diffusion coefficient and volumetric heat capacity. The experiment was conducted in accordance with the German ISO22007–2 "The Standard Test Method of Thermal Conduction" and used the hot disk thermal constant analyzer for measurement. The specific test results are summarized in Table 4. In order to intuitively reflect the effect of different graphite additive amounts on the thermal properties of cement, Figure 3 shows the thermal conductivity coefficient and Figure 4 shows the volumetric heat capacity.

Table 4.　Effects of different additive amounts of graphite on the thermal properties of cement.

Number	Graphite/ cement (wt%)	Thermal conductivity coefficient/ $[W \cdot (m \cdot K)^{-1}]$	Thermal diffusion coefficient/ $(\times 10^{-7}\,m^{-2} \cdot s^{-1})$	Volumetric heat capacity/ $[MJ \cdot (m^3 \cdot K)^{-1}]$
S0	0	0.9325	5.1261	173.52
S1	4	1.1525	6.7979	169.54
S2	8	1.4300	8.6268	165.71
S3	12	1.8937	11.7477	161.20
S4	16	2.3292	15.2139	153.10
S5	20	2.5114	17.0866	146.98

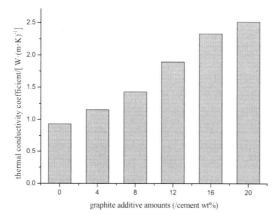

Figure 3.　Effects of different additive amounts of graphite on the thermal conductivity coefficient.

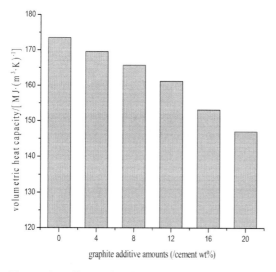

Figure 4.　Effects of different additive amounts of graphite on the volumetric heat capacity.

It can be seen from Figure 3 that with the increase in graphite additive amounts, the samples' thermal conductivity coefficient increased gradually. When the graphite additive amount was 4 wt%, 8 wt%, 12 wt%, 16 wt%, and 20 wt%, the samples' thermal conductivity coefficient, respectively, increased by 25.59%, 53.35%, 102.68%, 149.78%, and 169.32% when compared with the blank samples. It can be seen that with the increase in graphite additive amounts, the thermal conductivity coefficient first increased and then reduced. From the analysis on the microstructure, because at a low temperature, most of the inorganic nonmetallic materials' electrons were bound, and could not become the carrier of thermal conductivity, heat transfer was found to rely on the lattice vibration. According to the quantum theory, the energy of lattice vibration is quantized, usually the quanta of lattice vibration are called phonons (Huang Kun and Han Ruqi, 1998). Each carbon atom in the graphite forms only three covalent bonds with the surrounding three carbon atoms, the remaining one electron is not bound and constrained, and can move freely with respect to the free electron in the metal. So, in addition to the phonon heat transfer, the inter-action and collision between electrons is also the main way that graphite conducts heat (Castro Neto. A. H et al., 2009). From the analysis on the macrostructure, the cement's thermal conductivity coefficient was about 0.93 $W \cdot (m \cdot K)^{-1}$ far less than graphite (about 135 $W \cdot (m \cdot K)^{-1}$), so the addition of graphite changed the material's composition and greatly increased the material's thermal conductivity coefficient.

In addition, when the graphite additive amount was low, the graphite particles existed separately in the cement matrix and could not form good thermal conductive loops. At this time, the material's thermal conductivity coefficient increased insignificantly. With the increase in graphite additive amounts, the chances of forming thermal conductive loops increased gradually and the material's thermal conductivity coefficient increased significantly. It can be seen from formula

$$\lambda = \frac{1}{3} \cdot C_v \cdot \overline{v} \cdot l \qquad (1)$$

that the main effect factor of the thermal conductivity coefficient is the phonon mean free path. The size of the phonon mean free path is mostly decided by the scattering caused by the collision between phonons and the scattering caused by interaction between phonons and grain boundaries, defects, and impurities. Under the condition of low temperature, the short wavenumber affecting the interaction between phonons reduced sharply. The scattering caused by the interaction between phonons weakened rapidly. Now, the size of the phonon mean free path is mainly decided by the scattering caused by the interaction between phonons and grain boundaries, defects, and impurities (Hu Peng and Chen Zeshao, 2009). So, when the graphite additive amount was large, the massive material defects such as holes and cracks shortened the phonon mean free path and the increase in the thermal conductivity coefficient reduced.

Material heat capacity has two main representation modes, namely volumetric heat capacity and mass heat capacity, because the material has many defects such as holes and cracks, so there is a more practical significance to choose the volumetric heat capacity in this experiment. Volumetric heat capacity refers to the heat absorption by rising per 1 k or the heat release by reducing per 1 k. Its main effect factors are material composition, internal porosity and bulk density. It can be seen from Figure 4 that with the increase in graphite additive amounts, the samples' volumetric heat capacity reduced significantly. When the graphite additive amount was 20 wt%, the samples' volumetric heat capacity reduced by 15.29% when compared with the blank samples. This is because the addition of graphite not only changed the material composition but also increased the material internal porosity and reduced the bulk density, which caused the volumetric heat capacity to reduce greatly. The decrease in volumetric heat capacity made the temperature of the material to rise greatly when absorbing the same heat, contributing to the heat gathering and transfer.

3.3 *SEM analysis*

This experiment studied the effects of graphite on the cement stone microstructure of different hydration ages by using the Scanning Electron Microscope (SEM). The SEM photographs of the blank samples and the additive 12 wt% graphite samples of different hydration ages are shown in Figure 5,

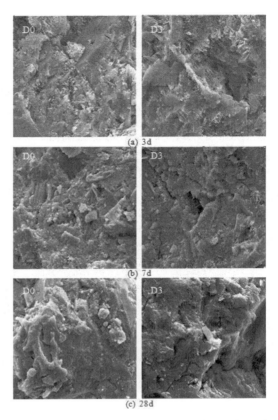

Figure 5. SEM photographs of the blank samples and the additive 12 wt% graphite samples at different hydration ages.

where D0 is the blank sample and D3 is the additive 12 wt% graphite sample.

It can be seen from Figure 5 that in the 3d hydration age, the two sets of samples formed lots of needle bar AFt and part of six-party slab Ca(OH)2, but the amount of AFt in D3 was much greater, the crystals were much thicker and longer, clusters grew in a crisscross pattern and the continuity between the hydration products in D3 was poor, many lattice defects such as holes and cracks existed, and the cement stone structure was very loose. At the 7 d hydration age, the amount of Ca(OH)2 in the two sets of samples greatly reduced and formed part of gels, the cement stone structures were more compact than that at 3 d age, a few holes and cracks existed in D0, but part of the holes and cracks still existed in D3. At the 28d hydration age, the amount of C-S-H gels in the two sets of samples greatly increased and the hydration products were almost all of AFt and C-S-H gels, the cement stone density greatly increased, the lattice defects such as holes and cracks were almost not seen in D0, but a few holes and cracks still existed in D3, so the compressive strength of D3 was significantly lower than that of D0.

4 CONCLUSIONS

1. With the increase in graphite additive amounts, the water requirement of normal consistency increased gradually. The samples' absolutely dry bulk density reduced gradually, and the samples' 3d,7d,28d compressive strength reduced significantly.
2. With the increase in graphite additive amounts, the samples' thermal conductivity coefficient increased gradually and the thermal conductivity coefficient first increased and then reduced, and the samples' volumetric heat capacity reduced significantly.
3. When the graphite additive amount was low, the thermal properties of cement did not increase significantly. When the graphite additive amount was large, the mechanical properties of cement reduced significantly, so the comprehensive analysis showed that the graphite's suitable additive amount was 8%~20%.

ACKNOWLEDGMENTS

This research was supported by the Science and Technology Development Project of Shandong Province (No. 2013GGB01156).

REFERENCES

[1] Castro N.A.H., Guinea F., Peres N.M.R. et al. (2009) The electronic properties of graphene. Reviews of Modern Physics (81): 109–162.

[2] Ferriere A., Chaussavoine C., Leyris J.P. et al. (2003) Numerical simulation of the cooling of a Hot disk rapidly subjected to combined convective and radiant heat losses. International Journal of Heat and Mass Transfer 46(13): 2485–2493.

[3] Huang K. and Han R.Q. (1998) *Solid State Physics.* Higher Education Press, Beijing, China. (in Chinese).

[4] Hu P. and Chen Z.S. (2009) *Calorimetry techniques and thermal physical properties determination.* China Science and Technology University Press, Hefei, China. (in Chinese).

[5] Jan O.D. (1996) Solar energy in building renovation. *Energy and Buildings* 24(1): 39–50.

[6] Jonas N. and Fredrik H. (2012) Concrete vs wood in buildings energy system approach. *Building and Environment* (51): 361–369.

[7] Melissa A.S. and Melissa E. (2009) Technology S-curves in renewable energy alternatives: Analysis and implications for industry and government. Energy Policy (37): 1767–1781.

[8] Parameshwaran R., Kalaiselvam S. et al. (2012) Sustainable thermal energy storage technologies for buildings. *Renewable and Sustainable Energy Reviews* (16): 2394–2433.

[9] Philippe M., Dominique F. and Marie L.L. (2003) Prices versus quantities: choosing policies for promoting the development of renewable energy. *Energy Policy* (31): 799–812.

[10] Wang F., Bennett A.M. et al. (2012) A feasibility study on solar-wall systems for domestic heating—An affordable solution for fuel poverty. Solar Energy (86): 2405–2415.

[11] Xiao J.Z., Huang Y.B. and Ren H.M. (2008) Influence Factors of Thermal Properties of Concrete and Theoretical Analysis. Coal Ash China (5): 17–20. (in Chinese).

[12] Yuan H.W., Lu C.H. et al. (2012) Mechanical and thermal properties of cement composite graphite for solar thermal storage materials. Solar Energy (86): 3227–3233.

Material Science and Engineering – Chen (Ed.)
© 2016 Taylor & Francis Group, London, ISBN 978-1-138-02936-1

Research on the preparation and properties of foamed cement with metakaolin

N. Li, Q. Wang & P. Song
School of Materials Science and Engineering, University of Jinan, Jinan, China

ABSTRACT: Incorporation of metakaolin gives useful enhancements to the cement properties. This paper obtained foamed cement with metakaolin by using metakaolin, replacing part of cement, as binding material, and adding admixture and foam, which possess a more excellent performance compared with foamed cement. We studied the effects of different replacements of cement with metakaolin on cement strength and preparation technology on dry density, strength and water absorption by using the orthogonal test. The results indicate that the optimum substitution by metakaolin was found to be 10 wt%; the optimized combination was 0.8 water/cement ratio, 0.35% foaming agents and 0.15% foam stabilizer, namely G1. The dry density, bending and compressive strength, water absorption and thermal conductivity were 479 kg/m^3, 0.56 and 1.72 MPa, 61.2% and 0.077 W/(m·K), respectively. More C4AH13 existed in G1 as characterized by the XRD analysis. Holes were spherical and regular in G1 as revealed by the SEM analysis.

Keywords: foamed cement; metakaolin; dry density; strength; water absorption

1 INTRODUCTION

Over the past century, cement-based concrete with its versatility and generally highly reliable performance has become the highest volume manufactured product worldwide. Meanwhile, it leads to a high sector-wide consumption of raw materials [1], energy [2–3], emission of CO_2 [4], and dust pollution. With the policies of saving energy and reducing emissions, building insulation materials has been developing rapidly. At present, building insulation materials on the market are divided into two groups: one is the organic insulation materials, mainly XPS and EPS, with the characteristics of low self-weight and thermal insulation, but their durability and fireproof ability are poor. The other is the inorganic insulation materials, mainly expansion perlite and aerated concrete, which also possess the excellent abilities of durability and fireproof except for low self-weight and thermal insulation.

Foamed cement is a cement-based lightweight porous material, which consists of binding material with a high degree of void space. It is one kind of inorganic insulation material. Although there are lots of advantages, there exist defects such as lower strength and higher thermal conductivity. The effective way to improve the performance of foamed cement is to add pozzolans into cement.

According to Awal and Hussin [5], adding pozzolans into concrete can enhance the concrete properties in addition to its economic advantages. Metakaolin with stable components, extensive sources and high pozzolanic activity, as a pozzolanic material in cement, has been paid much attention by researchers since the early 1980s [6–9]. Metakaolin is prepared with kaolin by heat-treated technique. It contains a great deal of amorphous SiO_2 and Al_2O_3, which react with $Ca(OH)_2$ that is produced during cement hydration, producing additional products (secondary C-S-H gel, AFt and C_4AH_{13}) that can improve the performance of cement [10–14]. It has been reported that the replacement of cement with 5–20% of metakaolin increases the compressive strength for concrete and mortar at 28 days [15]. Kostuch et al. [16] reported that $Ca(OH)_2$ was significantly reduced over time at all replacement levels (0, 10, and 20%), and fully removed for 20% replacement in concrete at 28 days. The degree of the pozzolanic reaction of metakaolin was higher by 5% than by 10% and 20%. This is because a lower replacement level can be attributed to the higher concentration of $Ca(OH)_2$ [17]. Frías and Cabrera [18] determined the degree of hydration of metakaolin and cement. There are two hydration mechanisms that influence the $Ca(OH)_2$ content: increase in $Ca(OH)_2$ amounts in the hydration of cement, and decrease in values in the pozzolanic reaction of metakaolin.

The objective of this investigation was to improve the matrix strength of foamed cement through a replacement of cement by metakaolin. On this basis, foamed cement with metakaolin was prepared by adding admixture and foam, which possess a more excellent performance compared with foamed cement.

Table 1. Chemical compositions of cement (P.O 42.5).

Compound	SiO$_2$	Al$_2$O$_3$	Fe$_2$O$_3$	CaO	MgO	f-CaO
Content (%)	21.54	4.81	3.20	64.7	2.81	0.86

Table 2. Chemical analysis of kaolin (k) and metakaolin (mk).

Composition	SiO$_2$	Al$_2$O$_3$	Fe$_2$O$_3$	CaO	MgO	SO$_3$	K$_2$O	Na$_2$O	TiO$_2$	P$_2$O$_5$
k (%w/w)	44.95	37.80	0.38	0.23	0.24	0.72	0.15	0.25	1.04	0.15
mk (%w/w)	52.17	43.98	0.46	0.33	0.30	0.41	0.19	0.31	1.21	0.17

2 EXPERIMENTAL WORK

2.1 Materials

Ordinary Portland cement (P.O 42.5) used in this study was produced by Sunnsy Group. The chemical compositions are given in Table 1. Metakaolin was prepared by firing kaolin at 800°C for 2 h, supplied from Lingshou County, Hebei Province. The chemical analysis of kaolin (k) and metakaolin (mk) is given in Table 2. The phase composition of metakaolin was studied by means of X-Ray Diffraction (XRD), and the XRD pattern shown in Figure 1 indicates the presence of mainly amorphous alumina and silica in metakaolin. Foaming agents were prepared by dodecyl sodium sulfate and saponin in accordance with a proportion of 1:1. Both of them were supplied from Shanghai China. The foam stabilizer was Hydroxypropyl Methylcellulose (HPMC); the coagulant was calcium chloride Anhydrous (AR).

2.2 Formula design

During the foamed cement with the metakaolin process, some parameters were taken as the factors of the orthogonal experiment, i.e. water/cement ratio, foaming agents and foam stabilizer. In the present investigation, the levels of factors were restricted in order to obtain the specimen with optimal performance. The selected process parameters and their levels in the experiment are listed in Table 3.

2.3 Preparation process of specimens

Cement and metakaolin were mixed with water for four minutes. When the slurry was homogeneous, foam was obtained by stirring the foam stabilizer solution mixed with foaming agents. Mixing of the fresh paste was continuously done for 3 minutes and then molded into 40 mm × 40 mm × 160 mm molds. The paste samples were first cured in a 50 °C oven for up to 24 h, then demolded and dried at the temperature of 60°C for 24 h, and then at 80°C

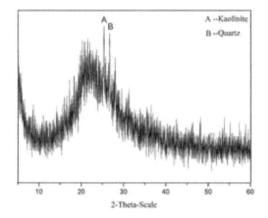

Figure 1. XRD pattern of metakaolin.

for 24 h and at 100°C until the weight was constant in the electrothermal blowing dry box.

2.4 Test methods

Density test: a group of three specimens was dried at the temperature of 60°C for 24 h, then at 80°C for 24 h and at 100°C until the weight was constant in the electrothermal blowing dry box. The weight-to-volume ratio was density. Strengths test: after the density test, strength was tested immediately at the loading rate of 10 mm/min. Water absorption test: it was conducted in accordance with GB/T 11970–1997. It was usually measured by drying the specimen until the weight was constant, and immersing it in water and measuring the increase in weight as a percentage of dry weight. Thermal conductivity test: it was conducted in accordance with GB/T 10294–2008. The mineral phases, morphology and microstructure of the hardened neat foamed cement with metakaolin paste were studied using X-Ray Diffraction (XRD) and Scanning Electron Microscopy (SEM).

3 RESULTS AND DISCUSSION

3.1 Effects of different replacements of cement with metakaolin on cement strengths

As can be seen from Figure 2, the strength of the sample first increased and then decreased with the increased replacement of cement with metakaolin. The optimum substitution of cement with metakaolin was found to be 10 wt%. The compressive strengths of 3 d, 7 d and 28 d were 6.2%, 28.8% and 50.1%, being higher than the blank sample. The above results indicated that metakaolin contained a great deal of amorphous SiO_2 and Al_2O_3, which reacted with $Ca(OH)_2$ that was produced during cement hydration, producing additional products (secondary C-S-H gel and C_4AH_{13}), which improved the strengths of cement, on the one hand, and promoted the hydration of cement on the other hand. However, the amount of cement reduced when the replacement with metakaolin was too high. $Ca(OH)_2$ produced during cement hydration was insufficient, which attenuated the pozzolanic reaction and the cement hydration rate.

3.2 Range analysis

The results of the range analysis of foamed cement with metakaolin are shown in Figure 3, which indicated that the effect of the water/cement ratio was greater than the other two factors, the secondary material was foaming agents and foam stabilizer, respectively, for the bending strength/dry density. The effect of the water/cement ratio was greater than the other two factors, and the secondary material was foam stabilizer and foaming agents, respectively, for the compressive strength/dry density. The effect of foaming agents was greater than the other two factors. The secondary material was foam stabilizer

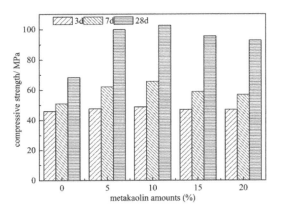

Figure 2. Effects of different replacements of cement with metakaolin on cement strength.

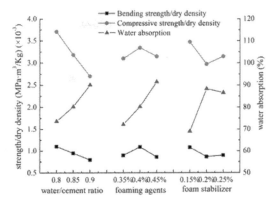

Figure 3. Range analysis of foamed cement with metakaolin.

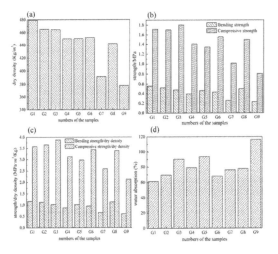

Figure 4. The physical properties of foamed cement with metakaolin: dry density (a), strengths (b), strength/dry density (c) and water absorption (d).

and water/cement ratio, respectively, for water absorption. The requirements of new wall materials are lightweight, high strength and low water absorption. The key is to find a balance point. Through the range analysis (Figure 2), the optimal combination was 0.8 water/cement ratio, 0.15% foam stabilizer, and 0.35% foaming agents, namely G1.

3.3 Effect of preparation technology on dry density, strengths and water absorption of foamed cement with metakaolin

The physical properties of the samples are shown in Figure 4. The strength variations are commensurate with changes in the dry density of samples, with lower dry density samples exhibiting lower strengths, which is related to porosity. There are many formulas to describe the relationship between strength and porosity (formulas (1)-(4)) [19–20]:

$$f_c = f_{c,0}(1-p)^n \quad \text{(Balshin)} \tag{1}$$

$$f_c = f_{c,0}\,e^{-k_r p} \quad \text{(Ryshkeviteh)} \tag{2}$$

$$f_c = k_s \ln\left(\frac{p_0}{p}\right) \quad \text{(Sehiller)} \tag{3}$$

$$f_c = f_{c,0} - k_H p \quad \text{(Hassclmann)} \tag{4}$$

where

f_c—compressive strength of concrete with the porosity of p;
$f_{c,0}$—compressive strength of concrete with the porosity of 0;
n—coefficient, but it is not necessarily constant;
p_0—porosity when the compressive strength is 0;
k_r, k_s and k_H are empirical constants.

There are two mechanisms that influence water absorption:capillary permeability and interconnected pore permeability [21]. Capillary pores are generated in the early stage of cement hydration. Liquid film produces uneven diffusion under the double action of gravity and surface tension and slurry extrusion, leading to incomplete holes. Bleeding paths are generated by bleeding when the water/cement ratio is large. Both of them result in an interconnected pore formation. Obviously, the water absorption variations are in contrast to the changes in the dry density of samples, with lower dry density samples exhibiting higher water absorption. The main reason is that the porosity of the samples rises with the decrease in dry density; meanwhile, the wrapping layer of the liquid film becomes thinner. It is indicated that the probability of interconnected pore formation greatly improves during the hydration of cement. In addition, foams extract large amounts of water during the hydration of cement, resulting in the increase of the water/cement ratio to a certain extent, which makes capillary permeability to relatively increase, thus increasing water absorption.

The bending strength and the bending strength/dry density are highest and the compressive strength is second only to G3, although the dry density of G1 is maximum. Besides, water absorption is minimum; the dry density of G9 is minimum, but its strength and strength/dry density are smallest, and water absorption is maximum. Through the comprehensive analysis, G1 conforms to the characteristics of low weight, high strength and low water absorption.

3.4 *Analysis and discussion of the levels of factors*

As it can be seen from Figure 3 and Figure 4, the influence of the water/cement ratio on the strength/density and water absorption is due to an increase in the porosity. The strength is reduced further compared with the dry density. The strength/density of the samples first increases and then decreases, and water absorption increases with the increase in foaming agents. This is because an increase in the foam volume is observed as foaming agents increase. The porosity increases, but the decline range of the strength and dry density is not the same; the strength/density of the samples first decreases and then increases, and water absorption first increases and then decreases with the increase in foam stabilizer. This is because the stability of foam increases and the breakage of foam reduces. The porosity increases and the strength is reduced further compared with the dry density. However, when the content of the foam stabilizer continues to increase, it hinders the development of foam and reduces the foam volume. The porosity decreases, and the strength increases more than the dry density.

4 MICROSTRUCTURE ANALYSIS

4.1 *Mineral composition*

As it can be seen from Figure 5, the hydration products in the samples are mainly composed of C_4AH_{13} and $Ca(OH)_2$ crystals. The diffraction peak of the C-S-H gel is not obvious because it is a flocculent gel rather than a crystalline substance. The absence of new diffraction peaks indicates that metakaolin does not have a significant influence on the composition of hydration products of cement.

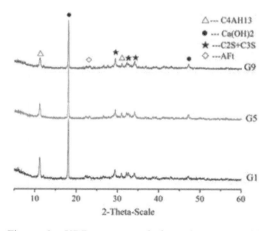

Figure 5. XRD pattern of foamed cement with metakaolin.

The highest diffraction peak intensity of C_4AH_{13} and $Ca(OH)_2$ crystals appearing in G1 indicates that metakaolin reacts with $Ca(OH)_2$, producing more C_4AH_{13}, on the one hand, and promoting the hydration of C_3S and C_2S, and increasing the nucleation and precipitation of $Ca(OH)_2$ on the other hand. On the contrary, the lowest diffraction peak intensity of C_4AH_{13} and $Ca(OH)_2$ crystals appearing in G9 indicates that the pozzolanic reaction is not sufficient.

4.2 Morphological and microstructural analysis

As it can be seen from Figure 6 (a) and (c), the holes in G1 are spherical and regular. Most holes closed independently can reduce the stress concentration and improve the strength of the samples. In contrast, the holes in G9 are flat and irregular. The formation of interconnected pores make the strength of the samples worse. Comparatively speaking, many flocculent products are generated in G1. $Ca(OH)_2$ crystals and other hydration products' stagger is grown more closely, which forms the stable structure of foamed cement with metakaolin (Figure 6 (b)). Many $Ca(OH)_2$ crystals exhibiting a hexagonal morphology are generated in G9. $Ca(OH)_2$ crystals show increased orientation (Figure 6 (d)). The spectrum analysis of G1 is shown in Figure 7. The high contents of C, O, Al and Si indicate that the product is a mixture of HPMC, C-S-H gel, metakaolin and cement. HPMC molecules forms the membrane and reticular structure through entangling with each other and wrapping cement, metakaolin and water through the adsorption of hydroxyl groups on the macromolecular chains in the hydration process, which guaranteed the stable structure of foamed cement with metakaolin.

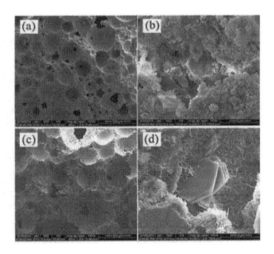

Figure 6. SEM analysis of G1 ((a) and (b)) and G9 ((c) and (d)).

5 CONCLUSIONS

1. The strength of the samples first increased and then decreased with the increased replacement of cement with metakaolin. The optimum substitution of cement by metakaolin was found to be 10 wt%.
2. The orthogonal test results show that the effect of the water/cement ratio on strength/dry density was more dominant than the other two factors. The effect of foaming agents on water absorption was more dominant than the other two factors.
3. Confirmation experiments were conducted under optimal conditions: the dry density, bending and compressive strengths, water absorption and thermal conductivity were 479 kg/m³, 0.56 and 1.72 MPa, 61.2% and 0.077 W/(m·K), respectively.

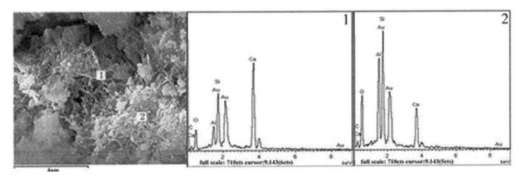

Figure 7. Spectral analysis of G1.

4. Amorphous SiO_2 and Al_2O_3 in metakaolin reacted with $Ca(OH)_2$ that was produced during cement hydration, producing more C_4AH_{13}, on the one hand, and promoting the hydration of C_3S and C_2S, increasing the nucleation and precipitation of $Ca(OH)_2$, on the other hand. The holes that were spherical and regular and closed independently can reduce the stress concentration and improve the strength of the samples.

ACKNOWLEDGMENTS

This project was supported by the Science and Technology Development Plan of Shandong Province [2013GGB01156].

REFERENCES

[1] UNIDO. 2006. Cement production in Vertical Shaft kilns in China- status and opportunities for improvement. *A report to the United Nations Industrial Development Organization.*

[2] Schneider M., Romer M., Tschudin M., Bolio H. 2011. Sustainable cement production-present and future. *Cem Concr Res* 41(7): 642–50.

[3] Damtoft J.S., Lukasik J., Herfort, Sorrentino D., Gartner E.M. 2008. Sustainable development and climate change initiatives. *Cem Concr Res* 38: 115–27.

[4] Olivier J.G.K., Janssens-Maenhout G., Peters J.A.H.W. 2012. Trends in global CO_2 emissions; 2012 report. *PBL Netherlands Environmental Assessment Agency.* The Hague, Netherlands.

[5] Awal A.S.M., Hussin M.W. 1997. Some aspects of durability performances of concrete incorporating palm oil fuel ash. *In: Proceeding of 5th International conference on structural failure, durability and retrofitting, Singapore* 210–7.

[6] Siddique R., Klaus J. 2009. Influence of metakaolin on the properties of mortar and concrete: a review. *Appl Clay Sci* 43: 392–400.

[7] Güneyisi E., Gesoglu M., Mermerdas K. 2008. Improving strength, drying shrinkage, and pore structure of concrete using metakaolin. *Mater Struct* 41: 937–49.

[8] Sabir B.B., Wild S., Bai J. 2001. Metakaolin and calcined clays as pozzolans for concrete: a review. *Cem Concr Compos* 23: 441–54.

[9] Güneyisi E., Gesoglu M., Karaoglu S., Mermerdas K. 2012. Strength, permeability and shrinkage cracking of silica fume and metakaolin concretes. *Constr Build Mater* 34: 120–30.

[10] Ding J.T., Li Z. 2002. Effects of metakaolin and silica fume on properties of concrete. *ACI Mater J* 99(4): 393–8.

[11] Badogiannis E., Papadakis V.G., Chaniotakis E., Tsivilis S. 2004. Exploitation of poor Greek kaolins: strength development of metakaolin concrete and evaluation by means of k-value. *Cem Concr Res* 34: 1035–41.

[12] Siddique R., Klaus J. 2009. Influence of metakaolin on the properties of mortar and concrete: a review. *Appl Clay Sci* 43: 392–400.

[13] Badogiannis E., Kakali G., Dimopoulou G., Chaniotakis E., Tsivilis S. 2005. Metakaolin as a main cement constituent. Exploitation of poor Greek kaolins. *Cem Concr Compos* 27: 197–203.

[14] Badogiannis E., Tsivilis S. 2009. Exploitation of poor Greek kaolins: durability of metakaolin concrete. *Cem Concr Compos* 31: 128–33.

[15] F. Curcio, B.A. Deangelis, S. Pagliolico. 1998. Metakaolin as a pozzolanic microfiller for high-performance mortars, *Cem Concr Res* 28 (6): 803–809.

[16] Kostuch, J.A., Walters, G.V., Jones, T.R., 1993. High Performance Concrete Incorporating Metakaolin-a review. *Concrete 2000.* University of Dundee, pp. 1799–1811.

[17] Poon, C.S., Lam, L., Kou, S.C., Wong, Y.L., Wong, R., 2001. Rate of pozzolanic reaction of metakaolin in high-performance cement pastes. *Cem Concr Res* 31: 1301–1306.

[18] Frías, M., Cabrera, J. 2000. Pore size distribution and degree of hydration of MK-cement pastes. *Cem Concr Res* 30: 561–569.

[19] M. Roler, I. Oder. 1985. Investigations on the relationship between porosity, structure and strength of hydrated Portland cement Pastes: 1. Effect of Porosity. *Cem Concr Res* 15: 320–330.

[20] G. Fagerlund. 1973. Strength and porosity of concrete, Proc. Int. RILEM SymP. *Pore Structure,* Prague 51–73 (Part2).

[21] He B., Huang H.K., Yang J.J. 2007. Investigation on the water absorption of foamed concrete. *New wall materials* 12: 24–26. [Chinese].

Material Science and Engineering – Chen (Ed.)
© 2016 Taylor & Francis Group, London, ISBN 978-1-138-02936-1

Deformation tracking method of asphalt mixture particles based on cross-calibration

H. Ying, Y. Liu, H. Chen, Q. Wu, X.X. Li & X.Y. Lu
College of Architecture and Traffic Engineering, Guilin University of Electronic Technology, Guilin, China

ABSTRACT: In order to detect the deformation of particles within the asphalt mixture, a method for deformation tracking of particles based on cross-calibration was proposed. This method can realize the deformation calculation at any point within the particle by tracking the displacements of the target center, which printed on the particle before and after deformation as well as the corner of the long axis. In the process of capturing the target, the noise jamming, such as the target color fading, abrasion, stains and grain boundary, can be eliminated by using a rectangular-ambulatory-plane gradient and the gradient histogram quantile method combined with the aggregate boundary mask. Then, the target image noise suppression can be further enhanced by using the Hough transform, segment splicing threshold and segment filtering suppression. Finally, the geometric characteristics of the target will be used to complete target identification, location and parameter calculation. The result shows that this method can realize the deformation tracking of the particles, which provides an effective method to analyze the motion of the internal particles of the asphalt mixture under the action of load and the link between macroscopic deformation and microstructural morphological parameter.

Keywords: road engineering; asphalt mixture; deform tracking; cross-calibration; digital image technology

1 INTRODUCTION

From the mesoscopic structure of the asphalt mixture, excavated mesoscopic characteristics of materials find the key factors that influence the properties of materials, such as aggregate shape, texture, and pore [1–4], and have pertinence to improve the research ideas of the mixture performance that has gradually attracted the attention of the academic circle, which presents a new trend of the hybrid material design. Gradation has a significant effect on the rutting macroscopic morphology [5]. In the final analysis, it is due to the different types of mixtures caused by the mesoscopic deformation form different under the vehicle load, considering the smaller skeleton type mixture rut deformation, especially distance loading area more distant locations with little deformation, so it is very necessary for the research particle deformation detection method of high precision understanding of gradation to the influences of rut deformation.

Many researchers have performed studies on tracking technology to explore the mixture particle swarm deformation, such as Huang Longsheng [6], Wu Wenliang [7], and Ying Hong [8] who used the template match method, to the board of rut specimen section aggregate particles are pattern recognition in loading before, after, through the position correlation, get deformation of aggregate particles swarm. However, in the course of loading, especially in the nearby of the loading area, asphalt is extrusion from the gap of the aggregate particles, to the surface of the aggregate cause pollution, this situation to template matching bring greatly distressed; and when the particles rotate after, use the template matching rules is incapable of action. Qiao Yingjuan [9] captured the corner of block target that affixed to the test surface, in order to detect the deformation of particles, but the method of the target feature is too simple, and relies only on the four corner features of diamond, in the interference consists of a large corner of aggregate boundaries, the target easily submerged in noise, thus resulting greatly reduced in automatic recognition of degree. This paper presents the cross of target printed on the aggregate surface, by tracking the displacement, rotation of the target realization deformation calculation of any point on particle, cross of target has obviously linear, vertical intersection, size ratio geometric features, both for the image recognition and conducive to the deformation calculation, in order to provide an effective way to solve the mixture internal granular swarm deformation tracking problem.

2 TRACKING METHOD OF PARTICLE SWARM DEFORMATION

Cross of target by two orthogonal line segments constituted the cross, longer line is the long axis, shorter line for the short axis, the longest radius as long radius, intersection is the target center, the definition long radius left the negative x axis counterclockwise rotation angle is the target angle. The cross-target makes a stamp, the use of photosensitive ink is printed on the aggregate particle surface, particle happen displacement or rotation, by tracking displacement of the target center and the corner of the target can be calculated displacement of any point on aggregate. As shown in Figure 1, the center point of target (xt,yt), after deformation coordinates for (xt′,yt′), initial angle of target is a, after the deformation angle for a′, any point of the aggregates (x,y), after the deformed coordinate (x′,y′) for:

$$x' = x_t' + (x - x_t)\cos(a' - a) - (y - y_t)\sin(a' - a)$$
$$y' = y_t' + (x - x_t)\sin(a' - a) + (y - y_t)\cos(a' - a) \quad (1)$$

The use of crosses of the target to mixture particle swarm conduct deformation tracking, first, the mixture is cut, ground, make the section aggregate particles surface as far as possible smooth, flat, then the target is printed on the surface of the aggregate particles, air dry after acquisition initial position image of particle swarm; and then the test piece is assembled, loaded, acquisition image of particles group in after loaded; by capturing the particle swarm target in two images, obtain the target center point and the target angle of each target, can according to the above formula aggregate calculated displacement of each point, making the displacement map of particle swarm. The main difficulty in the cross of target recognition is binarization of the target and positioning of the target.

3 BINARIZATION METHOD OF CROSSES OF THE TARGET

In the image of the mixture section, the noise has three main types in aggregate particles and nearby regions:

1. Wear, fade: target printed on the aggregate particle surface, liquid ink through osmosis, air drying, the solidification on the particle surface, as a result of the aggregate absorbent, bright and clean degree of the surface is different, the target color is not uniform, the part of the target

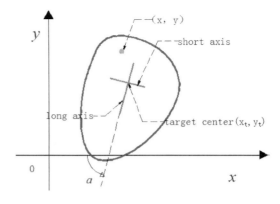

a . Before deformation

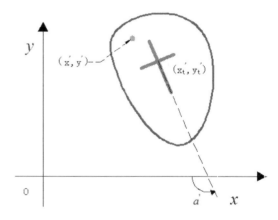

b . After deformation

Figure 1.   Deformation tracking principle.

color dim, in addition, printed with the target of mixture section in the loading process, the particles surface of the friction is also easy to cause target color fuzzy;

2. Asphalt pollution: in the process of loading, asphalt has been pushed out, leading to the pollution of the aggregate surface;

3. Particle boundary and colloidal interference: general particle boundary area color changed drastically, the so-called strong edge area, some particles because of the relationship of the shape, there are large areas of asphalt mortar within the external rectangular of the particles, so some rely on the algorithm of the gray-scale statistical characteristics that caused a serious interference.

Figure 2 shows the three main types of noise of the illustrations, and respectively with classical threshold segmentation method of Otsu, edge

a. Wear, fade

b. Asphalt pollution

c. Particle boundary and colloidal interference

Figure 2. Particle external rectangle region mainly noise types.

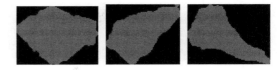

Figure 3. Particles' area map.

cles' area map of the three types of interference image.

2. Reference edge detection method, using back to shape template calculation the gray gradient of image, set the outer circle radius of back to shape for N, the inner circle of radius is M, then (i,j) of back to the shape gray gradient $G(i,j)$ for:

$$G(i,j) = \frac{1}{(2N+1)^2} \sum_{x=i-N}^{x=i+N} \sum_{y=j-N}^{y=j+N} g(x,y)$$

$$- \frac{1}{(2M+1)^2} \sum_{x=i-M}^{x=i+M} \sum_{y=j-M}^{y=j+M} g(x,y) \quad (2)$$

In the formula, g(x,y) for image (x,y) gray level.

3. Use the back to shape template to scan the whole image, computing the whole image back to the shape gradient histogram. The histogram indicates the probability density distribution of image gray gradient, which is defined as follows:

$$h(r_k) = n_k \quad k = 0, 1, 2, \dots L-1 \quad (3)$$

In the formula, $h(rk)$ is the gradient histogram of the gray level gradient grade in [0, L-1]; rk is the K gray level gradient.

nk is the number of pixels that the gray level gradient is rk in the image, usually by the normalized histogram represented as follows:

$$P(r_k) = \frac{n_k}{n} \quad k = 0, 1, 2, \dots L-1 \quad (4)$$

In the formula, n is the total number of pixels in the image.

4. By gray gradient histogram calculation of the cumulative histogram of gray gradient $D(rk)$, the graph is defined as follows:

$$D(r_k) = 1 - \sum_{m=0}^{m=k} P(r_m) \quad k = 0, 1, 2, \dots L-1;$$
$$m = 0, 1, 2, \dots L-1 \quad (5)$$

In the formula, $D(rk)$ is the pixel scale that gray gradient greater than rk in the gray gradient histogram.

5. The calculation complete of ratio γi of cross-target and the particle image, whose size of

detection method Sobel edge processing, got their respective threshold image and edge image, dim targets in a figure, in the threshold graph, target to blend together with the texture of aggregate, in the edge graph of the target boundary is not obvious; b in figure targets are polluted by noise, the threshold image and edge image which are asphalt stains pollution; c figure due to the glue sand area is complex, plastic sand and stone has a clear bimodal in the gray-level histogram, therefore targets disappeared in the threshold figure, be mistaken for stone, and the edges in the graph, there was too much the interference of other particles edge, which brings great difficulties to the subsequent straight line detection.

To realize the binarization of the cross-target, overcome above the noise interference, the design algorithm is as follows:

1. Using the multi-stage adaptive boundary segmentation method [10] for the specimen cross-section image into binarization, the completion of the various particle boundary tracking, we calculate the area of each particle, denoted as Ai; and to each trace of the particle conduct number, requires each particle has a unique number, and then we use this number to the particle area pixel conduct assignment and make number map of the particle area. Thus, we make use of external rectangular to a single particle conduct interception, taking in the map and pixel regions of the current particle the same number, so the other particle is excluded, as shown in Figure 3. Figure 2 lists the parti-

477

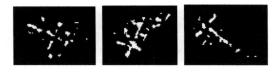

Figure 4.   Cross-target image of binarization.

cross-target has been fixed, as the known quantity, denoted as Across, then $\gamma i = Across/Ai$, which the Ai boundary tracking complete the calculation in step (1).

To draw lessons from the classical p points a gray-level threshold segmentation method of thought, find the segmentation threshold of Ti in gray gradient accumulative histogram, Ti is just to make the $D(rk)$ more than γi back to the shape of the gray gradient magnitude, using of Ti conduct segmentation in the back to the shape of the gradient map, get the target binary diagram, as shown in Figure 4.

4   TARGET POSITIONING

To particle images conduct binarization after, found pattern of cross-target is intermittent of several lines segments in the binary image, and with noise segments are mixed together, it is still difficult to achieve the extraction of the target, therefore, in this paper, using the Hough transform idea design, a stitching algorithm of image line segment, realizations two intersecting line segment detection in the cross-target.

The traditional Hough transformed to a point (x, y) in Cartesian coordinates transformation for a curve $\rho = x\cos\theta + y\sin\theta$ in ρ-θ the space, on a straight line of a series points (xi, yi) in Cartesian coordinates transformation to the ρ-θ space for a series intersections of curves $\rho = x\cos\theta + y\sin\theta$. Therefore, Hough transformation is widely used in line detection in image processing, but not for line segment detection, in order to overcome this shortcoming, and able to connect the target segment after binarization, filter out noise segment, the design improved the algorithm as follows:

1. Using the traditional Hough transform will wait for inspection point in the images the transform for a series of curves of ρ-θ space, as shown in Figure 5 a. In the image, curve in Table 1 for family of curves of various points in ρ-θ space, the curve in the graph has many intersection points, easy to set up filter conditions, will be more than three common intersection points curve find out, such as curve A, B, C, D common intersect to X1 point, curve R, M, N, P, Q common intersect at point

X2, by the intersection point coordinate can obtain two lines $\rho X1 = x\cos\theta X1 + y\sin\theta X1$ and $\rho X2 = x\cos\theta X2 + y\sin\theta X2$.

2. Each wait for the test point of the image into two line equations, easy to judge A, B, C, D four points in the line X1, R, P, Q, M, N five points in line X2, as shown in Figure 5b; then to each point on the straight line conduct ordering and distance calculation, get the line AC and DE on line X1, line MN, RQ on line X2, and can get each line spacing dCD and dMQ on the same line.

3. Set the line segment splicing threshold x, when the distance of each line segment on the line between less than this value, considering the line segment is close enough, spacing may be due to noise formed, can be on both ends of the line segment conduct stitching in this distance. As shown in Figure 5c, we assume that Tseg < dCD, Tseg > dMQ, AC and DE segments, which will be stitching into the line AE.

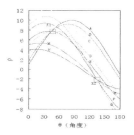

a Hough transform

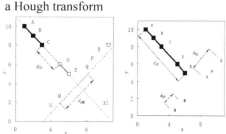

b. Line search   c. Line filter, stitching

Figure 5.   Based on Hough transform of line splicing, filter the graph of principle.

Table 1.   Hough transform wait for the test point coordinate.

Pixel coordinates	A	B	C	D	E	N	M	Q	P	R
x		1	2	3	5	6	4	5	8	9	10
y		10	9	8	6	5	1	2	5	6	7

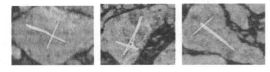

a .The noise segment

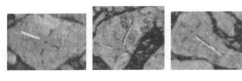

b. The recognition effect

Figure 6.   Cross-target positioning.

4. After splicing of the length of each line segment conduct calculated, as shown in Figure 5c, we obtain three line segment lengths, and then set the line filter threshold Trvd into the smaller segment MN, RQ filter, which reserves the final results AE.

After line segment splicing, the cross-targets relatively show a complete display, but in addition to the target, there is still some noise segment, according to geometric features of the target eliminate excessive lines, according to the following rules to realize the target orientation:

1. The two lines' segment intersection, and the crossing point in two lines' segments internal;
2. The two lines' segment angle of φ is in the range of $90 \pm 10°$;
3. $|\varphi - 90|$ is minimum.

Recognition effects of the target are shown in Figure 6, in which figure a shows post-Hough transformation, line segment splicing, the filtered results, and figure b shows according to the geometric characteristics of targets to complete the target location, targets' cross-center has used circle markers in the figure.

5   PARTICLE SWARM DEFORMATION TRACKING EXPERIMENT

When using cross-target tracking particle swarm deformation, target the smallest size can do that the long axis was 5 mm, the short axis was 3 mm, and this size completely can print on the surface of that the size of particles above 4.75. Although a small amount of the target pattern exceeded the aggregate particle boundaries, it can still be recognized, in order to guarantee the recognition rate, when the target pattern printed, as far as possible will be

printed targets the internal of particle boundary, and away from the boundary is far better. As shown in Figure 6, using rutting specimens, it will cut to print the targets, using a high-resolution scanner collected before the deformation section image a, then it will reload specimens in the test module conduct rut test, loading accomplish after, remove the specimen, acquisition deformation after of the section image b; respectively to capture targets in the figure of a and b, getting the center point and the rotation angle of the each target, according to formula (1) to calculate the displacement of the grid point on the aggregate, production a deformation mapping, as shown in figure C. In order to highlight circumstances of deformation of the particle, the displacement is magnified 5 times, as can be seen from the graph, in the wheel loading region, particles mainly downward movement, and along the depth direction spread to the sides; when this diffusion being restrain by the left and right boundaries of the test mode, the morphology of particle displacement gradually becomes upward uplift, but the deformation of the left and right sides is asymmetric, such as the right side of the particles eminence is large, due to the space distribution of the mixture internal granular and particle movement trend are closely related, the right side sliding line of the particle is the smallest part of motion resistance of particles. From these analyses, it can be seen that the deformation tracking of material inside particles swarm, it is not only internal reflect of the macroscopic deformation of test pieces, and with the mixture internal aggregate of contact form, aggregate shape, and spatial distribution pattern, the micro-parameter has a natural close contact, therefore it has the ability for particle swarm

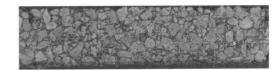

a Cross section image before deformation

b Cross section image after deformation

Figure 7.

deformation as a bridge, and the macro-mechanical properties and mesoscopic structure shape parameters link together, profoundly to reveal the mixture macroscopic behavior and improve the performance of the mixture to provide an effective analysis tool.

6 CONCLUSION

1. This paper presents the use of the mixture cross-section internal particle group deformation tracking method of cross-target, by capturing the cross-target that is printed on the surface of the aggregate particle group, by the displacement of the center point of the target before and after loading and long axis rotation amount calculating the deformation of aggregate internal at any point, drawing deformation map, the shape is simple, easy to calculate, applicable to small particles.
2. Aiming at the target color faded, wear and stains and grain boundary interference, using the back to shape gradient, gradient histogram fractile method, combined with the masking of the grain boundaries design a cross-target binarization methods, the method can better eliminate noise above a few kinds, realize the target of binarization, to lay the foundation for the location of subsequent target.
3. On the basis of binarization in the cross-target, and the traditional Hough transform, by line splicing threshold Tseg and line filter threshold Tseg to the target line conduct enhancement noise suppression. and combining with the geometrical feature of the target, to accomplish the cross-target positioning.

ACKNOWLEDGMENT

This work was supported by the Natural Science Foundation of China under Grant No. 51208130, and the Natural Science Foundation of Guangxi Zhuang Autonomous Region under Grant No. 2013GXNSFBA019258.

REFERENCES

[1] Kutay M Emin, Ozturk Hande I, Abbas Ala R, et al. Comparison of 2D and 3D image-based aggregate morphological indices [J]. International Journal of Pavement Engineering, 2011, 12(4): 421–431.
[2] Eyad Masad and Dana Olcott, Correlation of Imaging Shape Indices of Fine Aggregate with Asphalt Mixture Performance [C], TRB 80th Annual Meeting, Washington, D.C, 2001, Paper No. 01–2132.
[3] Tan Yi-qiu, Chen Guo-ming, Shi kun-lei, Wang Zheren, Asphalt Mixtures performance Research Based On Coarse Aggregate Surface Textures [A], TRB [C], 2006.
[4] Pei Jianzhong, Zhang Jialin, Chang Mingfeng. Influence of Mineral Aggregate Gradation on Air Void Distribution Characteristic of Porous Asphalt Mixture [J]. China Journal of Highway and Transport, 2010, 23(1):1–6.
[5] Li Lihan, Su Zhou, Chen Jianjun. Research on Uplift Coefficient of Asphalt Mix by Loaded Wheel Tracker [J]. Journal of Highway and Transportation Research and Development, 2007, 24(12):42–45.
[6] Huang Longsheng. Asphalt Concrete Macroscopic Rutting and Behavior Mechanism of Microscopic Track of Analysis [D]. Doctoral Dissertation of National Cheng Kung University, 2003.
[7] Wu Wenliang, Zhang Xiaoning, Lizhi. Analysis of Coarse Particle Trajectory in Asphalt Mixture Rutting Test [J]. Journal of South China University of Technology, 2009, 37(11):27–30.
[8] Ying Hong. Study and Application of Digital Image Processing Technique on Asphalt Concrete [D]. Master Dissertation of Chongqing Jiaotong University, 2008.
[9] Qiao Yingjuan. Asphalt mixture displacement field determination and Analysis on flow rutting [D]. Doctoral Dissertation of Dalian University of Technology, 2008.
[10] Huang Qing-ming, Gao Wen, Cai Wen-jian. Thresholding technique with adaptive window selection for uneven lighting image [J]. Pattern Recognition Letters, 2005, 26(1):801–808.

Material Science and Engineering – Chen (Ed.)
© 2016 Taylor & Francis Group, London, ISBN 978-1-138-02936-1

Discussion on the development and application of new environmental protection building materials

A. Wang & M.M. Li

Education School, Jiangxi Science and Technology Normal University, Nanchang, China

ABSTRACT: Building materials, an important part of building, have a close relationship with people's material life and spiritual life. With the development of construction industry, people's demand for building materials has become much higher. Traditional building materials cannot meet the people's needs, but new environmental protection building materials are able to meet the demand, which is the development trend of modern building materials. This paper focuses on the new environmental protection building materials and its application.

Keywords: environmental protection building materials; development; application

1 INTRODUCTION

With the development of science and technology, people's material and cultural life level has grown increasingly, while the deterioration of the ecological environment has also caused problems to human health and life. Therefore, in the use of building materials, such as wall and floor tiles, sanitary ceramics, and glass, people are not only satisfied with their performance, but also pursue conditions that provide comfort and benefits to the physical and mental health, which is the development direction of world building materials. The current building materials, with a single function, only pay attention to the structure and decoration, while neglecting their comprehensive function, especially that some decorative materials will release the gas or ray that are harmful to human health. Therefore, new high technology, such as SOL-GET technology, new ceramic preparation technology, photoelectric catalysis technology, and controlled release technology are adopted, to improve people's living environment. Healthy environmental protection materials have become the urgent task of building materials' scientific and technical workers.

2 THE MEANING OF NEW ENVIRONMENTAL PROTECTION BUILDING MATERIALS

New environmental protection building materials consist of five points. First, high-performance building materials are produced with relatively low resource, energy consumption and environment pollution. For example, first, modern advanced science and technology is used and processed to produce high-quality cement. Second, it can relatively reduce building energy consumption in the process of production; for example, new wall materials are widely used at present, with the characteristics of lightweight, heat insulation, and sound insulation. Third, it involves high use efficiency and excellent performance, which can effectively reduce the consumption of materials, for example, high-performance concrete. Fourth, it can improve people's living environment and is beneficial to people's health, for example, the latest multifunctional glass, ceramic and coatings. Fifth, it can be reused or be made of industrial waste; for example, cement materials can purify sewage and cure toxic and harmful industrial waste slag.

3 THE STATUS QUO OF DEVELOPMENT AND APPLICATION OF NEW ENVIRONMENTAL PROTECTION BUILDING MATERIALS

3.1 *Waste plant fiber*

Waste plant fiber, a kind of renewable biological resource with many uses, mainly refers to the crop straw, waste wood, and bamboo. Compared with other building materials, the processing process of block with the straw plant fiber as raw materials is simple, efficient, and non-environmental pollution, which can be regarded as green environmental protection and energy-saving materials. Therefore, the building made of those materials can also be regarded as 100% ecological building. However, because of the characteristics of the materials itself, it still has some limitation in some building.

3.2 *Gypsum building materials*

Gypsum building materials have the following advantages: 1) the energy consumption of gypsum calcinations is relatively low (only 1/4 of the cement and 1/3 of the lime), so using the gypsum as building materials can greatly save energy; 2) gypsum building materials can save more materials than solid brick and concrete; 3) gypsum building materials can be recycled and they do not produce construction waste; 4) gypsum is non-toxic and harmless and has a good heat resistance and fire resistance.

3.3 *Fly ash*

Fly ash is a kind of industrial waste discharged from the pulverized coal boiler in the thermal power. In 2001, the emission of fly ash was 0.16 billion t in China, and has been constantly increasing every year. Fly ash stacks area and threatens the ecological environment seriously. Thus an effective way of handling fly ash is to apply it in building materials. Fly ash can be used instead of clay to fire hollow bricks, wall and floor tiles, or ceramsite. The pottery production of fly ash is a promising new energy-saving building material.

3.4 *Foam glass*

Foam glass is a new building material with the advantages of environmental protection, insulation, flame retardancy, moisture insulation, and sound absorption. The raw material is a waste of flat or bottle glass fragments, which belongs to the waste utilization. Its production has significant environmental benefits. Foam glass, with the characteristics of lightweight, high compressive strength, low thermal conductivity, good fire resistance, waterproof ability, and high chemical stability, is a kind of building material of thermal insulation and cold insulation. At present, foam glass is widely used in various places, such as roof, wall and ceiling materials, and components of thermal insulation instead of brick and block.

3.5 *Film material*

The composite building membrane material has a good light transmittance, low density, high mechanical strength, durability, fire resistance, anti-UV and other good properties, thus becoming the new environmental friendly and energy-saving building materials. There are a variety of film materials used in building, which can be divided into two catego-

ries according to different materials: the first is the PTFE film material, whose resin content is greater than 90%; the second is the PVC film material.

4 THE APPLICATION OF NEW ENVIRONMENTAL PROTECTION BUILDING MATERIALS

4.1 *The application of new environmental friendly wall materials*

Environmental protection wall material is a kind of new aerated concrete block. The development of new wall materials, conducive to the ecological balance, environmental protection and energy conservation, not only meets the requirement of national industrial policy, but can also improve the function of building. At the same time, it can make full use of local resources, and comprehensively use fly ash and other industrial waste to produce wall materials, so as to speed up the pace of developing new wall materials. For example, the clay can be replaced with the rich fly ash, coal gangue, and slag to produce fly ash brick, clay-sintered, coal gangue-sintered brick, and slag brick.

4.2 *The application of new-style environmental protection wall coverings*

The environmental protection wall coverings include grass, linen wallpaper, yarn silk cloth and other mildew wallpaper materials with the characteristics of moisturizing, health care and insecticide function. The green wall decoration materials, after chemical treatment, have a smooth surface and a good air permeability, so as to effectively solve the problem of mildew, foaming and breeding fungi that the decorating material is prone to in humid air.

4.3 *The application of new-style environmental protection flooring materials*

The environmental protection flooring materials mainly refer to the glass planting brick road. It is one kind of colored porous paving material products and mainly made up of recycled high-density polyethylene. The new environmental friendly building materials, mainly used in public facilities, can not only effectively reduce storm water runoff and prevent pollution of surface water, but also discharge the ground water efficiently.

4.4 *The application of new environmental protection lighting*

The environmental protection lighting is the lighting system saving energy and reducing the pollution on the environment, including the lighting circuit, lighting tools, and lighting switch in the practical application. The entire lighting system should be designed scientifically to create a comfortable and healthy lighting environment through safe, high-quality lighting electrical products.

4.5 *The application of new environmental protection paint*

The environmental protection paint mainly refers to the biological emulsion paint. Not only a variety of colors can be chosen, but also the construction process is relatively simple. If the wall is damaged or soiled, it can also be painted or cleaned by a detergent, so as to reduce a lot of procedures. In addition, the environmental protection paint can also emit fragrance smell. So, it can not only bring colors to people's living, but also effectively prevent fungus growth. Therefore, the environmental-friendly paint is widely used in architectural decoration.

4.6 *The application of new antibacterial materials*

Antibacterial materials mainly fix the antimicrobial component on the building materials through the methods of exchanging, melting, and adsorption, so that it can break down environmental pollutants effectively and play the role of cleaning air. For example, in the irradiation of indoor fluorescent lamp, experimental surface lighting catalytic sanitary ceramics can kill *Escherichia coli*, *Pseudomonas aeruginosa*, and bacteria attaching on the top effectively. If used in the bathroom, it can not only kill bacteria, but also play the role of anti-slip and anti-fouling. So, the new environmental protection materials are generally used in the kitchen and bathrooms, which are prone to produce bacteria, as well as in hospitals and other public places.

5 THE DEVELOPMENT DIRECTION OF NEW ENVIRONMENTAL PROTECTION BUILDING MATERIALS

In our country, building materials have a huge market. The annual construction area is more than 0.8 billion m^2, requiring a large number of building materials and decorative materials. Green building materials are the new requirement of humans over the ancient field of building materials, and are also the inevitable road to the sustainable development of building materials. Although some achievements have been made in the development of building materials, it is still at the initial stage. Compared with the foreign developed countries,

there is still a large gap. Therefore, we must learn from the foreign advanced experience of green building materials and work for the development of building materials.

5.1 *Saving building materials*

It includes resource conservation and energy conservation. Land resources are seriously few, but the traditional building material resource consumption is great, for example, soil resources should be consumed to produce 10 thousand pieces of clay bricks, as well as considerable energy. Therefore, the method that uses new materials to replace traditional materials can save resources and energy effectively.

5.2 *Environmental protection building materials*

The construction industry is not only the major energy consumption, but also the environment pollution producer. In the traditional process, the production of ordinary Portland cement always discharges large amounts of carbon dioxide, sulfur dioxide and dust, causing a very bad influence on the environment. The building materials containing formaldehyde and aromatic compounds not only pollute the environment, but also influence human health and life. Therefore, developing environmental-friendly green building materials is the main development direction in the future.

5.3 *Function building materials*

Green building is to provide a safe and comfortable living space for humans, which should not only optimize the environment, but also be beneficial to human health. To meet these requirements, green building materials must be versatile, beautiful, and durable.

6 CONCLUSION

Briefly, bad building decoration materials pollute the home environment, and the damage caused to the people's health cannot be estimated. So, people should always choose pollution-free, non-radioactive environmental protection materials as decoration materials. In this way, poor building materials will have no market value inevitably. In addition, relevant national laws and regulation should restrict their use, and building materials manufacturers should choose pollution-free, non-radioactive materials and produce green environmental protection decoration materials with clean technology. So new environmental protection building materials are the inevitable trend of future development of building materials products.

REFERENCES

[1] Huang Haiyan. 2011. On the application of environmental protection building materials [J]. city construction theory research (electronic version), 2011(34):112–113.

[2] Yao Lei, Jia Kaiwu, Li Xiaozhi. 2008. Saving society and green building materials [J]. Shanxi building materials, 2008, 34(9):11–12.

[3] Shang Mei, Shao Mingshuang. 2011. Discussion on the sustainable development of residential building promoted by environmental protection building materials [J]. Liaoning building materials, 2011(4):86–87.

[4] Ding Xiaoling. 2008. Energy saving application of ecological building [J]. Building energy conservation, 2008(12):46–48.

[5] Yang Xiaohong. 2008. Study on the performance of new thermal building materials [J]. New building materials, 2008(8):50–52.

[6] Cui Yanqi. 2008. The foreign green building materials and its enlightenment to China [J]. New building materials, 2008(10):37–39.

Material Science and Engineering – Chen (Ed.)
© *2016 Taylor & Francis Group, London, ISBN 978-1-138-02936-1*

A new type of filter wall

H. Wen

Southwest Jiaotong University, Chengdu, Sichuan, China

ABSTRACT: Waste mining residue piled up in the open air can cause soil and groundwater pollution by heavy metal ions. This is because the ions flow into the earth during rainfall and due to the wet air around the ground. So, we design a new type of filter wall with five layers that can be set in the soil at the area where waste residues are piled up, in order to wash the water before it goes into the groundwater system. The layers of the wall are symmetrical, except the one in the middle of the five layers. The two outermost layers are pervious concrete; they are used to support the whole wall. Two layers that are under the concrete are made up of microfiltration membrane, which can prevent the entry of large particulate matter. The very middle layer contains diatomaceous earths and polyaluminum as the flocculent, which can precipitate heavy metal ions.

Keywords: diatomaceous earths; microfiltration membrane; pervious concrete; subtend arrangement; heavy metal ion; Polyaluminium (PAC)

1 INTRODUCTION

1.1 *Overview*

Application background: Waste mining residue piling up in the open air can easily weather and oxidize. After washed and soaked by rainfall, harmful elements are dissolved and flow into the earth, and then into the groundwater system, causing soil and ground water pollution. Curing these kinds of pollution will require a large number of money and energy, so we want to design a device or a system to prevent the pollution from the source.

1.2 *Materials and construction*

The filter wall can be divided into five layers; the layers of the wall are symmetrical, except the one in the middle of the five layers. The two outermost layers are pervious concrete, the inner two are microfiltration membranes, and the middle layer contains diatomaceous earths and polyaluminum as the flocculent. The arrangement is shown in Figure 1.

1.3 *The middle layer: diatomaceous earths with polyaluminum*

Description. The middle part of the wall is designed to provide reaction and adsorption, so we use polyaluminum as the flocculent to convert heavy metal ions into large particles and allow diatomaceous earths to adsorb them [1].

Diatomaceous earth. It is a kind of siliceous sedimentary rock that is mainly composed of opal. As a kind of clay mineral, it has a high adsorption ability, wide sources and low price, so it is suitable for the adsorption of heavy metal ions. The combination of diatomaceous earths and flocculation has a good effect on cleaning water with Cd^{2+}, Pb^{2+}, Cr^{3+}, and Cu^{2+}.

Working principle of diatomaceous earths and the inner packing. The adsorption of the heavy ions of diatomaceous earths lies in its carboxyl. The hydrogen on the carboxyl group can be free, making the surface of the diatomite carry a negative charge. This enhances the diatomaceous earth surface's

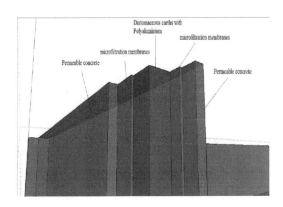

Figure 1. Five layers of the filter wall.

ability to attract positively charged heavy metal ions. On the other hand, silicon carboxyl still can make heavy metal ion complexing adsorption on the surface of the diatomaceous earth, which can be thought of as a form of complexation reaction [2]. In addition, the use of Polyaluminum (PAC) as a flocculating agent can also make the ions flow into the sediment so as to clean the water, making the residual heavy metal ions at or near the standard [3]. The chemical structure of diatomaceous earth is shown in Figure 2.

Diatomaceous earth adsorption compared with the other methods of waste water treatment. Facing the increasingly serious pollution of heavy metals, many methods have been used to remove heavy metal ions in the water. Common ways are the chemical precipitation, membrane filtration method, ion exchange method and activated carbon adsorption method. One of the most effective methods is active the carbon adsorption method. However, its recycling requires high costs and the other traditional craft is also expensive. Besides, the pollutant removal rate of traditional ways is low. So, compared with the traditional process, the wide sources, low cost and no secondary pollution, environment mineral materials, diatomaceous earth has attracted the attention of researchers. Diatomaceous earth has a unique microporous structure. As a typical natural microporous material, it has good adsorption properties. We can not only make an effective use of mineral resources, but also reduce the cost if we use it to deal with waste water; what's more important is that it does no harm to the environment.

1.4   *The outermost layers: permeable concrete*

Description. Permeable concrete is also called the porous concrete, which consists of aggregate, cement and water and becomes a kind of porous lightweight concrete. It does not contain a fine aggregate, and the coarse aggregate coated on the surface of a thin layer of cement can bond with each other to form a cavity with an evenly distributed cellular structure.

Advantages
High water permeability. Waterproof concrete has 15%-15% of the pore, making a permeable rate

of 31–52 l/m/hour, far more than the most heavy rainfall in the most excellent drainage under the configuration of the discharge rate.

High bearing capacity. By the testing authority appraisal, the bearing capacity of the permeable floor is completely capable of achieving the C20-C25 concrete bearing standard, which is higher than the bearing capacity of the permeable brick in general.

High durability. The durability and abrasion performance of permeable concrete is superior to asphalt, avoiding the short usage time [4].

1.5   *The inner layers: microfiltration membranes*

Description. The inner layers that are under the concrete are microfiltration membranes. We use them to prevent the larger particles from going through the wall and passing into the soil outside the wall or blocking the permeable pore of the concrete.

The mechanism of the microfiltration membrane. Microporous is a kind of filter medium. Under the impetus of the water pressure, it can block particles between 0.1 and 1 micron such as gravel, silt and clay particles, allowing a large amount of solvents and small molecules, and a small amount of solute molecules passing through. There are three kinds of microfiltration filtering mechanism according to the location of particles in the process of microfiltration intercept: screening, filter and bridge. Generally, we think that the mechanism of microfiltration is screening. In addition, factors such as adsorption and electric properties may also affect the rejection rate. The three mechanisms are shown in Figure 3.

Screening. Microfiltration membranes block the particles bigger than or equal to the membrane pore size. It is also known as mechanical screening.

Filter. Particles can be filtered through physical and chemical adsorption. Particle size that is less than the film hole can be trapped.

Bridge. Particle accumulation jostled each other, because many particles cannot enter the film hole or card in the hole, and then complete the intercept.

This design mainly uses the first kind of filtration principle. From the perspective of the mechanical and physical block, the larger particles are at the inner side of the membrane.

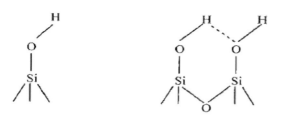

Figure 2.   Chemical structure of diatomaceous earth.

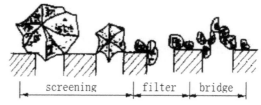

Figure 3.   Three kinds of filtering mechanisms.

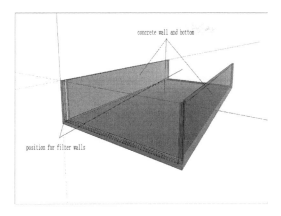

Figure 4. Position of the filter walls.

2 CONSTRUCTION STEPS

Concrete can be used for laying the bottom of the pit and casting a pair of side walls.

SBS-modified asphalt waterproof coiled material is set at the surface of the bottom to ensure that water will not go through the bottom.

A pair of filter walls is set at the other pair of directions, forming an available path for water to pass through. The relative position of the concrete walls and filter walls is shown in Figure 4.

Later period maintenance
Because of the fact that the diatomaceous earth is a kind of consumable, it means that the diatomaceous earth should be capable of changing during the period of use. So, we design the middle layer of the wall, a replaceable core with diatomaceous earths and PAC in it. When the materials inside are used, the core can be directly extracted out and changed.

3 CONCLUSION

By using a five-layered wall structure and reasonable anti-side arrangement, the water have to flow through the filter wall. The heavy metal ions will be adsorbed or become flocculated to be blocked in the reaction center.

This device will provide an available method to eliminate heavy metal pollution from the source, avoiding the post-pollution treatment.

The two outermost layers are pervious concrete; they are used to support the whole wall. Two layers that are under the concrete are made up of microfiltration membrane, which can prevent large particulate matter from going through. The very middle layer contains diatomaceous earths and polyaluminum as the flocculent; they can precipitate heavy metal ions. Faced with the growing problem of heavy metal pollution, much of the processing methods have been used for the removal of heavy metal ions. Comparison of commonly used methods such as chemical precipitation, membrane filtration, ion exchange and activated carbon adsorption method is made. The most effective method is activated carbon adsorption, but the manufacturing and recycling of activated carbon requires a relatively high cost. Other traditional crafts are also costly, and involve less removal of pollutants. Compared with the traditional process, a wide variety of sources, low cost and no secondary pollution of the environment mineral material, diatomaceous earth has attracted the attention of researchers. Diatomite has a unique microporous structure, as a typical natural porous material having good adsorption properties. Using it to treat waste water can not only effectively utilize mineral resources, but also reduce costs and, more importantly, further environmental protection role.

REFERENCES

[1] Information on http://baike.baidu.com.
[2] Materialized modification of diatomite and Cd_2_ and Pb_2_ adsorption performance research-Ye Zhang, 2012.5 (In Chinese).
[3] Adsorption performance of Diatomite for Cd_2_ and Pb_2_ in waste water-Jing Ling, 2013.5 (In Chinese).
[4] Waterproof concrete construction technology-Zhanqiu Zhou, Yang Zhou, Junfeng Xu, 2007.7-TU755 (In Chinese).
[5] Object and materials Science of Germany, No. Oct-14th-2014.

Fiber material

Material Science and Engineering – Chen (Ed.)
© 2016 Taylor & Francis Group, London, ISBN 978-1-138-02936-1

Adsorption of Pb(II) from aqueous solutions by immobilized persimmon tannins on collagen fiber

H.-J. Liang & Z.-M. Wang
School of Materials Science and Engineering, Guilin University of Electronic Technology, Guilin, China

G.-Y. Li
School of Life and Environmental Sciences, Guilin University of Electronic Technology, Guilin, Guangxi, China

C.-Y. Ma & P.-B. Dai
School of Materials Science and Engineering, Guilin University of Electronic Technology, Guilin, China

ABSTRACT: In this work, a novel adsorbent (PTCF) was synthesized by immobilized persimmon tannin on collagen fiber. Various adsorption parameters such as pH, solid/liquid ratio, initial concentration of Pb(II) and temperature were determined. The maximum adsorption capacity reached 140 mg/g at 303 K and pH 4.5 when the initial concentration of Pb (II) was 200 mg/L and the solid/liquid ratio was 1.0. The Freundlich model and the pseudo-second-order model can well fit to explain its adsorption isothermal and kinetic data, respectively. All these results indicated that the PTCF is a promising bioadsorbent of Pb (II) in waste water treatment.

Keywords: persimmon tannins; biosorption; Pb(II); adsorption capacity; collagen fiber

1 INTRODUCTION

Heavy metal ions such as Pb, Cd, Hg, Cr, Ni, Zn and Cu are non-biodegradable, toxic and carcinogenic even at very low concentrations. Hence, they usually pose a serious threat to the environmental and public health (Liu et al. 2008). The permissible levels of Pb in drinking water and waste water are $0.05 \, mg \cdot L^{-1}$ and $0.005 \, mg \cdot L^{-1}$, respectively. When Pb is accumulated at high levels, Pb can generate serious health problems such as liver and kidney damage, mental retardation, infertility and abnormalities in pregnant women (Singh C.K et al. 2008 & Singh V. et al. 2007).

At present, precipitation, ion exchange, extraction, ultrafiltration, reverse osmosis, electrodialysis, and adsorption are conventional methods for the removal of heavy metal ions from aqueous solutions. Adsorption technology has received increasing attention due to its simplicity, high efficiency and low cost in the removal of heavy metal ions from aqueous solutions. Activated carbon, synthetic resin (Wang et al. 2010) and other adsorbents have been used for adsorptive Pb(II) from waste water. However, these adsorbents not only have a low adsorption capacity but also release environmentally unacceptable chemicals in the river. Therefore, some kind of environmental

friendly and high-adsorption capacity adsorbent is needed.

It has been reported that vegetable tannins could recover metal ions from aqueous solutions, such as bayberry, valonia and persimmon (Takeshi et al.). Tannin is a kind of natural polyphenol compound that contain abundant adjacent phenolic hydroxyl groups that are capable of chelating with metal ions (Huang et al. 2010). Hence, tannin can be utilized as a new kind of environmental protection material for heavy metal disposal. However, tannin cannot be directly used as an adsorbent for metal extraction because it is water-soluble. To overcome this disadvantage, many attempts have been made to immobilize tannins on various water-insoluble matrices such as cellulose, SiO_2 and collagen fiber (He et al. 2012). Collagen fiber is an abundant natural biomass and has many functional groups including $-NH_2$, $-COOH$ and $-OH$, which can react with other chemicals. So, collagen fiber is selected as matrices in our study.

In this paper, persimmon tannin was immobilized on collagen fiber by the glutaraldehyde cross-linking reaction to prepare a novel adsorbent (PTCF). The sorption capacity of PTCF for the removal of Pb (II) ions from the aqueous solution was evaluated.

2 EXPERIMENTAL PROCEDURE

2.1 *Materials*

Persimmon tannin and collagen fiber were purchased from Guangxi Huikun Company of Agricultural Products. Glutaraldehyde and $Pb(NO_3)_2$ were from Guangdong Xilong Chemical Co., Ltd. All other chemicals used were of analytical grade without further purification.

2.2 *Preparation of PTCF by immobilized persimmon tannin on collagen fiber*

A 30 g sample of persimmon tannin was dissolved in 1000 mL of deionized water and mixed with 20 g of collagen fiber. The mixture was stirred at room temperature for 20 h. Then, the suspended matter was filtered and washed with distilled water. Afterwards, 100 mL of glutaraldehyde (25 wt%) were added into the suspended matter. The mixture was first stirred at 298 K for 4 h and then stirred at 313 K for 4 h. Finally, the product was washed with distilled water and filtered. The filter cake was dried in vacuum at 323 K for 24 h. Then, the adsorbent was crushed and sieved to 100 mesh prior to use.

2.3 *Effect of initial pH on adsorption capacity*

A 100 mg sample of PTCF was suspended in 100 mL of 200 mg/L of $Pb(NO_3)_2$ solutions. The pH of the initial solution was adjusted to 2.0–5.5 by using 1M HNO_3 or 1M NaOH solutions. The adsorption process was conducted at 303 K with constant stirring for 24 h to reach equilibrium. After adsorption equilibrium, the suspension was filtered and Pb (II) ion concentration in the filtrate was analyzed by an atomic adsorption spectrometer (AAS, Zeenit 700P, Germany).

Adsorption capacity was calculated according to Equation (1):

$$Q_e = (C_o - C_e) \cdot V / W \qquad (1)$$

where Q_e is the adsorption capacity; V is the volume of the solution (L); W is the weight of dry PTCF (g); C_o is the initial concentration of Pb(II) ions; and C_e is the equilibrium concentration.

2.4 *Effect of the solid/liquid ratio on the adsorption capacity*

A 100 mL solution of 200 mg/L $Pb(NO_3)_2$ with varying mass of the adsorbent (solid/liquid ratio 0.5, 1.0, 2.0, 4.0, 6.0, 8.0) was used, and the solution with the initial pH was adjusted to 4.5 by 1M HNO_3 or 1M NaOH solutions. After 24 h, the samples were filtered and the concentration of Pb(II) ions in the filtrate was determined by AAS. Adsorption capacity was calculated according to Equation (1).

2.5 *Adsorption isothermal studies*

Isothermal studies were carried out by shaking a 100 mg adsorbent with the initial concentration of $Pb(NO_3)_2$ solution ranging from 40 mg/L to 200 mg/L, and the initial pH of the solution was adjusted to 4.5 by 1M HNO_3 or 1M NaOH solutions. The adsorption process was conducted with constant stirring for 24 h at 303 K, 313 K and 323 K. After 24 h, the samples were filtered and the concentration of Pb(II) in the filtrate was determined by AAS. Adsorption capacity was calculated according to Equation (1).

2.6 *Adsorption kinetics*

A 100 mg sample of adsorbent was added to 100 mL of 200 mg/L Pb $(NO_3)_2$ solution, and the initial pH of the solution was adjusted to 4.5 by 1M HNO_3 or 1M NaOH solutions. The adsorption process was conducted with constant stirring at 303 K, 313 K and 323 K. The concentration of Pb(II) in the filtrate was determined by AAS at a regular interval during the adsorption process. Adsorption capacities at time t (min) were calculated according to Equation (1).

3 RESULTS AND DISCUSSION

3.1 *Effect of initial pH on the adsorption capacity*

The effect of pH on Pb adsorption by PTCF results is shown in Figure 1. The pH solution has a significant effect on the adsorption capacity.

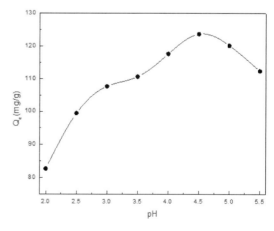

Figure 1. Effect of initial pH on the adsorption capacity of Pb(II).

In the pH range of 2.0–4.5, the adsorption capacity increased with the pH increasing until reaching an optimum pH value around 4.5 and then decreased with a further increase in pH.

The highest biosorption efficiencies were observed at pH 4.5. These facts indicated that the adsorption of Pb(II) onto PTCF was highly pH-dependent. It has been reported that a high ionization degree of phenolic hydroxyls and a high cationic charge of metal species favored the chelating reaction during the adsorption process (Li et al. 2012). When the pH was increased, electrostatic repulsion between cations and surface sites and the competing effect of hydrogen ions decreased and, consequently, the metal uptake was increased. Above the pH value of 5.5, Pb(II) starts precipitating as $Pb(OH)_2$ and hence studies in this range are not conducted. Therefore, the adsorption capacity sharply decreased when the pH exceeded 5.0. Additionally, the phenolic hydroxyls of persimmon tannin could be easily oxidized at a higher pH value.

3.2 Effect of the solid/liquid ratio on the adsorption capacity

As it can be seen from Fig. 2, the adsorption capacity of Pb(II) gradually decreased from 138 mg/g to 19.7 mg/g as the solid/liquid ratio increased from 0.5 to 8.0. This may be due to the adsorbents containing a large number of active groups, which led to competitive adsorption between hydrated ions and Pb(II) on the surface of the adsorbent (Zhang et al. 2012). On the other hand, when the solid/liquid ratio was increased, the surface of the adsorbent easily produced the overlapping of adsorption sites, so the adsorption capacity reduced sharply. However, the adsorption extent of Pb(II) in the solution increased sharply with the increasing solid/liquid ratio, which can be attributed to the fact that the adsorption sites of the adsorbent were insufficient at a low solid/liquid ratio (Oladoja et al. 2011).

3.3 Adsorption isothermal studies

Figure 3 shows the adsorption isothermal curve at 303 K, 313 K and 323 K, respectively. The adsorption capacity of Pb(II) affected by temperature was not obvious, indicating that the adsorption of Pb(II) on PTCF was dominated by chemical adsorption. Langmuir and Freundlich models were the most commonly used isotherms to explain solid-liquid adsorption systems.

The general form of the Freundlich and Langmuir model (Huang et al. 2010) could be expressed by Equation (2) and Equation (3) as follows:

$$\ln(q_e) = 1/n\ln(c_e) + \ln(k) \qquad (2)$$
$$c_e/q_e = c_e/q_{max} + 1/bq_{max} \qquad (3)$$

where k and 1/n are the Freundlich isotherm constant; ce is the equilibrium concentration (mg/L); qe is the adsorbed value of metal ions at an equilibrium concentration (mg/g); b is the Langmuir model constant; and qmax is the maximum adsorption capacity (mg/g).

The parameters as well as the correlation coefficients (R^2) are summarized in Table 1. It is shown

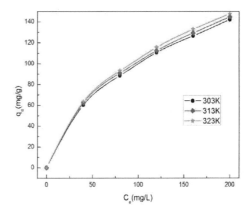

Figure 3. The adsorption isotherms of Pb(II) on PTCF.

Table 1. The Freundlich and Langmuir model parameters of the adsorption of Pb(II) on PTCF.

T	Langmuir fitting			Freundlich fitting		
	q_m(mg/g)	b	R^2	k	1/n	R^2
303 K	142.26	34.625	0.981	47.5	0.954	0.990
313 K	144.77	47.412	0.980	36.4	0.891	0.995
323 K	147.78	60.437	0.978	63.1	0.793	0.991

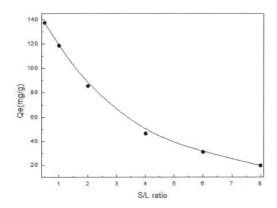

Figure 2. Effect of solid/liquid ratio on the adsorption capacity of Pb(II).

that the Freundlich isotherm is suitable for characterizing the experimental adsorption isotherms since all of the correlation coefficients are larger than 0.99. The Langmuir model exhibited a poor linear fitting to the adsorption isotherm data, which had low correlation coefficients.

3.4 *Adsorption kinetics*

The kinetics of metal ion adsorption was considered to be an important parameter for designing the adsorption process. In this study, the sorption curve shown in Figure 4 exhibited a similar trend, that is, qe increased quickly in 60 min, then slowly reached equilibrium, and finally remained constant after 300 min.

In order to examine the mechanism of the adsorption process, the adsorption rates for adsorbents at different temperatures were analyzed by the pseudo-first-order and pseudo-second-order models. The equations for the pseudo-first-order model (Ramesh et al. 2008) and the pseudo-second-order model (Yu et al. 2011) can be expressed as follows:

$$\log(q_e - q_t) = \log q_t - k1/2.303 \qquad (4)$$
$$t/q_t = 1/k2qe^2 + t/q_e \qquad (5)$$

where q_c (mg/g) and q_t (mg/g) are the amounts of the adsorbed metal at equilibrium and at time t (min), respectively. k1 (min^{-1}) and k2 (g/mg·min) are the rate constants.

Table 2 lists the parameters obtained from the two models. This showed that the adsorption of Pb(II) on PTCF was more appropriately described by the pseudo-second-order model. The correlation coefficients (R2) for the pseudo-second-order model were higher than 0.99, and the adsorption capacities calculated by the model were in agreement with the experimental data. This suggests

Table 2. Adsorption kinetic parameters of PTCF to Pb(II).

T (K)	qe.exp (mg/g)	pseudo-first-order model			pseudo-second-order model		
	Test	qt.cal (mg/g)	k1	R2	qt.cal (mg/g)	k2	R2
303	142.33	67.01	0.11	0.944	139.44	0.003	0.999
313	143.01	43.33	0.37	0.932	141.18	0.037	0.993
323	143.68	92.37	0.16	0.895	142.01	0.013	0.997

that the adsorption velocity of Pb(II) on PTCF was dominated by the surface adsorption rate (Y.S HO & G McKay, 1999), and should not exist internal diffusion resistance.

4 CONCLUSIONS

A novel adsorbent was prepared through immobilizing persimmon tannin on collagen fiber by the glutaraldehyde cross-linking reaction. PTCF exhibited a high adsorption capacity of Pb(II) in the aqueous medium. The adsorption capacity reached 140 mg/g at 303 K and at the pH value 4.5 when the initial concentration of Pb(II) in the aqueous solution was 200 mg/L and the solid-liquid ratio was 1.0. The adsorption behavior fitted well with the Freundlich model and the pseudo-second-order model. Thus, the novel adsorbents can be practically useful for removing Pb(II) from the waste water solution.

ACKNOWLEDGMENTS

This work was supported by the Guangxi Experiment Center of Information Science (20130113), the Guangxi Postgraduate Association Training program (20121225–06-Z), and the fund of Guangxi Key Laboratory of Metabolic Diseases Research (181H2011-01).

REFERENCES

[1] He Li, Gao Si-ying & Wu Hao. 2012. Antibacterial activity of silver nanoparticles stabilized on tannin-grafted collagen fiber. *Mater. Sci. Eng* 32: 1050–1056.
[2] Huang Xin, Liao Xuepin & Shi Bi, 2010. Tannin-immobilized mesoporous silica bead (BT-SiO$_2$) as an effective adsorbent of Cr(III) in aqueous solutions. *J. Hazard. Mater* 173: 33–39.
[3] Huang Xin, Wang Yanpin, Liao Xuepin. 2010. Adsorptive recovery of Au^{3+} from aqueous solutions using bayberry tannin-immobilized mesoporous silica. *J. Hazard. Mater* 183: 793–798.

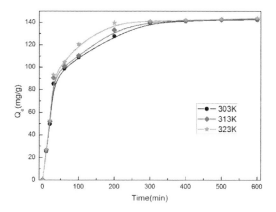

Figure 4. The kinetic of Pb(II) adsorption on PTCF.

[4] Li Weiguang, Gong Xujin & Li Xin. 2012. Removal of Cr(VI) from low-temperature micro-polluted surface water by tannin acid immobilized powdered activated carbon. *Bioresource Technology* 113: 106–113.

[5] Liu Changkun, Bai Renbi & Ly Quan San. 2008. Selective removal of copper and lead ions by diethylenetriamine functionalized adsorbent: behaviors and mechanisms. *Water Res* 42:1511–1522.

[6] Oladoja N.A., Alliu Y.B. & Ofomaja A.E. 2011. Synchronous attenuation of metal ions and colour in aqua stream using tannin–alum synergy. *Desalination* 271: 34–40.

[7] Ogata Takeshi & Nakano Yoshio. 2005. Mechnisms of gold recovery from aqueous solutions using a novel tannin gel adsorbent synthesized from natural condensed tannin. *Water Research* 39(18): 4281–4286.

[8] Ramesh A., Hasegawa H. & W. Sugimoto. 2008. Adsorption of gold(III), platinum(IV) and palladium(II) onto glycine modified crosslinked chitosan resin, *Bioresource Technology* 99: 3801–3809.

[9] Singh C.K., Sahu J.N. & Mahalik K.K. 2008. Studies on the removal of Pb(II) fromwastewater by activated carbon developed from tamarind wood activated with sulphuric acid. *J. Hazard. Mater* 153:221–228.

[10] Singh V., Tiwari S. & Sharma A. Kumar. 2007. Removal of lead from aqueous solutions using Cassia grandis seed gum-graft-poly(methylmethacrylate). *J. Colloid Interface Sci.* 316: 224–232.

[11] Wang Lin, Xing Ronge & Yu Huahua. 2010. Recovery of silver (I) using a thiourea-modified chitosan resin. *J. Hazard. Mater* 180: 577–582.

[12] Yu Dan., Wang Wei & Wu Jianwen. 2011. Preparation of conductive wool fabrics and adsorption behavior of Pd(II) ions on chitosan in the pretreatmen. *Synthetic Metals* 161: 124–131.

[13] Y.S Ho, G. McKay. 1999. Pseudo-second order model for sorption processes. *Process Biochemistry* 34(5): 451–465.

[14] Zhang Zhanhua, Liu Shouxin & Zhang Bin. 2012. Equilibrium and kinetics studies on the adsorption of Cu(II) from the aqueous phase using a β-cyclodextrin-based adsorbent, *Carbohydrate Polymers* 88: 609–617.

Material Science and Engineering – Chen (Ed.)
© *2016 Taylor & Francis Group, London, ISBN 978-1-138-02936-1*

Thermal stability of cotton cellulose modified with terbium complexes

Y.L. Yang, C.B. Guo & M. Gao
*School of Environmental Engineering, North China University of Science and Technology, Yanjiao,
Beijing, China*

ABSTRACT: Complexes of cell-THPC-thiourea-ADP with Tb^{3+} have been prepared. The thermal stability and smoke suspension of the samples are determined by TG, DTA and cone calorimetry. The activation energies for the second stage of thermal degradation have been obtained by following Broido equation. Experimental data show that for the complexes of cell-THPC-thiourea-ADP with Tb^{3+}, the activation energies and thermal decomposition temperatures are higher than those of cell-THPC-thiourea-ADP, which shows the metal ion can increase the thermal stability of cell-THPC-thiourea-ADP.

Keywords: cotton cellulose; thermal stability; flame retardant; smoke suspension

1 INTRODUCTION

Cotton cellulose is used extensively to make life pleasant, comfortable and colorful. Unfortunately, it is flammable and causes a fire hazard. According to fire statistics, about 50% of fires are caused by textiles in the world (Gaan & Sun, 2007), and the cotton cellulose is one of important components in textiles. So the emphasis on reducing combustibility has centered on its chemical modification. There are also many studies on thermal degradation of cotton cellulose treated with flame retardants (Lessan et al. 2011). However, with environmental sustainability required, the effects of flame-retardants on both smoke generation and the toxicity of combustion products have become special important, as flame retardant cellulose has been reported to produce denser smoke than pure cellulose (Grexa & Lubke 2001).

In previous papers (Tian et al. 2003), compounds of transition metals have been found to be effective smoke retarders. However, there is no information about the effects of Tb^{3+} on smoke suspension and thermal degradation of cotton cellulose treated with flame retardants. So the main objective of the work reported here is to investigate the effects of Tb^{3+} on the thermal degradation and smoke suspension of cotton cellulose modified with flame retardant.

In this paper, complexes of cell-THPC-thiourea-ADP with Tb^{3+} were prepared. The thermal degradation of samples was studied from ambient temperature to 800°C by TG, DTA.

2 EXPERIMENTAL

2.1 *Materials*

Cotton cellulose of commercial grade (Hebei province, China) was selected for flame-retardant treatment. The cotton cellulose was immersed in 24% NaOH solution at room temperature for 24 h (mercerization process). The alkali was then filtered off and the sample was washed repeatedly with distilled water. The sample was dried in an oven at 60°C and then stored in a desiccator.

2.2 *Instrumentation*

The elemental analysis was carried out using a Carlo Eroa 1102 Elemental Analyzer. LOI values were determined in accordance with ASTM D2863-70 by means of a General Model HC-1 LOI apparatus. Thermogravimetry (TG) was carried out on a DTA-2950 thermal analyzer (Dupont Co. USA) under a dynamic nitrogen (dried) atmosphere at a heating rate of 10°C min^{-1}.

2.3 *Cotton cellulose treatment*

The preparation of the samples is corresponded to references (Tian et al. 2003). THPC (Shanghai, China) was neutralized with NaOH to give a pH value equal to 6.5 and its 45% solution was mixed with 22.5% thiourea solution. The pH value was adjusted to 6.5 and a small amount of ADP was added. The resulting mixture was used as the treating solution. The mercerized cotton cellulose was immersed in the treating solution for 30 min at room temperature. The treated cotton cellulose was dried at 60°C in an oven for 60 min. Curing of these treated cellulose was carried out by heating at 160°C for 5 min in the oven. After cooling, the sample was thoroughly washed with distilled water for an hour and dried in an oven at 60°C. Tb^{3+} complexes of cell-THPC-thiourea-ADP were prepared by treating 6 g of cell-THPC-thiourea-ADP in each instance with 5% aqueous solutions of $Tb(NO_3)_3$ at room temperature for 72 h under constant stirring. Each product was washed repeatedly with water until the filtrate was free from metal salt and dried overnight in an oven at 60°C then stored in a desiccator.

3 RESULTS AND DISCUSSION

The DTA, TG curves of (1) cotton cellulose, (2) cell-THPC-thiourea-ADP, (3) Tb^{3+} complexes of cell-THPC-thiourea-ADP were obtained in a dynamic air atmosphere from ambient temperature to 800°C and are shown in Figure 1.

3.1 *Differential thermal analysis*

From the DTA curves of samples 1–3, the initiation temperatures (T_i), peak temperatures (T_p) and termination temperatures (T_t) of the various endotherms and exotherms were investigated and are given in Table 1. The DTA curve of cotton cellulose shows two large exotherms with their respective peak maxima at 363 and 459°C. Before 350°C, decomposition and dehydration occur to form some flammable volatile products, and the first exotherm peaking at 363°C is due to the oxidation of these volatile products. Another exotherm,

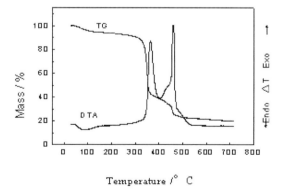

Figure 1. Caption of a typical figure. Photographs will be scanned by the printer. Always supply original photographs.

Table 1. Peak temperatures in DTA curves of samples 1–3.

No.	Compound	DTA curve			Nature of peak
		T_i	T_p	T_t	
1	Cellulose	320	363	403	Exo(large)
		403	459	539	Exo(large)
2	Cell-THPC-thiourea-ADP	255	326	414	Exo(large)
		414	485	614	Exo(large)
3	Tb^{3+} complex of cell-THPC-thiourea-ADP	79	174	214	Endo(small)
		214	326	401	Exo(large)
		401	498	627	Exo(large)

peaking at 459°C, represents oxidation of the charred residues. The dehydration process dominates at low temperatures and ultimately leads to a carbonaceous residue. At higher temperatures, cleavage of glycosyl units by intra-molecular transglycosylation starts, forming ultimately a tarry mixture with levoglucosan as the major constituent (Kandola et al. 1996). Levoglucosan decomposes into volatile and flammable products and therefore plays a key role in the flammability of cellulose.

The DTA curve of cell-THPC-thiourea-ADP is quite distinct from that of pure cotton cellulose. The treated cotton cellulose seemed to decompose in two steps (Mostashari, S.M. & Mostashari, S.Z. 2008). A breakdown or depolymerization of the THPC-thiourea-ADP finish, a catalyzed dehydration of the cellulose, and some bond formation occurred during the first step. Two large exotherms with peak maxima at 326 and 485°C are shown in Figure 2, respectively. The second step involved a breakdown of the cellulose chain, evolution of gases from both the cellulose and the finish polymer, and continuation of bond formation. The bond formation was probably due to a phosphorylation reaction at the C-6 hydroxyl group of the anhydroglucose unit as suggested. Phosphorylation at this position would inhibit the formation of levoglucosan and prevent further breakdown to flammable gases. This would account for the increased amount of char formed over that for untreated cotton cellulose. The last large exotherm peaking at 485°C is due to the combustion of the char (Janowska et al. 2008).

In the DTA curves of the metal complexes of cell-THPC-thiourea-ADP (sample 3), the first peak, a new endotherm with peak maxima in the range 79–214°C, represents depolymerization, a catalyzed dehydration of the cellulose and some bond formation. For the metal complexes (sample 3), there are also two large exotherms in each case. The decomposition stage, which is represented by

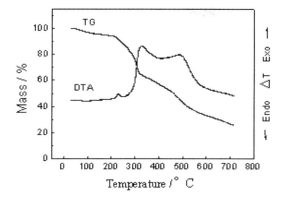

Figure 2. Thermal analysis curves of cell-THPC-thiourea-ADP in air.

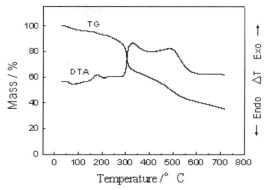

Figure 3. Thermal analysis curves of Tb3+ complex of cell-THPC-thiourea-ADP in air.

the first exotherm, is very different in the decomposition temperature in the complexes. The peak temperature of the complexe is 326°C. The last exotherm, which is due to oxidative decomposition of the residual products, also shows considerable variation in the complexes.

For cotton cellulose, the two exotherms are sharp and narrow, which shows a large rate of heat release. For the samples 2–3, the two exotherms become small and broad. Heat release is distributed between two broad peaks covering a wide area, resulting in a major reduction in rate of heat release and flammable products which fuel the flaming combustion reaction. In the other hand, the second exotherms become much smaller for the samples 2–3, which indicates that oxidation of the charred residues becomes more difficult due to the existence of flame retardants.

3.2 Thermogravimetry

From Figure 1, it can be seen that the second stages in thermal decomposition of the samples, decompose mainly and quickly, play a key role attributed to the combustibility. So we mainly discuss this stage. Temperature Range (TR), Mass Loss (ML) at the second stage (quick mass loss rate) in TG are listed in Table 2. Generally, at lower temperatures, the thermal degradation of cellulose includes dehydration, depolymerization, oxidation, evolution of carbon monoxide, carbon dioxide, and formation of carbonyl and carboxyl groups and ultimately a carbonaceous residue; at higher temperatures, cellulose decomposes into a tarry mixture (mainly levoglucosan) which further decompose into volatile and flammable products. The main role of flame retardants containing phosphorus is to minimize the formation of levoglucosan by lowering the decomposition temperature of cellulose

Table 2. Thermal degradation and analytical data of samples 1–3.

No.	TR (°C)	ML (%)	Ea (kJ/mol)	CY (%)	LOI (%)	P (%)	N (%)	M (%)
1	326–365	46	198.3	20.2	18.0	–	–	–
2	214–318	28	78.1	26.5	29.6	1.69	3.78	–
3	290–340	21	120.5	36.0	28.5	1.50	3.00	0.65

and enhancing char formation by catalyzing the dehydration and decomposition reaction. However, lowering the decomposition temperature of cellulose is to decrease its thermal stability, which is not favorable. So the two points must be considered simultaneously.

From Figure 1 and Table 2, it can be seen that for the cotton cellulose (sample-1), initial decomposition temperature is 326°C, the second stage is in a range 326–365°C, and the mass loss is 46%. For cell-THPC-thiourea-ADP, initial decomposition temperature is 214°C, the second stage is in a range 224–318°C, all these much decreased compared with those of cotton cellulose, which shows that the thermal stability of the cell-THPC-thiourea-ADP is much decreased because of the catalyzing dehydration and decomposition reaction. For sample 3, the decomposition temperature range is 290–340°C, higher than that of cell-THPC-thiourea-ADP, which shows that the thermal stability of samples is increased.

In order to understand the flame retardant properties of these samples, LOI of samples is measured, given in Table 2. From Table 2, we can see that samples 2–3 show high values of LOI. This suggests that the combustibility of cotton cellulose treated with flame retardants decreases.

Moreover, the sample-2 containing highest content of phosphorus and nitrogen shows highest value of LOI. The second stage of decomposition for sample-2 occurs at lower temperatures (224–318°C) and produces less flammable volatile products, resulting in higher flame retardancy.

The kinetic parameters for the second stage were determined using the following equation, given by Broido (Broido, 1969):

$$\ln\left(\ln\frac{1}{y}\right) = -\frac{Ea}{R}\cdot\frac{1}{T} + \ln\left(\frac{R}{Ea}\cdot\frac{Z}{|\mathring{A}}\cdot T_m^2\right)$$

where y is the fraction of the number of initial molecules not yet decomposed, T_m the temperature of the maximum reaction rate, β the rate of heating and Z the frequency factor.

For the second stage, the major degradation and mass loss stage, the energy of activation for sample-2 is 78.1 kJ/mol, is much decreased compared to cotton cellulose (198.3 kJ/mol). The reason is that the flame retardant catalyzes decomposition reaction. The lower decomposition temperatures (224–318°C) also support this. The energy of activation for sample-3 is 120.5 kJ/mol, are much higher than that of sample-2, which shows the thermal stability of sample-3 is increased.

4 CONCLUSIONS

For complexes of cell-THPC-thiourea-ADP with metal ions, the activation energies and thermal decomposition temperatures are higher than those of cell-THPC-thiourea-ADP. The metal ions (Tb^{3+}) can increase the thermal stability of cell-THPC-thiourea-ADP. However, the two exotherms in DTA curves are very different in the decomposition temperature in all the complexes.

REFERENCES

[1] Gaan, S. & Sun, G. 2007. Effect of phosphorus flame retardants on thermo-oxidative decomposition of cotton. Polymer degradation and stability, 92(6), 968–974.
[2] Lessan, F. Montazer, M. & Moghadam, M.B. 2011. A novel durable flame-retardant cotton fabric using sodium hypophosphite, nano TiO$_2$ and maleic acid. Thermochimica Acta, 520, 48–54.
[3] Grexa, O. & Lubke, H. 2001. Flammability parameters of wood tested on a cone calorimeter. Polymer Degradation and Stability, 74, 427–432.
[4] Tian, C.M. Xie, J.X. Guo, H.Z. & Xu, J.Z. 2003. The effect of metal ions on thermal oxidative degradation of cotton cellulose ammonium phosphate. J. Therm. Anal. Cal., 73, 827–834.
[5] Kandola, B.K. Horrocks, A.R. Price D. & Coleman, G.V. 1996. Flame-retardant treatments of cellulose and their influence on the mechanism of cellulose pyrolysis. J. Macromal Sci, Rev Macromol Chem Phys., C36 (4), 721–794.
[6] Mostashari, S.M. & Mostashari, S.Z. 2008. Combustion pathway of cotton fabrics treated by ammonium sulfate as a flame-retardant studied by TG. J. Therm. Anal. Cal. 91, 437–441.
[7] G. Janowska, T. Mikołajczyk & M. Olejnik. 2008. Effect of montmorillonite content and the type of its modifier on the thermal properties and flammability of polyimideamide nanocomposite fibers. J. Therm. Anal. Cal. 92, 495–503.
[8] Broido, A. 1969. A simple, sensitive graphical method of treating thermogravimetric analysis data. Journal of Polymer Science Part A-2: Polymer Physics. 7(10), 1761–1773.

Material Science and Engineering – Chen (Ed.)
© 2016 Taylor & Francis Group, London, ISBN 978-1-138-02936-1

Identification of textile fiber components using Thermo-Gravimetric Analysis

J. Luo, J.C. Hu & D.Y. Qi
Guangzhou Fibre Product Testing and Research Institute, Guangzhou, China

ABSTRACT: This work studies the identification of textile fiber components using Thermo-Gravimetric Analysis (TGA). Optimal conditions for TGA in the identification of the fiber components were investigated. Under the optimal conditions, the feature library of thermal stability of both conventional fiber and high-performance fiber has been established. The results indicated that TGA is suitable for the identification of textile fiber components, in particular for high performance fiber.

Keywords: fiber component; thermogravimetric analysis; textile

1 INTRODUCTION

Fibre component is one of the most fundamental properties of textile. How to detect fiber component and to quantify its content in yarn, fabric or textile is of significant importance for product, design, analysis and management. And it's also the basis for government to carry out the supervision of quality verification. With the advent of new materials and their scale-up of production, a variety of novel fibers, such as environmentally friendly fibers, high-performance fibers, functional fibers, have been exploited. Also, there is a tendency to produce differential fibers. The development of novel fibers has undoubtedly put forward new challenging in determining fiber content qualitatively.

An accurate identification of fiber components is necessary before making a quantitative analysis of fiber content (Langley & Kennedy, 1981). Currently, several testing methods are frequently used in identifying textile fibers, such as burning behavior method, microscopy method, chemical dissolution method, qualitative observation of colour-production for chlorine and nitrogen, fibre melting point, etc. (Cox & Flanagan, 1997), most of which are described in detail on the textile industry standard FZ/T 01057 series. The disadvantage of burning behavior method and microscopy method is that they are inefficient and error-prone require a lot of manpower to operate. The usage of large amount of inorganic acid and organic solvents in chemical dissolution method results in highly corrosive waste, causing serious pollution to the environment.

For an unknown sample, it is usually necessary to use several methods at the same time to identify fiber components. Consequently, the textile industry standard FZ/T 01057 series have a series of methods for identification of fiber component, which need to be updated constantly due to the development of novel fibers.

Thermal analysis is a branch of material science which studies the properties of materials as they change with temperature (Wendlandt, 1974). According to the property which is measured, the commonly used methods of thermal analysis are classified as: Differential Thermal Analysis (DTA) which measures temperature difference, Differential Scanning Calorimetry (DSC) which measures heat difference, and Thermo-Gravimetric Analysis (TGA) which measures mass difference(Ehrenstein, Riedel & Trawiel, 2012; Hatakeyama & Liu, 1998).

TGA is a method of thermal analysis in which changes in mass are measured as a function of increasing temperature. It uses the recording thermobalance to automatically measure and record weight changes when the sample is heated or cooled under a controlled program. The resulting weight change versus temperature curve, or its derivative, gives information concerning the thermal stability and decomposition of the original sample. Therefore, TGA is an ideal method for identification of textile fiber component (Crighton & Holmes, 1998; Manich, Bosch, Carilla, Ussman, Maillo & Gacén, 2003).

In this study, the usage TGA in identifying fiber components has been tested. The results indicated that TGA is suitable for identification of textile fiber components, especially with the application of high performance fiber.

2 MATERIALS AND METHODS

All textile samples were purchased in the market. The mass of each sample was ~10 mg. The components of each textile sample were analyzed following the textile industry standard FZ/T 01057.

The thermogravimetric analysis is performed using a TA Instrument (TGA Q500). Thermal stability of textile is observed during the thermogravimetric analysis. Multiple TGA experiment conditions such as environmental atmosphere, heating rate and temperature range were investigated to find out the optimal conditions for the qualitative identification of textiles. In order to better obtain the temperature at the peak of decomposition, the Derivative Thermogravimetric curves (DTG) was used.

3 MATERIALS AND METHODS

3.1 *The optimum conditions*

In order to find out the optimal conditions for the qualitative identification of textiles, three factors including environmental atmosphere, heating rate and temperature range have been invested. Figure 1 shows the TG (left) and DTG (right) curves of a cotton/polyester blended sample under air or nitrogen condition, while the rest of test conditions are fixed. The temperature range is from 50°C to 800°C with a heating rate of 10°C/min. The results show two different thermal

diagram of decomposition for the same sample, which is due to the difference of environment atmosphere. In the air atmosphere, the sample can be oxidized and decomposed, which is different from the simple under inert atmosphere. In the air condition, there is another peak appeared at 500°C, which is generally considered as a further oxidation of polyester after the first stage degradation. Therefore, the better environmental atmosphere condition for identification of fiber component is nitrogen.

The effect of different heating rate on the thermogravimetry analysis of identification of fiber component is then studied. TG (left) and DTG (right) curves of a cotton/polyester blended sample under different heating rate are shown in Figure 2. Nitrogen is selected for environment atmosphere and the temperature range is from 50°C to 800°C. Different heating rate (2°C/min, 5°C/min, 10°C/min, 20°C/min) is tested, while other conditions are fixed. The dextral peak of the DTG curve shown in Figure 2 is the characteristic peak of polyester fiber. The curve of the heating rate of 2, 5, 10°C/min shows the same shape and quality of this peak, which is better than the curve of heating rate of 20°C/min. In the same quality of data, the higher heating rate means the less test time. Thus, 10°C/min is the fittest heating rate for the TG test.

We also considered the best temperature range between the different samples. Figure 3 displayed the DTG curve of conventional fibers and high-performance fiber. The results show that the temperature range from 100°C to 600°C is meeting the need of conventional fiber. But most of the high-performance fibers have a good thermal stability. Therefore, the highest temperature of TG test is no less than 800°C.

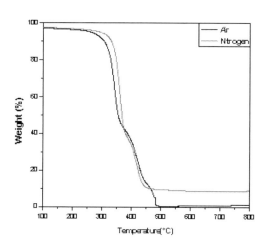 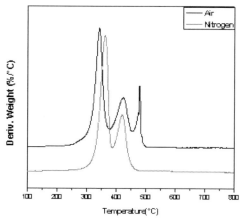

Figure 1. TG (left) and DTG (right) curve of a cotton/polyester blended sample under different environment atmosphere.

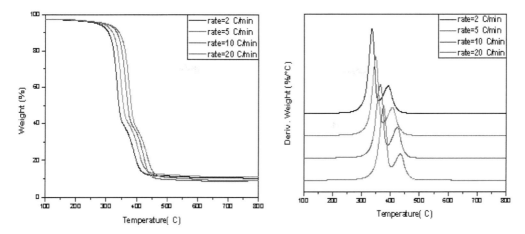

Figure 2. TG (left) and DTG (right) curve of a cotton/polyester blended sample under different heating rate.

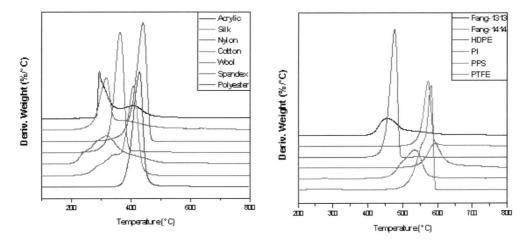

Figure 3. DTG curve of conventional fibers (left) and high-performance fiber (right).

Based on the above research, we select the following conditions for the TG test on identification of textile fiber component: nitrogen environmental atmosphere, temperature range from 100°C to 800°C with a heating rate of 10°C/min.

3.2 Standard thermal stability information of textile

Table 1 shows the thermal stability data of textile, including decomposition index (n) and the temperature of the maximum decomposition rate (Td). All the data of Table 1 are the average values over 10 of the similar single component samples. In the data of Td, the temperatures in the brackets stand for the maximum peak value in DTG curves.

3.3 Application

In order to investigate the validity of the identification method of textile fiber component by TG, a series experiments have been carried out. Identification of fiber components for single component samples and multicomponent samples by TG have the same principle, which is based on the characteristic peaks of the standard thermal stability. Thus, we used an unknown two components sample to illustrate the availability of the method.

The DTG curves of the sample and the corresponding standard fibers are shown in Figure 4, with the analysis result listed in Table 2. The decomposition index of sample is the sum of that of standard cotton fiber and standard

Table 1. The thermal stability data of textile.

Component	n	Td/°C
Cotton	1	363
Nylon	1	440
Acrylic	2	(295)/408
Polyester	1	429
Wool	1	319
Silk	2	(318)/377
Spandex	3	283/346/(409)
HDPE	1	476
Nomex	1	455
Kevlar	1	573
PI	1	592
PPS	1	535
PTFE	2	557/(581)

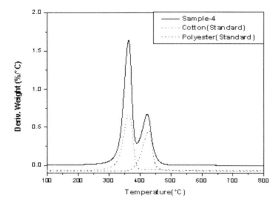

Figure 4. DTG curve of textile sample and the corresponding standard fibers.

Table 2. The analysis result of unknown sample by TG.

Sample	Two component unknown sample	
Corresponding standard	Cotton	Polyester
Decomposition index of sample	2	
Decomposition index of standard	1	1
The temperature of the maximum decomposition rate of sample (°C)	364	424
The temperature of the maximum decomposition rate of standard (°C)	363	429
Difference of sample values and standard values (°C)	1	−5

polyester fiber. And the peak-fitting results show that the Td of each peak is close to the Td of the corresponding standard fiber. Therefore, according to the TG identification method, the sample is cotton/polyester blended product, which is in agreement with the result analyzed from the methods following the textile industry standard FZ/T 01057.

4 SUMMARY

In this study, TGA was used to identify textile fiber components. Optimum conditions for TGA analysis in identifying fiber components are nitrogen environmental atmosphere and a temperature range from 100°C to 800°C with a heating rate of 10°C/min. Under the optimal conditions, the feature library of thermal stability of both conventional fibers and high-performance fiber can be established. The results indicate that TGA is suitable for identification of textile fiber components, in particular for high performance fiber.

ACKNOWLEDGEMENTS

This work was financially supported by Science Foundation of Guangzhou Bureau of Quality and Technical Supervision, Science Foundation of General Administration of Quality Supervision, Inspection and Quarantine of P.R.C.

REFERENCES

Cox, B.N., & Flanagan, G. (1997). Handbook of analytical methods for textile composites. National Aeronautics and Space Administration.
Crighton, J., & Holmes, D. (1998). The Application of Thermal Analysis to Textiles. Part I: Identification of Fibre-blend Components by Using Thermogravimetry. Journal of the Textile Institute, 89(2), 198–207.
Ehrenstein, G.W., Riedel, G., & Trawiel, P. (2012). Thermal analysis of plastics: theory and practice. Carl Hanser Verlag GmbH Co KG.
Hatakeyama, T., & Liu, Z. (1998). Handbook of thermal analysis.
Langley, K.D., & Kennedy, T.A. (1981). The identification of specialty fibers. Textile Research Journal, 51(11), 703–709.
Manich, A.M., Bosch, T., Carilla, J., Ussman, M., Maillo, J., & Gacén, J. (2003). Thermal analysis and differential solubility of polyester fibers and yarns. Textile Research Journal, 73(4), 333–338.
Wendlandt, W.W. (1974). Thermal methods of analysis. Wiley-Interscience. New York.

Material Science and Engineering – Chen (Ed.)
© 2016 Taylor & Francis Group, London, ISBN 978-1-138-02936-1

Effect of card taker-in speed on fiber length in flat strips

D.L. Shi & X. Guo
School of Fashion and Textile, Eastern Liaoning University, Dandong, Liaoning, China

X.Z. Yu
Liaoning Province Key Laboratory of Functional Textile Materials, Eastern Liaoning University, Dandong, Liaoning, China

ABSTRACT: The paper studied the influence of taker-in speed on fiber length and fiber length distribution in flat strips. The fiber length in flat strips produced under different taker-in speeds was tested by USTER AFIS single fiber tester. In point of fiber length parameters in flat strips, such as mean length by weight and by number, mean length of the 25% longest fibers by weight, mean length of the 5.0% longest fibers by number, short fiber content by weight and by number, the taker-in speed of 1400 r/min is most conducive to the improvement of flat strips quality, and the flat strips quality is the worst at that of 800 r/min as a whole. In point of fiber length distribution in flat strips, the flat strips quality is the best at the taker-in speed of 800 r/min and the worst at that of 1000 r/min. It may not be complete only using fiber length parameters to judge the carding quality, and it will be more objective to use fiber length distribution to evaluate the carding quality.

Keywords: taker-in speed; flat strips; fiber length parameter; fiber length distribution

1 INTRODUCTION

The taker-in speed has an impact on yarn quality, and thus has been one of the important problems in the carding study (2012). Fujino et al. (1962) pointed out that with the increase of taker-in speed, the opening ability to fibers increased and nep level in cotton web reduced significantly. Artzt et al. (1973) had shown that the taker-in speed of 800 r/min was conducive to nep reduction. Harrison et al. (1986) confirmed by experiment that the taker-in speed had little effect on neps in cotton web. Göktepe et al. (2003) researched the effect of taker-in speed on fiber properties on modern carding machines with a triple taker-in, and found that there was no significant difference in nep count and nep size, but the amount of trash, dust, and foreign material was higher when a higher taker-in speed was applied. They also indicated that the fibers in the waste had higher elongation values at a higher taker-in speed except for the taker-in droppings, and the fibers in the waste taken from the back part of the cylinder were longer and stronger at a lower taker-in speed.

Sun (2005) systematically discussed the taker-in speed selection of carding machine, and presented that the taker-in speed should be chosen according to different process conditions, blowing state and equipment, and the selection range of the taker-in speeds under different process conditions is also given. Yu et al. (2009) discussed the effect of cylinder-taker-in speed ratio on card sliver quality. The results confirmed that too big and too small speed ratio were all not conducive to the improvement of card sliver quality, and the speed ratio of 1.79 and 1.98 were more advantageous to the improvement of card sliver quality. Zhang et al. (2013) studied the effect of the speed change of different taker-in on yarn quality, and the results showed that the quality of cotton yarns produced by gill pin taker-in was better than those produced by saw-tooth taker-in as a whole. But the study on fiber length parameter change in flat strips under different taker-in speeds has rarely been reported. So the fiber length in flat strips produced under four-grade taker-in speeds was tested by USTER AFIS single fiber tester and the results analysis is as follows.

2 EXPERIMENTAL

2.1 Materials

Cotton from Xinjiang of China was used in the experiment. Its AFIS test results were as follows:

mean length by weight (L_w) was 25.7 mm, short fiber content by weight (SFC_w, < 16 mm) was 10.6%, mean length of the 25% longest fibers by weight was 30.3 mm, mean length by number (L_n) was 21.7 mm, short fiber content by number (SFC_n, < 16 mm) was 25.9%, mean length of the 5.0% longest fibers by number was 34.3 mm, nep content was 225 cnt/g, nep mean size was 675 μm, Seed Coat Neps (SCN) were 4 cnt/g, SCN mean size was 1136 μm. Impurity content was 173 cnt/g, dust content was 142 cnt/g, trash content was 31 cnt/g, and Visible Foreign Matter (VFM) content was 0.69%.

2.2 Experimental process flow

A002 bale plucker→A006B mixer→A034 step opener→A036 porcupine opener→A092 tandem feeder→A076A single beater lap machine→A186F carding machine.

2.3 Experimental conditions

1. The card laps used in the experiment were processed by the same picker and the lap quantity was 425 g/m.
2. In the experiment, a modified A186F carding machine that is equivalent to FA series carding machine was used. Speeds of cylinder and doffer were 360 and 30 r/min, respectively. The taker-in speed was 800, 1000, 1200 and 1400 r/min, respectively. Gauge of taker-in under casing was 6 mm at entrance, 1 mm at the fourth point and 1 mm at exit; while gauge of mote knife was 0.36 mm. There were 4 back stationary flats with the gauge of 0.89, 0.76, 0.64, 0.51 mm from bottom to top. Tooth density was 90 teeth/(25.4 mm)^2 and heel-toe difference was 0.42 mm. The card sliver quantity was 5 g/m.

2.4 Experimental methods

Two laps for each experiment plan were processed. In order to ensure high accuracy of test results (2010), 30 flat strips samples for each experiment plan were randomly taken and each sample was about 0.5 g, then their length was tested by USTER AFIS single fiber tester.

3 EXPERIMENTAL RESULTS

The test results of fiber length parameters and fiber length distribution in flat strips under different experiment plans are shown in Tables 1 and 2.

Table 1. Test results of fiber length parameters in flat strips under different experiment plans.

Taker-in speed/r·min^{-1}	L_w/mm	SFC_w, <16 mm/%	25% L_w/mm
800	22.5	21.7	28.3
1000	22.5	21.8	28.3
1200	22.5	22.1	28.3
1400	22.3	22.7	28.2
Taker-in speed/r·min^{-1}	L_n/mm	SFC_n, <16 mm/%	5% L_n/mm
800	16.5	47.8	31.9
1000	16.5	47.7	31.8
1200	16.4	48.2	31.8
1400	16.1	49.5	31.6

4 DISCUSSION OF EXPERIMENTAL RESULTS

4.1 Effect of taker-in speed on fiber length parameters in flat strips

Usually the following factors should be considered for evaluating flat strips quality (2015): ① less of the amount of flat strips would be better under the condition of ensuring yarn quality; ② the head is thick and the tail is thin for each flat strips and there is little longer bridge fiber between two flats; ③ there are less spinnable fibers and more short fibers in flat strips; ④ flat strips contain more impurities and neps. Therefore, the length parameters to evaluate flat strips are opposite to those to evaluate card slivers, viz. the shorter the fiber length parameters (L_w, L_n, 25% L_w and 5% L_n) in flat strips are, the greater SFC_w and SFC_n are, the better the flat strips quality is.

As shown in Table 1:

1. L_w in flat strips is the shortest at the taker-in speed of 1400 r/min, followed by that of 1200 r/min, 1000 r/min and 800 r/min, which are 0.2 mm longer than that of 1400 r/min.
2. L_n in flat strips is the shortest at the taker-in speed of 1400 r/min, followed by that of 1200 r/min, which is 0.3 mm longer than that of 1400 r/min; that of 1000 r/min and 800 r/min are the longest, which are 0.4 mm longer than that of 1400 r/min.
3. 25% L_w in flat strips is the shortest at the taker-in speed of 1400 r/min, followed by that of 1200 r/min, 1000 r/min and 800 r/min, which are 0.1 mm longer than that of 1400 r/min.
4. 5% L_n in flat strips is the shortest at the taker-in speed of 1400 r/min, followed by that of 1200 r/min and 1000 r/min, which are 0.2 mm longer than that of 1400 r/min; that of 800 r/min

Table 2. Test results of fiber length distribution in flat strips under different experiment plans.

Fiber length/mm	Length component content under different taker-in speed (%)			
	$800/\text{r} \cdot \text{min}^{-1}$	$1000/\text{r} \cdot \text{min}^{-1}$	$1200/\text{r} \cdot \text{min}^{-1}$	$1400/\text{r} \cdot \text{min}^{-1}$
0–2	2.7	2.6	2.6	2.5
2–4	11.7	11.0	11.7	11.4
4–6	7.7	7.4	7.6	7.6
6–8	6.0	5.8	6.0	6.0
8–10	5.2	5.1	5.1	5.3
10–12	4.8	4.9	4.7	4.8
12–14	4.8	4.9	4.7	4.8
14–16	5.3	5.3	5.0	5.2
16–18	5.5	5.6	5.6	5.6
18–20	6.0	6.0	6.0	6.0
20–22	6.5	6.6	6.3	6.4
22–24	6.9	7.0	6.8	6.9
24–26	7.0	7.0	7.1	7.1
26–28	6.4	6.7	6.5	6.4
28–30	5.2	5.4	5.4	5.3
30–32	3.7	3.9	3.8	3.9
32–34	2.1	2.3	2.3	2.3
34–36	1.1	1.2	1.1	1.2
36–38	0.6	0.5	0.6	0.5
38–40	0.2	0.2	0.3	0.3
40–42	0.1	0.1	0.2	0.1
42–44	0.1	0.1	0.1	0.1
44–46	0.1	0.1	0.1	0.1
46–48	0.1	0.1	0.1	0.1
48–50	0.0	0.1	0.1	0.1
50–52	0.0	0.0	0.0	0.1
Super-short fiber ≤ 6	22.1	21	21.9	21.5
Short fiber 6–12	16	15.8	15.8	16.1
Medium-short fiber 12–18	15.6	15.8	15.3	15.6
Medium-length fiber 18–24	19.4	19.6	19.1	19.3
Medium-long fiber 24–30	18.6	19.1	19	18.8
Long fiber 30–36	6.9	7.4	7.2	7.4
Super-long fiber > 36	1.2	1.2	1.5	1.4

is the longest, which is 0.3 mm longer than that of 1400 r/min.

5. SFC_{w}, <16 mm in flat strips is the greatest at the taker-in speed of 1400 r/min, followed by that of 1200 r/min, 1000 r/min and 800 r/min, which are respectively 2.64%, 3.96% and 4.41% smaller than that of 1400 r/min.

6. SFC_{n}, <16 mm in flat strips is the greatest at the taker-in speed of 1400 r/min, followed by that of 1200 r/min, 800 r/min and 1000 r/min, which are respectively 2.63%, 3.43% and 3.64% smaller than that of 1400 r/min.

According to the above analysis on fiber length parameters in flat strips, the taker-in speed of 1400 r/min is most conducive to the improvement of flat strips quality; and the flat strips quality is the worst at that of 800 r/min as a whole.

4.2 Research significance of fiber length distribution in flat strips

Short fiber content and mean length were used to indicate fiber damage during carding, the study results in literature (2006) showed that it might not be complete only using short fiber content to judge fiber damage during the carding and fiber length distribution should be used to evaluate fiber damage in the carding process. Thus, fiber length distribution in flat strips was used to study the effect of taker-in speed on fiber length distribution. According to literature (2006), the floating zone length of modern drafting system can be reduced to below 10.5 mm. The results of literature (2012) showed that the floating zone length of modern drafting system can reach 6–10 mm. Therefore, following definitions are made in this paper: define fiber length ≤ 6 mm as super-short fibers, which will affect yarn quality because they cannot be controlled by modern drafting system; define fiber length of 6–12 mm as short fibers, which can only be partially controlled by modern drafting system, so too many this kind of fibers will have an adverse effect on yarn quality; define that of 12–18 mm as medium-short fibers; define that of 18–24 mm as medium-length fibers; define that of 24–30 mm as medium-long fibers; define that of 30–36 mm as long fibers; define 36 mm and above as super-long fibers.

4.3 Effect of taker-in speed on fiber length distribution in flat strips

As shown in Table 2:

1. Super-short fiber content in flat strips is the most at the taker-in speed of 800 r/min, followed by that of 1200 r/min, 1400 r/min and 1000 r/min, which are respectively 0.9%, 2.71% and 4.98% less than that of 800 r/min. This shows that the removal of super-short fibers in flat strips is the most at the taker-in speed of 800 r/min.
2. Short fiber content in flat strips is the most at the taker-in speed of 1400 r/min, followed by that of 800 r/min, which is 0.62% less than that of 1400 r/min; that of 1000 r/min and 1200 r/min are the least, which are 1.86% less than that of 1400 r/min. This shows that the removal of short fibers in flat strips is the most at the taker-in speed of 1400 r/min.
3. Medium-short fiber content in flat strips is the most at the taker-in speed of 1000 r/min, followed by that of 800 r/min and 1400 r/min, which are 1.27% less than that of 1000 r/min; that of 1200 r/min is the least, which is 3.16% less than that of 1000 r/min. This shows that the removal of medium-short fiber in flat strips is the most at the taker-in speed of 1000 r/min.
4. Medium-length fiber content in flat strips is the most at the taker-in speed of 1000 r/min, followed by that of 800 r/min, 1400 r/min and 1200 r/min, which are respectively 1.02%, 1.53% and 2.55% less than that of 1000 r/min. This shows that the removal of medium-length fiber in flat strips is the most at the taker-in speed of 1000 r/min.
5. Medium-long fiber content in flat strips is the most at the taker-in speed of 1000 r/min, followed by that of 1200 r/min, 1400 r/min and 800 r/min, which are respectively 0.52%, 1.57% and 2.62% less than that of 1000 r/min. This shows that the removal of medium-long fiber in flat strips is the most at the taker-in speed of 1000 r/min.

6. Long fiber content in flat strips is the most at the taker-in speed of 1000 r/min and 1400 r/min, followed by that of 1200 r/min, which is 2.70% less than that of 1000 r/min and 1400 r/min; that of 800 r/min is the least, which is 6.76% less than that of 1000 r/min and 1400 r/min. This shows that the removal of long fiber in flat strips is the most at the taker-in speeds of 1000 r/min and 1400 r/min.
7. Super-long fiber content in flat strips is the most at the taker-in speed of 1200 r/min, followed by that of 1400 r/min, which is 6.67% less than that of 1200 r/min; that of 800 r/min and 1000 r/min are the least, which are 20% less than that of 1200 r/min. This shows that the removal of super-long fiber in flat strips is the most at the taker-in speed of 1200 r/min.

According to the above analysis on fiber length distribution in flat strips, short fiber (2–12 mm) content in flat strips at the taker-in speed of 800 r/min is obviously higher than that of 1200 r/min, 1400 r/min and 1000 r/min. The sum of medium-long fiber and long fiber content in flat strips at the taker-in speed of 800 r/min is also evidently lower than that of 1200 r/min, 1400 r/min and 1000 r/min. It shows that, in point of fiber length distribution in flat strips, the flat strips quality at the take-in speed of 800 r/min is significantly higher than the other three plans and the flat strips quality is the worst at that of 1000 r/min. This conclusion has a certain difference with that based on fiber length parameters, so it will be more objective to use fiber length distribution to evaluate the carding quality.

5 CONCLUSIONS

1. In point of fiber length parameters in flat strips, the take-in speed of 1400 r/min is most conducive to the improvement of flat strips quality, and the flat strips quality is the worst at that of 800 r/min as a whole.
2. In point of fiber length distribution in flat strips, the flat strips quality is the best at the take-in speed of 800 r/min and the worst at that of 1000 r/min.
3. It may not be complete only using fiber length parameters to judge the carding quality, and it will be more objective to use fiber length distribution to evaluate the carding quality.

ACKNOWLEDGMENTS

The authors would like to express their sincere appreciation to the financial support in this study from Key Project Plan of Science &

Technology Department of Liaoning Province
(No. 2012219021).

REFERENCES

[1] Artzt, P. & Schreiber, L.O. 1973, 1974. Correlation between card sliver nep and yarn nep. *Textil Praxis (international)* (11): 608–611; (6): 754–762.

[2] Cao, J.P., Lu, Q., Sun, P.Z., Liu, H.P. 2010. Test stability of Uster advanced fiber information system (AFIS). *Journal of Donghua University (Eng. Ed.)* 27(3): 412–418.

[3] Cao, J.P., Zhang, Z.D., Sun, P.Z. 2015. Effect of cylinder speed on fiber length distribution in flat strips. *Journal of Textile Research* 36(3): 24–26.

[4] Fujino, K. & Itani, W. 1962 Effects of carding on fibre orientation. *J. Text. Mach. Soc. Japan* 8(1): 1–8.

[5] Göktepe, F., Göktepe, Ö., Süleymanov, T. 2003. The effect of the licker-in speed on fiber properties on modern carding machines with a triple ticker-in. *Journal of the Textile Institute* 94(3/4): 166–176.

[6] Harrison, R.E. & Bargeron, J.D. 1986. Comparison of several nep determination methods. *Textile Research Journal* 56(2): 77–79.

[7] Sun, P.Z. 2005. Research and choice of licker-in speed of carding machine. *Cotton Textile Technology* 33(10): 15–19.

[8] Sun, P.Z. 2012. *Study on technology of carding machine*. Beijing: China Textile Press.

[9] Tang, W.H. & Zhu, P. 2012. The theory and practice of modern cotton spinning drafting. Beijing: China Textile Press.

[10] Wang, L., Sun, P.Z., Zhang, M.G. 2006. Effect of feed plate gauge on fiber length. *Journal of Textile Research* 27(4): 70–73.

[11] Wu, Y.Q. & Wei, H.X. 2006. Technology advancement of cotton spinning draft system (The First Part). *Cotton Textile Technology* 34(11): 22–25.

[12] Yu, X.Z. & Sun, P.Z. 2009. Test and discussion of relationship between cylinder taker-in speed ratio and card sliver quality. *Cotton Textile Technology* 37(3): 18–20.

[13] Zhang, Z.D., Yu, X.Z., Sun, P.Z. 2013. Study on the effect of gill pin taker-in speed on yarn quailty. *The 12th Asian Textile Conference* (10): 108.

Material Science and Engineering – Chen (Ed.)
© 2016 Taylor & Francis Group, London, ISBN 978-1-138-02936-1

Inter-ONU backup fiber based protection for Passive Optical Network against multiple link failures

H.J. Tan, X. Guo & Q. Gao
Sichuan Engineering Technical College, Deyang, P.R. China

Y.M. Xie
College of Information Science and Engineering, Northeastern University, Shenyang, P.R. China

ABSTRACT: Survivability is one of the key issues in the planning of Passive Optical Network (PON), because network component failure (e.g., fiber link cut) may cause huge data loss. Some standardized protection approaches have been widely studied to protect the fiber links by duplicating them. However, the fiber duplication approaches have to suffer from the challenge of optical signal attenuation and the risk of simultaneous failure of primary and duplicated fiber links. In this paper, we focus on the survivability of PON against multiple link failure that is a failure scenario less touched in previous works. A protection approach is proposed to first divide all Optical Network Units (ONUs) in the network into some clusters and then deploy backup fibers between different ONUs in the same cluster. Thus, the disconnected ONUs can transfer their traffic into other normal ONUs in the same cluster along the backup fibers between them. Aiming to minimize the cost of backup fibers, we put emphasis on the clusters of ONUs and deployment of backup fiber. Simulation results demonstrate that the proposed approach is effective in reducing deployment cost and enhancing network survivability.

Keywords: Passive Optical Network; survivability; backup fiber

1 INTRODUCTION

Due to the technological merits in bandwidth capacity and stability, Passive Optical Network (PON) has been widely acknowledged as a promising candidate for next-generation broadband access networks. Due to the tree topology, PON is vulnerable to network component failure (e.g., fiber link cut) which would cause the interruption of many traffic flows. Therefore, network survivability is always considered as one of the key issues in the planning of PON [1]. Some approaches have been standardized (e.g., ITU-T G.983.1) to protect PON by duplicating fiber links [2]. When the primary fiber link cuts, the interrupted connections can be recovered by switching them into the duplicated fiber link. However, the duplication of fiber links usually requires double splitting ratio, which may weaken the reach of optical signal. Furthermore, due to the geographical limitation and technological complexity, the primary and duplicated fiber links sometimes have to be deployed in the common physical resource [3]. Thus, they may cut simultaneously when the common physical resource fails.

In this paper, a protection approach called Inter-ONU Backup Fibers based Protection (IOBFP) is proposed to protect PON against multiple fiber-link failures that is a notable failure event in disaster scenario but less studied before. Here, the "fiber link" refers to the distribution fiber between Optical Network Unit (ONU) and splitter. As shown in Figure 1, we first divide all ONUs in the network into some clusters. Each cluster guarantees that, in case of the simultaneous failures of X fiber links, the normal ONUs should have enough residual capacity to carry the traffic from the disconnected

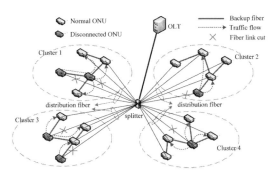

Figure 1. Illustration of the proposed IOBFP approach.

ONUs. Then, we deploy the backup fibers between different ONUs (i.e., inter-ONU backup fibers) such that at least one backup-optical-path is available between each pair of ONUs in the same cluster. Once the simultaneous failures of multiple fiber links, the disconnected ONUs can transfer their traffic into other normal ONUs in the same cluster along the backup-optical-paths between them. In the IOBFP approach, we aim to jointly optimize the ONU clusters and deployment of backup fibers with the minimum deployment cost. Integer Linear Programming (ILP) is employed to formulate the joint optimization problem and obtain the optimal solution for small-scale network. A heuristic algorithm is also proposed for the near-optimal solution in the large-scale network scenario.

The rest of paper is organized as follows. In Section 2, we employ ILP to formulate the joint optimization problem of ONU clusters and backup fibers deployment. In Section 3, we propose the heuristic algorithm to support the application of the proposed protection approach in large-scale network. The performance evaluation and results analysis are shown in Section 4. Finally, we conclude this paper in Section 5.

2 PROBLEM FORMULATION

For brevity of problem formulation, we first introduce the notations as follows:

- N_O: the number of ONUs in the network;
- o_i: the ONU indexed by i;
- $l_{i,j}$: the length of backup fiber between $\lambda_{i,j}$ and o_j;
- $\lambda_{i,j}$: a binary variable taking 1 if a backup fiber is deployed between o_i and o_j, and 0 otherwise;
- $N_C^{\max}$: the maximum number of ONU clusters;
- S_k: the set of ONUs in the kth cluster;
- τ_k: a binary variable taking 0 if the kth cluster is empty and 0 otherwise;
- $w_{i,j}^{p,q}$: a binary variable taking 1 if the backup-optical-path between o_i and o_j traverses over the backup fiber between o_p and o_q, and 0 otherwise;
- $\varepsilon_{i,j}$: a binary variable taking 1 if o_i and o_j are located in the same cluster, and 0 otherwise;
- $l_{i,j}$ $\forall i,j$: the traffic demand of o_j;
- H: the maximum length of backup-optical-path;
- C: the ONU capacity;
- I: an integer large enough.

– Objective:
It is clear that larger length of backup fibers will result in higher deployment cost [4]. Thus, the objective of minimizing the *deployment cost* of backup fibers can be transformed into minimizing the *length* of backup fibers, as follows.

Minimize $\sum_{i=1}^{N_O}\sum_{j=i+1}^{N_O} l_{i,j} \cdot \lambda_{i,j}$ (1)

For the ILP formulation of the optimization problem in IOBFP, we introduce the following linear constraints:

– Constraint on configuration of ONU clusters:

$$|S_k| = \sum_{i=1}^{N_O} \delta_i^k \qquad \forall k \tag{2}$$

$$|S_k| \geq \tau_k \cdot (X+1) \quad \forall k \tag{3}$$

$$\sum_{k=1}^{N_C^{\max}} \delta_i^k = 1 \qquad \forall i \tag{4}$$

$$\delta_i^k \leq \tau_k \qquad \forall k,i \tag{5}$$

The constraints in Eqs. (2) and (3) ensure that at least one ONU is normal in each non-empty cluster upon the simultaneous failures of X fiber links. The constraints in Eqs. (4) and (5) guarantee that each ONU is located in only one non-empty cluster.

– Constraint on flow conservation of backup-optical-path:

$$\sum_{q=1}^{N_O} w_{i,j}^{p,q} - \sum_{q=1}^{N_O} w_{i,j}^{q,p} = \begin{cases} \varepsilon_{i,j}, & \text{if } p=i \\ -\varepsilon_{i,j}, & \text{if } p=j \quad \forall i<j \\ 0, & \text{otherwise} \end{cases} \tag{6}$$

$$\varepsilon_{i,j} \leq \delta_i^k \qquad \forall i<j \tag{7}$$

$$\varepsilon_{i,j} \leq \delta_j^k \qquad \forall i<j \tag{8}$$

$$\varepsilon_{i,j} \geq \delta_i^k + \delta_j^k - 1 \quad \forall i<j \tag{9}$$

The constraint in Eq. (6) formulates the flow conservation of backup-optical-path. The constraints in Eqs. (7)–(9) formulate the definition of binary variable $\varepsilon_{i,j}$.

– Constraint on length of backup-optical-path:

$$\lambda_{p,q} \geq \frac{1}{I} \sum_{i=1}^{N_O} \sum_{j=i+1}^{N_O} w_{i,j}^{p,q} \qquad \forall p<q \tag{10}$$

$$\lambda_{p,q} \leq \sum_{i=1}^{N_O} \sum_{j=i+1}^{N_O} w_{i,j}^{p,q} \qquad \forall p<q \tag{11}$$

$$\sum_{p=1}^{N_O} \sum_{q=p+1}^{N_O} w_{i,j}^{p,q} \leq H \cdot \varepsilon_{i,j} \qquad \forall i<j \tag{12}$$

The constraints in Eqs. (10) and (11) formulate the definition of binary variable $\lambda_{p,q}$. The constraint in Eq. (12) specifies that the length of backup-optical-path should be no more than H hops.

– Constraint on full protection:

$$\sum_{i=1}^{N_O} (d_i \cdot \delta_i^k) \leq (|S_k| - X) \cdot C + (1 - \tau_k) \cdot I \ \forall k. \tag{13}$$

The constraint in Eq. (13) ensures that each non-empty cluster should have enough residual capacity to carry the traffic affected by the simultaneous failures of X fiber links even in the same cluster.

3 HEURISTIC ALGORITHM

In view of the time-consuming nature of ILP, we also propose a two-stage heuristic algorithm for the joint optimization problem. In the first stage, we employ the Genetic Algorithm (GA) to optimize the ONU clusters. The nth individual in the mth population is defined as a binary matrix $I_m^n = [\delta_i^k]_{N_O \times N_C^{\max}}$ to represent a feasible solution for ONU clusters. We evaluate the fitness of individual I_m^n according to the fitness function F_m^n in Eqs. (14) and (15).

$$F_m^n = \sum_{i=1}^{N_O} \sum_{j=i+1}^{N_O} l_{i,j} \cdot \varepsilon_{i,j} \qquad \forall\, m,n \qquad (14)$$

$$\varepsilon_{i,j} = \delta_i^k \cdot \delta_j^k \qquad \forall\, i,j,k \qquad (15)$$

Thus, a fitter individual I_m^n with larger F_m^n can yield the better solution which brings about less distance among the ONUs in each cluster. The evolution is carried out to create the new population of individuals by implementing the genetic operators including selection, crossover and mutation on the current population of individuals. In the proposed GA, we use the combination of roulette wheel selection, uniform crossover and random mutation [5][6] to create the new populations until the required number of populations is reached. Finally, we can obtain the best solution for ONUs clusters from the fittest individual. Generally, the proposed GA has a similar flowchart to the typical genetic algorithm. We omit its pseudo-code procedure for the purpose of saving space.

In the second stage, we optimize the deployment of backup fibers in each cluster. For any one cluster, e.g., the kth cluster, the backup fibers will be deployed one-by-one in a greedy manner until each pair of ONUs in the kth cluster have the backup-optical-path satisfying the length constraint. Particularly, when deploying the wth backup fiber in the kth cluster, we expect that the deployment of the wth backup fiber can bring about more ONU pairs whose backup-optical-path satisfies the length constraint, while the length of the wth backup fiber is minimized. Thus, the objective of deploying the wth backup fiber can be represented as minimizing the cost-efficiency $E_{i,j}^w = N_{i,j}^w / l_{i,j}$, where $N_{i,j}^w$ denote the increased number of ONU pairs whose backup-optical-paths satisfy the length constraint if the wth backup fiber is deployed between the ONUs o_i and o_j. The pseudo-code procedure of deploying backup fibers is summarized in Table 1.

Table 1. Pseudo-code procedure of deploying backup fibers.

Algorithm: Deploying backup fibers for the kth cluster

Input: $\delta_i^k \;\; \forall k$, $l_{i,j} \;\forall i,j, H$.
Output: $\lambda_{i,j}$.
1: Initialize $w \leftarrow 1$;
2: $\forall o_i, o_j \in S_k$, initialize $\lambda_{i,j} \leftarrow 0$;
3: Compute the number of ONU pairs in the kth cluster: $T_k = |S_k| \cdot (|S_k| - 1)/2$;
4: while $T_k \neq 0$ do
5:     $\forall o_i, o_j \in S_k$, compute $E_{i,j}^w = N_{i,j}^w / l_{i,j}$;
6:     Determine $o_{\bar{i}}$ and $o_{\bar{j}}$ with the maximum $E_{\bar{i},\bar{j}}^w$ $o_{\bar{i}}, o_{\bar{j}} \in S_k$;
7:     Deploy the wth backup fiber between $o_{\bar{i}}$ and $o_{\bar{j}}$: $\lambda_{\bar{i},\bar{j}} \leftarrow 1$;
8:     $\forall o_i, o_j \in S_k$, update the length of backup-optical-path between them;
9:     Update $T_k \leftarrow T_k - N_{i,j}^w$;
10:    Update $w \leftarrow w + 1$;
11: end while
12: Output $\lambda_{i,j} \; \forall o_i, o_j \in S_k$.

---

4 PERFORMANCE EVALUATION

In simulation, we set a PON in a 35km×35km square area. OLT and splitter are randomly placed in the network area with the 15km feeder fiber between them. There are 64 ONUs uniformly distributed in the network [7, 8]. For the purpose of generality, we define the normalized traffic demand $\bar{d}$ for each ONU as the ratio of the traffic demand to the capacity. Accordingly, the capacity of each ONU is normalized as 1. We reserve 10% guard capacity in each ONU to avoid the traffic congestion [9–11]. Thus, each ONU has the normalized traffic demand $\bar{d}$ with the upper bound of 90%. The length constraint of backup-optical-path H is set to 4 hops [12]. ILOG CPLEX 12.1 is used to solve the ILP on a computer with Intel Core i5 2.30 GHz CPU and 2GB RAM.

In Table 2, we compare the heuristic-based IOBFP and ILP-based IOBFP in the total length of backup fibers. To solve the ILP model in reasonable time, we set a small-scale network for this comparison including 16 ONUs and survivability level $X = 2$. All ONUs equally have the normalized traffic demand 30%, 40%, 50%, and 60% in different cases, respectively. The results are shown in Table 2. We observe that the heuristic-based IOBFP has the total length of backup fibers much close to the ILP-based IOBFP. Particularly, the gap between them is just 4.6% in the case of normalized traffic demand taking 30%. Furthermore, we compare the running time of both approaches

Table 2. Heuristic-based IOBFP vs. ILP-based IOBFP in total length of backup fibers.

Normalized traffic demand	Total length of backup fibers (km)	
	ILP-based IOBFP	Heuristic-based IOBFP
30%	53.62	56.21
40%	76.87	76.87
50%	76.87	76.87
60%	98.63	100.16

Table 3. Heuristic-based IOBFP vs. ILP-based IOBFP in running time.

Normalized traffic demand	Running time (s)	
	ILP-based IOBFP	Heuristic-based IOBFP
30%	21028.94	187.42
40%	4578.20	152.93
50%	4582.47	163.40
60%	9602.49	175.37

in Table 3 to highlight the advantage of the proposed heuristic approach in time-efficiency. We observe that, compared to the ILP-based IOBFP, the heuristic-based IOBFP can reduce the running time by more than 96.43% when the normalized traffic demand varies from 30% to 60%. Therefore, the heuristic-based IOBFP is highly efficient in producing the near-optimal solution. This motivates us to apply the proposed heuristic algorithm instead of the ILP-based IOBFP to larger-scale network.

In Figure 2, with the normalized traffic demand randomly set for each ONU, we compare the total length of backup fibers between Heuristic-based IOBFP, ILP-based IOBFP and the previous MCMP (Minimum Cost Maximum Protection) approach [4]. Since the MCMP approach considers only single fiber-link failure, we make a fair comparison with the survivability level $X = 1$. We observe that the total length of backup fiber of Heuristic-based IOBFP is much near to that of ILP-based IOBFP, which further demonstrates the near-optimality of the proposed heuristic algorithm. More importantly, the proposed heuristic algorithm requires the running time of only several seconds, while the ILP solution is obtained after some hours. Thus, the heuristic algorithm is much more time-efficient and applicable to the large-scale network compared to ILP. Furthermore, the length of backup fibers of IOBFP is much less than that of MCMP, especially in the case of higher traffic demand. The reason can be explained as follows. In MCMP, each disconnected ONUs can transfer their traffic into only the neighboring ONUs, while our IOBFP approach allows the traffic to be transferred into the remote ONUs more than one hop far away. Thus, IOBFP can improve the utilization of backup fibers and requires fewer backup fibers to be deployed.

In Figure 3, we apply the heuristic solution to the IOBFP approach with different X and examine its advantage in enhancing network survivability. We use the total length of backup fibers in MCMP as an upper bound to represent the

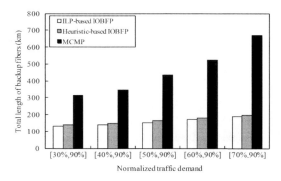

Figure 2. Analysis of reducing backup fibers length in the scenario of different normalized traffic demand.

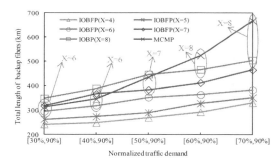

Figure 3. Analysis of enhancing network survivability in the scenario of different normalized traffic demand.

maximum affordable deployment cost, and then we analyze the number of simultaneous fiber-link failures that IOBFP can tolerate. For example, in case of normalized traffic demand $\bar{d} \in [30\%, 90\%]$, the IOBFP($X = 6$) has the total length of backup fibers less than MCMP, while the IOBFP($X = 7$) has the total length of backup fibers more than MCMP. Thus, in this case, IOBFP can guarantee the network survivability level $X = 6$, that is, the network can survive the simultaneous failures of 6 fiber links with enough residual capacity for traffic recovery. The results in Figure 3 demonstrate that, when the maximum affordable deployment cost is given, IOBFP can provide stronger network survivability than the previous MCMP approach.

In Figure 4, we show the comparison of the total length of backup fibers between Heuristic-based IOBFP, ILP-based IOBFP and MCMP with $X = 1$ in the scenario of different number of ONUs. As expected, with the number of ONUs increasing, more backup fibers need to be deployed for the full protection of all ONUs, thus the total length of backup fibers increases. However, since higher utilization of backup fibers in our IOBFP scheme, the total length of backup fibers of Heuristic-based IOBFP and ILP-based IOBFP is much lower than MCMP. Therefore, our IOBFP scheme has better performance than MCMP in reducing the deployment cost of backup fibers in the scenario of different number of ONUs. Specifically, the performance gains of both Heuristic-based IOBFP and ILP-based IOBFP remain higher than 68.82% when the number of ONUs ranges from 36 to 100.

In Figure 5, with the different number of ONUs, we investigate the impact of X on the total length of backup fibers of the IOBFP scheme. We can observe that IOBFP always has larger length of backup fibers when X is higher. We use the total length of backup fibers of MCMP as the upper bound and compute the number of simultaneous fiber-link failures that IOBFP can tolerate. We observe that the network survivability can reach to 8 in most cases. Namely, our IOBFP scheme can tolerate the simultaneous failures of at most 8 fiber links with less deployment cost of backup fibers than MCMP.

5 CONCLUSIONS

In this paper, we focus on the survivability of PON against multiple link failures. A protection approach called IOBFP has been proposed to first divide all ONUs into several clusters and then deploy backup fibers between different ONUs in the same cluster. Both ILP and heuristic algorithm have been proposed to optimize the ONU clusters and backup fibers deployment with the minimum deployment cost. Simulation results have demonstrated that the proposed approach is effective in reducing deployment cost and enhancing network survivability.

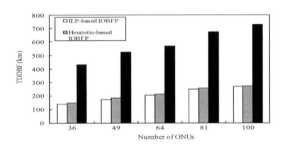

Figure 4. Analysis of reducing backup fibers length in the scenario of different number of ONUs.

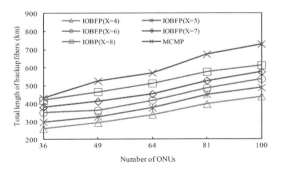

Figure 5. Analysis of enhancing network survivability in the scenario of different number of ONUs.

REFERENCES

[1] Y. Liu, L. Guo, X.T. Wei. Optimizing Backup Optical-Network-Units Selection and Backup Fibers Deployment in Survivable Hybrid Wireless-Optical Broadband Access Networks, *IEEE/OSA Journal of Lightwave Technology*, 2012, 30(10): 1509–1523.
[2] Y. Liu, L. Guo, C. Yu et al. Planning of Survivable Long-Reach Passive Optical Network (LR-PON) against Single Shared-Risk Link Group (SRLG) Failure, *Optical Switching and Networking*, 2014, 11(B): 167–176.
[3] Y. Luo, F. Effenberger. Optical network unit (ONU) power saving in time division multiplexed passive optical networks (TDM-PONs), in Proc. *OFC*, 2014, pp. 1–3.
[4] Y. Yu, Y. Liu, Y. Zhou, P. Han. Planning of survivable cloud-integrated wireless-optical broadband access network against distribution fiber failure, *Optical Switching and Networking*, 2014, 14(3): 217–225.
[5] L. Guo, Y. Liu, F. Wang, W. Hou, B. Gong. Cluster-based protection for survivable fiber-wireless (FiWi) access network, *IEEE/OSA Journal of Optical Communications and Networking*, 2013, 5(11): 1178–1194.
[6] N.U. Kim, H.S. Park, H.S. Lim. Adaptive packet transmission scheduling using multicast service efficiency in TDM-PON, *IEEE/OSA Journal of Lightwave Technology*, 2014, 32(9): 1759–1769.
[7] L. Guo, J. Cao, H Yu, L. Li. Path-based routing provisioning with mixed shared protection in WDM mesh networks, *IEEE/OSA Journal of Lightwave Technology*, 2006, 24 (3): 1129–1141.
[8] C.W. Chow, C.H. Yeh. Using downstream DPSK and upstream wavelength-shifted ASK for Rayleigh back-scattering mitigation in TDM-PON to WDM-PON migration scheme, *IEEE Photonics Journal*, 2013, 5(2): 1–3.

[9] M. Ruffini, D. Mehta, B.O. Sullivan, L. Quesada, L. Doyle, and D. Payne. Deployment strategies for protected Long-Reach PON, *IEEE/OSA Journal of Optical Communications and Networking*, 2012, 4(2): 118–129.

[10] M. Maier, M. Levesque. Dependable Fiber-Wireless (FiWi) access networks and their role in a sustainable third industrial revolution economy, *IEEE Transactions on Reliability*, 2014, PP(99): 1–15.

[11] M. Mahloo, J. Chen, L. Wosinska, A. Dixit, B. Lannoo, D. Colle, C.M. Machuca. Toward reliable hybrid WDM/TDM passive optical networks, *IEEE Communications Magazine*, 2014, 52(2): S14–S23.

[12] M.P.I. Dias, B.S. Karunaratne, E. Wong. Bayesian Estimation and Prediction-Based Dynamic Bandwidth Allocation Algorithm for Sleep/Doze-Mode Passive Optical Networks, *IEEE/OSA Journal of Lightwave Technology*, 2014, 32(14): 2560–2568.

Material Science and Engineering – Chen (Ed.)
© 2016 Taylor & Francis Group, London, ISBN 978-1-138-02936-1

A real-time current measuring system for fibers and textiles based on LabVIEW

G.P. Jia
College of Textile and Clothing, Yancheng Institute of Technology, Jiangsu, China

C.M. Zeng
Department of Light Chemical Engineering, Yancheng Textile Vocational Technology College, Jiangsu, China

ABSTRACT: Textile material electrical resistance testers measured the resistance of a bundle of oriented fibers, random fiber assembly, yarns, or fabrics after the test readings were stable, which took a few seconds, but the direct current of textile fiber was not absolutely stable with time. This makes the test results inconsistent, and these tests cannot be sufficient to reflect fiber current variation in detail. In order to research the textile electrical properties and fiber moisture regain, a real-time textiles electrical current testing system based on LabVIEW was developed to test a bundle of oriented fibers, yarns, random fiber assemblies and fabrics. The result showed that the relative error of the measuring current was no more than 3.01%, and the range of measurement was about 2.44×10^{-17}~5×10^{-3} A. This testing system integrated the traditional measuring function of testing electrical conductance of fibers, yarns, random fiber assembly and fabrics, and was more portable, functional, highly accurate, of wide range and low cost than the traditional meters. It provides a brand new technical method for textile electrical research.

Keywords: textile; current; measurement; LabVIEW

1 INTRODUCTION

Textile electrical properties are one of the important fields in researching textile performances. Their electrical properties have a strong relationship with fiber construction and assembling status of fibers. Electrical conductivity or resistivity is the common indices to characterize the textile material conductance (Mu 2009, Hearle 1953a, b, Hersh & Montogomery 1952). At present, many textile electrical resistance testing meters or raw cotton moisture regain meters can measure the electrical resistance after the data are stable, which will take a few seconds (BS EN 1997, ASTM 1990, GOST 1999, ASTM 1987, GB 1985, GB/T 2006, Reddick et al. 1959, Cusick & Hearle 1955) or within no more than one second (Ismail & Alyahya 2003, Byler 2004). This makes the measured results inconsistent; meanwhile, a higher applied voltage was often used for these meters. At a higher voltage, textile fibers can be polarized effectively, and can also produce a local electrical breakdown. This has a serious influence on measuring fibers and textile resistance. With the development of weak signal measuring techniques, lower voltage has been started to use for measuring fiber resistance in order to reduce the influence of polarization

and to avoid the local electrical breakdown. Fiber's varying current has a strong relationship with fiber micro-construction, fiber aging, and fiber moisture regain. Therefore, the development of the real-time textile current testing system has great importance to research textile material performances.

2 MEASURING PRINCIPLE

Fiber electrical current can be indirectly measured by measuring the voltage on the standard resistor, which is with fiber or textiles in series at certain conditions. Here, multi-range dividing voltage circuit was selected to expand the range of measurement, because the resistance of some hygroscopic fiber material from a low moisture regain to a high value can be of several orders of magnitude. The basic measuring principle circuit is shown in Figure 1.

Here R represents the textile resistance and R_i represents the standard resistance. V_i means the input voltage before amplifying and V_o is the output voltage after amplifying. The temperature coefficient and thermal noise of the standard resistor should be low in order to reduce the noise disturbance.

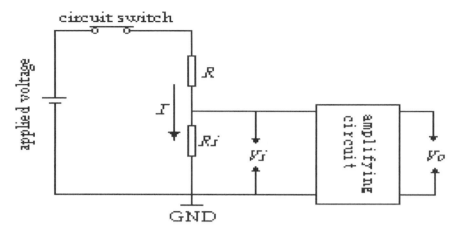

Figure 1.   Schematic of dividing voltage circuit of measuring textile current.

In order to maintain the current on R, which is basically equal to that of R_i, it means reducing the dividing current effect of input terminals of the amplifying circuit, which requires that the input resistance be at least two orders of magnitude compared with the standard resistance R_i. The current on fiber can also be calculated by Equation (1):

$$I = \frac{V_o}{R_i \times K} \tag{1}$$

where V_o is the amplified voltage; K is the amplification coefficient; and I is the textile current.

3   DESIGN OF DEVICES AND ELECTRODES FOR TEXTILES

3.1   *Random fiber assembly compression device and electrode design*

Current of fiber random assembly is associated not only with fiber itself, but also with the tightness between fibers. The increasing fiber filling ratio can make the fiber current adding. Therefore, in order to measure the fiber block current, it is necessary to control the density of the fiber assembly. The fiber compression device design is shown in Figure 2.

The specimen cell is made of Teflon. The compression pole is fixed, and the supporting plate can be moved with the movement of the piston. The density of the random fiber assembly can be controlled by the piston of the jack pushing the specimen cell upwards or downwards. Bronze can be used as electrodes. A pair of rectangular electrodes that are face to face is located in the specimen cell. The schematic diagram of the electrodes in the specimen cell is shown in Figure 2.

3.2   *Oriented fiber clipping device and electrode design*

In order to test the fiber current along the fiber axial direction or yarn, two ends of the fibers' bundle or yarn should be clipped by metal clip holders and the two clip holders have a certain distance; meanwhile, the clip holders act as the electrodes. The metal clip holders and clipping device are shown in Figure 3.

3.3   *Design of fabric holding device and electrodes*

Figure 4 shows the fabric holding device and electrodes. The electrode can use a three-circular flat plate electrodes system. It can test the fabric's volume current and surface current, respectively. The three-electrode system of the ZC36 high-resistance meter can be used in this real-time measuring system directly, which does not need to be modified. The construction in detail of the three-electrode system can be seen from the ZC36 high-resistance meter manual and the Chinese standard method for solid electrical insulating materials (GB/T 2006).

4   DESIGN OF THE MEASURING CIRCUIT

4.1   *Design of the amplifying circuit*

Generally, fiber polymer construction makes itself to conduct weakly. At certain temperature and moisture conditions, fiber conductivity can be of several orders of magnitude. For example, cotton fiber has a good hydroscopicity. Its moisture regain is in the range of 3% to 15%, and the order of magnitude of fiber current can be in the range of about 10^{-5} to 10^{-9} A. First, standard resistors with different resistance should be used in order to expand the measuring range; second, the current of fibers

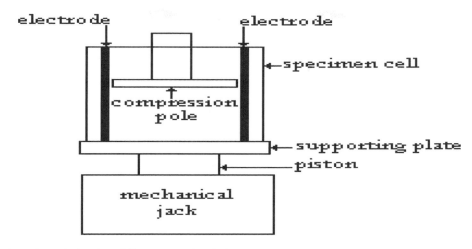

Figure 2. Schematic drawing of fiber compression device.

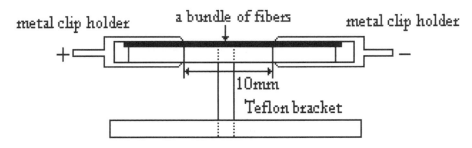

Figure 3. Schematic of oriented fibers/yarn metal clip holders and device.

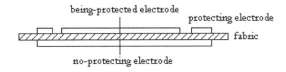

Figure 4. Schematic of fabric holding device and electrodes.

can be obtained indirectly by measuring the voltage of standard resistors. The amplification coefficient should not be too high. The very high coefficient will make the noise signal increasing, which has an influence on the measurement of the fiber current signal. Therefore, the amplification coefficient was set to 1000 or so. The two-stage amplification circuit is used for the circuit, because more stages of circuit can bring more circuit noise signals. In order to make the current flowing through fiber pass though the standard resistor as much as possible, the input resistance of the amplifying circuit is at least 100 times more than that of standard resistors. The OPA2111 instrument amplifier is used for this circuit because of lower input bias current, improved input offset voltage, and wider common mode voltage range, and because it is extremely stable with respect to time and temperature. It processes a high input resistance of 10^{13} Ω, so the resistance of the standard resistor is no more than 10^{11} Ω. In order to reduce the temperature drift, the same two amplifiers are composed of differential amplifying circuit, which can also obtain a high input resistance. The circuit is shown in Figure 5.

4.2 Design of anti-interference

Fiber current is so weak that it can be easily interfered by high frequency and electrical power frequency. In order to reduce the influence, it is necessary to make a special design on the circuit. At first, the circuit is needed to be electro-magnetic shielding. The whole circuit is put into an electro-magnetic shielding box in order to restrain it from electro-magnetic

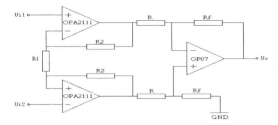

Figure 5.   Schematic of differential amplifying circuit.

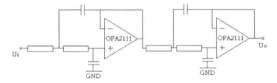

Figure 6.   Schematic of the fourth-order low-pass filtering circuit.

interference. The shielding cable is used as the signal transporting wire, and the grounding terminal of the circuit is grounded quite well. Signal input terminals of the amplifying circuit need to be used as the shielding ring in order to reduce the leaking current and contact noise from the printing circuit board. Meanwhile, the layout of the circuit should be wide to reduce the influence of stray capacitance.

In order to restrain the interfering signal, the low-pass filtering circuit is needed to cut off high-frequency signals. The fourth-order low-pass filter circuit is added to the measuring circuit. The gain is a unit and the cut-off frequency is 40 Hz. The filtering circuit is shown in Figure 6.

5   DESIGN OF DEVICES AND ELECTRODES FOR TEXTILES

5.1   Design of the testing program

The measuring function is developed based on LabVIEW, which is a famous virtual instrument platform (Long & Gu 2008). LabVIEW is a kind of graphic programming language. LabVIEW uses the structure diagram to construct the programming code. It is basically unnecessary to program the code, while using the icon, connecting line and block chart composed of the program. The program can be divided into front panel and back panel. The front panel is the interface of the application program, and consists of controls and indicators. The user can control the data input and program running by controlling controls, and indicators mainly indicate the results that the user wants to know on the PC monitor.

By software controlling, data acquisition card puts the voltage signal in the designated channel into the computer, and the voltage data signal is processed in the computer and transformed into the current signal and displayed by the indicators, and thus the user can observe the variation in real-time current intuitively. With the acquired voltage signal, the data is saved into the files for subsequent processing. In this real-time measuring system, the electrodes in the random fiber assembly compression device, the metal clip holders on the fiber clipping device and the electrodes on fabric holding device are connected to the channel zero of data acquisition card PCI1713 by wires, and all of them are connected with the selecting toggle switch. When testing the current of random fiber assembly, oriented bundle of fibers or fabrics, it can be chosen by the selecting toggle switch.

A 12-bit, 32-channel isolated analog input card PCI 1713 is used for acquiring the voltage signal. The PCI-1713 provides a sampling rate up to 100 K samples/s, and its conversion time is 2.5 us. At the ±5 V output, the card minimum resolution of is 2.44 mV. There are three trigger models for A/D conversion. Here, the software triggering is used. Part of the graphic programming codes of the voltage acquisition program and the real-time fiber electrical current tester front plane is shown in Figures 7 and 8, respectively.

5.2   Estimation of measuring error and measurement range

According to Equation (1), the current relative error can be calculated by

$$\frac{\Delta I}{I} = \pm \frac{\Delta U}{U} \pm \frac{\Delta K}{K} \pm \frac{\Delta R}{R} \pm \frac{\Delta n}{n} \qquad (2)$$

where $\Delta U/U$ is the relative error of PCI1713 card with gain 1, 0.01%FSR ± LSB; $\Delta K/K$ is the relative error of amplification coefficient, 1%; $\Delta R/R$ is the relative error of the standard resistor, 1%; and $\Delta n/n$ is the relative error of the dividing current, less than 1%. Therefore the relative error of measuring current is no more than 3.01%.

The amplification circuit can detect the voltage V_i ranging from 2.44 uV to 5 mV, and eleven standard resistors from 10 Ω to 10^{11}Ω. With $R \gg R_i$, the current can be calculated by

$$I = \frac{V_i}{R_i} = \frac{V}{R} \qquad (3)$$

For the description of the symbols I, V_i, R_i, V and R refer to Equation (1) and Figure 1. Therefore, the range of the current measurement is from about 2.44×10^{-17} A to 5×10^{-3} A.

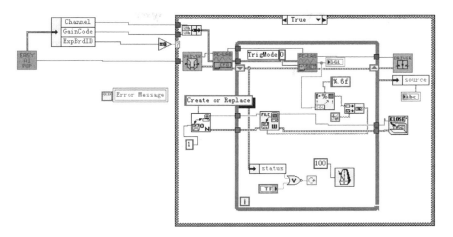

Figure 7.   Part of the graphic programming code of voltage acquisition.

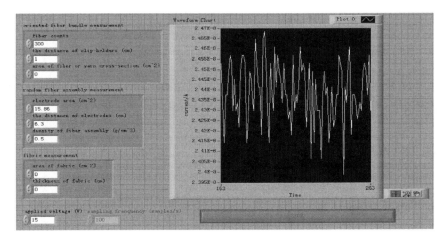

Figure 8.   The real-time textiles electrical current front plane.

5.3   *Measuring effect*

Cotton fiber grade 3 from Shandong Province, its moisture regain 6%, at atmosphere temperature 15°C, put it 50 g into the fiber compression device and compress it to density 0.5 g/cm³, then rotate the selecting toggle switch to random fiber assembly shift. Figure 9 shows that at 15 V, the direct current of the random cotton fiber assembly changed with time, which is from the circuit switch off as the zero second to 26.3 seconds.

From Figure 9, it can be clearly seen that the current of the random cotton fiber assembly at the above laboratory condition changes with the time, and when the circuit switch is off, the peak current appears at 5.19×10^{-8} A, and the current

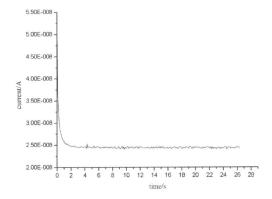

Figure 9.   Variation of random cotton fiber assembly current with time.

521

falls quickly with the time prolonging within about 4 seconds. The figure more intuitively shows the variation of fiber current with time than do the traditional meters.

6 CONCLUSIONS

The real-time textiles current measuring system was developed based on the LabVIEW platform, and it has a higher accuracy and wide measuring range. The relative error of the measuring current is no more than 3.01% and the measuring range is $2.44 \times 10^{-17} \sim 5 \times 10^{-3}$ A. By designing the real-time measuring system, it can be seen that this textiles measuring system based on the virtual instrument technology compared with the traditional measuring technology is more flexible, more functional and more vivid to reflect the variation of material current with time, and the traditional fibers, yarns, random fiber assembly or fabrics resistance meters can be integrated into one measuring system. Meanwhile, the current can be easily transformed to resistance, conductivity or resistivity, which can be shown on the PC monitor for the user by adding some simple graphic programming code. The real-time textiles current testing system based on the virtual instrument technology provided a brand new technical means to research textiles electrical performances.

ACKNOWLEDGMENT

The authors are grateful to the Textile Engineering Laboratory of Yancheng Institute of Technology and the Academician Research Center of Xi'an Polytechnic University, and acknowledge the financial support from the General Project of Jangsu Province University Natural Science Fund No. 13 KJD430006.

REFERENCES

[1] ASTM D 4238. 1990. Standard Test method for Electrostatic Properties of Textiles.
[2] ASTM STP926. 1987. Engineering Dielectric, Volume 2 B. Electrical Properties of Solid Insulating material.
[3] Byler, R.K. 2004. A device for rapidly predicting cotton lint moisture content. Trans. ASAE 47: 1331–1335.
[4] BS EN 1149-2. 1997. Protective clothing-Electrostatic Properties Part 2. Test method for measurement of the electrical resistance through a material (vertical resistance).
[5] Cusick, G.E. & Hearle, J.W.S. 1955. The effect of voltage and time on the electrical resistance of cotton. Text. Res. J. 25: 563–566.
[6] GB 6102.2. 1985. Test Method for Moisture Regain in Raw Cotton by Electrical Moisture Meter.
[7] GB/T 1410. 2006. Methods of test for volume resistivity and surface resistivity of solid electrical insulating materials.
[8] GOST 6433.2-71. 1999. Measurement of Electrical Resistance of Textile Materials, textile assemblies, in: Proceedings of the 27 Annual Meeting of ESA.
[9] Hearle, J.W.S. 1953. The electrical resistance of textile materials: I. the influence of moisture content. J. Text. Inst. 44: 117–143.
[10] Hearle, J.W.S. 1953. The electrical resistance of textile materials: III miscellaneous effects. J. Text. Inst. 44: 155–170.
[11] Hersh, S.P. & Montogomery, D.J. 1952. Electrical resistance measurement on fibers and fiber assemblies. Text. Res. J. 22: 805–818.
[12] Ismail, K.M. & Alyahya, S.A. 2003. A quick method for measuring date moisture content. Trans. ASAE 46: 401–405.
[13] Long, H.W. & Gu, Y.G. 2008. LabVIEW 8.2.1 and DAQ. Beijing: Tsinghua University Press.
[14] Mu, Yao. 2009. Textile material science. Beijing: China Textile & Apparel Press.
[15] Reddick, J.A. & Mayne, S.C. & Berkley, E. 1959. Improved instrument to measure the moisture content of lint cotton, seed cotton and cottonseed. Text. Res. J. 29: 219–222.

Material Science and Engineering – Chen (Ed.)
© 2016 Taylor & Francis Group, London, ISBN 978-1-138-02936-1

Thermal degradation of cotton cellulose modified with magnesium complexes

X.J. Ren, H. Wang, C.G. Song & M. Gao
School of Environmental Engineering, North China University of Science and Technology, Yanjiao, Beijing, China

ABSTRACT: Complexes of cell-THPC-thiourea-ADP with Mg^{2+} were prepared. The thermal stability and smoke suspension of the samples were determined by TG and DTA. The activation energies for the second stage of thermal degradation were obtained by following the Broido equation. Experimental data showed that for the complexes of cell-THPC-thiourea-ADP with Mg^{2+}, the activation energies and thermal decomposition temperatures were higher than those of cell-THPC-thiourea-ADP, suggesting that the metal ion can increase the thermal stability of cell-THPC-thiourea-ADP.

Keywords: cotton cellulose; thermal stability; flame retardant; smoke suspension

1 INTRODUCTION

Cotton cellulose is used extensively to make life pleasant, comfortable and colorful. Unfortunately, it is flammable and causes a fire hazard. According to fire statistics, about 50% of fire accidents are caused by textiles in the world (Gaan & Sun, 2007), and cotton cellulose is one of the important components in textiles. So, the emphasis on reducing combustibility has centered on its chemical modification. There are also many studies conducted on the thermal degradation of cotton cellulose treated with flame retardants (Lessan et al. 2011). However, with the requirement of environmental sustainability, the effects of flame retardants on both smoke generation and the toxicity of combustion products have become critical, as flame-retardant cellulose has been reported to produce denser smoke than pure cellulose (Grexa & Lubke 2001).

In previous papers (Tian et al. 2003), compounds of transition metals have been found to be effective smoke retardants. However, there is no information about the effects of Mg^{2+} on the smoke suspension and thermal degradation of cotton cellulose treated with flame retardants. So, the main objective of this work is to investigate the effects of Mg^{2+} on the thermal degradation and smoke suspension of cotton cellulose modified with flame retardants.

In this paper, complexes of cell-THPC-thiourea-ADP with Mg^{2+} were prepared. The thermal degradation of the samples was studied from ambient temperature to 800°C by TG and DTA.

2 EXPERIMENTAL DESIGN

2.1 *Materials*

Cotton cellulose of commercial grade (Hebei province, China) was selected for flame-retardant treatment. The cotton cellulose sample was immersed in a 24% NaOH solution at room temperature for 24 h (mercerization process). The alkali was then filtered off and the sample was washed repeatedly with distilled water. The sample was dried in an oven at 60°C and then stored in a desiccator.

2.2 *Instrumentation*

The elemental analysis was carried out using a Carlo Eroa 1102 Elemental Analyzer. LOI values were determined in accordance with ASTM D2863-70 by means of a General Model HC-1 LOI apparatus. Thermogravimetry (TG) was carried out on a DTA-2950 thermal analyzer (Dupont Co., USA) under a dynamic nitrogen (dried) atmosphere at a heating rate of 10°C min⁻¹.

2.3 *Cotton cellulose treatment*

The preparation of the samples has been described previously (Tian et al. 2003). THPC (Shanghai, China) was neutralized with NaOH to give a pH value equal to 6.5, and its 45% solution was mixed with a 22.5% thiourea solution. The pH value was adjusted to 6.5 and a small amount of ADP was added. The resulting mixture was used as the treating solution. The mercerized cotton cellulose was immersed in the treating solution for 30 min at room temperature. The treated cotton cellulose was dried at 60°C in an oven for 60 min. Curing of these treated cellulose samples was carried out by heating at 160°C for 5 min in the oven. After cooling, the sample was thoroughly washed with distilled water for an hour and dried in an oven at 60°C. Mg^{2+} complexes of cell-THPC-thiourea-ADP were prepared by treating 6 g of cell-THPC-thiourea-ADP in each instance with 5% aqueous solutions of $MgSO_4$ at room temperature for 72 h under constant stirring. Each product was washed repeatedly with water until the filtrate was free from the metal salt and dried overnight in an oven at 60°C, and then stored in a desiccator.

3 RESULTS AND DISCUSSION

The DTA and TG curves of (1) cotton cellulose, (2) cell-THPC-thiourea-ADP, and (3) Mg^{2+} complexes of cell-THPC-thiourea-ADP were obtained in a dynamic air atmosphere from ambient temperature to 800 °C, and are shown in Figure 1.

3.1 Differential thermal analysis

From the DTA curves of samples 1–3, the initiation temperatures (T_i), peak temperatures (T_p) and termination temperatures (T_t) of the various endotherms and exotherms were investigated, and are given in Table 1. The DTA curve of cotton cellulose shows two large exotherms with their respective peak maxima at 363 and 459°C. Before 350°C, decomposition and dehydration occur to form some flammable volatile products, and the first exotherm peaking at 363°C is due to the oxidation of these volatile products. Another exotherm, peaking at 459°C, represents the oxidation of charred residues. The dehydration process dominates at

low temperatures and ultimately leads to a carbonaceous residue. At higher temperatures, the cleavage of glycosyl units by intra-molecular transglycosylation starts, forming ultimately a tarry mixture with levoglucosan as the major constituent (Kandola et al. 1996). Levoglucosan decomposes into volatile and flammable products, and it therefore plays a key role in the flammability of cellulose.

The DTA curve of cell-THPC-thiourea-ADP is quite distinct from that of pure cotton cellulose. The treated cotton cellulose seemed to decompose into two steps (Mostashari, S. M. & Mostashari. S. Z. 2008). A breakdown or depolymerization of the THPC-thiourea-ADP finish, a catalyzed dehydration of the cellulose, and some bond formation occurred during the first step. Two large exotherms with peak maxima, respectively, at 326 and 485°C are shown in Figure 2. The second step involved a breakdown of the cellulose chain, evolution of gases from both the cellulose and the finish polymer, and the continuation of bond formation. The bond formation was probably due to a phosphorylation reaction at the C-6 hydroxyl group of the

Table 1. Peak temperatures in DTA curves of samples 1–3.

No.	Compound	DTA curve			Nature of peak
		T_i	T_p	T_t	
1	Cellulose	320	363	403	Exo (large)
		403	459	539	Exo (large)
2	Cell-THPC-thiourea-ADP	255	326	414	Exo (large)
		414	485	614	Exo (large)
3	Mg^{2+} complex of cell-THPC-thiourea-ADP	230	254	280	Endo (small)
		280	320	400	Exo (large)
		400	490	560	Exo (large)

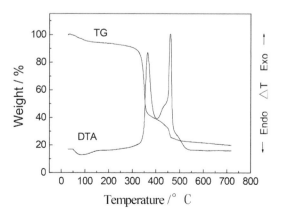

Figure 1. Thermal analysis curves of cotton cellulose.

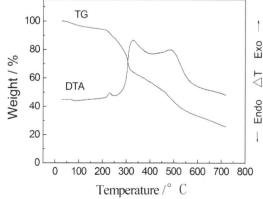

Figure 2. Thermal analysis curves of cell-THPC-thiourea-ADP.

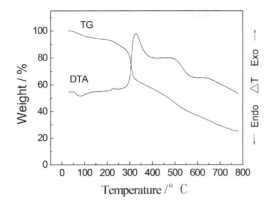

Figure 3. Thermal analysis curves of the Mg²⁺ complexe of cell-THPC-thiourea-ADP.

anhydroglucose unit, as suggested. Phosphorylation at this position would inhibit the formation of levoglucosan and prevent further breakdown into flammable gases. This would account for the increased amount of char formed over that for untreated cotton cellulose. The last large exotherm peaking at 485°C is due to the combustion of the char (Janowska et al. 2008).

In the DTA curves of the Mg^{2+} complexes of cell-THPC-thiourea-ADP shown in Figure 3, the first peak, a new endotherm with peak maxima in the range of 230–280°C, represents depolymerization, a catalyzed dehydration of the cellulose and some bond formation. For the metal complexes (sample 3), there are also two large exotherms in each case. The decomposition stage, which is represented by the first exotherm, is very different in the decomposition temperature in the complexes. The peak temperature of the complexes is 320°C. The last exotherm, which is due to the oxidative decomposition of the residual products, also shows considerable variation in the complexes.

For cotton cellulose, the two exotherms are sharp and narrow, which shows a large rate of heat release. For samples 2–3, the two exotherms become small and broad. Heat release is distributed between two broad peaks covering a wide area, resulting in a major reduction in the rate of heat release and flammable products, which fuel the flaming combustion reaction. On the other hand, the second exotherms become much smaller for samples 2–3, which indicates that the oxidation of the charred residues becomes more difficult due to the existence of flame retardants.

3.2 Thermogravimetry

From Figures 1–3, it can be seen that the second stages in the thermal decomposition of the samples,

decompose mainly and quickly, playing a key role in combustibility. So, we mainly discuss this stage. Temperature Range (TR) and Mass Loss (ML) at the second stage (quick mass loss rate) in TG are listed in Table 2. Generally, at lower temperatures, the thermal degradation of cellulose includes dehydration, depolymerization, oxidation, evolution of carbon monoxide, carbon dioxide, and formation of carbonyl and carboxyl groups and ultimately a carbonaceous residue. At higher temperatures, cellulose decomposes into a tarry mixture (mainly levoglucosan), which further decomposes into volatile and flammable products. The main role of flame retardants containing phosphorus is to minimize the formation of levoglucosan by lowering the decomposition temperature of cellulose and enhancing char formation by catalyzing the dehydration and decomposition reaction. However, lowering the decomposition temperature of cellulose decreases its thermal stability, which is not favorable. So, the two points must be considered simultaneously.

From Figures 1–3 and Table 2, it can be seen that for the cotton cellulose (sample-1), the initial decomposition temperature is 326°C, the second stage is in the range of 326–365°C, and the mass loss is 46%. For cell-THPC-thiourea-ADP, the initial decomposition temperature is 214°C and the second stage is in the range of 224–318°C. All these much decreased compared with those of cotton cellulose, which shows that the thermal stability of the cell-THPC-thiourea-ADP is much decreased because of the catalyzing dehydration and decomposition reaction. For sample 3, the decomposition temperature range is 280–310°C, higher than that of cell-THPC-thiourea-ADP, which shows that the thermal stability of samples is increased.

In order to understand the flame-retardant properties of these samples, the LOI of the samples is measured, which are given in Table 2. From Table 2, we can see that samples 2–3 show high values of LOI. This suggests that the combustibility of cotton cellulose treated with flame retardants decreases.

Table 2. Thermal degradation and analytical data of the samples.

No	1	2	3
TR (°C)	326–365	214–318	280–310
ML (%)	46	28	18
Ea (kJ/mol)	198.3	78.1	87.2
CY (%)	20.2	26.5	25.4
LOI (%)	18.0	29.6	28.0
P (%)		1.69	1.40
N (%)		3.78	3.05
M (%)			0.73

Moreover, the sample 2 containing the highest content of phosphorus and nitrogen shows the highest value of the LOI. The second stage of decomposition for sample 2 occurs at lower temperatures (224–318°C) and produces less flammable volatile products, resulting in higher flame retardancy.

3.3 *Activation energies*

The kinetic parameters for the second stage were determined using the following equation, given by Broido (Broido, 1969):

$$\ln\left(\ln\frac{1}{y}\right) = -\frac{Ea}{R}\cdot\frac{1}{T} + \ln\left(\frac{R}{Ea}\cdot\frac{Z}{|\hat{A}|}\cdot T_m^2\right)$$

where y is the fraction of the number of initial molecules not yet decomposed; T_m is the temperature of the maximum reaction rate; β is the rate of heating; and Z is the frequency factor.

For the second stage, the major degradation and mass loss stage, the energy of activation for sample 2 is 78.1 kJ/mol, which is much decreased compared with cotton cellulose (198.3 kJ/mol). The reason is that the flame retardant catalyzes the decomposition reaction. The lower decomposition temperatures (224–318°C) also support this. The energy of activation for sample 3 is 85.1 kJ/mol and is higher than that of sample 2, which shows that the thermal stability of sample 3 is increased.

4 CONCLUSIONS

For complexes of cell-THPC-thiourea-ADP with metal ions, the activation energies and thermal decomposition temperatures are higher than those of cell-THPC-thiourea-ADP. The metal ions (Mg^{2+}) can increase the thermal stability of cell-THPC-thiourea-ADP. However, the two exotherms in DTA curves are very different in the decomposition temperature in all the complexes.

ACKNOWLEDGMENT

The work was supported by the fundamental research funds for the Central Universities (3142013102).

REFERENCES

[1] Gaan, S. & Sun, G. 2007. Effect of phosphorus flame retardants on thermo-oxidative decomposition of cotton. Polymer degradation and stability, 92(6), 968–974.
[2] Lessan, F. Montazer, M. & Moghadam, M.B. 2011. A novel durable flame-retardant cotton fabric using sodium hypophosphite, nano TiO$_2$ and maleic acid. Thermochimica Acta, 520, 48–54.
[3] Grexa, O. & Lubke, H. 2001. Flammability parameters of wood tested on a cone calorimeter. Polymer Degradation and Stability, 74, 427–432.
[4] Tian, C.M. Xie, J.X. Guo, H.Z. & Xu, J.Z. 2003. The effect of metal ions on thermal oxidative degradation of cotton cellulose ammonium phosphate. J. Therm. Anal. Cal., 73, 827–834.
[5] Kandola, B.K. Horrocks, A.R. Price D. & Coleman, G.V. 1996. Flame-retardant treatments of cellulose and their influence on the mechanism of cellulose pyrolysis. J. Macromal Sci, Rev Macromol Chem Phys., C36 (4), 721–794.
[6] Mostashari. S.M. & Mostashari. S.Z. 2008. Combustion pathway of cotton fabrics treated by ammonium sulfate as a flame-retardant studied by TG.J. Therm. Anal. Cal. 91, 437–441.
[7] G. Janowska, T. Mikołajczyk & M. Olejnik. 2008. Effect of montmorillonite content and the type of its modifier on the thermal properties and flammability of polyimideamide nanocomposite fibers. J. Therm. Anal. Cal. 92, 495–503.
[8] Broido, A. 1969. A simple, sensitive graphical method of treating thermogravimetric analysis data. Journal of Polymer Science Part A-2: Polymer Physics. 7(10), 1761–1773.

had no significant effects on the molecular conformation of SF. However, the absorption peaks at 1320 cm⁻¹ and 770 cm⁻¹ in spectra b, c and d were disappeared compared to spectrum a, indicating that the sericin was removed by 0.1, 0.25 or 0.4 wt.% Na_2CO_3 solution.

According to studies on the crystalline structures of wild SF, with Cu Kα radiation, the main XRD peaks of α-helix structure are appeared at 11.8° and 22.0°, and those of β-sheet structure are at 16.5°, 20.2°, 24.9°, 30.90°, 34.59°, 40.97° and 44.12° (Tao et al. 2007).

As shown in Figure 2a, there were strong absorption peaks around 16.5° and 20.2° and medium peaks around 24.2°, indicating that undegummed *A. y* silk fibers have high crystallinity and the main molecular conformations of them were β-sheet structure. These diffraction peaks of undegummed fibers at 14.88°, 24.32°, 29.8° and 38.19° were observed. The presence of three bands at 14.88°, 24.32° and 29.8° were attributed to the characteristic peaks of SiO_2. The medium peak around 29.8° was the characteristic peaks of Na_2CO_3 and $CaCO_3$, and peak at 38.19° was assigned to the characteristic peak of Al_2O_3. The reason is that dust grains in the air were deposited on the sericin during spinning cocoons in wild (Li et al. 2012; Chun et al. 2014; Vielle et al. 2003, Min et al. 2014). By contrast, all fibers (undegummed and degummed) exhibited peaks at 16.5°, 20.2° and 24.2°. The results showed that the crystalline structure of SF remained unaffected by the concentrations of Na_2CO_3 solution. However, the diffraction peaks at 14.88°, 24.32°, 29.8° and 38.19° in spectra b, c and d disappeared compared to spectrum a.

The change indicated that 0.1, 0.25, or 0.4 wt.% Na_2CO_3 solution can remove the sericin surrounding silk fibers. The results were consistent with data from FTIR spectra.

3.2 Dissolution of A. y SF fibers

The *A. y* SF fibers were hardly dissolved in 7 M $CaCl_2$ solution at 40 °C, 70 °C, 95 °C for 1 h, indicating $CaCl_2$ solution had no obvious effect on dissolution of *A. y* SF fibers.

As shown in Table 1, when the molar ratios of $CaCl_2$ to EtOH were 1:1, 1:2 and 1:3, the solubility of silk fibroin fibers were 12.2, 11.5 and 9.3 wt.% at 70 °C, respectively. The results showed the *A. y* SF fibers cannot be totally dissolved by the mixed solution of $CaCl_2$ and EtOH.

As seen in Table 2, the ability of $Ca(NO_3)_2$ solution to dissolve the *A. y* SF fibers was obviously stronger than that of $CaCl_2$ solution. As the dissolution temperature was increased to 90 °C, the solubility was 99.7 wt.%.

Table 3 showed the dissolution of *A. y* SF fibers in LiSCN solution. The solubility of SF fibers in 9 M LiSCN solution at 35 °C reached only approximately 12.7 wt.%. When the concentration of LiSCN solution was increased to 11 M, the solubility was 99.0 wt.%.

The dissolving temperature can affect the dissolution of *A. y* SF fibers. As shown in Table 3, the solubility of *A. y* SF fibers in 11 M LiSCN solution at 20 °C was only about 55.1 wt.%. When the temperature was increased to 35 °C, the solubility was 99.0 wt.%.

The solubility was also affected by dissolution time. When *A. y* SF fibers were added in 11 M LiSCN solution for 20 min at 35 °C, the solubility can only reach to 93.8 wt.%. As the time prolonged to 60 min, solubility reached to 99.7 wt.%.

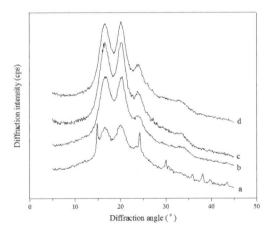

Figure 2. X-ray diffraction curves of *A. y* silk fibers: (a) undegummed *A. y* silk fibers; degummed *A. y* SF fibers by (b) 0.1 wt.%, (c) 0.25 wt.%, and (d) 0.4 wt.% Na_2CO_3 solution.

Table 1. Influence of the molar ratio of $CaCl_2$ to EtOH on solubility of *A. y* SF fibers.

$CaCl_2$: EtOH	Solubility wt.%
1:1	12.2
1:2	11.5
1:3	9.3

Table 2. Influence of dissolving temperature on solubility of *A. y* SF fibers in $Ca(NO_3)_2$.

Temperature [°C]	60	75	90	105
Solubility [%]	24.8	40.0	99.7	99.7

Table 3. Solubility of *A. y* SF fibers in LiSCN solution.

Condition	Temperature 35 °C Time 1 h				Time 1 h concentration 11 M			Temperature 35 °C concentration 11 M				
	Concentration (M)				Temperature (°C)			Time (min)				
	5	7	9	11	20	35	50	20	40	60	80	100
Solubility [%]	10.4	10.0	12.7	99.0	55.1	99.0	99.2	93.8	98.5	99.0	99.0	99.1

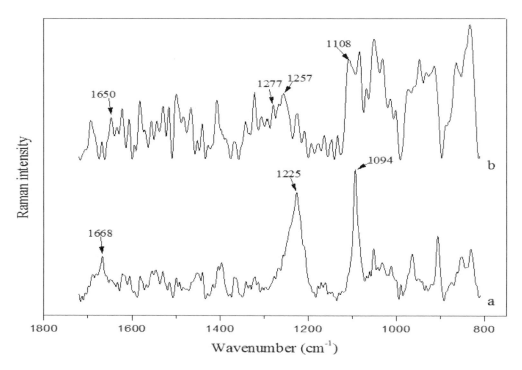

Figure 3. Raman spectra of *A. y* SF: (a) *A. y* SF fibers, (b) regenerated *A. y* SF.

3.3 Laser Raman spectroscopy of regenerated A. y SF

The Raman spectrum is sensitive to the molecular conformation of *A. y* SF. It was reported previously that, wild SF showed strong absorption bands at 1657 cm^{-1}, 1263 cm^{-1} and 1106 cm^{-1}, all assigned to α-helix structure. The bands at 1668 cm^{-1}, 1230 cm^{-1}, 1095 cm^{-1} and 1073 cm^{-1} were attributed to β-sheet structure (Zheng et al. 2010, Tao et al. 2007, Tsukada et al. 1995). The Raman spectrum of *A. y* silk fibers (Fig. 3a) displayed peaks at 1667 cm^{-1}, 1226 cm^{-1} and 1095 cm^{-1}, which were the characteristic bands of β-sheet structure. These meant that the molecular conformation of *A. y* silk fibers was primarily characterized by β-sheet structure. The results were consistent with that of the FTIR (Fig. 1) and XRD (Fig. 2). The Raman spectrum of *A. y* SF fibers dissolved by 11 M LiSCN solution for 1 h showed a presence of strong peaks at 1650 cm^{-1}, 1277 cm^{-1}, 1257 cm^{-1} and 1108 cm^{-1}. These specific peaks were related to α-helix structure. Results showed that β-sheet structure inside the *A. y* SF was destroyed during the dissolution, and the molecular conformation of *A. y* SF spontaneously aggregate into α-helix structure in aqueous solution.

4 DISCUSSION

The sericin surrounding *B. m* fibers can be removed by enzymes, acid and alkaline aqueous solution. The sericin in *B. m* fibers was able to be effectively degummed by 0.02–0.05 wt.% Na$_2$CO$_3$ (Cao et al.

2013, Wray et al. 2011, Sumana et al. 2013, More et al. 2013). However, sericin of *A. y* silk fibers was different from that of *B. m* silk fibers. The reason is that, *A. y* silk fibers were made during spinning cocoons in the wild, *A. y* must resist worse environment than *B. m*. The ability of *A. y* silk sericin to resist chemical reagents was stronger than that of *B. m* silk sericin. In this study, the degumming and dissolving methods of *A. y* silk fibers were investigated. As shown in Figure 1 and Figure 2, *A. y* silk sericin was removed by 0.1–0.4 wt.% Na_2CO_3.

B. m SF fibers were readily dissolved by concentrated neutral salts solution such as LiBr, $Ca(NO_3)_2$, LiSCN, $Ca(Cl)_2$, $Ca(SCN)_2$, etc. However, the molecular conformation was different between *A. y* SF fibers and *B. m* SF fibers. The conformation of *A. y* SF fibers is mainly antiparallel β-sheet. Hydrophobic interaction occurs in -$(Ala)_n$-. This intermolecular bonding is sufficiently strong to prevent the separation of molecular, so it is difficult to dissolve *A. y* SF fibers in various chemical solvents (Kundu et al. 2014, Hayashi et al. 1999). The results in this study indicated that *A. y* SF fibers were solubilized in $Ca(NO_3)_2$ solution at 90 °C for 1 h, and *A. y* SF fibers were also dissolved by 11 M LiSCN solution at 35 °C for 1 h. However, *A. y* SF fibers remained undissolved in $CaCl_2$ solution. Even if ethanol can promote the penetration of solvents, the solubility only can reach to 12.2 wt.% in the mixed solution of $CaCl_2$ and ethanol.

5 CONCLUSION

The sericin surrounding *A. y* silk fibers was able to be removed by 0.1–0.4 wt.% Na_2CO_3 at 98–100 °C. Degumming process had no obvious effects on the molecular conformation of *A. y* SF fibers. *A. y* SF fibers were successfully dissolved by aqueous $Ca(NO_3)_2$ at temperatures above 90 °C. Meanwhile, the fibers also can be totally dissolved by LiSCN solution at temperatures over 35 °C. However, high concentration of $CaCl_2$ solutions and the mixed solution of $CaCl_2$ and ethanol could not dissolve the silk fibroin fibers. The molecular conformation of *A. y* SF obtained by dissolved in LiSCN solution was mainly α-helix structure.

ACKNOWLEDGEMENTS

This work was supported by the National Nature Science Foundation of China (31370968), Natural Science Foundation of Jiangsu Province (BK20131177), College of Natural Science Research Project of Jiangsu Province (12KJA430003) and Jiangsu Province Science and Technology Support Program (BE2013734).

REFERENCES

[1] Cao, T.T., Wang, Y.J. & Zhang, Y.Q. 2013. Effect of strongly alkaline electrolyzed water on silk degumming and the physical properties of the fibroin fiber. PloS one 8(6): e65654.

[2] Chen, J.P., Chen, S.H. & Lai, G.J. 2012. Preparation and characterization of biomimetic silk fibroin/chitosan composite nanofibers by electrospinning for osteoblasts culture. Nanoscale research letters 7(1): 1–11.

[3] Chun, T.J., Zhu, D.Q., Pan, J. & He, Z. 2014. Recovery of Alumina from Magnetic Separation Tailings of Red Mud by Na_2CO_3 Solution Leaching. Metallurgical and Materials Transactions B 45(3): 827–832.

[4] Germershaus, O., Werner, V., Kutscher, M. & Lorenzl, M. 2014. Deciphering the mechanism of protein interaction with silk fibroin for drug delivery systems. Biomaterials 35(10): 3427–3434.

[5] Hayashi, C.Y., Shipley, N.H. & Lewis, R.V. 1999. Hypotheses that correlate the sequence, structure, and mechanical properties of spider silk proteins. International Journal of Biological Macromolecules 24(2): 271–275.

[6] Kar, S., Talukdar, S., Pal, S., Sunita, N., Pallavi, P. & Kundu, S.C. 2013. Silk gland fibroin from indian muga silkworm Antheraea assama as potential biomaterial. Tissue Engineering and Regenerative Medicine 10(4): 200–210.

[7] Kundu, B., Kurland, N.E., Yadavalli, V.K. & Kundu, S.C. 2014. Isolation and processing of silk proteins for biomedical applications. International journal of biological macromolecules 70: 70–77.

[8] Li, J.N., Chen, C.Z. & Wang, D.G. 2012. Surface modification of titanium alloy with laser cladding RE oxides reinforced Ti3Al–matrix composites. Composites Part B: Engineering 43(3): 1207–1212.

[9] Liu, G.F., Wang, X.L. & Hu, C. 1993. Thermal analysis and molecular structure of antheraea yamamai silk. Journal of Zhejiang Institute of Silk Textiles 10(1): 1–5.

[10] Masuhiro, T., Giuliano, F., Yoko, G. & Nobutami, K. 1994. Physical and chemical properties of Tussah silk fibroin fibers. Journal of Polymer Science Part B: Polymer Physics 32(8): 1407–1412.

[11] Mhuka, V., Dube, S. & Nindi, M.M. 2013. Chemical, structural and thermal properties of Gonometa postica silk fibroin, a potential biomaterial. International journal of biological macromolecules 52: 305–311.

[12] Min, Y.J., Hong, S.M., Kim, S.H., Lee, K.B. & Jeon, S.G. 2014. High-temperature CO_2 sorption on Na_2CO_3-impregnated layered double hydroxides. Korean Journal of Chemical Engineering 31(9): 1668–1673.

[13] More, S.V., Khandelwal, H.B., Joseph, M.A. & Laxman, R.S. 2013. Enzymatic degumming of silk with microbial proteases. Journal of Natural Fibers 10(2): 98–111.

[14] Sumana,D., Sudarshan, M., Thakur, A.R. & Chaudhuri, S.R. 2013. Degumming of raw silk fabric with help of marine extracellular protease. American Journal of Biochemistry and Biotechnology 9(1): 12.

[15] Tao, W., Li, M.Z. & Zhao, C.X. 2007. Structure and properties of regenerated Antheraea pernyi silk fibroin in aqueous solution. International journal of biological macromolecules 40(5): 472–478.

[16] Tsukada, M., Freddi, G., Monti, P., Bertoluzza, A. & Kasai, N. 1995. Structure and molecular conformation of tussah silk fibroin films: Effect of methanol. Journal of Polymer Science Part B: Polymer Physics 33(14).

[17] Vieille, L., Rousselot, I., Leroux, F., Besse, J.P. & Guého, C.T. 2003. Hydrocalumite and its polymer derivatives. 1. Reversible thermal behavior of Friedel's salt: a direct observation by means of high-temperature in situ powder X-ray diffraction. Chemistry of materials 15(23): 4361–4368.

[18] Wray, L.S., Hu, X., Gallego, J., Georgakoudi, I., Omenetto, F.G., Schmidt, D. & Kaplan, D.L. 2011. Effect of processing on silk-based biomaterials: Reproducibility and biocompatibility. Journal of Biomedical Materials Research Part B: Applied Biomaterials 99(1): 89–101.

[19] Yang, B.S., Li, J. & Wang, H. 2013. Research progress in sequences comparison and crystal structure of silk fibroin. Advanced Materials Research 664: 443–448.

[20] Zhang, F., Lu, Q., Yue, X.X., Zuo, B.Q., Qin, M.D., Li, F., Kaplan, D.L. & Zhang, X.G. 2015. Regenerated silk fibroin fiber with high quality wet spinning from CaCl2 formic acid solvent. Acta biomaterialia 12: 139–145.

[21] Zheng, Z.H., Wei, Y.Q. & Yan, S.Q. 2010. Preparation of regenerated Antheraea yamamai silk fibroin film and controlled-molecular conformation changes by aqueous ethanol treatment. Journal of applied polymer science 116(1): 461–467.

Material Science and Engineering – Chen (Ed.)
© 2016 Taylor & Francis Group, London, ISBN 978-1-138-02936-1

Effects of fractured cell-walls on the creep of low density open-cell foams

Z.G. Fan

Institute of Systems Engineering, CAEP, Mianyang, P.R. China

ABSTRACT: Open-cell foams have excellent mechanical and physical properties and are widely used in structural components, energy adsorption, heat transfer, sound insulation, and so on. When their in-service temperature is high, time dependent creep may become significant. By taking the mass at strut nodes into account, and based on the three dimensional Voronoi models and finite element method, the creep deformations of metal foams having fractured cell-walls morphological imperfection are calculated. Numerical results show that the morphological imperfection has a significant effect on foam creep performance; with the increasing of the fraction of struts broken, the secondary foam creep rates increase rapidly.

Keywords: open-cell foams; micromechanical model; creep; finite element method

1   INTRODUCTION

Low density open-cell foams made of high thermal conductive materials, such as aluminum and copper alloys, have excellent heat exchange performance and could be used as heat exchanger in autos, trains, aircrafts and micro-electronics [Ashby et al. 2000, Banhart 2001, Evans et al. 1998]. The in-service temperature of these applications could be well above $0.3T_m$, where T_m is the melting temperature. Under such circumstances, time-dependent creep deformation in metals is usually deemed to be significant and cannot be neglected [Gibson & Ashby 1997, Andrews et al. 1999a, Andrews et al. 1999b]. It is thus necessary to study the creep responses of foams carrying loads for long period of time and at higher temperatures [Andrews et al. 1999a].

Based on a bending deformation mechanism dominated cubic model, Gibson and her coworkers [Gibson & Ashby 1997, Andrews et al. 1999a] derived an equation for the uniaxial secondary creep rate of low density open-cell foams. While Hodge and Dunand (2003) received another creep rate expression for axial stretching deformation mechanism dominated foam models. Besides, Oppenheimer and Dunand (2007) constructed four different architectures which represented four types of foam models having various deformation mechanisms, and calculated the uniaxial secondary creep rates of them. Andrews and Gibson (2001) simulated the secondary creep rates of three two-dimensional cellular solids under uniaxial stress. And Huang and Gibson (2003) analyzed the uniaxial secondary creep rates of a

three-dimensional open-cell Voronoi model created with 27 random seeds. All of the Finite Element Analysis (FEA) results [Oppenheimer & Dunand 2007, Andrews & Gibson 2001, Huang & Gibson 2003] indicated that the changes of microstructure of foam models could affect the creep deformation remarkably. Moreover, in real metal foams, several morphological imperfections, including curved and wrinkled cell walls, non-uniform wall thickness, non-uniform cell size distribution, fractured cell-walls, and missing cells are inevitable, which may improve the creep rates remarkably. So a numerical study on the creep deformations of open-cell foams based on a accurate model is necessary.

In the work reported here, three dimensional (3D) Voronoi models are constructed to investigate the creep deformations of low density open-cell foams under uniaxial compressive and tensile loadings, and the effects of morphological imperfection, fractured cell-walls, on the creep of foams are focused.

2   FINITE ELEMENT MODEL

2.1   *3D Voronoi model*

A series of 3D Voronoi models with 128 cells are constructed using the software compiled by Gan et al. (2005), FEA code ABAQUS are adopted to simulate the time-dependent creep behavior of Voronoi foams. The constituent struts in the models are assumed to have a constant circular cross section and are meshed by Timoshenko beam elements (B32). According to the advices given by Chen et al. (1999) and Zhu et al. [Zhu et al. 2001,

Zhu & Windle 2002], periodic boundary condition is applied. One of the random Voronoi models is shown in Figure 1.

Power law equation (Eq.1) without primary creep is employed, as [Ashby et al. 2000, Gibson & Ashby 1997]

$$\dot{\varepsilon}_{ss}^{U} = A\sigma_s^n \exp\left(-\frac{Q_s}{RT}\right) = \dot{\varepsilon}_{0s}\left(\frac{\sigma_s}{\sigma_{0s}}\right)^n \qquad (1)$$

where the superscript U represents the uniaxial loading condition; $\dot{\varepsilon}_{ss}^{U}$ is the uniaxial strain rate;

A, $\dot{\varepsilon}_{0s}$ and σ_{0s} are the creep constants; σ_s is the uniaxial stress; n is the stress exponent; Q is the creep activation energy; R is the ideal gas constant and T is the absolute temperature. Creep properties of Al-6101 in T6 condition are introduced to model the properties of the solid making up the struts [Andrews et al. 1999a, Oppenheimer & Dunand 2007]: creep constant $A = 1.95 \times 10^3$ MPa^{-n}·s^{-1}, stress exponent $n = 4.0$ and activation energy $Q = 173$ KJ·mol^{-1}. Other properties of solid needed during FEA: the ideal gas constant $R = 8.314$ J·mol^{-1}·K^{-1}, Young's modulus $E_s = 69$ GPa and Poisson ratio $\nu_s = 0.35$.

2.2 Relative density

The foam relative density ρ is defined as the volume fraction of the solid material of foams. And in this study, all the struts are assumed to have the same and constant circular cross section with area S, while the area of the cross-section is variable along the strut length in real foams.

In general, for simplicity, the foam relative density is determined by the cross-section area S and the total length of struts $\sum l_i$,

$$\bar{\rho} = \frac{S\sum l_i}{V_0} = \frac{\pi r^2 \sum l_i}{L^3} \qquad (2)$$

where r is the radius of circular cross section of struts, and L is the edge length of Voronoi tessellations.

Figure 1. Three-dimensional open-cell Voronoi model.

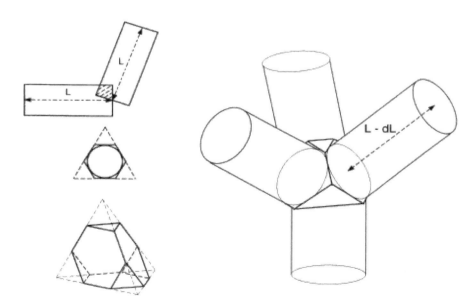

Figure 2. Foam solid volume consists of one truncated tetrahedron and four truncated cylindrical members. [Wang and Cuitino, 2000].

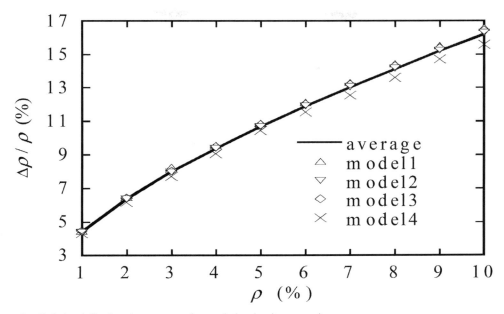

Figure 3.   Relative deflections between two foam relative density expressions.

As Gan et al. (2005) and Wang and Cuitino (2000) noted that the volume of intersection points of struts is calculated repeatedly in Eq. 2 (Fig. 2 shows the geometric structures on the intersection points of tetrakaidecahedral cell model). And the simplification is effective when the foam relative density is low enough, whereas the errors will be increasing with the increment of foam relative density. So a modified expression for the relative density of Voronoi foams was presented by Gan et al. (2005),

$$\rho = \frac{27\pi r^2\left(\sum l_i - 2\sqrt{2}rN\right) + 46\sqrt{6}Nr^3}{27L^3} \quad (3)$$

Here, N is the number of vertices of a Voronoi structure.

Figure 3 shows the relative deflections between two foam relative density expressions. It could be seen that with the increasing of foam relative density, the deviation increases quickly; for instance, the value of relative deviation can be up to 10% when the foam relative density equals 5%. So the mass at strut nodes should be taken into account, and Eq. 3 is used in this study.

3   NUMERICAL ANALYSIS

A damage parameter β, represented the fraction of struts fractured, is defined to quantify the imperfection, as the ratio of the number of fractured

struts N_{damage} to the total number of struts N_{intact} involved in the model [Gan et al. 2005].

$$\beta = N_{damage}/N_{intact} \quad (4)$$

The effect of randomly fracturing struts on the secondary creep rate of open-cell Voronoi foams is shown in Figure 4. With the increasing of the fraction of struts broken, the secondary foam creep rates increase rapidly; the creep rates of the damaged Voronoi foam, normalized by that of the intact foam, raise to −14.3 power of the damage parameter $1−\beta$, namely,

$$\frac{\dot{\varepsilon}_{damage}^{U}}{\dot{\varepsilon}_{intact}^{U}} = (1-\beta)^m \quad (5)$$

where $\dot{\varepsilon}_{damage}^{U}$ and $\dot{\varepsilon}_{intact}^{U}$ represent the secondary foam creep rates of open-cell foams with and without fractured cell-walls, respectively. The parameter m equals −14.3.

So fractured cell-walls morphological imperfection has a significant effect on the creep rate of open-cell Voronoi foams. Huang and Gibson (2003) also studied the effect of morphological imperfection, and found order influence of removal of a small fraction of the struts on the creep of open-cell Voronoi foams. Compared with the FEA results in this study, Huang and Gibson (2003) over-predicted the degree of the influence of fractured cell-walls morphological imperfection

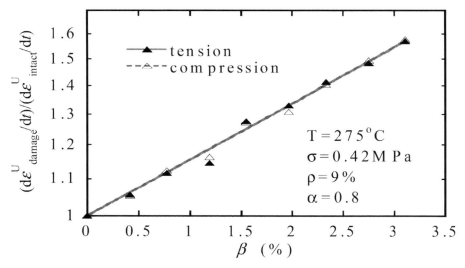

Figure 4. The creep rate of the damaged Voronoi foam, normalized by that of the intact foam, plotted against the fraction of struts fractured under uniaxial stress. (stress: $\sigma = 0.42$ MPa; temperature: $T = 250°C$; relative density: $\rho = 9\%$).

4 CONCLUSIONS

Using 3D Voronoi models and by FEA, the average secondary creep rates of foams having fractured cell-walls morphological imperfection under uniaxial compressive and tensile stresses are calculated. Effect of morphological imperfection on the creep rates of Voronoi models reveals that with the increasing of the fraction of struts fractured, foam creep rates increase quickly.

ACKNOWLEDGMENTS

This work is supported by the National Natural Sciences Foundation of China (No. 11372295).

REFERENCES

[1] Ashby, M.F. et al. 2000. *Metal Foams: A Design Guide.* Butterworth Heinemann: Oxford.
[2] Evans, A.G. et al. 1998. Multifunctionality of cellular metal systems. *Progress in Materials Science* 43(3): 171–221.
[3] Banhart, J. 2001. Manufacture, characterisation and application of cellular metals and metal foams. *Progress in Materials Science* 46(6): 559–632.
[4] Gibson, L.J. & Ashby, M.F. 1997. *Cellular Solids: Structure and Properties.* Cambridge: Cambridge University Press.
[5] Andrews, E.W. et al. 1999a. The creep of cellular solids. *Acta Materialia* 47(10): 2853–2863.
[6] Andrews, E.W. et al. 1999b. Creep behavior of a closed-cell aluminum foam. *Acta Materialia* 47(10): 2927–2935.
[7] Hodge, A.M. & Dunand, D.C. 2003. Measurement and modeling of creep in open-Cell NiAl foams. *Metallurgical and Materials Transactions A* 34(10): 2353–2363.
[8] Oppenheimer, S.M. & Dunand, D.C. 2007. Finite element modeling of creep deformation in cellular metals. *Acta Materialia* 55(11): 3825–3834.
[9] Andrews, E.W. & Gibson, L.J. 2001. The role of cellular structure in creep of two-dimensional cellular solids. *Materials Science and Engineering A* 303(1–2): 120–126.
[10] Huang, J.S. & Gibson, L.J. 2003. Creep of open-cell Voronoi foams. *Materials Science and Engineering A* 339(1–2): 220–226.
[11] Gan, Y.X. et al. 2005. Three-dimensional modeling of the mechanical property of linearly elastic open cell foams. *Int J Solids and Structures* 42(26): 6628–6642.
[12] Chen, C. et al. 1999. Effect of imperfections on the yielding of two-dimensional foams. *J Mech Phys Solids* 47(11): 2235–2272.
[13] Zhu, H.X. et al. 2001. Effects of cell irregularity on the elastic properties of 2D Voronoi honeycombs. *J Mech Phys Solids.* 49(4): 857–870.
[14] Zhu, H.X. & Windle, A.H. 2002. Effects of cell irregularity on the high-strain compression of open-cell foams. *Acta Mater* 50(5): 1041–1052.
[15] Wang, Y. & Cuitino, A.M. 2000. Three-dimensional nonlinear open-cell foams with large deformations. *J Mech Phys Solids* 48(5): 961–988.

on the foam creep rates as only 27 random seeds were used to construct Voronoi models by them.

Material Science and Engineering – Chen (Ed.)
© 2016 Taylor & Francis Group, London, ISBN 978-1-138-02936-1

BFRP-based FBG strain sensor

H.Y. Yu, Q. Wei, X.H. Qin, M. Wang & Y.D. Liu
School of Information and Electrical Engineering, Shandong Jianzhu University, Jinan, China

ABSTRACT: This paper presents a FBG strain sensor based on BFRP, which can monitor the surface strain of steel building structure and has a better sensing performance than most of the existing sensors. The structure of the BFRP-FBG sensor and its preparation method are described in this paper. The sensing properties of BFRP-FBG are tested preliminarily. This paper proves a preliminary proof of sensor-wise better performance.

Keywords: BFRP; FBG; strain sensor

1 INTRODUCTION

Optical fiber sensor in strain monitoring has better performance and obvious advantages than the traditional metal electronic ones [1]. Optical fiber strain sensor has the advantages of simple structure, strong corrosion resistance, immune to electromagnetic interference, distant signal transmission, low sensitivity to temperature [2], good insulation performance of sensors and sensing circuits [3], and suitable for use in the strain of building structure monitoring. In recent years, the Basalt Fiber Reinforced Plastics is more widely used. BFRP exhibits a good performance, and is more cheaper than the widely used carbon Fiber Reinforced Plastics. It has a better performance in insulation and heat insulation in addition to other advantages [4].

This paper presents a kind of optical fiber strain sensor, using fiber Bragg grating as the sensing original components, using Basalt fiber composite material [5] as packaging components, and used in building structure surface strain monitoring. This paper first introduces the basic principle of the optical fiber strain sensor, and then specifically explains the structural design and the manufacturing method of the optical fiber strain sensor. Finally, by adopting a method of simulating the surface strain of building structure using cantilever whose intensity is equal to steel [6], the optical fiber strain sensor is introduced and strain sensing performance is tested, concluding with the relationship between building structural strain and the maximum power and wavelength of the optical fiber sensor. The test result proves that the optical fiber strain sensor, introduced in this paper, demonstrates a satisfied capability of strain sensing.

2 THE PRINCIPLE OPTICAL FIBER STRAIN SENSOR

Based on the optical wave guide theory, when the phase matching condition is satisfied, the Bragg wavelength of fiber grating is given by

$$\lambda_B = 2n_{eff}\Lambda \qquad (1)$$

where λ_B is the Bragg wavelength; n_{eff} is the effective refractive index of optical fiber transmission mode; and Λ is the grating period. Figure 1 shows the principle of the fiber Bragg grating sensor.

The Bragg wavelength of optical fiber grating is affected by both the outside environment temperature and the axial force. When the environment temperature remains constant and optical fiber generates an axial strain along the axial direction only by an external force, the relationship between the variation and the strain of Bragg wavelength can be expressed as follows:

$$\frac{\Delta\lambda_B}{\lambda_B} = (1 - P_e)\Delta\varepsilon \qquad (2)$$

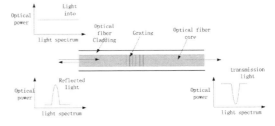

Figure 1. Fiber Bragg grating sensor principle.

According to formula (2), when FBG generates a strain variation $\Delta\varepsilon$ by an axial external force, the variation $\Delta\lambda_B$ of wavelength is linearly proportional to the axial strain $\Delta\varepsilon$.

When applied in reality, because strain is transferred from the measured object to optical fiber grating, during the transmission process, the strain through the interface will suffer an unavoidable loss. If the packaging material of the optical fiber grating sensor is excessively thick, there would be a moment of force leading to a lag of strain transfer [7]. Anyway, the strain will unavoidably suffer a loss in the real situation. The features of measured surface structure, properties of adhesive and forms of packaging [8] will directly influence the outcome of measurement of strain loss, thereby affecting the sensitivity of the measurement. If an unpacked optical fiber grating sensor is directly attached to the surface of the measured object, due to its small radial size that causes the contact area to be very small, it will lead to a strain transmission loss as well [9]. In order to optimize the performance of the optical fiber strain sensor, perfect package and less thickness are necessary.

3 DESIGN AND MANUFACTURING OF BFRP-FBG STRAIN SENSOR

3.1 Design of the optical fiber strain sensor

According to the features of the optical fiber sensor and the specificity of the strain measurement of tower components, a packaging form of unidirectional fiber mixed with epoxy resin is introduced in Figure 2.

Optical fiber strain sensor is based on fiber Bragg grating, which has an equal cycle and is able to realize a quasi-distributed measurement and easy to compose the sensor network [10]. Its peak wavelength signal is the sensor signal as well as the sensor address.

The main material used in the packaging of optical fiber is mainly bilateral basalt fiber-reinforced material, which enjoys a better anti-shear and anti-tensile capacity and insulation performance [11]. It has been used widely in building structure reinforcement. The structure of the reinforced material is similar to the optical fiber, so it can be combined well with each other. This design can

provide the optical fiber grating a good protection, which solves the problem that fiber grating is easily damaged. Reinforced material is able to form a wrapped structure around the optical fiber, which increases the contact area of the optical fiber grating along the axial direction. The design can not only prevent the detachment of fiber grating from the packaging material, but also increase the transmission area of the axial force, reducing the strain loss during the course of transmission.

In the application, if the fiber material used as the packaging material is oversize, then the elasticity modulus is excessive, which causes that the strain of the part, which is attached to the sensor, is less than the other parts in the whole structure; therefore, the measurement data cannot reflex the real situation. The size of the sensing zone of the unpackaged optical fiber sensor is usually 10 mm. Fiber material on the upper and lower layers can fully wrap around the fiber Bragg grating.

Basalt fiber material is infiltrated by epoxy resin. Epoxy resin enjoys good mechanical properties, and its elasticity modulus is large enough after solidification [12]. Its function of strain transmission is better, and able to combine the fiber grating and fiber material with good adhesion, as well as of excellent insulation and anti-corrosion. Fiber Bragg grating can insulate external air, preventing from aging equipment.

3.2 BFRP-FBG strain sensor preparation

The first step involves the cutting of basalt fiber cloth according to the design requirements, and ensures the fabric consistency of fiber cloth and smoothness.

The second step involves the fixing of the grating and then the fixing of optical fiber with a rapid curing adhesive. After the bottom of the fiber, the cloth is fixed and the adhesive is applied onto the grating along the direction axially parallel to the fiber.

The third step involves the pressing of the infiltrated optical fiber strain sensor with heavy and flexible stuff, so that the optical fiber strain sensor becomes as thin as possible and avoids breaking fiber grating.

The steps are shown in Figure 3.

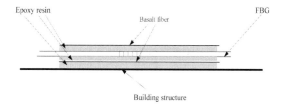

Figure 2. Optical fiber strain sensor package.

Figure 3. BFRP-FBG strain sensor preparation.

4 PERFORMANCE TEST AND ANALYSIS ON THE BFRP-FBG STRAIN SENSOR

Different packaging materials and different packaging forms will have a various impact on the sensing performance of the optical fiber strain sensor. Therefore, it is of great significance to test the performance of the packaged optical fiber strain sensor.

The sensor is attached to the surface of the steel structure in order to monitor its strain. Cantilever whose intensity is equal to steel is used to simulate the strain in the real situation. As shown in Figure 4, when the initial section of the cantilever is applied with the vertical force, the surface strain of the cantilever can approximately simulate the strain of the steel tower. The optical fiber sensor is attached to the surface of the cantilever, so the performance can be tested.

In the experiment, fiber demodulation equipment displays the wavelength-power distribution. The change in the wavelength of maximum power should be observed related to different strains. Every time when the strain changes, the wavelength of maximum power also changes. While changing the strain by changing the force applied to the cantilever and observing the wavelength of maximum power, the variation $\Delta\lambda_B$ related to different strains is recorded and the relationship between the amount of strain ε- the variation $\Delta\lambda_B$ of wavelength of maximum power is established.

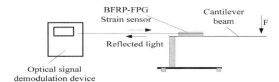

Figure 4. Optical fiber strain sensor test.

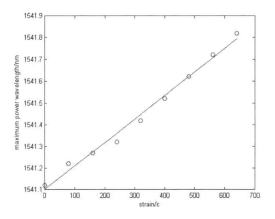

Figure 5. Part of the sensor performance curve.

During the test, with the strain linearly increasing, the corresponding change in the maximum power wavelength approximately linearly increases.

By fitting at least squares principle, strain variation ε- maximum power wavelength $\Delta\lambda B$ proportionality coefficient is 0.00117 nm/$\mu\varepsilon$. As shown in Figure 5, the table is part of the performance of the optical fiber strain sensor. Dots are the maximum power wavelength related to different strains, and the straight line stands for the sensing property. In the whole testing process, the optical fiber strain sensor shows a good linearity.

The test demonstrates that the optical fiber strain sensor is capable to measure the strain of the building structure.

5 CONCLUSIONS

After the manufacturing and preliminary test of the proposed optical fiber strain sensor, the performance is demonstrated to be satisfied.

According to the test, the BFRP-FBG strain sensor has a good linearity. It increases the contact area between the sensor and the building structure. It improves the protection performance of fiber Bragg grating.

In the future experiment, some further step can be taken to test the performance of the sensor. The variation can be magnified with meeting the requirements of the sensor to test the performance under the negative strain condition. The sensitivity to the temperature of the packaged sensor should also be examined so as to realize the temperature compensation.

ACKNOWLEDGMENTS

This research was supported by the Ministry of Housing and Urban Rural Development Project: Study on optical fiber sensor used for illumination of building energy monitoring (2013 K145); the Ministry of Housing and Urban Rural Development Project: Study on two dimensional intelligent Basalt fiber sheet used in building strengthening project (2013 K806); and Shandong Province Science and Technology Development Plan: Study on the coupling performance of high power laser (2009GG20003017).

REFERENCES

[1] Zhang Kuangwei, Zhang Shaojie, Zhao Xiaoxia, et al. 2014. Application of FBG strain sensors in bridge structure monitoring system. Optical Instruments 36(1):15–19.

[2] Zhao Hongxia, Bao Jilong, Chen Ying. 2005. The Polymer Package Craft and Performance Tests of FBG. Optoelectronic Technology & Information 18(5): 39–42.

[3] Li Jun, Bian Chao, Wu Jun. Power cable with optical fiber grating temperature on-line monitoring system [J] Jiangsu Electrical Engineering, 2005, 24 (1):6–7 (in Chinese).

[4] Hu Xianqi, ShenTu Nian, The Applications of the CBF in War Industry & Civil Fields [J]. Hitech Fiber & Application 2005, 30(6):8–13. (in Chinese).

[5] Tian Gaojie, Li Chuan, You Jing, et al. 2010. Fiber Bragg grating packaged by Basaltfiber0reinforced plastic. Dam & Safety (1):28–30.

[6] Gao Yufei, Liu Chao, Mou Haiwei. 2014. Pressure response of fiber Bragg grating based on plate diaphragm and equal-strength cantilever. Optical Instruments 36(4):333–336.

[7] Li Hong, Zhu Lianqing, Liu Feng, et al. Surface mounting structure of fiber Bragg grating strain analysis and experimental study of transfer [J]. Journal of optical instrument 2014(8):1744–1749 (in Chinese).

[8] Guo Wei, Li Xinliang, Song Hao. Transfer analysis strain is stuck on the surface of fiber Bragg grating sensor [J]. Measurement technology, 2011 (4):1–4 (in Chinese).

[9] Wu Jun, Chen Weimin, Zhang Peng, Liu Li, Liu Hao. Influence of bond layer characteristics on strain sensing properties of FBG [J]. Sensors Optics and Precision Engineering, 2011, 19(12):2941–2946 (in Chinese).

[10] Zhou Guangdong, Li Hongnan, Ren Liang. A parametric study of the effects of strain transfer of fiber Bragg grating sensor [J] Engineering Mechanics, 2007 (6):169–173 (in Chinese).

[11] Yashiro S, Takeda N, Okabe T, et al. A new approach to predicting multiple damage states in composite laminates with embedded FBG sensors [J]. Composites Science and Technology, 2005, 65(3): 659–667.

[12] Huang Guanglong, Lian Xu, Chen Qiao, xu Hong-zhong. Sensing properties of FBG in fiber reinforced composite bars, Journal of Nanjing University of Technology (Natural Science Edition) 2010, 32(1):23–27 (in Chinese).

Material Science and Engineering – Chen (Ed.)
© *2016 Taylor & Francis Group, London, ISBN 978-1-138-02936-1*

Thermogravimetric analysis on the combustion of hemicellulose, cellulose and lignin

L.K. Li
Advanced Materials R&D Center of WISCO, Beijing, China
Materials Science and Engineering, Huazhong University of Science and Technology, Wuhan, China

Y. Zhou, J.Y. Hwang, J.X. Liu, X.P. Mao & G.F. Zhou
Advanced Materials R&D Center of WISCO, Beijing, China

J.P. Suo
Materials Science and Engineering, Huazhong University of Science and Technology, Wuhan, China

ABSTRACT: The thermal behavior of typical biomass and its three components (hemicellulose, cellulose and lignin) during combustion was studied by means of thermogravimetric analysis. Combustion kinetic of the three components and their blends were also studied. The result indicated that biomass, hemicellulose and cellulose have a large amount of volatile matter, which means that fuel is easy to ignite and tends to burn quickly. Meanwhile, lignin has a large amount of fixed carbon. The reaction order of biomass, hemicellulose, cellulose, lignin and their blends are close to 1. The activation energy of fixed-carbon combustion is lower than that of volatile matter combustion when two components of the major components blended. However, a little change is observed when the three components blended because the activation energy of the different components shows distinctly. This work reports that the synergistic effect exists during the co-combustion of the major components.

Keywords: biomass; combustion; kinetic

1 INTRODUCTION

Nowadays, energy crisis problems and global environment problems have become a serious issue. Biomass as a clean and renewable energy has been paid more attention (Fernández, G.F. et al. 2012, Qu, T.T. et al. 2011, Nasrin, A. et al. 2011, Munir S. et al. 2009). In this paper, biomass is defined as "all kinds of materials directly or indirectly produced, not too long ago, from photosynthesis reactions from vegetable matter and its derivates: wood-fuel, wood derived fuels, fuel crops, agricultural and agro-industrial by-products and annual by-products" (Loo, S.V. and Koppejan J. 2008). Currently, biomass is one of the major energy sources, which provides 14% of the world's needs (Shen, D.K. 2009).

Biomass contains three major components: hemicellulose, cellulose and lignin. The thermal behavior of these major components during pyrolysis has been investigated by many researchers (Yang H.P. 2006). Previous studies indicated that it is an independent parallel decomposition of the major components when they blended (Wang, G. 2008,

Han, J. and Kim, H 2008). As there are only a few studies on the combustion of the three components, which is important to biomass as well, the study on the combustion of the major components becomes critical.

Meanwhile, the kinetic properties of biomass combustion have been investigated by many researchers. Such knowledge is needed for the design of conversion systems (Sanchez, M.E. 2009, Xiao, H. M 2009). Some researchers have developed a two-stage reaction kinetic model according to different temperature regions (Fang, X. 2013).

Thermogravimetric Analysis (TGA) is one of most commonly used methods to study thermal events, which has been widely used to investigate the thermal process and kinetics of the combustion and pyrolysis of biomass, coal, refuse-derived fuel and other fuel (Li, X.G. 2011, Fangxian, L. 2009, Otero, M. 2008, Varol, M. 2010, Shen, B.X. and Qin, L. 2006).

This paper aimed to investigate the thermal behavior of typical biomass and its three major components (hemicellulose, cellulose and lignin) during combustion. The kinetic model of the three components and their blends were also obtained.

2 EXPERIMENTAL DESIGN

2.1 Materials and methods

Four kinds of as-prepared samples were studied in this research, including sawdust, hemicellulose, cellulose and lignin. Sawdust is one of the typical biomass materials, which was collected from a lumber mill. Because hemicellulose can hardly be purchased, xylan was used as a representative of hemicellulose in the combustion process (Wang, G. W 2008). Xylan was purchased from Beijing xinao Biological Technology Corporation. Cellulose was purchased from Beihu Company. The as-prepared samples were dried in an oven at 373 K for 3 h. Then, hemicellulose and cellulose, hemicellulose and lignin, cellulose and lignin were blended with weight percentages of 20 wt%, 40 wt%, 50 wt%, 60 wt%, and 80 wt%. Meanwhile, hemicellulose, cellulose and lignin were blended with weight ratios of 1:1:1, 1:1:2, 1:2:1, and 2:1:1.

The combustion characteristics of sawdust, hemicellulose, cellulose, lignin and the blends were determined on a discovery thermogravimetric analyzer. In an air flux of 25 mL/min, materials were heated from ambient temperature to 1173 K under air atmosphere with a heating rate of 20 K/min. The same mass of each material (about 10 mg) was loaded onto an Al_2O_3 ceramic pan.

2.2 Theoretical estimation

From previous studies, kinetic parameters from thermogravimetric (TG) data can be obtained from the following rate expression (Salin, J.M. and Seferis, J.C. 1993):

$$\frac{d\alpha}{dt} = A\exp\left(\frac{-E_a}{RT}\right)(1-\alpha)^n \qquad (1)$$

Equation (1) shows the relationship time and the fraction of the material consumed in the form of Arrhenius expression, where α is the fractional conversion; t is the time; A is the pre-exponential; E_α is the activation energy; T is the temperature; and n is the reaction order.

But

$$\beta = \frac{dT}{dt} \qquad (2)$$

where β is the heating rate, which is 20 K/min in this study.

Hence

$$\frac{d\alpha}{dT} = \frac{A}{\beta}\exp\left(\frac{-E_a}{RT}\right)(1-\alpha)^n \qquad (3)$$

Taking the natural log on both sides, we obtain the following equation:

$$\ln\left(\frac{d\alpha}{dT}\right) = \ln\frac{A}{\beta} - \frac{E_a}{RT} + n\ln(1-\alpha) \qquad (4)$$

Assuming $x = \ln\frac{A}{\beta}$, $b = -\frac{1}{RT}$, $c = \ln(1-\alpha)$, $y = \ln\left(\frac{d\alpha}{dT}\right)$, the equation can be translated into the following expression:

$$y = x + bE_a + cn \qquad (5)$$

From TG data, a coefficient matrix can be obtained, in which id is used to solve x, E_a and n, and we denote the coefficient matrix as G. According to the least-squares method, the coefficient matrix should satisfy the following equation:

$$[x, E_a, n]^T = (G^T G)^{-1} G^T Y \qquad (6)$$

Hence, x, E_a, n can be determined directly from the TG data, and A can be obtained from x.

3 RESULTS AND DISCUSSION

3.1 Kinetic analysis of biomass and the major components

The combustion process of sawdust and the major components using the TG/DTG profile are shown in Figure 1a and b. The kinetic parameters of sawdust and the major components according to the temperature regions are listed in Table 1. As cellulose has a small amount of fixed carbon and lignin has a small amount of volatile matter, the kinetic parameters of cellulose and lignin are shown at one temperature.

The combustion of sawdust and the three major components can be divided into three stages: water evaporation, combustion of volatile matter and combustion of fixed-carbon (Yang, Y.B. 2004). Based on the TG curve shown in Figure 1, it can be observed that sawdust, hemicellulose and cellulose contain a large amount of volatile matter as the sample mass decreases, which mainly occurs in the second stage, which means that fuel is easy to ignite and tends to burn quickly. Meanwhile, lignin contains a large amount of fixed carbon as the sample mass decreases, which mainly occurs in the third stage, which means that fuel is hard to ignite. It can be seen that the reaction order of sawdust and the three major components are close to 1.

3.2 Kinetic analysis of the blends of two components

The kinetic analyses of the blends of hemicellulose and lignin, cellulose and lignin, hemicellulose and

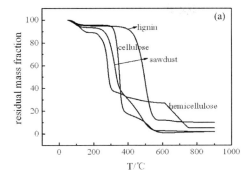

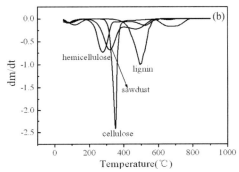

Figure 1. TG-DTG curves of sawdust and the major components: (a) TG curve (b) DTG curve.

Table 1. The kinetic parameters of sawdust and the major components.

	Temperature region	lnA	E	n	R^2
Sawdust	200°C–400°C	17.55	91.34	1.04	0.95
	400°C–600°C	10.09	69.01	0.90	0.99
Cellulose	280°C–400°C	47.83	246.35	1.38	0.96
Lignin	400°C–600°C	19.68	134.99	1.00	0.98
Hemicellulose	180°C–350°C	28.17	130.57	1.32	0.95
	580°C–800°C	31.01	253.20	1.25	0.96

*A, pre-exponential; E, activation energy; n, reaction order; R^2, correlation coefficients.

cellulose were performed, and the result is summarized in Table 2, Table 3 and Table 4. The temperature regions were divided according to the peaks of DTG curves.

The combustion of the blends of hemicellulose and lignin mainly occurs at 200°C~340°C and 340°C~600°C. The blends of cellulose and lignin combustion occur at 280°C~380°C and 380°C~600°C, and those of hemicellulose and cellulose combustion occur at 200°C~400°C and

Table 2. The kinetic parameter of the blends of hemicellulose and lignin.

	Temperature region	lnA	E	n	R^2
20%H	200°C–340°C	18.84	92.53	0.75	0.96
80%L	340°C–600°C	6.18	50.55	0.56	0.95
40%H	200°C–340°C	20.25	98.00	0.85	0.96
60%L	340°C–600°C	7.24	55.77	0.67	0.97
50%H	200°C–340°C	21.34	102.18	0.94	0.98
50%L	340°C–600°C	4.47	40.26	0.49	0.95
60%H	200°C–340°C	20.01	96.46	0.90	0.96
40%L	340°C–600°C	9.26	66.08	0.82	0.97
80%H	200°C–340°C	20.51	99.69	0.95	0.98
20%L	340°C–600°C	9.41	66.38	0.85	0.96

Table 3. The kinetic parameter of the blends of cellulose and lignin.

	Temperature region	lnA	E	n	R^2
20%C	280°C–380°C	54.83	279.01	1.43	0.96
80%L	380°C–600°C	6.56	52.57	0.57	0.95
40%C	280°C–380°C	56.75	289.59	1.33	0.97
60%L	380°C–600°C	7.33	56.90	0.63	0.97
50%C	280°C–380°C	51.08	262.33	1.14	0.98
50%L	380°C–600°C	8.78	64.98	0.72	0.97
60%C	280°C–380°C	58.80	300.40	1.28	0.96
40%L	380°C–600°C	9.12	66.70	0.74	0.98
80%C	280°C–380°C	55.33	283.28	1.20	0.96
20%L	380°C–600°C	9.01	67.45	0.78	0.95

200°C~550°C. It is evident from the data that the reaction order of the blends is close to 1. The activation energy of the blends hemicellulose and lignin, cellulose and lignin, hemicellulose and cellulose at a lower temperature region is, respectively, in the range of 92.534~102.177 kJ/mol, 262.332~300.402 kJ/mol and 72.354~83.498 kJ/mol, while at a higher temperature region, they are, respectively, in the range of 40.264~66.375 kJ/mol, 52.571~67.452 kJ/mol and 14.261~49.243 kJ/mol. The activation energy at a higher temperature is lower than that at a lower temperature. The main reason can be attributed to the combustion of volatile matter, which results in the increase of the area of internal pores and the external surface and the heat release from volatile matter that makes the reaction more easily.

3.3 *Kinetic analysis of the blends of the three components*

The kinetic parameters of the blends of hemicellulose, cellulose and lignin are listed in Table 5.

Table 4. The kinetic parameter of the blends of hemi-cellulose and cellulose.

	Temperature region	lnA	E	n	R^2
20%H	200°C–400°C	15.14	83.50	0.70	0.98
80%C	400°C–550°C	5.53	49.24	0.77	0.95
40%H	200°C–400°C	13.90	74.95	0.82	0.96
60%C	400°C–550°C	2.64	23.61	0.62	0.95
50%H	200°C–400°C	13.80	74.61	0.81	0.97
50%C	400°C–550°C	2.44	27.33	0.64	0.95
60%H	200°C–400°C	13.70	72.35	0.91	0.96
40%C	400°C–550°C	0.95	14.26	0.62	0.97
80%H	200°C–400°C	16.02	80.80	1.15	0.95
20%C	400°C–550°C	4.85	18.71	0.72	0.96

Table 5. The kinetic parameters of the blends of hemi-cellulose, cellulose and lignin.

	Temperature region	lnA	E	n	R^2
1H1C1L	200°C–400°C	15.68	83.45	0.90	0.98
	400°C–600°C	10.48	73.64	0.83	0.98
1H2C1L	200°C–400°C	15.83	85.51	0.80	0.95
	400°C–600°C	12.30	84.29	0.90	0.97
2H1C1L	200°C–400°C	13.78	73.00	0.92	0.98
	400°C–600°C	11.94	80.81	0.93	0.97
1H1C2L	200°C–400°C	14.36	78.03	0.79	0.95
	400°C–600°C	13.85	94.07	0.92	0.98

The combustion occurs at 200°C~400°C and 400°C~600°C, while the temperature regions are the same as that for sawdust combustion. As the main components of biomass are hemicellulose, cellulose and lignin, biomass can be regarded as the blends of hemicellulose, cellulose and lignin, which reduces the combustion of biomass mainly occurring at 200°C~400°C and 400°C~600°C. The combustion of the blends is almost the first-order reaction. The activation energy changes to a little extent whether at a higher temperature region or a lower temperature region, ranging from 73.001 kJ/mol to 85.508 kJ/mol at a lower temperature region and from 73.642 kJ/mol to 94.072 kJ/mol at a higher temperature region.

4 CONCLUSIONS

In summary, biomass, hemicellulose and cellulose have a large amount of volatile matter, which means that fuel is easy to ignite and tends to burn quickly. Meanwhile, lignin has a large amount of fixed carbon.

1. The reaction order of biomass, hemicellulose, cellulose, lignin and their blends are close to 1.
2. The activation energy of fixed-carbon combustion is lower than that of volatile matter combustion when two components of the major components blended, which has a little change when the three components blended.
3. With respect to the difference in the components, the activation energy is distinct, which indicates that the synergistic effect exists during the co-combustion of the major components.

REFERENCES

[1] Aghamohammadi, N. et al. 2011. Combustion characteristics of biomass in South East Asia. *Biomass And Bioenergy* 35(9): 3884–3890.
[2] Fang, X. et al. 2013. A weighted average global process model based on two–stage kinetic scheme for biomass combustion. *Biomass and Bioenergy* 48: 43–50.
[3] Fernández, R.G. et al. 2012. Study of main combustion characteristics for biomass fuels used in boilers. *Fuel processing technology* 103: 16–26.
[4] Han, J. & Kim, H. 2008. The reduction and control technology of tar during biomass gasification/pyrolysis: an overview. *Renewable and Sustainable Energy Reviews* 12(2): 397–416.
[5] Li F.X. et al. 2009. Thermal analysis study of the effect of coal-burning additives on the combustion of coals. *Journal of Thermal Analysis and Calorimetry* 95(2): 633–638.
[6] Li X.G. et al. 2011. Thermogravimetric investigation on co-combustion characteristics of tobacco residue and high-ash anthracite coal. *Bioresour Technol* 102(20): 9783–9787.
[7] Loo, S.V. & Koppejan, J. 2008. The handbook of biomass combustion and co-firing, Earthscan Publications Ltd (eds). biomass combustion & co-firing: 108–201. London.
[8] Munir S. et al. 2009. Thermal analysis and devolatilization kinetics of cotton stalk, sugar cane bagasse and shea meal under nitrogen and air atmospheres. *Bioresource Technolgy* 100(3): 1413–1418.
[9] Otero, M. et al. 2008. Co-combustion of different sewage sludge and coal: a non-isothermal thermogravimetric kinetic analysis. *Bioresource Technology* 99: 6311–6319.
[10] Qu, T. et al. 2011. Experimental study of biomass pyrolysis based on three major components: hemicellulose, cellulose, and lignin. *Industrial & Engineering Chemistry Research* 50(18): 10424–10433.
[11] Salin, J.M. & Seferis, J.C. 1993. Kinetic analysis of high-resolution TGA variable heating rate data. *Journal of Applied Polymer Science* 47(5): 847–856.
[12] Sanchez, M. et al. 2009. Thermogravimetric kinetic analysis of the combustion of biowastes. *Renewable Energy* 34: 1622–1627.
[13] Shen, B. & Qin, L. 2006. Study on MSW catalytic combustion by TGA. *Energy Conversion and Management* 47: 1429–1437.

[14] Shen, D. et al. 2009. Kinetic study on thermal decomposition of woods in oxidative environment. *Fuel* 88: 1024–1030.

[15] Varol, M. et al. 2010. Investigation of co-combustion characteristics of low quality lignite coals and biomass with thermogravimetric analysis. *Thermochimica Acta* 510: 195–201.

[16] Wanga, G. et al. 2008. TG study on pyrolysis of biomass and its three components under syngas. *Fuel* 4–5: 552–558.

[17] Xiao, H. et al. 2009. Isoconversional kinetic analysis of co-combustion of sewage sludge with straw and coal. *Applied Energy* 86: 1741–1745.

[18] Yang, H. et al. 2005. In-depth investigation of biomass pyrolysis based on three major components: hemicellulose, cellulose and lignin. *Energy Fuels* 20(1): 388–393.

[19] Yang, Y.B. et al. 2004. Effect of air flow rate and fuel moisture on the burning behaviours of biomass and simulated municipal solid wastes in packed beds *Fuel* 83: 1553–1562.

Material Science and Engineering – Chen (Ed.)
© 2016 Taylor & Francis Group, London, ISBN 978-1-138-02936-1

Dyeing of silk/nylon blend with heterobifunctional reactive dyes in one bath

M. Xiong & R.C. Tang

National Engineering Laboratory for Modern Silk, College of Textile and Clothing Engineering,
Soochow University, Suzhou, P.R. China

ABSTRACT: In this work, three heterobifunctional reactive dyes (Everzol Orange ED, Rubine ED and Navy ED) with sulphatoethylsulphone and monochlorotriazinyl reactive groups were applied to dye silk/nylon blends with the aim of obtaining high wet color fastness, and the silk and nylon fabrics with the same weights were immerse into one dyebath in order to imitate the dyeing of silk/nylon blend. The dyeing was first performed under an acidic condition for dye sorption, and then under an alkaline condition for dye fixation. The influence of the initial pH values of dyebath on the exhaustion as well as the effects of the dyeing temperature and the dosage of sodium carbonate used during the dye fixation process on the color depth of dyed fabrics was investigated. Moreover, the union dyeing property of silk and nylon was also discussed. It was found that trichromatic reactive dyes exhibited high exhaustion at low initial pH values, and the dyed fabrics showed high color depth and high color fastness to washing and rubbing in the case of appropriate sodium carbonate dosage and a temperature of 70 °C.

Keywords: silk; nylon; blend; reactive dyes; dyeing

1 INTRODUCTION

Silk is a natural protein fiber excreted by the moth larva Bombyx mori, and a fine continuous monofilament fiber. The prestige silk fiber has some excellent properties such as lightness, soft luster, good touch, high strength, good hygroscopicity, etc. Because of its shortcomings such as high cost, creasing, yellowing and poor dimensional stability, it finds very limited use in textiles. Nylon is one of three synthetic fibers and has good abrasion resistance and flexibility but poor hygroscopicity. The silk/nylon blend combines the advantages of silk and nylon fibers, providing good dimensional stability, good hygroscopicity and low cost of production, and opens a broader development. Both silk and nylon fibers can be dyed with acid dyes, but the wet fastness of dyed fabrics is unsatisfactory [1]. Though it is possible to improve the wet fastness by choosing appropriate acid dyes and fixing agents for aftertreatment, it is hard to meet the high wet fastness requirements [2]. Silk fiber can be dyed with lots of reactive dyes [3]. Some investigations reported the reactive dyeing and printing of nylon fiber, and revealed that the dyed nylon fibers had high wet color fastness [1, 2, 4]. In this work, three heterobifunctional reactive dyes (Everzol Orange ED, Rubine ED and Navy ED) with sulphato-ethylsulphone and monochlorotriazinyl

reactive groups were applied to dye silk/nylon blends with the aim of obtaining high wet color fastness. In order to perform the investigations, the dyeing of the silk and nylon fabrics with the same weights which were immerse into one dyebath was used to imitate the dyeing of silk/nylon blend. The dyeing was first performed under an acidic condition for increasing the extent of dye sorption, and then under an alkaline condition for increasing the extent of dye fixation. Some important dyeing parameters were discussed.

2 EXPERIMENTAL

2.1 Materials

The silk (11160 habotai) and nylon (55.6 dtex/48 F FDY in warp, and 50.0 dtex/34 F FDY in weft) fabrics as well as the silk/nylon blend containing 50% silk and 50% nylon were purchased from Wujiang Zhiyuan Co., Ltd., China. Trichromatic reactive dyes (Everzol Orange ED, Rubine ED, Navy ED) were kindly provided by Everlight Chemical Industrial Co., Ltd., Taiwan.

2.2 Dyeing method

All dyeing experiments were carried out in the sealed and conical flasks immersed in the

XW–ZDR low-noise oscillated dyeing machine (Jiangsu Jingjiang Xinwang Dyeing and Finishing Machinery Factory, China). All of the dye solutions contained Levelling Agent O (0.5 g/L). The liquor ratio was kept at 1: 50. At the end of dyeing, the fabric samples were washed in distilled water, soaped in the solution of 2 g/L Soaping Agent SW ECO at 95 °C using a liquor ratio of 1: 50 for 15 min, and finally dried in the open air.

Three main factors (pH, temperature and sodium carbonate concentration) were discussed. An initial dyeing was carried out at 2% dye owf to determine the effect of pH on dyeing. Five initial pH values (3.7, 4.4, 5.5, 9.6 and 10.4) were obtained by using McIlvaine buffers; the fabrics were immersed into the dye solutions at 30 °C, and then the temperature was raised to 70 °C at 1 °C/min, then sodium carbonate was added after dyeing 20 min and the dyeing continued for 40 min. The effect of temperature on dyeing was assessed by using the 2% owf dye solution whose initial pH was adjusted to 3.7; after the dyebath was heated to the required temperature (50–90 °C) and the dyeing was conducted for 20 min, sodium carbonate (10 g/L) was added and the dyeing continued for 40 min. The effect of sodium carbonate concentration was also assessed by using the 2% owf dye solution whose initial pH was adjusted to 3.7, and the same dyeing procedure as ascribed above; after the sorption process of dyes, the different sodium carbonate dosages ranging from 5 to 30 g/L was added into dyebath. The union dyeing property was carried out by using the same dyeing procedure as ascribed above, and by applying an initial pH of 3.7 for dye sorption and 10 g/L sodium carbonate for dye fixation.

2.3 Measurements

The percentage of dye exhaustion (E) was determined spectrophotometrically, using a Shimadzu UV-1800 UV-vis spectrophotometer (Shimadzu Co., Japan), according to Eq. 1, where A_0 and A_1 are the absorbance of dye solution before and after dyeing, respectively. The unfixed and hydrolyzed dyes in the samples were removed by soaping and then measured spectrophotometrically. The apparent color depth (K/S) of the dyed fabrics was measured using a HunterLab UltraScan PRO reflectance spectrophotometer at illuminant D65 and 10° standard observer. The balance value of the union dyeing property (K) was characterized by Eq. 2, where $(K/S)_S$ and $(K/S)_N$ are the color depth of the dyed silk and nylon fabrics, respectively. The wash fastness of the fabrics was carried out by a WashTec-P fastness tester (Roaches International, England), using the standard test method ISO 105-C06. The rubbing fastness was measured on a

Model 670 crockmaster (James H. Heal, England) according to ISO 105-X12.

$$\%E = 100 \times [(A_0 - A_1)/A_0] \qquad (1)$$

$$K = (K/S)_S/(K/S)_N \qquad (2)$$

3 RESULTS AND DISCUSSION

3.1 Effect of pH of dye solution

As the pH of dye solution affects the charge properties of silk and nylon fibers and the corresponding sorption and reaction of dyes on fibers, it is important to discuss the effect of initial pH values of dye solution on dyeing. Figure 1 shows that the exhaustion of dyes for silk/nylon blend increased obviously with decreasing initial pH values. This is due to the fact that the protonation extent of the amino groups of silk and nylon fibers increased with decreasing initial pH values, resulting in the increasing ion-ion interaction between fibers and reactive dyes and the improved sorption. In an alkaline medium, there is almost no ionic bonding between fibers and dyes because the fibers are negatively charged, and the sorption of dyes on the fibers occurs by means of hydrogen bonding and van der Waals force [4], and thus the exhaustion of dyes was very low.

3.2 Effect of dyeing temperature

As shown in Figure 2, the color depth of nylon fabric increased greatly with increasing temperature as well, reaching a maximum at 80 °C and 90 °C for three dyes. For the nylon fiber with a compact structure, when temperature increased, its swelling, the chain segment motion of its molecules and the diffusion rate of dye molecules into its

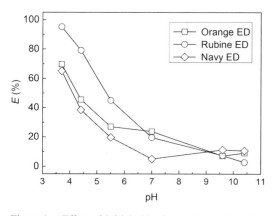

Figure 1.   Effect of initial pH values on the exhaustion of reactive dyes.

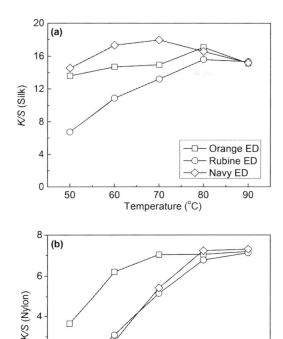

Figure 2. Effect of dyeing temperature on the color depth of silk and nylon fabrics.

existence form of dyes, the charge nature of fibers, and the ionization extent of amino and hydroxyl groups in fibers. The low ionization extent of amino groups in silk and nylon fibers, the high ionization extent of hydroxyl groups in silk fiber would increase the fixation of dyes. In this work, we tried to carry out the sorption of dyes under acidic conditions followed the fixation of dyes under alkaline conditions. Here the application of acidic dyeing conditions was to increase the sorption of dyes, whereas the application of alkaline fixation conditions was to increase the fixation of dyes. After the sorption process of dyes, the added sodium carbonate neutralizes acids, and then provides the alkaline condition for the fixation of reactive dyes. The influence of alkali concentration on the color depth of silk and nylon fabrics dyed with trichromatic reactive dyes is shown in Figure 3. The color depth of silk fabric increased with increasing alkali concentration in the range of 5–20 g/L, then decreased as the alkali concentration further increased. However, the color

interior increased [2]. The factors contribute to an increase in the color depth of nylon with increasing temperature. Meanwhile, the color depth of silk fabric increased with increasing temperature, reaching a maximum at 70 °C for Navy ED, and at 80 °C for Orange ED and Rubine ED, and then decreased as the temperature further increased. In a certain temperature range, the increased color depth of silk should be caused by the enhanced diffusion of the dye molecules into silk fiber with increasing temperature. At the higher temperature, the decreased color depth of silk resulted from the increased uptake of dyes by nylon, and the increased hydrolysis of dyes.

3.3 *Effect of alkali concentration*

For the dyeing of silk and nylon fabrics with heterobifunctional reactive dyes, the exhaustion of dyes should mainly depend on the pH of dyebath, while the fixation of dyes or the reaction of dyes with fibers would be related to the pH of dyebath, the

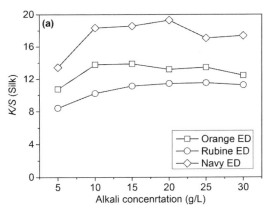

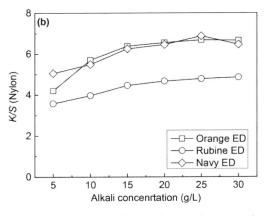

Figure 3. Effect of sodium carbonate dosage on the color depth of silk and nylon fabrics.

depth of nylon fabric increased with increasing alkali concentration. The reason is that there are more covalent bonds between dyes and fibers in an alkaline medium. Taking into consideration the color depth of silk and nylon fabrics as well as the damage of silk fiber at high alkali concentrations, the optimum dosage of sodium carbonate was 10 g/L.

3.4 Color effects and fastness of dyeings

In the case of the dyeing of silk and nylon fabrics in one bath, the color depth of silk was obviously higher than that of nylon (Table 1). Therefore some measures should be taken to improve the union dyeing property of silk and nylon fibers after further research. Three dyes were also used to dye silk/nylon blends, and the color fastness was determined. Table 2 shows that the dyed silk/nylon blends exhibited very high color fastness to washing and rubbing.

Table 1. Union dyeing property of silk and nylon fabrics.

Dyes	K/S, silk	K/S, nylon	K
Orange ED	14.9725	7.0325	2.1290
Rubine ED	13.2125	5.1625	2.5593
Navy ED	17.9550	5.4325	3.3051

Table 2. Color fastness of dyed silk/nylon blends to washing and rubbing.

Dyes	Washing			Rubbing	
	Color change	Staining		Dry	Wet
		Cotton	Silk		
Orange ED	4	5	5	5	4–5
Rubine ED	5	5	5	5	4–5
Navy ED	4	5	5	5	4–5

4 CONCLUSIONS

Three heterobifunctional reactive dyes with sulphatoethylsulphone and monochlorotriazinyl reactive groups were applied to dye silk/nylon blends, and the dyeing was first performed under an acidic condition for increasing the extent of dye sorption, and then under an alkaline condition for increasing the extent of dye fixation. The trichromatic reactive dyes exhibited high exhaustion at low initial pH values, and the dyed fabrics showed high color depth and high color fastness to washing and rubbing in the case of appropriate sodium carbonate dosage and a temperature of 70 °C. The dyed silk showed higher color depth than the dyed nylon, and therefore the union dyeing property of silk/nylon blend should be improved through further research.

ACKNOWLEDGEMENTS

This study was funded by the Priority Academic Program Development (PAPD) of Jiangsu Higher Education Institutions.

REFERENCES

[1] Johnson, H.L. 1965. Artistic development in autistic children. *Child Development* 65(1): 13–16.
[2] Burkinshaw, S.M., Chevli, S.N. & Marfell, D.J. 2000. Printing of nylon 6,6 with reactive dyes part I: preliminary studies. *Dyes Pigments* 45(3): 235–242.
[3] Son, Y.A., Hong, J.P. Lim H.T. & Kim T.K. 2005. A study of heterobifunctional reactive dyes on nylon fibers: dyeing properties, dye moiety analysis and wash fastness. *Dyes Pigments* 66(3): 231–239.
[4] Wu Z. 1998. Recent developments of reactive dyes and reactive dyeing of silk. *Rev. Prog. Coloration* 28: 32–38.
[5] Soleimani-Gorgani, A. & Taylor J.A. 2006. Dyeing of nylon with reactive dyes. Part 1. The effect of changes in dye structure on the dyeing of nylon with reactive dyes. *Dyes Pigments* 68 (2–3): 109–117.

Mechanical property of materials

Material Science and Engineering – Chen (Ed.)
© 2016 Taylor & Francis Group, London, ISBN 978-1-138-02936-1

Moisture sorption isotherm and thermodynamic property of *Litopenaeus vannamei*

W.H. Cao

Guangdong Provincial Key Laboratory of Aquatic Product Processing and Safety, Zhanjiang, Guangdong Province, China
Key Laboratory of Advanced Processing of Aquatic Products of Guangdong Higher Education Institution, Zhanjiang, Guangdong Province, China
College of Food Science and Technology, Guangdong Ocean University, Zhanjiang, Guangdong Province, China

S. Tian

College of Food Science and Technology, Guangdong Ocean University, Zhanjiang, Guangdong Province, China

C.H. Zhang

Guangdong Provincial Key Laboratory of Aquatic Product Processing and Safety, Zhanjiang, Guangdong Province, China
Key Laboratory of Advanced Processing of Aquatic Products of Guangdong Higher Education Institution, Zhanjiang, Guangdong Province, China
College of Food Science and Technology, Guangdong Ocean University, Zhanjiang, Guangdong Province, China

ABSTRACT: Moisture contents, water activity, freezing point and heat melting enthalpy were determined to research the relationship between the frozen water and unfrozen water of *Litopenaeus vannamei*. Modeling of the isotherms was done using the GAB (Guggenheim-Anderson-de Boer) model and the BET (Brunauer-Emmett-Teller) model. The results showed that the GAB model had a better fit quality than the BET model in the whole period. The determination coefficient was 0.98 and 0.90 when the water activity was observed between 0.40 and 0.90, respectively. The monolayer moisture contents calculated by the BET model and the GAB model were 0.0532 and 0.0868 g g^{-1} on the dry basis, respectively. There was a linear relationship between the frozen water and the melting enthalpy. When the moisture content was 74.8%, the freezing point was −4.49°C. A linear relationship between the unfrozen moisture content and the hot melting enthalpy was also observed. The unfrozen moisture content was 11.1% in *L. vannamei* when the melting enthalpy was zero.

Keywords: *L. vannamei*; sorption isotherms; water activity; freezing point; melting enthalpy

1 INTRODUCTION

L. vannamei, one of the three highest yield breeding shrimps worldwide, has been an important commercial species for almost 30 years in China. Shrimp industries in China produced approximately 1,560,000 t of shrimp in 2011, in which *L. vannamei* accounted for a total output of about 70% [1]. *L. vannamei* is delicious and rich in nutrients. They are consumed fresh, frozen, canned, or smoked. Protein degradation generally causes the spoilage of shrimp products during transportation, storage, retail display and consumption. In this cellular process, the structural organization is destroyed by ice or the presence of microorganisms. Thus, the moisture sorption isotherm and thermophysical property of *L. vannamei* must be

investigated because they provide a theoretical basis for the process and storage of *L. vannamei*.

Moisture sorption isotherm is the relationship between the equilibrium moisture content (X) and the water activity (a_w) of food at a constant temperature (T) [2]. This relationship is an essential tool in the dehydration process to predict shelf-life stability, packaging and drying of a desired product. Water activity of food is an important parameter of spoilage and quality. Food stability is different as the water activity changes. Controlling the moisture content during the processing of foods is an ancient method of preservation. This is achieved by either removing water or binding it, such that the food becomes stable with respect to both microbial and chemical deterioration [3]. For this reason, much attention has been given to the sorption properties of foods. Sorption characteristics have and are currently being examined in light of their influence on the storage stability of dehydrated products, as well as their effect on the diffusion of water vapor [4]. Moisture sorption isotherms represent the equilibrium relationship between the water activity and the moisture content of foods at a constant pressure and temperature [5]. A sound knowledge of the relationship between moisture content and equilibrium relative humidity is essential in the formulation of foods and in their storage stability [6].

The water in the food is divided into frozen water and unfrozen water, and frozen water can be further divided into free water and intermediate water. Free water is frozen at a freezing point temperature, and the transition among phases and the heat melting enthalpy in the intermediate water are lower than the pure water. Given the strong relationship between the unfrozen water and the water–gel of an organism, the unfrozen water cannot be frozen at a freezing point temperature. Thus, water in *L. vannamei* and other aquatic products is an important factor in storage time and processing [7]. Moisture sorption isotherms are mainly studied in plants, such as in chestnut and wheat flours [8], citrus reticulate leaves [9], prickly pear cladode [10] and jasmine rice crackers [11]. Some studies have also been conducted in livestock, poultry [12] and aquatic animals [13]. However, this relationship is not investigated in *L. vannamei*.

Despite the theoretical limitations of the BET adsorption analysis, the BET monolayer concept has been found to be a reasonable guide with respect to various aspects of interest in dried foods [14]. In more recent years, the GAB isotherm model has been widely used to describe the sorption behavior of foods [15]. Having a reasonably small number of parameters, the GAB model has been found to represent adequately the experimental data in the range of water activity of most practical interest in foods, i.e., 0.10 ± 0.90. This study determines the moisture sorption isotherms of *L. vannamei* throughout the changes in moisture content, water activity, freezing point and water content of unfrozen moisture. The objectives of this study were to determine the moisture sorption isotherms and thermodynamic property of *L. vannamei* using the GAB and BET models to investigate the holding capacity of *L. vannamei*, and provide the basic data for food processing and storage.

2 MATERIALS AND METHODS

2.1 *Materials*

Fresh shrimp (*L. vannamei*) weighing approximately 15 g to 20 g were purchased from Zhanjiang Dongfeng Market. They were placed on ice and transported within approximately 1 h to the Department of Key Laboratory, Guangdong Ocean University. Upon arrival at the laboratory, the shrimp were washed with clean water, de-headed and stripped, and their gut strings were removed. Then, these shrimp were pulped, packed in ventilated cellophanes and pressed into a 1 cm slice length. The slice was embedded in allochroic silica gels to de-water at 20°C. Sampling was performed every 6 h to 10 h within 10 days. Moisture content, water activity, freezing point and heat melting enthalpy (ΔH_f) were determined.

Allochroic silica gel was obtained from Weihai Pearl Silica Gel Company (Weihai, Shangdong, China). Ventilate cellophanes were obtained from Guangzhou Sea Source Packaging Materials Company (Guangzhou, Guangdong, China).

2.2 *Determination of moisture content*

The moisture content of shrimp samples was determined by drying the meat in an oven at 105°C until a constant weight was obtained [16]. Following drying, the samples were removed, placed in a dessicator and weighed. Percent moisture was calculated using the following formula: Percentage moisture = (wet weight - dry weight) / wet weight × 100.

2.3 *Determination of water activity*

Shrimp samples at 25°C were prepared and kept overnight in a constant temperature heating plate (Thermal Temperature Plate; Decagon Devices, Pullman, Washington, USA) at those temperatures. The water activity of these samples was then determined at 25°C using the water activity meter, Hygrolab3 (Rotronic, Switzerland), which allowed temperature control during the measurement. Both a_w measuring devices were calibrated and tested with shrimp samples within the a_w range of interest in this work. For each a_w determination, three replicates were obtained and the average is reported.

2.4 *Determination of the freezing point and enthalpy of water fusion (ΔHf)*

Freezing point and enthalpy of water fusion were determined according to the method of Wu et al. [17]. Samples (5 g) were placed in a special aluminum miniature cup and sealed. A differential scanning calorimeter (Pyris1 DSC-7, PE, USA) was used to plot the heat flux curve. The conditions were set as follows: initial temperature, −40°C; final temperature, 25°C. The ascending temperature speed was 2.0°C/min. The enthalpy of water fusion (ΔH_f) and freezing point were calculated from the heat flux curve.

2.5 Establishment of moisture sorption isotherm models

Several models have been proposed to describe the relationship between moisture content and water activity. The GAB model and the BET model [18–21] are commonly used models to describe the moisture sorption isotherm [22]. The equations of the BET and GAB models are as follows:

BET model:

$$m_t = \frac{C_B a_w m_0}{(1 - a_w)(1 + C_B a_w - a_w)} \quad (1)$$

Equation (1) can be transformed to

$$\frac{a_w}{(1 - a_w)m_t} = \frac{C_B - 1}{m_0 C_B} a_w + \frac{1}{m_0 C_B} \quad (2)$$

GAB model:

$$m_t = \frac{K_G C_G a_w m_0}{(1 - K_G a_w)(1 - K_G a_w + C_G K_G a_w)} \quad (3)$$

Equation (3) can be transformed to

$$\frac{a_w}{m_t} = \frac{K_G(1 - C_G)}{m_0 C_G} a_w^2 + \frac{C_G - 2}{m_0 C_G} a_w + \frac{1}{m_0 C_G K_G} \quad (4)$$

The nomenclature of the models is shown below.

Nomenclature	
a_w	The water activity
m_t	The total moisture content (%)
m_0	The monolayer moisture content (%)
C_B	The surface thermal constants of the BET model
C_G	The surface thermal constant of the GAB model
K_G	The constant of the GAB model

2.6 Statistical analysis

All the experiments were performed in triplicate. The analysis of all data and comparison was carried out using JMP 7 data analysis. To evaluate the ability of each model to fit the experimental data, the mean relative percent Error (E) and the Standard Error (S.E.) between the experimental and predicted data were determined by the following equations:

$$E = \frac{100}{N} \sum_{i=1}^{n} \frac{|m_e{}^i - m_p{}^i|}{m_e{}^i} \quad (5)$$

$$S.E. = \sqrt{\frac{\sum_{i=1}^{n}(m_e{}^i - m_p{}^i)^2}{N}} \quad (6)$$

where m_e and m_p are the experimental and calculated values of moisture content, respectively; and N is the number of the trial value. These statistical parameters were also used by other authors [23, 24]. The mean relative percent error was the first criterion used to evaluate the quality of the fit.

3 RESULTS AND DISCUSSION

3.1 Relationship between mt and aw of L. vannamei

The initial m_t and a_w of *L. vannamei* were 74.8% and 0.98, respectively. The decrease in m_t resulted in the decrease of a_w. The a_w value was located between 0.86 and 0.98 when the water content was higher than 31% (Figure 1). When a_w was higher than 0.40, the moisture sorption isotherm obtained by the experiment corresponded to the moisture sorption isotherm characteristics of livestock, poultry [12], aquatic animals and plants [8, 9].

From Equations (1) and (3), we used excel to make a curve fitting and obtain the following regression equations:

BET model:

$$\frac{a_w}{(1 - a_w)m_t} = 23.940272a_w - 4.118569 \quad (7)$$

GAB model:

$$\frac{a_w}{m_t} = -11.0394a_w^2 + 10.6068a_w + 1.5300 \quad (8)$$

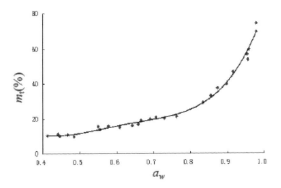

Figure 1. The relationship between water content and water activity for *L. vannamei*.

a_w of *L. vannamei* was observed between 0.40 and 0.98, and the BET model showed a bad fit quality. The determination coefficient (R^2) was only 0.60 (Figure 2 A). However, when a_w of *L. vannamei* was below 0.90, a very good fit quality ($R^2 = 0.90$) in the BET model was observed (Figure 2 B), and a much better fit quality ($R^2 = 0.98$) existed in the GAB model during the whole determination (Figure 3). The results are the same as those reported by Siripatrawan and Jantawat [11], who studied the moisture sorption isotherms of jasmine rice crackers. The present study reported that the experimental data followed the BET model when a_w was below 0.6 ($R^2 = 0.81$), but the data followed the GAB model between 0.10 and 0.95 ($R^2 = 0.98$).

Figure 4 shows the trial and calculated values by the BET and GAB models in the moisture sorption isotherm of *L. vannamei*. The moisture sorption isotherm and the saturated water content in a single molecular layer of food are important factors in food processing. The monolayer water contents in *L. vannamei* were 0.0532 g g^{-1} (dry basis) and 0.0868 g g^{-1} (dry basis) as determined by the BET and GAB models, respectively.

In general, the smaller the E-value and S.E. of the BET and GAB models, the higher the degree of fitting in the models. The E-value below 10% indicates an adequate fit for practical purposes [25, 26]. Table 1 lists the parameters calculated by the BET and GAB models in the moisture sorption isotherm of *L. vannamei*.

We observed that the mean relative percent errors of the BET and GAB models were all less than 10%. This result indicates that these models were adequately fit for practical purposes. The calculated values of both models in the moisture sorption isotherms of *L. vannamei* were analyzed using the F test to verify whether the models were fit for the experimental data. The results are summarized in Table 2. The F values of the BET and GAB models were extremely significant, indicating that the fitting effect of the two models is very good.

3.2 Relationship between mt and freezing point in L. vannamei

Different freezing points at various m_t in *L. vannamei* were determined using a differential scanning calorimeter, and then a relational schema was

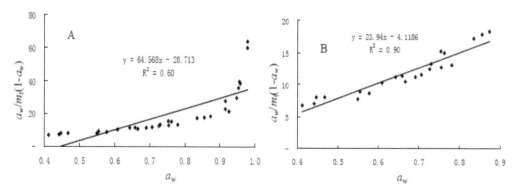

Figure 2. BET model between $a_w/m_t(1-a_w)$ and a_w (A 0.40~0.98; B 0.40~0.90).

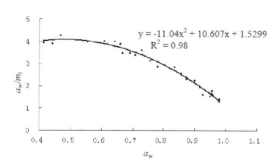

Figure 3. GAB model between a_w/m_t and a_w.

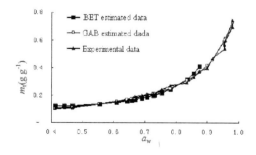

Figure 4. Trial values and calculated values by the BET and GAB model in the moisture *sorption* isotherms of *L. vannamei*.

Table 1. Parameters calculated by the BET and GAB model in the moisture sorption isotherms of *L. vannamei*.

Model	Parameters calculated	g g^{-1} (dry basis)	R^2	$E(\%)$	S.E.
BET	m_0	0.0505	0.0532	0.8999	5.0082	0.0085
	C_B	−4.8127				
GAB	m_0	0.0799	0.0868	0.9795	0.6196	0.0137
	C_G	13.1083				
	K_G	0.6241				

Table 2. Analysis of variance of the BET and GAB models in the moisture sorption isotherm of *L. vannamei*.

Model	Sum of squares	Mean square	F-value
BET	225.84470	225.845	179.7669**
Error	25.12639	1.256	
Summation	250.97109		
GAB	24.217247	12.1086	548.7268**
Error	0.507534	0.0221	
Summation	24.724781		

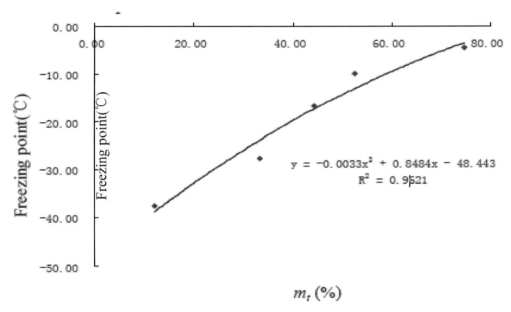

Figure 5. Relationship between m_t and freezing point of *L. vannamei*.

created between them (Figure 5). It is a second-degree polynomial equation. When the highest m_t was 74.8%, the freezing point was −4.49°C, but the freezing point reduced as m_t in *L. vannamei* decreased. When m decreased to 12.1%, the freezing point decreased to −37.52°C. The coefficient of determination (R^2) was 0.9621, which showed a very good degree of fitting. Thus, the freezing point was observed to decrease as m_t also decreased. This finding provides a theoretical basis for the parameters of the *L. vannamei* process.

3.3 Relationship between m_u and enthalpy of water fusion in L. vannamei

The enthalpy of water fusion of adsorption of *L. vannamei* ranged from 1.781 to 147.73 J g^{-1}

on the dry basis for the range of m_u varying from 0.2128 to 2.8139 g^{-1} on the dry basis (Figure 6). A linear relationship between the unfrozen m_u and the enthalpy of water fusion was observed. By fitting between the unfreezing m_u and the hot melt of *L. vannamei*, we can obtain the linear equation as follows:

$$\Delta H_f = 54.979 m_u - 6.8498 (R^2 = 0.9955) \qquad (9)$$

where m_u is the unfrozen moisture content (g/g on the dry basis) in *L. vannamei* and ΔH_f is the enthalpy of water fusion calculated by DSC in *L. vannamei*.

When $\Delta H_f = 0$, m_u was 0.1246 (g g^{-1} on the dry basis) from Equation (9), m_u in *L. vannamei* was 11.1%. This result indicates that only unfrozen

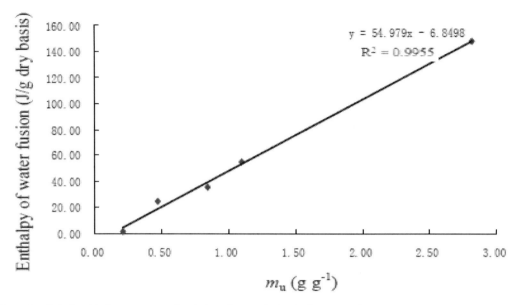

Figure 6.   Relationship between m_u and enthalpy of water fusion in *L. vannamei*.

water in *L. vannamei* was observed at this time. When m_u was below 11.1%, the hot melt cannot be determined by DSC in *L. vannamei*. The water combined with protein or lipid by the adsorption of static electricity and hydrogen bonding, where a much stronger force existed. Thus, the water content was generally higher than the saturated water content in a single molecular layer. Tsami *et al.* [27] suggested that the rapid increase in the isosteric heat of sorption at low m_u is due to the existence of highly active polar sites on the surface of the material. The surface is covered with water molecules forming a mono-molecular layer. Takizawa and Nakata [28] studied the moisture characteristics of a larger biological molecule–water gel system. The present study showed that the change of enthalpy was not only the phase translation among the water molecule, but also between the water molecule and the larger biological molecule when DSC was used to determine it in a larger biological molecule–water gel system.

4   CONCLUSIONS

This paper investigated the relationships between a_w, freezing point, water content, unfreezing water content and hot melt to analyze the moisture characteristics in the dehumidification process. a_w also decreased along with the decrease of m_t. The moisture sorption isotherm of *L. vannamei* belonged to type II (sigmoid curve). a_w changed between

0.855 and 0.98 when the water content was higher than 31%.

We used the BET and GAB models to fit the adsorption isothermal curve. The result exhibited a much better fit ability in the GAB model, which was suitable for a_w between 0.40 and 0.98. The saturated water content in a single molecular layer calculated by the GAB model was 0.0868 g g^{-1} on the dry basis. This study showed that the GAB model can estimate the change in water content and a_w in *L. vannamei*.

A linear relationship was observed between the water content and the hot melt in *L. vannamei* through the DSC analysis. The results of this study indicated that the freezing *point* of fresh shrimp can be estimated from the fitted curve between water content and the freezing point. The unfreezing water content can be calculated by the linear equation between the water content and the hot melt in *L. vannamei*.

ACKNOWLEDGMENTS

This work was supported by the Science and Technology Planning Project of Guangdong Province, China (No. 2012B020312005), the Guangdong Natural Science Foundation of China (S2013010012459) and the Innovation Experiment Program for College Students of Guangdong Ocean University & Guangdong Province (No. 1056610028).

REFERENCES

[1] H. Cui, L. Xiao, Current situation of China shrimp industry in 2011–2012 and its prospect. China Fisheries. 39(2012) 85–87.

[2] H.A. Iglesias, J. Chirife, Prediction of the effect of temperature on water sorption isotherms of food materials. Int. J. Food. Sci. Tech. 11(1976) 109–116.

[3] T.P. Labuza, The effect of water activity on reaction kinetics of food deterioration. Food Technol. 44(1980) 36–59.

[4] T.P. Labuza, Sorption phenomena in foods. Food Technol. 2(1968) 15–24.

[5] F. Kaymak-Ertekin, M. Sultanoglu, Moisture sorption isotherm characteristics of peppers. J. Food. Eng. 47(2001) 225–231.

[6] E. Tsami, M.K. Krokida, A.E. Drouzas, Effect of drying method on the sorption characteristics of model fruit powders. J. Food. Eng. 38 (1999) 381–392.

[7] N. Hamdami, J.Y. Monteau, A.L. Bail,. Transport properties of a high porosity model food at above and subfreezing temperatures (Part I: Thermophysical properties and water activity). J. Food Eng. 62 (2004) 373–380.

[8] R. Moreira, F. Chenlo, M.D. Torres, D.M. Prieto, Water adsorption and desorption isotherms of chestnut and wheat flours. Ind. Crops Prod. 32(2010) 252–257.

[9] N. Bahloul, N. Boudhrioua, N. Kechaou. Moisture desorption–adsorption isotherms and isosteric heats of sorption of Tunisian olive leaves (Oleaeuropaea L.). Ind. Crops Prod. 28(2008) 162–176.

[10] S. Lahsasni, M. Kouhila, M. Mahrouz, M. Fliyoua. Moisture adsorption–desorption isotherms of prickly pear cladode (Opuntia ficus indica) at different temperatures. Energy Convers. Manage. 44 (2003) 923–936.

[11] U. Siripatrawan, P. Jantawat, Determination of moisture sorption isotherms of jasmine rice crackers using BET and GAB models. Food Sci. Tech. Int.12(2006) 459–466.

[12] R.B. Singh, K.H. Rao, S.R. Anjaneyulu, G.R. Patilc, Moisture sorption properties of smoked chicken sausages from spent hen meat. Food Res. Int. 34 (2001) 143–148.

[13] E.S. Monterrey-Quintero, P.J.A. Sobral. Sorption isotherms of edible films of Tilapia myofibrillar proteins. Proceedings of 12th International Drying Symposium. Noordwijkerhout, The Netherlands, 2000.

[14] H.A. Iglesias, J. Chirife, Handbook of food isotherms. Academic Press, NY, USA, 1982.

[15] H.A. Iglesias, J. Chirife, An alternative to the GAB model for the mathematical description of moisture sorption isotherms of foods. Food Res. Int. 28 (1995) 317–321.

[16] AOAC, Official methods of analysis,16th ed., Association of Official Analytical Chemists, Washington, DC, 1990.

[17] G.H. Wu, H.B. Huang, M.Q. Zhang, Y. Chen, Study on moisture desorption isotherm and thermophysical property for Red sea bream Chrysophrys major Temminck et Schlegel. Prog. in Fish. Sci. 30 (1999) 115–121.

[18] S. Brunauer, P.H. Emmett, E. Teller, Adsorption of Gases in multimolecular layers. J. Am. Chem. Soc. 60(1938) 309–319.

[19] R.B. Anderson, Modifications of the BET equation. J. Am. Chem. Soc. 68(1946) 686–691.

[20] J.H. de Boer, The Dynamical Character of Adsorption, Clarendon Press, Oxford, U.K., 1953.

[21] E.A. Guggenheim, Applications of Statistical Mechanics, Clarendon Press Oxford, U.K., 1966.

[22] C. Van den Berg, S. Bruin, Water activity and its estimation in food systems: theoretical aspects. In: Rockland, L.B., Stewart, G.F. (Eds.), Water Activity: Influence on Food Quality, Academic Press, New York, 1981.

[23] M.P. Tolaba, M. Peltzer, N. Enriquez, M.L. Pollioc, Grain sorption equilibrium of quinoa grains. J. Food Eng. 61(2004) 365–371.

[24] A. Iguaz, P. Vírseda, Moisture desorption isotherms of rough rice at high temperatures. J. Food Eng. 79(2007) 794–802.

[25] R.J. Aguerre, C. Suarez, P.Z. Viollaz, New BET type multi-layer sorption isotherms-Part II: Modelling water sorption in foods. Lebensmittel-Wissenchaft und Technology, 22(1989) 192–195.

[26] K.J. Park, Z. Vohnikova, F.P.R. Brod. Evaluation of drying parameters and desorption isotherms of garden mint leaves (Mentha crispa L.). J. Food Eng. 51(2002) 193–199.

[27] E. Tsami, Z.B. Maroulis, D. Morunos-Kouris, G.D. Saravacos, Heat of sorption of water in dried fruits. Int. J. Food Sci. Tech. 25(1990) 350–359.

[28] T. Takizawa, Y. Nakata, New small endothermic peaks with hysteresis commonly observed in the differential scanning calorimetric study of biopolymer-water systems. Thermochimica. Acta. 7 (2000) 352–353.

Material Science and Engineering – Chen (Ed.)
© 2016 Taylor & Francis Group, London, ISBN 978-1-138-02936-1

The effect of Carbon Black on dynamic mechanical properties of Nitrile Butadiene Rubber vulcanizates

L. Liu, B.G. Zhang, X.R. Xie & X. Geng
Institute of High Performance Polymer, Qingdao University of Science and Technology, Qingdao, China

W. Li & Z.S. Chen
College of the Environmental Science and Engineering, Peking University, Beijing, China

ABSTRACT: The effect of the loads, types and structures of Carbon Black (CB) on the dynamic mechanical properties of NBR vulcanizates was studied. The results showed that the dependence of the storage modulus (G') and loss modulus (G'') on strain amplitude increased with the increasing CB specific surface area. There was a positive relationship between (G_0'-G_∞') and CB specific surface area. The Payne effect becomes increasingly conspicuous with the increase in carbon black N330 loads over 40 phr loads. On the contrary, the G' and G'' of vulcanizates with less than 40 phr loads were hardly influenced by strain amplitude in the strain range. The contribution to G' and G'' of the CB structure was larger than their specific surface area.

Keywords: Nitrile Butadiene Rubber; dynamic performance; Carbon Black

1 INTRODUCTION

Nitrile Butadiene Rubber (NBR), a copolymer of acrylonitrile and butadiene monomers, has a good property of oil resistance. Therefore, it is widely used in a range of oil applications under several conditions of temperature, frequency, and deformation (Buford, P.R. 1995) [1].

Carbon Black (CB) is widely used as a reinforcing filler in polymeric matrices, causing considerable changes in dynamic properties. Da Costa et al. (2002) [2] found that the dependence of the modulus on temperature for different concentrations of CB in the NR vulcanizates decreases with the increasing temperature. Sirisinha (2001) [3] reported that particle sizes influenced on the distribution of CB in each phase of BR/NBR blends, and the degree of decreases in tanδ max at a given CB particle size is more significant in BR than in NBR. Xiu Feng et al. (2012) [4] proved that there was the Payne effect with the increment of CB content. Trexler et al. (1985, 1986) [5,6] studied the effects of the types of CB with an equal filler loading on the dynamic properties of natural rubber and bromobutyl rubber. Chuayjuljit et al. (2002) [7] investigated the effects of particle size and amount of CB and calcium carbonate on the dynamic mechanical properties of vulcanized natural rubber. Gabriel. (1996) [8] reported the effects of CB loadings, temperature, and shear frequency on the dynamic mechanical properties of EPDM. Tripathy & Dutta (1989, 1992) [9,10] studied the effects of the

types of fillers on the physical and dynamic properties of bromobutyl vulcanizates. According to previous work, the dynamic mechanical properties of polymers are largely influenced by factors such as type, volume fraction, irregular shape and distribution of the aggregate shape of CB (Maiti, S. et al. 1992, Sirisinha, C. et al. 1998) [11, 12].

However, there are only a few studies reporting on the dynamic mechanical properties of NBR vulcanizates filled with CB. In this article, the changes in the dynamic mechanical properties of NBR vulcanizates related to the loading, particle size, and structure of CB on the temperature and strain amplitude were compared.

2 EXPERIMENTAL PROCEDURE

2.1 *Materials*

All materials were used as received. NBR 3445 (33.4 wt% acrylonitrile, Mooney viscosity 47) was supplied by Lanxess Germany Company. CB was supplied by Evonik Degussa Company, Qingdao. Other compounding ingredients, such as zinc oxide, and stearic acid were obtained from the market.

2.2 *Formulation of the NBR compound*

The quantity of ingredients is as follows (phr): NBR 100, zinc oxide 5, stearic acid 1, sulfur 1.5, dibenzothiazyl disulfide (DM) 1, and CB 40

Table 1. Physical properties of CB.

Properties	220	326	330	375	550	660	774	990
BET [m²/g]	121	83.2	82	90.4	43	36	29	9
DBP [cm³/100 g]	114	71.9	102	113.8	121	90	72	43

(physical properties of CB used in this study are presented in Table 1).

2.3 Dynamic mechanical property testing

Dynamic mechanical property testing was performed on a rubber processing analyzer RPA 2000 by Alpha Company (American) in two modes, namely temperature sweeps and strain sweeps. In the case of strain sweeps, strain amplitudes ranging from 0.1 to 140% were applied with a frequency of 1 Hz at constant temperatures of 60°C. The temperature sweep measurements were carried out at temperatures ranging from 50 to 150°C with a strain of 7% and a frequency of 1 Hz.

3 RESULTS AND DISCUSSION

3.1 Effect of CB types

The influence of CB types on the dynamic mechanical properties of NBR vulcanizates is shown in Figure 1. A dramatic decrease in all curves and a typical nonlinear behavior with increasing dynamic stain amplitude were observed, which is called the Payne effect. The Payne effect is influenced by the filler-filler interaction and the filler dispersion. CB with a high specific surface area and a high surface energy is prone to congregate. The higher the CB specific surface area, the more the formation of CB aggregates and agglomerates, furthermore, the more the network was formed in the polymer matrix. The contribution of the CB network to G' can be characterized by $(G_0' - G_\infty')$. Figure 1c shows the comparison of $(G_0'-G_\infty')$ for NBR vulcanizates filled with different CB types. It was seen that $(G_0'-G_\infty')$ decreased almost linearly when the CB specific surface area decreased. The amount of the filler network increased with the increasing specific surface area of filler (Chih-Cheng, P. et al. 2005.)[13], and G' was higher as well. Then, as the strain amplitude increased, the breakdown rate of the filler-filler network was greater than the reforming rate of physical bonds between the filler aggregates, which resulted in a decrease of G'. When the strain amplitude was higher enough, the whole CB network was destroyed and G' decreased to almost the same level, which was only contributed by the rubber matrix.

The dependence of G'' on the strain amplitude increased with increasing CB specific surface area

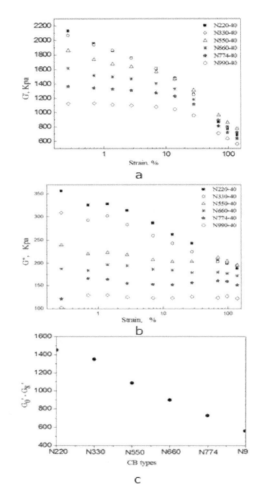

Figure 1. Dependence of G' and G'' on strain amplitude of NBR vulcanizates filled with different CB types.

(Figure 1b). N220 and N330, both with a smaller particle size and a greater specific surface area (Table 1), were easy to aggregate and could form a strong network structure at the 40 phr load. As the strain amplitude increased, we observed a steady decrease in the curves. The other CBs with large particle sizes can only form small agglomeration, which was difficult to form a strong filler network at the 40 phr load; thus, the G'' maintained a plateau over the range of strain amplitudes tested.

3.2 Effect of CB contents

The dependence of G′ and G″ on the strain amplitude of NBR vulcanizates filled with different N330 loads is shown in Figure 2. When the N330 load was below 40 phr, the G′ of NBR vulcanizates was similar and almost independent of the strain amplitude, and higher than the unfilled NBR vulcanizates. The reason may be that no filler agglomerates existed below that level of loading. Then, as the N330 loads increased from 40 phr to 80 phr, a significant augmentation at low strain amplitudes was observed, and decreased dramatically with the increasing strain amplitude. That is, the higher the CB load added, the more the CB networks formed in the polymer matrix, and the Payne effect was more pronounced for a greater filler load.

The G″ increased with the increase in N330 loads, but the G″ of vulcanizates filled with N330 with less than 40 phr kept a balance value similar to G′ during the range of the strain amplitude. Furthermore, when the load of N330 was more than 40 phr, there was a dramatic drop in G″ with the increasing strain amplitude (Fig. 2b). It could be interpreted that the load of 40 phr was the critical concentrations of N330 to form filler-filler networks in the NBR matrix. Because the degree of the filler-filler network was inversely proportional to the degree of filler dispersion, the N330 were added over 40 phr with a bad dispersion, and the friction of filler-polymer interactions increased. Under the effect of strain, the destruction and reconstruction of the filler network produced a lot of hysteresis and lead to the increase in G″.

3.3 Effect of CB structure

In Figure 3, it could be observed that the G′ of NBR vulcanizates filled with N330 was similar to N375, being higher than N326. Compared with the characteristic parameters of N375, N330, and N326 (Table 1), N375 had the highest structure while N326 had the lowest. The order of the specific surface area of three CB types was N326 ≈ N330 < N375. The contribution to G′ of the CB structure was greater than the specific surface area. From the fluid dynamics point of view, the higher the degree of the CB structure with the same load, the more the filler networks formed in the polymer matrix, which increased the rubber trapped in the filler networks or agglomerates. The rubber being trapped or caged lost its identity as an elastomer and, instead, behaved as a filler. Therefore, the effective volume of the rubber bearing the stress imposed upon the sample was reduced, which resulted in an increased modulus that was governed by the CB structure. The breakdown of the filler network would release the trapped rubber with the increasing strain amplitude, so the modulus would decrease.

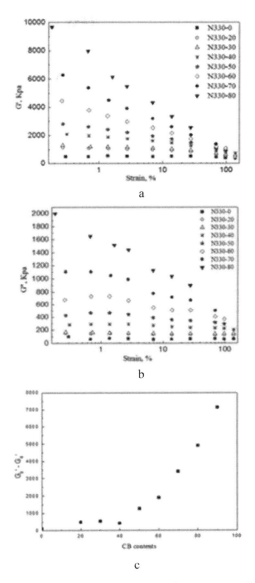

Figure 2. Dependence of G′ and G″ on strain amplitude of NBR vulcanizates filled with CB loads.

In Figure 3b, the G″ exhibited a maximum stain amplitude of 1.38%, and then there was a reduction with the increasing strain amplitude. This may be explained by the fact that at the strain amplitude less than 1.38%, the reform of the filler-filler and filler-rubber networks could be able to keep up with their breakdown, so that the friction and hysteresis loss increased, and G″ increased. Then, as the stain amplitude increased, the breakdown of networks was faster than their reform, so G″ decreased. The contribution of the CB structure was larger than their specific surface area, which is similar to G′.

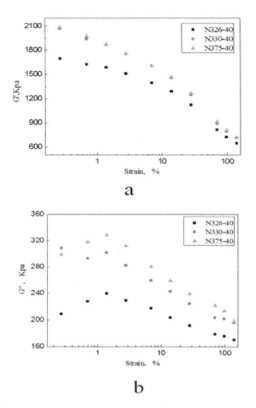

a

b

Figure 3. Dependence of G' and G″ on strain amplitude of NBR vulcanizates filled with different CB structure.

4 CONCLUSIONS

The main objective of this study was to highlight the effect of the loads, types and structures of CB on the dynamic mechanical properties of NBR vulcanizates. The dependence of the G' and G″ on the strain amplitude increased with the increasing CB specific surface area. There was a positive relationship between the $(G_0' - G_\infty')$ and the CB specific surface area. The Payne effect becomes increasingly conspicuous with the increase in carbon black N330 loads over 40 phr loads. On the contrary, the G' and G″ of vulcanizates with less than 40 phr loads were hardly influenced by the strain amplitude in the strain range. The contribution to G' and G″ of the CB structure was larger than their specific surface area.

REFERENCES

[1] Buford, P.R. 1995. Nitrile rubber—past, present and future. J. Rubber Chem. Technol. 68:540–546.
[2] Da Costa, H.M. et al. 2002. Mechanical and dynamic mechanical properties of rice husk ash-filled natural rubber compounds. J. Appl Polym Sci. 83:2331–2346.
[3] Sirisinha, C. & Prayoonchatphan, N. 2001. Study of carbon black distribution in BR/NBR blends based on damping properties: influences of carbon black particle size, filler, and rubber polarity. J. Appl Polym Sci. 81:3198–3203.
[4] Xiu Feng, Lv, et al. 2012. Preparation and properties of carbon black filled EPDM/POE thermoplastic vulcanizates. J. Appl Polym Sci. 125:3794–3801.
[5] Harold E, Trexler & Michael, C.H. Lee. 1985. Rubber Division Meeting.
[6] Harold E, Trexler & Michael, C.H. Lee. 1986. Effect of types of carbon black and cure conditions on dynamic mechanical properties of elastomers. J. Appl Polym Sci. 32: 3899–3912.
[7] Chuayjuljit, S. et al. 2002. Effects of particle size and amount of carbon black and calcium carbonate on curing characteristics and dynamic mechanical properties of natural rubber. J. Materials and Minerals.12:51–57.
[8] Gabriel, J.O. 1996. Effects of temperature and strain amplitude on dynamic mechanical properties of EPDM gum and its carbon black compounds. J. Appl Polym Sci. 59:567–575.
[9] Tripathy, D.K. & Dutta, N.K. 1989. Strain-dependent dynamic mechanical properties of black-loaded vulcanizates. J. Kautsch Gummi Kunstst. 42:665–671.
[10] Tripathy, D.K. & Dutta, N.K. 1992. Effects of types of fillers on the molecular relaxation characteristics, dynamic mechanical, and physical properties of rubber vulcanizates. J. Appl Polym Sci. 44:1635–1648.
[11] Maiti, S. et al. 1992. Quantitative estimation of filler distribution in immiscible rubber blends by mechanical damping studies. J. Rubber Chem Technol. 65: 293–302.
[12] Sirisinha, C. et al. 1998. Study of carbon black distribution in butadiene rubber-nitrile/butadiene rubber blends based on damping properties. Effect of some grades of carbon black. J. Plast Rubber Compos Process Appl. 27:373–375.
[13] Chih-Cheng, P. et al. 2005. "Smart" silica-rubber nanocomposites in virtue of hydrogen bonding interaction. J. Polym. Adv. Technol. 16:770–782.

Material Science and Engineering – Chen (Ed.)
© 2016 Taylor & Francis Group, London, ISBN 978-1-138-02936-1

Numerical simulation and lift loss assessment for a scaled S/VTOL aircraft proximity of the ground

S. Liu, Z.X. Wang & L. Zhou
School of Engine and Energy, Northwestern Polytechnical University, Xi'an, China

ABSTRACT: During the S/VTOL aircraft taking off proximity of the ground, both suckdown and fountain effects occur which will induce the lift loss. Moreover, the hot gas ingestion can also make the temperature at the front of the compressor distortion thus reducing the lift produced by the cruise engine. The effects above-mentioned were studied in this paper. The flow field was calculated with FLUENT, using unstructured grid for a scaled aircraft which has two impinge jets proximity of the ground and also the lift loss was assessed. The flow field and the lift loss characteristic were got by the numerical simulation. The temperature distribution on the ground and the pressure distribution under the fuselage and the wing at various heights, headwind speeds were studied for the scaled S/VTOL aircraft model.

Keywords: S/VTOL; suckdown; ingestion; numerical simulation; lift loss

1 INTRODUCTION

When the S/VTOL aircraft hovers in proximity of the ground, the lift jets entrain the surrounding fluid causing a negative pressure under the fuselage leading to lift loss. Otherwise, the lift jets impinging onto the ground meet at a stagnation line and turn upwards to form an upwash fountain which may interact with the fuselage. The fountain flow can offset part of the lift loss caused by the jet entrainment and also flow along the surface under the fuselage and into the engine inlet which may cause Hot Gas Ingestion (HGI). When the S/VTOL aircraft operates in crossflow, the jet spreading forward along the ground stagnates and forms horseshoe vortex which can also be ingested into the engine inlet [1–4]. In addition, the crossflow can also affect the pressure field on the fuselage. It is necessary to study the complicated flow field of the S/VTOL aircraft in proximity of the ground to develop efficient S/VTOL aircraft technology. The complicate flow field study over past years includes characteristic study of the lift loss caused by the jet entrainment and flow field phenomenon study for a certain concept aircraft model [5–9]. Most of the former research object is similar to the Harrier lift layout which has four hot and subsonic hot jets [10–13]. The recent development of the F-35B fighter has made people pay more attention to this successful supersonic aircraft capable of short takeoff and vertical landing with two lift jets. The purpose of this paper is to analyze the environment around a numerical model of a scaled S/VTOL aircraft which is similar to the F-35B lift layout in ground proximity.

2 NUMERICAL APPROACH

2.1 Numerical model and calculation domain

The jet induced pressure reduction in this relatively simple flow geometry can be modeled or predicted from inviscid potential flow theory to simulate the entrainment effect of the jet and also capture most of the physics of jet/ground interaction problems [14–15]. The time-averaged, inviscid, compressible, three dimensional method for ideal gas are employed for the present computations.

The fuselage of the aircraft is similar to F-35B fighter, which has two jets, side inlet configuration. The forward jet is supersonic cold flow and the aft nozzle jet is supersonic hot flow. The calculation domain is shown in Figure 1. Because the flow field is symmetric about the aircraft center plane, only half of the flow field was calculated. The calculation domain is 2.7 meters high, 7.2 meters wide, 15 meters long. The half width of the aircraft is 0.67 meters, and the diameter of the nozzle is 0.11 meters (D). The domain inlet is 71D upstream from the engine inlet. The distance between the engine inlet and the axis of the forward nozzle is nearly equal to D in x direction. The distance between the axis of the forward and aft jets is 9.2D, and the axis of the aft jet is 56D upstream form the exit boundary.

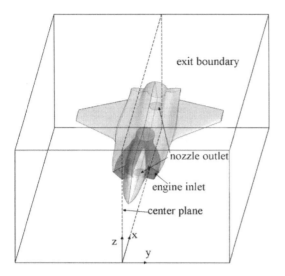

Figure 1.   Computational domain.

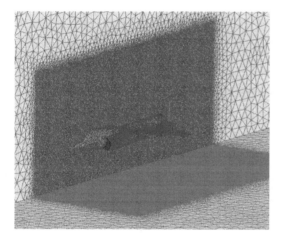

Figure 2.   Mesh structure of the model.

2.2 Mesh and boundary

The inviscid unstructured mesh of the calculation domain is generated by ICEM. Based on the characteristic of the flow field, the mesh is generated in two parts which has both finer and coarse mesh as shown in Figure 2. The transition ratio between the coarse and finer mesh does not exceed 5. The finer mesh is applied to the region with a larger gradient in velocity and temperature which is near the fuselage and the coarse mesh is used for the rest region. The grid has 2 million nodes, as shown in Figure 3.

The temperature and velocity of the upstream boundary were specified to be uniform. The free

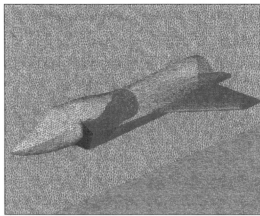

Figure 3.   Finer mesh around the aircraft.

stream temperature was 300 K, and the headwind velocity was 3 m/s, 10 m/s and 30 m/s. The mean velocity of the forward nozzle exit is 405 m/s, and a uniform temperature of 300 K is specified for the forward lift jet. The mean velocity of the aft nozzle exit is 788 m/s, and a uniform temperature of 800 K is specified for the aft lift jet. The pressure outlet boundary condition is imposed at the exit of the engine inlet duct and at the flow field downstream exit. A symmetric boundary condition is specified at the aircraft center plane. The mass of the engine inlet exit maintains the same with that of the aft vertical lift jet in FLUENT pressure outlet boundary settings.

Open your old file and the new file.

3   RESULTS

3.1 Features of flowfield

The flowfield of the S/VTOL aircraft in proximity of the ground has two main phenomenons: hot gas ingestion and suckdown forces. Both of them have adverse effect on aircraft operating in S/VTOL mode proximity of the ground. The temperature distortion may cause the engine to stall. The suckdown forces may cause lift loss and aircraft pitch moment unsteady.

The fountain occurs when the adjacent jet impinge onto the ground and meet each other. The high temperature fountain flow impinge onto the under surface of the fuselage and flow into the vicinity of the engine inlet. The hot gas ingested causes the temperature distortion of the engine inlet. Compared with the aircraft with hot jets, the forward cold jet plus the aft hot jet aircraft suffer lower temperature distortion. The forward cold jet spreads along the ground and meets the aft spreading hot jet forming a stagnation line to prevent the engine inlet from hot gas ingestion from the surrounding environment. Figure 4 is the inlet side cut plane at y = 0.13 m position which is appeared in Figure 5. Figure 5 and 6 is the result calculated for U = 3 m/s, and H/D = 3. The fountain is shown in Figure 5 in the streamlines and temperature contours plot of a vertical plane at y = 0.13 m position through the engine inlet. The stagnation line which is in the streamlines and temperature contours plot of the ground plane is shown in Figure 6.

When the lift jets expand, they entrain the surrounding air flow to the high speed jets creating negative pressures under the aircraft and causing lift loss. This feature can be seen in Figure 7, which is the result calculated for U = 3 m/s, and H/D = 3. The flowfield is shown as streamlines and velocity contours in plot of the aircraft center plane. In Figure 7, the streamlines show the air flow around the forward and aft jet is induced to the high velocity jets.

3.2 Effect of variations of height and headwind speed on hot gas ingestion

The effect of variations in the strength of the headwind and height on hot gas ingestion can be seen by

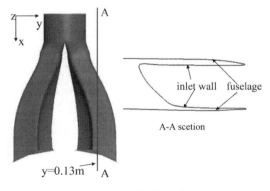

Figure 4.   Inlet section profile of the fountain result.

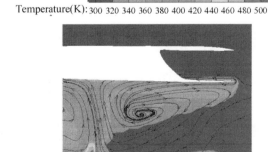

Figure 5.   Hot gas ingestion from the fountain flow.

Temperature(K): 350   490   630   770   910   1050

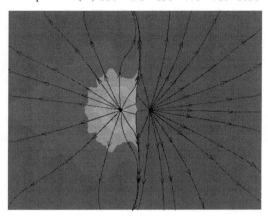

Figure 6.   Stagnation line distribution between the cold and hot jet on ground.

Velocity(m/s): 50   150   250   350   450   550   650   750

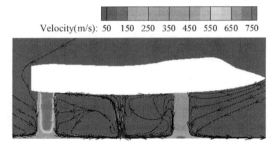

Figure 7.   Jet entrainment under the fuselage.

comparing calculated results for U = 3 m/s, 10 m/s and 30 m/s, with H/D = 3 and 7. Figure 8 shows temperature distributions in the aircraft center plane and the aircraft under surface which is corresponding to the center plane.

Comparing between Figure 8(a) and 8(d) or 8(b) and 8(e), in the aircraft center plane, the temperature at the top of the fountain region drops gradually as the aircraft height increasing with the same headwind strength. As the height increasing, the fuselage under surface suffers lower temperature from the fountain flow shown in Figure 8. As the fountain temperature is lower, they spread along the fuselage with temperature dropping and finally reach the lowest around the engine inlet under the fuselage. So as the height increasing, the temperature of hot gas ingestion from the fountain will be lower.

As the aircraft height is fixed, the effect of the variations in the strength of the headwind on the fountain is calculated for U = 3 m/s, 10 m/s and

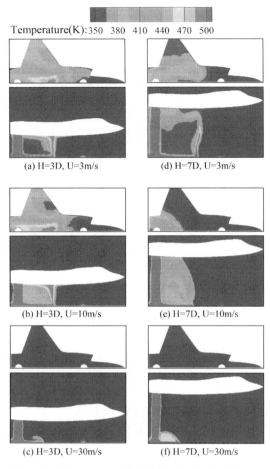

Temperature(K): 350 380 410 440 470 500

(a) H=3D, U=3m/s   (d) H=7D, U=3m/s

(b) H=3D, U=10m/s   (e) H=7D, U=10m/s

(c) H=3D, U=30m/s   (f) H=7D, U=30m/s

Figure 8.   Temperature distribution in the aircraft center plane and surface under the aircraft.

30 m/s for the same aircraft configuration. For U = 3 m/s condition, the fountain main region is almost perpendicular to the under surface of the aircraft in weak headwind case at H = 3D and H = 7D shown in the aircraft center plane. Compared with U = 3 m/s condition, the fountain high temperature region trends to the aft jet at U = 10 m/s condition, the temperature between the fountain and aft jet is higher than that at U = 3 m/s condition. The comparison of temperature distribution under the aircraft surface between Figure 8(a) and 8(b) or Figure 8(d) and 8(e) show that the high temperature distribution under the aircraft surface is greatly affected by the strength of the headwind. As the high temperature fountain region flowing to the backward of the aircraft under surface affected by stronger headwind, the gas temperature spreading to the aircraft forward becomes lower and lower until be ingested into the engine inlet. So the

stronger headwind may cause the high temperature fountain region moving backward and lead to lower temperature gas be ingested into the engine. For U = 30 m/s condition, in the aircraft center plane of Figure 8(c) and 8(f), the fountain region is greatly affected by the strong headwind flowing backward as vortex attached to the ground. The fountain front section inclines to the aft jet and at a certain angle to the vertical plane between the forward and aft jets. Especially, the fountain upwash flow has no effect on the aircraft flow which can be seen evidently in both aircraft center plane and under surface. So for too strong headwind, the fountain upwash may be flowed away from the fuselage by the crossflow.

Figure 9 shows the temperature distribution in the vertical plane through the inlet duct for cases with various heights and headwind strength calculated in this paper. The highest temperature appears in the lower outside corner of the duct corresponding to the temperature distributions under the fuselage.

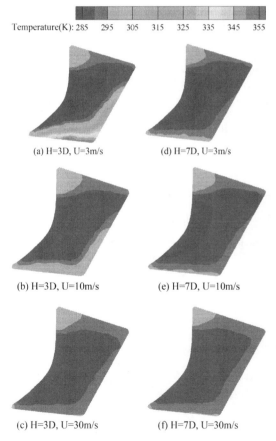

Temperature(K): 285   295   305   315   325   335   345   355

(a) H=3D, U=3m/s   (d) H=7D, U=3m/s

(b) H=3D, U=10m/s   (e) H=7D, U=10m/s

(c) H=3D, U=30m/s   (f) H=7D, U=30m/s

Figure 9.   Temperature distribution of the inlet duct vertical section.

The high temperature fountain flows forward along the fuselage meeting and flowing around the forward cold jet to the outer side of the inlet duct. So the high temperature distribution of the inlet duct is due to the special configuration of the aircraft. As the height and headwind strength increasing, the inlet duct highest temperature decreases, thus suffers weaker temperature distortion.

3.3 *Variation of height and headwind speed effect on lift loss*

The effect of variations in the strength of the headwind and height on gas entrainment can be seen by

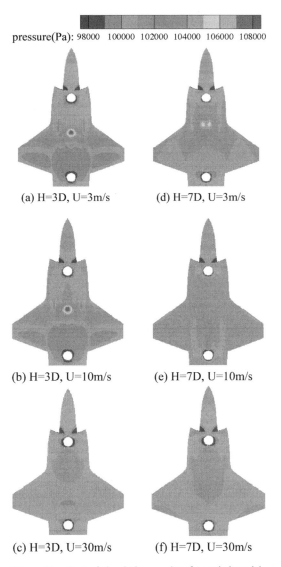

pressure(Pa): 98000 100000 102000 104000 106000 108000

(a) H=3D, U=3m/s   (d) H=7D, U=3m/s

(b) H=3D, U=10m/s   (e) H=7D, U=10m/s

(c) H=3D, U=30m/s   (f) H=7D, U=30m/s

Figure 10. Part of simulation results of coupled model.

Table 1. Lift loss with various headwind speeds and heights.

	ΔL (H = 3D)	ΔL (H = 7D)
U = 3 m/s	21.6%	13.4%
U = 10 m/s	19.3%	8.6%
U = 30 m/s	23%	15.2%

comparing calculated results for U = 3 m/s, 10 m/s and 30 m/s, with H/D = 3 and 7. Figure 10 shows the pressure distribution in the aircraft under surface.

Comparing the pressure distribution under the aircraft in Figure 10(a) and 10(d) or 10(b) and 10(e), as the height increasing, the low pressure region caused by jet entrainment around the forward and aft jets becomes smaller, the pressure of the fountain flow acting on the aircraft under surface reduces.

The effect of the variations in the strength of the headwind on jet entrainment is calculated for U = 3 m/s, 10 m/s and 30 m/s for the same aircraft configuration. For U = 3 m/s condition, comparing the pressure distribution under the aircraft in Figure 10(a) and 10(b) or 10(d) and 10(e), as the strength of the headwind increasing, the low pressure region caused by jet entrainment around the forward and aft jets becomes smaller. For U = 30 m/s condition, as aforementioned in this paper, the fountain upwash flow has no effect on the aircraft flow. So in too strong headwind condition, the fountain upwash may be flowed away from the fuselage by the crossflow, thus the lift loss increases. The lift loss for various headwind speeds and heights is shown in Table 1.

4 CONCLUSION

Compared with the all hot jets aircraft, the forward cold jet plus the aft hot jet aircraft suffer lower temperature distortion. Due to the special configuration of the aircraft, the highest temperature appears in the lower outside corner of the duct corresponding to the temperature distributions under the fuselage. As the height and headwind strength increasing, the inlet duct highest temperature decreases, thus suffers weaker temperature distortion. As the height increasing, the strength of the fountain is reducing gradually. For U = 10 m/s condition, the fountain makes bigger force to the fuselage than the other two headwind strength cases for the aircraft configuration in this paper. The too strong headwind may make the fountain out of the fuselage attached to the ground causing bigger lift loss.

REFERENCES

[1] Louisse, J. & Marshall, F.L. 1974. Prediction of ground effects for VTOL aircraft with twin lifting jets. AIAA-1974–1167.

[2] Webber, H.A. & Gay, A. 1975. VTOL Reingestion model testing of fountain control and wind effects. AIAA 1975–1217.

[3] Smith, C.W. & Karemaa, A. 1978. Induced effects of multijet surfaces fountains on flat-plate. AIAA-1978–1516.

[4] Foley, W.H. 1978. Methodology for Prediction of VSTOL Propulsion Induced Forces in Ground Effect. AIAA-1978–1281.

[5] Tafti, D. 1990. Hot gas environment around STOVL aircraft in ground proximity, Part II numerical study. AIAA-1990–2270.

[6] Barata, J.M.M. 1991. Fountain flows produced by multi jet impingement on a ground plane. AIAA-1991–1806.

[7] Barata, J.M.M. 1995. Fountain flows produced by multiple impinging jets in a crossflow. AIAA-1995–0190.

[8] Behrouzi, P. & McGuirk, J.J. Particle image velocimetry for intake ingestion in short takeoff and landing aircraft. Journal of aircraft, 37(6), 2000:994–1000.

[9] Page, G.J. & Qinling, L. LES of impinging jet flows relevant to vertical landing aircraft. AIAA-2005–5226.

[10] Smith, M. & Chawla, K. Numerical simulation of a complete STOVL aircraft in ground effect. AIAA-1991–3293.

[11] VanOverbeke, T.J. & Holdeman, J.D. Three dimensional turbulent flow code calculations of hot gas ingestion, Journal of aircraft, 27(7), 1990:577–582.

[12] Fricker, D.M. & Holdeman, J.D. Calculations of hot gas ingestion for a STOVL aircraft model, Journal of aircraft, 31(1), 1994:236–242.

[13] VanOverbeke, T.J. & Holdeman, J.D. A numerical study of the hot gas environment around a STOVL aircraft in ground proximity. AIAA-1988–2882.

[14] Kotansky, D.R. The influence of VTOL vehicle configuration variables on vectored jet induced flow fields. AIAA-1978–1021.

[15] Agarwal, R.K. & Deese, J.E. Euler solutions for air foil jet ground-interaction flowfields, Journal of aircraft, 23(5), 1985:376–381.

Material Science and Engineering – Chen (Ed.)

The synthesis of the amidosilane-based surfactants and the investigation of their surface tension

X.L. Sun, F.L. Xing, Q. Xu, Y.P. Wang, X. Li & L.Y. Wang

The Key Laboratory of Fine Chemicals of Qiqihar University, Heilongjiang, China

ABSTRACT: A series of novel organic silicone surfactants with amide bond were synthesized by the reaction of aminopropyl trisiloxane with succinic anhydride and phthalic anhydride, 4-((3-(1,1,1,3,5,5,5-heptamethyltrisiloxan-3-yl)propyl)amino)-4-oxobutanoatesodium,2-((3-(1,1,1,3,5,5,5-heptamethyltrisiloxan-3-yl)propyl)carbamoyl)cyclohexa-2,4-dienecarboxylatesodium. -N-(3-(1,1,1,3,5,5,5-heptamethyltrisiloxan-3y l)propyl) acetamide was prepared by the reaction of aminopropyl trisiloxane with glutaryl dichloride. All synthesized surfactants were characterized with IR and ^{1}H-NMR spectroscopy. Their surface tension was measured. These surfactants were found to reduce the surface tension of water less than 34 mN/m at a level of 10–5–10–4 mol/L, and their surface activities were compared. A conclusion was obtained on how the different hydrophilic groups affect the surface tension of the organic silicone surfactants with amide bond. The super surface activities of compound **1** is attributable to the structure of the hydrophilic group, which is an ionic group and relatively small in size.

Keywords: trisiloxane; surface activity; amides bond; hydrophobic groups

1 INTRODUCTION

In recent years, with environmental pressures increasing, green efficient surfactants [1–3] have become the mainstream of research in the development of surfactants. However, organic silicone surfactants with amide bond were studied less than others. In the USA, the study on surfactants was carried out earlier; for instance, America Union Carbide Company first synthesized nonionic silicone surfactants containing polyether. In China, the study of this kind on surfactants was started relatively late. In 2004, a family of glucosamide-based trisiloxane gemini surfactants was reported by Fu Han [4]. The members of this family reduced the surface tension of water to 21 mN/m at concentration levels of 10–5 mol/L. In 2014, a novel trisiloxane gemini surfactant was synthesized by Fenglan Xing [5]. The surface activities of the surfactant were compared with each other, and a discussion was reported on how the hydrophobic groups attached on the silicone atoms affect their surface activities.

Organic silicone surfactants are special surfactants with a silicone group as the hydrophobic side and organic groups attached on it as the hydrophilic side. Their uncommon surface properties were determined by their special structures, such as a hydrophobic behavior with high flexibility and easy orientation at the interface. Therefore, the widely used surfactants were used as detergents and agents in agricultural chemicals, coating material, products of cosmetics and textiles [6–10].

There are many factors that affect the surface activity, such as the carbon chain length of the surfactant, the type of the surfactant, the properties of the adsorbent, the pH of the solution and the temperature [11–15]. In this paper, the same hydrophobic group was connected with a different hydrophilic group. We first used succinic anhydride, phthalic anhydride and glutaryl dichloride to link 3-(diethoxy(methyl)silyl)pro-pan-l-amine to get three surfactants, namely compound **1**, compound **2** and compound **3**. The three chemical compounds are organic silicone surfactants with the amide bond bearing the same hydrophobic group but a different hydrophilic group. The surface activities were measured, and how the hydrophobic groups attached on the silicone atoms affect their surface activities is discussed. The three chemical compounds were prepared, as shown in Figure 1.

2 MATERIALS AND METHODS

2.1 *Materials*

Hexamethyldisiloxane, tetramethylammonium hydroxide (TMAH) and 3-(diethoxy(methyl) silyl)pro-pan-l-amine were obtained from Aladdin Industrial Corporation; glutaryl dichloride,

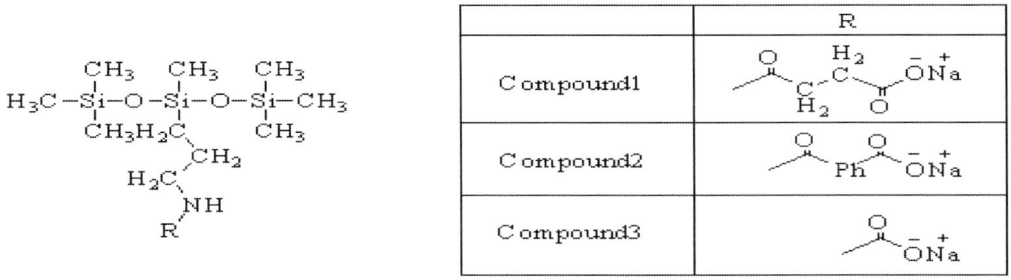

		R
Compound 1		(structure: acetyl-CH₂-CH₂-C(=O)-O⁻Na⁺)
Compound 2		(structure: acetyl-Ph-C(=O)-O⁻Na⁺)
Compound 3		(structure: acetyl-C(=O)-O⁻Na⁺)

Figure 1. Molecular structure of the amidosilane-based surfactant.

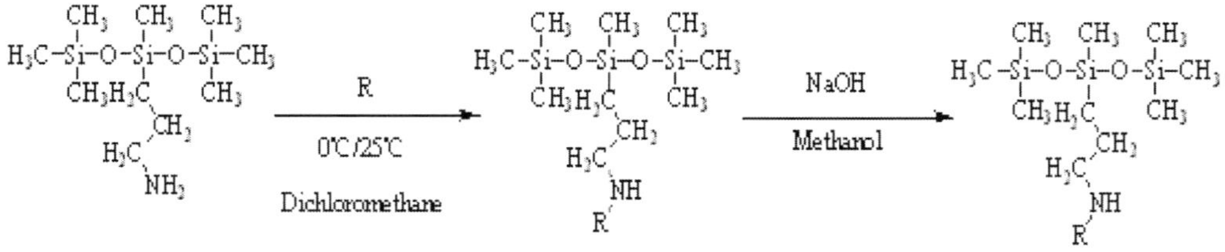

Figure 2. Synthesis of the amidosilane-based surfactant.

succinic anhydride and phthalic anhydride were ordered from Tianjin Fuyu Fine Chemical Co. Ltd. Water was doubly distilled. Other chemicals, methanol and dichloromethane were used as received without further purification.

2.2 Methods

The structure of the prepared products was confirmed through infrared (IR) and ^{1}H NMR spectroscopy. IR spectra were measured on a Perkin-Elmer Spectrum one spectrometer. ^{1}H NMR spectra were performed using a Bruker Ascend 600 MHz spectrometer, and the products were recorded as ppm in CDCl$_3$, with TMS as the internal standard. Surface tension was determined at 25°C using an Auto Surface Tensiometer JK99C (Zongchen Digital Technical Equipment Co. Ltd).

2.3 Surface tension

The samples of the surfactant solution were prepared with different concentrations. The method described in Ref. [16] was used to measure the samples of the surfactant solutions with different concentrations at 25°C. The products of the surfactants were rinsed with double-distilled water. When necessary, the mixture was processed by ultrasonic vibration at 25°C until limpid mixture was obtained. The double-distilled water and all samples were tested three times. The mean value of the three measurements was recorded and used

as its surface tension. The curve of the surface tension was generated. The break point of the curve and the logarithm of the concentration curve were used as the cmc and the γ-cmc of the surfactant solution.

2.4 Synthesis of amidosilane-based surfactants

2.4.1 Preparation of compound 1

Hexamethyldisiloxane (50.00 mmol), 3-(diethoxy (methyl)silyl)pro-pan-l-amine (10.00 mmol) and tetramethylammonium hydroxide (TMAH) (0.250 mmol) were added to 125 mL of a three-neck flask with a magnetic stirrer and cohobation under nitrogen. The mixture was heated to reflux for 2h, and then the temperature was raised to 130°C for 30 min to remove TMAH. The remaining compounds were evaporated in vacuo, washed with distilled water (10 mL × 15) and dried in vacuo for 4 h to obtain the material. Succinic anhydride (10.00 mmol) and dichloromethane were added to 125 mL of the three-neck flask with a magnetic stirrer under nitrogen. The mixture was stirred at 0 °C until the limpid mixture was obtained. A solution of the material (10.00 mmol, in 5 mL of dichloromethane) was added dropwise for 30 min in an ice bath. After the addition was finished, stirring was continued in the ice bath for 2 h. Then, the reaction was stopped at room temperature and stirring was continued for 2 h. The mixture was charged with NaOH (10.00 mmol) and the reaction was removed from the ice bath for 2 h. After the reaction, the mixture was dried by evaporation under reduced pressure. The crude product achieved was dried at 60°C *in vacuo* for 8 h. A 4.400 g orange-reddish viscous solid was obtained (95.65% yield). IR: 3,286–3,084 (N-H), 2,958 (CH$_2$), 1,576–1,650 (C = O), 1,078 (Si-O-C), 841 (Si-C). ^{1}H NMR (600 MHz, CDCl$_3$, ppm) δ: 0.06 (s, 3H, CH$_3$), 0.09 (s, 18H, CH$_3$), 1.54 (m, 2H, CH$_2$), 2.53 (t, J = 13.2, 2H, CH$_2$), 2.69 (t, J = 7.8, 2H, CH$_2$), 3.26 (m, 2H, CH$_2$), 7.76 (s, H, NH).

2.4.2 *Preparation of compound 2*

The same synthesis method for preparing compound **1** was used to prepare compound **2**. A 2.070 g orange-reddish viscous solid was obtained (96.28% yield). IR: 3,285–3,084 (N-H), 2,955 (CH2), 1,576–1,650 (C = O), 1,078 (Si-O-C), 841 (Si-C). ^{1}H NMR (600 MHz, DMSO, ppm) δ: 0.08 (s, 3H, CH3), 0.09 (s, 18H,CH3), 1.54 (m, 2H, CH2), 2.74 (m, 2H, CH2), 3.14 (m, 2H, CH2), 7.47–7.49 (m, 4H, CH), 8.15 (m, H, NH).

2.4.3 *Preparation of compound 3*

The same synthesis method for preparing compound **1** was used to prepare compound **3**. A 2.039 g of golden viscous liquid was obtained (91.7% yield). IR: 3,286–3,084 (N-H), 2,958 (CH2), 1,654 (C = O), 1,046 (Si-O-C), 842 (Si-C). ^{1}H NMR (400 MHz, CDCl3, ppm) δ: 0.08 (s, 3H, CH3), 0.12 (s, 18H, CH3), 0.45 (t, J = 4.2 Hz, 2H, CH2), 1.50 (m, 2H, CH2), 1.94 (s, 3H, CH3), 3.21(m, 2H, CH2), 5.57 (s, H, NH).

3 RESULTS AND DISCUSSION

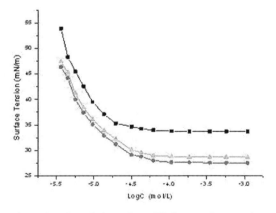

Figure 3. Correlation of equilibrium surface tension of aqueous solutions of the three surfactant molecules and their log mole concentrations. Compound 1 (filled circle), compound 2 (filled triangle) and compound 3 (filled square).

4 SURFACE OF THE THREE SURFACE MOLECULES

The method described in Ref. [16] was used to measure the surface tension of the three surfactants with different concentrations at room temperature. The data are shown in Figure 1. There are many factors of hydrophilic groups that affect the surface activity of amidosilane-based surfactants, including the hydrophilic type of the surfactants, the hydrophilic size of the surfactants and the effect of a mixed hydrophilic group. These molecules of the three surfactants have the same hydrophobic groups while linking to different hydrophilic groups. The surface tension (γ-cmc) at the critical micelle concentration (cmc) of the three surfactants is 27.51 mN/m, 28.65 mN/m and 33.68 mN/m. It was found that compound **3** reduced the surface tension of water less than the others, and the surface activity of compound **1** was better than compound **3**. When the hydrophilic group was the formyl group, its γ-cmc at cmc was slightly higher than the others. The main reason for this result is probably the fact that compound **3** is a nonionic surfactant and the others are ionic surface-active agents. The ionic surfactant is easily deflected by the inorganic electrolyte. The surface activity can be enhanced while the inorganic electrolyte is added to the surfactant. However, the effect of the inorganic electrolyte is not good at the nonionic surfactant of polar molecules without electric field. The surface activity of compound **1** was better than that of compound **2**. This may be due to the size of hydrophilic groups. When the size of the hydrophilic group is increasing, this may affect the area of the surfactant molecules adsorbed on the surface layer. The hydrophilic groups of compound **2** has a benzene ring and compound **1** has none, and the larger hydrophilic groups of compound **2** affect the molecular arrangement of the mixture to reduce the surface activity [17–20].

5 CONCLUSIONS

Of the three synthesized amidosilane-based surfactants, compound **1** has a super surface activity, and can reduce the surface tension of water to 27.51 mN/m; at the critical micelle concentration, the compound can be easily obtained. This is attributable to the structure of the hydrophilic group, which is an ionic group and relatively small in size. Compound **3** has a high surface tension of water, because it is a nonionic surfactant, which is not easily deflected by the inorganic electrolyte. Compound **2** has a low surface activity. The reason for this is that the larger hydrophilic groups affect the molecular arrangement of the mixture to reduce the surface activity.

This kind of surfactant has a better surface activity, a simple production technology and a lower-cost surfactant performance. So, the surfactants can be used as detergents and agents in the textile industry, petrochemical engineering, agriculture and cosmetics industry.

ACKNOWLEDGMENTS

The authors gratefully acknowledge the financial support from the Fund for the Heilongjiang Province High Education Key Laboratories Fine Chemicals (JX201207).

AUTHOR BIOGRAPHIES

Xiaolong Sun is a master's degree student at Qiqihar University, Qiqihar, P.R. China. His main research interest is the synthesis of amidosilane-based surfactants.

Fenglan Xing is currently a professor at Qiqihar University, Qiqihar, China. Her research interest is synthesis of surfactants and the investigation of their physico-chemical properties.

Qun Xu is a researcher at Qiqihar University, China. His research interest is the synthesis and the applications of surfactants.

Yuping Wang is a master's degree student at Qiqihar University, P.R. China. Her main research interests are the synthesis and evaluation of surfactant organosilicon surfactants.

Xu Li is a master's degree student at Qiqihar University, P.R. China. Her main research interests are the synthesis and evaluation of gemini surfactants.

Liyan Wang received her PhD in 2008 at Niigata University in Japan. She worked at the Key Laboratory of Fine Chemicals of Qiqihar University, China. Her research interest involves the applications and synthesis of surfactants.

REFERENCES

[1] Peng Z., Lu C. 2009. Synthesis and properties of novel double-tail trisiloxane surfactant. *J. Surf Deterg* 12(2): 331–336.

[2] Snow S.A. 1993. Synthesis, characterization, stability, aqueous surface activity, and aqueous solution aggregation of the novel, cationic siloxane surfactants $(Me_3SiO)_2Si(Me)(CH_2)_XNM$ $e^{2+}(CH_2)^2ORX^-(R = H, C(O)Me, C(O)NH(Ph);$ X = Cl, Br, I, NO_3, $MeOSO_3$). *Langmuir* 9(3):424–430.

[3] Hill R.M., Silicone surfactants new developments. *Curr Opin Colloid Interface Sci* 7(2): 255–261.

[4] Han F., Zhang G. 2004. Synthesis and characterization of glucosamide-based trisiloxane gemini surfactants. *J. Surf Deterg* 7(2004): 175–180.

[5] Xing F., Gao Y. 2013. The Effects of the Organic Groups Attached at the Silicone Atoms of the Organosilane-Based Gemini Nonionic Surfactants on Their Surface Activities. *Journal of Surfactants and Detergents* 7(3): 1–7.

[6] Hill R.M. ed. 1999. Silicone surfactants. Marcel Dekker: NewYork.

[7] Riess G., Hurtrez G. 1999. Encyclopedia of polymer science and engineering. Wiley: New York.

[8] Perry D. 2005. Silicone surface-active agents, Dow Corning Corporation.

[9] Schlachter I. 1998. Silicone surfactants. Marcel Dekker: New York.

[10] Wagner R., Richter L. 1996. Silicon-modified carbohydrate surfactants II: siloxanyl moieties containing branched structures. *Appl Organomet Chem* 10 (1996): 437–450.

[11] Hüsing, Nicola. 2003. Silicone containing surfactants as templates in the synthesis of mesostructured silicates. *Journal of solgel science and technology* 26(2003): 609–613.

[12] Seyferth D., Rochow. 1955. EGOrganopoly siloxanes containing group IV organometallic substituents in the side chain. *J. Polym. Sci* 18(1): 543–558.

[13] Zhou T., Yang H. 2008. Synthesis, surface and aggregation properties of nonionic polyethylene oxide gemini surfactants. *Colloids. Surf.* A317(2): 339–343.

[14] Yoshimura T., Esumi K. 2004. Synthesis and surface properties of anionic gemini surfactants with amide groups. *J. Colloid Interface Sci* 276(4): 231–238.

[15] Yoshimura T., Ichinokawa T. 2006. Synthesis and surface-active properties of sulfobetaine-type zwitterionic gemini surfactants. *Colloids. Surf.* A 273(2): 208–212.

[16] Harkins W.D., Jordan H.F. 1930. A method for the determination of surface and interfacial tension from the maximum pull on a ring. *J. Am. Chem. Soc* 52(1): 1751–1772.

[17] Klaus Lunkenheimer, Anna Lind. 2003. Surface tension of surfactant solutions. *J. Phys. Chem. B*107 (3): 7527–7531.

[18] Wagner R., Richter L. 1997. Silicon-modified carbohydrate surfactants III: cationic and anionic compounds. *Appl Organomet Chem* 11(1): 523–538.

[19] Julian Eastoe, Alison Paul. 2002. Effects of fluorocarbon surfactant chain structure on stability of water-in-carbon dioxide microe-mulsions. links between aqueous surface tension and microemulsion stability. *Langmuir* 18(2): 3014–3017.

[20] Wagner R., Richter L., Wersig R. 1996. Silicon-modified carbohydrate surfactants I: synthesis of siloxanyl moieties containing straight-chained glycosides and amides. *Appl Organomet Chem*10 (1): 421–435.

Material Science and Engineering – Chen (Ed.)
© 2016 Taylor & Francis Group, London, ISBN 978-1-138-02936-1

Airflow analysis of the web cleaner with one fully closed end

Y.P. Zhao & X.G. Han
School of Mechanical and Electrical Engineering, Eastern Liaoning University, Dandong, P.R. China

P.Z. Sun
School of Textiles and Fashion, Eastern Liaoning University, Dandong, P.R. China

J. Ning
School of Information Technology, Eastern Liaoning University, Dandong, P.R. China

ABSTRACT: With web cleaner being a fundamental device in a high production carding machine, less attention has been paid to its theoretical analysis of the applicability behavior of eliminating neps and impurities. Hence, a three-dimensional model of web cleaner with one fully closed end was established by using FLUENT CFD. The static pressure test was conducted to examine the airflow pattern and the pressure distribution at different suction rates at its outlet. CFD calculation results indicate that the airflow in the web cleaner is in a state of negative pressure and forms a strong circumferential fluid. The airflow flows along the axial direction of the web cleaner when the suctioned air is exhausted at the outlet. Because one end of the web cleaner is fully closed, the strong rotary fluid has a very little axial speed at this end, which affects the removal of impurities and its efficiency.

Keywords: web cleaner; suction rate; airflow pressure; CFD

1 INTRODUCTION

In the carding machine, web cleaners on the plate of the cylinder play a role in removing impurities and eliminating neps with the help of its air suction effect. Unfortunately, the aerodynamic performance of the web cleaner has not reported because it is difficult to install sensors on its structure for measurement. Therefore, a new method is used to modify the structure of the web cleaner for investigating its performances such as pressure distribution and flow path.

The airflow in the carding process has been one of the important research directions in the carding theory. Bayes et al. (1951) qualitatively researched the airflow around the flat strip and the cylinder with the titanic chloride smoke. Lauber et al. (1995) studied the speed and direction of airflow between the cylinder and the doffer using a high-speed photography and Laser Doppler Velocimeter (LDV) for analyzing the fiber motion in the carding process. Jehle (2001) recommended the velocity distribution of airflow on the wedge area below the gage point of the cylinder-doffer setting under the actual production condition using Particle Image Velocimetry (PIV). Dehghani et al. (2004) found turbulence under the gage area of the cylinder-doffer by using a high-speed camera to capture the movement track of smoke. Mhlmann et al. (2007) used PIV for measuring the airflow of the mixed gas in the carding area between the cylinder and the doffer, and used CFD to quantitatively analyze the mixed gas of air and fibers below the gage point. Lee et al. (2006) built the motion model of single fiber in the airflow and used a high-speed image recording for verification.

Wang (1987) qualitatively analyzed the airflow around the cylinder and flat strip. Yu (1981) discussed the influence of the front knife plate on airflow around the cylinder. Han et al. (2009) used CFD to analyze the static pressure of the airflow on the cover of licker-in and the back plate of the cylinder. Zhang et al. (1998) qualitatively researched the effect of the web cleaner on the front plate on the airflow, and found that the web cleaner is advantageous to fiber transfer from the cylinder to the doffer, and is able to reduce the airflow pressure on the front plate and to decrease turbulence generation. Sun et al. (2006) made an experimental research on the influence of the airflow in the web cleaner installed on the back plate on the silver quality. Shao (2007) carried out a theoretical analysis of the action of the airflow in the web cleaner. However, the above researchers did not make quantitative research on the airflow and the detailed function mechanism of the web cleaner.

Web cleaner is an essential assembly for removing short fibers and eliminating neps. This paper used the 3D CFD simulation to analyze the airflow in the web cleaner with one fully closed end on the back plate, and found out the characteristics of the airflow and the principle of the removal of short fibers and neps.

2 WEB CLEANER AND ITS AIRFLOW MODEL

Three-dimensional steady airflow in the web cleaner with one fully closed end was established and calculated by using the 3D CFD simulation. The pressures tested at the inlet and outlet were used as the initial boundary conditions for calculation. The flow configuration geometry model of the web cleaner was created by the actual measurement of dimensions shown in Figure 1. In the carding process, a modified web cleaner was installed on the back plate of the cylinder, as shown in Figure 2. The Navier-Stokes equations were solved using

FLUENT with the standard k-ε turbulence model. In the calculation of the airflow field in the web cleaner, the distributions of the airflow speed, static pressure and turbulence areas were analyzed.

2.1 Eddy viscosity model for the web cleaner

Despite much advances in the development of CFD models, selecting an appropriate turbulence model for a certain fluid problem to get results in accordance with experimental measurements is still an important process for researchers and engineers. Considering the balance of computation costs and accuracy, the realizable k-ε model is chosen and employed in CFD simulations of the web cleaner herein. Generally speaking, the eddy viscosity model with k-ε double transport equations presents three similar forms of models such as the standard, the improved RNG, and the realizable k-ε models for calculating the turbulence field. Moreover, the standard k-ε model is commonly used at a large strain rate, which results in the negative normal stress.

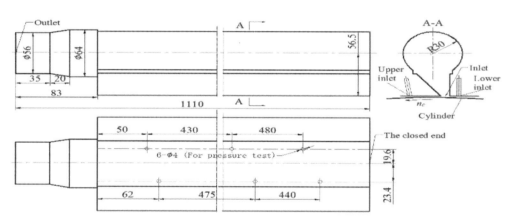

Figure 1. Structure diagram of the web cleaner.

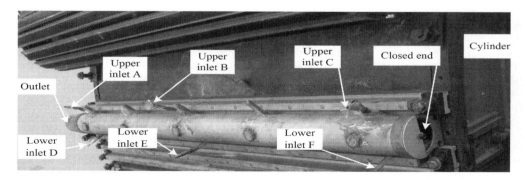

Figure 2. The set-up of the web cleaner on the back plate.

2.2 Boundary condition

Boundary condition is a critical component of FLUENT simulation, which it specifies the airflow variables on the initial boundaries of the physical model of the web cleaner. In addition to the above transport equations, it is necessary to specify the appropriate initial boundary conditions and walls for determining the airflow inside the web cleaner. According to the airflow rate tested in the web cleaner, the airflow is assumed as an incompressible fluid. The boundary conditions of the web cleaner for the model calculation consist of three different elements: the inlet of its rectangular opening as a pressure inlet, the outlet of its suction as a tested airflow outlet, and its inner surface as the wall with a certain roughness, in which the standard wall function is chosen. These boundary conditions are used to implement the physical model of the web cleaner in the FLUENT calculation.

2.3 Establishment model of the web cleaner

The pressure distribution and velocity profile of the airflow in the web cleaner are calculated and analyzed by the CFD method in FLUENT 6.3, which is a special CFD software for calculating from incompressible fluid to moderately compressible fluid, and to a high compressible fluid. In the software, FLUENT solvers, which use the fully unstructured grid and the control volume method, are able to achieve an optimal convergence precision due to the adoption of various solution methods and multi-grid convergence acceleration techniques. Moreover, the solvers, which have a flexible unstructured grid, an adaptive mesh resolution and a wealth of physical fluid models, are capable of calculating many types of fluids accurately.

In order to obtain high-precision solutions and make a three-dimensional model of the web cleaner as closer as possible to its real shape, the geometric model is built in Pro/E software and then is imported into pre-processing software Gambit, which is the software dealing with geometry modeling and CFD mesh generation. Moreover, excellent mesh generation strategies can significantly reduce the preprocessing time to ensure the best grid generation. One of the key steps in pretreatment related directly to the accuracy of the calculation results is to establish an appropriate model of the web cleaner and to generate mesh. It is important to find a balance between the computational cost and accuracy in the numerical simulation. Hence, in order to ensure the accuracy of the boundary condition, the technical route is that mesh-refinement for the boundary surface is first made, and computational mesh generation of the

Figure 3. Mesh generation of a cavity in the web cleaner.

Table 1. Carding technology parameters.

Levels	Cylinder speed (rev/min)	Licker-in speed (rev/min)	Air volume (m³/h)
1			126.6
2	350	1000	85.7
3			56.5

internal body is finally performed using the meshing code integrated in the FLUENT package, as shown in Figure 3.

3 AIRFLOW TEST IN THE WEB CLEANER

3.1 Technology parameters

The initial boundary data at the inlet and outlet were obtained from the carding test for the CFD calculation of the web cleaner. The tests were conducted on a carding machine type A186 with a specially modified web cleaner. In order to measure the airflow pressure in the web cleaner, technology parameters of the carding machine were chosen, as listed in Table 1. A large power vacuum cleaner TC3080D type, in which three levels of air volume were quantitatively pre-calibrated, was used to control the air volume of the outlet.

3.2 Positions of static pressure test in the web cleaner

The tests were carried out for measuring static pressures and air volume in the web cleaner with one fully closed end. The six specially made drainage pipe connectors were installed on both sides of the inlet in the web cleaner at a certain distance for the static pressure test, as shown in Figures 1–2, in which installation positions are labeled as inlets A, B, C at the upper position and inlets D, E, F at the lower position. In the static pressure test, the connectors at six inlet positions were connected with six Peter tubes while the right end of the web cleaner was fully closed. The tested static pressures are represented as the height of the water column in Table 2. At the same time, 6332D Kanomax type hotwire anemometer was also used for measuring the pressures at inlet gages for an additional calibration of the pressure data tested by Peter tubes for the CFD calculation.

Table 2. Static pressure in the web cleaner.

No.	Air volume (m³/h)	Outlet pressure (Pa)	Pressure at inlets (Pa)					
			A	B	C	D	E	F
1	126.6	340	121	172	83	407	306	289
2	85.7	220	59	113	49	240	206	206
3	56.5	180	39	91	37	186	17	167

4 ANALYSIS OF THE CFD CALCULATION

A three-dimensional steady single phase flow model was made for analyzing the airflow in the web cleaner with one fully closed end, in which their good following performances with the airflow were not taken into account due to the smaller densities of short fibers and impurities. The airflow in the web cleaner was merely computed in the model. The airflow at the outlet was under the manual control as the air volume was 56.5 m³/h, 85.7 m³/h and 126.6 m³/h, respectively.

Using the CFD calculation in FLUENT software, the calculation results of the airflow in the web cleaner are shown in Figures 4–7, in which the air volume at the outlet is 56.5 m³/h. In Figure 4, the airflow flows into the inlet and produces the strong vortex flow, which is affected by the wall surface and forms the velocity distribution in the states of higher speed at the outer ring and of

Figure 4. Distribution of velocity vectors at the positions of inlets along the cross-sections (m/s).

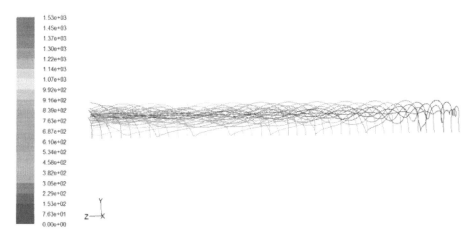

Figure 5. Pathlines of the airflow (particle ID).

Figure 6. Static pressure on the wall surface (Pa).

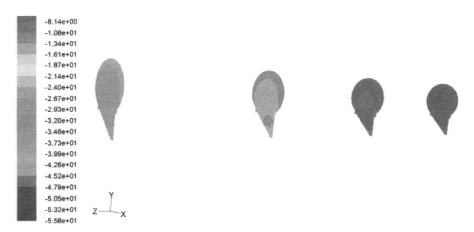

Figure 7. Static pressure at the positions of the inlets along the cross-sections (Pa).

lower speed at the inner ring. Moreover, the vortex flow is conducive to impurity fibers collection while the vortex flow is under the action of the outlet suction made by the vacuum cleaner and produces the axial movement, in which the axial velocity increases along the axial direction of the web cleaner and is up to a maximum of 5 m/s at the outlet but down to 0 m/s at the fully closed end. Hence, the impurity of fibers around the fully closed end has a more circumferential rotation, and cannot be easily removed out of the web cleaner.

Figure 5 clearly shows the pathlines of the airflow in the web cleaner, in which the distribution of the airflow speed at the inlet is not uniform and its average speed is 4.2 m/s, but the speed at the fully closed end is 2.4 m/s. So, the non-uniformity of the airflow speed in the web cleaner is about $\delta v = (8.0-2.4)/4.2 = 133.3\%$, which will make the

dust removal efficiency of the whole web cleaner produce a great difference.

In Figure 6, the inner side of the web cleaner is in the state of negative pressure, which is beneficial to taking the impurities of fibers into the web cleaner for the goal of cleaning, because the inhalation force at the outlet of the web cleaner is made by the vacuum cleaner. The static pressure distribution gradually reduces from the fully closed end to the outlet, in which the static pressure is the maximum negative pressure −3.2 Pa at the fully closed end and the minimum pressure is −55.8 Pa at the outlet, respectively. It is the axial distribution of the static pressure that makes the airflow speed increase along the axial direction from the fully closed end to the outlet. The static pressure distribution along each cross-section is basically uniform. The outer ring of the pressure distribution is slightly higher and the inner ring of the pressure

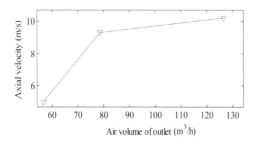

Figure 8. Axial velocity along the cross-section.

distribution is slightly lower at each cross-section, as shown in Figure 7.

When the air volume of the outlet expands from 56.5 m³/h to 85.7 m³/h and 126.6 m³/h, the CFD calculation results show that distribution manifolds of the airflow have obviously no changes, but their values of pressure and speed of the airflow have some changes. The axial velocity in the outlet increases from 5 m/s to 9.3 m/s, which is conducive to the removal of impurity, as shown in Figure 8. When the air volume at the outlet increases to 126.6 m³/h, the airflow speed and vacuum in the web cleaner increase further and the axial speed of the airflow at the outlet increases to 10.2 m/s, as shown in Figure 8.

5 CONCLUSION

This paper studied the mechanics and characteristics of the web cleaner, and introduced its inner airflow under three suction conditions, from which the below conclusions can be drawn.

The interior of the web cleaner is in the state of a negative static pressure, and the vacuum at the outlet is higher than that at fully closed end. The static pressure distribution along the cross-section of the web cleaner is uniform. The static pressure at the center of the cross-section is slightly higher than that at the outer ring of the cross-section. With the airflow increasing at the outlet, the vacuum in the web cleaner increases gradually. When the airflow comes into the inlet, it produces a strong circumferential swirl fluid. Moreover, the vortex fluid turns stronger with the air volume increasing. Because the right end of the web cleaner is fully closed, the airflow has a strong swirl flow and a small axial flow, which makes impurities difficult to be removed and then affects the cleaning efficiency.

Distributions of both airflow velocity at the inlet and axial speed are uniform. The airflow

speed is maximum at the outlet and is minimum at the fully closed end. Non-uniformity, which affects the cleaning efficiency of the web cleaner, is more than 130% under various suction conditions.

These conclusions imply that the web cleaner with one fully closed end needs to be modified to make a half circular hole at the closed end in order to improve its cleaning efficiency. In addition, this study also provides a mathematical model to predict the effective mechanism and the structure option of the web cleaner.

ACKNOWLEDGMENT

This work was supported by the National Natural Science Foundation of China under Grant No. 11152002 and the Liaoning Provincial Natural Science Foundation of China under Grant No. 2102079.

REFERENCES

[1] Bayes, A.W., Cheng R., & Morton W.E. 1951. Control of flat strip in carding. *J. Text. Inst. Proc.* 42(8): 442–456.
[2] Dehghani, A., Lawrence, C.A., Mahmoudi, M., et al. 2004. Aero-dynamics and fiber transfer at the cylinder doffer interface. *J. Text. Inst.* 95(1): 35–49.
[3] Han, X.G., Sun, P.Z., Zhao, Y.P. 2009. Analysis of static pressure in area between back plate and cylinder of a carding machine with CFD. *J. Donghua Univ.* 26(3): 242–245.
[4] Jehle, V. 2001. Visualization and quantitative recording of airflows in the card. *Melliand Engl.* 83(6): 117–118.
[5] Lauber, M., Wulfhorst, B. 1995. Non-contact gauging of the fibre flow during carding and drafting of cotton. *Melliand Textilberichte Engl.* 76(5): 77–78.
[6] Lee, M.E.M., Ockendon, H. 2006. The transfer of fibres in the carding machine. *J. Eng. Maths.* 54(2): 261–271.
[7] Mhlmann, I., Seide, G., Mathis, P., & et al. 2007. CFD analysis of the airflow in the card. *Melliand Engl.* 89(7–8): 96–97.
[8] Sun, P.Z., & Shao, J.D. 2006. Influence of wind's flow of web cleaner in a carding machine on the quality of card sliver. *J. Text. Res.* 27(3): 74–76.
[9] Shao, J.D. 2007. Theoretical discussion on installation of additional carding equipment at the back sheet of a card. *J. Text. Res.* 28(6): 115–117.
[10] Wang, X.X. 1987. The function of airflow around cylinder. *Text. Bulletin* 11: 22–25.
[11] Yu, W.Y. 1981. Analysis on the functions of front upper plate of the carding machine. *J. Text. Res.* 2(4): 44–48.
[12] Zhang, X.C., Yu, X.Y., Zhou, Y., et al. 1998. Application of Trex apparatus in C4 carding machine. *Cotton Text. Technol.* 26(6): 11–13.

Material Science and Engineering – Chen (Ed.)
© 2016 Taylor & Francis Group, London, ISBN 978-1-138-02936-1

Research of different clamping modes on stretch forming twisted part

Y.Y. Cheng
Dieless Forming Technology Center, Jilin University (Nanling Campus), Changchun, P.R. China
Department of Chemical Machinery, Jilin Vocational College of Industry and Technology, Jilin, P.R. China

M.Z. Li
Dieless Forming Technology Center, Jilin University (Nanling Campus), Changchun, P.R. China

J. Xing
Dieless Forming Technology Center, Jilin University (Nanling Campus), Changchun, P.R. China
Engineering Training Center, Northeast Dianli University, Jilin, P.R. China

ABSTRACT: Multi-Gripper Flexible Stretch Forming (MGFSF) is a new forming process of sheet metal, which can flexibly stretch forming regular part. Irregular twisted part was selected as the research object. Two finite element models of different clamping modes were built, which were the integral gripper and multi-gripper clamping modes. The strain, thickness and springback were comparatively analyzed. The simulation results show that strain, thickness and springback distribution are well-proportioned; strain, thickness thinning and springback value are smaller; forming quality with the multi-gripper clamping mode gets better than with the integral gripper clamping mode. The stretch forming experiment for twisted part was conducted with MGFSF machine developed by Jilin University. The experimental results are in good agreement with simulation results. That proves that the MGFSF machine could stretch forming irregular twisted part and provides bases for the theories and practice to expand application of MGFSF machine.

Keywords: plasticity processing; numerical simulation; stretch forming; flexible forming technology

1 INTRODUCTION

With the improvement of science and technology, the demands and production of three-dimensional parts increase rapidly, the shape gets more complex in military and civil manufacturing fields. Quality and cost of three-dimensional parts are mainly determined by the level of advanced manufacturing technologies [1–4]. In the manufacturing field there have been some flexible forming technologies to meet the need of surface forming [5–8], MGFSF is one of them. MSFSF is a new processing method combined with flexible manufacturing and computer technology.

2 MGFSF

MGFSF machine [9] is shown in Figure 1. Multi-point digital devices substitute for the entity die [10–12], and it can effectively reduce production cost and time in skin parts manufacture process. The integral gripper is replaced by multiple discrete grippers, multiple discrete grippers are controlled by the hydraulic mechanism and each gripper can move independently. In clamping sheet metal, multiple grippers are arranged into a line state; in stretch forming, each gripper can make real-time

movement adjustment leading to displacements and rotations to the variety of the die curvature by the autoregulation of stretch forming device. After stretch forming the part load distributes uniformly. That avoids stress concentration and poor moldability, and improves the forming quality. So MGFSF machine can stretch forming more complex three-dimensional parts, which contribute to extensive use of stretch forming technology. At present, regular parts which include cylindrical, saddle, spherical parts and so on were selected as

Figure 1. MGFSF machine.

the research object of stretch formed parts, but irregular twisted parts were studied less. Therefore, in this paper twisted part was studied and the law of stretch forming part was analyzed.

3 FE MODEL

The finite element technology is widely used to simulate stretch forming processes of sheet metal, predicts the forming results and reduces experimental costs. In this paper, finite element software ABAQUS/explicit, an explicit dynamic code, is chosen to simulate multi-gripper stretch forming process for twisted part. In the numerical investigation, the material of sheet metal is aluminum alloy 5083 P-O. The size is 2440 mm × 1220 mm and thickness 3 mm. The mechanical properties of 5083P-O are: density ρ = 2780 kg/m^3, Elastic modulus E = 66.57 GPa, Poisson's coefficient μ = 0.33, yield strength σ_y = 145 MPa. The stress σ—strain ε curve of 5083P-O is presented in Figure 2.

To save the calculation time and reduce the difficulty of the model, the entity die substituted for multi-point die and the gripper was simplified as a rigid body. Twisted part is asymmetric and three-dimensional model of the die is seen in Figure 3. The FE model of MGFSF was composed of three parts: multiple grippers, stretch die and sheet

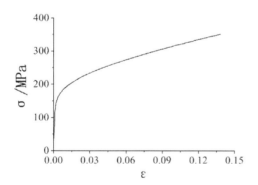

Figure 2.   The stress σ—strain ε curve of 5083P-O.

Figure 3.   Three-dimensional model of the die.

Figure 4.   The finite element model of MGFSF.

metal in Figure 4. Stretch die and the multiple grippers were modeled with complicated rigid element R3D4. The element S4R modeled the sheet metal. The R3D4 is bilinear quadrilateral three-dimensional rigid element with four nodes, and the S4R is doubly curved quadrilateral shell element with four nodes. During the simulation, the sheet metal was assumed to be isotropic and followed Mises yield criterion. General contact algorithm was adopted. The friction at the interfaces between the sheet metal and the die was assumed to follow Coulomb's model, the friction coefficient was considered as 0.15. The tangential behavior at the interfaces between the grippers and the sheet metal was modeled with the Rough friction formulation.

4 NUMERICAL RESULTS

4.1 Strain distribution

The stain distribution of formed parts in the different clamping modes is shown in Figure 5. Apparently, the strain value of the multi-gripper clamping mode is less than that of the integral clamping mode. The strain unevenly distributes in integral gripper clamping mode and the strain value of transition area is the maximum from Figure 5a. The integral gripper remains flat in stretch forming, the curvature of transition area changes large. So it easily generates crack defects in transition area. The strain distributes uniformly and the strain value of clamping area is the maximum in multi-gripper clamping mode from Figure 5b. Variety of multiple grippers' curvature can make variety of the dies' curvature and the area between forming and clamping area transites smoothly. Therefore the strain distribution is uniform and forming quality gets well.

4.2 Thickness distribution

The thickness distribution of formed parts in the different clamping modes is shown in Figure 6. In forming area the thickness of formed parts ranges from 2.825 mm to 2.995 mm in integral gripper clamping mode from Figure 6a; the thickness

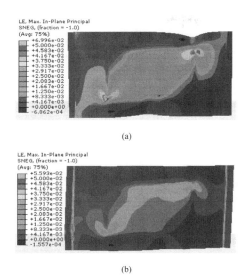

(a)

(b)

Figure 5. The strain distribution of formed parts. (a) The integral gripper; (b) The multi-gripper.

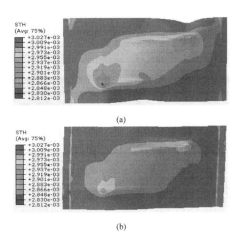

(a)

(b)

Figure 6. The thickness distribution of formed parts (m). (a) The integral gripper; (b) The multi-gripper.

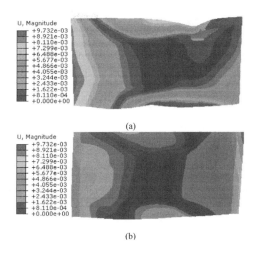

(a)

(b)

Figure 7. The springback distribution of formed parts (m). (a) The integral gripper; (b) The multi-gripper.

Figure 8. The part of MGFSF experiments.

ranges from 2.955 mm to 2.996 mm in multi-gripper clamping mode from Figure 6b. Thus the maximum thinning ratio and distribution range have decreased 74.284% and 75.882% respectively in forming area of MGFSF part contrasting to integral gripper stretch forming part. This is the reason that strain distribution is uniform and value is small in multi-gripper clamping mode.

4.3 Springback

Springback is an inevitable phenomenon in sheet metal stretch forming, and it greatly affects the geometrical accuracy of forming parts. Hence the prediction of springback is necessary. The springback distribution of formed parts is shown in Figure 7. The springback value increases with distance between sheet metal and forming part center. The maximum value appears at sheet metal edge. From Figure 7a and Figure 7b the springback distribution is more uniformly, and the value is smaller in forming area of the multi-gripper clamping mode contrasting to the integral gripper clamping mode. That is mainly owing to small elastic deformation and uniform strain distribution.

5 EXPERIMENT

Stretch forming experiments for 5083P-O were done with MGFSF machine which was developed by Jilin University. In the experiment, the size of sheet metal was consistent with that of numerical simulation. It is described the part of MGFSF experiments in Figure 8. The formed parts are no wrinkling and cracking, and can completely conform to the forming die from Figure 8. Forming effect is excellent. The experiment results agree with the simulation results. The above validates the feasibility that the irregular part can be stretched forming with MGFSF machine.

6 CONCLUSIONS

The influences of different clamping modes on strain, thickness and thickness distribution law have been conducted through finite element method and experiments to understand the affecting on the forming quality. Strain, thickness and springback distribution are well-proportioned; strain, thickness thinning and springback value are smaller; forming accuracy with the multi-gripper clamping mode gets better than with the integral gripper clamping mode. The results of finite element method and experiments test that MGFSF machine is highly flexible. It is feasible to stretch forming twisted part. That provides the reference for other irregular parts by MGFSF.

ACKNOWLEDGMENTS

This work was supported by the EU Seventh Research Framework (FlexiTool-CT-2010-261925).

REFERENCES

[1] Jeswiet, J. Geiger, M. Engel, U. and et al. Metal forming progress since 2000. *CIRP Journal of Manufacturing Science and Technology*, 2008(1): 2–7.

[2] Zhang, S.H. Wang, Z.R. Wang, Z.T. and et al. Some new features in the development of metal forming technology. *Journal of Materials Processing Technology*, 2004, 151(3): 39–42.

[3] Li, M.Z. Cai, Z.Y. Sui, Z. and et al. Multi-point forming technology for sheet metal. *Journal of Materials Processing Technology*, 2002, 129(1–3): 333–338.

[4] Li, M.Z. Liu, Y.H., Su, S.Z. and et al. Multi-point forming: a flexible manufacturing method for a 3-d surface sheet. *Journal of Materials Processing Technology*, 1999, 87: 277–280.

[5] Wang, T. Platts, M.J. Levers, A. A process model for shot peen forming. *Journal of Materials Processing Technology*, 2006, 172: 159–162.

[6] Cai, Z.Y. Wang, S.H. Xu, X.D. and et al. Numerical simulation for the multi-point stretch forming process of sheet metal. *Journal of Materials Processing Technology*, 2009, 209(1): 396–407.

[7] Wang, S.H. Cai, Z.Y. Li, M.Z. Numerical investigation of the influence of punch element in multi-point stretch forming process. *International Journal of Advanced Manufacturing Technology*, 2010,49(5/6/7/8): 475–483.

[8] Wang, S.H. Cai, Z.Y. Li, M.Z. FE simulation of shape accuracy using the multi-point stretch forming process. *International Journal of Materials & Product Technology*, 2010, 38(2/3): 223–236.

[9] Li, M.Z. Han, Q.G. Fu, W.Z. and et al. (2011) Study of a flexible stretch forming machine. Proc 10th Int Techn Plast (ICTP2011) *Aachen, Germany*: 655–658.

[10] Li, M.Z. Cai, Z.Y. and et al. *Journal of Materials Processing Technology*, Vol. 129(1/3) (2002): 333–338.

[11] Papazian, J.M. Anagnostou, E.L and et al. The 6th Joint FAA/DoD/NASA Conf.on Aging Aircraft, *San Francisco,CA,USA, September*(2002): 16–19.

[12] Cai, Z.Y. Wang, S.H. and et al. *Journal of Materials Processing Technology*, Vol. 209(2009): 396–407.

Material Science and Engineering – Chen (Ed.)
© 2016 Taylor & Francis Group, London, ISBN 978-1-138-02936-1

The influence of friction coefficient on forming results in Multi-Gripper Flexible Stretch Forming

J. Xing
Dieless Forming Technology Center, Jilin University (Nanling Campus), Changchun, P.R. China
Engineering Training Center, Northeast Dianli University, Jilin, P.R. China

Y.Y. Cheng
Dieless Forming Technology Center, Jilin University (Nanling Campus), Changchun, P.R. China
Department of Chemical Machinery, Jilin Vocational College of Industry and Technology, Jilin, P.R. China

M.Z. Li & Y. Wang
Dieless Forming Technology Center, Jilin University (Nanling Campus), Changchun, P.R. China

ABSTRACT: Multi-Gripper Flexible Stretch Forming (MGFSF) is a recent technological innovation of sheet metal forming process. Straight jaws in traditional stretch forming machine are replaced by a pair of opposed clamping mechanisms which can significantly improve the conformability of sheet metal to the die and the material utilization. In this paper, spherical part was selected as the study object and FE models of MGFSF were established by a commercial finite element software ABAQUS. Four levels of friction coefficients are selected to study their influences on strain distribution in the forming zone and forming precision of the simulated parts. The simulation results reveal that the strain in the forming zone as a whole decreases with the friction coefficient increased. It is also found that a friction coefficient equal to 0.1 would result in a simulated part with the highest forming precision.

Keywords: stretch forming; numerical simulation; spherical part; forming precision

1 INTRODUCTION

The sheet stretch forming is used mainly in aeronautics to produce skin parts, which has the following advantages: low tooling cost (only male tools are needed), short manufacturing cycle, simple processing and low elastic return. Currently there are two main types of stretch forming machines used for this process: the transverse machine and the longitudinal machine. When a transverse machine is used to produce a large curvature part, the displacements of the material near the jaws zone are almost at the same level, easily lead to a product with uneven strain distribution, wrinkle or even material failure. In addition, a large processing allowance named the transition zone is required which would cause a low material utilization. A longitudinal stretch machine commonly has two jaw assemblies, and each jaw assembly includes a jaw having an array of adjacent grippers or groups of gripper. Prior to forming, it's necessary to move the grippers or groups of grippers relative to each other to form the opposite edge portions of a sheet metal into a predetermined configuration manually or automatically, therefore, it's less efficient. In addition, the transition zone is still necessary, but the material utilization is increased to a certain extent.

Currently, a tremendous variety of sheet metal parts both in size and complexity are employed in various industries such as automobile, aerospace and architecture. Under such backgrounds, various flexible forming methods have been proposed and deeply investigated since the innovational idea of Nakajima [1]. Walczyk and Hardt addressed the design and analysis issues involved with movable die pins, turned a matrix of die pins into a rigid tool and the pin matrix containment frame [2]. Papazian developed a production prototype system with reconfigurable tools for stretch forming sheet metal parts of aircraft [3]. Anagnostou revealed that a set of reconfigurable die could replace hundreds of fixed tools that are currently in use, and the cost analysis showed that the use of reconfigurable tooling can reduce the cycle time for parts production from weeks to hours [4]. Koc and Thangaswamy introduced a new reconfigurable tooling for the fabrication of three-dimensional free-form objects under process conditions of thermoplastic molding [5]. Multi-Point Forming (MPF), named by Li, is a matched-die forming process for three-dimensional parts, in which the conventional stamping dies are replaced by a pair of matrices of closely stacked discrete elements,

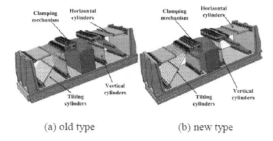

(a) old type   (b) new type

Figure 1. Schematics of two types of MGFSF apparatus.

and the height of each element is controlled by an electric motor [6]. Till now, Li and co-workers have developed several MPF apparatus and applied this technique into the manufacturing of locomotives, aircraft panels and steel structure parts of architectures successfully [7]. Later, Li and his co-workers developed a type of Multi-Gripper Flexible Stretch Forming (MGFSF) apparatus (shown in Fig. 1a), and conducted systematic investigations. Li et al. addressed the mechanism design and forming characters, and pointed out that MGFSF have several advantages over the traditional stretch forming, such as a simpler servo control system, a higher material utilization and an easier conformability to the desired shape [8]. Chen et al. investigated influences of the transition length, gripper number and clamping mode on the forming results [9–11]. Then, Li and his co-workers developed a new type of MGFSF apparatus (shown in Fig. 1b) to further improve the conformability of sheet metal and material utilization. Peng et al. conducted comparisons of two types of MGFSF apparatus on the von Mises stress, thickness and springback of simulated parts [12]. However, process parameters of the new type of MGFSF apparatus on the quality of formed parts are not deeply investigated.

Through numerical and experimental analyses, the authors have proved that the sheet metal can be formed with a very small transition zone or even without a transition zone by utilizing the new type of MGFSF, which significantly improve the material utilization and the quality of formed parts. So the length of the transition zone on the forming results is not considered in the study. The main purpose of this paper is study the influences of friction coefficient on strain, thickness and forming precision of the simulated part.

2 FORMING APPARATUS

MGFSF is a novel technological innovation derived from traditional stretch forming and MPF technol-

ogy, in which the Pascal's Law of multi-cylinder hydraulic system, strain hardening and minimal resistance force of the material are employed. Figure 2 shows the newly developed MGFSF apparatus. A stretch forming apparatus of the type mainly consists of a pair of opposed discrete clamping mechanisms, a set of Multi-Point Die (MPD), three rows (horizontal, inclined, vertical) of hydraulic cylinders installed on each side of the MPD, and a frame. The hydraulic cylinders in the same row are controlled by a common electromagnetic direction valve to simplify the control system, therefore, they will produce the same tensile force during the forming process. During the forming operation, the discrete clamping mechanisms would make real-time position adjustment to assume a contour roughly in the shape of the curved surface of the die, which results in a better conformability of sheet metal to the desired curvature.

Figure 3a shows schematics of the clamping mechanism and the gripper. The clamping mechanism mainly consists of a clamping cylinder, a piston, two grippers, and a clamping part, which is a vital component of MGFSF machine and has a great influence on the forming quality. The main dimensions of the gripper are shown in Figure 3b. It can be seen that the side edges and the end edge are rounded with a fillet radius of 10 mm and 30 mm respectively, in order to weaken local stress concentration. The clamping face of the gripper is

Figure 2. Newly developed MGFSF apparatus.

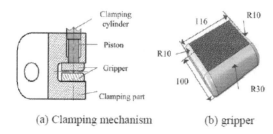

(a) Clamping mechanism   (b) gripper

Figure 3. Schematics of clamping mechanism and the gripper.

designed into staggered saw-tooth configuration to prevent sliding of sheet metal, and also with surface quenching to improve its wear resistance.

3 FE MODEL OF MGFSF

3.1 Material model

A deep-drawing cold-rolled steel (ST14) was considered in the current study. The relevant mechanical properties are: density $\rho = 7.85$ g/cm³; elastic modulus E = 207 GPa; Poisson's coefficient $v = 0.28$; Initial yield strength $\sigma y = 176.3$ MPa; Swift coefficients: K = 596 MPa and n = 0.247.

The material is modeled as elastic-plastic where the elasticity is taken to be isotropic and the plasticity is modeled as anisotropic using the Hill quadratic anisotropic yield criterion. In ABAQUS, anisotropic yield behavior is modeled through the use of yield stress ratios, Rij. However, the sheet anisotropy is commonly defined by the plastic strain ratios of width strain to thickness strain (i.e. r-value) in the cases of sheet metal forming applications. Therefore, mathematical relationships are then necessary to convert the strain ratios to stress ratios that can be input into Abaqus by the following expressions.

$$R_{11} = 1$$

$$R_{22} = \sqrt{\frac{r_{90}(r_0 + 1)}{r_0(r_{90} + 1)}}$$

$$R_{33} = \sqrt{\frac{r_{90}(r_0 + 1)}{(r_0 + r_{90})}}$$

$$R_{12} = \sqrt{\frac{3(r_0 + 1)r_{90}}{2(r_{45} + 1)(r_0 + r_{90})}}$$

where $r_0 = 1.88$, $r_{45} = 1.4$, $r_{90} = 2.23$.

3.2 FE model discritization

The numerical simulation of MGFSF with a MPD is more sophisticated than that with a continuous surface of the die because of the discontinuous contact generated by the MPD, which would also result in an unacceptable computational time. Therefore, the MPDs are replaced by equivalent continuous surfaces and the interpolators are removed to simplify the FE model and cut the simulation time. The die with a sphere-shaped surface with a radius of 2000 mm is selected as the study object, and the dimensions in the stretch and transverse directions are 1660 mm and 1300 mm, respectively. The fillet radii on the edges of the die are 30 mm. The initial

size of the blank sheet is 2030 × 1210 mm with a thickness of 1.5 mm.

In view of the symmetry of the model and the computational time, only a quarter of the sheet, hydraulic cylinders, forming die and clamping mechanisms were modeled. Figure 4a illustrates a finite element model of MGFSF simulations. The clamping parts and the grippers were modeled with C3D4 elements and translated into rigid elements in the assembly models. The die was modeled with R3D4 elements, and the sheet is modeled with S4R elements with five Simpson integration point along the thickness. The hydraulic cylinders were simplified with connectors. General contact algorithm was adopted and the frictions at the interfaces were assumed to follow Coulomb's model. The friction coefficient between the blank and die was considered as 0.1. As there was no relative slip, the tangential behavior at the interfaces between the grippers and the blank sheet was modeled with the Rough friction formulation.

3.3 Boundary conditions

The blank sheet was separated into three distinct zones, which are the forming zone, the transition zone and the clamping zone, as shown in Figure 4b. OA and OB are denoted by x- and y-, respectively, are the symmetry axes of the sheet metal. The y-axis agrees with the stretch direction of the blank sheet. The symmetry axes of the sheet metal were

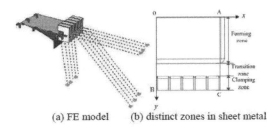

(a) FE model   (b) distinct zones in sheet metal

Figure 4. Finite element model of MGFSF (1/4 symmetry).

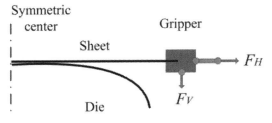

Figure 5. Schematic of the loading sequence of hydraulic cylinder.

constrained with symmetry boundary conditions and the die was fixed in the FE simulations. In practical application, the loading sequences are very complicated. In order to simplify the simulation process, a representative loading sequence of the hydraulic cylinders as shown in Figure 5 is selected to investigate the influences of friction coefficient on strain and forming precision of the simulated part.

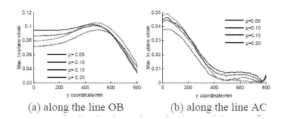

(a) along the line OB   (b) along the line AC

Figure 7. Strain distributions along the denoted lines. Influence on thickness.

4 NUMERICAL SIMULATION ANALYSES

4.1 *Influence on strain distribution*

Figure 7 shows the Max In-plane principal strain contour plots of the simulated parts with different coefficients. As shown in the figure, the strain in the middle of the part decreases with the coefficient increased (the color in the middle of the part changes from red to yellow). In addition, it is found that the largest strain in the simulated part is not located right in the middle of the part. This is because when forming a part with a large length in the stretch direction, the middle region of the material complies with the die first and then its deformation would be highly limited by the increasing contact force with the forming process continues. However, as the contact force in the region near the transition zone is smaller than that in the middle region of the forming zone, a larger deformation would be occurred in the region near

the transition zone. In order to study the influence of friction coefficient on strains in the simulated part quantitatively, the strain distributions along the lines OB and AC are given in Figure 7. As shown in the Figure 7a, the strain in the middle of the forming zone decreases with the coefficient increased while the strain in the forming zone near the transition zone increases with the coefficient increased. As shown in the Figure 7b, the strain is mainly manifested as a decreasing trend with the coefficient increased in general.

4.2 *Influence on thickness*

The thickness distributions of the simulated parts are shown in Figure 8. As shown in the figure, the uniformity of thickness decreases with the fiction coefficient increased. However, it doesn't show significant difference in the thickness of simulated part from the overall perspective.

4.3 *Influence on forming precision*

As known, springback is multi-factor controlled variable, including the strain magnitude, strain change, contact condition, wrapping angle between the sheet and die and so on. However, springback is only studied in the perspective of friction coefficient in the study. In order to investigate the friction coefficient on the forming precision, springback analyses have been conducted by importing numerical results from Abaqus/Explicit into Abaqus/Standard. Then the error calculations between the simulated and the target parts were conducted in the software of "Geomagic Qualify 7". Figure 9 illustrates the normal error distributions of the simulated spherical parts under different friction coefficients. As shown in the figure, the normal error is mainly distributed in the green area whose value ranges between −1.50 mm and 1.50 mm. In addition, it is clearly seen that the green area of the simulated part under a efficient equal to 0.10 is the largest which demonstrates that a efficient equal to 0.10 would results in the highest forming precision of the simulated part. As the strain mag-

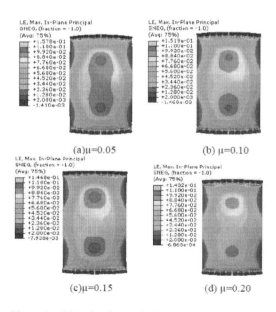

Figure 6. Max. in-plane principal strain contour plots of the simulated parts under different friction coefficients. (a) μ = 0.05, (b) μ = 0.10, (c) μ = 0.15, (d) μ = 0.20.

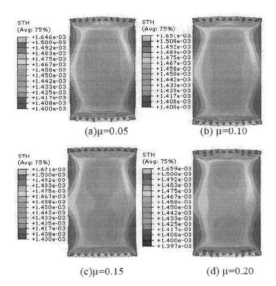

Figure 8. Thickness distributions of the simulated parts under different friction coefficients. (a) μ = 0.05, (b) μ = 0.10, (c) μ = 0.15, (d) μ = 0.20.

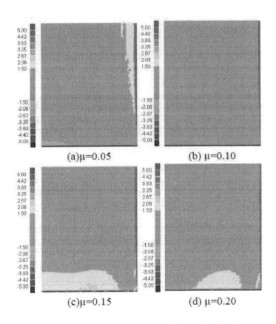

Figure 9. Normal error (mm) of the simulated parts under different friction coefficients.

nitude and springback are sensitive to the forming force and there is no pressure sensor installed on the newly self-developed MGFSG apparatus, only qualitative experiments are conducted. Figure 10 shows the photographs of the simulated and the

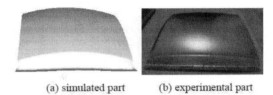

(a) simulated part   (b) experimental part

Figure 10. Photographs of the simulated part and the experimental part.

experimental parts. It is clearly seen that the contour of the clamping zone in the simulated part is similar to that in the experimental result, which indicates that the experiment result shows a good agreement with the simulated result.

5 CONCLUSIONS

This paper studied on the influences of friction coefficient on strain and forming precision of the simulated part by using commercial finite element software of ABAQUS and reverse engineering software of Geomagic qualify. The simulation results show that the strain in the forming zone of the simulated part decreases with friction coefficient increased. In addition, the uniformity of thickness decreases with the friction coefficient increased. It is also found that the sheet metal can be formed with the best forming precision with the friction coefficient equal to 0.10. Finally, experimental validation is conducted on self-developed forming apparatus, and the experimental result shows a good agreement with the simulation result.

ACKNOWLEDGMENTS

This work was supported by the EU Seventh Research Framework (FlexiTool-CT-2010-261925).

REFERENCES

[1] Nakajima, N., 1969. A newly developed technique to fabricate complicated dies and electrodes with wires. *Bull of JSME* 12, 1546–1554.
[2] Walczyk, D.F., Hardt, D.E., 1998. Design and analysis of reconfigurable discrete dies for sheet metal forming. *J. Manuf. Syst.* 17, 436–454.
[3] Papazian, J.M., 2002. Tools of change: reconfigurable forming dies raise the efficiency of small-lot production. *Mech. Eng.* 124, 52–55.
[4] Anagnostou. E.L., 2002. Optimized tooling design algorithm for sheet metal forming over reconfigurable compliant tooling. *PhD Thesis, State University of New York, USA.*

[5] Koc, B., Thangaswamy, S., 2011. Design and analysis of a reconfigurable discrete pin tooling system for molding of three-dimensional free-form objects. *Robotics and Computer-Intergrated Manufacturing* 27, 335–348.

[6] Li, M.Z., Nakamura, K., Watanabe, S., et al, 1992. Study of the basic principles (1st report: research on multi-point forming for sheet metal). *In: Proceedings of the Japanese Spring Conference for Technology of Plasticity, Japan*, pp. 519–522.

[7] Li, M.Z., Cai Z.Y., Sui Z., et al, 2008. Principle and applications of multi-point matched-die forming for sheet metal. *P. I. Mech. Eng B-J. Eng*. 222, 581–589.

[8] Li, M.Z., Han, Q.G., Fu, W.Z., et al., 2011. Study of a flexible stretch forming machine. *In: proceedings of the international conference on technology of plasticity, Aachen, Germany*, pp. 655–658.

[9] Chen, X., Li, M.Z., Fu, W.Z., et al, 2010. Research on multi-head stretch forming technology of sheet metal. *In: proceedings of 2010 international conference on electrical and control engineering, Wuhan, China*, pp. 2952–2955.

[10] Chen, X., Li, M.Z., Zhang, H.H., et al, 2012. The effect of discrete grippers on sheet metal flexible-gripper stretch forming. *Advanced Materials Research* 121–126, 488–492.

[11] Chen, X., Li, M.Z., Fu, W.Z., et al, 2012. Numerical simulation of different clamping modes on stretch forming parts. *Advanced Materials Research* 189–193, 1922–1925.

[12] H.L. Peng, M.Z. Li, Q.G. Han, et al, Design of flexible multi-gripper stretch forming machine by FEM, *Adv. Mater. Res*. 328–330 (2011) 13–17.

Mechanical engineering and manufacturing technology

Material Science and Engineering – Chen (Ed.)
© 2016 Taylor & Francis Group, London, ISBN 978-1-138-02936-1

The influence of delay time on the dynamic characteristics of milling system

B. Rong & W.F. Xu
Shanghai Aircraft Manufacturing Co. Ltd., Shanghai, P.R. China

Y.F. Yang & N. He
College of Mechanical and Electrical Engineering, Nanjing University of Aeronautics and Astronautics, Nanjing, P.R. China

ABSTRACT: The paper studies the influence of delay time between inner and outer waves on the dynamic characteristics of milling system. A method based on Routh criterion is proposed to analyze the variations of damping and stiffness of milling system caused by delay time and to predict stability limitation. The simulation and experimental results indicated that the method was effective and the root cause of occurrence of chatter could be revealed through the method. Additionally, it can be concluded that the method is useful in designing variable pitch cutters to avoid chatter according to existing experiments. Finally, a new viewpoint is proposed that the concepts of process damping could be developed.

Keywords: delay time; dynamic characteristics; Routh criterion; root cause; variable pitch cutters; process damping

1 INTRODUCTION

Productivity and surface quality in milling processes have direct effects on cost and quality of machined parts. Chatter is one of the most common limitations for productivity and part quality in milling operations. Poor surface finish with reduced productivity and decreased tool life are the usual results of chatter. Additional operations, mostly manual, are required to clean the chatter marks left on the surface. Thus, chatter vibrations result in reduced productivity, increased cost and inconsistent product quality. The importance of modeling and predicting stability in milling has further increased within last couple of decades due to the advances in high speed milling technology. At high speed, the stabilizing effect of process damping result from plowing effect diminishes and the possibility of chatter increases. The mathematical model of the stability leads to better machine tool and cutting tool design.

Chatter develop due to dynamic interactions between the cutting tool and workpiece. Tlusty [1] and Tobias [2] were the leading pioneers in the research of cutting dynamic, they identified the most powerful source of self-excitation, *regeneration*, which is associated with the structural dynamics of the machine tool and the feedback between the subsequent cuts on the same cutting surface. These and the following other fundamental studies are applicable to orthogonal cutting where the direction of the cutting force, chip thickness and system dynamics do not change with time. On the other hand, the stability analysis of milling is complicated due to the rotating tool, multiple cutting teeth, periodical cutting forces and multi-degree-of-freedom structural dynamics.

In the early milling stability analysis, Tlusty [3] used his orthogonal cutting model considering an average direction for the cut. Later, Tlusty et al [4] presented the numerical simulation of the milling dynamics including nonlinear behavior such as the tool jumping out of cut. Sridhar et al [5, 6] performed a comprehensive analysis of milling stability which involved numerical evaluation of the dynamic milling system's state transition matrix. Minis et al [7, 8] used Floquet's theorem and the Frourier series for the formulation of the milling stability, and numerically solved it using the Nyquist criterion. Altintas and Budak [9] presented the first analytical solution that led to the prediction of stability lobes directly in the frequency domain. Budak and Altintas [10] also demonstrate a higher order solution which provides improved prediction when the process is highly intermittent at small radial immersions. Stepan et al [11] solved the stability in discrete time domain, which allowed for inclusion of periodically varying system parameters directly. Bayly [12] applied time finite method to analyze the stability of interrupted cutting.

Generally, lots of methods have been proposed to generate stability diagrams which have been extensively used in high speed milling machining. However, few researchers have analyzed the variations of dynamic characteristics of milling system before and after the chatter, which is just the key problem in vibration analysis. The paper propose a method based on Routh criterion to analyze the variations of damping and stiffness of milling system caused by delay time between inner and outer waves. The method can also predict the stability limitation accurately, which is presented by examples. Additionally, it is concluded that the method based on Routh criterion can be also helpful to design variable pitch cutters. Variable pitch tools can change system damping indirectly to avoid chatter through variable delay time. Finally, a new viewpoint is proposed that the concepts of process damping could be developed, the system damping caused by delay time can also been seen as a type of process damping.

2 DYNAMIC EQUATION OF MILLING

A milling cutters with N teeth is considered to have flexibility in two orthogonal directions (x, y) as shown in Figure 1. The vibrations are projected to rotating tooth number j in the radial or chip thickness direction at instantaneous angular immersion φ_j of tooth j measured anticlockwise from negative y-axis. The dynamic equations can be shown as

$$\begin{cases} m_x \ddot{x} + c_x \dot{x} + k_x x = \sum_{j=1}^{N} F_{xj} = F_x(t) \\ m_y \ddot{y} + c_y \dot{y} + k_y y = \sum_{j=1}^{N} F_{yj} = F_y(t) \end{cases} \quad (1)$$

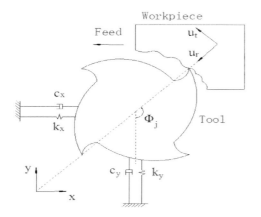

Figure 1. Dynamic milling system.

where the m_x, c_x, k_x are modal parameters in x direction. m_y, c_y, k_y are modal parameters in y direction. These dynamic parameters can be obtained by modal experiments. F_{xj}, F_{yj} are the dynamic cutting force in x and y direction. F_x, F_y are the total dynamic cutting force, which are given by [9]

$$\{F(t)\} = \frac{1}{2} a_p K_t [A_0]\{\Delta(t)\} \quad (2)$$

where

$$\{F(t)\} = \begin{Bmatrix} F_x(t) \\ F_y(t) \end{Bmatrix} \quad [A_0(t)] = \frac{N}{2\pi} \begin{bmatrix} \alpha_{xx} & \alpha_{xy} \\ \alpha_{yx} & \alpha_{yy} \end{bmatrix}$$

$$\{\Delta(t)\} = \begin{Bmatrix} \Delta x(t) \\ \Delta y(t) \end{Bmatrix} = \begin{Bmatrix} x(t) - x(t-T) \\ y(t) - y(t-T) \end{Bmatrix} \quad (3)$$

where, a_p are the axial depth, K_t is the cutting coefficients in tangential direction. $[A_0(t)]$ are directional coefficient matrix, the elements in matrix are

$$\alpha_{xx} = \frac{1}{2}[\cos 2\phi - 2K_r\phi + K_r \sin 2\phi]_{\phi_{st}}^{\phi_{ex}}$$

$$\alpha_{xy} = -\frac{1}{2}[-\sin 2\phi - 2\phi + K_r \cos 2\phi]_{\phi_{st}}^{\phi_{ex}}$$

$$\alpha_{yx} = -\frac{1}{2}[-\sin 2\phi + 2\phi + K_r \cos 2\phi]_{\phi_{st}}^{\phi_{ex}} \quad (4)$$

$$\alpha_{yy} = \frac{1}{2}[-\cos 2\phi - 2K_r\phi - K_r \sin 2\phi]_{\phi_{st}}^{\phi_{ex}}$$

where, ϕ_{st} and ϕ_{ex} are the start and exit immersion angles of the cutter to and from the cut, respectively. K_r is the cutting coefficients in radial direction. The differenced vibrations are $\Delta x(t) = x(t) - x(t-T)$, $\Delta y(t) = y(t) - y(t-T)$, T is the tooth passing interval and also the delay time for uniform pitch cutter. The delay time T is the essential factor for the occurrence of chatter.

3 EFFECT OF DELAY TIME

Lots of solutions for Eq. (1) in frequency and time domain have been presented. The paper proposes one idea to analyze the influence of delay time T on the system stability from the essence of vibration problems.

The Eq. (1) is converted from time to frequency domain,

$$\left(-\begin{bmatrix} m_x & 0 \\ 0 & m_y \end{bmatrix} \omega^2 + \begin{bmatrix} c_x & 0 \\ 0 & c_y \end{bmatrix} j\omega + \begin{bmatrix} k_x & 0 \\ 0 & k_y \end{bmatrix}\right) \begin{Bmatrix} X(\omega) \\ Y(\omega) \end{Bmatrix}$$

$$= \frac{1}{2} a_p K_t (1 - e^{-j\omega T})[A_0] \begin{Bmatrix} X(\omega) \\ Y(\omega) \end{Bmatrix}$$

$$(5)$$

(5) can be written simply

$$\left(\left[Z(\omega)\right]-\frac{1}{2}a_pK_t\left(1-e^{-j\omega T}\right)[A_0]\right)\left\{\begin{matrix}X(\omega)\\Y(\omega)\end{matrix}\right\}=\{0\} \quad (6)$$

where, $[Z(\omega)]$ is the dynamic stiffness matrix of milling system. The characteristic equation of milling system can be obtained from Eq. (6)

$$\det\left(\left[Z(\omega_c)\right]-\frac{1}{2}aK_t\left(1-e^{-j\omega_cT}\right)[A_0]\right)=0 \quad (7)$$

The expansion of Eq. (7) is shown as follows

$$\det\left(-\begin{bmatrix}m_1 & 0\\0 & m_2\end{bmatrix}\omega_c^2+\begin{bmatrix}c_1+c_{11} & c_{12}\\c_{21} & c_2+c_{22}\end{bmatrix}j\omega_c\right.$$
$$\left.+\begin{bmatrix}k_1+k_{11} & k_{12}\\k_{21} & k_2+k_{22}\end{bmatrix}\right)=0 \quad (8)$$

where, $m_1=m_x$, $m_2=m_y$, $c_1=c_x$, $c_2=c_y$, $k_1=k_x$, $k_2=k_y$.

Additional damping elements:

$$c_{11}=-\frac{N}{4\pi}a_pK_t\frac{\sin(\omega_cT)}{\omega_c}a_{xx}$$

$$c_{12}=-\frac{N}{4\pi}a_pK_t\frac{\sin(\omega_cT)}{\omega_c}a_{xy}$$

$$c_{21}=-\frac{N}{4\pi}a_pK_t\frac{\sin(\omega_cT)}{\omega_c}a_{yx} \quad (9)$$

$$c_{22}=-\frac{N}{4\pi}a_pK_t\frac{\sin(\omega_cT)}{\omega_c}a_{yy}$$

Additional stiffness elements:

$$k_{11}=-\frac{N}{4\pi}a_pK_t(1-\cos(\omega_cT))a_{xx}$$

$$k_{12}=-\frac{N}{4\pi}a_pK_t(1-\cos(\omega_cT))a_{xy}$$

$$k_{21}=-\frac{N}{4\pi}a_pK_t(1-\cos(\omega_cT))a_{yx} \quad (10)$$

$$k_{22}=-\frac{N}{4\pi}a_pK_t(1-\cos(\omega_cT))a_{yy}$$

In Eq. (4), $\alpha_{xy}>0$, $\alpha_{yx}<0$, in general. For regenerative chatter, the phase $\beta=\omega_cT\in[\pi,2\pi)$, thus, $\sin(\omega_cT)<0$. For additional damping terms, $c_{12}>0$, $c_{21}<0$, for additional stiffness terms, $k_{12}<0$, $k_{21}>0$. It is the non-diagonal elements that cause chatter, which can be verified by Routh criterion. In Eq. (8), $j\omega_c$ can be replaced by s, and the equation can be transformed into Laplace domain.

$$\det\left(-\begin{bmatrix}m_1 & 0\\0 & m_2\end{bmatrix}s^2+\begin{bmatrix}c_1+c_{11} & c_{12}\\c_{21} & c_2+c_{22}\end{bmatrix}s\right.$$
$$\left.+\begin{bmatrix}k_1+k_{11} & k_{12}\\k_{21} & k_2+k_{22}\end{bmatrix}\right)=0 \quad (11)$$

Eq. (11) can be expanded

$$a_4s^4+a_3s^3+a_2s^2+a_1s+a_0=0 \quad (12)$$

Where

$$\begin{cases}a_4=m_1m_2\\a_3=m_1(c_2+c_{22})+m_2(c_1+c_{11})\\a_2=m_1(k_2+k_{22})+(c_1+c_{11})(c_2+c_{22})\\\quad+m_2(k_1+k_{11})-c_{12}c_{21}\\a_1=(c_1+c_{11})(k_2+k_{22})+(c_2+c_{22})(k_1+k_{11})\\\quad-k_{12}c_{21}-k_{21}c_{12}\\a_0=(k_1+k_{11})(k_2+k_{22})-k_{12}k_{21}\end{cases} \quad (13)$$

The first row of Routh criterion Table is

$$Ls=\left[a_4 \quad a_3 \quad \frac{a_3a_2-a_4a_1}{a_3} \quad a_1-\frac{a_3a_0}{Ls(3,1)} \quad a_0\right]^T \quad (14)$$

According to Routh criterion, the system becomes unstable when there are negative elements in Eq. (14). In Eq. (9)–(10), $k_{12}<0$, $c_{21}<0$, thus $k_{12}c_{21}>0$, then $k_{21}>0$, $c_{12}>0$, then $k_{21}c_{12}>0$, as a result, a_1 decrease. And $k_{12}k_{21}<0$ results in a_0 increase. The coefficient Ls (4,1) decrease obviously as a result of a_1 decreasing and a_0 increasing. Meanwhile, the other elements in Eq. (14) have no obvious trends to decrease. Thus, the chatter occurrence and critical cutting depth can be predicted when the $Ls(4,1)$ become negative as a_p increases, the prediction can be verified by milling experiments.

4 EXPERIMENTAL VERIFICATION

For verification of the presented method, numerical simulations and extensive experiment are carried out. The cutting conditions are given as below

Tool: carbide end mill, overhang length: 70 mm
Overall length: 125 mm, 4 Flutes, helix angle: 35°
Diameter: $D=12$ mm
Workpiece: aluminum alloy T6061
Cutting conditions: cutting width $a_e=2$ mm federate $f_t=0.1$ mm/z
Cutting constants: $K_t=781.7$ MPa, $K_r=0.3$

The structural modal parameters of tool:
Natural Frequency: $\omega_{nx} = \omega_{ny} = 1155$ (Hz)
Modal Stiffness: $k_x = k_y = 1.044 \times 10^7$ (N/m)
Modal Damping Ratio: $\xi_x = \xi_y = 0.046$

The simulation and experimental results are shown in Figure 2. For the simulation method based on Routh criterion, the critical axial depth can be predicted according to Eq. (14), when the spindle speed n and chatter frequency ω_c are given. The chatter frequency ω_c and phase β can

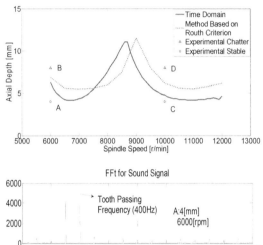

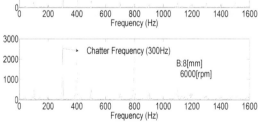

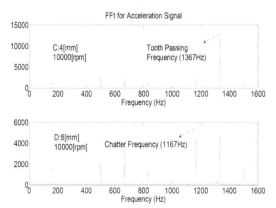

Figure 2. Comparisons of the zero order solution against numerical and experimental results.

be obtained through semi-discrete method. The semi-discrete time domain solution require to be searched by scanning a range of trial spindle and depths of cuts, however the characteristic multipliers cross unit circle almost linearly and the phase changes little in the process. Thus, much fewer numbers of discrete steps are required to obtain chatter frequency and phase, large amounts of computing time are saved.

It can be seen from the Figure 2 that the method based on Routh criterion show good agreement with the experimental results. The experimental results also indicate that the stability has been predicted accurately. The peak value in chatter frequency becomes apparent with increasing a_p, and chatter occurs consequently. Relevant parameters are listed in Table 1. Vibration marks on the machined surface can be seen from Figure 3, which can represent variable types of vibration.

As shown in Figure 3, the shapes of vibration marks can be classified into two types: straight lines and slant lines. The straight lines correspond to forced vibrations. The slant lines include left-hand and right-hand, and the latter is the symbol of regenerative chatter. The chatter frequency ω_c is listed in Table 1 which is correspond to right-hand marks.

Table 1. Chatter frequency and phase in experiments.

n (rpm)	ω_c (Hz)	T (s)	β/π
6000	300	0.0025	1.5
10000	1167	0.0015	1.5

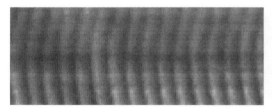

(a)(n=10000r/min, a_p=8mm)

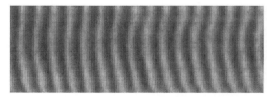

(b)(n=6000r/min, a_p=8mm)

Figure 3. Chatter marks on the machined surface.

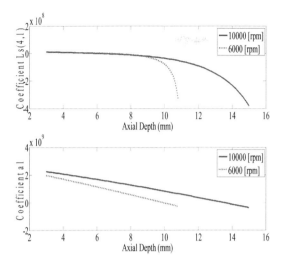

Figure 4. Variation trend of coefficient in Eq. (14).

From Table 1, the chatter phase β are all 1.5π in experiments at the unstable point B and D, which indicate the experimental chatter can be regarded as typical Hopf bifurcation. According to Eq. (14), the non-diagonal damping elements c_{12} and c_{21} increase and coefficient a_1 decrease, the characteristic multipliers cross unit circle as conjugate complex numbers. The variation trend of Coefficient Ls (4,1) and a_1 in Eq. (14) can be seen from Figure 4, which indicate the stability criterion proposed in paper is valid.

Another instability type, *double period bifurcation*, has not been observed from experiment. In this case, phase is about π, the non-diagonal stiffness elements k_{12} and k_{21} increase and a_0 coefficient increase, the characteristic multipliers cross unit circle along negative x-axis. This instability type is not common as Hopf bifurcation in practice. It can be seen that the additional damping (c_{12}, c_{21}) plays more important role than additional stiffness (k_{12}, k_{21}) for the occurrence of chatter.

5 DISCUSSIONS

The method in the paper reveals the root cause of the occurrence of chatter, which are the variations of dynamic characteristics of milling system caused by delay time between inner and outer waves. The viewpoint has been verified by simulations and experiments. Many methods have been proposed to avoid chatter through changing delay time, such as variable speed cutting and variable pitch cutters, however the action mechanism of these methods still lack comprehensive explanations. [13] Optimized the variable pitch based on numerical

methods and spent much more time to compute. In fact, the optimization of variable pitch for regenerative chatter mitigation is not one complex task if the mechanism of chatter is well learned.

Budak [14, 15] designed variable pitch cutters to avoid vibration in milling. An optimization objective for variable tooth pitch $f = \sum_{i=1}^{N} \sin(\omega_c T_i)$ is proposed based on zero order solution in [14] and the tooth pitch designed satisfied $f \approx 0$, which has been verified experimentally in [15]. This objective can be proved from the essence of the vibration problems through the method based on Routh criterion.

For variable pitch cutters, the delay time and phase for each tooth are different, thus $N \sin(\omega_c T)$ in Eq. (9) can be replaced by $\sum_{i=1}^{N} \sin(\omega_c T_i)$. If the f is equal to about 0, the phase β can return to 2π, and the additional damping (c_{12}, c_{21}) in Eq.(14) return to zero so as to the Hopf bifurcation are eliminated. Besides, the shape of vibration marks can transform from slant lines to straight lines. Variable pitch cutters can change delay time and thus to change damping indirectly.

Further, it is important to utilize process damping for low machinability materials [16]. The process damping can be increased by designing different tool geometries. A.R. Yusoff et al [17] evaluated the performance of process damped milling considering different tool geometries (edge radius, rake and relief angles and variable pitch). Besides it indicated that variable pitch angles most significantly increase process damping performance through experiments, however, the concepts of process damping require the further discussions.

The process damping arises from the machining process itself [18]. Most researchers [16–19] think that the process damping is caused by plowing effect. However, the damping caused by delay time as shown in the paper differ from the ones caused by plowing effect, which may can also been seen as another type of process damping. Increased tool edge radius and decreased relief angles result in increased indentation volume and resistances, however, variable pitch cutters adopt another way. The concepts of process damping could be developed probably.

6 CONCLUSIONS

A great deal of research has been carried out on the chatter problem in milling, and lots of significant advances have been made over the years. The paper accomplishes some new work in this research direction.

1. For studying the physical essence of regenerative chatter in milling, the method based on Routh criterion is proposed to analyze the dynamic characteristics of milling system and to solve the stability equations. Experimental results prove that the method can predict stability limitation accurately. Most importantly, the method provides powerful explanations for the action mechanism of chatter suppressing measures such as variable pitch cutters, which is very helpful to the design of chatter mitigation.

2. During the analytical process, a new viewpoint was proposed that the concepts of process damping could be developed. Theoretical analysis and existing experiments indicate that variable pitch cutters can change the system damping through variable delay time to suppress chatter. The damping caused by the delay time can also be seen as a type of process damping, which differs from the process damping result from plowing effect.

REFERENCES

[1] J. Tlusty, M. Polacck,. The stability of machine tools against self-excited vibration. *ASME International Research in Production Engineering*, 465–474, 1963.

[2] S.A. Tobias. *Machine tool vibration*. Blackie and Sons Ltd, 1965.

[3] F. Koenigsberger, J. Tlusty. *Machine tool structures*. Pergamon Press, 1967.

[4] J. Tlusty, F. Ismail. Basic Nonlinearity in machine chatter. *Annals of the CIRP*, 30:21–25, 1981.

[5] R. Sridhar, R.E. Hohn, G.W. Long. General formulation of the milling process equation. *Transaction of ASME Journal of Engineering for Industry*, 90:317–324, 1968.

[6] R. Sridhar, R.E. Hohn, G.W. Long. A stability algorithm for the milling process. *Transaction of ASME Journal of Engineering for Industry,* 90:330–334, 1968.

[7] I. Minis, R. Yanushevsky, A. Tembo. Analysis of linear and nonlinear chatter in milling. *Annals of the CIRP*, 39(1):459–462, 1990.

[8] I. Minis, R. Yanushevsky. A new theoretical approach for the prediction of machine tool chatter in milling. *Transaction of ASME Journal of Engineering for Industry*, 115:1–8, 1993.

[9] Y. Altintas, E. Budak. Analytical prediction of stability lobes in milling. *Annals of the CIRP*, 44(1):357–362, 1995.

[10] E. Budak, Y. Altintas. Analytical prediction of chatter stability conditions for multi-degree of systems in milling. *Transactions of ASME Journal of Dynamic Systems Measurement and Control*, 120:22–30, 1998.

[11] T. Insperger, G. Stepan. Updated semi-discretization method for periodic delay-differential equations with discrete delay. *International Journal for Numerical Methods in Engineering*, 61:117–141, 2004.

[12] P.V. Bayly, J.E. Halley, B.P. Mann, et al. Stability of interrupted cutting by temporal finite element analysis. *Journal of Manufacturing Science and Engineering*, 125:220–225, 2003.

[13] A.R. Yusoff, N.D. Sims. Optimisation of variable helix tool geometry for regenerative chatter mitigation. *International Journal of Machine Tools & Manufacture*, 51:133–141, 2011.

[14] E. Budak. An analytical design method for milling cutters with nonconstant pitch to increase stability, part1: theory. *Journal of Manufacturing Science and Engineering*, 125:29–31, 2003.

[15] E. Budak. An analytical design method for milling cutters with nonconstant pitch to increase stability, part 2: application. *Journal of Manufacturing Science and Engineering*, 125:35–38, 2003.

[16] E. Budak, L.T. Tunc. A new method for identification and modeling of process damping in machining. *Journal of Manufacturing Science and Engineering*, 131:1–10, 2009.

[17] A.R. Yusoff, M.S. Taloy, N.D. Sim. The role of tool geometry in process damped milling. *The International Journal of Advanced Manufacturing Technology*, 50:883:895, 2010.

[18] Chao-Yu Huang. *Analysis of process damping and system dynamics in milling*. Ph.D. Dissertation, National Cheng Kung University, 2006.

[19] L.T. Tunc, E. Budak. Effect of cutting conditions and tool geometry on process damping in machining. *International Journal of Machine Tools and Manufacture*, 57:10–19, 2012.

Material Science and Engineering – Chen (Ed.)
© 2016 Taylor & Francis Group, London, ISBN 978-1-138-02936-1

Research on the screw machine for the teaching aid

F. Feng, J.G. Ling, J.M. Zhou, F.Q. Bai & R.Q. He
Ningbo Dahongying University, Ningbo, Zhejiang, China

ABSTRACT: This design is a "multi-functional" in the mechanical and gas-electric display aid. This comprehensive teaching aid can be used as a future display aid in the classroom. In addition to making our mechanical and gas-electric display aid play a greater role and use of sense, our work can also be seen as an automatic screw machine, whose biggest feature is a set of electronic controls, mechanical, and pneumatic, which are the three major areas of knowledge of machinery in one modern product design. The perfect combination of knowledge through electronic control, mechanical, pneumatic transmission into automatic screw machines abandons the old tradition by artificially screwing the inefficient feature. Furthermore, it can serve as a comprehensive teaching aid to provide a convenient way to teachers and their students, one mechanical and gas-electric can be used as a teaching display aid in the mechanical courses of electronic design, mechanical, and pneumatic, which are the three categories of knowledge to explain, specifically linking the knowledge and production of life.

Keywords: teaching aid; mechanical; screw machine

1 INTRODUCTION

Nowadays, with the rapid development of manufacturing industry, the scale and output of China's mechanical industry has surpassed Japan and has become the second in the world, and exports surpass German to be the first. Thus, China has already become a global manufacturing country. However, why has not China been a world machinery superpower? Although we have a large amount of mechanical products, the core technology comes from abroad, so that we are controlled by others. As for colleges, a comprehensive visual training aid for engineering classes has the advantage of low cost, which will provide a more vivid class to the students.

Therefore, we designed a kind of comprehensive training aid, which is mainly composed of mechanical structure, electronic control and pneumatic transmissions. Besides, for the following purposes, we made this automatic filature: (1) With the effective combination of 3 mechanical principles, namely mechanical structure, electronic control and pneumatic transmissions, our work can achieve automation to help teachers show the relevant motion principles and laws more vividly. (2) Traditional filatures require someone to operate it and a screwdriver to tighten the screws, which leads to low efficiency. On the contrary, this automatic filature can solve this, improve efficiency and reduce labor costs.

2 CONSTRUCTION AND PERFORMANCE INDEX

1. This handheld filature achieves three kinds of mechanical field knowledge, namely machine construction, pneumatic transmission and electronic control. In addition, it provides aid with the deep combination of machine theory and production life. Moreover, it provides a comprehensive aid to engage mechanical teaching teachers. At the same time, it is a good example for students to learn the knowledge of machine construction, pneumatic transmission and electronic control, and it is also convenient for the connection with production life.
2. The basic size of machine construction and its gas electric display aid: length 365 mm, width 253 mm, height 157 mm.
3. The number of screws (guide rail) can carry for one time: 32 trunks.
4. The number of screws (screw gun) can carry for one time: 1 trunk.
5. Screw specification: M1.7X6.

3 DESIGN SCHEME

3.1 *The mechanical structure and the basic structure of the handheld filature.*

Handheld filature mainly includes screw conveying mechanism, screw mechanism and handheld electric lock screw batch. Screw conveying mechanism involves the gear drive, setting of the inner cavity with a guide rail, and combining the linear vibrator forward so as to realize discharge. Screw institution involves the push-pull type material, screw lined up until the institution, and the cylinder for translational motion of a single screw, which will be moved to the position so as to realize points. Handheld electric lock screw batch is modified on the basis of the traditional electric group, and a device with a similar collet chuck realized function of screw.

This handheld filature shows the materials that have the advantages of compact structure and preciseness, with a convenient operation and a precise locking screw.

3.2 *The workflow of the handheld filature*

The workflow of the handheld filature: the feeder puts the qualified screws into the feeding mechanism with an orderly arrangement, as shown in Figure 3–1. The action of the cylinder achieves the dosing of screws, as shown in Figure 1, and separates the screws waiting to be locked, and then puts them inside the tube for sending the nail, as shown in Figure 2, and finally it presses down the electric screwdriver to lock the screws.

3.3 *The design of the dosing mechanism*

The dosing mechanism is also the key structure of this device. During the design process, we select two programs; one is under the pressure, as shown in Figure 3. But after the experiment, this structure very easily gets stuck.

After several tests, the under-pressure dosing mechanism uses a way of push-pull. The schematic of this dosing mechanism is shown in Figure 4.

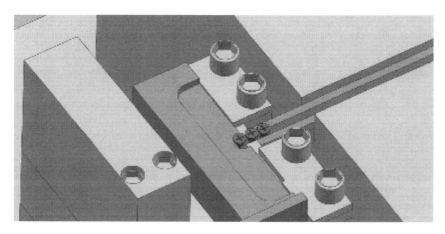

Figure 1. Nesting schematic.

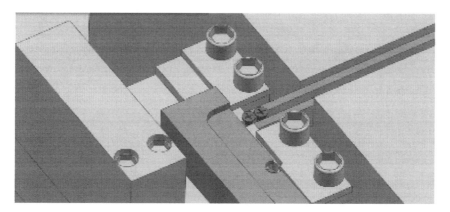

Figure 2. Nesting schematic.

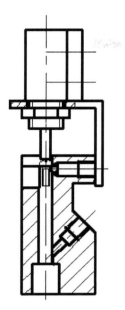

Figure 3. The schematic of under-pressure dosing mechanism.

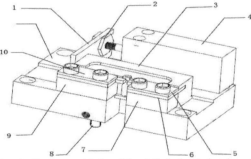

1-webs 2-screw nut 3-the slider 4-Cylinder 5-skateboard 6-right plate 7-right guide block 8-tube of cutting 9-left guide block 10-left plate

Figure 4. The design schematic of the dosing mechanism of the screw machine.

4 OPERATING PRINCIPLE

The following is the concrete analysis of this product that involves the analysis of the basic structure, the working principle and the introduction of the running performance.

1. This work is controlled by the structure of the single chip microcomputer of the circuit part, in which the single chip connects to the circuit, through the calculation of main control points of the material and feeding mechanism. Micro-computer control of the motor: the motor running gear is driven to rotate the feeding wheel, in which there are six internal arc plastic sheet feeding wheels, through the rotation of the spare screw hopper in the drive and then falls on the aggregate structure.

2. Use the aggregate structure to collect screws, and put a part of vertical distribution of the screws on the straight rail, and leave the rest on the lag screw hopper.

3. Let the single chip microcomputer control linear vibrators, make the screw fall vertically aligned straight rail-way through a linear vibrator slow vibrations, and make the screw motion to divide the material structure.

4. When the screw reaches a certain amount, the sensor sensing microprocessor controls the feeding wheel to stop turning.

5. When the screw goes through a filter structure, it can intercept screw, which does not conform to the sport rules on the straight guide rail.

6. When the screw goes to the end of the straight guide,it will be divided into a round hole, which is stuck on the material structure, single chip microcomputer control cylinder screws into the screw gun conduit to pressure to force the screw into the feed tube nails.

7. The screw head screw puts the screw into leather mouth, screw position to be locked position, under pressure locking screwdriver, screws automatically carve into the screw holes.

8. Pressure screw gun's control switch, SCM, cylinder automatic operation, screw rails continue straight into the feed screw nail gun tube, and repeat the cycle.

9. When the number of screws straight rails less than a certain value, the MCU control feed rollers continue to run, and repeat the above action.

5 INNOVATIONS AND APPLICATIONS

5.1 *Innovations*

1. With the effective combination of 3 mechanical principles, namely mechanical structure, electronic control and pneumatic transmissions, our work can achieve automation to help teachers show the relevant motion principles and laws more vividly.

2. Mechanical structure of the swing of sliders and gears drive can apply to mechanical principle courses: analysis of mechanism composition, analysis of motion and force, and design of linkage mechanisms. Besides, the mechanical structure of the radiotube and the circuit can apply to SCM, which belongs to the electrician electronic technology. in addition, the combination of cylinders and the swing of sliders can apply to hydraulic and pneumatic transmission courses.

3. As a mechanical engineering visual training aid, our work provides visual learning experience. Furthermore, as a practical mechanical product, it has a more innovative design and more ideal performance than any existing filature.
4. Structurally, our work makes full use of the features of the mechanical structure. By combining the drawing-enroller with a motor, together with a swing mechanism composed of a small cylinder, we realize the auto-feeding mechanism.

5.2 Applications

This work can be applied to all kinds of engineering colleges as a visual training aid, and operates as a comprehensive visual training aid in mechanical design manufacturing and automation classes, electronic design and its automation classes. By solving the problem of lacking visual product, our work provides the students a more vivid class.

This work is suitable for mass production and a joint-school project as well. Considered as a marketing product, this innovative training aid has the advantage of low-cost and high-cost performance, and deserves good market prospects.

ACKNOWLEDGMENTS

This work was financially supported by the National College Students' Innovation and Entrepreneurship Training Plan for 2014 (201413001004) and the College Students' Scientific Research of Ningbo DaHongYing University.

REFERENCES

[1] <Machinerys Handbook> (NEW). China Machine Press, 2004.
[2] Krzyst of Mianowski. Simple and very low cost remote systems for tele-manipulation, 1996.
[3] Dan Zhang ZhenGao. Hybrid head mechanism of the groundhog-like mine rescue robot [J]. Robotics and Computer-Integrated Manufacturing, 2011, (27): 460–470.

Material Science and Engineering – Chen (Ed.)
© 2016 Taylor & Francis Group, London, ISBN 978-1-138-02936-1

Research on the vibration-damping technology of some underwater vehicle

K. Liu, X. Zhang, J. Sun, S.C. Ding & B.Y. Zhang
Scientific Research Department, Naval University of Engineering, Wuhan, China

ABSTRACT: As for some underwater vehicle, the high vibration noise exerts great influences on its stealth performance and sonar detection capability. For the shell structure of this underwater vehicle, we use the Spectral Finite Element Method (SFEM) to establish the spectral element equations, respectively, for two cases, namely with and without damping materials, and analyze the damping effect by adopting different constrained-layer damping schemes through the simulation. Then, we carry out the shell vibration-damping experimental study based on the above analysis. Finally, an optimal vibration-damping scheme is gained under similar constraint conditions.

Keywords: shell; vibration; damping; vibration-reduction; underwater vehicle

1 INTRODUCTION

In modern naval battles, discovering and being discovered and detecting and being detected have become a decisive factor for the outcome of a war. Therefore, the "acoustic performance", which is an invisible fighting force, has aroused the attention of various naval powers [1]. All the new types of underwater vehicles take "acoustic performance" as one of the important examination indices [2]. The "noise problems" of some underwater vehicle are mainly reflected in two aspects: one is radiation noise, which has a close relation to its stealth performance; the other is self-noise from sonar parts, which affects the sonar detection distance of the underwater vehicle. How to effectively reduce the vibration and noise levels of underwater vehicles in China has become an important problem to be considered by many experts, scholars and engineers.

The damping technology is a kind of widely used vibration and noise reduction technique [3], easy to operate, and has the following advantages for solving the vibration and noise problems of underwater vehicles:

1. The damping technology can easily achieve a good noise reduction effect for underwater vehicles and other similar thin-shell structures.
2. The damping materials are especially suitable for reducing vibration and noise at a medium-high frequency, which coincides with the main frequency associated with the underwater vehicle;
3. Few studies have been involved in the use of damping measures, and no influence is exerted on the design of other structures.

However, it is not easy to apply the damping technology to reduce the vibration and noise of underwater vehicles. The vehicle has a complicated internal structure and a narrow space, so the thickness and weight of damping materials and components available are limited. Additionally, the influence exerted by high temperature and oil pollution should also be considered. Under such harsh conditions, it is really difficult to obtain a good effect of vibration and noise reduction. This paper adopts SFEM to establish the spectral element equations, respectively, for two cases, namely with and without damping materials, and analyzes the damping effect by adopting different constrained-layer damping schemes through the simulation. Then, it executes the shell vibration-damping experimental study based on the above analysis. Finally, an optimal vibration-damping scheme is gained under similar constraint conditions.

2 MODELING OF VIBRATION-DAMPING DIGITAL SHELL

The SFEM is a new type of spectral domain modeling method, which combines the finite element analysis with parsing methods and expresses the nodal displacement and nodal forces with frequency domains. The dynamic-stiffness matrix is also related to the frequency, so the corresponding structural unit is called spectral elements and dynamic-stiffness matrices. This paper adopts the SFEM to establish a shell model for the underwater vehicle.

The shell structure of some underwater vehicle is shown in Figure 1, which is mainly composed

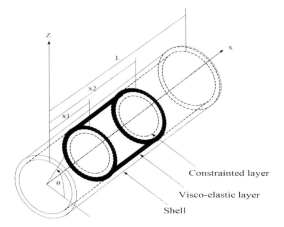

Figure 1. Schematic diagram of the engine compartment shell.

of the base shell, viscoelastic layer and constrained layer. During modeling, the underwater vehicle shell, which is partially laid with constraint and damping materials, can be divided into one subshell laid with damping materials and two subshells laid without damping materials. First, we analyze the one with damping materials and ignore its constrained and damping layers, and then we gain a relevant equation [4] for the sub-shell without damping materials. In this way, each subshell can be regarded as an independent spectral element.

2.1 Spectral element equation of the sub-shell with damping materials

In the SFEM, the derivation of a spectral element equation for structures is generally started with the differential equation for structural movement without an external force, i.e. the homogeneous differential equation [5]. The spectral element equation for the cylindrical shell with n circumferential modes is as follows:

$$S_e^{(n)}(\omega) \cdot d_e^{(n)} = F_e^{(n)} \qquad (1)$$

where, $d_e^{(n)}$ refers to the spectral displacement component corresponding to n circumferential modes, and $F_e^{(n)}$ refers to the spectral external force vector corresponding to n circumferential modes, $S_e^{(n)}(\omega)$:

$$S_e^{(n)}(\omega) = K_e^{(n)}(\omega) - \omega^2 M_e^{(n)}(\omega) \qquad (2)$$

where $K_e^{(n)}(\omega)$ and $M_e^{(n)}(\omega)$, respectively refer to spectral element stiffness matrix and the spectral element quality matrix [6].

2.2 Derivation of the spectral element equation of the sub-shell without damping materials

After ignoring the constrained and damping layers in the spectral element equation of the subshell with damping materials, we gain the spectral element equation of sub-shells without damping materials, which is as follows:

$$S_e^{(n)}(\omega) \cdot d_e^{(n)} = F_e^{(n)} \qquad (3)$$

$$S_e^{(n)}(\omega) = \begin{bmatrix} N_{12}^{-1} N_{11} & -N_{12}^{-1} \\ N_{21} - N_{22} N_{12}^{-1} N_{11} & N_{22} N_{12}^{-1} \end{bmatrix} \qquad (4)$$

When the number of circumferential modes is n, the spectral transfer matrix of sub-shells without damping materials is as follows:

$$e^{A(\omega) \cdot L_1} = \{N_{ij}\}, \quad (i, j = 1, 2), \qquad (5)$$

3 SIMULATED ANALYSIS OF THE VIBRATION-DAMPING EFFECT

The total thickness of the entire constrained damping structure should be within 4 mm due to the limitation of some underwater vehicle. Considering the production and construction factors, the constrained and damping layers with 3 different thicknesses are selected for analysis and research. The thicknesses h_v of damping materials are, respectively, 2 mm, 2.5 mm and 3 mm, and the thicknesses h_c of aluminum-alloy constrained layers are respectively 1 mm, 1.5 mm and 2 mm. The schematic diagram of the underwater vehicle shell is shown in Figure 1: length $L = 585$ mm, radius $R = 267$ mm, density $\rho_s = 2710$ kg/m^3, Poisson's ratio $\mu_s = 0.34$, elastic modulus $E_s = 70$ GPa, $x_1 = 170$ mm, $x_2 = 415$ mm visco-elastic layer length $L_v = 245$ mm, radius $R = 267$ mm, density $\rho_s = 1328$ kg/m^3, Poisson's ratio $\mu_v = 0.49$, complex shear modulus $G_v = (8 + 3j)$ MPa; constrained layer length $L_c = 245$ mm, radius $R = 267$ mm, density $\rho_c = 2710$ kg/m^3, Poisson's ratio $\mu_c = 0.34$, and elastic modulus $E_c = 70$ GPa.

Assuming that both ends of the underwater vehicle shell are free and the clamped connection is adopted between the sub-shell without damping materials and the sub-shell with damping materials, we use the SFEM to calculate the inherent frequency and the modal loss factor under different modes, and study the influences caused, respectively, by the thicknesses of constrained and damping layers. This paper only takes transversal displacement and vibration into account since the vibration of the engine compartment shell is dominated by the bending wave.

Assuming that the thickness of damping layers is kept 2 mm and those of constrained layers are 1 mm, 1.5 mm and 2 mm, respectively, the change in the inherent frequency and modal loss factor at different numbers of modes in the three cases are, respectively, shown in Figure 2 and Figure 3. The results show that the increase in the thickness of constrained layers has a little influence on the inherent frequency, but will cause an increase in the modal loss factor; thus, increasing thickness of constrained layers is good for improving the damping effect.

In order to further study the thickness influence of damping layers on the inherent frequency and modal loss factor, assuming that the thickness of constrained layers is kept 1 mm and those of damping layers are 2 mm, 2.5 mm and 3 mm,

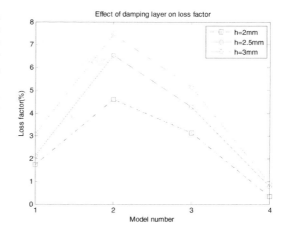

Figure 4. Influence of the thickness of damping layers on modal loss factor.

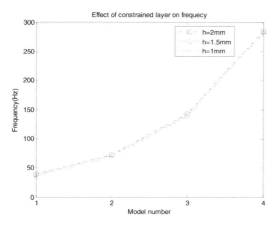

Figure 2. Influence of the thickness of constrained layers on inherent frequency.

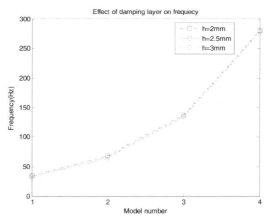

Figure 5. Influence of thickness of damping layers on inherent frequency.

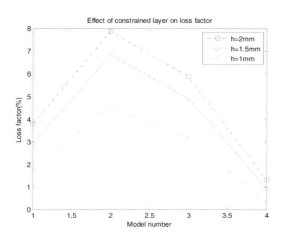

Figure 3. Influence of the thickness of constrained layers on modal loss factor.

respectively, the change in the inherent frequency and modal loss factor at different numbers of circumferential modes in the three cases is compared, as shown in Figure 4 and Figure 5. The results show that the increase in the thickness of damping layers may cause a slight decrease in the inherent frequency, but will cause an obvious increase in the modal loss factor.

4 EXPERIMENTAL STUDY

It can be seen from the simulation results, the difference between the thicknesses of damping and constrained layers exerts a great influence on the vibration-damping effect of some underwater vehicle. To achieve the best damping effect from the constrained and damping layers, an experimental study using the 3 different thicknesses of damping and constrained layers is carried out to analyze their damping effect (see Figure 6), so as to offer a basis for the final scheme.

4.1 *Experimental scheme*

For the first cylindrical shell used in the experiment, the paste thicknesses of damping materials and aluminum-alloy constrained layers are, respectively, 2.5 mm and 1.5 mm; for the second cylindrical shell, the thicknesses are, respectively, 3 mm and 1 mm; for the third shell, the thicknesses are, respectively, 2 mm and 2 mm. The location and width of the damping and constrained materials are consistent with those of the actual damping materials inside the engine compartment. The experimental schematic diagram is shown in Figure 7.

4.2 *Analysis of the experimental results*

The vibration-level difference is considered as an evaluation index of the vibration-damping effect if the calculation results are not comparable due to

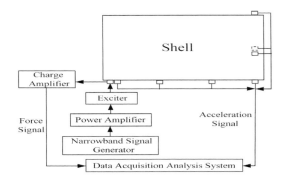

Figure 7. Schematic diagram of the vibration-damping performance test.

Table 1. Difference in vibration levels at response points with and without damping.

Designation	4 kHz	5 kHz	6.3 kHz	8 kHz
No. 1 shell	13.5 dB	18.7 dB	19.8 dB	23.4 dB
No. 2 shell	5.7 dB	5.4 dB	13.3 dB	14.3 dB
No. 3 shell	14 dB	16 dB	21.7 dB	14 dB

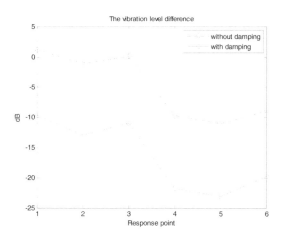

Figure 8. Comparison of vibration-level differences at 6 response points with and without damping at the central frequency of 4 kHz.

a different exciting force before and after the test. When the exciting signal is the 1/3 octave random signal with the central frequency of 4 kHz, 5 kHz, 6.3 kHz and 8 kHz, the difference in vibration levels at 4 response points, respectively, for the shells with and without damping is given in Table 1, and detailed testing results are shown in Figures 8–11.

As it can be seen from the above table, the damping structure adopting the first cylindrical shell has

a) Shell after Pasted with Damping Layer

b) Shell after Pasted with Constrained Layer

Figure 6. Structural radiation of constrained and damping layers.

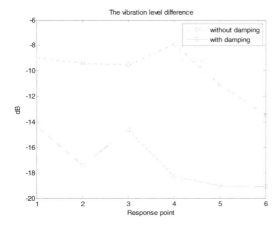

Figure 9. Comparison of vibration-level differences at 6 response points with and without damping at the central frequency of 5 kHz.

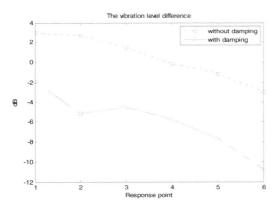

Figure 10. Comparison of vibration-level differences at 6 response points with and without damping at the central frequency of 6.3 kHz.

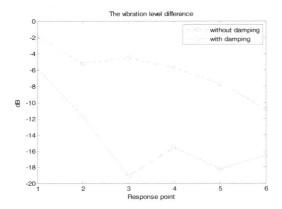

Figure 11. Comparison of vibration-level differences at 6 response points with and without damping at the central frequency of 8 kHz.

a better damping effect at 1/3 octave when the central frequency is 5 kHz, and there is also a better damping effect at other similar frequency bands. Through the above analysis, the best constrained and damping structure scheme is to adopt 2.5 mm-thick damping materials and 1.5 mm-thick aluminum-alloy constrained layers.

5 CONCLUSION

As for vibration-damping problems of some underwater vehicle, this paper adopts the SFEM to establish the spectral element equations, respectively, for two cases, namely with and without damping materials, and analyzes the vibration-damping effect by adopting different constrained-layer damping schemes through the simulation, and masters the influence of different thicknesses of damping and constrained layers on the inherent frequency and modal loss factor of the cylindrical shell that is partially laid with PCLD. Besides, a shell vibration-damping experimental study is carried out to finally get an optimal vibration-damping scheme under similar constraint conditions. The test results show that when the vibration-damping scheme designed in this paper adopts random signals at the 1/3 octave of a single central frequency as the exciting force, the vibration-damping effect of all the structures laid with damping and constrained layers is higher than 5dB, which effectively reduces the vibration of the shell of some underwater vehicle.

REFERENCES

[1] Shijian Zhu, Lin He. Control on Ship and Machinery Vibration [M]. Beijing: National Defense Industry Press, 2006.
[2] Yingfu Zhu, Guoliang Zhang. Ship Stealth Technology [M]. Harbin: Harbin Engineering University Press, 2003.
[3] China Aeronautical Materials Handbook (Volume 10) [S]. Beijing: Standards Press of China.
[4] Bagley R.L, Torvik P.J. On the Fractional Calculus Model of Viscoelastic Behavior [J]. Journal of Rheology, 1986, 30:133–135.
[5] Lesieutre G.A, Mingori D.L. Finite Element Modeling of Frequency-dependent Material Damping Using Augmenting Thermodynamic Fields [J]. Journal of Guidance and Control, 1990, 13(6):1040–1050.
[6] Golla D.F, Hughes P.C. Dynamic of Viscoelastic Structure-A Time Domain Finite Element Formulation [J]. Journal of Applied Mechanics, 1985, 52:897–960.

Material Science and Engineering – Chen (Ed.)
© 2016 Taylor & Francis Group, London, ISBN 978-1-138-02936-1

Research on the dynamic characteristics of hydraulic shock absorber

X. Wang, P.B. Wu & D.X. Yang
State Key Laboratory of Traction Power, Southwest Jiaotong University, Chengdu, China

ABSTRACT: Hydraulic shock absorber is one of the important damped components in a mechanical field. The dynamic characteristics of one anti-yaw shock absorber used in high-speed EMU are studied based on the test-bed, and two experimental factors, loading frequency and loading amplitude, are considered in the test. The dynamic characteristics are found by analyzing the test results. Compared with Maxwell model analysis results, the results of the test indicates that the Maxwell model is appropriate to represent the actual dynamic characteristics of the phase angle, but not appropriate for the dynamic damping coefficient and dynamic stiffness. The test results also show that the dynamic damping coefficient increasingly varies with the increasing loading frequency at a low frequency, and then decreases at a high frequency. The dynamic stiffness increasingly varies with the increasing loading frequency at a low frequency, and then tends to a steady state at loading frequencies above 4 Hz.

Keywords: hydraulic shock absorber; Maxwell model; dynamic characteristics; test

1 INTRODUCTION

Hydraulic shock absorber is used to absorb vibration energy, which is produced when the train is running, by the liquid viscous drag. Its advantage is that the resistance is the function of the velocity, and its characteristic is that the larger the vibration amplitude is, the more the attenuation is [1]. At present, the hydraulic shock absorber has been widely used in the railway vehicle suspension system, and its performance directly affects the safety and stability of the operation of the train [3].

Studies [1, 4] have discussed the Maxwell model of the hydraulic shock absorber in detail, and analyzed its dynamic characteristics. A study [3] has described the working principle of the hydraulic shock absorber, and established and verified a mathematical model. Another study [6] has proposed some kinds of models to analyze the hydraulic shock absorber. In these articles, the hydraulic shock absorber has not been tested to obtain the dynamic characteristics, in order to compare with the results of the models. Based on one anti-yaw shock absorber used in high-speed EMU as the research object, according to the standard BS EN 13802:2004 "Railway Applications Suspension components—Hydraulic dampers" regulation, we test the dynamic characteristics of an anti-yaw shock absorber, and analyze the results and compare with the results of the Maxwell model.

2 THE WORKING PRINCIPLE OF THE HYDRAULIC SHOCK ABSORBER

Hydraulic shock absorber performs a reciprocating motion in the external incentive operation, generating pressure loss when the hydraulic oil operates in the cylinder through the damping valve, and then the shock absorber generates the damping force. Additionally, parts of the mechanical energy of vehicle vibration are converted into heat energy and dissipating, so as to achieve the purpose of reduced vibration [3]. The physical model of an anti-yaw shock absorber is shown in Figure 1.

In the process of compression, as shown in Figure 1 (a), the piston moves down, the check valve of the piston opens, the check valve of the bottom valve closes, and the top and bottom of the pressure cavity interlink. Then, the oil of the bottom of the pressure cavity passes into the top through the check valve of the piston, due to the piston rod that occupies a certain space on the top of the pressure cavity, forcing the oil into the oil cylinder through the damping valve. So, a transient pressure is produced in the pressure cavity, because the role played by the area of the top and bottom of the pressure cavity is different, which can produce a pressure to hinder the movement of the piston.

In the process of extension, as shown in Figure 1 (b), the piston moves up, the check valve of the piston closes, and the check valve of the bottom valve opens. The oil on the top of the pressure

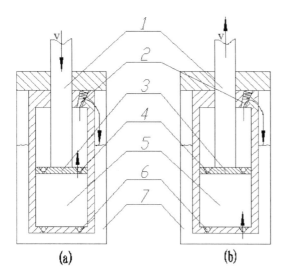

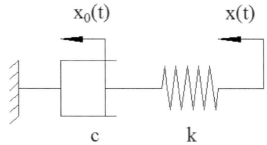

Figure 2. Maxwell model of the hydraulic shock absorber.

Figure 1. Structure and oil flow diagram of the shock absorber. (1) piston rod, (2) damping valve, (3) the piston, (4) check valve of piston, (5) pressure cavity, (6) check valve of bottom valve, (7) oil cylinder.

cavity is forced into the oil cylinder through the damping valve. Due to the reduced pressure in the bottom of the pressure cavity, the oil in the cylinder is sucked into the bottom of the pressure cavity. On the top of the pressure cavity, a transient pressure is produced to hinder the movement of the piston.

3 MAXWELL MODEL OF THE HYDRAULIC SHOCK ABSORBER

Due to the rubber joint on both ends of the hydraulic shock absorber and some compressed air in the oil cylinder, the stiffness of the hydraulic shock absorber occurs in the axial direction. When considering the joint stiffness and liquid stiffness of the hydraulic shock absorber, the phase change between the damping force and the piston velocity will produce, and this feature is called dynamic damping characteristics [4]. Using the Maxwell model to perform the theoretical research of the hydraulic shock absorber, the shock absorber is designed as a combination of components with damping and spring in series, as shown in Figure 2.

Transient pressure hinders the movement of the piston.

When the shock absorber end receives the sine excitation x with frequency ω and amplitude A, if the quality of the piston is m and the displacement is x_0, the mathematical model of this system is given by

$$m\ddot{x}_0 = k(x - x_0) - c\dot{x}_0 \tag{1}$$

The inertial force can be ignored since the weight is too small. So, Equation (1) can be written as follows:

$$c\dot{x}_0 + k(x_0 - x) = 0 \tag{2}$$

The displacement of the excitation is assumed as follows:

$$x(t) = A\sin(\omega t) \tag{3}$$

So, Equation (4) can be obtained as follows:

$$x_0 = \frac{k^2 A}{k^2 + c^2 \omega^2}\sin(\omega t) - \frac{c\omega k A}{k^2 + c^2 \omega^2}\cos(\omega t) \tag{4}$$

Therefore, the shock absorber damping force can be expressed as follows:

$$F = \frac{c\omega k^2 A}{k^2 + c^2 \omega^2}\cos(\omega t) + \frac{c^2 \omega^2 k A}{k^2 + c^2 \omega^2}\sin(\omega t) \tag{5}$$

By simplifying Equation (5), we can obtained the following equation:

$$F = \frac{c\omega k A}{\sqrt{k^2 + c^2 \omega^2}}\sin(\omega t + \varphi) \tag{6}$$

where $\varphi = \arctan\dfrac{k}{c\omega}$ (7)

In this case, the damping force F is no longer the same phase with the excitation speed, and there is a phase lag, and the damping force amplitude is reduced accordingly.

To put the displacement of the excitation (Equation (3)) into Equation (5), the damping force can be described as follows:

$$F = \frac{c\omega k^2 A}{k^2 + c^2\omega^2}\sqrt{1 - \left(\frac{x}{A}\right)^2} + \frac{c^2\omega^2 k}{k^2 + c^2\omega^2}x \qquad (8)$$

Equation (8) can obtain the indicator diagram feature, which is a deflected oval because of the phase angle between the damping force and the excitation speed.

Equation (7) can obtain the dynamics damping coefficient of the hydraulic shock absorber as follows:

$$c = \frac{k}{\omega\tan\varphi} \qquad (9)$$

where ω is the external excitation frequency; φ is the phase angle between the damping force of the hydraulic shock absorber and the displacement of the external excitation; and k is the dynamic stiffness of the hydraulic shock absorber:

$$k = \frac{F_{max}}{x_{max}}\sqrt{1 + \tan^2\varphi} \qquad (10)$$

where F_{max} is the force amplitude of the hydraulic shock absorber in the axial direction, and x_{max} is the displacement amplitude of the external excitation.

4 THE DYNAMIC CHARACTERISTICS ANALYSIS OF THE HYDRAULIC SHOCK ABSORBER

The anti-yaw shock absorber with rubber joints is a dynamic characteristic test object. In this experiment, one end of the anti-yaw shock absorber is fixed and the other is connected to the actuator, using displacement sensors and force sensors to collect the data. In the test process, a single test and a parallel test were conducted via changing tooling, as shown in Figure 3.

Figure 3.   Test-bed of the shock absorber.

The stiffness of the end of the shock absorber fixed in the test device is big enough, with each part of the whole system being connected to the solid and reliable, to ensure that it does not absorb the loading displacement during the test. Before the test, the center of the shock absorber and the actuator are guaranteed at the same level. In the test process, the data are collected by the shock absorber module in the MTS system, and the collected data must be enough to guarantee that the indicator diagram is a smooth curve.

In this experiment, using the sine wave as the loading waveform, the anti-yaw shock absorber moves 10 cycles under the reciprocating motion at each frequency and amplitude. Gathering the last four cycles, each cycle collects 400 points, so that we can get the curve of the force and displacement, and then obtain the dynamic damping coefficient and the dynamic stiffness of the shock absorber.

The content of the test is as follows: loading excitation amplitudes were 0.5 mm, 1 mm, 2 mm, loading excitation frequencies were 1 Hz, 2 Hz, 4 Hz, 6 Hz, 8 Hz, 10 Hz, and total of 18 test conditions.

In Figure 4, the indicator diagram of both the test and the Maxwell model is deflected oval, but with the increase in the excitation amplitude, the indicator diagram of the test will lead to the occurrence of deformation and is no longer completely oval, but the indicator diagram of the Maxwell model does not lead to the occurrence of deformation. Under the condition of the same amplitude, with the increase in the frequency, the indicator diagram of the test and the Maxwell model is counterclockwise deflection, which is caused by the phase angle between the damping force and the excitation speed.

In Figure 5, the phase angle calculated by the Maxwell model decreases gradually with the increase in the frequency, which is consistent with the test results. However, the phase angle calculated by the Maxwell model on the value does not change with the varying amplitude, and the phase angle achieved by the test is slightly changed. At the same frequency, under the condition of different amplitudes, the phase angle calculated by the Maxwell model is smaller than that calculated by the test.

In Figure 6, the dynamic damping coefficient calculated by the Maxwell model decreases with the increase in the frequency. The dynamic damping coefficient achieved by the test increasingly varies with the increasing frequency at a low frequency, and then decreases at a high frequency. The Maxwell model cannot reflect the inflection point of the dynamic damping coefficient. The dynamic damping coefficient calculated by the Maxwell model on the value does not change with the varying amplitude. The dynamic damping coefficient achieved by the test exhibits a large change, and

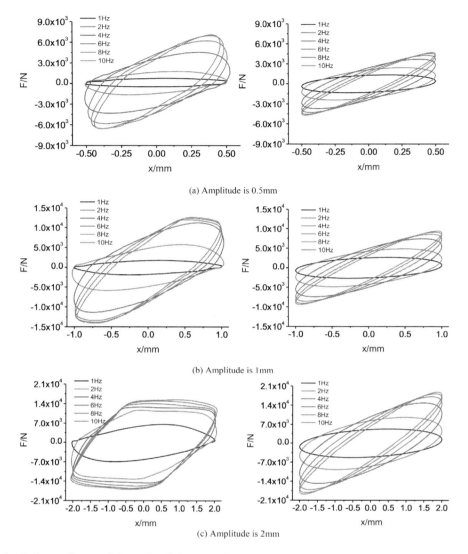

(a) Amplitude is 0.5mm

(b) Amplitude is 1mm

(c) Amplitude is 2mm

Figure 4.   Indicator diagram (left: results of the test; right: results of the Maxwell model).

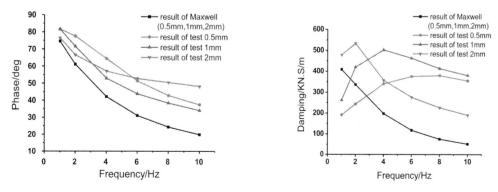

Figure 5.   Phase angle.

Figure 6.   Dynamic damping coefficient.

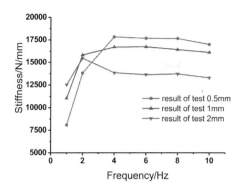

Figure 7. Dynamic stiffness of the test.

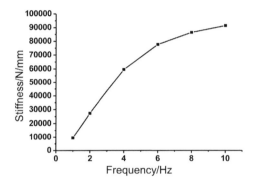

Figure 8. Dynamic stiffness of the Maxwell model.

the Maxwell model on the numerical computation changes over the amplitude, and with the increase in the amplitude, the frequency of the location of the inflection point decreases.

In Figure 7, the dynamic stiffness achieved by the test increasingly varies with the increasing frequency at a low frequency, and then tends to a steady state at the frequency above 4 Hz, resulting in a large change of the value over the amplitude. In Figure 8, the dynamic stiffness calculated by the Maxwell model increases with the increase in the frequency, and does not change in the value over the amplitude. The dynamic stiffness calculated by the Maxwell model is close to the test results at 1 Hz, and then becomes larger at frequencies above 1 Hz.

Through the above analysis, it can be found that the Maxwell model cannot well reflect the dynamic characteristics of the hydraulic shock absorber.

5 SUMMARY

The following conclusions can be drawn by analyzing the dynamic characteristic test data of the hydraulic shock absorber:

1. The phase angle calculated by the Maxwell model decreases gradually along with the increase in the frequency, which is close to the experimental test results.
2. The dynamic damping coefficient calculated by the Maxwell model decreases with the increase in the frequency, but the test results show that the dynamic damping coefficient increasingly varies with the increasing frequency at a low frequency, and then decreases at a high frequency. The Maxwell model cannot reflect the inflection point of the dynamic damping coefficient.
3. The dynamic stiffness calculated by the Maxwell model increases with the increase in the frequency. The test results show that the dynamic stiffness increasingly varies with the increasing frequency at a low frequency, and then tends to a steady state at frequencies above 4 Hz.
4. The Maxwell model cannot well reflect the dynamic characteristics of the hydraulic shock absorber.

ACKNOWLEDGMENTS

This paper was supported by the National High-technology Research and Development Program ("863"Program) of China (2012AA112001–02), the Project of China Railway Corporation (2014J012-C), and the Railway Ministry Science and Technology Research and Development Program [2012G002].

REFERENCES

[1] Y. li. 2007. Applied research of railway hydraulic shock absorber [D]. *Chengdu: Southwest Jiaotong University*.
[2] BS EN13802:2004. Railway applications—Suspension components—Hydraulic dampers [S].
[3] Y.D. Zhang & L.J. Dong. 2004. Simulation of railway locomotive vehicle hydraulic shock absorber system [J]. *Machine Tool with Hydraulic*: 108–109.
[4] C.H. Huang. 2012. Study on Vibration Reduction Technologies for High Speed Cars [D]. *Chengdu: Southwest Jiaotong University*.
[5] G.Z. Yang & F.T. Wang. 2003. Hydraulic Dampers for Railway Rolling Stock [M]. *China Railway Publishing House, Beijing*.
[6] Z.H. lv & S.M. Li. 2002. The development of dynamic characteristics simulation technology of drum liquid resistance of shock absorber [J]. *Journal of Tsinghua University* 42 (11): 1532–1536.
[7] G.D. Lu. 2006. The application of anti-yaw shock absorber in the high-speed trains [J]. *Railway Vehicle* 44 (8): 6–8.
[8] H.B. Gong, G.Z. Yang & F.T. Wang. 2002. Damping characteristic analysis of hydraulic shock absorber [J]. *Railway Vehicle* 40 (12): 6–8.

Material Science and Engineering – Chen (Ed.)
© 2016 Taylor & Francis Group, London, ISBN 978-1-138-02936-1

The diagnosis of fuel injection quantity through the fuel pressure sensor of the common rail in diesel engines

C.L. Xu
China North Engine Research Institute, Tianjin, China

G.Z. Yue
Institute of Engineering (Baotao), College of Engineering Peking University, Baotou, China

Y. Lei
College of Environmental and Energy Engineering, Beijing University of Technology, Beijing, China

ABSTRACT: The diagnosis of fuel injection quantity plays an important role in the common rail fuel system of diesel engines. This paper researches a method to diagnose the fuel injection quantity by analyzing the signal of the pressure sensor. It was concluded that the fuel injection quantity causes the fuel pressure drop in the common rail. The experiment results showed that the rail pressure drop has a linear approximation with the injection quality.

Keywords: diesel engines; common rail; fuel injection quantity; pressure drop

1 INTRODUCTION

With increasing interest in energy saving and environment protection, the improvement of engine performance and pollutant emissions is required [1–2]. One of the key components is fuel injection equipment that determines engine torque, fuel consumption and emissions. Nowadays, the common rail fuel system has been widely applied due to its precise control of rail pressure and injection timing [3–5]. The injector is an important component of the fuel system, delivering the fuel into the cylinder precisely [6]. The injector operates in a harsh environment. On the one hand, high pressure fuel washes the needle valve, holes and sealing cone of the needle, resulting in the change in the structural parameters of the injector. On the other hand, the injector is located inside the combustion chamber, experiencing the high temperature. The thermal stress in the cylinder may influence the abnormal function of the injector. Additionally, injector holes wear and the hole diameter becomes larger because of the erosion of fuel impurities, causing increasing fuel injection quantity and the deterioration of emission [7]. Therefore, the diagnosis of fuel injection quantity plays an important role in the fuel system [8–9].

This paper researches a method to diagnose the fuel injection quantity by analyzing the signal of the pressure sensor on the common rail. A model of fuel pressure in the common rail was developed. Theoretical analysis indicates that impact factors of the fuel rail pressure volatility are engine speed, aim rail pressure and fuel injection quantity. In the engine test bench, the output signal of the fuel rail pressure sensor was analyzed. It was proved that the pressure drop resulted from the fuel injection.

2 THEORETICAL ANALYSIS

The main elements of a common rail fuel system are a low pressure circuit, including the fuel tank and a low pressure pump, a high pressure pump with a delivery valve, a common rail and injectors. The low pressure pump sends the fuel coming from the tank to the high pressure pump. The delivery valve delivers the fuel from the low pressure pump to the high pressure pump, supplying the common rail. Injectors send the fuel from the common rail to the cylinder for combustion.

In this model, the common rail is taken separately to analyze the fuel pressure in the common rail (see Figure 1). The fuel inlet of the fuel rail is the high pressure pump. The fuel outlet of the fuel rail is the injectors and the fuel leakage of the high pressure circuit. The fuel leakage of the high pressure circuit is much less than the fuel injecting amount. Therefore, the fuel leakage of the high pressure circuit can be ignored. The pressure limit valve only opens if the common rail pressure exceeds the maximum pressure. As a result, the pressure of the fuel rail is affected by the high pressure pump and injectors.

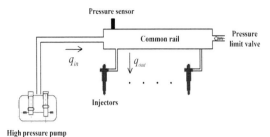

Figure 1. A block scheme of the fuel inlet and outlet in the common rail.

In this model, considering that the diesel fuel is compressible and the mechanical deformation of the fuel rail is negligible, the fuel in the rail can be expressed by the bulk modulus of elasticity as follows:

$$K_f = \frac{dp}{d\rho/\rho} \qquad (1)$$

where ρ is the fuel density, and $d\rho$ is the density increase of a unit volume of liquid due to a dp pressure increase. K_f and p have the same units. In normal operating conditions, K_f is set equal to 12000 bar and can be related to the fuel pressure p by the following expression.

$$K_f = 12000 \cdot \left(1 + \frac{p}{1000}\right) \qquad (2)$$

The fuel pressure changes in the rail can be expressed as follows:

$$\frac{dp}{dt} = \frac{K_f}{V} \cdot (q_{in} - q_{out}) \qquad (3)$$

where V is the volume of the fuel rail; q_{in} is the fuel inlet from the high pressure pump and q_{out} is the fuel outlet of the injectors.

At the steady engine speed and the fuel rail pressure, the fuel inlet from the high pressure pump is determined by a delivery valve that has a different opening degree at a different driving current. The fuel outlet of the injectors includes fuel injection and fuel leakage. Because the fuel leakage is much less than the fuel injection, it is not considered in this model.

Therefore, we get Equation (4) as follows:

$$\frac{dp}{dt} = \frac{12000}{V} \cdot \left(1 + \frac{P_{aim}}{1000}\right) \cdot (q_{pump} - q_{inj}) \qquad (4)$$

where q_{pump} is the fuel inlet from the high pressure pump; q_{inj} is the fuel outlet of the injectors; and q_{aim} is the aimed fuel pressure.

From Equation (4), it is concluded that the fuel pressure in the fuel rail is affected by the aimed fuel pressure, the high pressure pump and the injectors. Because the response time of the fuel injector solenoid valve is much smaller than the delivery valve on the high pressure pump, the fuel injection could cause the fuel rail pressure to drop, which would be compensated by the pressure pump with a time delay.

3 EXPERIMENTAL AND TEST RESULTS

To verify that the fuel injection can cause a pressure drop of the fuel in the rail, a series of the test under various operating conditions were conducted at the engine test bench. The rail pressure was measured by the pressure sensor in the rail. Wei Chai WP10 diesel engine was used. Table 1 outlines the detailed engine specifications.

At the engine test bench, the injectors of cylinder 1 to 5 are shut off and the aimed fuel injection of cylinder 6 is 46 mm³/cycle. The engine speed is 2000 r/min. As shown in Figure 2, the period of an operating cycle is 60 ms. From 10 ms to 12 ms, the rail pressure is 1200 bar at the beginning and then declines sharply to 1155 bar. The pressure drop of 45 bar is caused by the fuel injection. In order to verify it, the injection rate was measured and calculated (see Figure 3). It was found that the rail pressure dropped during the injection process. Therefore, the fuel pressure drop from 10 ms to 12 ms resulted from the fuel injection. Then, the rail pressure was kept around 1155 bar, and the fluctuation was less than 5 bar from 12 ms to 22 ms. The fuel pressure rose up after 22 ms due to the increasing fuel supply amount in the high pressure pump. The fuel injector responded much more quickly than the delivery valve of the fuel pump. Therefore, fuel injection caused the rail pressure drop, and the high pressure pump compensated the pressure drop in a delay time.

In Figure 2, by recording the rail pressure drop caused by the fuel injection, the average pressure drop of 10 continuous operating cycles was taken. Figure 4 shows the relationship between the rail pressure drop and the injection quantity. It was found that the rail

Table 1. Engine specification.

Property	Value
Number of cylinder	6
Displacement (L)	9.726
Rated power (kW/(rpm))	247/2200
Maximum torque ((Nm)/(rpm))	1250/1200–1600
Air intake	Turbocharged, intercooled

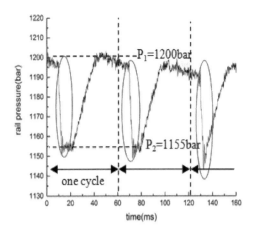

Figure 2. Rail pressure under engine speed of 2000 r/min and single cylinder fuel quantity of 45 mg/cycle.

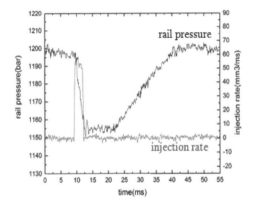

Figure 3. Rail pressure and fuel injection rate under the engine speed of 2000 r/min and the single cylinder fuel quantity of 45 mg/cycle.

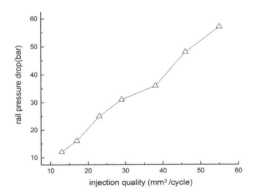

Figure 4. Rail pressure drop against injection quantity.

pressure drop increased as the injection quantity was greater. The linearity of the rail pressure drop and the injection quantity was good, and it could be used as a method to diagnose the fuel injection.

4 CONCLUSIONS

This paper concluded theoretically that the pressure in the common fuel rail is affected by the aimed fuel pressure, the high pressure pump and the injectors. It is proposed that the linearity of the rail pressure drop and the injection quantity was good based on the experimental results.

ACKNOWLEDGMENT

This research was financially supported by the National Science Foundation of the Inner Mongolia Autonomous Region (2014MS0501).

REFERENCES

[1] Salvador F.J., Martínez-López J., Caballer M., et al. Study of the influence of the needle lift on the internal flow and cavitation phenomenon in diesel injector nozzles by CFD using RANS methods [J]. Energy Conversion and Management, 2013, 66: 246–256.

[2] Qiu T., Li X., Liang H., et al. A method for estimating the temperature downstream of the SCR (selective catalytic reduction) catalyst in diesel engines [J]. Energy, 2014, 68: 311–317.

[3] Xu-Guang T., Hai-Lang S., Tao Q.I.U., et al. The Impact of Common Rail System's Control Parameters on the Performance of High-power Diesel [J]. Energy Procedia, 2012, 16: 2067–2072.

[4] Chen P.C., Wang W.C., Roberts W.L., et al. Spray and atomization of diesel fuel and its alternatives from a single-hole injector using a common rail fuel injection system [J]. Fuel, 2013, 103: 850–861.

[5] Guan L., Tang C., Yang K., et al. Effect of di-n-butyl ether blending with soybean-biodiesel on spray and atomization characteristics in a common-rail fuel injection system [J]. Fuel, 2015, 140: 116–125.

[6] Asi O. Failure of a diesel engine injector nozzle by cavitation damage [J]. Engineering Failure Analysis, 2006, 13(7): 1126–1133.

[7] Lowe D.P., Wu W.L., Tan A.C.C. Experimentally induced diesel engine injector faults and some preliminary acoustic emission signal observations. In: World congress on engineering asset management, Cincinnati, USA, 3–5 October, 2011.

[8] Qiu T., Li X., Lei Y., et al. The prediction of fuel injection quantity using a NOx sensor for the on-board diagnosis of heavy-duty diesel engines with SCR systems [J]. Fuel, 2014.

[9] Luján J.M., Bermúdez V., Guardiola C., et al. A methodology for combustion detection in diesel engines through in-cylinder pressure derivative signal [J]. Mechanical Systems and Signal Processing, 2010, 24(7): 2261–2275.

Material Science and Engineering – Chen (Ed.)
© *2016 Taylor & Francis Group, London, ISBN 978-1-138-02936-1*

Novel friendly hydrophobic-associated hydrogel AM-co-AEO-AC: Particular behavior and extraordinary mechanical properties

T.T. Gao, H. Huang, G. Gao & F.Q. Liu
College of Chemistry, Jilin University, Changchun, China

ABSTRACT: A new hydrophobic-associated hydrogel with controllable mechanical strength and environmental-friendly properties was successfully prepared by micelle co-polymerization. AM-co-AEO-AC-gel exhibits exceptional mechanical and quick self-healing properties (AEO: fatty alcohol ethylates). Three new hydrophobic monomers AEO-AC were synthesized by the reaction of chloride: (1) heterogeneous AEO-AC-10-05 (number of carbon in the carbon chain 10; number of poly-ethylene oxide in the PEG chain 5), (2)heterogeneous AEO-AC-13-10 (as (1)), and (3) straight AEO-AC-13-05 (proportion of straight AEO-AC-12-05 and AEO-AC-14-05 1:1). All the above compounds were characterized by Fourier transform infrared spectroscopy. In this paper, hydrophilic monomer acrylamide was AM (10 wt%) and hydrophobic monomer was AEO-AC (1%~5% mol/AM). Their mechanical properties and swelling behaviors were studied. The results show that the 13-05-gel exhibits a high mechanical strength and non-degradable in water, which is due to their relatively homogeneous compact hydrophobic association network structure. The other dissolved in water within two months.

Keywords: hydrophobic association; controllable; self-healing

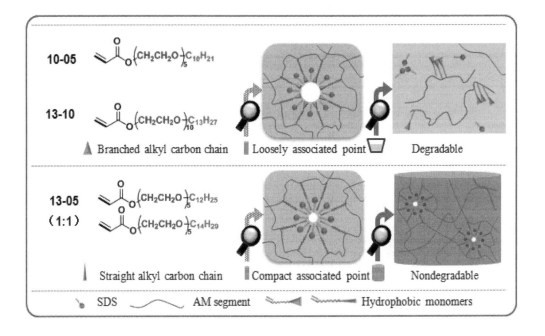

1 INTRODUCTION

Hydrogels have received significant attention for their challenging application in drug delivery, tissue engineering, and high-performance liquid chromatography, intelligent sensors such as temperature pH-sensitive. Furthermore, the regulation of the subtle structure and the suppression of network heterogeneity can improve the fragile of hydrogels, and provide them with multifunctionality. Thus, it is a worthwhile issue to develop the structure and network of the hydrogel as we know: tree-branching gel, NC gel, π-π stacking interaction, certain metal-ligand coordination bonds, hydrogen bonding, host-guest monomers and ion interaction.

Hydrophobic structure was first referred by Miguel F. Refojo in 1967. He found a secondary structure in the poly(2-hydroxyehtyl methacrylate) homogeneous hydrogel. The author assumed *a priori* that the hydrophobic interaction between the methyl groups in the position may also be the principal factors contributing to the stability of the secondary structure of the PHEMA hydrogel. There are three major network factors: (a) hydrophobic group content; (b) the distribution of the hydrophobic group in the hydrophilic chain; and (c) the length of the hydrophobic micro-block in the hydrophilic chain.

Based on these findings, our group designed a kind of hydrophobic-associated hydrogel, and successfully synthesized via micellar co-polymerization. The first OP-AC (OP: nonylphenol polyoxyethylene ether) gel has almost ideal properties: transparent, outstanding mechanical properties. The associated micelles acted as physical crosslinking points in the network of HA gels, so the mechanical properties could be controlled, particularly the stress and elastic modulus, as well as the lifetime in water. Moreover, OP-7-AC has pH-sensitive properties.

Here, we first use AEO (fatty alcohol ethylates) replaced OP (nonylphenol ethylates) because it is a non-toxic and non-polluting widespread non-ionic surfactant (without benzene). Then, we synthesize three new hydrophobic monomers by the reaction of chloride: AEO-AC-10-05, AEO-AC-13-05, AEO-AC-13-10. The mechanical properties were obtained through tensile and swelling test by the control variate method.

2 EXPERIMENTAL PROCEDURE

2.1 *Materials*

Polyoxyethylene ether fat (AEO-A-E, A = 10, 13, E = 05, 10, where A is the number of alkyl group, E means the total number of ethoxy units in a molecule, CP grade) was provided by Jiangsu Haian Petrochemical Plant, China. Sodium Dodecyl Sulfate (SDS, CP grad) was provided by Tianjin Guangfu Fine Chemical Research Institute. Acryloyl Chloride (AC, CP grad) was purchased from Shanghai Haiqu Chemical Co., China. These reagents were used without further purification. Acrylamide (AM) and potassium persulfate (KPS) was provided by Tianjin Fuchen Chemical Reagent Factory, China, and recrystallized with distilled water before use. Triethylamine (TEA) was provided by Tianjin Fuyu Chemical Reagent Factory, China. Dichloromethane CH_2Cl_2 was provided by Beijing Chemical works. These agents were purified by distillation.

2.2 *Synthesis of hydrophobic monomers*

The AEO-A-E-AC gels were synthesized as follows: 0.10 mol AEO-A-E, 0.12 mol TEA, 0.15 mol AC, and 160 mL CH_2Cl_2 were added to a three-neck flask with an electromagnetic stirrer in an ice-salt water bath. When a homogeneous solution was achieved, 0.12 mol AC in 27 mL CH_2Cl_2 was added dropwise to the flask under stirring, and the temperature of the water bath maintained between −5 and 0°C. After the dropwise addition was complete, the reaction was carried out for 5 h, and CH_2Cl_2 was removed by rotary evaporation. Then, to separate TEA hydrochloride that was the sediment of the reaction system, the appropriate amount of acetone was added into the mixture, and the upper clear liquid containing AEO-A-E-AC was distilled under reduced pressure to remove diethyl ether. Finally, the residual TAC in the product was separated by centrifugation and the final product AEO-A-E-AC was dried to constant weight in vacuum at 50°C.

2.3 *Synthesis of HA gels*

The HA gels were synthesized by micellar co-polymerization. The reaction system generally consisted of water-soluble monomers, hydrophobic monomers, surfactants and water. As a typical example, AM (1.00 g), AEO-AC (varied), SDS (varied), and distilled water were added to a beaker, in which the total mass of the reaction solution was 10 g. The mixture was treated ultrasonically until a homogeneous solution was achieved. Then, 0.50 mL of the KPS solution (0.01 g/mL) was added to the above solution. Afterwards, the solution was put into a test tube and removed dissolved oxygen with N_2 for 10 min under normal pressure, and reduced the pressure to dispel the air bubbles. Immediately, the test tube was sealed. After being placed at room temperature for 1 h, the test tube containing the reactant solution was heated to 50°C in a water bath for 12 h, and then the HA gels with high mechanical properties were prepared.

Table 1. The major carbon chain composition of three AEO surfactants.

AEO-AC-10-05		AEO-AC-13-10		AEO-AC-13-05	
DB*2.1		DB 2.9		DB 0	
CC**	Content	CC	Content	CC	Content
C9	4%	C12	21%	C12	50%
C10	89%	C13	70%	C14	50%
C11	3%	C14	6%		

*DB, degree of branching.
**CC, major carbon chain length.

Table 2. The content of compositions in the initial reaction solution for the HA gels and naming.

	AM	AEO	
Group	%wt.	%mol	SDS
AEO-A-E-AM-X% (Group-AM for short X = 8, 10, 12, 14, 16)	X*	2	3% wt.
AEO-A-E-AEO-Y% (Group-AEO for short Y = 1, 2, 3, 4, 5)	10	Y**	3% wt.
AEO-A-E-SDS-Y%-R (Group-SDS for short R = 0.5, 1.0, 2.0, 3.0, 4.0)	10	1	R***

*X wt% of hydrophilic monomer AM to total solution.
**Y mol% of AEO-AC-A-E to AM.
***R is the mole percentage of the surfactant SDS to hydrophobic monomer AEO-AC.
****KPS is the initiating agent, 10 mg/mL.
*****Each HA-gel prepared solution is 15 g, in addition to compositions on the table, the rest are distilled water.

2.4 Characterization

Fourier transform infrared spectra were recorded in the wavenumber range of 4000~500 cm^{-1} by using a Nicolet 360 FT-IR spectrophotometer. The samples were categorized into three kinds of AEO samples and corresponding AEO-AC samples. They were prepared by drying under ultraviolet light.

2.5 Measurements of mechanical properties

The tensile stress-strain measurements were performed by using a universal tester at 25°C (AG-I 1KN, Shimadzu, Japan). The tensile experiment was carried out at a cross-head speed of 100 mm/min. The cylindrical gel samples were 6 mm in diameter and the original length between the top and foot clamps was 20 mm.

2.6 Measurements of swelling

Swelling experiments were performed by immersing as-prepared AEO-A-E gels (initial size Φ6 mm × 5 mm length) in a large excess of distilled water at room temperature, and distilled water was replaced every day. In each measurement, the samples were removed from distilled water and weighed after removing excess distilled water, and then these samples were dried to constant weight in oven at 50°C. The swelling degree was expressed by the ratio of the weight of the swollen hydrogel to its corresponding dried gel weight, and the gel fraction was expressed by the ratio of the dried gel weight after a certain swelling time to its theoretical dried gel weight. Here, the theoretical dried gel weight means the theoretical solid weight based on the experimental recipe.

We used AEO-SDS-1.0%-1.0 gels to determine their swelling properties. The Swelling degree (Sr) and the remaining Gel Fraction (GF) were calculated by the following equations:

$$Sr = \frac{W_t}{W_d} \times 100\% \tag{1}$$

$$GF = \frac{W_d}{\frac{\Sigma(W_A + W_B + W_C + W_D + W_E)}{\Sigma(W_a + W_b + W_c + W_d + W_e)} \times W_0} \times 100\% \tag{2}$$

where W_t is the weight of the swollen hydrogel sample at the regular time interval t; W_d is the weight of the corresponding dry gel; W_0 is the ordinary weight of the hydrogel. Five unswelled gels in each group were dried directly after being cut, where W_A ~W_E is the weight of the unswelled gel after drying and W_a ~W_e is the ordinary weight of these gels.

3 RESULTS AND DISCUSSION

3.1 Structural characterization

In the IR spectrum of AEO and AM-co-AEO-AC-gel, (1) a broad peak appeared at 3473 cm^{-1} intermolecular hydrogen bond; this peak was found in AEO but not in AEO-AC, thus—OH was destroyed; (2) the peak at 1731 cm^{-1} was the carbonyl stretching vibration for ester; this peak was found in AEO-AC but not in AEO; thus, eater construction was created; (3) the peak at 1410 cm^{-1} contained a substituent carbon double bond in-plane vibration peak shear deformation, only in AEO; (4) the peak at 1195 cm^{-1} corresponded to the asymmetrical stretching vibration of C-O-C. Steps (2)~(4) show the evidence for AEO-AC.

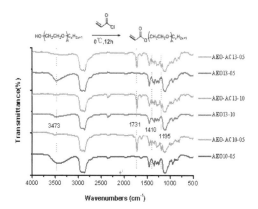

Figure 1. FT-IR spectra of AEO-AC and AEO.

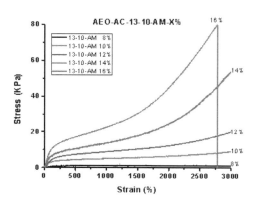

Figure 2. The stress-strain curves of the gels with varied AM content.

3.2 Varying the content of AM-co-AEO-AC

Figure 2 shows the stress-strain curves of the Group AM. We fixed the ratio of hydrophobic monomer AEO-AC to hydrophilic monomer (2% mol to AM) and kept the content of SDS (3% wt). Obviously, as the amount of the two monomers increased proportionately, the stress of the gels also increased. In our testing range, the elongations of these gels were totally stretched to 3000%, but the AEO-13-10-AM-16% had a shorter elongation 2773%.

First, by increasing proportionately these two monomer contents, the network would be more tight and strong. Then, the stress was increased proportionately by adding the AM content. In the test, the Max_Stress had 80.00 KPa of AEO-13-10-AM-16%, as the Max_Stress was 1.61 KPa of AEO-13-10-AM-8%. From 8% to 16%, twice the AM content, the stress improved 50~60 times.

3.3 Varying hydrophobic monomer AEO-AC content

First, by increasing the content of the hydrophobic monomer as SDS was full dose, the amount of the hydrophobic monomer in each SDS micelle increased. In addition, the content of AM was the same as in this Group-AEO. The length of the AEO-AC micro-block in the AM carbon chain was longer. In Figure 3b and 3c, we could see that the stress was stronger and the utmost elongation decreased as the AEO-AC monomer increased from 1% to 5%.

The similar tendency is also shown in Figure 3a, but did not last when raising the content of AEO-AC

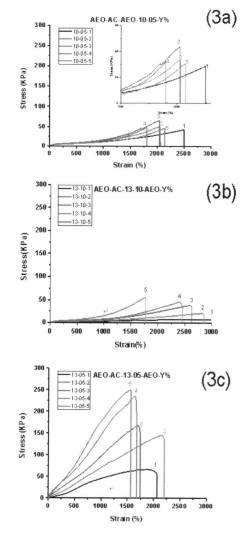

Figure 3. Stress–strain curves of Group-AEO.

above 3%. In fact, even though we added the content from 3% to 5%, the stress declined and the utmost elongation increased. AEO-AC-10–05 had a shorter hydrophilic polyoxyethylene chain than 13–10, though 13–10 had a longer alkyl chain. Finally, 10–05 showed a higher hydrophobicity and higher mechanical properties, similar to the AEO-AC mole content. However, 13–05 had a straight carbon chain and a similar alkyl carbon chain as the surfactant sodium dodecyl sulfate. As a result, 13–05 had the highest mechanical properties.

3.4 Varying the molecule ratio of SDS to AEO-AC

Figure 4 shows the stress-strain curves of Group-SDS. As the ratio R (SDS: AEO mol%) increased from 0.5 to 1.0, the Max_Stress also increased but the utmost elongation decreased. In the range of R from 1.0 to 3.0, the Max_Stress declined but the utmost elongation increased. When the R value was 0.5, the SDS was not sufficient enough to incorporate all the hydrophobic monomers into SDS micelles, leading

to reduced associated points and a weak network of the 0.5-gel. The gel had an ideal property when the R value was equal to 1.0. The hydrophobic monomers were dissolved in SDS micelles, so the hydrophobic micro-segment length was longer and the segment distribution was good. Continuous addition of more SDS in the system may lead to more free SDS in the solution. These free SDS and SDS empty micelles in the system, providing a negative effect on the mechanical strength of the gels, but had a positive effect on the elongation of the gels.

3.5 Swelling test

The AEO-13–05-SDS-1.0%-1.0 gels maintained their shape in water and had a little small Sr value

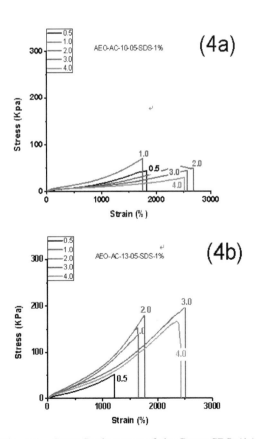

Figure 4. Stress-Strain curves of the Group-SDS: (4a) AEO-AC-10–05, (4b) AEO-AC-13–05.

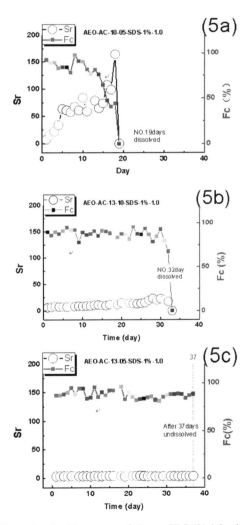

Figure 5. Swelling curves of Group-SDS-1%-1.0 gels.

in the range of 3~5, as the Fc changed to a small extent. They had a tight well-distributed network because the straight carbon chain had little space steric hindrance and could provide a more stable and strong AEO-SDS micelle. Finally, these strong micelles became strong associated points, so that the gel cannot be dissolved in water.

In Figure 5a and 5b, we could see that the gel had a branched alkyl-chain hydrophobic monomer AEO-AC dissolved after 19days (10–05) and 32days (13–10).

Compared with the above gels in Figure 5c, due to steric hindrance, the branched alkyl-chain would form a loose micellar structure with SDS. So, the associated points were not compact and became weaker. As a result, the "branching gel" will be dissolved.

3.6 *Self-healing*

The period of hydrogel rod ($\Phi6$ mm $\times$ 4 cm) was cut into four even parts, dyeing two of them in blue with methylene blue to observe clearly.

We put four parts onto the horizontal table closely without the external force. After 1 minute, they were self-healed quickly and could support their weight, so that we could pick up "the one rod" by a tweezer (see Figure 6a). After 24 h, two parts were almost totally healed, which were wrapped together closely by plastic wrap on a horizontal table (see Figure 6b).

The associated points of the HA gel were not immutable. Under the action of free surfactant SDS, there was an association-disassociation equilibrium in the gel system. Thus, it could heal by building new associated points in two close cross-sections.

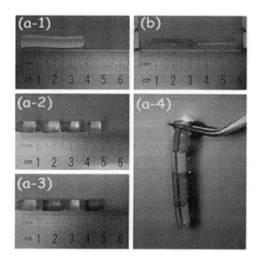

Figure 6. Self-healing gels.

4   CONCLUSIONS

We successfully prepared a new kind of HA-gel with self-healing and environmental-friendly properties. It has a certain strength and excellent ductility, and we can control the strength of the hydrogels by adjusting the hydrophobic-associated points.

Keeping other conditions unchanged, the concentration of the hydrophobic monomer, which can be soluble in this scope, was only increased. The strength of hydrophobic-associated domains increases. However, with the increase in the number of hydrophobic groups, and the number of hydrophilic backbone being unchanged, both of these result in an uneven distribution of the hydrophobic portion, more likely to agglomerate. Finally, we obtain the gels with an unbalancing network. Although the modulus increased, the mechanical properties declined. If we increased the amount of AM and the AEO in the same proportion, the content of the hydrophobic group will increase uniformly, and can be uniformly distributed as well. Then, the stress of the gel will grow exponentially. With the same and sufficient amount of SDS, AEO-AC-13-10 gels exhibited better mechanical properties. In this case, the length of the microblocks plays a major role; while the molar ratio of SDS and AEO is no more than 4, AEO-AC-10-05 gel exhibits a more greater stress. Then, the uniform distribution of hydrophobic domains plays a leading role. As for heterogeneous AEO-AC gels, it is obvious that the linear AEO-AC has a more uniform gel network, such gels will definitely exhibit more higher mechanical properties.

Under normal circumstances, those materials can be soluble in water, which are called sols, while nonsoluble materials are called as gels. This gel, nonetheless, can survive in water for about two months, and finally it can be completely degraded in the water. So, it can be called "pseudo—gel". We studied their mechanical tensile properties and swelling properties, and estimated their network structure simulation parameters according to the Mooney curve. We tried to establish a preparation model from the micro to the macro level. All of these results will open up new ideas for our future sol-gel research.

REFERENCES

[1] Arıca, M.Y., Gülsu, A.Ö., Denizli, A. 2001. Novel hydrophobic ligand-containing hydrogel membrane matrix: preparation and application to γ-globulins adsorption. *Colloids and Surfaces B: Biointerfaces* 21: 273–283.

[2] Burnworth, M., Tang, L., Kumpfer, J.R., Duncan, A.J., Beyer, F.L., Fiore, G.L., Rowan, S.J., Weder, C. 2011. Optically healable supramolecular polymers. *Nature* 472: 334–338.

[3] Burattini, S., Colquhoun, H.M., Fox, J.D., Friedmann, D., Greenland, B.W., Harris, P.J.F., Hayes, W., Mackay, M.E., Rowan, S.J. 2009. A self-repairing, supramolecular polymer system: heal ability as a consequence of donor-acceptor pi-pi stacking interactions. *Chemical Communications* 44: 6717–6719.

[4] Chen, L. 2012. *Study on the synthesis and properties of isomeric fatty alcohol polyoxyethylene ether and their derivatives.* Jiangnan University in Wuxi: master thesis in China National Knowledge Internet.

[5] Gou, M.L., Li, X.Y., Dai, M., Gong, C.Y., Wang, X.H., Xie, Y., Deng, H.X., Chen, L.J., Zhao, X., Qian, Z.Y., Wei, Y.Q. 2008. A novel injectable local hydrophobic drug delivery system: biodegradable nanoparticles in thermo-sensitive hydrogel. *International Journal of Pharmaceutics* 359: 228–233.

[6] Haraguchi, K., Takehisa, T. 2002. Nanocomposite hydrogels: a unique organic-inorganic network structure with extraordinary mechanical, optical, and swelling/de-swelling properties. *Advanced Materials* 14: 1120–1124.

[7] Jiang, G.Q., Liu, C., Liu, X.L., Zhang, G.H., Yang, M., Chen, Q.R., Liu, F.Q. 2009. Construction and properties of hydrophobic association hydrogels with high mechanical strength and reforming capability. *Macromolecular Materials and Engineering* 294: 815–820.

[8] Jiang, G.Q., Liu, C., Liu, X. L., Zhang, G.H., Yang, M., Chen, Q.R., Liu, F.Q. 2010a. Self-healing mechanism and mechanical behavior of hydrophobic association hydrogels with high mechanical strength. *Journal of Macromolecular Science, Part A: Pure and Applied Chemistry* 47: 335–342.

[9] Jiang, G.Q., Liu, C., Liu, X.L., Zhang, G.H., Yang, M., Chen, Q.R., Liu, F.Q. 2010b. Network structure and compositional effects on tensile mechanical properties of hydrophobic association hydrogels with high mechanical strength. *Polymer* 51: 1507–1515.

[10] Karinoa, T., Masuib, N., Hiramatsuc, M., Yamaguchic, J., Kuritac, K., Naito J. 2002. Stabilization of hydrophobic domains in hydrogel by intermolecular hydrogen bonds between carboxylic groups at the distal end of alkyl side-chain. *Polymer* 43: 7467–7475.

[11] Liu, C., Yu, J.F., Liu, X.L., Li, Z.Y., Gao, G., Liu, F.Q. 2013. *Journal of Material Science* 48: 774–784.

[12] Refojo, M.F. 1967. *Journal of Polymer Science: Part A-1,* 5: 3103–3113.

[13] Varshosaz, J., Falamarzian, M. 2001. Drug diffusion mechanism through pH-sensitive hydrophobic/polyelectrolyte hydrogel membranes. *European Journal of Pharmaceutics and Biopharmaceutics* 51: 235–240.

[14] Volpert, E., Selb, J., Candau, F. 1998. Association behavior of poly-acrylamides hydrophobic ally modified with dihexylacrylamid. *Polymer* 39: 1025–1033.

[15] Wang, Q., Mynar, J.L., Yoshida, M., Lee, E.J., Lee, M.S., Okuro, K., Kinbar, K., Aida, T. 2010. High-water-content mouldable hydrogels by mixing clay and a dendritic molecular binder. *Nature* 463: 339–343.

[16] Wu, C.J., Gaharwar, A.K., Chan, B.K., Schmidt, G. 2011. Mechanically tough pluronic F127/laponite nanocomposite hydrogels from covalently and physically cross-linked networks. *Macromolecules* 44: 8215–8224.

[17] Xing, B.G., Yu, C.W., Chow, K.H., Ho, P.L., Fu, D.G., Xu, B. 2002. Hydrophobic interaction and hydrogen bonding cooperatively confer a vancomycin hydrogel: a potential candidate for biomaterials. *Journal of the American Chemical Society* 124: 14846–14847.

[18] Yang, M., Liu, C., Li, X.L., Gao, G., Liu, F.Q. 2010. Temperature-responsive properties of poly (acrylic acid-co-acrylamide) hydrophobic association hydrogels with high mechanical strength. *Macromolecules* 43: 10645–10651.

[19] Zheng, Y., Hashidzume, A., Takashima, Y., Yamaguchi, H., Harada, A. 2012. Switching of macroscopic molecular recognition selectivity using a mixed solvent system. *Nature Communications* 3: 813.

Material Science and Engineering – Chen (Ed.)
© 2016 Taylor & Francis Group, London, ISBN 978-1-138-02936-1

Optimization design for the driving shaft of walking beam cooling bed transmission system

W.Y. Yu & Y. Shang
Institute of Mechanical Engineering, Inner Mongolia University of Science and Technology, Inner Mongolia, Baotou, China

ABSTRACT: This paper focuses on the transmission system of cooling bed used by the rolling production line in some steel plants as the research object. The supporting force of many fixed bearing supports, which exist in the transmission system, cannot be manually calculated. So, the application of the dynamics simulation software (ADAMS platform) can simulate the motional characteristics of the transmission system, and calculate the bearing force and torque on the transmission shaft on the basis of the simulation result. In addition, according to the calculation result, it can be found that the journal value is less than the original journal value. Therefore, the similar principles and methods proposed in this paper can solve the statically indeterminate problem, save the raw materials of transmission shaft, and provide a theoretical basis for the optimum design of similar transmission shafts.

Keywords: walking beam cooling bed; statically indeterminate structure; operation cycle cooling bed; kinematic analysis

1 INTRODUCTION

Transmission system plays a vital role in cooling bed equipment, in which the kinematic performance will directly affect the safety and performance of the cooling bed. The transmission form of cooling bed is that the motor actuates two turbine reducers through the connecting shaft and then the double output shafts of the turbine reducer that drives the long shaft rotating. Also, the eccentric wheel installed in the transmission long shaft can finally realize the step motion of the cooling bed.

The maximum journal value of the dangerous section of the transmission shaft is 180 mm. The previous design for the transmission shaft, at home and abroad, is by the manual calculation with a higher safety factor, by which the journal value is relatively larger. It can satisfy the using requirement but is not economical. Now the application of the dynamics simulation software to simulate the working process of the transmission shaft can calculate the bearing force at each point of the transmission shaft and the journal value of the dangerous section of the transmission shaft. This method can reap certain economic benefits by saving the raw material of the transmission shaft.

2 MOTION ANALYSIS

2.1 *Technical specifications of cooling bed*

The total length of the transmission system is 120 meters in length and 6 meters in width. Besides, two sets of arrangement are, respectively, installed at the position −30 m away from two ends of the transmission system. There are two forms of transmitting long shaft, namely single idler roller and dual idler roller. Each of the long shafts is composed of 20 segments connected by rigid coupling, supported by 40 fixed bearings. A three-dimensional model is shown in Figure 1.

Cooling bed is driven by a variable frequency motor, which can control the cooling bed starting with a constant acceleration and stopping with a constant deceleration. The action cycle of cooling bed is

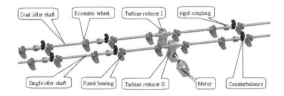

Figure 1. Partial schematic diagram of the transmission system.

Table 1. Technical specifications of the cooling bed.

Name	Value	Unit
The mass of each segment of the active system M_r	3899	kg
The maximum mass of the metal on the rack M_m	25000	kg
Rack of a working trip time	5	s
Rack of teeth	51	
The distance between the cogging	100	mm
The distance between the rack	1000	mm

5 s and the stepping eccentricity is 50 mm, with bar specifications ø of 12~45 mm. The technical specifications of the cooling bed are outlined in Table 1.

2.2 Action cycle of the cooling bed

In controlling the cooling bed starting, the transmission shaft deviates from the lowest point at an angle A of $\pi/12$; at this moment, counterbalance provides additional starting torques, thus reducing the motor's starting torque. As figure 2 shows, from "0" bit to "1" bit is the angle A_1 that the cooling bed runs from the beginning to a constant speed. When the eccentric wheel moves to "2" bit, the dynamic rack and the static rack coincide to the point of the towed steel point of the cooling bed. From "1" bit to "5" bit is the uniform motion and "4" bit is the cooling bed release steel point; "5" bit is the brake initiation and the brake stops when reaching "0" bit, and the brake angle is A_2.

2.3 Motion calculation of the cooling bed

$$\begin{cases} \frac{1}{2}\alpha_1 t_1^2 + \alpha_1 t_1(t_3 + t_4 + t_5) + \frac{1}{2}\alpha_2 t_2^2 = 2\pi \\ t_1 + t_2 + t_3 + t_4 + t_5 = 5 \\ \frac{1}{2}\alpha_1 t_1^2 = A_1 \\ \frac{1}{2}\alpha_2 t_2^2 = A_2 \\ \alpha_1 t_1 \cdot t_3 = A_3 \end{cases} \quad (1)$$

In formula (1), α_1 is the angular acceleration; α_2 is the angular deceleration and A is the start angle of $\pi/12$, $A_1 = A_2 = \pi/3$, $A_3 = \pi/12$.

Movement results calculated according to formula (1) are summarized in Table 2. Transmission system suffers the greatest impact at the towed steel point where the transmission shaft bears the maximum support force and torque. Within the rotation period, the transmission shaft ranges from uniform acceleration to uniform to uniform

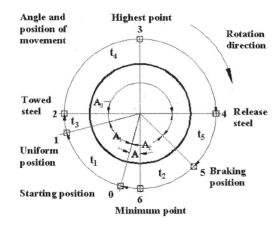

Figure 2. Cooling bed action cycle.

Table 2. Movement calculation results.

Term	Results	Unit	Remark
$\alpha_1 = \alpha_2$	1.34	rad/s2	
t_1	1.25	s	Uniform acceleration
t_3	0.156	s	Uniform
t_4	1.875	s	Uniform
t_2	1.25	s	Uniform deceleration
n	16	r/min	Uniform rotation speed

deceleration, and the towed steel point time $t_1 + t_3 = 1.406$ s, reaching the highest point time $t_1 + t_3 + t_4/2 = 3.344$ s, time for release steel point $t_1 + t_3 + t_4 = 3.281$ s. The impact coefficient at the point of the tow steel is K = 1.5; with a total of 80 eccentric wheels supporting the weight of 20 group activities racks and load in the two sets of parallel transmission shafts. Maximum load and resistance moment can be calculated for each eccentric wheel as follows:

$$\begin{cases} F_{max} = \frac{(20M_r + K \cdot M_m) \cdot g}{80} \\ M_{max} = F_{max} \cdot r \end{cases} \quad (2)$$

The values of M_r and M_m are given in Table 1; K is the impact coefficient, K = 1.5; and r is the distance from the tow steel point to the center of rotation, r = 245 mm.

3 THE ESTABLISHMENT OF THE ENTITY MODEL

This paper uses Solidworks to establish the three-dimensional entity model of the demanding parts, to set material properties and then to assemble them according to the constraint condition. The assembled model was then imported into the kinematic analysis software-ADAMS to add constraints and loads. According to the movement of the transmission shaft in a rotation period of time, we use the IF function compiling the function rotational speed during the operation, and the direction of rotation is consistent with the eccentric wheel initial deflection direction. We select acceleration as the type of motion and then add the driver functions as IF (time −1.25: 76.8 d, 76.8 d, if (time −3.75: 0 d, 0 d, if (time −5: −76.8 d, −76.8 d, 0))). The simulation time is set to a campaign period T = 5 s, step = 50 and the model is validated to ensure the accuracy of the operation.

4 ANALYSIS OF THE RESULTS

There are two transmission drives, in which configuration and installation are perfectly symmetrical with each other. By comparing the simulation results of both inside and outside of the segment shaft in the two transmission devices, the larger bearing force and torque on the shaft are selected as the analysis objects. The outside shaft represents the shafts between the transmission device and the adjacent end portion, and the inside shaft represents the shafts between the two sets of transmission devices. The transmission shaft force diagrams are shown in Figure 3 and Figure 4, and the torque graphs between the transmission shaft are shown in Figure 5 and Figure 6.

As can be seen from Figure 3 and Figure 4, the difference in the value between the biggest support force of the inside and outside transmission shafts is very small, both occurring near the transmission device position. The difference in the middle part support force is not very significant.

It is known from Figure 5 and Figure 6 that the maximum torque values occur in the t = 1.4 s position, that is, the time of the tow steel point. The torque of both sides of the segment shaft changes regularly, and the closer the transmission device, the larger the

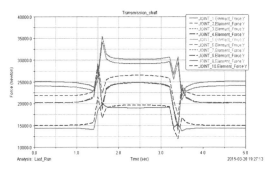

Figure 3.   Force diagrams of the inside shaft.

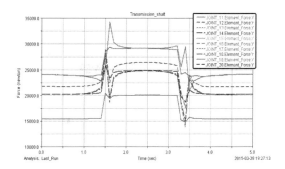

Figure 4.   Force diagrams of the outside shaft.

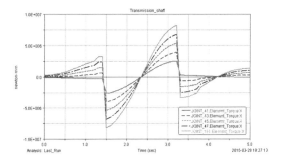

Figure 5.   Torque curves of the inside shaft.

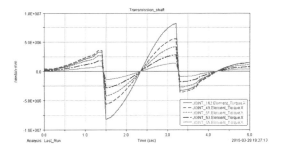

Figure 6.   Torque curves of the inside shaft.

value of torque. Comparing the torque value of the outside transmission shaft with the inside, the maximum torque value occurs in the inside transmission shaft near the transmission device.

5 CONCLUSION

Force and torque values of the transmission shaft can be calculated according to the simulation results. After an analysis of the force bored on the shaft, we can calculate the journal value of the dangerous section of the transmission shaft as 150 mm and 30 mm less than the value of the original dangerous section journal.

This paper uses the dynamic analysis software ADAMS to solve the force of the transmission shaft, which further solves the statically indeterminate problem existing in the transmission system. The application of this method can save the raw materials of the transmission shaft, and provide a theoretical basis for the optimum design of similar transmission shafts.

REFERENCES

[1] Chen, H.J. & Hu, Z.Q. & Yang. C.L. 2006. Machinery Design & Manufacture. *Design and apply of the new walking beam cooling bed.* 06 (6): 12–13.

[2] Zhang, C.L. & Lang, Y.C. 1999. Coal Technology. *Design and usage of the step cold bed.* 18 (4): 33–34.

[3] Chen, B.N. & Song, K. & Liu, R. 2008. Journal of Chongqing Institute of Technology (Natural Science). *Optimization Design for the Lifting System of Walking Beam Cooling Bed.* 22 (7): 30–33, 72.

[4] Kuz'menko, A.G. & Ermolenko, A.F. & Pozdnyakov, M.A. & Baibuzenko, V.N. & Kleshchenko, D.A. & Smiyanenko, I.N. & Sheremet, V.A. 2001. Metallurgist. *New Equipment for Levelling and Facing Rolled Products on the Carryover-Type Cooling Beds of Light-Section Mills.* 45 (7), 277–280.

[5] Li, Y.H. 2005. Inner Mongolia Science Technology & Economy. *Institutional Analysis and Calculation of Walking Beam Cooling Bed.* (22): 23–24.

[6] Liang, H. & Wang, R.Q. & Su, L. 2011. *Walking Beam Cooling Bed Drive Unit Optimization.* Eighth China Steel Annual Symposium.

[7] Lijun Zhangr & D Center Shanghai Hitachi Electrical Appliances Co. *Stiffness analysis for the shaft of double rotator compressor with CAE technology.* Proceedings of the 3 ~ (rd) International Compressor Technique Conference.

[8] Yuan, A.F. & Lu, Y. 2013, Journal of Mechanical Transmission, *Virtual Prototype Design and Simulation Analysis of the Cutting Machine based on the Solidworks and ADAMS.* 37 (04): 60–63.

[9] Li, Z.G. 2006. *Detailed Examples of Entry and ADAMS*: 1–5. Beijing: Defense Industry Press.

[10] Zheng, K. & Hu, R.X. & Chen, L.M. 2006. *ADAMS 2005 Senior Mechanical Design Applications.* Beijing: Machinery Industry Press.

[11] Cheng, D.X. 2010. *Mechanical Design Handbook.* Beijing: Chemical Industry Press.

[12] Wu, W.Q. 2012. Mechanical Engineering and Automation. *Design of Two-stage Bevel and Cylindrical Gear Reducer Shaft* 174 (5): 174–176.

Material Science and Engineering – Chen (Ed.)
© 2016 Taylor & Francis Group, London, ISBN 978-1-138-02936-1

The design and analysis of the whole drilling rig moved device

H.B. Zhao & X.H. Shen
School of Mechanical Engineering, Xi'an Shiyou University, Xi'an, Shaanxi, China

X.W. Sun
Products Research Institute, Lanzhou LS Petroleum Equipment Engineering Co. Ltd., China

ABSTRACT: Taking the whole drilling rig moved device as the research object, this article designs a kind of device by improving the peripheral equipment removed with the actual environment and the requirements of different move distances. The device has a number of advantages, such as safe, reliable, good bearing. Besides, the overall layout is reasonable, and the connection and detachment are rapid and firm. The finite element modeling and analysis are done on the clamping support of the whole drilling rig moved device according to the related theory and method of the finite element analysis software. The results show an overall stress distribution and deformation of the brackets, and then the structure is optimized and improved.

Keywords: drilling rig; moved device; finite element analysis software

1 INTRODUCTION

Weight is mainly borne by the top rise liquid cylinder and connection rolling device in the process of the whole drilling rig moved. So, the four wheels of the traditional moved devices are equipped with individual driving in the pre—and post-axle shaft. In other words, it always keeps the form of four wheel driving. It also can distribute the traction according to the actual situation of field to make sure the passing ability and stability.

Each wheel is hanged on the rolling device bracket alone through the elastic suspension, to reduce the stress of the tire, improve the wheel ground adhesion, reduce removed devices focus, through the uneven road surface, left and right wheels beating alone, disconnected, reduce the tilt of moved device and vibration [1, 2]. However, its complex structure, high cost, maintenance inconvenience will increase the weight and overall dimensions of the moved units. The moved device has only its own weight, a part of the rig and the traction effect, but the main load is ignored. This paper designs a new type of the whole drilling rig moved device by fully considering the advantages and disadvantages of traditional moved devices.

2 OVERALL STRUCTURE SCHEME

The principal structures of the drilling rig include crown block, swim vehicle, hook, swivel, rotary table, independent drive, winch, derrick substructure, wire rope, driller control atrial, and right and left folk prescription. The total weight of the drill machine is about 800 tons, which includes the principal structures and all the pipes (250 tons). It does not include the ramp, slides and ladders that are needed to be removed in the moving process [3]. The whole drilling rig moved device is composed of the power system, removed system, braking system, hydraulic system and other parts. The power system is equipped with motor, transmission shaft, differential, driving gear train, and controller. The removed system is composed of roll board, chain plate, shaft, roller and bracket [4]. Obviously, roller and roll board are the main parts in the moving. One side of the roller contacts with the ground, but the other side of the roller contacts with the roll board.

The roller can make the drilling rig continue to move through its movement from top to bottom and from front to behind. The traction needed by the roller is decided by the friction in the roller and the roller board, as well as the friction in the roll and the ground. The roller and the roll board are for steel to steel friction, so the friction coefficient is small. Actually, it is good for removing the traction with less than 0.06 when it is working. The bracket is connected directly to the top of the cylinder. It also can connect with the rotary device. What's more, brackets can be made into various shapes by the working environment, so that it is convenient to connect with other peripheral equipments.

Overall arrangements for the whole drilling rig moved device are thought of step by step to move.

That is to say, a set of stepper device should be added in the front and end of a bracket's right and left. When the rig hoisting operation condition should be moved, it just needs to jack the main body of the machine with four sets of stepping device of the jack-up cylinder. Then, the main of the drilling rig moves to the front or back driven by the agency. The brake system brake after reaching the designated position shrinks the lifting oil cylinder. The body of the drilling rig moves to the target location and peripheral equipment with corresponding moved device to the designated place.

3 THE MODEL OF THE BRACKETS STRUCTURE

Brackets arrange averagely at the four corners of the drilling rig. So, only 1/4 of the whole drilling rig moved device model is set up, whose shape size is 1500 mm × 600 mm. The structure is shown in Figure 1. Brackets are welded by the steel plate called Q345. Its thickness is 30 mm and the elastic modulus is $2.06 × 10^5$ Mpa. Its Poisson's ratio is 0.3 and the density is 7850 kg/m^3. The bracket finite element model is set up in the corresponding parameters, applying the vertical load. The front plane is selected as the bearing surface and specified constraint area.

This paper adopts the number of shell 63 unit. This unit has both bending capacity and membrane force. It also can take the load and the method of the plane load. Besides, each unit node has six degrees of freedom. We consider not only the stress effect and large deformation capacity, but also the direction translation of XYZ node coordinates and the axis rotation of XYZ node coordinates. This article adopts the artificial mesh method to ensure the quality of the grid. The grid is divided

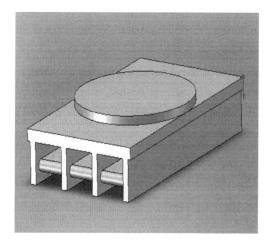

Figure 1.   3 D model of the bracket.

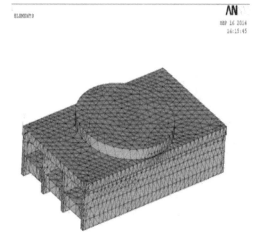

Figure 2.   Bracket meshing model.

into the reference unit with a single side length of 20 mm. The connection of each panel uses a continuous fillet weld, and it also can be dealt with in a node sharing way. All units in the boundary of the model are divided by the method of equal fraction. This method can make its node superposition, and then merge nodes. All meshes of the whole brackets are shown in Figure 2.

4 THE FINITE ELEMENT ANALYSIS OF BRACKETS

Brackets are only under the quality of the device when the moved device is stationary. However, drill is mainly under its own weight and the overall weight of the drill rig. Moved devices have symmetry, so loads also have symmetry. The maximum equivalent contour of brackets is shown in Figure 3. We can infer that the maximum stress value is 25537.6 MPa in the connecting plate with the bracket and the cylinder. The max stress is not beyond the allowable stress values, but it is related to the welding quality of the plate. The second max stress value is 17029 Mpa, which occurs in the plate and top panel joint area. The stress in the rest part of the bracket is far less than the allowable stress values, and the value shows that it is relatively safe. The maximum stress of the bracket can be seen from the displacement nephogram appearing in the middle of the plate connected with the panel.

From the above static finite element analysis of the bracket, the largest stress occurs in the welded parts except some stress concentration phenomenon parts with some displacement constraints and load action. All the above results are in line with the general engineering facts. In practice, the welding qualities directly decide the strength of the part size. The actual stress value may be less than the simulation value, because

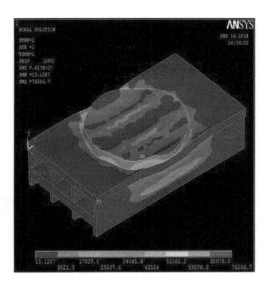

Figure 3. Brackets equivalent displacement nephogram.

of it being in the form of node shared connection without any other special processing in terms of modeling. Overall, a larger stress mainly appears in the roller contact surface and the cross-section mutation position. Other parts of it is safe and reliable, with their stress being far less than the allowable stress values. Otherwise, subsequent optimization can be conducted in view of the local stress.

5 THE OPTIMIZATION OF BRACKETS

Under the premise of meeting the structure strength requirement, the lightweight optimization and reliability optimization are performed for the brackets. The fimincon function provided by the MATLAB optimization toolbox is applied to solve nonlinear optimization. The thickness of the panel and the lower height of the support plate on the brackets are considered as design variables, while the minimum anchorage quality is the optimization goal. The constraint condition is only considered as the maximum stress value, which is less than the allowable stress of each design variable value. Through establishing the optimization model, the lightweight optimization design results can be obtained as follows: the reliability index is 3.54, the failure probability is 2.3411×10^{-4}, the maximum stress value is 35987 Mpa, the panel thickness is 25 mm, and the quality is reduced by 4.68%. It improves the dynamic performance of the moved device and saves material, but reducing the strength, stiffness and safety coefficient of the brackets. What's more, it makes the structure reliability to reduce. Reliability optimization design also can get the results indicating that the reliability index is 7.24, the fail-

ure probability is 1.9671×10^{-4}, the maximum stress value is 19054 Mpa, the panel thickness is 30 mm, and the quality is reduced by 1.95%. How to coordinate lightweight and reliability are the key to the structure optimization design of brackets. To ensure a safe and reliable structure under the premise of lightweight components as far as possible is the ultimate goal. Thus, it can both control the design stage cost and get a higher ratio of economic products.

The above two models of optimization are performed under the premise of meeting the design requirements in structure strength. The difference between them is that the former seems to reduce the quality as the optimization goal, and reads the parameters in the optimization process to determine the amount, while the latter designs itself by aiming at safety and reliability and considering the influence of random factors. From the above results, it can be seen that the reliability optimal design is safer, but this may increase the material usage. So, most time, it uses a lightweight design to reduce the material cost; when only a higher reliability is needed, the latter is used to ensure structural safety.

6 CONCLUSION

This paper designs a new side of the drilling rig moved device by analyzing the traditional moved device. The finite element analysis software ANSYS is used to build modeling for the main stress components of brackets. The finite element analysis software ANSYS is used for grid tumble and the analysis of stress and displacement. The result of the software shows that stress is larger in the roller contact surface and the cross-section mutation position. At the same time, the strength of the welding is determined directly by the quality of the parts' size. The analysis results show that the strength and stiffness of brackets meet the design requirements, but there are also local stress concentrations that need to be further improved.

REFERENCES

[1] Guoqing Liu, Qingdong Yang. ANSYS engineering application tutorial [M]. Beijing: China railway publishing house, 2003.
[2] Changjin Wei, Xiaolin Wang, Anjin Yang. Artificial field drill rig dedicated mobile platform development. Science and technology information. 2009, 02.
[3] Jian Sun. To develop the design of convenient mobile device [J]. Coal minemachinery. 2012, 33(5):172~173.
[4] Meiling Huang. The design of the whole mobile device [J]. Journal of Shengli oil field worker University. 2009, 23(3):40~42.
[5] HanZhao, Demeng Zhao. Automobile structure lightweight design based on ANSYS [J]. Journal of agricultural machinery. 2005, 36(6):12~15.

Material Science and Engineering – Chen (Ed.)
© 2016 Taylor & Francis Group, London, ISBN 978-1-138-02936-1

Effect of parameters on flying wing UAV buzz responses

J. Xu
School of Aeronautics, Northwestern Polytechnical University, Xi'an, P.R. China

X.P. Ma
UAV Research Institute, Northwestern Polytechnical University, Xi'an, P.R. China

ABSTRACT: The effect of parameters on flying wing UAV buzz was studied based on the CFD/CSD coupling method. The RANS equations and finite element methods were established, in which the aerodynamic dynamic meshes used the combination of spring-based smoothing and local remeshing methods. The interfaces between structural and aerodynamic models were built with an exact match surface where load transferring was performed based on 3D interpolation. The flying wing UAV transonic buzz responses were studied, and the rudder parameter buzz response and aileron, elevator and flap vibration responses caused by rudder motion were also provided.

Keywords: flying wing UAV; buzz; CFD/CSD; parameter; rudder

1 INTRODUCTION

The flying wing UAV has a high aspect ratio. It is important to accurately calculate the nonlinear aeroelastic responses. The literature about buzz is scarce. This paper mainly focuses on the detailed flying wing UAV structural model buzz responses analysis, and the rudder motion constraints at the end of the flying wing UAV rudder structural model.

2 CFD/CSD METHOD

The aeroelastic equations and boundary conditions can be represented as follows:

$$M\ddot{U} + C\dot{U} + KU = F(t) \tag{1}$$

$$\frac{\partial}{\partial t}\iiint_\Omega \bar{Q}d\Omega + \iint_S \left(\bar{G} - \bar{Q}\vec{q}_b\right)\cdot d\vec{S} = \frac{1}{Re}\iint_S \bar{F}^V\cdot d\vec{S} \tag{2}$$

$$\sigma_s \cdot \boldsymbol{n} = -p\boldsymbol{n}, \boldsymbol{u}_s = \boldsymbol{u}_F \tag{3}$$

Equation (1) is the structural dynamic. Equation (2) is the integral form of the unsteady Navier-Stokes equation. Equation (3) is, respectively, the normal vector equilibrium and displacements compatibility conditions of the interface surface.

3 FLYING WING UAV MODEL

There were eight boxes and four root ribs in the fuselage; the inside wing structure was composed of wall plates, girders, wing ribs; there were four beams and six ribs in the inside wing. The outside wing was composed of three beams and sixteen ribs; there were three rubs behind the outside wing, namely rudder, aileron and elevator. The geometric model of the flying wing UAV is shown in Figure 1.

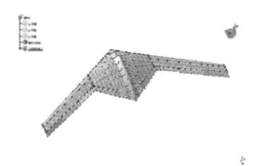

Figure 1. Geometry model of the flying wing UAV.

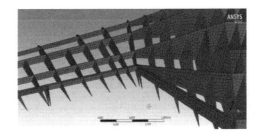

Figure 2. Local structural grid of the half model.

4 RUDDER BUZZ RESPONSE

The constraints of the rudder are shown in Figure 3. The rudder motion conditions were constrained on the flying wing UAV structural model. The motion equation of the rudder is as follows:

$$\beta(t) = \beta_0 + \beta_m \sin(wt) \qquad (4)$$

The RANS N-S equations with the SST turbulence model, and the finite volume method were used to discretize the N-S equation. The second-order implicit time marching method was also used. The unstructured dynamic meshes used the combination of the spring-based smoothing and local remeshing methods.

The CFD/CSD loosely coupling method was developed to calculate the flying wing UAV rudder response. The coupling calculation time step was 0.001s, and the aerodynamic convergence error was 1e-6. The RANS N-S equation with the SST turbulence model was used to simulate the aerodynamic models.

The rudder buzz responses are shown in the figure, indicating the outside and inside monitoring point of the rudder. The vibration responses of the rudder, aileron, elevator and flap induced by the rudder motion are shown in Figure 5. The surface pressure coefficient of the flying wing UAV is shown in Figures 6 and 7.

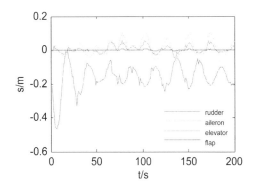

Figure 5. The vibration time responses.

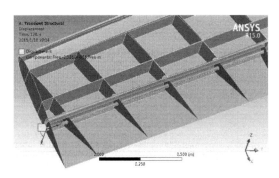

Figure 3. Constraints of the rudder buzz.

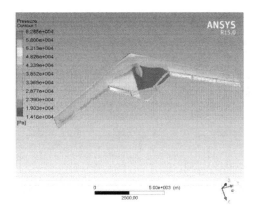

Figure 6. Surface pressure coefficient.

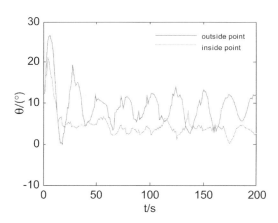

Figure 4. The rudder angle responses.

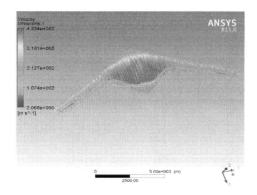

Figure 7. Flying wing UAV streamline.

5 BUZZ PARAMETER ANALYSIS

The effect of the rotational frequency on the rudder angle is shown in Figure 8. The rotational frequency had a greater effect on the buzz responses. In addition, the effect of the rotational frequency on the angle acceleration is shown in Figure 9.

The effect of the rotational frequency on the rudder, aileron, elevator and flap displacement is shown in Figures 10–13. It can be seen from these figures that the rotational frequency had a greater effect on the vibration displacement.

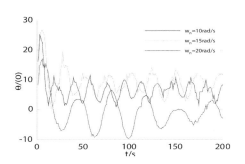

Figure 8.   Effect of rotational frequency on the rudder angle.

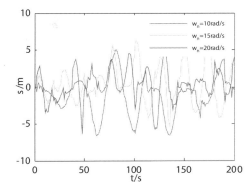

Figure 11.   Effect of rotational frequency on the aileron displacement.

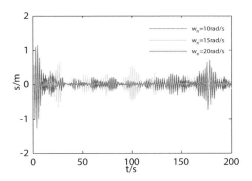

Figure 9.   Effect of rotational frequency on the angle acceleration.

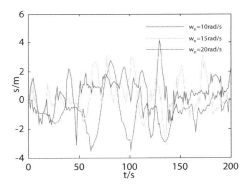

Figure 12.   Effect of rotational frequency on the elevator displacement.

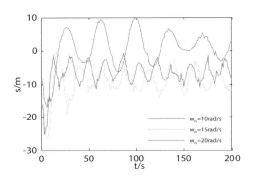

Figure 10.   Effect of rotational frequency on the rudder displacement.

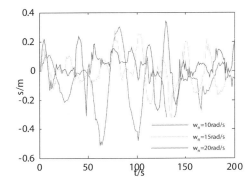

Figure 13.   Effect of rotational frequency on the flap displacement.

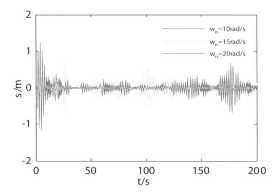

Figure 14.   Effect of rotational frequency on the rudder acceleration.

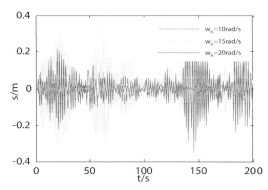

Figure 15.   Effect of rotational frequency on the aileron acceleration.

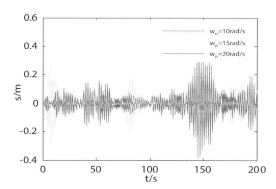

Figure 16.   Effect of rotational frequency on the elevator acceleration.

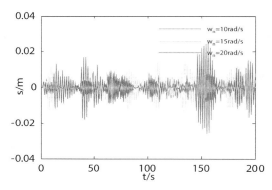

Figure 17.   Effect of rotational frequency on the flap acceleration.

The effect of the rotational frequency on the rudder, aileron, elevator and flap acceleration is shown in Figures 14–17. It can be seen from these figures that the rotational frequency had a greater effect on the vibration acceleration.

6   CONCLUSIONS

A detailed structural and aerodynamic model was developed to study the rudder buzz responses of the flying wing UAV, and the effect of the rotation angular was found to have a greater influence on the buzz response frequency.

REFERENCES

[1] Earl Dowell, John Edwards, Thomas Strganac. Nonlinear aeroelasticicy [J]. Journal of Aircraft 2003, 40(5): 857–874.
[2] Chen-gi Pak, Myles L, Baker. Control surface buzz analysis of a generic NASP wing [J]. AIAA Paper 2001–1581, 2001.
[3] Xu Jun, Ma Xiaoping. Transonic rudder buzz on a tailless flying wing UAV [J]. Transactions of Nanjing University of Aeronautics & Astronautics, 2015, 32(1): 61–69.
[4] Jun Xu, Xiaoping Ma. Effects of parameters on flutter of a wing with an external store [J]. Advanced Materials Research, 853(2014): 453–459.

Material Science and Engineering – Chen (Ed.)
© *2016 Taylor & Francis Group, London, ISBN 978-1-138-02936-1*

Determination method for rutting specimen cross-section deformation curve based on line-structured light

H. Ying, Q. Wu, Y. Liu, H. Chen, W. Wang, H.J. Xu & J.Q. Gao
College of Architecture and Traffic Engineering, Guilin University of Electronic Technology, Guilin, China

ABSTRACT: The difference of the material can result in the changes of mixture rut deformation pattern, and the rut cross-sectional deformation curve can reflect the characteristics of the internal structure of the mixture. To overcome the shortcomings of detection methods for the rut deformation pattern, a new determination method for the rut cross-sectional deformation curve based on line-structured light vision is proposed. In this method, the camera and line laser, which installed into constant temperature box of the rut test machine, combined with the rut sample are used to set up a structure light vision inspection system. Secondly, the parameters of the camera model will be calibrated. Then construct the light-plane-equation control points that are not collinear by using 2D coordinates of the target and calculate the coefficient of the light-plane-equation based on bi-plane grid target, morphology thinning and Hough transform, and the calibration of detection system will be realized. Finally, the determination for cross-sectional deformation curve of rut will be realized by capturing the light stripe image of the rut before and after loading and calculating the 3D coordinates of the light stripes from the structure—light model.

Keywords: road engineering; rut; line-structured light; vision measurement; hot mixture

1   INTRODUCTION

The rut cross-sectional deformation curve is an important characteristic of rut deformation pattern, and it is the accumulation of the internal structure of mixture under the repeated application of traffic loads. Because the diversity of the material can lead to the difference of the cross-sectional curve, the relationship between cross-section characteristics and material structure can be studied from the rut cross-sectional deformation curve [1] [2]. It is provide a technical means to reveal the behavior of the material under load and assess the high temperature performance of the materials, thus it is of great importance to research a test method for rut cross-sectional deformation curve with high precision and high efficiency. Now, there are three major classes on the methods to get the rut cross-sectional deformation curve: the first, manual measurement with the vernier caliper, but the testing precision is difficult to guarantee; the second, dot laser array is used to obtain the deformations of several points of the cross-section, but limited by the number of sensors and the size of the constant temperature box of the rut test machine, it is difficult to describe the cross-sectional deformation of the rut sample with finite sampling points, so this method is mainly used in automotive rut

rapid acquisition for physical engineering [3]; the third, line laser is used to describe the rut cross-section [4], but it is almost rely on artificial measurement in the distance measurement model and the model calibration, so the degree of automation is very low. In this paper, the line-structured light is used to make sure the system can automatically capture the reference points, laser light stripe and light control points by using grid target in different planes in terms of calibration. In this way, the system calibration will be fulfilled instantly, thus the system becomes more efficient and convenient.

2   COMPONENT OF MEASUREMENT SYSTEM

In this paper, the line laser light stripe, which project onto the surface of asphalt mixture which have rut deformation, forms a light stripes curve. Then the light stripe image will be captured by CCD camera, so we can get the 3D information of the light stripe according to the principle of structure light vision to realize the rut deformation curve detection. Figure 1 illustrates the system structure. This system uses Sentech STC-Pocl232 A area-array camera which parameters include the following aspects: resolution 1600*1200, frame rate 30fps,

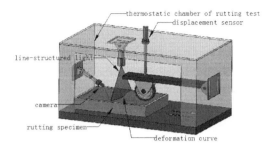

Figure 1. Rut deformation detection model based on structure light vision.

Cameralink output; Fujinon 50 mm optical lens, DALSA PC2-CameraLink image acquisition card and red-lined line laser which power is 100 mW. The laser is fixed at the top of the constant temperature box of the rut test machine through a movable joint which can adjust the angle between the laser light and the specimen. The camera is fixed at the side wall of the constant temperature box of the rut test machine through threes movable joints which allow the camera can up and down move, horizontal move and rotary move, which make the position and orientation adjustment of the camera be convenience.

3 MODEL OF MEASUREMENT SYSTEM

The positional information of the line-structured light stripe is captured by the camera and the image is formed on the CCD sensor. The space position relation from the stripe to the image is the structure light model which is composed of the camera model and the light-plane-equation. Suppose that the world coordinate system are comprised of $X_w = [X_w, Y_w, Z_w]^T$ and (u, v) is the image coordinate in pixels. The following formulas are the relationships between the coordinates of point P $[X_w, Y_w, Z_w]^T$ in terms of the world coordinate system and the image coordinate of point P(u, v) [5]:

$$s[u,v,1]^T = \mathbf{M}(\mathbf{R}\mathbf{X_w} + \mathbf{t}) \tag{1}$$

$$Z_w = AX_w + BY_w + C \tag{2}$$

where, s is the scale factor; $\mathbf{M} = \begin{bmatrix} a_x & 0 & u_0 \\ 0 & a_y & v_0 \\ 0 & 0 & 1 \end{bmatrix}$ is the

parameter within the camera, which determined by a_x, a_y, u_0, v_0 (u_0, v_0) is the coordinates of the projection in coordinate system subject to the center of the lens. a_x and a_y stands for the internal parameters in connection with the physical size and focal

length of the camera; R is the rotation matrix of the camera relative to the world coordinate system and t is the translation matrix of the camera relative to the world coordinate system; A, B, C are the coefficients of the structured light plane.

Equation (1) is the camera model while equation (2) is the light plane equation. The structured light model is consisted of equation (1) and equation (2). After the structured light system is calibrated, which means that s, R, t and A, B, C, D are determined, the world coordinate of the light stripe will be calculated according to the model as long as the coordinate (u, v) of any point on the modulated light stripe is captured by the camera, thus 3D coordinates of the rut deformation curve can be calculated.

4 MODEL CALIBRATION

The intrinsic and extrinsic parameters of the camera and the coefficient of the light-plane-equation need to be determined before the structured-light vision system can be used, and this process is the calibration process of the visual system. The process to seek the intrinsic and extrinsic parameters of the camera is called camera calibration which adopts the camera calibration method based on plane grid points presented by Zhengyou Zhang [6]. The plane grid points which we have known their coordinates are used in this method, that is, using the target consisted of reference points to solve the intrinsic and extrinsic parameters of the camera from different perspectives. Because the targets are simply manufactured and the calibration is highly automated, this method is applied widely.

When determining the light plane, we can get the light-plane-equation as long as three or more non-collinear points are obtained by the properties of plane that a plane is determined by two intersecting lines. However, it is very difficult to capture the precise 3D coordinates of the space points in road engineering laboratory. Thus, in this paper, we use two plane grid targets to construct the 3D space points, which need for the structured light plane calibration, as the control points, and then calculate their 3D coordinates to complete the calibration work. Figure 2 shows that the 3D control points calibration system is constituted by two grid targets A and B that in different planes, and the light from the line laser project onto the target A, forming the light lines LA. Similarly, the light lines LB is formed on the target B. $A1$, $A2$, $A3$ and $A4$ are the arbitrary points on LA, while $B1$, $B2$, $B3$ and $B4$ are the arbitrary points on LB. As long as figure out the 3D coordinates of the eight control points, the light-plane-equation are available from regression analysis.

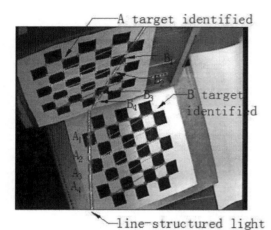

Figure 2. Structured light plane calibration system constituted by two grid targets.

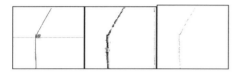

(a)Binarization (b) Light stripes thinning (c) Hough transform straight line detection of light stripe

Figure 3. Light stripe detection.

The light plane calibration method is shown as follows:

1. *Extract the centerlines of the light strips:* Because the image of the stripe is very bright, close to white, and the target is also white, they have the figure of low contrast. By using the domain division method, the result is shown in Figure 3a. The light strips in the binary image is coarse, so the further thinning treatments are necessary. The method of thinning treatments consist of erosion, expansion, Hilditch, Deutch, model refinement, morphology thinning and so on, and then detect the liner equation of the light stripe by using Hough transform, but it is difficult to fit the straight line of the light stripe just rely on simple erosion and expansion. After the image are processed like Hilditch, Deutch, model refinement and morphology thinning, the centerlines of the light strips are almost the same by means of Hough transform detection method. Considering the complexity of the algorithm, this paper choose morphology thinning to process the binary image, and then fit the straight line of the light stripe by using

Hough transform. The morphology thinning effects are shown in Figure 3b while the centerlines of the light strips detection effects are shown in Figure 3c.

2. *Select the control points:* The two straight lines that form on the target which in different planes by the light plane are detected, apparently, they are all on the light plane, and randomly picking a set of points ($A1\sim A4$, $B1\sim B4$) as for control points for the light-plane-equation regression.

3. *Calculate the target pose parameters:* Because they are in their own target plane ($Z_w = 0$ plane), for the control *points* on the LA, their image coordinates and the 3D coordinates meet the following equation when take the upper-left corner of the target A as the origin of the world coordinate system:

$$s\begin{bmatrix} u \\ v \\ 1 \end{bmatrix} = \begin{bmatrix} a_x & 0 & u_0 & 0 \\ 0 & a_y & v_0 & 0 \\ 0 & 0 & 1 & 0 \end{bmatrix}\begin{bmatrix} \mathbf{R}_A & \mathbf{T}_A \\ 0^T & 1 \end{bmatrix}\begin{bmatrix} X_w \\ Y_w \\ 0 \\ 1 \end{bmatrix} \quad (3)$$

where, $\mathbf{R}_A$ and $\mathbf{T}_A$ is the rotation matrix and the translation matrix of the camera optical center relative to the plane of the target A, respectively.

For the control points on the LB, their image coordinates and the 3D coordinates meet the following equation when take the upper-left corner of the target B as the origin of the world coordinate system:

$$s\begin{bmatrix} u \\ v \\ 1 \end{bmatrix} = \begin{bmatrix} a_x & 0 & u_0 & 0 \\ 0 & a_y & v_0 & 0 \\ 0 & 0 & 1 & 0 \end{bmatrix}\begin{bmatrix} \mathbf{R}_B & \mathbf{T}_B \\ 0^T & 1 \end{bmatrix}\begin{bmatrix} X_w \\ Y_w \\ 0 \\ 1 \end{bmatrix} \quad (4)$$

where, $\mathbf{R}_B$ and $\mathbf{T}_B$ is the rotation matrix and the translation matrix of the camera optical center relative to the plane of the target B, respectively. $\mathbf{R}_A$, $\mathbf{T}_A$, and $\mathbf{R}_B$, $\mathbf{T}_B$ are the pose parameters of the target which can be calibrated by 36 inside corners on the target A and B, which also adopts the camera calibration method based on plane grid points presented by Zhengyou Zhang [6].

4. *Coordinate-transformation of target B:* The coordinate *systems*, which take the upper-left corner of the target A as the origin of the entire system, can be established by formula (3) and (4), therefore, the coordinate-transformation formula of target B as following:

$$\begin{bmatrix} X_A \\ Y_A \\ Z_A \end{bmatrix} = \mathbf{R}_A^{-1}\left(\mathbf{R}_B\begin{bmatrix} X_B \\ Y_B \\ 0 \end{bmatrix} + \mathbf{T}_B - \mathbf{T}_A \right) \quad (5)$$

where, X_B and Y_B are the coordinates which take the upper-left corner of the target B as the origin, while X_A, Y_A and Z_A are the coordinates on the target B which take the upper-left corner of the target A as the origin.

5. *Return to the light plane:* The 3D coordinates of the *control* points which on the target A and B of the structure light stripe can be calculated, and can return to the light plane, then complete the parameter calibration of the structured light system.

5 RUT DEFORMATION CURVE CALCULATION

Rut deformation curve reflect the deformation on the vertical slice (outline) of the specimen, thus it is necessary to convert the curvilinear coordinate into the vertical plane when calculate the rut cross-sectional deformation curve. To keep things simple, firstly, the target A should be placed on the top of the specimen when we need to calibrate the light-plane-equation and make the x-y plane of the target A parallel to the top of the specimen; secondly, make sure that the y-axis of the target A parallel to the movement direction of the wheel; thirdly, the light stripe should be parallel to the x-axis of the target by adjusting the light plane; finally, we should adjust the camera to make sure that the light stripe in the camera's view is parallel to the abscissa of the image, and the center of the image should be positioned as close as possible to the intersection of the light stripe and the center of the wheel path. Thus, in the detection system coordinates which take the upper-left corner of the target A as the origin, the z-axis directly reflect the cross-sectional deformation of the rut tests, while the x-axis represents the horizontal position of the cross-section on the specimen.

Limited by the condition such as the size of the constant temperature box of the rut test machine, the camera working distance and the angle between the camera and the light plane, the sampling width of the rut specimen must be controlled within about 26 cm in the center of the specimen to ensure the computation precision of the rut deformation curve. Figure 4a shows the light stripe image of the specimen before the test, while Figure 4b shows the light stripe image of the specimen after the test. After the 3D coordinates of the two light stripes are calculated respectively, the rut deformation curve can be obtained when performed subtraction on two 3D coordinates.

According to Figure 4, we plug the coordinates of the light stripe image before and after loading into the formula (1) to get the world coordinates of the light stripe. Because target A is parallel to

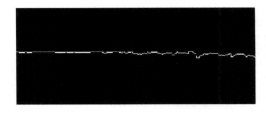

(a) Beforeloading

(b) After loading

Figure 4. Light stripe image.

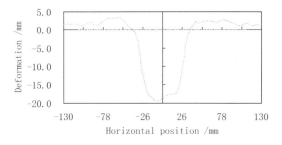

Figure 5. Rut cross-sectional deformation curve.

the top of the specimen before loading, the rut cross-sectional deformation continuous curve can be obtained by using coordinate interpolation on the Z_w-axis, as shown in Figure 5. From this curve we can calculate the parameters of the curve, like notch depth, notch area, uplift height, uplift area and the position of the uplift peak and so on, which provide an effective analysis tool for describing the rut deformation pattern, establishing the quantity relationship between the internal structure of the mixture and the rut deformation pattern, and revealing the deformational characteristics of the internal particles of the mixture.

6 CONCLUSIONS

1. This paper presents a determination method for rut cross-sectional deformation curve based on

line-structured light vision. Through the light stripe modulated by the mixture rut specimen from the line-structured light, the 3D coordinates of the thinning light stripe are obtained by using structure light vision principle and image processing technology, and then the cross-sectional deformation curve can be calculated from these coordinates.

2. During the light plane calibration in the structured-light vision system, this paper proposes a method for calibration of the light plane by using two grid targets that in different planes. The information, such as the reference points of the two targets, the pose parameters and the light stripe image, can be captured by the computer, then using binarization, thinning and Hough transform to preprocess the light stripe image to get the intersecting straight line on the light plane and the two targets. The 3D coordinates of any point on the two straight lines can be got by the spatial relationships between the two targets, and then return to the plane equation to complete the calibration. The whole process is fully controlled by the computer, and this method has the advantages of convenience and rapidity, with high practicality.

ACKNOWLEDGEMENT

This work is supported by the Natural Science Foundation of China under No 51208130, Natural Science Foundation of Guangxi Zhuang Autonomous Region under No 2013GXNSFBA019258.

REFERENCES

[1] Li Li-han, Chen Jian-jun, Su Zhou, etc. Analysis of Uplift Deformation and Loose Phenomena of Asphalt Layer [J]. Journal of Tongji University (Nature Science), 2009, 37(1):52–57.

[2] Li Li-han, Su Zhou, Chen Jian-jun. Research on uplift coefficient of asphalt mix by loaded wheel tracker [J]. Journal of Highway and Transportation Research and Development, 2007(7):42.

[3] Ma Rong-gui. Study on the Principle and Method of Road Three-Dimension Measurement System [D]. Chang'an University, 2008, 6.

[4] Li Li, Sun Li-jun, Tan Sheng-guang, etc. Line-Structured Light Image Processing Procedure for Pavement rut Detection [J]. Journal of Tongji University (Nature Science), 2013, 41(5):710–716.

[5] Zhang Guang-jun. Vision Measurement [M]. Beijing: Science Press, 2008, 33–36.

[6] Zhang Zheng-you. A flexible new technique for camera calibration. IEEE Trans. on Pattern Analysis and Machine Intelligence, 2000, 22(11):1330–1334.

Material Science and Engineering – Chen (Ed.)
© 2016 Taylor & Francis Group, London, ISBN 978-1-138-02936-1

Virtual prototype simulation of a hybrid quad-rotor aircraft based on RecurDyn and Simulink

G. Tong
Liaoning Academy of General Aviation, Shenyang Aerospace University, Liaoning, Shenyang, China

X.J. Zhang
College of Mechanical and Electrical Engineering, Shenyang Aerospace University, Liaoning, Shenyang, China

ABSTRACT: In order to avoid the tedious process of deriving the kinetic equations of a hybrid four-rotor aircraft, and improve the efficiency of the mechanical and control systems, a physical 4-rotor aircraft model was created on the software CATIA, and then imported the model into the multi-body dynamic software RecurDyn in the format of STP. Using RecurDyn/Control and Matlab for data exchange, a four-rotor aircraft control system was created based on Matlab/Simulink. Finally, the simulation of the four-rotor aircraft joint virtual prototype was realized based on RecurDyn and Simulink. The simulation results can provide a theoretical basis for further development of aircrafts, and reflect the co-simulation feasibility of RecurDyn and Matlab.

Keywords: quad-rotor aircraft; RecurDyn; virtual prototype; simulation

1 INTRODUCTION

Quad-rotor aircraft is a type of non-coaxial multi-rotor butterfly aircraft, which is capable of vertical take-off and landing and autonomous hovering. Currently, the quad-rotor aircraft basically belongs to the micro-unmanned aircraft in the market. Since 1970, with the Breguet brothers first realizing the launch of a four-rotor helicopter, there has not been too much progress. In recent years, with the development of new materials, such as MEMS, control technology, and four-rotor aircraft research, people's enthusiasm has been stimulated again. Currently, the autonomous flight control technology is the focus of quad-rotor aircraft research. The research is mainly divided into three aspects: aeromodeling quad-rotor aircrafts, small quad-rotor aircrafts, and micro quad-rotor aircrafts [1], among which the more mature application is remote control aeromodeling quad-rotor aircrafts [2];

Due to navigation and load limits, the usage of these aircrafts is very limited. In order to solve these problems, the Liaoning General Aviation Research Institute developed a hybrid quad-rotor aircraft. This paper will study this new type of quad-rotor aircraft based on the RecurDyn and Simulink co-virtual prototype technology.

2 MODELING OF THE 3D STRUCTURE SYSTEM OF THE QUAD-ROTOR AIRCRAFT

Quad-rotor aircraft, also known as the four-rotor aircraft, produces 6 degrees of freedom in the sky while it only has 4 inputs; as a consequence, it belongs to the under-actuated motion system with the complex characteristics of multivariables, strong coupling and non-linear and uncertainty [3–5]. The quad-rotor aircraft has two pairs of positive and negative rotor and an adjacent propeller rotating in the opposite direction to counteract the spin force generated by the rotation of the propeller; as a result, it does not need a special tail propeller to counteract the anti-twist torque, and the carrying capacity can be improved by using a multi-propeller [6]. The top view of the four-rotor aircraft structure is shown in Figure 1.

As it can be seen in Figure 1, the aircraft has two pairs of mutually symmetrical rotors: rotor 1 and rotor 3 are one pair, rotor 2 and rotor 4 are the other pair. Also, the altitude of the aircraft can be achieved by adjusting the rotational speed of the rotors. The pitching motion can be realized by increasing (decreasing) the speed of rotor 1 and, at the same time, reducing (increasing) the speed of rotor 3. The roll motion can be realized

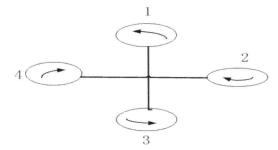

Figure 1.   Top view of four-rotor aircraft structure.

Figure 2.   Hybrid quad-rotor aircraft model drawn by using CATIA.

by increasing (decreasing) the speed of rotor 2 and, at the same time, reducing (increasing) the speed of rotor 4. The yaw motion can be realized by increasing (decreasing) the speed of rotor 1 and rotor 3 and correspondingly reducing (increasing) the speed of rotor 2 and rotor 4 [7]. In this paper, using the hybrid 4-rotor aircraft developed by the Liaoning General Aviation Research Institute, the virtual prototype simulation can be realized based on RecurDyn and Simulink software.

The aircraft is an oil and electric hybrid aircraft, which consists of engine, motor, steering gear, bevel gear, rack, rotor, battery and electric equipment. Every rotor aircraft is required to configure an engine, a battery, a steering gear and other auxiliary components, with each part being controlled independently. The engine provides the rotor with the rotational kinetic energy and the servos control the rotors up and down deflection to achieve the pitch function. Using the three-dimensional software CATIA to draw the quad-rotor aircraft entity model, the entity model is shown in Figure 2.

Because the aircraft has hundreds of parts, in order to simplify the difficulties of simulation, we considered only the chassis, steering, power-train, rotors and other components.

3   HYBRID QUAD-ROTOR AIRCRAFT VIRTUAL PROTOTYPE MODEL

Virtual prototyping technology is a digital design method based on computer modeling and simulation, which enables the engineers to develop a computer model and virtual moving parts before building the physical prototype model, and then simulates the motion behavior of the prototype [8–9]. Virtual prototype consists of three parts: functional analysis of the simulation model, simulation model of expressing the geometry of the CAD, and model for visualization and expression of the virtual environment [10]. RecurDyn is a virtual prototyping software that can be used for modeling, solving and simulating the dynamic behavior of mechanical systems, which can well solve the traditional kinematics and dynamics problems. However, before establishing the co-simulation model, which is primarily based on RecurDyn and additionally based on Simulink, we must ensure the correctness of the dynamic model of the aircraft. The hybrid quad-rotor aircraft model is comprised of the following three parts: 3D entity model of the aircraft, dynamics model of the aircraft and co-simulation control model of the aircraft.

The model was created in CATIA and imported directly into RecurDyn software in the format of STP. The virtual prototype model was created after editing the properties of every component, such as defining quality, materials, driving force and other related attributes. The virtual prototype model is shown in Figure 3.

The freedom of a multi-rigid body is calculated by using the following: $D = 6n - \sum_{i=1}^{m} p_i$, where n is the number of the rigid body; m is the number of kinematic pairs; and p_i is the i-th kinematic pair of constraint. The total degrees of freedom is 84 by calculation.

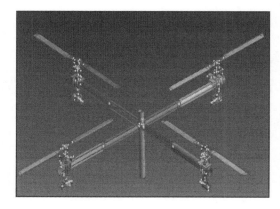

Figure 3.   Four-rotor aircraft virtual prototype model.

4 CO-SIMULATION BASED ON RECURDYN AND SIMULINK

In order to achieve the co-simulation based on RecurDyn and Matlab/Simulink, we must import the aircraft model into Matlab, which was built in RecurDyn, and then establish the joint virtual prototype model in Matlab/Simulink. Using the sub-module "control" of RecurDyn generated the relevant files after setting the parameters. In order to achieve data transfer, we need to set parameters variables in RecurDyn, including four speed control variables and three Euler angle variables, as given in Table 1.

In RecurDyn, the rotation speed of the 4 rotors is defined as the input variable "PINT", where they are used to store the rotational speed of Matlab's control system. The Euler angle is defined as the output variable "POUT", which is regarded as the angle feedback and imported into the control system. After setting the variable parameters, we click the sub-module of RecurDyn to export the quad-rotor aircraft. m file. We start Matlab and set the working path into the path where the quad-rotor aircraft. m is located. Also, we input the "quad-rotor aircraft. m "in the command window" in order to open the relevant file. Then, we enter "rdlib" in the command window where it will pop up an interface in which it contains the RecurDyn_Plant_Library. Finally, we establish a joint virtual prototype simulation model in Simulink.

5 SIMULATION ANALYSIS

The aircraft needs to have the following driving modes according to the requirements of navigation: engine drives alone, servo drives alone and hybrid drive of engine and servo. The driving torque is independently applied into the engine and servo, after setting the simulation time and step to realize the dynamics simulation, and we gain the curves of the rotors pitch angel and the roll angle that changes with time.

5.1 Engine drives alone

Just supply driving torque to engine and after setting the simulation time realizes the dynamics simulation of the aircraft. The aircrafts' pitch angel curves change with time, as shown in Figure 4. The variation in the pitch angles of rotor 1 and rotor 3 is the same and counterclockwise, and the variation in the pitch angles of rotor 2 and rotor 4 is the same and clockwise, and the angles of every rotor are the characteristic of linear motion, resulting in the adjacent rotor pitch in the opposite directions and offsetting the spin force produced by the pitch of the rotors. The aircrafts deflect 4 degrees within 0.4 minute, which corresponds with the experimental data.

5.2 Servo drives alone

By adjusting the servos' torque to control the flight posture of the aircraft, this paper analyzes only the linear motion of the rotors in order to simplify the calculation. Only applying the torque to the servos, the roll angle of the rotors is shown in Figure 5. The green curves represent the curves for rotor 1 and rotor 3, and the curves are the same. The red curves represent rotor 2 and rotor 4. We can draw the following conclusion by analyzing the different curves: the roll equal angle and the roll angle of rotor 1 and rotor 3 are the same; the roll equal angle and the roll angle of rotor 2 and rotor 4 are the same and the angle is just opposite to rotor 1 and rotor 3. The opposite roll angle of the adjacent rotors just offsets the torque produced by the rotors; as a consequence, it can guarantee the stability movement of the four-rotor aircraft.

5.3 Hybrid drive of engine and servo

Applying the driving torque to engines and servos, respectively, and observing the pitch angle and roll angle of the rotors, the angles change with time, and analyze the relationship between the curves. Because the quad-rotor aircraft is symmetrical in structure, we just need to analyze the movement of the adjacent rotor. Simulation curves are shown in Figure 6: one pitch angle increases in clockwise

Table 1. The setting parameter variables in RecurDyn.

Name	Variables	Function
Roll angle	Angelx	F(time...) = AY(.part.center.marker_30, part.ground.marker_45)
Pitch angle	Angely	F(time...) = AX(.part.center.marker_30, part.ground.marker_45)
Yaw angle	Angelz	F(time...) = AZ(.part.center.marker_30, part.ground.marker_45)
Speed of rotor aircraft 1	Speed 1	F(time...) = 10
Speed of rotor aircraft 2	Speed 2	F(time...) = 10
Speed of rotor aircraft 3	Speed 3	F(time...) = 10
Speed of rotor aircraft 4	Speed 4	F(time...) = 10

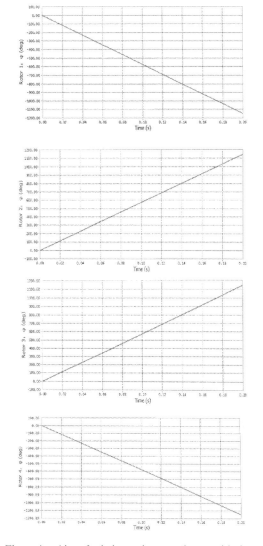

Figure 4. Aircraft pitch angel curves change with time.

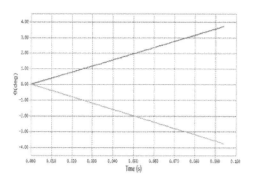

Figure 5. Four-rotor aircraft deflection angle curves.

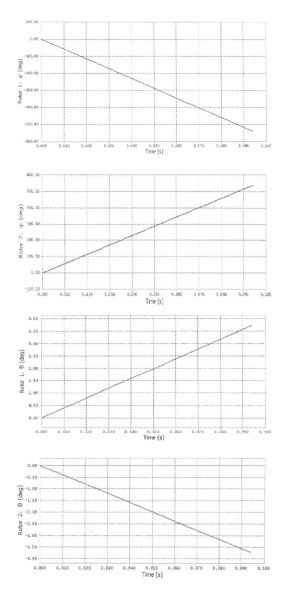

Figure 6. Adjacent rotor pitch angle curves and roll angle curves.

and the other pitch angle increases in anticlockwise, and the increasing trend is identical, and the roll angle is identical to the pitch angle, which can meet the need of the quad-rotor aircraft.

6 CONCLUSIONS

Using RecurDyn and Matlab for co-simulation, we can quickly build the hybrid quad-rotor aircraft virtual prototype model and avoid the tedious derivation of kinetic equations. In this paper, we establish the quad-rotor aircraft entity model and the virtual prototype model based on CATIA, RecurDyn and Matlab/Simulink, and realize the virtual prototype simulation of the aircraft and obtain the pitch and roll curves of the aircraft based on engines drive alone, servos drive alone and hybrid drive, and verify the operating characteristics of the quad-rotor aircraft, which was developed by the Liaoning General Aviation Research Institute. The simulation results can provide a theoretical basis for further development of the aircraft.

In subsequent studies, we will establish the precise mathematical model of the aircraft and study deeply the dynamics problems of the aircraft under the effect of low Reynolds numbers, and ensure that the virtual simulation environment is much closer to the actual working environment and further verify the correctness of the simulation by testing.

REFERENCES

[1] Huang Guo Hua, Zhu Qing Song. Target detection and recognition Summary of micro four-rotor aircraft, Mechanical and electrical products development and innovation, pp. 16–18, 2011.

[2] Zhen Hong Tao, Qi Xiao Hui etc. Four-rotor unmanned helicopter flight control technology overview, Flight Mechanics, pp. 295–299, 2012, 8.

[3] E Altug, J.P Ostrowski, R Mahony. Control of a Quadrotor Heli-copter Using Visual Feedback. IEEE Trans, on Robotics and Automation, 2002:72–77.

[4] E Altug, J.P Ostrowski, C.J Taylor. Quadrotor Control Using Dual Camera Visual Feedback. IEEE Trans, on Robotics and Automation' 2003: 4294–4299.

[5] R0 Saber. Nonlinear Control of Underactuated Mechanical Systems with Application to Robotics and Aerospace Vehicles. MIT, 2001, 2.

[6] Nie Bo Wen, Ma Hong Xu, Wang Jian etc. The research status and key technology of the four rotor Mini Rotorcraft, Electro-optical and Control, pp. 113–117, 2007.

[7] Gao Ying Jie, Chen Ding Xin, Li Rong Ming. Research on control Algorithm of small unmanned four rotor aircraft, pp. 4–7, 2011.

[8] Liu Yi, Chen Guo Ding, Li Ji Shun. Flexible Multibody Simulation Approach in the Analysis of FrictionWinder [J]. Advanced Materials Research Vols, pp. 2594–2597, 2010.

[9] Pan Teng. Research on Four-wheel Driving Carrier of Coal Cutter Based on Virtual Prototype, 2010, 6.

[10] Sun Qiang. The intelligent sewing equipment simulation based on Simulink and Recurdyn, 2009.

Material Science and Engineering – Chen (Ed.)
© 2016 Taylor & Francis Group, London, ISBN 978-1-138-02936-1

Experimental study on the micro-channel coil heat transfer performance of the current air-cooled scroll chiller

Y.R. Zhou
Mechanical and Electrical Engineering Institute, Shandong Institute of Commerce and Technology, Jinan, Shandong, China

P.J. Shang
Control Technology Institute, Wuxi Institute of Technology, Wuxi, Jiangsu, China

ABSTRACT: A method for testing design and test data analysis was proposed for comparing the performance of a condenser coil. The comparison between the performance of the current air-cooled scroll chiller with a 0.30 mm wall thickness flat tube of a micro-channel coil and a new design with a 0.36 mm wall thickness flat tube of a coil were analyzed through experimental data. The results show that the difference in the condensing temperature of the 0.36 mm wall thickness flat tube of the coil is around 0.36°C to 0.97°C higher than that of the 0.30 mm wall thickness flat tube at the 35°C ambient condition. The decrease in COP ranged from 1.2% to 11.4%.

Keywords: micro-channel coil; heat transfer performance; data analysis method; flat tube wall thickness

1   INTRODUCTION

The micro-channel heat exchanger, compared with the traditional tube-fin heat exchanger, has the advantages of small volume, lightweight, high thermal efficiency, and compact structure (Zhang Chao, 2011). They are widely applied in the field of automotive air conditioning, but are sparsely used in the field of household air conditioning. In recent years, with the cost and performance advantages of the micro-channel heat exchanger, the application has caused a wide public concern in household air conditioning and central air conditioning (Zhang Lei, 2010). Many colleges and enterprises at home and abroad have studied the application of the micro-channel in household air conditioning. Yan Ruidong (2013) studied the flow distribution of the micro-channel in household air conditioning. Wang Ying (2014) comparatively analyzed the whole machine performance and the filling volume of the household packaged chiller air conditioning instead of the tube heat exchanger.

In this paper, the performance difference in the air cooled micro-channel heat exchanger is tested and analyzed between the 0.30 mm and 0.36 mm wall thickness flat tubes, using an R134a refrigerant air-cooled eddy-type water chiller. We compare the test and analysis of the existing an air-cooled eddy-type water chiller, and put forward the performance test and data analysis methods for comparing the different heat transfer mechanisms. The similarities and differences in the design of

the micro-channel are as follows. The refrigerant channel number and channel hydraulic diameter of the flat tube are the same as the processing and the manufacturing tolerance requirements. All single heat exchangers have the same number of flat tubes. The position of the pass partition in the head tube is the same. The number of the flat tube

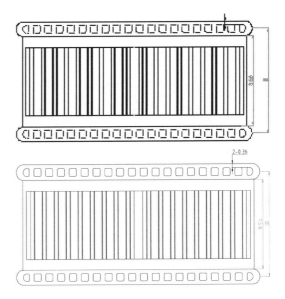

Figure 1.   Chart showing the comparison between the 0.30 mm and 0.36 mm flat tubes.

placed in the flow path is identical. Fin wing-type specifications are the same, and the designed lateral fin adopts the 21 FPI shutter.

Because of the different flat tube wall thicknesses, and the center distance between two flat tube phases at the same time, the fin height of the flat tube-fin area is different. The flat tube wall thickness is set to 0.36 mm and the flat fin height is reduced by 0.12 mm, causing the fin side heat transfer area to be reduced. At the constant fan design, the air-side resistance losses of the wall thickness of the 0.36 mm flat micro-channel heat exchanger would increase. The air volume would decrease through the micro-channel heat exchange. For the main thermal resistances in the air-side of the micro-channel heat exchanger, the heat transfer performance would be reduced, due to the wind speed, and the heat transfer area would be decreased. Figure 1 shows the chart of the comparison between the 0.30 mm and 0.36 mm flat tubes.

2 EXPERIMENTAL SET-UP

The performance contrast test of the micro-channel heat exchanger is conducted on the current air-cooled scroll chiller (Zhao Yu, 2009). The rated refrigerating capacity is 288 kW. The turbine is rated at 97.5 kW. The dual systems, systems # 1 and # 2, are configured with 2 pieces of heat exchanger. The unit adopts a dry-type shell-and-tube evaporator.

First, we ran the wall thickness of the 0.30 mm flat micro-channel heat exchanger in the test. According to the test plan, we replaced all 4 pieces of the flat tube wall thickness to 0.36 mm once the test is done. A complete set of test conditions was conducted. Hence, each test condition could be comparatively analyzed.

2.1 The operating condition

The test environmental temperature was 18°C, 25°C, 30°C, 35°C, 40°C and 43°C, respectively. In each test environment temperature, we changed the compressor operating frequency to 12 Hz, 20 Hz, 25 Hz, 30 Hz, 35 Hz, 40 Hz, 45 Hz and 50 Hz, as given in Table 1. The refrigerant side mass flow and the condensation heat transfer were changed. In this process, the air volume into the micro-channel heat exchanger was adjusted by using the frequency changer to control the fans' rotational speed.

2.2 *The testing process*

In each environment temperature of the testing point, all the variable frequencies of the compressor and the fan frequency condition, the refrigerant injection quantity, were not adjusted, but the outlet temperature of the condenser refrigerant remained

Table 1. The operating condition.

The test environment temperature		The compressor operating frequency	
	18°C		12 Hz
	25°C		20 Hz
	30°C		25 Hz
	35°C		30 Hz
	40°C		35 Hz
	43°C		40 Hz
			45 Hz
			50 Hz

basically the same under various test conditions by changing the electronic expansion valve opening. The outlet temperature of the condenser refrigerant decreased gradually when the frequency of the compressor operation reduced. When there was no guarantee that the environment temperature and the outlet temperature of the condenser refrigerant were exactly the same, we could modify the experimental data through the following methods. All the benchmark test data were the same.

2.3 *Temperature-revised method*

The temperature-revised method for the outlet temperature of the condenser refrigerant is as follows. The temperature difference is the inlet saturation temperature minus the inlet air temperature, $dT_{sat} = T_{sat,coil} - T_{air,on}$. The outlet temperature of the condenser refrigerant $T_{liq,out}$, the average of all the environmental temperature point test data $T_{liq,average}$, and the condensation heat transfer temperature difference were corrected according to the following formula: $dT_{sat,correct} = dT_{sat} - (T_{liq,average} - T_{liq,out})*0.3$. The difference in the condensation heat transfer temperature depends on the condensing temperature and the liquid temperature. If the liquid temperature is low at one test condition, the difference in the heat transfer temperature difference would reduce (Yang Haiming, 2008).

2.4 Tempeature curve drawing

The functional relationship between the transfer temperature difference and the condensation heat transfer graphics is plotted. It is the proportional or approximate quadratic curve relationship. We can analyze the heat transfer performance of different heat exchangers and boil down to all the factors affecting the heat transfer temperature difference and the condensation heat transfer of data.

3 RESULTS AND DISCUSSION

Figure 2 to Figure 5 show the test point at the 35°C environmental temperature. The actual test environmental temperature range is from 34.7°C to 35.5°C.

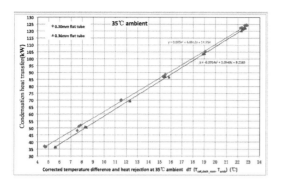

Figure 2. The relationship curve of the corrected temperature difference and heat rejection at the 35°C ambient condition.

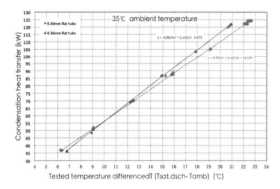

Figure 3. The relationship curve of the tested temperature difference and heat rejection at the 35°C ambient condition.

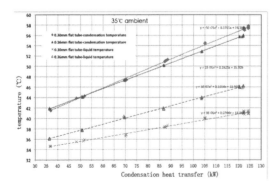

Figure 4. Comparison of the condensing temperature and the liquid temperature at the 35°C ambient condition.

In the testing process, the compressor frequency was reduced gradually from 50 Hz to a minimum of 12 Hz. When reducing the frequency of the compressor operation, the air displacement through the micro-channel heat exchanger decreased and the heat load reduced. The heat transfer temperature

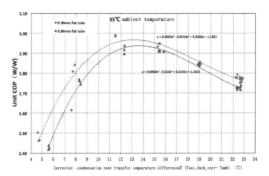

Figure 5. Unit COP at the 35°C ambient condition.

difference also decreased. From Figure 4, we can see that the outlet temperature of the condenser refrigerant with the 0.30 mm flat tube is about 34.6°C~41.5°C, and the outlet temperature with the 0.36 mm flat tube is about 35.6°C~46.3°C. Heat transfer is big at a full load and the liquid temperature is high. Conversely, heat transfer is small at a minimum load and the liquid temperature is low. The correct condensation heat transfer temperature difference in all the tests with the average liquid temperature of 40.2°C was observed. The influencing factors of temperature, pressure drop and heat transfer coefficient of the liquid phase can boil down to $dT_{sat,correct}$. The deviations of the inlet and outlet water temperature of chilled water, environment temperature, the frequency of the compressor operation were reflected in the experimental data points and the contrast of the fitted curve. We can see that the curve is basically parallel.

From Figure 2, we can see the condensation heat transfer of the 0.36 mm flat tube to be low around 2.0% to 10.2% than that of the 0.30 mm flat tube under the same revised condensation heat transfer temperature difference. Similarly, under the revised condensation heat transfer, the condensation heat transfer temperature difference was found to be low around 0.36°C to 0.97°C, at an average below 0.57°C. The heat transfer temperature difference is calculated according to the condensation temperature, environmental temperature and the outlet temperature of the condenser liquid phase.

Figure 3 shows that the outlet temperature of the condenser refrigerant, with the 0.30 mm flat tube being low around −0.5°C to 1.5°C than the 0.36 mm flat tube, and the liquid temperature are lower around 1.3°C to 5.4°C.

From Figure 5, it can be seen that the chiller COP with the 0.36 mm flat tube of the micro-channel heat exchanger is lower around 5.5%~11.4% than that with the 0.30 mm flat tube. Table 2 presents the COP with the 0.36 mm flat tube of the micro-channel heat exchanger at several different experimental conditions.

Table 2. COP with the 0.36 mm flat tube.

Environment temperature	COP
18°C	lower around 2.35%~3.3%, reduce by an average of 2.9%
25°C	lower around 0.28%~1.58%, reduce by an average of 0.79%
30°C	lower around 4.5%~7.4%, reduce by an average of 6.1%
35°C	lower around 5.5%~11.4%
40°C	lower around 1.2%~2.1%
43°C	lower around 1.25%~4.4%, reduce by an average of 3.3%

4 CONCLUSIONS

The article puts forward the test method and data analysis method to compare the different kinds of the heat transfer performance of heat exchanger. It can boil down to all the factors affecting the heat transfer temperature difference and condensation heat transfer data.

The test selected 35°C as the environmental temperature test point and compared the performance of different wall thicknesses of the flat microchannel heat exchanger. Using the wall thickness of 0.36 mm, the heat transfer temperature can be improved about 0.36°C to 0.97°C, and the COP can be reduced by about 1.2%~11.4%.

ACKNOWLEDGMENTS

This work was supported by Montreal Convention on the International Multilateral Fund Projects (C/III/S/14/358).

REFERENCES

[1] Wang Ying, Xu Bo, Chen Jiangping, et al. (2014). The Experimental Research of Microchannel Heat Exchanger on Packaged Air Conditioning System. *J. Journal of Refrigeration*.
[2] Yan Ruidong, Xu Bo, Chen Jiangping, et al. (2013). The impact on air condition system of two-phase distribution in microchannel heat exchanger. *J. Journal of Refrigeration*.
[3] Yang Haiming, Zhu Kuizhang, Zhang Jiyu, et al. (2008). Study on the flow and heat-transfer properties of microchannel heat-exchanger. *J. Cryogenics and Superconductivity*.
[4] Zhang Chao, Liu Ting, Zhou Guanghui. (2011). Application analysis for micro-channel heat exchanger in refrigeration and air-conditioning system. *J. Cryogenics & Superconductivity*.
[5] Zhang Lei. (2010). Analysis of performance of air conditioners with micro-channel heat exchangers. *J. Refrigeration Technology*.
[6] Zhao Yu, Qi Zhaogang, Chen Jiangping. (2009). Flow Configuration in Micro-channel Parallel Flow Evaporator. *J. Journal of Refrigeration*.

Material Science and Engineering – Chen (Ed.)
© 2016 Taylor & Francis Group, London, ISBN 978-1-138-02936-1

Effects of the production process on the properties of inorganic non-metallic material

F.Q. Zhang, X.T. Yue, L.N. Xu & Q.B. Tian
College of Material Science and Engineering, Shandong Jianzhu University, Jinan, China

ABSTRACT: Inorganic non-metallic material is composed of oxides, carbides, silicates and other substances of certain elements. It is primarily held in covalent bonds. Therefore, it has a high melting point, hardness, oxidation resistance; excellent electrical and optical properties. It includes cement, glass and ceramics. The production of cement is primarily determined by aluminum silicate materials, siliceous materials and calcareous materials, which provide the main components of cement including CaO, SiO_2 and Al_2O_3. Besides, with the improvement of production technology, the current output is over 10,000 tons per day. Furthermore, the quality of products is improved greatly. The 28-day intensity can reach up to 62.5 Mpa, which can fully meet the requirements of building applications.

Keywords: inorganic non-metallic materials; reforming production process; performance

1 INTRODUCTION

The production of inorganic non-metallic materials is mainly dependent on raw materials such as aluminum silicate, siliceous material, and calcareous material to provide CaO, SiO_2 and Al_2O_3. It is provided in the form of bulk material with hard texture. Fragmentation process is essential for homogenization and baking ingredients. Preparation of powder is one of the indispensable processing sectors for most inorganic non-metallic materials. Powder has a high specific area, which is essential for the preparation of ceramic ingredients, cement raw materials and cement products. Usually, we prepare the powder of ceramics in the wet process, and for cement and glass, we prepare it in dry process. Shaping process is due to the demand of use and further processing. Shaping is one of the processes, but it can come in different orders in the process. That is, for ceramics, it might come before the high-temperature hot working, while for glass it can come after that. And for cement, shaping means processing into concrete products. Drying involves the removal of free water from the material or stock. The clay and mixed materials need drying before the preparation of powder in cement, and the shaping of the green body of ceramics needs drying to be sintering. As to the high temperature treatment, for cement, it means obtaining a liquid phase and clinker by rapid cooling. For glass, it means obtaining a liquid phase and preventing the crystalization of glass. Besides, for ceramics, it is the process of acquiring a sintered body. The above methodology shows that all

the products have a similar production process as well as a significant difference.

Inorganic non-metallic material is composed of oxides, carbides, silicates and other substances of certain elements. It is primarily held in covalent bonds. Therefore, it has a high melting point, hardness, oxidation resistance, and excellent electrical and optical properties. It includes cement, glass and ceramics. So, there is generality and individuality in its production process, which generally consists of material crushing, powder preparation, shaping, drying and high-temperature treatment. The production process of different products is different, which will lead to diverse performance. However, the main processes are similar. So, the performance characteristics of inorganic non-metallic materials determine the similarity in its production process, and vice versa.

Inorganic non-metallic materials technology experiment is an experimental design that attaches to the basic technical course of inorganic non-metallic materials technology in materials science and engineering. The main content of this experimental design covers the experimental preparation of cement, glass and ceramic. In theoretical courses of inorganic non-metallic materials technology experiment, the description of the equipment, experimental procedures, experimental methods and principles is usually boring. It is hard for students to understand because of the abstract content, which, in turn, leads to boredom and a bad teaching effect.

Moreover, the planned experiments are usually needed to be verified. However, the relatively less comprehensive experiments and simple processing reduce the students' learning motivation. To overcome these problems, it is necessary to reform the experimental teaching of this course appropriately [1–3].

2 EXPERIMENTAL PROCEDURE

The component of inorganic non-metallic material is determined by the X-ray photoelectron spectrometer. The strength of cement is tested by the automation testing system in a universal testing machine.

3 RESULTS AND DISCUSSION

For the actual production of cement, glass and ceramic usually involve a relatively complicated large equipment, and it is not possible to carry out the experiments in the laboratory. Therefore, it is necessary to study the specific materials individually.

In the inorganic non-metallic materials technology experiment, the cement part mainly covers the production of cement, including the ratio of raw materials, mixing and the so-called "two grinding and one burning" process. However, for modern companies, they are new dry process kilns used to produce cement with a daily output of 5000–16000 tons. Figure 1 shows the daily average output of one cement plant in different years.

From the chart, we can find that with the development of technology, the daily average output of one cement plant increases across the years.

Therefore, it is impossible to simulate the actual production in a laboratory. Depending on this situation, it is necessary to develop related experimental animations using a modern teaching method to simulate cement production [4–6].

On the one hand, it makes the students to well understand the experimental equipment and production process. On the other hand, it can improve the learning interest and learning activity. There-fore, as for the experimental course of cement, we can integrate the experimental animations and small experiments. In small experiments, students can simulate mixing and sintering in a high-temperature furnace according to their knowledge learnt from the theory courses [7–9].

Because most of the students graduated from the School of Materials & Engineering, Shandong Jianzhu University would find a job in the construction organization associated with projects, the curriculum design and courses arrangement place emphasis on engineering. Specifically, many engineering courses are created, such as "Building Construction", "Cementitious Materials Science", and "Concrete Science", and in most of these courses, the experiments are related to cement and concrete. In view of the coherence of knowledge, it is necessary to integrate the "Inorganic Non-metallic Materials Technology Experiment" and "Building Construction Experiment". On the one hand, it is convenient to use the calcine clinker prepared in the courses of "Inorganic Non-metallic Materials Technology Experiment" to produce concrete.

After quenching, grinding and adding other ingredients, the as-prepared calcine clinker can be directly used to produce concrete, followed by testing various properties. This experimental system not only realizes the preparation of cement and subsequent tests including the cement initial setting time, hydration heat and specific surface area, but also realizes the production of concrete and successive tests, including the initial setting time, collapsibility, intensity of 3 days and 28 days, respectively. Table 1 presents the different cement strengths, with the curing time increasing and the strength of the cement being improved obviously.

In the inorganic non-metallic materials technology experiment, the ceramic part mainly covers the ceramic preparation and sintering. There is no fixed formulation for ceramic due to various sources of raw materials. Ceramic could be prepared when the formula is within a reasonable range.

There are two main types of experiments on ceramic, including compression molding and drawing. Compression molding is mainly realized by the simple tablet compressing machines, while the drawing is via a withdrawing machine. Usually,

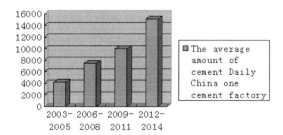

Figure 1.   Cement plant average daily output.

Table 1.   The cement strength.

Time	Strength of the cement
6 h	0.5 Mpa
24 h	2 Mpa
1 day	10 Mpa
3 day	25.5 Mpa
28 day	62.5 Mpa

the tablet compressing machines cannot gain much attention from the students due to its so-called boring and simple operation. However, for students, it is difficult to get a high-quality ceramic using the withdrawing machine. Therefore, here we propose the slip casting process experiments for ceramic preparation because this process has fixed mold that generally leads to high-quality artificial products and improves the students' interests and activity.

In the inorganic non-metallic materials technology experiment, the ceramic part mainly covers the preparation of glass. Similar to cement, there is also no fixed formula for the preparation of glass due to various sources of raw materials. Glass could be prepared when the formula is within a reasonable range. The experiments on glass that are offered in the School of Materials Science and Engineering, Shandong Jianzhu University are mainly about glass melting, which require the students to prepare a transparent glass without bubbles in muffle furnace by themselves.

However, many students cannot accomplish it. On the other hand, there is a big difference between the experimental preparation and the actual production in the factory. To solve this problem, multimedia can be used to simulate the online production of glass. Moreover, students can experience the production of glass in factories via the cooperation of schools and factories. Due to the limitations of experimental conditions in the laboratory, the lack of enough muffle furnace and a long experimental period, cooperation with factories becomes urgent and quite important.

4 SUMMARY

The production of inorganic non-metallic materials mainly depends on raw materials. For example, the production of cement is primarily determined by aluminum silicate materials, siliceous materials and calcareous materials, which provide the main components of cement including CaO, SiO_2 and Al_2O_3. Besides, with the improvement of production technology, the current output is over 10,000 tons per day. Furthermore, the, the quality of products is improved greatly. The 28-day intensity can reach up to 62.5 Mpa, which can fully meet the requirements of building applications.

ACKNOWLEDGMENTS

This work was supported by the Shandong Youth Education Science Foundation (15SC092).

REFERENCES

[1] Wang. Qi: Inorganic non-metallic materials (China Building Material Technology Press, China 2005).
[2] Chen. Shuguang, Xia. Qing and Ye. Chang: China Education Innovation Herald. Vol. 09B (2007), p. 18.
[3] Wu. Yin, Gong. Jianghong and Tang. Zilong: Experimental Technology and Management. Vol. 28 (6) (2011), p. 257.
[4] Ma. Quanshan, Liu. Guangjun: Experimental Technology and Management. Vol. 27(6) (2010), p. 150.
[5] Liu. Huiying, Wang. Yi: Research and Exploration in Laboratory, Vol. 25(6) (2006), p. 25.
[6] He. Feng, Xie. Junlin: Journal of Technology College Education, Vol. 26(5) (2007), p. 84.
[7] Chen. Guohua, Liu. Guizhong: Journal of Technology College Education, Vol. 26(5) (2007), p. 102.
[8] Chen. Yanwen, Niu. Wangyang, Pan. Wenhua and Sun, Xiaowei: Educational Research, Vol. 09(2012), p. 34.
[9] Wang. Yingling, Qi. Xiwei, Luo. Shaohua and Liu. Xuanwen: Experimental Technology and Management. Vol. 28(6) (2011), p. 300.

Other engineering topics

Material Science and Engineering – Chen (Ed.)
© 2016 Taylor & Francis Group, London, ISBN 978-1-138-02936-1

Evacuation strategy and simulation study of urban rail transit emergencies

L. Zhang & Y.B. Lv

School of Traffic and Transportation, Beijing Jiaotong University, Beijing, China

ABSTRACT: In recent years, urban rail transit has received much attention. When an emergency occurs, how to evacuate large passenger flow in a short time becomes an increasingly important topic. This paper considers the example of poisonous gas emergency evacuation simulation at Dongsishitiao Station of line 2, studies the characteristic of passenger flow and the inner construction of the station, calculates the largest number of passengers gathered, sets up four evacuation plans, and uses ANYLOGIC to simulate these plans. Through the analysis of these plans, the corresponding optimal evacuation plan is obtained under different environments.

Keywords: emergencies; urban rail transit; evacuation simulation; social-force model

1 INTRODUCTION

Urban rail transit emergency refers to the influence operational status of urban rail transit on a wide range of events, which contain natural disaster events, production events, and terrorism. In all emergencies in urban rail transit, terrorist events occur less, mainly, for example, the poison gas attack. However, although the frequency of its occurrence is less, it has caused serious consequences every time, such as the poison gas attack in Japan's subway. Therefore, it is necessary to carry out evacuation simulation under gas attack to ensure that at the time of the accident, we can take effective measures to reduce casualties.

Urban rail transit has been developing for more than one hundred years, and various scholars have carried out extensive research on the emergency evacuation of urban rail transit, making a series of achievements. Kikuji Togawa (1955) proposed the empirical formula that includes the factors of passageway and exit. Helbing (2000) established the social force model to simulate the Conformity Phenomenon. Wenlong Lu (2012) proposed a cellular automata model of the fine mesh in the rail transit station for modeling and simulating the pedestrian movement. R.L. Hughes (2000) revealed the movement of Hajj crowd based on the continuum theory successfully. Yanfen Zhang (2012) used a social force model and ANYLOGIC software to build the evacuation model of pedestrian, carried out the analysis of simulation, and evaluated the evacuation plans of the Beijingnan Railway Station.

2 THE CALCULATION OF EVACUATION NUMBERS

Before evacuation simulation, we need to know evacuation crowd. This paper calculates the largest number of passengers gathered, and selects it as the simulation object. The specific principle of calculating is presented below.

Maximum passengers means the internal number of all passengers, including stops traffic and outbound traffic. For the station, there are two entrances that influence the number inside the station: one is the entrance, through which passengers travel in and out of the station; another is the platform, by which passengers arrive or depart the station by train, as shown in Figure 1.

Station traffic is an instantaneous value at a certain moment. In order to calculate the largest number of passengers gathered, this paper assumes that a passenger spends 'a' riding away from the station, and a passenger spends 'b' going the outbound. In other words, a passenger stayed in the station for 'a', and an outbound passenger stayed in the station for 'b'. The assembling at the time 't_0' is both the group rides away from the station after entering into the station and the group leaves the train has not been the outbound. The crowded of entering into the station includes the person of '$t_0 - a$' to 't_0', and the outbound crowded includes the person of 't_0' to '$t_0 + b$' traffic. Station at time t_0 assembling is both combined, as shown in Figure 2.

Through the field survey, line 2 departure intervals, the value of a, and the value of b are given in Table 1.

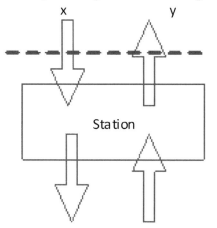

Figure 1. Sources of passenger flow at the station.

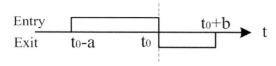

Figure 2. Passenger flow monitoring on time t_0.

Table 1. Dongsishitiao Station measurement data.

Period of time	a (min)	b (min)
16:00–17:00	7.5	5
17:00–18:00	6.5	5
18:00–19:00	5.5	5
19:00–20:00	6.5	5

After the investigation, the time of maximum passengers in the rail transit station is 18:45, and the gathered in the peak at 714. Therefore, the number of evacuation is 714.

3 PEDESTRIAN EVACUATION SIMULATION OF CURRENT DONGSISHITIAO STATION

Dongsishitiao Station lies away from the Workers' Stadium about 1 km. When the stadium has a game or events, most viewers will choose to take the subway forth and back, which can cause a large passenger concentration in the Dongsishitiao Station. In case of emergency, the casualties and social impacts are enormous. Dongsishitiao Station has four entrances that are built in the intersection of the four corners of the place. The station hall is divided into a north hall and a south hall, and has an island platform, as shown in Figure 3.

In this paper, in order to better study the evacuation plans, the same pedestrian source is set. The process of the simulation parameters are as follows:

We randomly set the pedestrian area of diameter between 0.4 m–0.5 m;

Pedestrian speed: we set the speed of the evacuation channel at 1.5 m/s, and the speed of evacuation stairs at 1 m/s;

Number of the total passenger evacuation: we select the largest number of total passenger traffic in Dongsishitiao Station as 714 persons.

This paper designed four kinds of evacuation plan for the evacuation simulation. The specific plans are as follows:

Plan A: passengers walk freely. According to the probability of choice, all the passengers on the platform are free to choose the north hall or the south hall, and the probability that all the passengers will choose to go to the north hall or the south hall is 50%. In this plan, according to the principle, the passengers cannot go back. That is, if a passenger enters the south hall, he cannot choose exit A or B; likewise, if a passenger enters the north hall, he cannot choose exit C or D. After passengers enter the station hall, they makes a choice according to the principle of 1:1 export options.

Plan B: controlling the pedestrian path. The station controls the pedestrian evacuation path. Passengers in Gate 1, Gate 2 and Gate 3 all move to the south hall. Passengers in Gate 1, Gate 2 and Gate 3 all move to the north hall. When passengers arrive the south hall, 50% choose exit C and 50% choose exit D. The same principle is applied for the north hall.

Plan C: expanding capacity. In this plan, the passengers on the platform walk freely. In addition, the probability that all the passengers will choose to go to the north hall or the south hall is 50%. When through the stair, the passenger has three

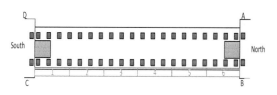

Figure 3. The platform structure.

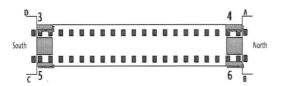

Figure 4. Expanding the stair capacity.

options, for the south of the hall: 60% more likely to choose the stairs, the possibility of 20% chooses emergency staircase through 3, and 20% probability of choosing 5 emergency by stairs. After entering the station hall, 50% of passengers are more likely to choose exit D and 50% are more likely to choose exit C, as shown in Figure 4.

Plan D: controlling the pedestrian path combined with expanding capacity. In this plan, it requires not only the pedestrian path for the whole of the control, but also the expansion of the capacity of platform stairs, and this plan is based on plan 3; therefore, two plans of evacuation numbers are the same. In the specific plan, 60% of pedestrians are in Gate 1 into 5 emergency staircase and 40% of the pedestrian into 1 main staircase. All pedestrians in Gate 2 are into the 1 main staircase, 60% of pedestrians in Gate 3 into the 3 emergency staircase, and 40% of pedestrian into 1 main staircase. In this plan, pedestrians in Gate 1 go to 5 emergency staircase in order to avoid the flow conflict with pedestrians in Gate 2. The same principle is followed by pedestrians in Gates 4, 5 and 6.

4 THE SIMULATION RESULTS

The ANYLOGIC simulation result is as follows: the evacuation time of plan A is 346.1 s, that of plan B is 267 s, that of plan C is 298.1 s, and that of plan D is 242.5 s.

Because there are no measures in plan A, this paper compares plan A with other plans. Through the comparison, we can draw the following conclusion:

Based on the evacuation efficiency of plan A, plan B improves the evacuation efficiency by 23%, plan C improves by 13%, and plan D improves by 30%.

Both the plan of controlling the pedestrian path and the plan of expanding capacity can improve the efficiency of evacuation, and the plan of using a combination of these plans can achieve the best effect.

The evacuation efficiency of plan B increases by 23% over the evacuation efficiency of plan A. The evacuation efficiency of plan D increases by 19% over the evacuation efficiency of plan C. The evacuation efficiency of plan C increases by 14% over the evacuation efficiency of plan A. The evacuation efficiency of plan D is over the evacuation efficiency of plan B. As we can see, the plan of controlling the pedestrian path is more efficient than the plan of expanding capacity by 10%. Therefore, the station should focus on strengthening the staff's ability to control large populations. In the case of high risk, the station can carry out the plan D.

5 CONCLUSION

This paper studies the characteristic of passenger flow, carries out the investigation on Dongsishitiao Station, sets up four evacuation plans, and uses ANYLOGIC to simulate these plans. Through the analysis of these plans, the corresponding optimal evacuation plan is obtained under different environments.

ACKNOWLEDGMENTS

This research was supported by the Center of Cooperative Innovation for Beijing Metropolitan Transportation; Science & Technology Evaluation and Statistics of Special (2014SE-0205).

REFERENCES

[1] Togawa K. 1995. "Study of Fire Escapes Basing on the Observations of Multiple Currents" *Japan. Report No. 14, Building Research Institute, Ministry of Construction.*

[2] Helbing D, Farkas I & Vicsek T. 2000. "Simulating dynamical features of escape panic" *Nature*, 407:487–490.

[3] Wenlong Lu. 2012. "Research of urban rail transit emergency evacuation" *China Academy of Railway Sciences.*

[4] R.L. Hughes. 2000. "The flow of large crowds of pedestrians" *Mathematics and Computers in Simulation*, 53(4–6):367–370.

[5] Yanfen Zhang. 2012. "Study of simulation method of Large passenger traffic emergency evacuation" *Beijing Jiaotong University.*

Material Science and Engineering – Chen (Ed.)
© *2016 Taylor & Francis Group, London, ISBN 978-1-138-02936-1*

Study on emergency evacuation strategies for unexpected large passenger flow in urban rail transit station

Q.W. Ye & Y.B. Lv
School of Traffic and Transportation, Beijing Jiaotong University, Beijing, China

ABSTRACT: In recent years, urban rail transit has become the backbone of the transport system of urban traffic, developing rapidly in several large and medium-sized cities in the country. However, the unexpected large passenger flow poses a serious impact on urban rail system, resulting in economic losses and damage to passengers' travel, and safety of property and life. Therefore, it is of extremely great importance to study issues related to emergency strategies for unexpected large passenger flow in urban rail transit. This paper worked out three emergency plans for the platform of Beijing subway Baishiqiao South Station, conducted a simulation study with the help of simulation software Anylogic, and found the optimal solution.

Keywords: urban rail transit; unexpected large passenger flow; emergency strategies; simulation

1 INTRODUCTION

Urban rail transit includes subway, light rail, tram and magnetic levitation trains, among which subway and light rail are most widely used in major cities. Urban rail transportation has become an effective means to alleviate urban transport for its advantages in traffic volume, speed, safety, comfort, punctuality, environmental protection, energy conservation, etc.

Due to expansion of urbanization and increasing population density in China, as well as the initial formation of the subway network operation pattern, travel by subway traffic is increasing. Currently, daily subway traffic of major cities such as Beijing, Shanghai, and Guangzhou is approaching 7 million people, among which the personnel density is really large in the world. During the subway operation, subway station platform gathers a lot of the public. The passenger flow of the line section, going out and into the station, and transferring is basically at its maximum volume especially at peak hours in the morning and the evening. Some lines reach a maximum long-term design flow as soon as they open. Such a large passenger flow generally will have a great impact on subway system operation.

2 CHARACTERISTIC ANALYSIS

Unexpected large passenger flow in rail transportation system means that a centralized, distributed large passenger flow is formed quickly within a certain period of time in the station which cannot be afforded by the normal passenger transport facilities and daily traffic organization programs.

There are four main reasons why urban rail transit passenger generates unexpected large passenger flow: (1) to hold large-scale cultural and sports activities; (2) tourist travel; (3) severe weather emergencies; (4) operational failures.

Unexpected large passenger flow is an unconventional mega fast passenger distribution, mainly manifesting as crowded or very crowded, or pedestrians moving slowly and the serious interference of station passenger crossover. Here are some basic characteristics of unexpected large passenger flow: (1) the large total passenger flow; (2) the relative concentration of passenger distribution time; (3) the uneven spatial and temporal distribution of passenger flow and obvious peak; (4) the relative dispersion of traffic sources, but more consistent trip purpose; (5) the transitivity of effect on traffic, but relatively shorter duration; (6) the extremely large security risk in generation of unexpected passenger flow.

3 EMERGENCY STRATEGIES

Developing emergency strategy for unexpected large passenger flow in urban rail transit station (hereinafter referred to as emergency strategy) is to choose the most reasonable management and regulation measures to improve the distribution efficiency and reduce operational risk while ensuring safe and efficient distribution of passenger flow.

Here are a few basic principles to follow in making emergency strategies with passenger flow control at its core for urban rail transit to ensure the accuracy and effectiveness of emergency programs.

1. Security
   Pedestrian flow lines in the subway station include: inbound, outbound and transfer. The mass passenger flow brings great impact on security, gates, escalators, etc., which can cause constant conflict, small personal space and the herd, etc., and there will be traffic safety risks.
2. Psychological effects
   A large number of pedestrians gather on station can cause many troubles such as intense oppression on or off the train, obvious bottlenecks, port traffic density, and the impact of unidirectional passenger flow in transfer channels. All of them can cause pedestrian panic and impulsive behavior. Historical experience shows that overcrowding and stampede pedestrian safety is the biggest risk.
3. Scientificity of communication and control systems
   For timely reporting of emergency situations, urban rail transit systems need to have scientific communications systems and equipment such as emergency evacuation and rescue communications system. To guarantee smooth and rescue information disclosure, stations and tunnels should be equipped with necessary materials such as staffing rescue, line rescue and vehicle maintenance, with some emergency communication such as walkie-talkies, telephones, and car radio.
4. Hierarchy
   Systematic emergency decision should be considered at all levels to ensure the integrity of the strategy, including a few basic elements: detection prediction layer, risk identification layer, planning and design layer, emergency decision-making layer, emergency command layer, measures implementation layer, security and rescue layer and so on.

4 SIMULATION STUDY

4.1 A brief introduction of the station

Located at the Capital Gymnasium South Road (north-south) and Chegongzhuang West (EW) intersection, Baishiqiao South Station is a transfer station of the Beijing Metro Line 6 and the Beijing Metro Line 9.

Figure 1.   Baishiqiao South Station location map.

Table 1.   Three plans.

	No. 1 stair	No. 2 stair	No. 3 stair	No. 4 stair
Plan A	Out	In	In	Transfer
Plan B	In	In	Out	Transfer
Plan C	In	Out	In	Transfer

Two lines are both underground line station. Line 6 island platform is located at the third underground and Line 9 island platform is located at the second underground. Line 9 is located above the line 6. Basement of the station is station hall. A joint AFC hall for both lines is set at the "L" shaped node, through which pedestrians walk in and out of the station.

4.2 Plan designation

1. Situation selection
   For some reasons, the road traffic may be paralyzed and become impassable, and thus causing continuing large passenger transport task for the urban rail transit stations to bear. In Baishiqiao South Station, there are too many pedestrians, both drawing out and in, which is the same for the platform of line 9. In this section, the simulation is conducted on the basis of the platform of line 9 in Baishiqiao South Station, where the subway train arrives in two directions.
2. Plan ideas
   Since the number of pedestrians both drawing out and in is big at the platform of Line 9 in the Baishiqiao South Station, a total of four stairs is so designed as two for coming in and the other two for out. No. 4 stair is used for the getting off pedestrians to transfer for line 6, which means that only two of the stairs are for coming-in-platform pedestrians, one for them to leave for the subway hall, and the last one for them to transfer to line 6. In this way, there are three plans. As shown in Table 1, the "in" means pedestrians coming into the platform

through the stairs, "out" represents the getting-off pedestrians leaving the platform of line 9 for the subway hall, and "transfer" means those getting-off pedestrians transferring for line 6 through the stairs.

3. Plan description

Plan A: with guidance, outward pedestrians can only get off from the first six pairs of doors, namely doors No. 1, 2, 3, 4, 5, 6, 1', 2', 3', 4', 5', and 6' to leave the station through No. 1 stair, pedestrians coming in the platform through No. 2 stair can only get on the train through the then six pairs of doors, namely doors No. 7, 8, 9, 10, 11, 12, 7', 8', 9', 10', 11', and 12', pedestrians coming in the platform through the No. 3 stair can only get on the train through the then six pairs of doors, namely doors No. 13, 14, 15, 16, 17, 18, 13', 14', 15', 16', 17' and 18', and passengers transferring for line 6, as with the Blizzard case, can only get off the train through the last six pairs of doors, namely doors No. 19, 20, 21, 22, 23, 24, 19', 20', 21', 22', 23' and 24'.

Plan B: pedestrians coming in the platform through No. 1 stair can only get on the train through the first six pairs of doors, namely doors No. 1, 2, 3, 4, 5, 6, 1', 2', 3', 4', 5', and 6', pedestrians coming in the platform through No. 2 stair can only get on the train through the next six pairs of doors, namely doors No. 7, 8, 9, 10, 11, 12, 7', 8', 9', 10', 11', and 12', all pedestrians leaving for the subway hall can only get off through the train's then six pairs of doors and go to the hall through No. 3 stair, namely doors No. 13, 14, 15, 16, 17, 18, 13', 14', 15', 16', 17' and 18', the designation of No. 4 stair is the same with that of Plan A.

Plan C: pedestrians coming in the platform through No. 1 stair can only get on the train through the first six pairs of doors, namely doors No. 1, 2, 3, 4, 5, 6, 1', 2', 3', 4', 5', and 6', all pedestrians leaving for the subway hall can only get off through the train's then six pairs of doors and go to the hall through No. 2 stair, pedestrians coming in the platform through No. 3 stair can only get on the train through the next six pairs of doors, the designation for No. 4 stair is the same with that of Plan A and Plan B.

4. Parameter setting

In order to set the arrival rate of each pedestrian entrance, passenger flow volume of four stairs

at peak hours was investigated. Data has been collected as follows: the total number of passengers coming in the platform through those stairs at peak hours is 6184, among which 4788 passengers are at No. 4 stair. In consideration of much more emergent passenger flow shared by subway, assuming passenger flow increasing by 50%, then a total of 9276 was resulted among which 7182 is for those who transferring for line 6. The arrival rate of stairs No. 1, 2, and 3 are set 9672/3 = 3092 persons per hour, and that of the last six pairs of doors at 7182/12 = 600 persons per hour.

Normally the speed of pedestrians is around 0.5–1 m/s. While in the case of emergent large passenger flow, pedestrian speed will increase, so the pedestrians' speed down stairs in this simulation is set at 0.5–1.5 m/s, and that of pedestrians getting off 0.5–1 m/s.

In the software simulation, pedestrians will automatically select the shortest queue to queue up. The plans will divide the pedestrians into different queues according to the width of the pedestrian stairway. In this scenario, an escalator can be divided into two queues, and ordinary stairs 4 queues, which means allowing two or four pedestrians up and down stairs simultaneously.

4.3 Simulation process and results

1. Simulation process

Plan A:

At the time of the 10th second, there are some pedestrians walking out one after another from the first six pairs of doors, namely No. 1, 2, 3, 4, 5, 6, 1', 2', 3', 4', 5', and 6', towards No. 1 stair, some pour out from No. 2 and No. 3 stairs, and some getting off from the train's last six pairs of doors, namely doors No. 19, 20, 21, 22, 23, 24, 19', 20', 21', 22', 23', and 24', towards No. 4 stair.

At the time of the 30th second, there are more and more pedestrians on the platform. The pedestrian density around No. 2, No. 3 and No. 4 stairs is large. The platform reaches a stable state and there is evidence that people queue automatically.

At the time of the 60th second, train's dwell time is up, and it leaves. As waiting for the

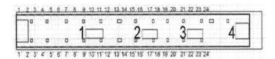

Figure 2. Platform diagram.

Figure 3. Platform condition (10ths, Plan A).

Figure 4.   Platform condition (30ths, Plan A).

Figure 5.   Platform condition (60ths, Plan A).

Figure 6.   Platform condition (10ths, Plan B).

Figure 7.   Platform condition (30ths, Plan B).

Figure 8.   Platform condition (60ths, Plan B).

next train, there is no congestion on the platform. A current-limiting process is conducted to prevent the platform from being too much overcrowded.

Plan B:

At the time of the 10th second, pedestrians out from No. 1 and No. 2 stairs walk towards the first 12 pairs of doors, namely doors No. 1, 2, 3, 4, 5, 6, 1', 2', 3', 4', 5', 6', 7, 8, 9, 10, 11, 12, 7', 8', 9', 10', 11', and 12'. Pedestrians getting off from the train's last 12 pairs of doors walk towards No. 3 and No. 4 stairs to get out of the platform and transfer for line 6.

At the time of the 30th second, the number of pedestrians on the platform increases as well as the density. It is evident that people began to queue up especially in the vicinity of No. 3 and No. 4 stairs.

At the time of the 60th second, there is no sign of congestion. Train's dwell time is up and it leaves the station with doors closed. As waiting for the next train, a current-limiting process is conducted to prevent the platform from being too much overcrowded.

Plan C:

At the time of 10th second, pedestrians pouring out from No. 1 stair walk towards the first six pairs of doors, namely doors No. 1, 2, 3, 4, 5, 6, 1', 2', 3', 4', 5', and 6'. Pedestrians getting off from the train's next six pairs of doors leave the platform through No. 2 stair. Pedestrians coming in through No. 3 stair walk towards the then six pairs of doors to get on the train, those getting off from the last six pairs of doors walk towards No. 4 stair to transfer for line 6.

At the time of 30th second, the number of pedestrians on the platform increases to reach a relatively large density. There is obvious sign of queuing up especially at No. 2 and No. 3 stairs.

At the time of 60th second, the density continues to increase, causing a congestion at No. 1 stair. Train's dwell time is up and leaves the station with doors closed. As waiting for the next train, a current-limiting is conducted to prevent from being too much overcrowded.

2.   Simulation results

In this paper, statistics about pedestrians getting on the train, transferring for another, leaving the station and the platform density in one minute is collected as follows.

It can be concluded that Plan A is the best one, which means No. 1 stair for pedestrians to come in the subway hall, No. 2 and No. 3 stairs for coming in the platform, and No. 4 stair for

Figure 9.   Platform condition (10ths, Plan C).

Figure 10.   Platform condition (30ths, Plan C).

Figure 11.   Platform condition (60ths, Plan C).

Table 2. Plan statistics.

	On (person)	Transfer (person)	Leave (person)	Platform density (person/m^2)
Plan A	430	336	56	0.166
Plan B	359	334	45	0.204
Plan C	446	345	27	0.179

transference for line 6. The biggest advantage of Plan A is to avoid interweaving of passenger "in" and "out" flow, thus improving the emergency efficiency, which manifests as "On" plus "Transfer" and "Leave" being maximum while the relevant platform density being minimum. Secondly, No. 1 stair consists of two columns of escalators which are more likely to get damage than normal stairs. No. 1 stair is organized for pedestrians to get off into the subway hall, thus increasing the security, because the "out" number is much smaller than the "in" number, which makes No. 1 stair shoulder a relatively small load. Finally, the 12 pairs of doors in the middle which are relatively close to No. 2 and No. 3 stairs are organized for pedestrians to access, making the walking distance of pedestrians coming in for a ride relatively short, thus increasing the rationality of the plan, considering that this part of pedestrian is large.

5 CONCLUSION AND EXPECTATION

This paper studies the contingency strategies for emergent large passenger flow in urban rail transit station, analyzes the passenger flow characteristics, flow line conditions and the platform structure on the basis of actual investigation of Beijing Metro Baishiqiao South Station, designs contingency plans, including two kinds of scenarios and four plans, conducts simulations on those plans with the help of AnyLogic software, and evaluates and selects the optimal solution according to the simulation results.

In this paper, the platform emergency is studied. However, as the author's capacity is limited and the study is not deep enough, further study is still urgent mainly in the following areas: simulation on the entire station, instead of merely the platform,

can make it more comprehensive and complete; future research in this area may be conducted from a macro point of view, such as train organizational operations in urban rail transit and the entire urban traffic connection.

REFERENCES

[1] Cao, Zhigang 2013. Unexpected passenger flow evolution mechanism analysis in urban rail transit network and emergency strategy study. *Beijing Jiaotong University.*
[2] Ling, Qiao 2013. Research on the adaptability of transit organization in urban rail transit under outburst mass massenger flow. *Southwest Jiaotong University.*
[3] Zhu, Xiaojie 2010. Countermeasures on high passenger flow at shanghai people square subway hub. *Urban Mass Transit.*
[4] Gao, Xuexiang 2006. A study of countermeasures on metro emergency event. *Urban Mass Transit.*
[5] Li, Mingbo & Shen, Jinsheng 2007. Countermeasures against metro outburst events. *Urban Mass Transit.*
[6] Kyriakidis, Miltos, Hirsch, Robin, & Majumdar Arnab 2012. Metro railway safety: An analysis of accident precursors. *Safety Science* 50: 1535–1548.
[7] Hoogendoom S.P. & Daamen W. 2004. Design assessment of Lisbon transfer stations using microscopic pedestrian simulation. *Proceedings of Computers in Railways* Ix[c]: 135–147.
[8] Liu, Lianhua & Jiang, Liang 2011. Research on passenger flow control method of urban rail transit network. *Railway Transport and Economy.*
[9] Wang, Hao, Hong, Ling & Xu, Ruihua 2012. Analysis of the emergency evacuation guide at metro station. *Urban Rail Transit.*
[10] Gao, Chang 2013. Research on the adaptability of transit organization in urban rail transit under outburst mass passenger flow. *Southwest Jiaotong University.*
[11] Wang, Weinan 2008. Study on operation organization of urban rail transit system under outburst mass passenger flow. *Beijing Jiaotong University.*
[12] Jiang, Yajun & Yang, Qixin 2004. Metro disasters prevention and rescue system. *Urban Rail Transit.*
[13] He, Ziqiang, Mao, Baohua & Song, Lily 2005. Study on passenger security assurance system in railway station under massive passenger flow. *China Safety Science Journal.*
[14] He, Wenkai, Liu, Jian & Wang, Jintao 2006. Safety management in urban rail transit on the pre-crisis management. *Urban Rail Transit.*

Material Science and Engineering – Chen (Ed.)
© 2016 Taylor & Francis Group, London, ISBN 978-1-138-02936-1

Research on the evaluation and selection of the returned logistics operation model of iron and steel enterprise

M.N. Li & Y.B. Lv
School of Traffic and Transportation, Beijing Jiaotong University, Beijing, China

ABSTRACT: This paper sums up three kinds of modes of returned logistics of iron and steel enterprises, analyzes the influencing factors that should be considered when the enterprise evaluates and chooses the suitable logistics mode, establishes the evaluation index system and an example based on the analytic hierarchy process and fuzzy comprehensive evaluation, and gives a detailed description of the evaluation process of the mode of returned logistics.

Keywords: returned logistics mode; selection and evaluation; analytic hierarchy process; fuzzy comprehensive evaluation

1   INTRODUCTION

With a large scale of production, iron and steel industry in China has developed rapidly. Over the years, China has been the largest steel-producing and steel consumption country in the world. However, as the iron and steel industry suffers from long-term overcapacity, this has led to a glut on the market. Domestic iron and steel enterprises are in the pattern of high-input, high-output and low-income circumstances, which become the bottleneck to the development of steel industry. Enterprises rely on the release of production capacity, and increased production capacity to dilute the costs and earn the benefits to a leading opponent, which has been unable to work.

Returned logistics as the enterprises' management function, the same with the other targets, aims to pursue maximum profits. As an important part of the logistics system of iron and steel enterprises, it is one of the important directions of enterprise logistics' development. So, it is crucial for iron and steel enterprises to choose a suitable returned logistics mode.

2   SUMMARY OF THE RETURNED LOGISTICS MODE OF IRON AND STEEL ENTERPRISES

Currently, there are three main kinds of operation mode of the returned logistics of iron and steel enterprises, including self-operating mode, outsourcing mode and enterprise-associating mode. This section mainly introduces the advantages and disadvantages of the three kinds of operation mode and the applicable conditions.

2.1   *Self-operating mode of returned logistics*

Self-operating mode of returned logistics for an iron and steel enterprise refers to the enterprise itself, establishes its independent returned logistics system, and manages the recycling processing business of waste generated during the course of production. Under the self-operating mode, the enterprise establishes the returned logistics network, in order to recycle the solid waste generated in the process of production and take all that waste material to the recycling centers to receive central treatment.

2.2   *Outsourcing mode of returned logistics*

Under the outsourcing mode, in order to focus on enhancing the core competitiveness, iron and steel enterprises will have a slag recycling business either wholly or partly run by enterprises specialized in the returned logistics service through making an agreement and bearing the cost.

2.3   *Enterprise-associating mode of returned logistics*

The enterprise-associating mode of returned logistics refers to some industry enterprises that produce the same products or similar products, and cooperate with each other to form an alliance and invest in an returned logistics system in a joint venture in order to achieve better results than the individuals. This system is responsible for all these iron and steel enterprises' returned logistics business, and this system is also available to other cooperative enterprises at the same time. Then, a partnership among all these enterprises that trust mutually, and share the risk and share profit, will be established.

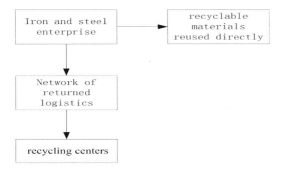

Figure 1. Self-operating mode of returned logistics..

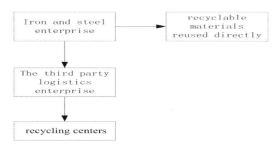

Figure 2. Outsourcing mode of returned logistics.

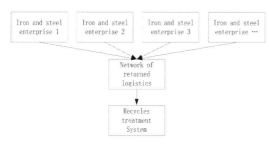

Figure 3. Enterprise-associating mode of returned logistics.

3 CONSTRUCTION OF THE EVALUATION INDEX SYSTEM OF THREE KINDS OF MODES OF RETURNED LOGISTICS

3.1 *Analysis of influencing factors for evaluation*

According to previous-related research, combined with the characteristics of iron and steel enterprises, operating characteristics of the various logistics operation model and their advantages and disadvantages, according to the design idea of the evaluation system, this paper presents the selection process of iron and steel enterprises' suitable returned logistics mode by using a multi-index layer overcoming the shortcomings of a single-index evaluation that emphasizes the selecting mode comprehensively. In the index design, it abandons the indices pursuing the blind profit and seeking economic benefits one-sidedly, increases and strengthens the indices reflecting returned logistics capability for iron and steel enterprises and information service ability, and focuses on the indices making the returned logistics' development balanced. At the same time, the index design fully considers the factors of returned logistics and operational aspects of the company, making the evaluation index to have a stronger practicability.

Therefore, considering various evaluation indices comprehensively, this paper mainly analyzes the selection factors of logistics mode for iron and steel enterprises from the following aspects.

3.1.1 *The strength and characteristics of iron and steel enterprises*

The strength and characteristics of iron and steel enterprises are the influencing factors that must be taken into consideration for logistics operation mode selection from the enterprise itself. It mainly includes the level of logistics facilities, the logistics information technology level, the scale of capital investment, and the control of the logistics activities.

Generally speaking, from the point of the enterprise scale and the strength, larger and stronger iron and steel enterprises usually own better logistics facilities and higher logistics information technology level, and are capable of obtaining funds through various channels for the construction of the logistics system. Besides, it is relatively easy for large-scale enterprises to find the right partner composition complementary logistics alliance. As for the small and medium iron and steel enterprises, with the limitation of staffs, funds and managing resource, it is difficult to improve the efficiency of logistics management. So, relying on the enterprise itself to build a completely self-operation logistics system and logistics alliance is almost impossible, and selecting a service to the good-quality third party logistics supplier will be the first choice.

From the point of enterprise investment scale, it is different among different modes of operation from the logistics aspects of different enterprises. In the self-management mode, the investment of returned logistics is mainly from the enterprise itself. However, for the outsourcing mode and enterprise-associating mode, enterprises only need to bear part of the investment or none of it.

From the point of recyclables' own characteristics, it is applicable to the self-operating mode for the products that can be used directly for the iron and steel enterprises and have a large number. For the smaller number of products or products that have more complicated reprocessing procedures, it is better to select the logistics outsourcing mode.

3.1.2 *Logistics management ability*

Returned logistics management ability refers to the operation and management ability on these three returned logistics operation modes for iron and steel enterprises. From the point of logistics management capability, it reflects the ability of business management, in which information confidential level and management concept are the main factors to be chosen for a logistics agency enterprise or a logistics alliance.

The better the enterprise logistics management services are, the more appropriate the completely self-operating logistics operation mode is in the case of ripe conditions. If some conditions are lacking, the rest of the modes can also be applied.

3.1.3 *Logistics cost*

Logistics cost refers to the monetary manifestation of labor cost in logistics activities, manifesting as the composite cost of recycling collection, transportation and reprocessing.

The level of enterprise logistics cost is one of the important factors to be considered in choosing the kind of logistics operation mode. In the long term, the completely self-operating logistics mode and logistics alliance is beneficial in reducing the logistics costs. However, not all the enterprises can afford the large capital inputs. Hence, it is just for big and abundant capital enterprises. Other medium-sized and small enterprises cannot afford abundant large capital inputs, so it is suitable to choose a third-party enterprise.

3.1.4 *The logistics service ability*

Although the cost is important, the ability to provide logistics services for the enterprises is also crucial. It is mainly reflected in two aspects, namely safety and timeliness.

3.2 *The establishment of the evaluation index hierarchical structure*

According to factors during the selection of the operation mode, the evaluation index hierarchical structure is determined as follows:

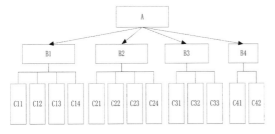

Figure 4. Index system.

A: modes of iron and steel enterprise
B1: strength and characteristics of iron and steel enterprises
B2: logistics management ability
B3: logistics cost
B4: logistics service ability
C11: investment fund scale
C12: level of logistics facilities
C13: level of logistics information
C14: characteristics of recycle materials
C21: quality of manager
C22: ability of communication and inter-personal
C23: level of logistics control
C24: efficiency of logistics process
C31: administration cost of logistics business
C32: transportation cost
C33: investment risk degree
C41: tameness of logistics transportation
C42: security during transportation

4 THE APPLICATION OF AHP AND FUZZY COMPREHENSIVE EVALUATION METHOD

4.1 *Design of index weight*

Several experts are invited to score the weights of each index value through multiple comparisons to obtain a fuzzy judgment matrix. The results of the calculation are as follows.

4.2 *Application of the fuzzy comprehensive evaluation method to evaluate and select the returned logistics mode*

According to the returned logistics model for iron and steel enterprises based on the AHP and fuzzy comprehensive evaluation, the membership matrix of three modes can be obtained through the expert

Table 1. Index weight.

	B1	C11	0.1395
		C12	0.0882
		C13	0.0510
		C14	0.0330
	B2	C21	0.0326
		C22	0.0395
A		C23	0.0791
		C24	0.0586
	B3	C31	0.0999
		C32	0.1816
		C33	0.0550
	B4	C41	0.1138
		C42	0.0379

Table 2. The final result of fuzzy evaluation.

Modes	Evaluation result			
	4	3	2	1
Mode 1	0.2422	0.3510	0.2423	0.1779
Mode 2	0.2416	0.3494	0.2985	0.1186
Mode 3	0.2145	0.3082	0.3145	0.1763

judgment method for the hierarchical fuzzy synthetic evaluation and the final evaluation results as follows.

The assignment of the fitness level and evaluation grade value can be transformed into a vector $Y = [4, 3, 2, 1]^T$, calculating the synthetic evaluation value P_N:

$$P = Y * X \qquad (1)$$

where X is the evaluation result of the three modes.

Finally, the score of the self-operating mode is $P_1 = 2.685$, that of the outsourcing mode is $P_2 = 2.730$, and that of the enterprise-associating mode is $P_3 = 2.5879$.

From the above synthetic evaluation score results, the outsourcing returned logistics mode is most suitable for the iron and steel enterprise.

5 CONCLUSION

Returned logistics is the new horizons in the field of logistics, and is also of great significance for enterprises to achieve resource conservation and environmental protection, which is an important part of the enterprise management strategy.

Selecting a suitable returned logistics mode is discussed in this paper based on the analytic hierarchy process and fuzzy comprehensive evaluation, which will be of practical value.

ACKNOWLEDGMENTS

This paper was supported by the Center of Cooperative Innovation for Beijing Metropolitan Transportation; Science & Technology Evaluation and Statistics of Special (2014SE-0205).

REFERENCES

[1] Minghui Sun, *Study on Establishing Returned Logistics System for Scrap Steel* (D), Beijing Jiaotong University.
[2] Hui Yue, Xueyan Zhong, Huaizhen Ye, *Fuzzy Evaluation Research for The Third Party Reverse Logistics Enterprises* (J), 2006, 19(5): 39–42.
[3] Chunlin Lu, Jiazhen Huo, *Research on scrap reverse logistics in steelworks* (J), 2007, 35(1): 1–4.

Material Science and Engineering – Chen (Ed.)
© 2016 Taylor & Francis Group, London, ISBN 978-1-138-02936-1

Research on the evaluation index system of the comprehensive competitiveness of coal mine projects under coal electricity joint operation conditions based on AHP

R.J. Ding & M. Li

School of Management, China University of Mining and Technology, Beijing, China

ABSTRACT: In this paper, we analyze the factors that influence the comprehensive competitiveness of coal mine projects under coal electricity joint operation conditions. For this purpose, we use the Analytic Hierarchy Process (AHP) method to establish the evaluation index system of the comprehensive competitiveness of coal mine projects. By solving the AHP model, we obtain the priority coefficient of each factor influencing the coal mine projects' comprehensive competitiveness under coal electricity joint operation conditions.

Keywords: comprehensive competitiveness; AHP method; evaluation index system; coal mine projects

1 INTRODUCTION

During 2014, the coal industry suffered industry depression, and coal production, sales and consumption suffered a rare negative growth. This article attempts to construct the evaluation index system of the comprehensive competitiveness of coal mine projects under coal electricity joint operation conditions with the AHP method. The results of this paper will provide theoretical support for the evaluation of the coal mine projects' comprehensive competitiveness scientifically, promoting coal mine projects sustainable development in China.

2 THE CONNOTATION OF COMPREHENSIVE COMPETITIVENESS OF COAL MINE PROJECTS

Coal mine projects can be divided into pre-projects, infrastructure projects and projects putting into production according to the project stage. Pre-projects refer to the stage which contains the project opportunities study until all the work before the construction. Infrastructure projects means the infrastructure construction project. Production projects refer to the complete coal mine trial production and passing the acceptance for use in the production [1].

Li Weiming, Li Weihong, Huo Tianxiang (2012) suggested that the comprehensive competitiveness of coal enterprises can be defined as follows:

in the competitive market conditions, decided by the coal enterprises' different resources, technology, management and other individual competitive advantages, the coal enterprises can continuously provide much more or better quality products or services than other similar enterprises in the market, so as to achieve the maximization of profit's comprehensive ability [2]. Zuomin (1997) constructed the evaluation index system of comprehensive competitiveness of coal enterprises, which include the enterprise resource utilization level, the economic efficiency of enterprises, the enterprise social benefit, enterprise safety level, enterprise growth, enterprise production and management level [3]. Wang Boan (2002) showed that the competition ability index system includes the influence enterprise survival capacity, the ability of enterprise development and enterprise external environment three parts, altogether 47 indices [4]. Zhang Jinchang (2002) put forward the evaluation index system of enterprise competitiveness based on earnings, with the index system mainly including the evaluation index, the competitiveness determinants, and determining factors of the evaluation index [5].

In this paper, we indicate that the comprehensive competitiveness of coal mine projects under coal electricity joint operation conditions can be defined as follows: based on the coal mine projects' internal resources, industry coordination ability, profit ability and the location advantage, and external environmental factors, the comprehensive ability is formed to interact and integrate dynamically.

3 USING THE AHP MODEL TO DETERMINE THE WEIGHT OF THE EVALUATION INDEX SYSTEM

Analytic hierarchy process is a method of the analysis system proposed by the American operations researcher T.L. Saaty [6] in the early 1970s. The AHP decision analysis usually consists of the following steps: (1) establishing the hierarchical structure model; (2) constructing the judgment matrix; (3) calculating the weight vector; and (4) conducting the consistency test.

We determine the comprehensive competitiveness evaluation index system of coal mine projects by the Delphi method, questionnaire method, expert consultation and enterprise research methods. This paper constructs the evaluation index system of coal mine projects' comprehensive competitiveness in different project phases, including industry cooperation, resources, location advantages, the occurrence condition of coal, the external environment and the enterprise profit ability factors. Here, we just take the production coal projects' calculation using the AHP method as an example.

3.1 Establishing the hierarchical structure model

The hierarchical structure model of coal mine projects' comprehensive competitiveness is shown in Figure 1.

3.2 Constructing the judgment matrix and consistency test

We obtain the relative weight index of each judgment matrix calculated by the AHP method. The calculation process is as follows.

① Industry synergy index weight calculation results are as follows.
② Resources condition's index weight is calculated as follows.
③ Location advantage index weight is calculated as follows.
④ The occurrence conditions' index weight is calculated as follows.
⑤ The external environment index weight is calculated as follows.
⑥ The profit ability index weight is calculated as follows.

3.3 Total taxis of hierarchy and consistency test

The process to determining the relative importance of priority weights of all factors of a layer for the overall goal is called the total taxis of hierarchy. This procedure is conducted sequentially from the top to the bottom layer. For the top layer, the single-level sequencing results are also the result of the total order.

With the same method, we can obtain the index weight of the infrastructure and pre-mine projects. The total taxis of hierarchy of the comprehensive competitiveness of coal mine projects under coal electricity joint operation conditions is shown as follows.

The consistency of the total sequencing result can be tested as follows:

Consistency test for pre-coal mine projects:

$$CR = \frac{\sum_{j=1}^{m} CI(j)a_j}{\sum_{j=1}^{m} RI(j)a_j} = \frac{\begin{pmatrix} (0,0.0194,0.0738,0.0035,0.0181) \\ (0.14,0.34,0.30,0.11,0.11) \end{pmatrix}^T}{\begin{pmatrix} (0.58,0.58,1.32,0.9,0.58) \\ (0.14,0.34,0.30,0.11,0.11) \end{pmatrix}^T}$$

$$= \frac{0.0311}{0.8372} = 0.0371 < 0.1$$

Consistency test for infrastructure coal mine projects:

$$CR = \frac{\sum_{j=1}^{m} CI(j)a_j}{\sum_{j=1}^{m} RI(j)a_j} = \frac{\begin{pmatrix} (0,0.0194,0.0738,0.0035,0.0046) \\ (0.14,0.34,0.30,0.11,0.11) \end{pmatrix}^T}{\begin{pmatrix} (0.58,0.58,1.32,0.9,0.58) \\ (0.14,0.34,0.30,0.11,0.11) \end{pmatrix}^T}$$

$$= \frac{0.0296}{0.8372} = 0.0353 < 0.1$$

Consistency test for production coal mine projects:

$$CR = \frac{\sum_{j=1}^{m} CI(j)a_j}{\sum_{j=1}^{m} RI(j)a_j} = \frac{\begin{pmatrix} (0,0.0194,0.0738,0.0035,0.0046,0) \\ (0.15,0.27,0.27,0.08,0.08,0.15) \end{pmatrix}^T}{\begin{pmatrix} (0.58,0.58,1.32,0.9,0.58,0.58) \\ (0.15,0.27,0.27,0.08,0.08,0.15) \end{pmatrix}^T}$$

$$= \frac{0.03}{0.8054} = 0.03 < 0.1$$

Accordingly, the consistency of the total order is acceptable.

3.4 Analysis of the results

Through the analysis, we conclude that the influencing factors of the comprehensive competitiveness of coal mine projects in different stages of the project are different. For pre—and infrastructure projects, the resources condition (weight 0.34) and the occurrence condition of coal (weight 0.30) are the most important factors that influ-

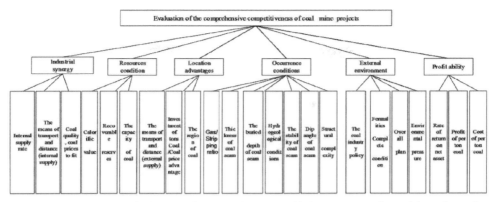

Note: The open-cast coal mine index without gas; the pre—and infrastructure projects without the profit-ability index.

Figure 1.   The hierarchical structure model.

Table 1.   The comprehensive competitiveness of the coal mine project's judgment matrix (production projects).

	Resources condition	Occurrence conditions	Industrial synergy	Profit ability	Location advantage	External environment	Weight
Resources condition	1	1	2	2	3	3	0.2695
Occurrence conditions	1	1	2	2	3	3	0.2695
Industrial synergy	1/2	1/2	1	1	2	2	0.1486
Profit ability	1/2	1/2	1	1	2	2	0.1486
Location advantage	1/3	1/3	1/2	1/2	1	1	0.0819
External environment	1/3	1/3	1/2	1/2	1	1	0.0819

$\lambda_{\max} = 6.018$, $CI = \dfrac{\lambda_{\max} - n}{n - 1} = 0.0037$, $CR = \dfrac{CI}{RI} = 0.00297 < 0.10$. The consistency of the judgment matrix is acceptable.

Table 2.   Industry synergy index weight calculation table.

	Internal supply rate	Means of transport and distance (internal supply)	Coal quality, coal prices to fit	Weight
Internal supply rate	1	3	3	0.60
Means of transport and distance (internal supply)	1/3	1	1	0.20
Coal quality, coal prices to fit	1/3	1	1	0.20

$\lambda_{\max} = 3$, $CI = \dfrac{\lambda_{\max} - n}{n - 1} = 0$, $CR = \dfrac{CI}{RI} = 0.0 < 0.10$. The consistency of the judgment matrix is acceptable.

Table 3.   Resource condition index weight calculation table.

	Calorific value	Recoverable reserves	Coal mine production capacity	Weight
Calorific value	1	2	3	0.5390
Recoverable reserves	1/2	1	2	0.2973
Coal mine production capacity	1/3	1/2	1	0.1638

$\lambda_{\max} = 3.0092$, $CI = \dfrac{\lambda_{\max} - n}{n - 1} = 0.0194$, $CR = \dfrac{CI}{RI} = 0.0334 < 0.10$. The consistency of the judgment matrix is acceptable.

Table 4. The location advantage index weight calculation table.

	Coal price advantage	The region of coal	Means of transport and distance (external supply)	Weight
Coal price advantage	1	2	3	0.5390
Region of coal	1/2	1	2	0.2973
Means of transport and distance (external supply)	1/3	1/2	1	0.1638

$\lambda_{max} = 3.0092$, $CI = \dfrac{\lambda_{max} - n}{n-1} = 0.0046$, $CR = \dfrac{CI}{RI} = 0.0079 < 0.10$. The consistency of the judgment matrix is acceptable.

Table 5. The occurrence condition index weight calculation table.

	Gas/stripping ratio	Stability of coal seam	Hydrogeological conditions	Buried depth of coal seam	Thickness of coal seam	Dip angle of coal seam	Structural complexity	Weight
Gas/stripping ratio	1	5	3	2	2	2	1/3	0.2034
Stability of coal seam	1/5	1	1/3	1/3	1/3	1/3	1/3	0.0463
Hydrogeological conditions	1/3	3	1	2	2	2	1/3	0.1416
Buried depth of coal seam	1/2	3	1/2	1	1/2	1/2	1/3	0.0823
Thickness of coal seam	1/2	3	1/2	2	1	1	1/3	0.1084
Dip angle of coal seam	1/2	3	1/2	2	1	1	1/3	0.1084
Structural complexity	3	3	3	3	3	3	1	0.3095

$\lambda_{max} = 7.4429$, $CI = \dfrac{\lambda_{max} - n}{n-1} = 0.0738$, $CR = \dfrac{CI}{RI} = 0.0559 < 0.10$. The consistency of the judgment matrix is acceptable.

Table 6. Exterior environment index weight calculation table.

	Coal industry policy	Overall planning of mining area	Formalities complete condition	Environmental pressure	Weight
Coal industry policy	1	2	2	3	0.4231
Overall planning of mining area	1/2	1	1	2	0.2272
Formalities complete condition	1/2	1	1	2	0.2272
Environmental pressure	1/2	1	1/2	1	0.1225

$\lambda_{max} = 4.010$, $CI = \dfrac{\lambda_{max} - n}{n-1} = 0.0035$, $CR = \dfrac{CI}{RI} = 0.0038 < 0.10$. The consistency of the judgment matrix is acceptable.

Table 7. Profit ability index weight calculation table.

	Rate of return on common stockholders' equity	Profit of per ton coal	Cost of per ton coal	Weight
Rate of return on common stockholders' equity	1	1	1	0.34
Profit of per ton coal	1	1	1	0.33
Cost of per ton coal	1	1	1	0.33

$\lambda_{max} = 3$, $CI = \dfrac{\lambda_{max} - n}{n-1} = 0$, $CR = \dfrac{CI}{RI} = 0 < 0.10$. The consistency of the judgment matrix is acceptable.

Table 8. Total taxis of hierarchy.

Index layer / Criteria layer	Industrial synergy	Resources condition	Location advantage	Occurrence condition	External environment	Profit ability	Pre-projects	Infrastructure projects	Production projects
Pre and infrastructure projects	0.1365	0.3388	0.1113	0.3008	0.1126	-	Total ordering index		
Production projects	0.1486	0.2695	0.0819	0.2695	0.0819	0.1486			
Internal supply rate	0.60	-	-	-	-	-	0.08	0.08	0.09
Means of transport and distance (internal supply)	0.20	-	-	-	-	-	0.03	0.03	0.03
Coal quality, coal prices to fit	0.20	-	-	-	-	-	0.03	0.03	0.03
Calorific value	-	0.54	-	-	-	-	0.18	0.18	0.15
Recoverable reserves	-	0.30	-	-	-	-	010	010	0.08
Coal production capacity	-	0.16	-	-	-	-	0.05	0.05	0.04
Region of coal	-	-	0.334/0.3	-	-	-	0.04	0.03	0.03
Means of transport and distance (external supply)	-	-	0.26/0.16	-	-	-	0.03	0.02	0.01
Tonnage expenditure/coal price advantage	-	-	0.40/0.54	-	-	-	0.04	0.06	0.04
Gas/stripping ratio	-	-	-	0.20	-	-	0.06	0.06	0.05
Stability of coal seam	-	-	-	0.05	-	-	0.02	0.02	0.01
Hydrogeological conditions	-	-	-	0.14	-	-	0.04	0.04	0.04
Buried depth of coal seam	-	-	-	0.08	-	-	0.03	0.03	0.03
Thickness of coal seam	-	-	-	0.11	-	-	0.03	0.03	0.03
Dip angle of coal seam	-	-	-	0.11	-	-	0.03	0.03	0.03
Structural complexity	-	-	-	031	-	-	0.09	0.09	0.08
Coal industry policy	-	-	-	-	0.4196	-	0.05	0.05	0.03
Overall planning of mining area	-	-	-	-	0.2300	-	0.03	0.03	0.02
Formalities complete condition	-	-	-	-	0.2300	-	0.03	0.03	0.02
Environmental pressure	-	-	-	-	0.1200	-	0.01	0.01	0.01
Rate of return on common stockholders' equity	-	-	-	-	-	0.003	-	-	0.05
Profit of per ton coal	-	-	-	-	-	0.003	-	-	0.05
Cost of per ton coal	-	-	-	-	-	0.003	-	-	0.05

ence the comprehensive competitiveness of coal mine projects. The second is industrial synergy ability (weight 0.14). The location advantage (weight 0.11) and external environment (weight 0.11) are at the end of the factors. For the production projects, the resources condition (weight 0.27) and the occurrence condition of coal (weight 0.27) are the most important factors that influence the comprehensive competitiveness of coal mine projects. The second is industrial synergy ability (weight 0.15) and profitability (weight 0.15), and the last is the location advantage (weight 0.08) and external environment (weight 0.08). Through the level of total ordering of all of these factors, the principal factors influencing the comprehensive competitiveness of coal mine projects include internal supply rate, the calorific value, recoverable reserves, and gas.

4 SUMMARY

The above analysis shows that, in order to increase the competitiveness of China's coal mine projects, obtaining high-quality coal resources is the prerequisite. On this basis, improving the industrial coordination level and profitability is the key factor. In addition, we need to pay close attention to the impact of the external environment and location condition on the coal mine's comprehensive competitive in the coal mine projects' development and operation process. The result can basically reflect the actual situation of the coal projects' comprehensive competitiveness. The results of this study also provides theoretical support for how to improve the comprehensive competitiveness of coal mine projects under coal electricity joint operation conditions.

ACKNOWLEDGMENT

This research was supported by the Twelfth Five Years of National Science and Technology Support Program (No. 2013BAK04B01-03).

REFERENCES

[1] Qiao, Z., Qiao, Z.C., Cheng, W.F. 2008. The contents and steps of the preliminary work for coal mining projects. *Coal Engineering*, 3:112.
[2] Li, W.M., Li, W.H., Huo, T.X. 2012. Study on the evaluation index system of China's comprehensive competitiveness of coal enterprises. *Chinese coal magazine*, 5:15–17.
[3] Zuo, M. 1997. Study on evaluation index system of the comprehensive competitiveness of coal enterprises. *Coal Economic Research*, 5:29–30.
[4] Wang, B.A. 2002. Design of evaluation index system of enterprise competition ability. *Journal of industrial technology economy*, 2:35–37.
[5] Zhang, J.C. 2002. The theory and method of international competitiveness evaluation. Beijing: *Economic Science Press*, 75–88.
[6] Saaty, T.L. 1978. Modeling unstructured decision problems-the theory of analytical hierarchies. *Math Comput Simulation*, 20:147–158.

Material Science and Engineering – Chen (Ed.)
© *2016 Taylor & Francis Group, London, ISBN 978-1-138-02936-1*

Double image encryption based on multiple-parameter fractional Fourier transform

F.S. Song, H.Y. Mao & X.B. Meng
Department of Automation Control Engineering, Shenyang Institute of Engineering, Shenyang, China

ABSTRACT: A double image encryption scheme is proposed by using discrete multiple-parameter fractional Fourier transform. Therefore, the purpose of the paper is to improve the key space of the system. One of the two original images was encoded into the phase function of a complex signal after being scrambled, and the other original image was encoded into its amplitude. The method of encoding double images was used to multiply the complex signal by a pixel scrambling operation and random phase mask, and then to encrypt by a discrete multiple-parameter fractional Fourier transform. In conclusion, numerical simulations were done to prove the validity and security of the proposed method. In this paper, we present a novel double image encryption algorithm, and provide a much higher security level.

Keywords: double image encryption; discrete multiple-parameter fractional fourier transform; pixel scrambling

1 INTRODUCTION

Nowadays, image's security is becoming more and more important, so more researchers are working on image encryption and many image encryption methods have been proposed [1–15]. However, most of them can only encrypt one image simultaneously [1–7]. In many case, two images with strong relevance need to be encrypted at the same time. Zhong Z et al. [13] encoded one target image into the phase function of a complex signal, and encoded the other target image into its amplitude. This can improve the efficiency of the system greatly. For the image encryption algorithm, the conventional Double Random Phase Encoding (DRPE) is used most widely. The conventional double random phase encoding technique allows the target image to be multiplied in the random phase, and then performs a Fourier transformation, and then after a second random phase, it multiplies in the frequency domain and applies the Fourier transform to obtain the encrypted image with a stable white noise distribution. However, this algorithm is linear, and uses only two random phases as the system's key, and it is vulnerable to a variety of attacks and can crack easily [11–12]. Therefore, we change the Fourier transformation to the fractional Fourier transform, whose orders also serve as the keys. This method can increase the system's key space and improve the system's security [13–14]. However, four keys are not enough. Shan MG et al. [15] applied the Discrete Multiple-Parameter Fractional Fourier Transform (DMPFRFT) to image encryption. The discrete multiple-parameter

fractional Fourier transform has more number of parameters, and the larger system's key space can improve system performance significantly.

In this paper, a method using discrete multiple-parameter fractional Fourier transform is proposed to encrypt two images into one complex signal. This method has a larger key space and the performance is better. Numerical simulations show that the proposed scheme provides an efficient and secure way for double image encryption.

The rest of this paper is organized as follows. Section 2 introduces the proposed double image encryption method. Section 3 presents the numerical simulation results to demonstrate the performance of the method. Finally, Section 4 states the conclusions.

2 PRINCIPLE

The DMPFRFT has a transform kernel with multiple parameters. It gives us more choices to represent signals with extra degrees of freedom by vector parameters. We first explain the operations by using a one-dimensional case. For a transformable digital signal $\mathbf{X} = (x_0, x_1, \ldots, x_{N-1})^T$ with the length of an arbitrary integer N, discrete multiple-parameter fractional Fourier transform with order α, periodicity M and vector parameter $\mathbf{n}'$ can be defined [11] as follows:

$$\mathbf{X}_M^\alpha (\mathbf{n}') = F_M^\alpha (\mathbf{n}') \mathbf{X} \tag{1}$$

where $\mathbf{n}' = (n_0', n_1', \cdots n_{(M-1)}') \in \mathbb{Z}^M$, and the eigen-decomposition structure of **DMPFRFT** is given by

$$F_M^\alpha(\mathbf{n'}) = VD^\alpha V^T = \begin{cases} \sum_{k=0}^{N-1} \exp\left\{(-2\pi i/M)\left[\alpha\left(\text{mod}(k,M)+n'_{\text{mod}(k,M)}M\right)\right]\right\}v_k v_k^T & \text{for N odd} \\ \sum_{k=0}^{N-2} \exp\left\{(-2\pi i/M)\left[\alpha\left(\text{mod}(k,M)+n'_{\text{mod}(k,M)}M\right)\right]\right\}v_k v_k^T \\ \quad + \exp\left\{(-2\pi i/M)\left[\alpha\left(\text{mod}(N,M)+n'_{\text{mod}(N,M)}M\right)\right]\right\}v_{N-1}v_{N-1}^T & \text{for N even} \end{cases} \tag{2}$$

and T is the matrix transpose; V is a matrix with the eigenvectors as column vectors, i.e. $V = [\mathbf{v}_0|\mathbf{v}_1|\dots|\mathbf{v}_{N-2}|\mathbf{v}_{N-1}]$ for N odd and $V = [\mathbf{v}_0|\mathbf{v}_1|\dots|\mathbf{v}_{N-2}|\mathbf{v}_N]$ for N even; and D is a diagonal matrix with its diagonal entries corresponding to the eigenvalues for each column eigenvectors $\mathbf{v}_k$ in V.

Figure 1 shows the proposed encryption process. Let $f(x,y)$ and $g(x,y)$ represent the two target images. For encryption, one pixel scrambling operation is applied to the target image $g(x,y)$, and then the result is encoded in a full phase function. Next, the two images are multiplied by each other to obtain the complex signal $M(x,y)$, which can be expressed as follows:

$$M(x,y) = f(x,y)\exp\left\{i\pi J_1[g(x,y)]\right\} \tag{3}$$

Then, the complex signal $M(x,y)$ is applied to another pixel scrambling operation, and multiplied by the random phase mask $\exp[ip(x,y)]$, where $p(x,y)$ is statistically white sequence uniformly distributed in $[0,2\pi]$. Finally, by using 2D DMPFRFT with parameters of (ML,MR;αL,αR; mL,nL; mR,nR), the final distribution $E(x,y)$ of the output encrypted image can be expressed as follows:

$$\begin{aligned} E(x,y) &= F_{(M_L,M_R)}^{(\alpha_L,\alpha_R)}\left(\mathbf{n'_L},\mathbf{n'_R}\right) \\ &\quad \times \left\{J_2[M(x,y)]\exp[ip(x,y)]\right\} \\ &= F_{(M_L,M_R)}^{(\alpha_L,\alpha_R)}(x,y) \\ &\quad \times \left\{J_2\left\{f(x,y)\exp\left\{i\pi J_1[g(x,y)]\right\}\right\}\exp[ip(x,y)]\right\} \end{aligned} \tag{4}$$

Figure 2 shows the proposed decryption process. The encrypted image is applied by the inverse of 2D DMPFRFT, and then multiplied by the conjugate of RPM. After using the inverse pixel scrambling, the complex signal M(x,y) can be obtained as follows:

$$M(x,y) = J_2^{-1}\left\{F_{(M_L,M_R)}^{(-\alpha_L,-\alpha_R)}[E(x,y)]\exp[-ip(x,y)]\right\} \tag{5}$$

The two decrypted images can be expressed as follows:

$$f(x,y) = |M(x,y)| \tag{6}$$

$$g(x,y) = J_1^{-1}\left(|\arg\{M(x,y)\}/\pi|\right) \tag{7}$$

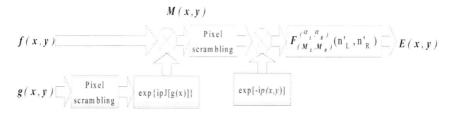

Figure 1. Schematic of encryption.

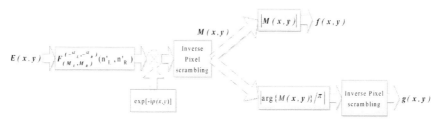

Figure 2. Schematic of decryption.

It can be seen that to retrieve the original image $f(x, y)$, the periodicities (M_L, M_R) transform orders (α_L, α_R) and vector parameters $(\mathbf{m}_L, \mathbf{n}_L; \mathbf{m}_R, \mathbf{n}_R)$ of DMPFRFT, and the pixel scrambling operation J_2 and RPM are essential. However, when retrieving the original image $g(x, y)$, not only the above keys but also the pixel scrambling operation J_1 is needed. They all serve as the keys of this algorithm, and the key space is improved significantly.

3 SIMULATION RESULTS

Numerical simulation experiments are carried out to verify the proposed encryption method. We take "Lena" and "camera man" with the size of 256×256 pixels, as shown in Figure 3(a) and (b), as the two target images. We let Lena as the amplitude-based image $f(x, y)$ and camera man as the phase-based image $g(x, y)$. After pixel scrambling J_1, we can get the scrambled $g(x, y)$, as shown in Figure 3(c). It can be seen that we cannot obtain any information of $g(x,y)$ in this result. The DMPFRFT parameters are $(M_L, M_R; \alpha_L, \alpha_R) = (15,20;0.34,0.73)$. The vector parameters $(\mathbf{m}_L, \mathbf{n}_L)$ and $(\mathbf{m}_R, \mathbf{n}_R)$ are 1×15 and 1×20 random vec-

tors whose values are independent integer values, respectively. Figure 3(d) shows the final encrypted image with stationary white noise. We also cannot obtain any useful information about the two target images. When the correct system keys are used to decrypt, the retrieved images "Lena" are shown in Figure 3 (e). "Cameraman" can be obtained as shown in Figure 3(f), which are exactly the same as the target images without any noise or distortion.

In the following analysis, the influence of the deviation of different keys on the decrypted images is considered. Figure 4(a) and (b) shows the decrypted images obtained without the correct inverse pixel scrambling operation J_2. Figure 4(c) and (d) shows the images obtained by the wrong RPM. Figure 4(e) and (f) shows the decrypted images obtained without the correct inverse pixel scrambling operation J_1. Figure 5 shows the frames of decrypted movies obtained by the wrong keys of DMPFRFT. It can be seen that when the RPM, the inverse pixel scrambling operation J_1 and J_2 and/or DMPFRFT parameters are not correct, the image "camera man" can hardly be retrieved even if the other keys are correct. However, when the RPM and the inverse pixel scrambling operation J_1 are wrong, the image "Lena" can be recognized and its key space is smaller. However, the whole system still has high security.

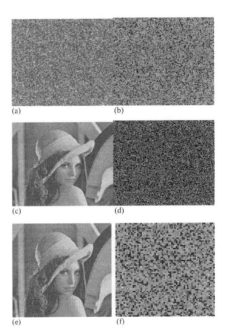

Figure 3. (a) Target image "Lena"; (b) target image "cameraman"; (c) scrambled image of "cameraman"; (d) the encrypted image; (e) decrypted "Lena"; (f) decrypted "cameraman".

Figure 4. (a)–(b) Decrypted images obtained without the correct inverse pixel scrambling operation J_2; (c)–(d) decrypted images obtained by the wrong RPM; (e)–(f) decrypted images obtained without the correct inverse pixel scrambling operation J_1.

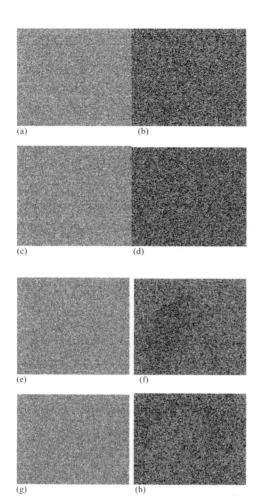

Figure 5. Decryption M(x, y) with different incorrect keys: (a)–(b) {(14,19); (−0.34, −0.73); $(\mathbf{m}_L, \mathbf{n}_L)$; $(\mathbf{m}_R, \mathbf{n}_R)$}, (c)–(d){(15,20); (−0.34+10^{-7}, −0.73+10^{-7}); $(\mathbf{m}_L, \mathbf{n}_L)$; $(\mathbf{m}_R, \mathbf{n}_R)$}, (e)–(f){(15,20); (−0.34, −0.73); $(\mathbf{m}_L+2, \mathbf{n}_L+1)$; $(\mathbf{m}_R, \mathbf{n}_R)$}, (g)–(h) {(15,20); (−0.34, −0.73); $(\mathbf{m}_L, \mathbf{n}_L)$; $(\mathbf{m}_R+1, \mathbf{n}_R+1)$}.

4 CONCLUSION

In this paper, we present a novel double image encryption algorithm based on the discrete multi-parameter fractional Fourier transform. The method enhances the key space of the system by using discrete multi-parameter fractional Fourier transform, and further provides a much higher security level. The computer simulation results show that the algorithm is sensitive to the keys. Naturally, as a novel method, there are still such many other questions to be further investigated, which will be our future task.

ACKNOWLEDGMENTS

This research was supported by the Scientific Study Project of Educational Commission in Liao Ning Province, China, No. L2014523, and Research on Cooperative Spectrum Detecting and Fusing Technology in Cognitive Radio.

REFERENCES

[1] Schwartz C. A new graphical method for encryption of computer data [J]. Cryptology, 1991, 15(1): 43–46.

[2] Fridrich J. Image encryption based on chaotic maps [C]. Proceedings of the IEEE International Conference on Systems, Man and Cybernetics, 1997, 2(1): 1105–1110.

[3] Hennelly B., Sheridan J.T. Optical Image Encryption by Random Shifting in Fractional Fourier Domains [J]. Opt. Exp., 2003, 28(4): 269–271.

[4] Liu S.T., Ren H.W., Zhang J.D., Zhang X.Q. Image-scaling problem in the optical fractional Fourier transform [J]. Applied Opt, 1997, 36(23): 5671–5674.

[5] Hennelly B.M., Sheridan I.T. Image encryption and the fractional Fourier transform [J]. Optik, 2003, 114(6): 251–265.

[6] Ozaktas H.M., Kutay M.A., Zalevsky Z. The fractional Fourier transform with Applications in Optics and Signal Processing [C]. John Wiley & Sons, 2000.

[7] Mendlovic, Ozaktas H.M., Lohmann. Fourier functions and fractional Fourier transform [J]. Opt. Commun., 1994, 42(11): 3084–3091.

[8] Meng X.F., Cai L.Z., He M.Z., Dong G.Y., Shen X.X. Cross-talk-free double-image encryption and watermarking with amplitude-phase separate modulations [J]. J. Opt. A: Pure Appl. Opt., 2005, 7: 624–631.

[9] Liu Z.J., Liu S.T. Double image encryption based on iterative fractional Fourier transform [J]. Optics Communications, 2007, 275: 324–329.

[10] Li H.J., Wang Y.R. Double-image encryption based on discrete fractional random transform and chaotic maps [J]. Optics and Lasers in Engineering, 2011, 49(7): 753–757.

[11] Refregier, Philippe, Javidi, Bahram. Optical image encryption based on input plane and Fourier plane random encoding [J]. Optics Letters, 1995, 20(7): 767–769.

[12] Javidi B., Sergent A., Zhang G., Guibert L. Fault tolerance properties of a double phase encoding encryption technique [J]. Optics Engineering, 1997, 36(4): 992–998.

[13] Zhong Z., Chang J., Shan M.G., Hao B.G. Fractional Fourier-domain random encoding and pixel scrambling technique for double image encryption [J]. Opt. Commun., 2012, 285: 18–23.

[14] Zhong Z., Chang J., Shan M.G., Hao B.G. Double image encryption using double pixel scrambling random phase encoding [J]. Opt. Commun., 2012, 285: 584–588.

[15] Shan M.G., Chang J., Zhong Z., Hao B.G. Double image encryption based on discrete multiple-parameter fractional Fourier transform and chaotic maps [J]. Opt. Commun., 2012, 285: 4227–4234.

Material Science and Engineering – Chen (Ed.)
© 2016 Taylor & Francis Group, London, ISBN 978-1-138-02936-1

An information service model based on "Cloud plus Terminal" for agriculture extension

L.F. Guo & W.S. Wang

Agricultural Information Institute of Chinese Academy of Agricultural Sciences, Beijing, China
Key Laboratory of Digital Agricultural Early-Warning Technology, Ministry of Agriculture, Beijing, P.R. China

ABSTRACT: Agriculture extension takes a very important role in the technology transformation. Information service for extension workers is necessary and urgent. Followed the development of ICTs, the model of information service is also changing quickly. Cloud computing and smart devices are the representative of the information technology of the next generation. Hence, this paper proposed a novel information service model based on "Cloud plus Terminal" and discussed what we have done. With the new information service model, extension workers can access the information service anywhere, anytime.

Keywords: agricultural extension; information service model; Cloud plus Terminal

1 INTRODUCTION

Chinese farmers, who have only 7% of the world's arable land, feed 22% of the world's population. Agriculture extension services are considered crucial for farmers to be able to improve their practice and livelihoods. The extension system in China is charged with spreading new agricultural technologies to rural inhabitants. China has the largest extension system in the world. In the last 50 years, great achievement has been accomplished in extension system in China and many lessons have also been accumulated. However, extension system still faces a lot of problems, especially the basic-level agro-extension system (Wang Wensheng, 2011). Firstly, the extension methods need urgent innovation to change the traditional model of "Two Legs and One Mouth". Secondly, the methods of managing agro-technique extension need improvement to optimize the performance evaluation of staffs in agro-technique extension. Thirdly, the quality and capacity of staffs require immediate improvement to update their unitary and stale knowledge.

As one extension work is responsible for several farmers and most extension works have developed a level of expertise in one or more specialized areas. What is more, extension works have high qualities and are much easier to master ICTs. To provide information service for agricultural extension team will be more effective for the transfer of agricultural technology.

Usually, extension workers spend most of their time in the fields and help farmers solve problems in zero distance. Agricultural extension team shows the following characteristics, nationwide coverage, unfixed workplace, strong liquidity and real-time information needs. The traditional information service is unable to meet the demand of extension worker.

Given this context, this paper proposed a novel information service model based the latest information technology, for example, cloud computing and 3G capable smart devices. On the one side, cloud provides a perfect platform for the accumulation of information resources; on the other side, smart devices enhances the working ability for grass-root agricultural extension workers.

2 RELATED WORKS

Followed the rapid development of information technology, agricultural extension information service can be divided into several stages.

1. Internet-based information service
   Internet is a global system of interconnected computer networks and it carries an extensive range of information resources and services. By the year of 2012, more than 2.4 billion people—over a third of the world's human population—have used the services of the Internet. Internet has become the main media of information exchange of people and is widely be used in the information service. Agricultural information institute has built the Chinese agricultural extension website since 2003 and served all national agricultural extension workers.

2. Mobile-based information service

Mobile phones are becoming an essential device for all types of users irrespective of the age group. Information services based on mobile phones are increasingly popular. Information can be disseminated with various forms like SMS based information services, voice based agro-advisory services, and videos though the mobile etc. In china, all three telecom operators launched mobile-based information service, such as agricultural hotline "110", "12316" and etc.

3. Television-based information service

Television is a telecommunication medium typically used in the 2000s for transmitting and receiving moving color images and sound. Agricultural TV program is also a vivid agricultural information dissemination mode. In 1987, Agriculture Film and CCTV co-produced a program: Agricultural Education and Science. Since 1995, Agriculture Film started making agricultural programs to be aired through CCTV 7 that was co-sponsored by MOA and CCTV. Now CCTV 7 has 11 programs being on air for 8 hours every day. All provinces have also launched television channels shot on the agricultural information services.

3 CONCEPTIONS

Information Services is to provide value-added information. Information service means that agency or department provides professional information on specific topics to its customers, or the general public. Information collection, processing, and provision are integrated into an industrial chain by the name of Information Services. The basic elements of information service include service content, service object, service provider, and service media.

Agricultural information service is an instance of information service which focuses on agricultural production and serving the farmers. According to the content of information service, it includes agroclimatic information service, agricultural market information service, agro-technology information service, agricultural advisory and etc.

As agricultural extension team plays a very important role in agriculture development, infor-

mation service for agricultural extension workers are increasing be studied. Information service for agricultural extension serves grass-root agricultural extension workers by providing special data, information resources and information applications, so as to promote the service level of agricultural extension team.

As for agricultural extension information service, service object is the agricultural extension worker; service provider is agricultural extension sector; service content covers special data, information resources, information applications, etc.; service media includes all kinds of information technology.

4 DESIGN

Information service is the combination of service consumers and service providers; hence most of information service just focus on the providers and pay a lot to the construction of a perfect information service platform based on the assumption that most service consumers have the capacity to consumer the information services. In case the service consumer lacks the basic ability to access the information service platform and service consumers are not in an equal position with the service providers, problems will arise and service won't yield good results.

Cloud computing is a type of computing that relies on sharing computing resources rather than having local servers or personal devices to handle applications, you can access the information from anywhere, where there is an Internet connection. In the meantime, smart devices are widely accepted in a very short time. A smart device is an electronic device, generally connected to other devices or

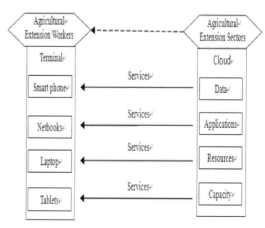

Figure 2.   Services based on "Cloud plus Terminal".

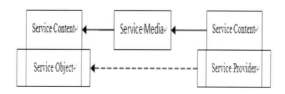

Figure 1.   Elements of information service.

networks via different protocols such as Bluetooth, NFC, WiFi, 3G, etc., that can operate to some extent interactively and autonomously. With a smart device, users can access the cloud, more convenient and more effective. Service based on "Cloud plus Terminal" are also introduced by many internet companies, such as Google, Microsoft etc.

This paper proposed a novel information service model based on "Cloud plus Terminal". Agricultural extension workers are connected with agricultural extension sectors with smart devices and cloud platform. Information services can be accessed by remote agricultural extension workers from anywhere. Agricultural extension sectors can concentrate on the construction of the cloud platform.

For the agricultural extension sectors, cloud platform is the foundation for providing the basic facilities for application deployment and data storage. Cloud platform provides various services to the user, such as data, application, resources and capacity. Data service is the most fundamental service. Users can get weather data, soil data, production figures etc. Application services are developed and deployed in the cloud according to user needs. Resource services provide the opportunity for self-learning and self-improvement and many videos and texts are stored in the cloud. Capacity for data computing or data storage is also provided.

For the agricultural extension workers, their capabilities are enhanced. With the help of smart devices, they can access the services in the cloud. Users can exchange information and share knowledge. They can also upload data to the cloud without concern of the security. Smart phones, netbooks, laptops, tablets and all other smart devices can be used to employ the agricultural extension workers.

5 IMPLEMENTATION

In 2009, the agricultural extension information platform was initially developed and used on-line. During the promotion process of this platform, we employed the cloud technology in the server side and reinforced the users' ability by adoption of the smart devices.

5.1 Agricultural extension cloud platform

Cloud computing platform provides the basic facilities for application deployment and data storage. We build a virtualized environment and agro-extension applications are deployed onto virtual servers. Resources allocated to agroextension application can be increased or decreased in real

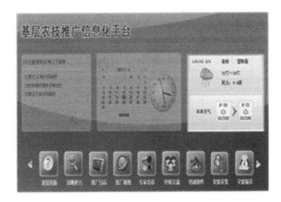

Figure 3. Agricultural extension cloud platform.

Figure 4. Agricultural extension information service.

time and on-demand. We build a cloud storage environment based on client-server model, providing storage capacity of more than 100 Terabytes.

Agricultural extension applications meet the demands of basic-level extension workers by providing information services with text, voice, video and other kinds. Applications can be divided into three major categories and there are more than fifty applications in our platform.

Data resources are very important part of our platform. Plenty of resources can be shared with agricultural extension workers. There are more than 100,000 data records and 5000 videos about field crops, livestock, disease prevention, aquaculture, fruit growing, vegetable growing, fish processing are provided in the platform.

5.2 Smart devices for agricultural extension

The term 3G refers to the third generation of cell phone networking. 3G is the latest mobile technology and is now the fastest growing host among mobile units and handsets. People can perform a lot of functions such as sending information and data and acquiring these via wireless access. You get to have data regardless of the time and location. An estimated 160 million active users of Android, and 80 million on iOS in mainland China at the end of 2012 are based on 3G.

As 3G is growing so well, all the smart devices are 3G capable in our program. With smart devices, agricultural extension workers can conduct high-speed data transfers which mean faster browsing, streaming, and downloading speeds. Corresponding applications are also built into the devices in advance, considering the compatibility.

6 CONCLUSIONS

Information service for extension workers is very important and urgent. Through constantly efforts, we found that better information service for agricultural extension put a great effect to further improve the extension level. 2010, we carried out demonstrations in Daxing Beijing and Xinghua Jiangsu. 2011, Luohe Henan and Turpan Xinjiang were included into the demonstration. 2012, 360 agricultural workers are full covered by the platform in Miyun Beijing. 2013, two other countys were joined into the demonstration. In 2013, the Ministry of Agriculture decided to promote the cloud platform to whole nation and provided agro-technique services to grass-root agro-technique extension team.

Although some achievements have been made, there is still a lot of work to do. In future, we will focus on the perfection of our platform, to further expand the functions, scale the demonstration scope and cooperate with friends both at home and abroad.

ACKNOWLEDGMENT

This paper is supported by National Science and Technology Support Program "Research and application of cloud storage and cloud computing technology for rural information service" (2013BAD15B02), Public sector (Agriculture) scientific research funding program "Integration and demonstration of agricultural extension service based on ICTs" (201303107).

REFERENCES

[1] Wang Wensheng. Build information platform with 3G technology and innovative grassroots agro-technical extension system. World Telecommunications. 2011:41–44. (in Chinese).

[2] Wang Wensheng. The Study and Integration of Key Technologies on Agricultural Extension Information based on 3G Technology [J]. China Science and Technology Achievements. 2011, 8:66. (in Chinese).

[3] Chen Jian long. A Study of Information Service Models [J]. Journal of Peking University, 2003, 40(3):124–132.

[4] Yang Shanlin, Luo He, Ding Shuai. Survey on multi-sources information service system based Oil cloud computing [J]. Journal of management sciences in China, 2012, 15(5):83–96.

[5] Hu Bohai. Construction and Informatization Direction of Service information network for Agricultural Technology Extension in China [J]. Agriculture Network Information, 2004, (1):3–6.

[6] Huang Zhaozhen, Zhang Ju, Feng Lianfang, Liu Xiaoying, Li Gang. Design and application of integrated service system of agrotechnical extension and information [J]. Acta Agriculturae Shanghai, 2010, 26(1):65–69.

[7] Zheng Huoguo, Hu Haiyan. On Agricultural Information Service Models and Their Role in "Agriculture, Rural Area, and Farmer" [J]. Journal of Library and Information Sciences in Agriculture, 2005, 17(2):137–139, 188.

[8] Liu Xiaoping, Huang He, Long Hai. Information Service Model in Less Developed Area of China [J]. Hubei Agricultural Sciences, 2011, 50 (23): 5000–5003, 5007.

[9] Zheng Huaiguo, Tan Cuiping, Li Guangda, Zhao Jingjuan. Agricultural Information Services Based on the Farmers' Perspective [J]. Journal of Anhui Agri. Sci. 2011, 39 (27):16978–16979.

[10] Gao Qijie, Zhang Chuanhong. Agricultural Technology Extension System In China: Current Situation And Reform Direction [J]. Management Science and Engineering, 2008, 2(4):48–58.

Material Science and Engineering – Chen (Ed.)
© *2016 Taylor & Francis Group, London, ISBN 978-1-138-02936-1*

The study of urban road intersection safety evaluation method based on the PPC model

Q. Zhou
School of Aeronautic Science and Engineering, Beihang University, Beijing, China

Z.S. Yang & H.B. Zhang
Naval Aeronautical Engineering Application, Beijing, China

ABSTRACT: Because of the increase in the vehicular traffic and the randomness of the system, the evaluation of the system will be affected when the issues of the city plane intersection safety evaluation are studied. There are many defects in the traditional evaluation methods, such as small sample size, large area, long cycle and low confidence, so the non-adaptive security evaluation is emerged for a small area. In order to solve the above problems, we propose a safety assessment model, which combines the technique of the traffic conflict and Projection Pursuit Classification Model (PPC). Then, an application is implemented in the traffic safety evaluation of urban intersection, and we apply the model to the city of Shanghai. A total of thirteen security-level plane crossing of Shanghai can be obtained. The experiment proved that the results of this model are reasonable, and this model has many advantages, such as a rapid, quantitative analysis. In summary, a new method and ideas of the safety evaluation for urban intersection are explored by the model.

Keywords: model of PPC; urban intersections; traffic conflict technique; accelerating genetic algorithm

1 INTRODUCTION

Recently, with the booming economical development and people's living standard continues to improve, the need for communal transportation is growing rapidly, leading to many problems, especially urban road transportation network in the city. The intersection of traffic accidents causes many problems that can be described as a top priority. Traffic safety evaluation is an assessment of the extent of traffic safety for a region, route, road or place (section), an objective description of the situation, and it also provides an objective analysis of road conditions that form a very important basis. Traffic evaluation can be used to express traffic safety. Through a variety of statistical indicators, it uses some arithmetic objective way to evaluate traffic safety. Urban road traffic safety evaluation is used to improve the urban road traffic safety and to study traffic management critical evaluation procedures. There are quality assessment method, multiple regression model, and grey evaluation fuzzy evaluation in the traditional traffic accident safety evaluation [1]. In addition, these traditional traffic accident safety evaluations are mainly based on statistical data to be implemented. The data of traffic accident are not exactly contained [2]. However, there are incompleteness and randomness in

traffic accident statistics, which contains a "small sample size, long cycle, large area, low reliability", and other features and defects. Furthermore, traditional data analysis methods cannot obtain good results for high-dimensional non-normal, non-linear data analysis.

For the above case, we propose a plane intersection safety evaluation model that combines the traffic conflict technique and PPC model. To meet the needs of the road safety evaluation method, the traffic conflict technique began to build up. It could be able to overcome many basic problems that the traditional road safety evaluation is difficult to overcome. For example, it could make some problems, which contain very small data sample for site-specific accident data and could affect the overall evaluation of the effectiveness and timeliness issues. Then, we combine the traffic conflict technique and PPC model to turn the high-dimensional problem down into a one-dimensional evaluation, which could avoid the "high-dimensional curse" problem and make it more simple and effective scientific.

The rest of this paper is organized as follows. A brief overview of the traffic conflict technique and related work is presented at Section 2. Section 3 provides an overview of the projection pursuit classification model. Section 4 makes an in-depth study on the above evaluation method by an application example. Finally, Section 5 concludes this paper.

2 INTERSECTION TRAFFIC CONFLICT TYPE

2.1 *Definition of traffic conflict*

Under observable conditions, when two or more road users for intersection are in the same space and time close to each other, any one of them takes a non-normal traffic behavior, such as steering, speed, and sudden braking, unless the other party has taken an appropriate hedging behavior, otherwise it will collide. This phenomenon is called intersection traffic conflicts [3]–[4].

2.2 *The relationship between traffic conflicts and accidents [5]–[6]*

The essence of traffic *conflict* is a form of insecurity travel behavior. Its development may lead to an even more serious accident. It may also avoid risk accidents by taking some related risk aversion. Thus, there is a very similar form of the accident and conflict. The only difference between them is whether it is a direct consequence of damage. The relationship between conflict and incidents of conflict is commonly made by qualitative analysis based on the degree of seriousness. The key studies of traffic conflict are how to determine that conflict is a serious conflict. So, we can determine the quantitative relationship between accidents and serious conflict.

Some conflicts and accidents show that there is a certain interchangeable relationship between them through correlation studies. This relationship is described by giving the replacement coefficient π, namely the value of π is expressed as a conflict caused by the size of the probability of an accident to occur, which is as follows:

$$\pi_1 = P_i \frac{C_i}{A_i} \qquad \pi_2 = P_i \frac{A_i}{C_i}$$

where A is the number of hours of accident records; C is the record for the number of hours the conflict; and P is the Poisson distribution, the value obtained by the maximum likelihood estimate as follows:

$$P_1 = \frac{\sum C_i}{\sum A_i} \qquad P_2 = \frac{\sum A_i}{\sum C_i}$$

$$P_1 = \frac{\sum C_i}{\sum A_i} \qquad P_2 = \frac{\sum A_i}{\sum C_i}$$

According to the statistical results of this paper, a conflict leads to the accident probability of 0.00001. Such an actual conversion factor of incidents and conflicts has a high level of confidence.

2.3 *The basic measurement of traffic conflict*

Based on accident research, the accident investigation mainly depends on the measurement T = S/V (where T = time, S = distance, V = velocity) of the basic relationship, namely we studied the correlation between the perpetrators and the point of accident by using V-S, T-V or T-S as the three measurement parameters. Traffic conflict is "quasi accident" that does not generate the damage consequences. The measurement parameters can be used for the following options:

Conflict distance T_s(m): when the conflictor take risk aversion, it is instantaneous distance from the accident contact point of effective location;
Conflict Speed C_s(m/s): when the conflictor take risk aversion, it is the instantaneous velocity;
Conflicts time C_s(s): when the conflictor take risk aversion, it is the time of contact accident;

Currently, the classification of the severity of the conflict is mainly as follows:
Spatial distance method: let us select a distance as the metric. The law, in practice, is very intuitive and very logical. In other words, the smaller the distance between the parties to the conflict, the greater the likelihood of collision. When the distance tends to infinity hours, the accident occurs.

Time distance method: let us select time as the metric. This method is based on the accident point projection of the actual speed and the distance of the approaching collision point time vector. To some extent, this reflects those involved to avoid the accident in the space requiring speed, distance, acceleration, and emergency steering capability. In the above two methods, there are advantages and disadvantages of each safety assessment. Under different circumstances, different metrics can be used, which can also be used in combination. No matter which parameters are used, the sole purpose is to quickly and accurately determine a serious conflict. In a variety of observation methods, the scene of the observer is the most common use of measurement. Conflict investigation time includes the period of surveys and the number of days with the investigation.

3 EVALUATION OF THE TRAFFIC CONFLIC PPC MODEL

The evaluation index is the number of conflicts with the time-averaged mixed traffic

(TC/MPCU) [7] ratio when we use the traffic conflict technique to make a traffic safety evaluation. MPCU represents an intersection traffic level, namely the distribution of traffic, traffic and traffic flow. Intersection traffic conflict represents a level of safety indicators, namely traffic management, legal awareness and distribution of transport facilities. There is an effectiveness of a high level of confidence, when the ratio of the two above-mentioned parameters relevant indicators is used as the grading evaluation of "intersection safety evaluation". This paper selected the TC/MPCU value of the morning peak and the evening peak period of peace as a safety evaluation. We established a traffic conflict technique and PPC model intersection safety evaluation model. Its model overcomes the problems of the traffic for multi-period analysis and conflict research data, which leads to consuming a variety of complex issues. The method has an objective more accurate than the previous time-averaged (TC/MPCU) ratio evaluation.

3.1 Basic principles and methods of PP

Projection pursuit model (referred to as the PP model) [8] is a high-dimensional data analysis, dimensionality reduction, method. It combines the overall spread level and the degree of local cohesion to form new indicators to make a new cluster analysis. The PP model not only overcomes the high-dimensional data of the "dimension curse", which causes difficulties for data analysis, but can also turn high-dimensional data into a one-dimensional space to project. We compare the results of different one-dimensional projections to find out the best projection direction through analyzing the projected one-dimensional data.

However, the standard genetic algorithm optimization efficiency also depends on the initial optimization variable change intervals to guarantee global convergence. Therefore, we choose a secondary evolutionary iterative algorithm to produce an excellent individuals change interval, which can be used as a new initial variable change interval, and then rerun the standard genetic algorithm to form an accelerated algorithm. The outstanding individuals will have a narrow range, and get closer with the most advantages, until the best individual optimization criterion function value is less than the pre-set value, or the algorithm runs to reach a predetermined acceleration times, and then the entire operation is completed. Furthermore, the best individual in the current population is to determine the best overall result of the accelerating genetic algorithm. This method of analysis is called the real coding Accelerating Genetic Algorithm (RAGA) [9].

3.2 Intersection evaluation of the PPC model modeling steps [10]–[11]

According to the basic principles and methods of PP, it can be reduced to a one-dimensional cube. There are local cohesion and the overall dispersion characteristics in the new indicators. We can perform cluster analysis according to the size of its optimal projection values. Using PP to perform the cluster analysis model is called the PPC model. The actual modeling process involves the following steps:

Step 1: evaluation of the normalized set

Let the sample set be the intersection index as follows:

$$X = (x_{ij})_{n \times p} \quad i = 1, 2, \cdots n; \quad j = 1, 2, \cdots, p$$

where z_{ij} is the i-th sample of the j-th index value and n, p are the number of samples and the number of indicators. In order to eliminate inter-dimensional indicators and harmonization of indicators of changes in scope, we use the following formula to normalize:

For bigger and more excellent indicators,

$$y(i, j) = \frac{x(i, j) - x_{\min}(j)}{x_{\max}(j) - x_{\min}(j)}$$

For the smaller and more excellent indicators,

$$y(i, j) = \frac{x_{\min}(j) - x(i, j)}{x_{\max}(j) - x_{\min}(j)}$$

where $x_{\max}(j)$, $x_{\min}(j)$ are, respectively, the maximum and minimum values of the j-th index, and $y(i, j)$ is the index characteristic value normalized sequences.

Step 2: building a projection index function $Q(a)$:

The PP model makes the P-dimensional data $\{y(i, j) \mid j = 1, 2, \cdots, p\}$ integrated into $a = \{a(1), a(2), \cdots, a(p)\}$ to project the direction of a one-dimensional projection value $z(i)$ as follows:

$$z(i) = \sum_{j=1}^{p} a(j) y(i, j) \ (i = 1, 2, \cdots, n)$$

where a is the unit length vectors. Then, according to $\{z(i) \mid i = 1, 2, \cdots, n\}$ of the one-dimensional distribution map, it can make a classification and evaluation. When we establish a comprehensive projection index value, the model requires the projection values $z(i)$. The characteristics of the local projection point spread as densely as possible. It is best to group together into a number of points. For the group as a whole, the projection point spreads as much as possible.

The projection objective function can be expressed as follows:

$$Q(a) = S_z * D_z$$

where a is the projection of the standard deviation value b and a is the projection of the local density value b:

$$D_z = \sum_{i=1}^{n} \sum_{j=1}^{n} (R - r(i,j)) * u(R - r(i,j))$$

$$S_z = \sqrt{\frac{\sum_{i=1}^{n} (z(i) - E(z))^2}{n-1}}$$

where $E(z)$ is the sequence $\{z(i) \mid i = 1, 2, \cdots, n\}$ of the average, and R is the radius of the local density of the window. The selection of the window not only contains the projection point of the average number that is not to a small extent, avoiding moving the average deviation too much, but also leads to the increase in the sample to a large extent. R is selected as 0. 1Sz is determined according to the test. $r(i,j)$ is the distance between the samples, $r(i,j) = |z(i) - z(j)|$; function $u(t)$ is the unit step function, if for $t < 0$, the function value is 0. When $t \geq 0$, the value of 1.

Step 3: optimization of the projection index function.

When the index value of each sample set is given, the projection index function Q (a) changes only with a change in the direction of projection. The different projection directions are reflected in the data structure of the type. The best projection direction is the maximum possible exposure to certain characteristics of high-dimensional data structure projection direction. Therefore, the projection index function can be maximized by solving the problem to determine the optimal projection direction. That is, the maximization objective function is given by

$$\max : Q(a) = S_z * D_z$$

Constraints:

$$s.t : \sum_{j=1}^{p} a^2(j) = 1$$

This is a complex nonlinear optimization problems based on $\{a(j) \mid j = 1, 2, \cdots, p\}$ variables that the conventional process could not deal with it. Therefore, this paper uses an internal simulation of biological survival of the fittest and group information exchange mechanism of chromosome Real-coded Accelerating Genetic Algorithm (RAGA) to get the best projection direction to resolve their high-dimensional global optimization problems.

Step 4: Classification and optimization of arrangement

The projection values of each sample point $z(i)$ can be obtained when we take the top projection direction as $z(i) = \sum_{j=1}^{p} a(j) y(i,j)$. The $z(i)$ values descending sort determine the extent of the sample traffic conflict. Setting the number of categories, we could use fast clustering to determine the cluster center. Through analyzing the distance from the cluster center, we could determine the category of each program.

4 APPLICATION EXAMPLES

Based on the safety assessment model, this paper selects sixteen intersections of Shanghai as an example. This paper studies the road intersections of Shanghai, and selects the intersection morning peak hours and evening peak hours of peace peak periods TC/MPCU as the evaluation index.

4.1 Selecting the parameter

This paper selects the parent initial population size of $N = 400$, crossover probability $P_c = 0.8$, the mutation probability of $P_m = 0.2$, and the number of 50 outstanding individuals, iterative twice into the acceleration cycle, $\alpha = 0.05$.

4.2 Searching for the final cluster centers

In this paper, we make a fast clustering for the projection values of sixteen regions in Shanghai by SPSS software. This paper classifies security issues into four categories: special safety; safe; criticality safety; unsafe [1]. The resulting final cluster centers and level are given in Table 1.

4.3 Making an evaluation

According to the projection values, this paper makes a descending order for Shanghai's sixteen regions to determine which situation the sixteen regional traffic safety belong to. The sorted results are given in Table 2.

From Table 2, sixteen intersection safety categories can be derived. Through the analysis, we

Table 1. The final cluster center.

Level	Special security	Security	Criticality safety	Unsafe
Projection values	0.1414	0.4704	0.9973	1.7269

Table 2. Shanghai 8 junctions TC/MPCU and the security level.

Intersection	Morning peak period	Evening peak period	Flat peak periods	Projection values	Category
1	0.0085	0.0131	0.0064	0.4951	Security
2	0.0156	0.0213	0.0115	1.0653	Criticality safety
3	0.0235	0.0311	0.0165	1.7269	Unsafe
4	0.0081	0.0091	0.0062	0.2334	Special security
5	0.0078	0.0083	0.0059	0.3038	Special security
6	0.0075	0.0079	0.0054	0.2646	Special security
7	0.0049	0.0054	0.0034	0.0489	Special security
8	0.004	0.0053	0.0031	0.0096	Special security
9	0.0043	0.0049	0.0033	0.0166	Special security
10	0.0104	0.0222	0.0079	0.7927	Criticality safety
11	0.0099	0.0101	0.0072	0.4591	Safety
12	0.0174	0.0217	0.012	1.1415	Criticality safety
13	0.0099	0.0136	0.0071	0.5387	Security
14	0.0056	0.0064	0.0039	0.1132	Special Security
15	0.0089	0.0099	0.0063	0.3887	Security
16	0.0156	0.0185	0.0113	0.9896	Criticality safety

Table 3. The 16 junctions safety sequence of Shanghai.

Intersection	1	2	3	4	5	6
Projection values	0.4951	1.0653	1.7269	0.2334	0.3038	0.2646
Sequence	7	3	1	12	10	11
Intersection	9	10	11	12	13	14
Projection values	0.0166	0.7927	0.4591	1.1415	0.5387	0.1132
Sequence	15	5	8	2	6	13

can get the first intersection, which is a more dangerous intersection consists of reality. Some measures are needed to be taken to make a investigation for the crossing of the traffic situation, to find out the root cause. We can make some improvements in the settings for the road intersection by using the traffic conflict technique, make decisions for the transport sector, and provide a basis for planning. Thus, this model gives a good solution for a specific place whose data are too small to affect the road safety evaluation of the timing and effectiveness.

Through the above research, we can not only obtain the safety level of the Shanghai's 16 junctions, but also give a safety sequence by the projection values. From Table 3, the sequence can be obtained. Of course, an in-depth analysis can provide the junction safety. For the different junctions, the diverse method can be adopted to make some improvements in settings. It just provides an evaluation for Shanghai, and some measures can be offered by the government by analyzing the reason for the safety level.

5 CONCLUSION

Compared with some of the traditional evaluation models of road traffic safety, traffic conflict technique is a non-accident statistics based on the indirect evaluation method, which has the advantage of rapid and quantitative analysis. The PPC model turns high-dimensional data into one-dimensional space by seeking the best projection direction. This model can not only make a comprehensive study of the sample, but also avoid the traditional method to determine the index weight objectivity. Combined with the PP model and the real code of the accelerating genetic algorithm, we solve global optimization problems on high-dimensional data, which greatly reduces the workload optimization and improves work efficiency, making the projection pursuit model to be used more widely in various fields. It is more conducive to make a rapid diagnosis intersection safety, research and analysis, and scientific evaluation by combining them. This evaluation model has more computational advantages, especially for small sample data, and the number of intersections evaluation index number of cases. This model is able not only to get samples sorted and classified by the degree of safety, but also to find the optimal projection direction indicators that reflect the overall evaluation of each evaluation of the contribution rate. Compared with the other safety evaluation method, the result is more rational and scientific, and provides a new avenue of research and ideas for a comprehensive evaluation of road traffic safety research.

REFERENCES

[1] Cheng Wei. Research on the Theory Model and Method of Traffic Accidents and Traffic conflict Technique in urban Streets [D], Jilin University doctoral dissertation. 2004. 3.

[2] Zhang Su. China Traffic Conflict Technique [D]. Southwest Jiaotong University doctoral dissertation, 1997. 3.

[3] Sayed, T., and Zein, S. Traffic Conflict Standards for Intersection, Transportation Planning and Technology, 1999, Vol. 22: pp. 309–323.

[4] Traffic Conflict Techniques for Safety and Operation-Observers Manual, U.S. Department of Transportaaion Federal Highway Administration, 1989. 1.

[5] Traffic Conflict Studies Report Working Paper 5, National Cooperative Highway Research Program Project Panel Members, 1990. 8.

[6] Kulmala, R. Measuring Safety Effect of Road Measures at Junctions. Accident Analysis and Prevention, 1994, 26(6):781–794.

[7] Cheng Wei. Wang Gui yong. Gray Cluster Evaluation of Safety at Intersections by Trafic Conflict Technique [J]. Joureal of Kunming University of Science and Technology, 2005(30):106–111.

[8] Friedman J.H, Tukey J.W. A projection pursuit algorithm for exploratory data analysis [J]. IEEE Trans. On Computer, 1974, 23 (9): 881–890.

[9] Cao Qing-pu, Dong Shu-fu, Luo Yun-qian etc. Evaluation of Network Performance Using Genetic Projection Pursuit Model Weighted Method [J]. Computer Simulation, 2010(27):103–107.

[10] Chen Guang-zhou, Wang Jia-quan, Xie Hua-ming. Application of Particle Swarm Optimization for Solving Optimization Problem of Projection Pursuit Modeling [J]. Computer Simulation, 2008(25): 159–161.

Material Science and Engineering – Chen (Ed.)
© *2016 Taylor & Francis Group, London, ISBN 978-1-138-02936-1*

Author index